Conversion Factors

Length

1 in. = 2.54 cm
1 ft = 0.3048 m
1 mi = 5280 ft = 1.609 km
1 m = 3.281 ft
1 km = 0.6214 mi
1 angstrom ($\mathring{A}$) = 10^{-10} m

Mass

1 slug = 14.59 kg
1 kg = 1000 grams = 6.852×10^{-2} slug
1 atomic mass unit (u) = 1.6605×10^{-27} kg
(1 kg has a weight of 2.205 lb where the
   acceleration due to gravity is 32.174 ft/s^2)

Time

1 day = 24 h = 1.44×10^3 min = 8.64×10^4 s
1 yr = 365.24 days = 3.156×10^7 s

Speed

1 mi/h = 1.609 km/h = 1.467 ft/s = 0.4470 m/s
1 km/h = 0.6214 mi/h = 0.2778 m/s = 0.9113 ft/s

Force

1 lb = 4.448 N
1 N = 10^5 dynes = 0.2248 lb

Work and Energy

1 J = 0.7376 ft·lb = 10^7 ergs
1 kcal = 4186 J
1 Btu = 1055 J
1 kWh = 3.600×10^6 J
1 eV = 1.602×10^{-19} J

Power

1 hp = 550 ft·lb/s = 745.7 W
1 W = 0.7376 ft·lb/s

Pressure

1 Pa = 1 N/m^2 = 1.450×10^{-4} lb/in.^{2}
1 lb/in.^{2} = 6.895×10^3 Pa
1 atm = 1.013×10^5 Pa = 1.013 bar =
   14.70 lb/in.^{2} = 760 torr

Volume

1 liter = 10^{-3} m^3 = 1000 cm^3 = 0.03531 ft^3
1 ft^3 = 0.02832 m^3 = 7.481 U.S. gallons
1 U.S. gallon = 3.785×10^{-3} m^3 = 0.1337 ft^3

Angle

1 radian = 57.30°
1° = 0.01745 radian

Standard Prefixes Used to Denote Multiples of Ten

Prefix	Symbol	Factor
Tera	T	10^{12}
Giga	G	10^{9}
Mega	M	10^{6}
Kilo	k	10^{3}
Hecto	h	10^{2}
Deka	da	10^{1}
Deci	d	10^{-1}
Centi	c	10^{-2}
Milli	m	10^{-3}
Micro	μ	10^{-6}
Nano	n	10^{-9}
Pico	p	10^{-12}
Femto	f	10^{-15}

Basic Mathematical Formulae

Area of a circle = πr^2

Circumference of a circle = $2\pi r$

Surface area of a sphere = $4\pi r^2$

Volume of a sphere = $\frac{4}{3}\pi r^3$

Pythagorean theorem: $h^2 = h_o^2 + h_a^2$

Sine of an angle: $\sin\theta = h_o/h$

Cosine of an angle: $\cos\theta = h_a/h$

Tangent of an angle: $\tan\theta = h_o/h_a$

Law of cosines: $c^2 = a^2 + b^2 - 2ab\cos\gamma$

Law of sines: $a/\sin\alpha = b/\sin\beta = c/\sin\gamma$

Quadratic formula:
   If $ax^2 + bx + c = 0$, then, $x = (-b \pm \sqrt{b^2 - 4ac})/(2a)$

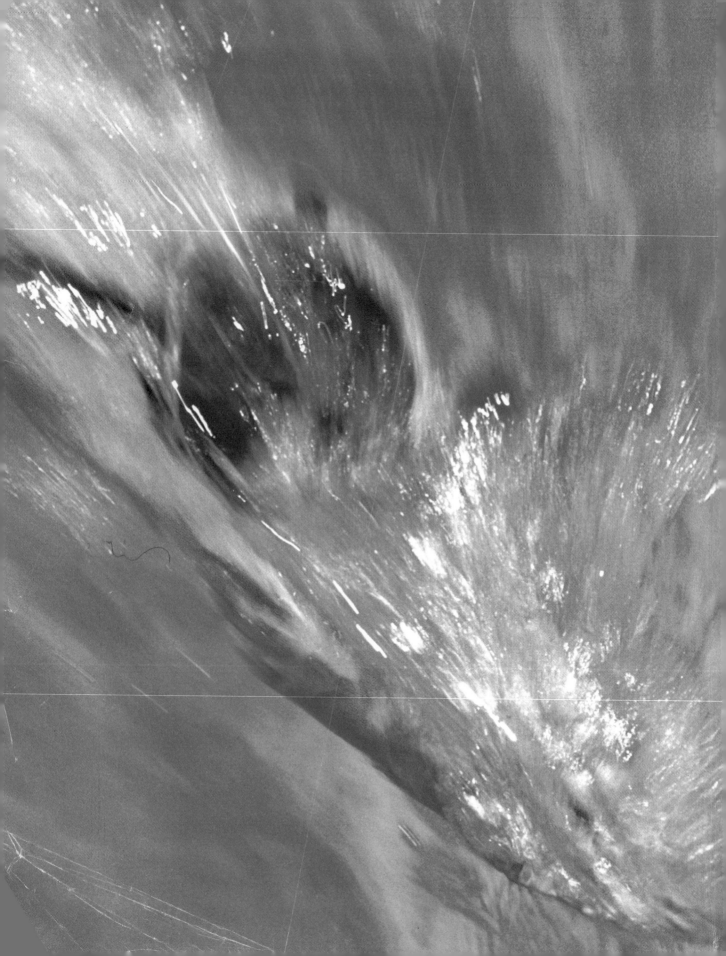

FOURTH EDITION • # PHYSICS

VOLUME 1

JOHN D. CUTNELL
KENNETH W. JOHNSON

Southern Illinois University at Carbondale

JOHN WILEY & SONS, INC.
NEW YORK • CHICHESTER • WEINHEIM • BRISBANE • SINGAPORE • TORONTO

ACQUISITIONS EDITOR   *Stuart Johnson*

DEVELOPMENTAL EDITOR   *Barbara Heaney*

SENIOR MARKETING MANAGER   *Catherine Faduska*

ASSOCIATE MARKETING MANAGER   *Catherine Beckham*

SENIOR PRODUCTION EDITOR   *Katharine Rubin*

COVER AND TEXT DESIGNER   *Karin Gerdes Kincheloe*

PHOTO EDITORS   *Ramón Rivera-Moret, Hilary Newman*

ASSOCIATE PHOTO RESEARCHER   *Elaine Paoloni*

ILLUSTRATION EDITOR   *Sigmund Malinowski*

ELECTRONIC ILLUSTRATIONS   *Precision Graphics, Poole Visual Communication Group*

COVER PHOTO   *Pete Saloutos/The Stock Market*

This book was set in 10/12 Times Roman by Progressive Information Technologies, and printed and bound by Von Hoffmann Press. The cover was printed by Lehigh Press.

Recognizing the importance of preserving what has been written, it is a policy of John Wiley & Sons, Inc. to have books of enduring value published in the United States printed on acid-free paper, and we exert our best efforts to that end.

This paper in this book was manufactured by a mill whose forest management programs include sustained yield harvesting of its timberlands. Sustained yield harvesting principles ensure that the number of trees cut each year does not exceed the amount of new growth.

Library of Congress Cataloging in Publication Data:

Cutnell, John D.

    Physics/John D. Cutnell, Kenneth W. Johnson. —4th ed.

      p. cm.

    Includes index.

    ISBN 0-471-15519-5 (set : cloth : alk. paper).—ISBN 0-471-19768-8

      (set : pbk. : alk. paper).—ISBN 0-471-19112-4 (v. 1 : pbk.).—

      ISBN 0-471-19113-2 (v. 2 : pbk.)

    1. Physics.   I. Johnson, Kenneth W.   II. Title.

QC23.C985   1997   97-21746

530—dc21   CIP

Printed in the United States of America

10 9 8 7 6 5

Volume 1 0471-19112-4 (pbk)

Volume 2 0471-19113-2 (pbk)

*T*his edition is dedicated to the memory of

Stella Kupferberg,

Director of the Photo Department,

a friend, a mentor, and a pillar

of dependability and excellence.

She had a strong and

enduring influence on our work

and taught us everything

we know about using photographs

in support of physics pedagogy.

We miss you, Stella,

and shall always remember

that a well-chosen photograph

should speak for itself,

without the need for lengthy

explanations.

BRIEF CONTENTS

CONTENTS

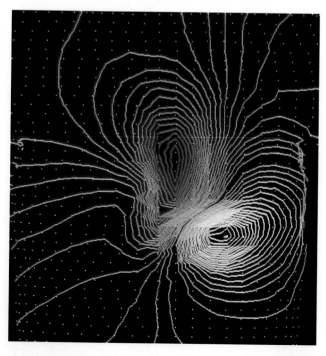

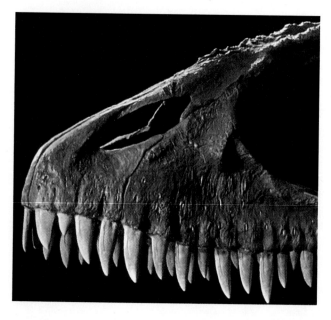

PREFACE

We have written this text for students and teachers who are partners in a one-year course in physics that uses algebra and trigonometry. Since each has only so much time in a given day, we have tried to produce a text that facilitates the learning and teaching processes. Considerable positive feedback from the many users of the third edition has encouraged us to make this fourth edition even more attuned to the needs of students and teachers. We have taken a fresh look at the goals and features of the third edition, adding refinements and new material that will help students learn and teachers teach. In addition, we have expanded our supplements package to include World Wide Web offerings and a CD-ROM that contains interactive problem-solving software and numerous simulations of physical phenomena.

GOALS

Our first goal is to help students develop conceptual understanding and use it in solving problems. One of the greatest challenges for teachers is to dispel the notion that physics is merely a large collection of equations that can be used to solve problems. Physicists know that the ability to reason is the cornerstone of problem solving and that this ability develops from a conceptual understanding of physical principles. The features of this edition that work toward this goal are: *Conceptual Examples*, *Explicit Reasoning Steps* in all examples, *Reasoning Strategies*, *Free-Body Diagrams*, and *Problem Solving Insights* placed in the margins.

Secondly, we want to help students see the interrelationships between the concepts of physics. In studying anything for the first time, it is easy to focus on the new details and lose sight of the overall picture. This is certainly true in physics, where the big picture consists of the relationships between basic concepts and the way in which they fit together to give a description of the physical world. To accomplish this goal, we have introduced a new feature called *Concepts at a Glance*, which consists of charts that illustrate the interrelationships diagrammatically. Each chapter also ends with a summary in the form of a condensed, but thorough, exposition of the chapter material, including equations. These reviews are intended to give the student an overview of how the chapter concepts have evolved.

Finally, we want to show students that physics principles come into play over and over again in their lives. It is always easier to learn something new when the learning has a direct relevance to our daily lives. Direct applications of physics principles are identified in the margins by a triangular-shaped icon, including the label *The Physics of.* . . . Many of these applications are biomedical in nature and deal with human physiology. We have also incorporated real-world situations into many of the worked-out examples, the line art, and the homework material at the end of each chapter.

ORGANIZATION AND COVERAGE

The text consists of 32 chapters and is organized in a fairly standard fashion according to the following sequence: (1) *Mechanics*, (2) *Thermal Physics*, (3) *Wave Motion*, (4) *Electricity and Magnetism*, (5) *Light and Optics*, and (6) *Modern Physics*. Within each chapter, material that is a likely candidate for omission is typically located in a subsection at the end of a main section or in a separate section near the end of a chapter. Sections marked with an asterisk can be omitted with little impact on the overall development of the material.

In addition to the one-volume hardbound edition, the book is available as a two-volume paperback version. Volume 1 consists of Chapters 1-17 (*Mechanics, Thermal Physics,* and *Wave Motion*), while Volume 2 consists of Chapters 18-32 (*Electricity and Magnetism, Light and Optics,* and *Modern Physics*).

Based on feedback from users and reviewers, we have added new material, made judicial deletions, and carefully tightened other material. The net result of these efforts is a more streamlined text that is about 50 pages shorter than the third edition. We have added new material on *The Center of Mass* (Section 7.5), *Conduction of Electrical Signals in Neurons* (Section 19.6), *The Doppler Effect and Electromagnetic Waves* (Section 24.5), *Medical Applications of the Laser* (Section 30.9), and *Cosmology* (Section 32.7). In Chapter 19, we have rewritten substantially the discussion of the electric potential, in order to strengthen the analogy to the discussion of gravitational potential energy presented in Chapter 6. We have updated our applications of physics principles and added a number of new ones. Many of these illustrate biomedical applications of physics and, together with the new section on the *Conduction of Electrical Signals in Neurons,* give the text an increased emphasis on the role of physics in biology and medicine.

FEATURES OF THE FOURTH EDITION

All of the features of the fourth edition have been designed to support the goals elaborated earlier in this preface. A list of the main features follows, including a description of each and selected illustrations.

Concepts at a Glance This new feature consists of flowcharts that occur in every chapter except Chapter 1. There are 66 of these charts, and they show in a vi-

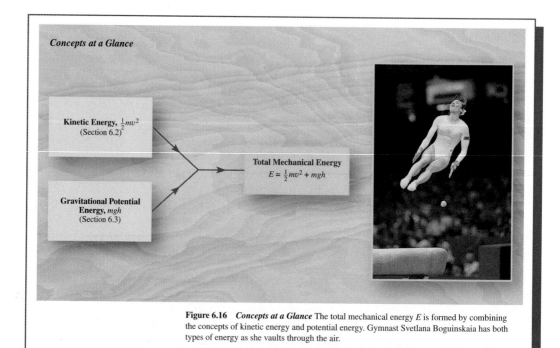

Figure 6.16 *Concepts at a Glance* The total mechanical energy *E* is formed by combining the concepts of kinetic energy and potential energy. Gymnast Svetlana Boguinskaia has both types of energy as she vaults through the air.

sual way the conceptual development of the new physics that is introduced in each chapter. Within a flowchart, the new physics concept being studied is placed in a gold panel, while the concepts presented earlier are placed in light blue panels. The flowcharts provide students with a coherent picture of how new concepts are built upon previous ones and, thus, reinforce fundamental unifying ideas. Included with each chart is at least one photograph, to help students connect the concepts being discussed with the real world.

Conceptual Examples Students often try to solve problems by searching for that elusive *right equation* and ignore the fact that equations are consequences of concepts, concepts that express physical ideas. We believe that good problem-solving techniques start with a foundation of conceptual understanding. Therefore, the text includes 93 examples that are entirely conceptual in nature. These examples are in addition to 282 standard quantitative examples. The *Conceptual Examples* are worked out in a rigorous, but qualitative fashion, with no (or very few) equations. The emphasis is on how to

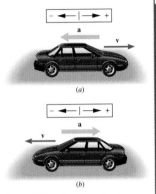

CONCEPTUAL EXAMPLE 9 • Deceleration Versus Negative Acceleration

A car is traveling along a straight road and is decelerating. Does the car's acceleration *a* necessarily have a negative value?

Reasoning and Solution We begin with the meaning of the term "decelerating," which has nothing to do with whether the acceleration *a* is positive or negative. The term means only that the acceleration vector points opposite to the velocity vector and indicates that the moving object is slowing down. When a moving object slows down, its instantaneous speed (the magnitude of the instantaneous velocity) decreases. One possibility is that the velocity vector of the car points to the right, in the positive direction, as Figure 2.14a shows. The term "decelerating" implies that the acceleration vector points opposite, or to the left, which is the negative direction. Here, the value of the acceleration *a* would indeed be negative. However, there is another possibility. The car could be traveling to the left, as in Figure 2.14b. Now, since the velocity vector points to the left, the acceleration vector would point opposite or to the right, according to the meaning of the term "decelerating." But right is the positive direction, so the acceleration *a* would have a positive value in Figure 2.14b. We see, then, that *a decelerating object does not necessarily have a negative acceleration.*

Related Homework Material: Problems 22 and 40

Figure 2.14 When a car decelerates along a straight road, the direction of the acceleration vector depends on the direction in which the car is traveling, as Conceptual Example 9 discusses.

apply a physics principle to arrive at a qualitative solution to a problem. The intent is to provide students with explicit models of how to "think through" a problem before attempting to solve it numerically. The *Conceptual Examples* deal with a wide range of topics, and in them we have addressed a large number of issues that often confuse students. Wherever possible, we have focused on real-world situations that have a direct relationship to physical principles and have structured the examples so that they lead naturally to homework material found at the ends of the chapters. At the end of most examples are explicit references to the related homework material, and that material contains a cross reference that encourages the student to review the pertinent *Conceptual Example*. Teachers can stress the importance of conceptual understanding by assigning the indicated homework material.

Explicit Reasoning Steps Since careful reasoning is the cornerstone of problem solving, we believe that students will benefit from seeing the reasoning stated explicitly. Therefore, the format in which examples are worked out includes an explicit reasoning step. In this step, we explain what motivates our procedure for solving the problem before any algebraic or numerical work is done. Teachers have applauded this feature of the third edition, and we have carefully reexamined each of the examples with a view toward improving the clarity of the reasoning steps. The *Conceptual Examples* and their associated homework material also strengthen our focus on reasoning as an essential part of problem solving.

REASONING STRATEGY

Applying the Equations of Kinematics

1. Make a drawing to represent the situation being studied. When solving kinematics problems, few of us can do without the aid of a drawing to help develop our reasoning and explain it to our coworkers.

2. Decide which directions are to be called positive (+) and negative (−) relative to a conveniently chosen coordinate origin. Do not change your decision during the course of a calculation.

3. In an organized way, write down the values (with appropriate plus and minus signs) that are given for any of the five kinematic variables (x, a, v, v_0, and t). Be on the alert for "implied data," such as the phrase "starts from rest," which means that the value of the initial velocity is $v_0 = 0$. The data summary boxes used in the examples in the text are a good way of keeping track of this information. In addition, identify the variables that you are being asked to determine.

Reasoning Strategies A number of the examples in the text deal with well-defined strategies for solving certain types of problems. In such cases, we have included summaries of the steps involved. These summaries, which are titled *Reasoning Strategies*, encourage frequent review of the techniques involved and help students focus on the concepts on which the techniques are based.

Important Definitions, Laws, and Theorems One of the tasks facing students is to distinguish between the basic concepts of physics and other, less fundamental, relations. To highlight the basic concepts, we have enclosed them within a gold panel. When the concept involves an equation, the meaning of each term in the equation is also explained. Since applying these concepts entails using correct units, the appropriate SI units have been included. The panels are used sparingly, only for the most important concepts.

The Physics of...
Applications of Physics
Principles

The Physics of . . . This edition contains over 200 real-world applications. These applications reflect our commitment to show students how prevalent physics is in their lives. We have added many new ones that are currently in the "headlines," such as Next Generation Weather Radar (NEXRAD), the Global Positioning Satellite system, digital satellite system TV, and the discovery of photoevaporation and star formation in the Eagle nebula. In addition, the number of biomedical applications has been increased to include Gamma Knife radiosurgery, cochlear implants, arthroscopic surgery, heart pacemakers, impedance plethysmography, endoscopy, defibrillators, and photorefractive keratectomy, and others. Each application is identified in the margin with the label **The Physics of . . .**, and those that illustrate a biomedical application are further marked with an icon in the shape of a microscope ⚲. A complete list of the applications can be found on pages xxi–xxiii.

Free-Body Diagrams Teachers are familiar with the importance of free-body diagrams when using Newton's laws of motion. We use free-body diagrams throughout the text, not just in the early chapters where Newton's laws are introduced. For instance, when the relation between pressure and depth in a fluid is developed in Chapter 11, a free-body diagram clarifies the discussion considerably. Free-body diagrams are also used in worked-out examples, as in Example 4 in Chapter 18, when we calculate the electrostatic forces that electric charges exert on each other.

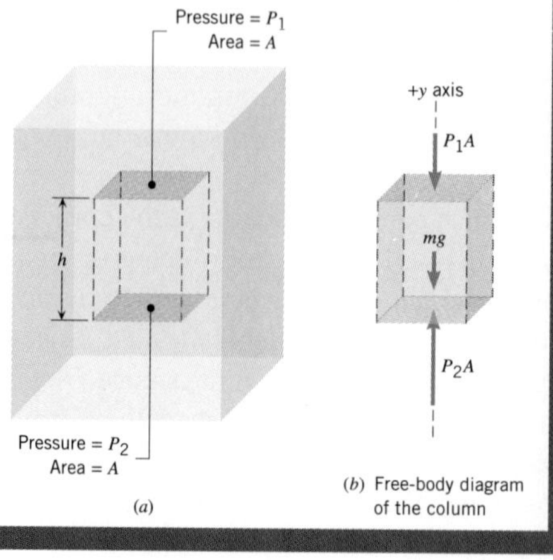

Pressure = P_1
Area = A

+y axis

P_1A

h

mg

P_2A

Pressure = P_2
Area = A

(b) Free-body diagram
of the column

(a)

Figure 11.7 (a) A container of fluid in which one particular column of the fluid is outlined. The fluid is at rest. (b) The free-body diagram, showing the vertical forces acting on the column.

Problem Solving Insights   To reinforce the problem-solving techniques illustrated in the carefully worked-out examples, we have included short statements in the margins, identified by a gold circle and the label **PROBLEM SOLVING INSIGHT**. These insights help students to develop good problem-solving skills by providing the kind of advice that a teacher would give when explaining a calculation in detail.

Homework Problems and Conceptual Questions   The fourth edition contains nearly 2400 problems and 600 conceptual questions for assignment as homework. About 25% of the problems are new or modified. In providing so many problems and questions, we have used a wide variety of real-world situations with realistic data. Building problem-solving skills involves the use of homework problems that progress from relatively easy to moderate to challenging levels of difficulty. In this spirit, we have ranked the homework problems according to difficulty. The most difficult are marked with a double asterisk (**), while those of intermediate difficulty are marked with a single asterisk (*). The easiest are unmarked. Those whose solutions appear in the *Student Solutions Manual* are identified with the label **ssm**. Those whose solutions are available on the World Wide Web (**http://www.wiley.com/college/cutnell**) are marked with the label **www**. Some of the problems are organized by section, whereas others are grouped without reference to any particular section under the heading **Additional Problems.** Problems and conceptual questions that are biomedical in nature are marked with an icon in the shape of a microscope ⚲.

• • •

In spite of our best efforts to produce an error-free book, there are no doubt errors that still remain. They are solely our responsibility, and we would appreciate hearing of any that you find. We hope that this text makes learning and teaching physics easier and more enjoyable, and we look forward to hearing about your experiences with it.

JOHN D. CUTNELL
KENNETH W. JOHNSON
Carbondale, Illinois, 1997

> • **PROBLEM SOLVING INSIGHT**
> When nonconservative forces are perpendicular to the motion, we can still use the principle of conservation of mechanical energy, because such "perpendicular" forces do no work.

SUPPLEMENTS

An extensive package of supplements to accompany *Physics*, 4th edition, is available to assist both the teacher and the student.

Student Study Guide, prepared by John D. Cutnell, Kenneth W. Johnson and Mark J. Comella. The Guide, which is designed to be used in close conjunction with the text, aids students with chapter previews, lists of important terms, discussions and explanations of commonly misunderstood topics, worked-out examples, practice problems, and chapter quizzes. Each chapter also has problems especially designed to help students study for the MCAT exam.

Student Solutions Manual, prepared by John D. Cutnell, Kenneth W. Johnson and Mark J. Comella. This manual contains carefully worked out solutions for approximately 600 of the odd-numbered problems located at the ends of the chapters. In the text, these problems are identified with the label **ssm**.

Instructor's Solutions Manual, Volume 1, chapters 1–17 and Instructor's Solutions Manual Volume 2, Chapters 18–32, for instructors only, prepared by John D. Cutnell, Kenneth W. Johnson and Mark J. Comella. This manual contains detailed solutions to all homework problems in the text. It also contains answers to the conceptual questions that are located at the ends of chapters.

Instructor's Solutions Disk, a computer disk version of the *Instructor's Solutions Manual*, for instructors only, is available in Microsoft Word for Windows and Macintosh.

Instructor's Resource Guide, prepared by David T. Marx of Southern Illinois University at Carbondale. The Guide contains an extensive listing of physics resources on the World Wide Web. It also includes teaching suggestions, lecture notes, demonstration suggestions, alternative syllabi for courses of different lengths and emphasis, strategies for incorporating supplements and materials from other texts, as well as conversion notes allowing the instructor to use class notes from other texts. For instructors who have homework problems from the third edition that they especially like to assign, there is a problem locator guide. This locator guide provides an easy way to correlate third-edition problem numbers with the corresponding fourth-edition numbers.

Test Bank, prepared by David T. Marx of Southern Illinois University at Carbondale. The test bank contains 2006 short-answer questions and problems, which represents about a 15% increase relative to those available in the third edition.

Computerized Test Bank PC and Macintosh versions of the entire *Test Bank* are available with full editing features to help you customize tests.

Homework Disk, for instructors only. Teachers of large classes often use a computer-graded, multiple-choice homework format. As part of the *Computerized Test Bank*, David T. Marx has converted nearly 1200 of the chapter-ending problems into a multiple-choice format, so teachers can generate their homework assignments in a convenient and effective way.

Four-color Transparency Acetates Nearly 300 four-color illustrations from the text have been resized and edited for maximum effectiveness, so that they can easily be projected in the classroom.

CD-ROM Contains the complete text, the *Student Solutions Manual*, the *Student Study Guide*, and *Learning Ware,* all connected with extensive hyperlinking. In addition, *Learning Ware* software guides students through solutions of a wide variety of problems. The solution process is developed interactively, with appropriate feedback and access to error-specific help for the most common mistakes. The *CD-ROM* also contains numerous simulations that allow students to explore various physics-related phenomena by varying parameters and observing the corresponding effects.

Learning Ware Software A separate software package containing the *Learning Ware* component of the *CD-ROM.*

Student's Pocket Companion A pocket-sized book that contains, on a section-by-section basis, a concise summary of all definitions, laws, theorems, reasoning strategies, and concepts. We believe students will find this quick reference extremely handy when doing homework assignments and studying for exams.

ACKNOWLEDGMENTS

It is a great pleasure to acknowledge the help of the team that brought this text to life. We are very fortunate in being associated with outstanding professionals, people who are talented, highly motivated, and take great pride in their work.

We are particularly grateful to our editor, Stuart Johnson. He came to Wiley at the onset of the fourth edition, quickly took the helm, and guided us through all phases of the book's development. We would also like to acknowledge our former editor, Cliff Mills, for the many important contributions that he made to this project over many years.

Barbara Heaney is our developmental editor, and, in a word, she's a gem. We have worked with her on two editions now, and greatly respect and admire her talents. She provided numerous ideas and continuous feedback, and many of the book's features have benefited from her help.

Hilary Newman and Ramón Rivera-Moret are the best photo researchers in the world, period. With a remarkable "sixth" sense, they selected photos that not only convey just the right physics but are beautiful as well. Sit back, relax, and just browse through the photographs (don't forget the cover photo!). We think you too will enjoy their work.

Our sincerest appreciation goes to Sigmund Malinowski and Ishaya Monokoff of the illustration department. All the illustrations were redrawn for this edition, which brought us into constant communication with Sigmund. He is professional, knowledgeable, cooperative and one of the nicest people we've ever met.

We are also indebted to Karin Kincheloe and Maddy Lesure of the design department. The beautiful design of the book and its cover is Karin's work. She united the wide array of features into a single whole that has a wonderful sense of color, balance, and openness.

Our thanks go to Katharine Rubin, Pam Kennedy, and Ann Berlin for coordinating the all-electronic production of the book and helping us achieve a result that adheres to the highest standards of production. It would not have been possible without their help.

Hats off to Lee Goldstein, master juggler. Page by page, she brought together the countless pieces of text, line art, photos, and margin elements to create a book that has a spacious look. We especially appreciate her patience and willingness to try out our many suggestions. It was a genuine pleasure working with Lee.

One of the most important aspects of producing a book is to have outstanding proofreaders. The eyes of Georgia Kamvosoulis Mederer and Betty Pessagno were absolutely invaluable in catching all types of errors, from typos, awkwardly worded sentences, wrong fonts, and, yes, even an incorrect equation or two. We are grateful for their conscientious work.

Catherine Faduska, Executive Marketing Manager, has been a friend of ours and of the book for many years. She's extremely knowledgeable about the marketing of college texts and has had a profound influence on the book's success. For all in the past and the future to come, thank you, Cathy.

Ethan Goodman and Catherine Beckham have also been a source of great help in keeping us abreast of the latest marketing developments. We appreciate their efforts in planning strategies and providing ideas on how to deal with the needs of an ever-changing market.

Our gratitude goes to Cynthia Rhoads for coordinating the extensive supplements package and to Virginia Dunn for copyediting the manuscript.

Jane Doub is a senior sales representative for John Wiley & Sons, Inc. and, fortunately, is the "rep" for our area. She stops by numerous times during the year and gives us valuable user feedback. Because of her great sales and marketing sense, we always take these opportunities to bounce new ideas off her. A true professional and a classy person, Jane is the best "rep" we've ever met.

The sales representatives of John Wiley & Sons, Inc. are in constant contact with physics departments throughout the country. They are a very knowledgeable and hard-working group, and we are most appreciative of their efforts.

To all of the physicists who have reviewed our work, both in this edition and in all previous editions, we owe a special debt. They have helped us to write more clearly and to remove ambiguities and inaccuracies. They have offered many suggestions that have influenced our decisions in producing this text. We have great respect for their work, and to each of them we extend our sincerest thanks. In particular, we thank the reviewers who helped us prepare the fourth edition:

Paul D. Beale, *University of Colorado at Boulder*
Roger Bland, *San Francisco State University*
Neal Cason, *University of Notre Dame*
Thomas Cobb, *Bowling Green State University*
Steven Davis, *University of Arkansas at Little Rock*
Lewis Ford, *Texas A&M University*
James B. Gerhart, *University of Washington*
Barry Gilbert, *Rhode Island College*
Larry Josbeno, *Corning Community College*
Randy Kobes, *University of Winnepeg*
Alfredo Louro, *University of Calgary*
Thomas P. Marvin, *Southern Oregon State College*
Paul Morris, *Abilene Christian University*
Vallabhaneni Rao, *Memorial University of Newfoundland*
R. S. Rubins, *University of Texas at Arlington*
Marc Sher, *College of William & Mary*
Rolf Vatne, *Portland Community College*

THE PHYSICS OF . . .

To show students that physics has a widespread impact on their lives, we have included a large number of applications of physics principles. Many of these applications are not found in other texts. The most important ones are listed below along with the page number locating the corresponding discussion. They are identified in the margin of the page on which they occur with a blue triangle and the title "**The Physics of...**" Biomedical applications are marked with an icon in the shape of a microscope. The discussions are integrated into the text, so that they occur as a natural part of the physics being presented. It should be noted that the list is not a complete list of all the applications of physics principles to be found in the text. There are many additional applications that are discussed only briefly or occur in the homework questions and problems.

TO THE STUDENT

Students like you have helped us prepare this 4th edition of *Physics*. Through focus groups held at several campuses with students using the 3rd edition, they have told us what aspects of the book work well for them and where improvements could be made. They also gave us excellent advice on the new features incorporated into this edition. As a result, the authors and John Wiley & Sons feel that this is the most user-friendly edition of the book that we have published, working equally well for instructors and students. The "Student to Student" letter printed below, for example, gives you advice from a fellow student on how well this book works and how best to study from it.

We invite you to share your comments and suggestions with us for the next edition. Please feel free to write us care of Physics Editor, College Division, John Wiley & Sons, Inc., 605 Third Avenue, New York, NY 10158-0012 or send us an e-mail at cutnell@wiley.com

STUDENT TO STUDENT

Annette Adams is a biology major, class of '98, at the College of William and Mary. She chose to attend the College of William and Mary because it's in such a beautiful part of Virginia, has a great dance program, and has a prestigious pre-medicine program. Her long-term career goal is to become a physician and possibly to pursue a Ph.D in biomedical sciences.

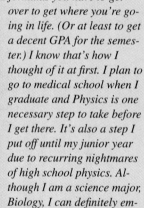

Hi there. *For many of you, Physics probably seems like a giant mountain of equations and formulas you have to get over to get where you're going in life. (Or at least to get a decent GPA for the semester.) I know that's how I thought of it at first. I plan to go to medical school when I graduate and Physics is one necessary step to take before I get there. It's also a step I put off until my junior year due to recurring nightmares of high school physics. Although I am a science major, Biology, I can definitely empathize with those of you who shy away from classes requiring long-drawn-out equations — and calculations. Let's face it, memorizing formulas and cranking out numbers on a calculator isn't too exciting. And that's exactly why I think this book will be one of your greatest assets as you tackle Physics this semester.*

You see, this book makes blatant, bland memorization unnecessary, because it explains the concepts in simple, easy-to-understand terms. It's so much easier to recall a formula when you understand how and why it is used. I found that the best way to use the text was to read through each section and try the Example problems. They have detailed explanations if you need help! Also, pay close attention to the Conceptual Examples and any "Problem Solving Insights" or "Reasoning Strategies." These can be key to understanding something that confused you in the text and/or in lecture. The Concept Charts that have been added to the fourth edition are a really neat idea. I didn't have them when I used the third edition, but I think that they will help you a lot in understanding the overall picture.

Surprisingly, I ended up really liking my Physics class. I was fortunate enough to have both a terrific, easy-to-read textbook AND a great professor. The whole point of a text for nonphysics majors is to demonstrate that physics doesn't apply just to physics — it applies to everything. This text definitely gets that point across. As a life-science major, I guarantee that random facts in chemistry and biology that you've accepted at face value will start to have a rhyme and reason behind them. And all of you, science and nonscience majors alike, will begin to look at the processes in the world around you with a new understanding. Who couldn't benefit from understanding a little bit more about what goes on around them?

Well, if you've been kind enough to humor me and read all the way to the bottom of "my" page, I thank you. I hope you get as much out of your Physics class as I did. Good luck to you!

Annette Adams

INTRODUCTION AND MATHEMATICAL CONCEPTS

The movie *Independence Day* is a tour de force of animation techniques, which rely heavily on computers and mathematical concepts. This chapter introduces some of the mathematical concepts—like trigonometry and vectors—that will be useful throughout this book in dealing with the laws of physics.

1.1 THE NATURE OF PHYSICS

The science of physics has developed out of the efforts of men and women to explain why our physical environment behaves as it does. These efforts have been so successful that the laws of physics now encompass a remarkable variety of phenomena, from planets orbiting the sun to lasers being used in eye surgery.

The laws of physics are equally remarkable for their scope. They describe the behavior of particles many times smaller than an atom and objects many times larger than our sun. The same laws apply to the heat generated by a burning match and the heat generated by a rocket engine. The same laws guide an astronomer in using the light from a distant star to determine how fast the star is moving and a police officer in using radar to catch a speeder. Physics can be applied fruitfully to objects as different as subatomic particles, distant stars, or speeding automobiles because it focuses on issues that are truly basic to the way nature works.

The strength of physics derives from the fact that its laws are based on experiment. This is not to say that intuition and educated guesses are unimportant. The great creative geniuses in science, as in art and music, work in leaps and bounds that no one can fully understand. In physics, however, a flash of insight never becomes accepted law unless its implications can be verified by experiment. This insistence on experimental verification has enabled physicists to build a rational and coherent understanding of nature.

The exciting feature of physics is its capacity for predicting how nature will behave in one situation on the basis of experimental data obtained in another situation. Such predictions place physics at the heart of modern technology and, therefore, can have a tremendous impact on our lives. Rocketry and the development of space travel have their roots firmly planted in the physical laws of Galileo Galilei (1564–1642) and Isaac Newton (1642–1727). The transportation industry relies heavily on physics in the development of engines and the design of aerodynamic vehicles. Entire electronics and computer industries owe their existence to the invention of the transistor, which grew directly out of the laws of physics that describe the electrical behavior of solids. The telecommunications industry depends extensively on electromagnetic waves, whose existence was predicted by James Clerk Maxwell (1831–1879) in his theory of electricity and magnetism. The medical profession uses X-ray, ultrasonic, and magnetic resonance methods for obtaining images of the interior of the human body, and physics lies at the core of all these. Perhaps the most widespread impact in modern technology is that due to the laser. Fields ranging from space exploration to medicine benefit from this incredible device, which is a direct application of the principles of atomic physics.

Because physics is so fundamental, it is a required course for students in a wide range of major areas. We welcome you to the study of this fascinating topic. You will learn how to see the world through the "eyes" of physics and to reason as a physicist does. In the process, you will learn how to apply physics principles to a wide range of problems. We hope that you will come to recognize that physics has important things to say about your environment.

1.2 UNITS

DEFINITION OF STANDARD UNITS

Physics experiments involve the measurement of a variety of quantities, and a great deal of effort goes into making these measurements as accurate and reproducible as possible. The first step toward ensuring accuracy and reproducibility is defining the units in which the measurements are made.

In this text, we will stress the system of units known according to the French phrase "Le Système International d'Unités," referred to simply as *SI units.* This system, by international agreement, employs the *meter* (m) as the unit of length, the *kilogram* (kg) as the unit of mass, and the *second* (s) as the unit of time. Two other systems of units are worth mentioning. The CGS system utilizes the centimeter (cm), the gram (g), and the second for length, mass, and time, respectively, whereas the BE or British Engineering system (the gravitational version) uses the foot (ft), the slug (sl), and the second. Table 1.1 summarizes the units used for length, mass, and time in the three systems.

Originally, the meter as a unit of length was defined in terms of the distance measured along the earth's surface between the north pole and the equator. Eventually, a more accurate measurement standard was needed, and by international agreement the meter became the distance between two marks on a bar of platinum-iridium alloy (see Figure 1.1) kept at a temperature of 0 °C. Today, to meet further demands for increased accuracy, the meter is defined as the distance that light travels in a vacuum in a time of 1/299 792 458 second. This definition arises because the speed of light is a universal constant that is defined to be 299 792 458 m/s.

The definition of a kilogram as a unit of mass has also undergone changes over the years. As Chapter 4 discusses, the mass of an object indicates the tendency of the object to continue in motion with a constant velocity. Originally, the kilogram was expressed in terms of a specific amount of water. Today, one kilogram is defined to be the mass of a standard cylinder of platinum–iridium alloy, like the one in Figure 1.2.

As with the units for length and mass, the present definition of the second as a unit of time is different from the original definition. Originally, the second was defined according to the average time for the earth to rotate once about its axis, one day being set equal to 86 400 seconds. The earth's rotational motion was chosen because it is naturally repetitive, occurring over and over again. Today, we still use a naturally occurring repetitive phenomenon to define the second, but of a very different kind. We use the electromagnetic waves emitted by cesium-133 atoms in an

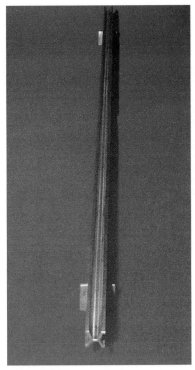

Figure 1.1 The standard platinum-iridium meter bar.

Figure 1.2 The standard platinum-iridium kilogram is kept at the International Bureau of Weights and Measures in Sèvres, France.

Table 1.1 Units of Measurement

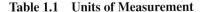

	System		
	SI	CGS	BE
Length	meter (m)	centimeter (cm)	foot (ft)
Mass	kilogram (kg)	gram (g)	slug (sl)
Time	second (s)	second (s)	second (s)

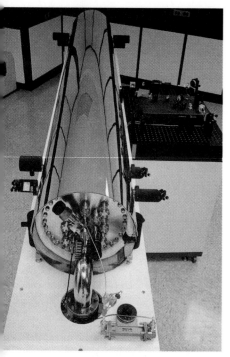

Figure 1.3   A cesium atomic clock.

Table 1.2   Standard Prefixes Used to Denote Multiples of Ten

Prefix	Symbol	Factor[a]
Tera	T	10^{12}
Giga[b]	G	10^{9}
Mega	M	10^{6}
Kilo	k	10^{3}
Hecto	h	10^{2}
Deka	da	10^{1}
Deci	d	10^{-1}
Centi	c	10^{-2}
Milli	m	10^{-3}
Micro	μ	10^{-6}
Nano	n	10^{-9}
Pico	p	10^{-12}
Femto	f	10^{-15}

[a] Appendix A contains a discussion of powers of ten and scientific notation.
[b] Pronounced jig′a.

atomic clock like that in Figure 1.3. One second is defined as the time needed for 9 192 631 770 wave cycles to occur.*

BASE UNITS AND DERIVED UNITS

The units for length, mass, and time, along with a few other units that will arise later, are regarded as ***base*** SI units. The word "base" refers to the fact that these units are used along with various laws to define additional units for other important physical quantities, such as force and energy. The units for these other physical quantities are referred to as ***derived*** units, since they are combinations of the base units. Derived units will be introduced as they arise naturally along with the related physical laws.

The value of a quantity in terms of base or derived units is sometimes a very large or very small number. In such cases, it is convenient to introduce larger or smaller units that are related to the normal units by multiples of ten. Table 1.2 summarizes the prefixes that are used to denote multiples of ten. For example, 1000 or 10^3 meters are referred to as 1 kilometer (km), and 0.001 or 10^{-3} meter is called 1 millimeter (mm). Similarly, 1000 grams and 0.001 gram are referred to as 1 kilogram (kg) and 1 milligram (mg), respectively. Appendix A contains a discussion of scientific notation and powers of ten, such as 10^3 and 10^{-3}.

1.3   THE ROLE OF UNITS IN PROBLEM SOLVING

THE CONVERSION OF UNITS

Since any quantity, such as length, can be measured in several different units, it is important to know how to convert from one unit to another. For instance, the foot can be used to express the distance between the two marks on the standard platinum-iridium meter bar. There are 3.281 feet in one meter, and this number can be used to convert from meters to feet, as the following example demonstrates.

EXAMPLE 1 • The World's Highest Waterfall

The highest waterfall in the world is Angel Falls in Venezuela, with a total drop of 979.0 m (see Figure 1.4). Express this drop in feet.

Reasoning   When converting between units, we write down the units explicitly in the calculations and treat them like any algebraic quantity. In particular, we will take advantage of the following algebraic fact: Multiplying or dividing an equation by a factor of 1 does not alter the equation.

Solution   Since 3.281 feet = 1 meter, it follows that (3.281 feet)/(1 meter) = 1. Using this factor of 1 to multiply the equation "Length = 979.0 meters," we find that

$$\text{Length} = (979.0 \text{ meters})(1) = (979.0 \text{ meters})\left(\frac{3.281 \text{ feet}}{1 \text{ meter}}\right) = \boxed{3212 \text{ feet}}$$

The colored lines emphasize that the units of meters behave like any algebraic quantity and cancel when the multiplication is performed, leaving only the desired unit of feet to describe the answer. In this regard, note that 3.281 feet = 1 meter also implies that (1 meter)/(3.281 feet) = 1. However, we chose not to multiply by a factor of 1 in this form, because the units of meters would not have canceled out.

---

* See Chapter 16 for a discussion of waves in general and Chapter 24 for a discussion of electromagnetic waves in particular.

A calculator gives the answer as 3212.099 feet. Standard procedures for significant figures, however, indicate that the answer should be rounded off to four significant figures, since the value of 979.0 meters is accurate to only four significant figures. In this regard, the "1 meter" in the denominator does not limit the significant figures of the answer, because this number is precisely one meter by definition of the conversion factor. Appendix B contains a review of significant figures.

In any conversion, if the units do not combine algebraically to give the desired result, the conversion has not been carried out properly. The next example also stresses the importance of writing down the units and illustrates a typical situation in which several conversions are required.

EXAMPLE 2 • Interstate Speed Limit

Express the speed limit of 65 miles/hour in terms of meters/second.

Reasoning As in Example 1, it is important to write down the units explicitly in the calculations and treat them like any algebraic quantity. Here, two well-known relationships come into play, namely, 5280 feet = 1 mile and 3600 seconds = 1 hour. As a result, (5280 feet)/(1 mile) = 1 and (3600 seconds)/(1 hour) = 1. Multiplying and dividing by these factors of unity does not alter an equation, a fact that will aid us in the conversions.

Solution By multiplying and dividing by factors of unity, we can find the speed limit in feet per second as shown below:

$$\text{Speed} = \left(65\,\frac{\text{miles}}{\text{hour}}\right)(1)(1) = \left(65\,\frac{\cancel{\text{miles}}}{\cancel{\text{hour}}}\right)\left(\frac{5280\text{ feet}}{1\,\cancel{\text{mile}}}\right)\left(\frac{1\,\cancel{\text{hour}}}{3600\text{ s}}\right) = 95\,\frac{\text{feet}}{\text{second}}$$

To convert feet into meters, we use the fact that (1 meter)/(3.281 feet) = 1:

$$\text{Speed} = \left(95\,\frac{\text{feet}}{\text{second}}\right)(1) = \left(95\,\frac{\cancel{\text{feet}}}{\text{second}}\right)\left(\frac{1\text{ meter}}{3.281\,\cancel{\text{feet}}}\right) = \boxed{29\,\frac{\text{meters}}{\text{second}}}$$

Figure 1.4 Angel Falls in Venezuela is the highest waterfall in the world.

A collection of useful conversion factors is given on the page facing the inside of the front cover. The reasoning strategy that we have followed in Examples 1 and 2 for converting between units is outlined as follows:

REASONING STRATEGY

Converting Between Units
1. In all calculations, write down the units explicitly.

2. Treat all units as algebraic quantities. In particular, when identical units are divided, they are eliminated algebraically.

3. Use the conversion factors located on the page facing the inside of the front cover. In your calculations, be guided by the fact that multiplying or dividing an equation by a factor of 1 does not alter the equation. For instance, the conversion factor of 3.281 feet = 1 meter might be applied in the form (3.281 feet)/(1 meter) = 1. This factor of 1 would be used to multiply an equation such as "Length = 5.00 meters" in order to convert meters to feet.

4. Check to see that your calculations are correct by verifying that the units combine algebraically to give the desired unit for the answer.

UNITS AS A PROBLEM-SOLVING AID

In addition to their role in guiding the use of conversion factors, units serve a useful purpose in solving problems. They can provide an internal check to eliminate certain kinds of errors, if they are carried along during each step of a calculation and treated like any algebraic factor.

Suppose, for instance, that the tank of a car contains 2.0 gallons of gas to start with and that gas is added at a rate of 7.0 gallons/minute. The total amount of gas in the tank 96 seconds later can be obtained by adding the amount put into the tank to the amount present initially. The amount put in can be calculated by multiplying the filling rate by the time the gas pump is on. But a lack of attention to the units in the calculation can lead to an erroneous result, as the following example shows.

$$\begin{aligned}\text{Total amount} \atop \text{of gas} &= {\text{Gas initially} \atop \text{present}} + {\text{Gas} \atop \text{added}} \\[2mm] &= 2.0 \text{ gallons} + \left(7.0 \frac{\text{gallons}}{\text{minute}}\right)(96 \text{ seconds}) \\[2mm] &= 2.0 \text{ gallons} + 672 \frac{\text{gallons} \cdot \text{seconds}}{\text{minute}}\end{aligned}$$

The answer cannot be $2.0 + 672 = 674$, because the units for the two added terms are not the same. ***Only quantities that have exactly the same units can be added (or subtracted).*** With the filling rate expressed as 7.0 gallons/minute, the correct answer can be obtained only if the time of 96 seconds is converted into minutes:

$$\text{Time} = (96 \ \cancel{\text{seconds}})\left(\frac{1 \text{ minute}}{60 \ \cancel{\text{seconds}}}\right) = 1.6 \text{ minutes}$$

$$\begin{aligned}\text{Total amount} \atop \text{of gas} &= 2.0 \text{ gallons} + \left(7.0 \frac{\text{gallons}}{\cancel{\text{minute}}}\right)(1.6 \ \cancel{\text{minutes}}) \\[2mm] &= 2.0 \text{ gallons} + 11 \text{ gallons} = 13 \text{ gallons}\end{aligned}$$

As indicated by the colored lines, the units of time now cancel algebraically when the multiplication is carried out, leaving only the desired unit of gallons. The procedure of "carrying along the units" serves as an automatic reminder to convert all data used in a calculation into a consistent set of units.

DIMENSIONAL ANALYSIS

We have seen that many quantities are denoted by specifying both a number and a unit. For example, the distance to the nearest telephone may be 8 meters, or the speed of a car might be 25 meters/second. Each quantity, according to its physical nature, requires a certain *type* of unit. Distance must be measured in a length unit such as meters, feet, or miles, and a time unit will not do. Likewise, the speed of an object must be specified as a length unit divided by a time unit. In physics, the term ***dimension*** is used to refer to the physical nature of a quantity and the type of unit used to specify it. Distance has the dimension of length (symbolized as [L]), while speed has the dimensions of length [L] divided by time [T], or [L/T]. Many physical quantities can be expressed in terms of a combination of fundamental dimensions such as length [L], time [T], and mass [M]. Later on, we will encounter certain other quantities, such as temperature, which are also fundamental. A fundamental quantity like temperature cannot be expressed as a combination of the dimensions of length, time, mass or any other fundamental dimension.

This scientist is using an automatic pipette to deliver a fixed volume of liquid into a sample cell.

Dimensional analysis is used to check mathematical relations for the consistency of their dimensions. As an illustration, consider a car that starts from rest and accelerates to a speed v in a time t. Suppose we wish to calculate the distance x traveled by the car, but are not sure whether the correct relation is $x = \frac{1}{2}vt^2$ or $x = \frac{1}{2}vt$. We can decide by checking the quantities on both sides of the equals sign to see if they have the same dimensions. If the dimensions are not the same, the relation is incorrect. For $x = \frac{1}{2}vt^2$, we write the dimensions as follows, using the dimensions for distance [L], time [T], and speed [L/T]:

$$x = \tfrac{1}{2}vt^2$$

Dimensions: $\qquad [L] \overset{?}{=} \left[\dfrac{L}{\cancel{T}}\right][T]^{\cancel{2}} = [L][T]$

Dimensions cancel just like algebraic quantities, and pure numerical factors like $\frac{1}{2}$ have no dimensions, so they can be ignored. The dimension on the left of the equals sign does not match those on the right, so the relation $x = \frac{1}{2}vt^2$ cannot be correct. On the other hand, applying dimensional analysis to $x = \frac{1}{2}vt$, we find that

$$x = \tfrac{1}{2}vt$$

Dimensions: $\qquad [L] \overset{?}{=} \left[\dfrac{L}{\cancel{T}}\right][\cancel{T}] = [L]$

The dimension on the left of the equals sign matches that on the right, so this relation is dimensionally correct. If we know that one of our two choices is the right one, then $x = \frac{1}{2}vt$ is it. In the absence of such knowledge, however, dimensional analysis cannot identify the correct relation. It can only identify which choices *may be* correct, since it does not account for numerical factors like $\frac{1}{2}$ or for the manner in which an equation was derived from physics principles.

• PROBLEM SOLVING INSIGHT
In problems that involve algebraic manipulations, you can check for errors that may have arisen during the manipulations by doing a dimensional analysis on the final expression.

1.4 TRIGONOMETRY

Scientists use mathematics to help them describe how the physical universe works, and trigonometry is an important branch of mathematics. Three trigonometric functions are utilized throughout this text. They are the sine, the cosine, and the tangent of the angle θ (Greek theta), abbreviated as $\sin\theta$, $\cos\theta$, and $\tan\theta$, respectively. These functions are defined below in terms of the symbols given along with the right triangle in Figure 1.5.

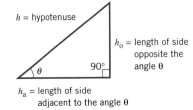

h = hypotenuse
h_o = length of side opposite the angle θ
h_a = length of side adjacent to the angle θ

Figure 1.5 A right triangle.

■ **DEFINITION OF SIN θ, COS θ, AND TAN θ**

$$\sin\theta = \frac{h_o}{h} \tag{1.1}$$

$$\cos\theta = \frac{h_a}{h} \tag{1.2}$$

$$\tan\theta = \frac{h_o}{h_a} \tag{1.3}$$

h = length of the **hypotenuse** of a right triangle
h_o = length of the side **opposite** the angle θ
h_a = length of the side **adjacent** to the angle θ

The sine, cosine, and tangent of an angle are numbers without units, because each is expressed as the ratio of the lengths of two sides of a right triangle. Example 3 illustrates a typical application of Equation 1.3.

EXAMPLE 3 • Using Trigonometric Functions

On a sunny day, a tall building casts a shadow that is 67.2 m long. The angle between the sun's rays and the ground is $\theta = 50.0°$, as Figure 1.6 shows. Determine the height of the building.

Reasoning Since we want to find the height of the building, we begin by identifying the height as the length h_o of the side opposite the angle θ in the colored right triangle in Figure 1.6. The length of the shadow is the length h_a of the side that is adjacent to the angle θ. The ratio of the length of the opposite side to the length of the adjacent side is the tangent of the angle θ, which can be used to find the height of the building.

Solution We use the tangent function in the following way, with $\theta = 50.0°$ and $h_a = 67.2$ m:

$$\tan \theta = \frac{h_o}{h_a} \qquad (1.3)$$

$$h_o = h_a \tan \theta = (67.2 \text{ m})(\tan 50.0°) = (67.2 \text{ m})(1.19) = \boxed{80.0 \text{ m}}$$

The value of $\tan 50.0°$ is found by using a calculator.

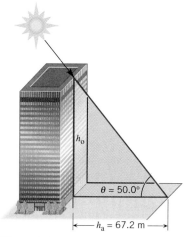

Figure 1.6 From a value for the angle θ and the length h_a of the shadow, the height h_o of the building can be found using trigonometry.

The sine, cosine, or tangent may be used in calculations such as that in Example 3, depending on which side of the triangle has a known value and which side is asked for. However, ***the choice of which side of the triangle to label h_o (opposite) and which to label h_a (adjacent) can be made only after the angle θ is identified.***

Often the values for two sides of the right triangle in Figure 1.5 are available, and the value of the angle θ is unknown. The concept of ***inverse trigonometric functions*** plays an important role in such situations. Equations 1.4–1.6 give the inverse sine, inverse cosine, and inverse tangent in terms of the symbols used in the drawing. For instance, Equation 1.4 is read as "θ equals the angle whose sine is h_o/h."

$$\theta = \sin^{-1}\left(\frac{h_o}{h}\right) \qquad (1.4)$$

$$\theta = \cos^{-1}\left(\frac{h_a}{h}\right) \qquad (1.5)$$

$$\theta = \tan^{-1}\left(\frac{h_o}{h_a}\right) \qquad (1.6)$$

The use of "-1" as an exponent in Equations 1.4–1.6 *does not mean* "take the reciprocal." For instance, $\tan^{-1}(h_o/h_a)$ does not equal $1/\tan(h_o/h_a)$. Another way to express the inverse trigonometric functions is to use arc sin, arc cos, and arc tan instead of $\sin^{-1}$, $\cos^{-1}$, and $\tan^{-1}$. Example 4 illustrates the use of an inverse trigonometric function.

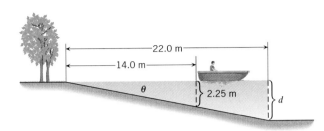

Figure 1.7 If the distance from the shore and the depth of the water at any one point are known, the angle θ can be found with the aid of trigonometry. Knowing the value of θ is useful, because then the depth d at another point can be determined.

EXAMPLE 4 • Using Inverse Trigonometric Functions

A lakefront drops off gradually at an angle θ, as Figure 1.7 indicates. For safety reasons, it is necessary to know how deep the lake is at various distances from the shore. To provide some information about the depth, a lifeguard rows straight out from the shore a distance of 14.0 m and drops a weighted fishing line. By measuring the length of the line, the lifeguard determines the depth to be 2.25 m. (a) What is the value of θ? (b) What would be the depth d of the lake at a distance of 22.0 m from the shore?

Reasoning Near the shore, the lengths of the opposite and adjacent sides of the right triangle in Figure 1.7 are $h_o = 2.25$ m and $h_a = 14.0$ m, relative to the angle θ. Having made this identification, we can use $\tan \theta = h_o/h_a$ to find the angle in part (a). For part (b) the procedure is similar, only farther from the shore the lengths of the opposite and adjacent sides become $h_o = d$ and $h_a = 22.0$ m. With the value for θ obtained in part (a), the tangent function can be used to find the unknown depth. Considering the way in which the lake bottom drops off in Figure 1.7, we expect the unknown depth to be greater than the value of 2.25 m that applies nearer the shore.

Solution

(a) Using Equation 1.3, we find that

$$\tan \theta = \frac{h_o}{h_a} = \frac{2.25 \text{ m}}{14.0 \text{ m}} = 0.161$$

Now that the value of $\tan \theta$ is known, the angle θ can be obtained by using the inverse tangent:

$$\theta = \tan^{-1}(0.161) = \boxed{9.15°}$$

(b) With $\theta = 9.15°$, the tangent function can be used to find the unknown depth farther from the shore, where $h_o = d$ and $h_a = 22.0$ m. Since $\tan \theta = h_o/h_a$, it follows that

$$h_o = h_a \tan \theta$$
$$d = (22.0 \text{ m})(\tan 9.15°) = \boxed{3.54 \text{ m}}$$

which is greater than 2.25 m, as expected.

The right triangle in Figure 1.5 provides the basis for defining the various trigonometric functions according to Equations 1.1–1.3. These functions always involve an angle and two sides of the triangle. There is also a relationship among the lengths of the three sides of a right triangle. This relationship is known as the **Pythagorean theorem** and is used often in this text.

■ **PYTHAGOREAN THEOREM**

The square of the length of the hypotenuse of a right triangle is equal to the sum of the squares of the lengths of the other two sides:

$$h^2 = h_o{}^2 + h_a{}^2 \qquad (1.7)$$

1.5 THE NATURE OF PHYSICAL QUANTITIES: SCALARS AND VECTORS

SCALARS

The volume of water in a swimming pool might be 50 cubic meters, or the winning time of a race could be 11.3 seconds. In cases like these, only the size of the numbers matters. In other words, *how much* volume or time is there? The "50" specifies the amount of water in units of cubic meters, while the "11.3" specifies the amount of time in seconds. Volume and time are examples of scalar quantities. A *scalar quantity* is one that can be described by a single number (including any units) giving its size or magnitude. Some other common scalars are temperature (e.g., 20 °C) and mass (e.g., 85 kg).

VECTORS

While many quantities in physics are scalars, there are also many that are not scalars, quantities for which magnitude tells only part of the story. Consider Figure 1.8, which depicts a car that has moved 2 km along a straight line from start to finish. When describing the motion, it is incomplete to say that "the car moved a distance of 2 km." This statement would indicate only that the car ends up somewhere on a circle whose center is at the starting point and whose radius is 2 km. A complete description must include the direction along with the distance, as in the statement "the car moved a distance of 2 km in a direction 30° north of east." A quantity that deals inherently with both magnitude and direction is called a *vector quantity.* Because direction is an important characteristic of vectors, arrows are used to represent them; *the direction of the arrow gives the direction of the vector.* The colored arrow in Figure 1.8, for example, is called the displacement vector, because it shows how the car is displaced from its starting point. Chapter 2 discusses this particular vector.

The length of the arrow in Figure 1.8 represents the magnitude of the displacement vector. If the car had moved 4 km instead of 2 km from the starting point, the arrow would have been drawn twice as long. *By convention, the length of a vector arrow is proportional to the magnitude of the vector.*

The practice of using the length of an arrow to represent the magnitude of a vector applies to any kind of vector. And in physics there are many important vectors, in addition to the displacement vector. All forces, for instance, are vectors. In common usage a force is a push or a pull, and the direction in which a force acts is just as important as the strength or magnitude of the force. The magnitude of a force is measured in SI units called newtons (N). An arrow representing a force of 20 newtons is drawn twice as long as one representing a force of 10 newtons.

The fundamental distinction between scalars and vectors is the characteristic of direction. Vectors have it, and scalars do not. Conceptual Example 5 helps to clarify this distinction and explains what is meant by the "direction" of a vector.

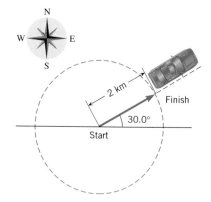

Figure 1.8 A vector quantity has a magnitude and a direction. The arrow in this drawing represents a displacement vector.

CONCEPTUAL EXAMPLE 5 • Vectors, Scalars, and the Role of Plus and Minus Signs

There are places where the temperature is +20 °C at one time of the year and −20 °C at another time. Do the plus and minus signs that signify positive and negative temperatures imply that temperature is a vector quantity?

Reasoning and Solution A vector has a physical direction associated with it, due east or due west, for example. The question, then, is whether such a direction is associated with temperature. In particular, do the plus and minus signs that go along with temperature imply this kind of direction? On a thermometer, the algebraic signs simply mean that the temperature is a number less than or greater than zero on the scale and have nothing to do with east, west, or any other physical direction. Temperature, then, is not a vector. It is a scalar, and scalars can sometimes be negative. *The fact that a quantity is positive or negative does not necessarily mean that the quantity is a scalar or a vector.*

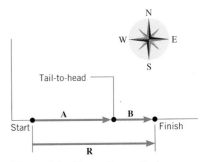

The velocity of this cyclist is another example of a vector quantity.

SYMBOLS USED FOR SCALARS AND VECTORS

Often, for the sake of convenience, quantities such as volume, time, displacement, and force are represented by symbols. This text follows the usual practice of writing vectors in boldface symbols* **(this is boldface)** and writing scalars in italic symbols (*this is italic*). Thus, a displacement vector is written as "**A** = 750 m, due east," where the **A** is a boldface symbol. By itself, however, separated from the direction, the magnitude of this vector is a scalar quantity. Therefore, the magnitude is written as "*A* = 750 m," where the *A* is an italic symbol.

1.6 VECTOR ADDITION AND SUBTRACTION

ADDITION OF COLINEAR VECTORS

Often it is necessary to add one vector to another, and the process of addition must take into account both the magnitude and the direction of the vectors. The simplest situation occurs when the vectors point along the same direction, that is, when they are colinear, as in Figure 1.9. Here, a car first moves along a straight line, with a displacement vector **A** of 275 m, due east. Then, the car moves again in the same direction, with a displacement vector **B** of 125 m, due east. These two vectors add to give the total displacement vector **R**, which would apply if the car had moved from start to finish in one step. The symbol **R** is used because the total vector is often called the *resultant vector.* With the tail of the second arrow located at the head of the first arrow, the two lengths simply add to give the length of the total displacement. This kind of vector addition is identical to the familiar addition of two scalar numbers (2 + 3 = 5), *and can be carried out here only because the vectors point along the same direction.* In such cases we add the individual magnitudes to get the magnitude of the total, knowing in advance what the direction must be. Formally, the addition is written as follows:

$$\mathbf{R} = \mathbf{A} + \mathbf{B}$$
$$\mathbf{R} = 275 \text{ m, due east} + 125 \text{ m, due east}$$
$$= 400 \text{ m, due east}$$

Figure 1.9 Two colinear displacement vectors **A** and **B** add to give the resultant displacement vector **R**.

---

* A vector quantity can also be represented without boldface symbols, by including an arrow above the symbol, e.g., $\vec{A}$.

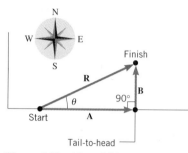

Figure 1.10 The addition of two perpendicular displacement vectors **A** and **B** gives the resultant vector **R**.

ADDITION OF PERPENDICULAR VECTORS

Perpendicular vectors are frequently encountered, and Figure 1.10 indicates how they can be added. This figure applies to a car that first travels with a displacement vector **A** of 275 m, due east, and then with a displacement vector **B** of 125 m, due north. The two vectors add to give a resultant displacement vector **R**. Once again, the vectors to be added are arranged in a tail-to-head fashion, and the resultant vector points from the tail of the first to the head of the last vector added. The resultant displacement is given by the vector equation

$$\mathbf{R} = \mathbf{A} + \mathbf{B}$$

The addition in this equation cannot be carried out by writing $R = 275$ m $+ 125$ m, because the vectors have different directions. Instead, we take advantage of the fact that the triangle in Figure 1.10 is a right triangle and use the Pythagorean theorem (Equation 1.7). According to this theorem, the magnitude of **R** is

$$R = \sqrt{(275 \text{ m})^2 + (125 \text{ m})^2} = 302 \text{ m}$$

The angle θ in Figure 1.10 gives the direction of the resultant vector. Since the lengths of all three sides of the right triangle are now known, either $\sin \theta$, $\cos \theta$, or $\tan \theta$ can be used to determine θ:

$$\tan \theta = \frac{B}{A} = \frac{125 \text{ m}}{275 \text{ m}} = 0.455$$

$$\theta = \tan^{-1}(0.455) = 24.5°$$

Thus, the resultant displacement of the car has a magnitude of 302 m and points north of east at an angle of 24.5°. This displacement would bring the car from the start to the finish in Figure 1.10 in a single straight-line step.

ADDITION OF VECTORS THAT ARE NEITHER COLINEAR NOR PERPENDICULAR

When two vectors to be added are not perpendicular, the tail-to-head arrangement does not lead to a right triangle, and the Pythagorean theorem cannot be used. Figure 1.11a illustrates such a case for a car that moves with a displacement **A** of 275 m, due east and then with a displacement **B** of 125 m in a direction 55.0° north of west. As usual, the resultant displacement vector **R** is directed from the tail of the first to the head of the last vector added. The vector addition is still given according to

$$\mathbf{R} = \mathbf{A} + \mathbf{B}$$

However, since the triangle in the drawing is not a right triangle, some means other than the Pythagorean theorem must be used to find the magnitude and direction of the resultant vector.

One approach uses a graphical technique. In this method, a diagram is constructed in which the arrows are drawn tail to head. The lengths of the vector arrows are drawn to scale, and the angles are drawn accurately (with a protractor, perhaps). Then, the length of the arrow representing the resultant vector is measured with a ruler. This length is converted into the magnitude of the resultant vector by using the scale factor with which the drawing is constructed. In Figure 1.11b, for example, a scale of one centimeter of arrow length for each 10.0 m of displacement is used, and it can be seen that the length of the arrow representing **R** is 22.8 cm. Since each centimeter corresponds to 10.0 m of displacement, the magnitude of **R**

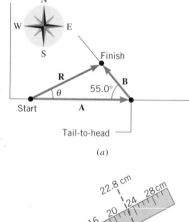

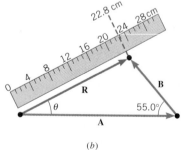

Figure 1.11 (a) The two displacement vectors **A** and **B** are neither colinear nor perpendicular but add to give the resultant vector **R**. (b) In one method for adding them together, a graphical technique is used.

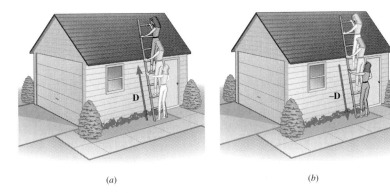

(a)   (b)

Figure 1.12 (*a*) The displacement vector for a woman climbing 1.2 m up a ladder is **D**. (*b*) The displacement vector for a woman climbing 1.2 m down a ladder is −**D**.

is 228 m. The angle θ, which gives the direction of **R**, can be measured with a protractor to be $\theta = 26.7°$.

SUBTRACTION OF VECTORS

The subtraction of one vector from another is carried out in a way that depends on the following fact. *When a vector is multiplied by −1, the magnitude of the vector remains the same, but the direction of the vector is reversed.* Conceptual Example 6 illustrates the meaning of this statement.

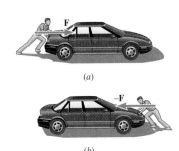

Figure 1.13 (*a*) The force vector for a man pushing on a car with 450 N of force in a direction due east is **F**. (*b*) The force vector for a man pushing on a car with 450 N of force in a direction due west is −**F**.

CONCEPTUAL EXAMPLE 6 • Multiplying a Vector by −1

Consider the two vectors described below:

1. A woman climbs 1.2 m up a ladder, so that her displacement vector **D** is 1.2 m, upward along the ladder, as in Figure 1.12*a*.

2. A man is pushing with 450 N of force on his stalled car, trying to move it eastward. The force vector **F** that he applies to the car is 450 N, due east, as in Figure 1.13*a*.

What are the physical meanings of the vectors −**D** and −**F**?

Reasoning and Solution A displacement vector of −**D** is (−1)**D** and has the same magnitude as the vector **D**, but is opposite in direction. Thus, −**D** would represent the displacement of a woman climbing 1.2 m down the ladder, as in Figure 1.12*b*. Similarly, a force vector of −**F** has the same magnitude as the vector **F**, but has the opposite direction. As a result, −**F** would represent a force of 450 N applied to the car in a direction of due west instead of due east, as in Figure 1.13*b*.

Related Homework Material: Question 15, Problem 66

In practice, vector subtraction is carried out exactly as vector addition, except that one of the vectors added is multiplied by a scalar factor of −1. To see why, look in Figure 1.14*a* at the two vectors **A** and **B**. These vectors add together to give a third vector **C**, according to **C** = **A** + **B**. Therefore, we can calculate vector **A** as **A** = **C** − **B**, which is an example of vector subtraction. However, we can also write this result as **A** = **C** + (−**B**) and treat it as vector addition. Figure 1.14*b* shows how to calculate vector **A** by adding the vectors **C** and −**B**. Notice that vectors **C** and −**B** are arranged tail to head and that any suitable method of vector addition can be employed to determine **A**.

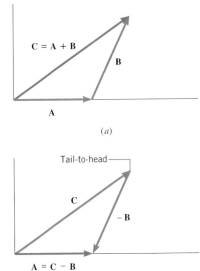

Figure 1.14 (*a*) Vector addition according to **C** = **A** + **B**. (*b*) Vector subtraction according to **A** = **C** − **B** = **C** + (−**B**).

1.7 THE COMPONENTS OF A VECTOR

VECTOR COMPONENTS

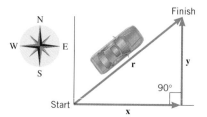

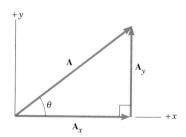

Figure 1.15   The displacement vector **r** and its vector components **x** and **y**.

Suppose a car moves along a straight line from start to finish in Figure 1.15, the corresponding displacement vector being **r**. The magnitude and direction of the vector **r** give the distance and direction traveled along the straight line. However, the car could also arrive at the finish point by first moving due east, turning through 90°, and then moving due north. This alternative path is shown in red in the drawing and is associated with the two displacement vectors **x** and **y**. The vectors **x** and **y** are called the *x* vector component and the *y* vector component of **r**.

Vector components are very important in physics, and two basic features of them are apparent in Figure 1.15. One is that the components add together to equal the original vector, as expressed by the following vector equation:

$$\mathbf{r} = \mathbf{x} + \mathbf{y}$$

The components **x** and **y**, when added vectorially, convey exactly the same meaning as does the original vector **r**, that is, they indicate how the finish point is displaced relative to the starting point. In general, *the components of any vector can be used in place of the vector itself in any calculation where it is convenient to do so.* The other feature of vector components that is apparent in Figure 1.15 is that **x** and **y** are not just any two vectors that add together to give the original vector **r**; they are perpendicular vectors.* This perpendicularity is a valuable characteristic, as we will soon see.

Any type of vector may be expressed in terms of its components, in a way similar to that illustrated for the displacement vector in Figure 1.15. Figure 1.16 shows an arbitrary vector **A** and its vector components $\mathbf{A}_x$ and $\mathbf{A}_y$. The components are drawn parallel to convenient *x* and *y* axes and are perpendicular. They add vectorially to equal the original vector **A**:

$$\mathbf{A} = \mathbf{A}_x + \mathbf{A}_y$$

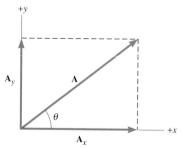

Figure 1.16   An arbitrary vector **A** and its vector components $\mathbf{A}_x$ and $\mathbf{A}_y$.

There are times when a drawing such as Figure 1.16 is not the most convenient way to represent vector components, and Figure 1.17 presents an alternative method. The disadvantage of this alternative is that the tail-to-head arrangement of $\mathbf{A}_x$ and $\mathbf{A}_y$ is missing, an arrangement that is a nice reminder that $\mathbf{A}_x$ and $\mathbf{A}_y$ add together to equal **A**.

The definition given below summarizes the meaning of vector components:

■ **DEFINITION OF VECTOR COMPONENTS**

In two dimensions, the vector components of a vector **A** are two perpendicular vectors $\mathbf{A}_x$ and $\mathbf{A}_y$ that are parallel to the *x* and *y* axes, respectively, and add together vectorially so that $\mathbf{A} = \mathbf{A}_x + \mathbf{A}_y$.

Figure 1.17   This alternative way of drawing the vector **A** and its vector components is completely equivalent to that shown in Figure 1.16.

The values calculated for vector components depend on the orientation of the vector relative to the axes used as a reference. Figure 1.18 illustrates this fact for a vector **A**, by showing two sets of axes, one set being rotated clockwise relative to the other. With respect to the black axes, vector **A** has perpendicular vector components $\mathbf{A}_x$ and $\mathbf{A}_y$; with respect to the colored rotated axes, vector **A** has different vector

---

* It is possible to introduce vector components that are not perpendicular, but, in general, they are not as useful as those introduced here.

components $\mathbf{A}_x'$ and $\mathbf{A}_y'$. The choice of which set of components to use is purely a matter of convenience.

SCALAR COMPONENTS

It is often easier to work with the *scalar components,* A_x and A_y (note the italic symbols), rather than the vector components $\mathbf{A}_x$ and $\mathbf{A}_y$. Scalar components are positive or negative numbers (with units) that are defined as follows. The component A_x has a magnitude that is equal to that of $\mathbf{A}_x$ and is given a positive sign if $\mathbf{A}_x$ points along the $+x$ axis and a negative sign if $\mathbf{A}_x$ points along the $-x$ axis. The component A_y is defined in a similar manner. The following table shows an example of vector and scalar components:

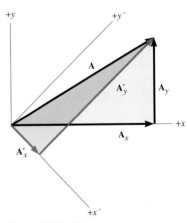

Figure 1.18 The vector components of the vector depend on the orientation of the axes used as a reference.

Vector Components	Scalar Components
$\mathbf{A}_x = 8$ meters, directed along the $+x$ axis	$A_x = +8$ meters
$\mathbf{A}_y = 10$ meters, directed along the $-y$ axis	$A_y = -10$ meters

In this text, when we use the term "component," we will be referring to a scalar component, unless otherwise indicated.

RESOLVING A VECTOR INTO ITS COMPONENTS

If the magnitude and direction of a vector are known, it is possible to find the components of the vector. The process of finding the components is called "resolving the vector into its components." As Example 7 illustrates, this process can be carried out with the aid of trigonometry, because the two perpendicular vector components and the original vector form a right triangle.

EXAMPLE 7 • Finding the Components of a Vector

A displacement vector $\mathbf{r}$ has a magnitude of $r = 175$ m and points at an angle of $50.0°$ relative to the x axis in Figure 1.19. Find the x and y components of this vector.

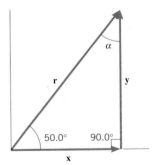

Figure 1.19 The x and y components of the displacement vector $\mathbf{r}$ can be found using trigonometry.

Reasoning and Solution 1 The y component can be obtained using the $50.0°$ angle and Equation 1.1, $\sin \theta = y/r$:

$$y = r \sin \theta = (175 \text{ m})(\sin 50.0°) = \boxed{134 \text{ m}}$$

In a similar fashion, the x component can be obtained using the $50.0°$ angle and Equation 1.2, $\cos \theta = x/r$:

$$x = r \cos \theta = (175 \text{ m})(\cos 50.0°) = \boxed{112 \text{ m}}$$

Reasoning and Solution 2 The angle α in Figure 1.19 can also be used to find the components. Since $\alpha + 50.0° = 90.0°$, it follows that $\alpha = 40.0°$. The solution using α yields the same answers as in Solution 1:

$$\cos \alpha = \frac{y}{r}$$

$$y = r \cos \alpha = (175 \text{ m})(\cos 40.0°) = \boxed{134 \text{ m}}$$

$$\sin \alpha = \frac{x}{r}$$

$$x = r \sin \alpha = (175 \text{ m})(\sin 40.0°) = \boxed{112 \text{ m}}$$

• **PROBLEM SOLVING INSIGHT**
Either acute angle of a right triangle can be used to determine the components of a vector. The choice of angle is a matter of convenience.

Since the vector components and the original vector form a right triangle, the Pythagorean theorem can be applied to check the validity of calculations such as those in Example 7. Thus, with the components obtained in Example 7, the theorem can be used to verify that the magnitude of the original vector is indeed 175 m, as given initially:

$$r = \sqrt{(112 \text{ m})^2 + (134 \text{ m})^2} = 175 \text{ m}$$

VECTORS THAT HAVE ZERO COMPONENTS

Depending on the orientation of the axes used as a reference, it is possible that one of the components of a vector can be zero. Figure 1.20 shows an example of this situation to emphasize that a vector is not zero merely because one of its components is zero. In this drawing, the y vector component is itself the vector **A**, the x vector component being zero. Vector **A** would be expressed as the sum of its vector components according to the following vector equation: $\mathbf{A} = 0 + \mathbf{A}_y$.

For a vector to be zero, every vector component must individually be zero. Thus, in two dimensions, saying that $\mathbf{A} = 0$ is equivalent to saying that $\mathbf{A}_x = 0$ and $\mathbf{A}_y = 0$. Or, stated in terms of scalar components, if $\mathbf{A} = 0$, then $A_x = 0$ and $A_y = 0$. This seemingly trivial fact plays an important role in physics. In particular, it will be used in Chapter 4 when we describe the equilibrium of an object by saying that the net force acting on the object is zero.

VECTORS THAT ARE EQUAL

Two vectors are equal if, and only if, they have the same magnitude and direction. Thus, if one displacement vector points east and another points north, they are *not* equal, even if each has the same magnitude of 480 m. In terms of vector components, two vectors, **A** and **B**, are equal if, and only if, each vector component of one is equal to the corresponding vector component of the other. In two dimensions, if $\mathbf{A} = \mathbf{B}$, then $\mathbf{A}_x = \mathbf{B}_x$ and $\mathbf{A}_y = \mathbf{B}_y$. Alternatively, using scalar components, we write that $A_x = B_x$ and $A_y = B_y$.

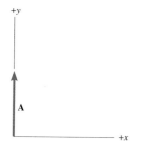

Figure 1.20 The x vector component of the vector **A** is zero, although the vector itself is not zero.

1.8 ADDITION OF VECTORS BY MEANS OF COMPONENTS

The components of a vector provide the most convenient and accurate way of adding (or subtracting) any number of vectors. For example, suppose that vector **A** is added to vector **B**. The resultant vector is **C**, where $\mathbf{C} = \mathbf{A} + \mathbf{B}$. Figure 1.21*a* illustrates this vector addition, along with the x and y vector components of **A** and **B**. In part *b* of the drawing, the vectors **A** and **B** have been removed, because we can use the vector components of these vectors in place of them. The vector component $\mathbf{B}_x$ has been shifted downward and arranged tail-to-head with the vector component $\mathbf{A}_x$. Similarly, the vector component $\mathbf{A}_y$ has been shifted to the right and arranged tail-to-head with the vector component $\mathbf{B}_y$. The x components are colinear and add together to give the x component of the resultant vector **C**. In like fashion, the y components are colinear and add together to give the y component of **C**. In terms of scalar components, we can write

$$C_x = A_x + B_x \quad \text{and} \quad C_y = A_y + B_y$$

The vector components $\mathbf{C}_x$ and $\mathbf{C}_y$ of the resultant vector form the sides of the right

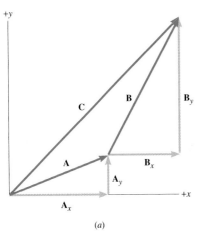
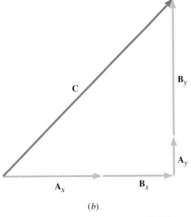
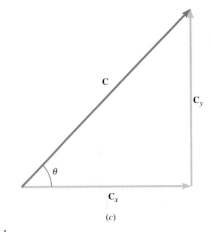

(a)   (b)   (c)

Figure 1.21 (*a*) The vectors **A** and **B** add together to give the resultant vector **C**. The *x* and *y* vector components of **A** and **B** are also shown. (*b*) The drawing illustrates that $C_x = A_x + B_x$ and $C_y = A_y + B_y$. (*c*) Vector **C** and its components form a right triangle.

triangle shown in Figure 1.21*c*. Thus, we can find the magnitude of **C** by using the Pythagorean theorem:

$$C = \sqrt{C_x^2 + C_y^2}$$

The angle θ that **C** makes with the *x* axis is given by $\theta = \tan^{-1}(C_y/C_x)$. Example 8 illustrates how to add several vectors using the component method.

EXAMPLE 8 • The Component Method of Vector Addition

A jogger runs 145 m in a direction 20.0° east of north (displacement vector **A**) and then 105 m in a direction 35.0° south of east (displacement vector **B**). Determine the magnitude and direction of the resultant vector **C** for these two displacements.

Reasoning Figure 1.22*a* shows the vectors **A** and **B**, assuming that the *y* axis corresponds to the direction due north. Since the vectors are not given in component form, we will begin by using the given magnitudes and directions to find the components. Then the components of **A** and **B** can be used to find the components of the resultant **C**. Finally, with the aid of the Pythagorean theorem and trigonometry, the components of **C** can be used to find its magnitude and direction.

Solution The first two rows of the table below give the *x* and *y* components of the vectors **A** and **B**. Note that the component B_y is negative, because **B**$_y$ points downward, in the negative *y* direction in the drawing.

Vector	*x* component	*y* component
A	$A_x = (145\text{ m})\sin 20.0° = 49.6\text{ m}$	$A_y = (145\text{ m})\cos 20.0° = 136\text{ m}$
B	$B_x = (105\text{ m})\cos 35.0° = 86.0\text{ m}$	$B_y = -(105\text{ m})\sin 35.0° = -60.2\text{ m}$
C	$C_x = A_x + B_x = 135.6\text{ m}$	$C_y = A_y + B_y = 76\text{ m}$

The third row in the table gives the *x* and *y* components of the resultant vector **C**: $C_x = A_x + B_x$ and $C_y = A_y + B_y$. Part *b* of the drawing shows **C** and its vector components. The magnitude of **C** is given by the Pythagorean theorem as

$$C = \sqrt{C_x^2 + C_y^2} = \sqrt{(135.6\text{ m})^2 + (76\text{ m})^2} = \boxed{155\text{ m}}$$

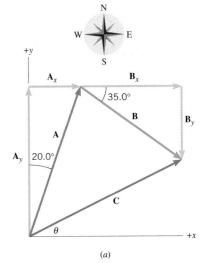

(a)

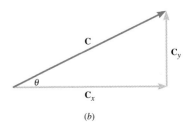

(b)

Figure 1.22 (*a*) The vectors **A** and **B** add together to give the resultant vector **C**. The vector components of **A** and **B** are also shown. (*b*) The resultant vector **C** can be obtained once its components have been found.

The angle θ that **C** makes with the x axis is

$$\theta = \tan^{-1}\left(\frac{C_y}{C_x}\right) = \tan^{-1}\left(\frac{76 \text{ m}}{135.6 \text{ m}}\right) = \boxed{29°}$$

In later chapters we will often use the component method for vector addition. For future reference, the main features of the reasoning strategy used in this technique are summarized below.

REASONING STRATEGY

The Component Method of Vector Addition

1. For each vector to be added, determine the x and y components relative to a conveniently chosen x, y coordinate system. Be sure to take into account the directions of the components by using plus and minus signs to denote whether the components point along the positive or negative axes.

2. Find the algebraic sum of the x components, which is the x component of the resultant vector. Similarly, find the algebraic sum of the y components, which is the y component of the resultant vector.

3. Use the x and y components of the resultant vector and the Pythagorean theorem to determine the magnitude of the resultant vector.

4. Use either the inverse sine, inverse cosine, or inverse tangent function to find the angle that specifies the direction of the resultant vector.

SUMMARY

Physics is an experimental science that uses precisely defined **units of measurement.** This text emphasizes SI (Système International) units, a system that includes the meter (m), the kilogram (kg), and the second (s) as base units for length, mass, and time, respectively. Units play an important role in solving problems, because the units on the left side of an equation must match the units on the right side. If the units on both sides do not match, either the equation is written incorrectly or the variables and constants in the equation are not expressed in a consistent set of units.

Trigonometry is used throughout physics. Particularly important are the sine, cosine, and tangent functions of an angle θ. These functions can be defined in terms of a right triangle that contains θ. The side of the triangle opposite θ is h_o, the side adjacent to θ is h_a, and the hypotenuse is h. In terms of these quantities $\sin\theta = h_o/h$, $\cos\theta = h_a/h$, and $\tan\theta = h_o/h_a$. Once the value of the sine, cosine, or tangent is known, the angle itself can be obtained using inverse trigonometric functions. The Pythagorean theorem, $h^2 = h_o^2 + h_a^2$, is useful when dealing with the sides of a right triangle.

A **scalar quantity** is described completely by its size, which is also called its magnitude. For a **vector quantity,** however, both magnitude and direction must be specified. Vectors are often represented by arrows, the length of the arrow being proportional to the magnitude of the vector and the direction of the arrow indicating the direction of the vector. The **addition of vectors** to give a resultant vector must account for both magnitude and direction. When the vectors are all colinear, the addition proceeds in the same way as the simple addition of scalar quantities. When the vectors are not colinear, one procedure for addition utilizes a graphical technique, in which the vectors to be added are arranged in a tail-to-head fashion. The subtraction of a vector is treated as the addition of a vector that has been multiplied by a scalar factor of -1. Multiplying a vector by -1 reverses the direction of the vector.

In two dimensions, the **vector components** of a vector **A** are two perpendicular vectors $\mathbf{A}_x$ and $\mathbf{A}_y$ that are parallel to the x and y axes, respectively, and add together vectorially so that $\mathbf{A} = \mathbf{A}_x + \mathbf{A}_y$. The **scalar component** A_x has a magnitude that is equal to that of $\mathbf{A}_x$ and is given a positive sign if $\mathbf{A}_x$ points along the $+x$ axis and a negative sign if $\mathbf{A}_x$ points along the $-x$ axis. The scalar component A_y is defined in a similar manner. Components provide the best way of adding any number of vectors. A vector is zero if, and only if, each of its vector components is zero. Two vectors are equal in two dimensions if, and only if, the x vector components of each are equal and the y vector components of each are equal.

CONCEPTUAL QUESTIONS

1. The table below lists four variables along with their units:

Variable	Units
x	meters (m)
v	meters per second (m/s)
t	seconds (s)
a	meters per second squared (m/s^2)

These variables appear in the following equations, along with a few numbers that have no units. In which of the equations are the units on the left side of the equals sign consistent with the units on the right side?

(a) $x = vt$

(b) $x = vt + \frac{1}{2}at^2$

(c) $v = at$

(d) $v = at + \frac{1}{2}at^3$

(e) $v^3 = 2ax^2$

(f) $t = \sqrt{\dfrac{2x}{a}}$

2. The variables x and v have the units shown in the table that accompanies question 1. Is it possible for x and v to be related to an angle θ according to $\tan\theta = x/v$? Account for your answer.

3. You can always add two numbers that have the same units. However, you cannot always add two numbers that have the same dimensions. Explain why not, and include an example in your explanation.

4. (a) Is it possible for two quantities to have the same dimensions, but different units? (b) Is it possible for two quantities to have the same units but different dimensions? In each case, support your answer with an example and an explanation.

5. In the equation $y = c^n at^2$ you wish to determine the integer value (1, 2, etc.) of the exponent n. The dimensions of y, a, and t are known. It is also known that c has no dimensions. Can dimensional analysis be used to determine n? Account for your answer.

6. Using your calculator, verify that $\sin\theta$ divided by $\cos\theta$ is equal to $\tan\theta$, for an angle θ. Try 30°, for example. Prove that this result is true in general by using the definitions for $\sin\theta$, $\cos\theta$, and $\tan\theta$ given in Equations 1.1–1.3.

7. $\sin\theta$ and $\cos\theta$ are called sinusoidal functions of the angle θ. The way in which these functions change as θ changes leads to a characteristic pattern when they are graphed. This pattern arises many times in physics. (a) To familiarize yourself with the sinusoidal pattern, use a calculator and construct a graph, with $\sin\theta$ plotted on the vertical axis and θ on the horizontal axis. Use 15° increments for θ between 0° and 720°. (b) Repeat for $\cos\theta$.

8. Which of the following quantities (if any) can be considered a vector: (a) the number of people attending a football game, (b) the number of days in a month, and (c) the number of pages in a book? Explain your reasoning.

9. Which of the following displacement vectors (if any) are equal? Explain your reasoning.

Vector	Magnitude	Direction
A	100 m	30° north of east
B	100 m	30° south of west
C	50 m	30° south of west
D	100 m	60° east of north

10. Are two vectors with the same magnitude necessarily equal? Give your reasoning.

11. A cube has six faces and twelve edges. You start at one corner, are allowed to move only along the edges, and may not retrace your path along any edge. Consistent with these rules, there are a number of ways to arrive back at your starting point. For instance, you could move around the four edges that make up one of the square faces. The four corresponding displacement vectors would add to zero. How many ways are there to arrive back at your starting point that involve *eight* displacement vectors that add to zero? Describe each possibility, using drawings for clarity.

12. (a) Is it possible for one component of a vector to be zero, while the vector itself is not zero? (b) Is it possible for a vector to be zero, while one component of the vector is not zero? Explain.

13. Can two nonzero perpendicular vectors be added together so their sum is zero? Explain.

14. Can three or more vectors with unequal magnitudes be added together so their sum is zero? If so, show by means of a tail-to-head arrangement of the vectors how this could occur.

15. In preparation for this question, review Conceptual Example 6. Vectors **A** and **B** satisfy the vector equation $\mathbf{A} + \mathbf{B} = 0$. (a) How does the magnitude of **B** compare with the magnitude of **A**? (b) How does the direction of **B** compare with the direction of **A**? Give your reasoning.

16. Vectors **A**, **B**, and **C** satisfy the vector equation $\mathbf{A} + \mathbf{B} = \mathbf{C}$, and their magnitudes are related by the scalar equation $A^2 + B^2 = C^2$. How is vector **A** oriented with respect to vector **B**? Account for your answer.

17. Vectors **A**, **B**, and **C** satisfy the vector equation $\mathbf{A} + \mathbf{B} = \mathbf{C}$, and their magnitudes are related by the scalar equation $A + B = C$. How is vector **A** oriented with respect to vector **B**? Explain your reasoning.

18. The magnitude of a vector has doubled. Can you conclude that the magnitude of each component of the vector has doubled? Explain your answer.

19. The tail of a vector is fixed to the origin of an x, y axis system. Originally the vector points along the $+x$ axis. As time passes, the vector rotates counterclockwise. Describe how the sizes of the x and y components of the vector compare to the size of the original vector for rotational angles of (a) 90°, (b) 180°, (c) 270°, and (d) 360°.

20. A vector has a component of zero along the x axis of a certain axes system. Does this vector necessarily have a component of zero along the x axis of another (rotated) axes system? Use a drawing to justify your answer.

PROBLEMS

Problems that are not marked with a star are considered the easiest to solve. Problems that are marked with a single star (*) are more difficult, while those marked with a double star (**) are the most difficult.

ssm Solution is in the Student Solutions Manual. **www** Solution is available on the World Wide Web at http://www.wiley.com/college/cutnell ⚕ This icon represents a biomedical application.

Section 1.3 The Role of Units in Problem Solving

1. ssm The mass of the parasitic wasp *Caraphractus cintus* can be as small as 5×10^{-6} kg. What is this mass in (a) grams (g), (b) milligrams (mg), and (c) micrograms (μg)?

2. The distance of the Boston marathon is 26 miles, 385 yards. What is the length of this race in meters?

3. How many seconds are there in (a) one hour and thirty-five minutes and (b) one day?

4. A 747 jetliner is cruising at a speed of 520 miles per hour. What is its speed in kilometers per hour?

5. ssm The largest diamond ever found had a size of 3106 carats. One carat is equivalent to a mass of 0.200 g. Use the fact that 1 kg (1000 g) has a weight of 2.205 lb under certain conditions, and determine the weight of this diamond in pounds.

6. A bottle of wine known as a magnum contains a volume of 1.5 liters. A bottle known as a jeroboam contains 0.792 U.S. gallons. How many magnums are there in one jeroboam?

7. The following are dimensions of various physical parameters that will be discussed later on in the text. Here [L], [T], and [M] denote, respectively, dimensions of length, time, and mass.

	Dimension		Dimension
Distance (x)	[L]	Acceleration (a)	$[L]/[T]^2$
Time (t)	[T]	Force (F)	$[M][L]/[T]^2$
Mass (m)	[M]	Energy (E)	$[M][L]^2/[T]^2$
Speed (v)	[L]/[T]		

Which of the following equations are dimensionally correct?

    (a) $F = ma$ **(d)** $E = max$

    (b) $x = \frac{1}{2}at^3$ **(e)** $v = \sqrt{Fx/m}$

    (c) $E = \frac{1}{2}mv$

8. The variables x, v, and a have the dimensions of [L], [L]/[T], and $[L]/[T]^2$, respectively. These variables are related by an equation that has the form $v^n = 2ax$, where n is an integer constant (1, 2, 3, etc.) without dimensions. What must be the value of n, so that both sides of the equation have the same dimensions? Explain your reasoning.

*9. ssm The depth of the ocean is sometimes measured in fathoms (1 fathom = 6 feet). Distance on the surface of the ocean is sometimes measured in nautical miles (1 nautical mile = 6076 feet). The water beneath a surface rectangle 1.20 nautical miles by 2.60 nautical miles has a depth of 16.0 fathoms. Find the volume of water (in cubic meters) beneath this rectangle.

*10. The CGS unit for measuring the viscosity of a liquid is the poise [P]: 1 P = 1 g/(s·cm). The SI unit is the kg/(s·m). The viscosity of water at 0 °C is 1.78×10^{-3} kg/(s·m). Express this viscosity in poise.

Section 1.4 Trigonometry

11. The gondola ski lift at Keystone, Colorado, is 2830 m long. On average, the ski lift rises 14.6° above the horizontal. How high is the top of the ski lift relative to the base?

12. A hill that has a 12.0% grade is one that rises 12.0 m vertically for every 100.0 m of distance in the horizontal direction. At what angle is such a hill inclined above the horizontal?

13. ssm www A highway is to be built between two towns, one of which lies 35.0 km south and 72.0 km west of the other. What is the shortest length of highway that can be built between the two towns, and at what angle would this highway be directed with respect to due west?

14. You are driving into St. Louis, Missouri, and in the distance you see the famous Gateway-to-the-West arch. This monument rises to a height of 192 m. You estimate your line of sight with the top of the arch to be 2.0° above the horizontal. Approximately how far (in kilometers) are you from the base of the arch?

15. The silhouette of a Christmas tree is an isosceles triangle. The angle at the top of the triangle is 30.0°, and the base measures 2.00 m across. How tall is the tree?

16. An observer, whose eyes are 1.83 m above the ground, is standing 32.0 m away from a tree. The ground is level, and the tree is growing perpendicular to it. The observer's line of sight with the treetop makes an angle of 20.0° above the horizontal. How tall is the tree?

*17. ssm What is the value of each of the angles of a triangle whose sides are 95, 150, and 190 cm in length? *(Hint: Consider using the law of cosines given in Appendix E.)*

*18. The drawing shows sodium and chlorine ions positioned at

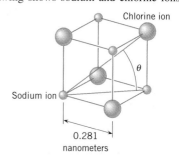

Chlorine ion

θ

Sodium ion

0.281 nanometers

the corners of a cube that is part of the crystal structure of sodium chloride (common table salt). The edge of the cube is 0.281 nm (1 nm = 1 nanometer = 10^{-9} m) in length. Find the distance (in nanometers) between the sodium ion located at one corner of the cube and the chlorine ion located on the diagonal at the opposite corner.

*19. What is the value of the angle θ in the drawing that accompanies problem 18?

**20. A regular tetrahedron is a three-dimensional object that has four faces, each of which is an equilateral triangle. Each of the edges of such an object has a length L. The height H of a regular tetrahedron is the perpendicular distance from one corner to the center of the opposite triangular face. Show that the ratio between H and L is $H/L = \sqrt{2/3}$.

Section 1.6 Vector Addition and Subtraction

21. **ssm www** One displacement vector **A** has a magnitude of 2.43 km and points due north. A second displacement vector **B** has a magnitude of 7.74 km and also points due north. (a) Find the magnitude and direction of **A** − **B**. (b) Find the magnitude and direction of **B** − **A**.

22. A chimpanzee sitting against his favorite tree gets up and walks 51 m due east and 39 m due south to reach a termite mound, where he eats lunch. (a) What is the shortest distance between the tree and the termite mound? (b) What angle does the shortest distance make with respect to due east?

23. A force vector F_1 points due east and has a magnitude of 200 newtons. A second force F_2 is added to F_1. The resultant of the two vectors has a magnitude of 400 newtons and points along the east/west line. Find the magnitude and direction of F_2. Note that there are two answers.

24. The drawing shows a triple jump on a checkerboard, starting at the center of square A and ending on the center of square B. Each side of a square measures 4.0 cm. What is the magnitude of the displacement of the colored checker during the triple jump?

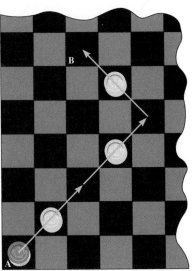

25. **ssm** Two ropes are attached to a heavy box to pull it along the floor. One rope applies a force of 475 newtons in a direction due west; the other applies a force of 315 newtons in a direction due south. As we will see later in the text, force is a vector quantity. (a) How much force should be applied by a single rope, and (b) in what direction (relative to due west), if it is to accomplish the same effect as the two forces added together?

26. A jogger travels due south, and in the process his displacement vector has a magnitude of 4.68 km. He then jogs due west. (a) What is the magnitude of his displacement vector in the due west direction, if the magnitude of his total displacement vector is 7.41 km? (b) What is the direction of his total displacement vector with respect to due south?

27. Vector **A** has a magnitude of 48.0 units and points due west, while vector **B** has the same magnitude but points due south. Determine the magnitude and direction of (a) **A** + **B** and (b) **A** − **B**. Specify the direction relative to due west.

*28. A basketball player runs a pattern consisting of three segments. The corresponding three displacement vectors **A**, **B**, and **C** have equal magnitudes of 7.0 m. Displacement **A** is directed forward and parallel to one side of the court, **B** is directed forward at a 45° angle with respect to the side of the court, and **C** is directed forward and parallel to the side of the court. With a scale drawing, use the graphical technique to find the magnitude and direction of the displacement vector for a straight-line dash between the starting and finishing points.

*29. **ssm www** A car is being pulled out of the mud by two forces that are applied by the two ropes shown in the drawing. The dashed line in the drawing bisects the 30.0° angle. The magnitude of the force applied by each rope is 2900 newtons. Arrange the force vectors tail to head and use the graphical technique to answer the following questions. (a) How much force would a single rope need to apply to accomplish the same effect as the two forces added together? (b) How would the single rope be directed relative to the dashed line?

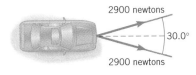

*30. In wandering, a grizzly bear makes a displacement of 1563 m due west, followed by a displacement of 3348 m in a direction 32.0° north of west. What are (a) the magnitude and (b) the direction of the displacement needed for the bear to *return to its starting point*? Specify the direction relative to due east.

*31. Vector **A** has a magnitude of 8.00 units and points due west. Vector **B** points due north. (a) What is the magnitude of **B** if **A** + **B** has a magnitude of 10.00 units? (b) What is the direction of **A** + **B** relative to due west? (c) What is the magnitude of **B** if **A** − **B** has a magnitude of 10.00 units? (d) What is the direction of **A** − **B** relative to due west?

Section 1.7 The Components of a Vector

32. A displacement vector has a magnitude of 177 m and points at an angle of 36.0° below the positive x axis. What are (a) the x scalar component and (b) the y scalar component of the vector?

33. **ssm** Vector **A** points along the $+y$ axis and has a magnitude of 100.0 units. Vector **B** points at an angle of 60.0° above the $+x$ axis and has a magnitude of 200.0 units. Vector **C** points along the $+x$ axis and has a magnitude of 150.0 units. Which vector has (a) the largest x component and (b) the largest y component?

34. A bicyclist is headed due east. A 5.00-m/s wind is blowing partially into the rider's face and is coming from a direction that is 35.0° south of east. The speed and direction of the wind constitute a vector quantity known as the velocity. In effect, then, the rider must "pump" against a component of the wind's velocity vector. What is the magnitude of this component?

35. An ocean liner leaves New York City and travels 18.0° north of east for 155 km. How far east and how far north has it gone? In other words, what are the magnitudes of the components of the ship's displacement vector in the directions (a) due east and (b) due north?

36. Your friend has slipped and fallen. To help him up, you pull with a force **F**, as the drawing shows. The vertical component of this force is 130 newtons, while the horizontal component is 150 newtons. Find (a) the magnitude of **F** and (b) the angle θ.

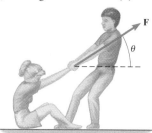

37. **ssm** The x vector component of a displacement vector **r** has a magnitude of 125 m and points along the negative x axis. The y vector component has a magnitude of 184 m and points along the negative y axis. Find the magnitude and direction of **r**. Specify the direction with respect to the negative x axis.

38. On takeoff, an airplane climbs with a speed of 180 m/s at an angle of 34° above the horizontal. The speed and direction of the airplane constitute a vector quantity known as the velocity. The sun is shining directly overhead. How fast is the shadow of the plane moving along the ground? (That is, what is the magnitude of the horizontal component of the plane's velocity?)

*39. The magnitude of the force vector **F** is 280 newtons. The x component of this vector is directed along the $+x$ axis and has a magnitude of 150 newtons. The y component points along the $+y$ axis. (a) Find the direction of **F** relative to the $+x$ axis. (b) Find the component of **F** along the $+y$ axis.

*40. The vector **A** in the drawing has a magnitude of 750 units. Determine the magnitude and direction of the x and y components

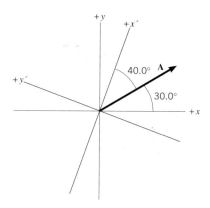

of the vector **A**, relative to (a) the black axes and (b) the colored axes.

41. **ssm www The drawing shows a force vector that has a magnitude of 475 newtons. Find the (a) x, (b) y, and (c) z components of the vector.

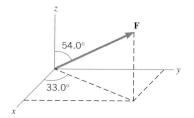

Section 1.8 Addition of Vectors by Means of Components

42. You are on a treasure hunt and your map says "Walk due west for 52 paces, then walk 30.0° north of west for 42 paces, and finally walk due north for 25 paces." What is the magnitude of the component of your displacement in the direction (a) due north and (b) due west?

43. A pilot flies her route in two straight line segments. The displacement vector **A** for the first segment has a magnitude of 243 km and a direction 50.0° north of east. The displacement vector **B** for the second segment has a magnitude of 57.0 km and a direction 20.0° south of east. The resultant displacement vector is **R** = **A** + **B**. What are the magnitude and direction of **R**? Use the component method and specify the direction relative to due east.

44. The force vector $\mathbf{F_A}$ has a magnitude of 45.0 newtons and points 30.0° north of east. The force vector $\mathbf{F_B}$ has a magnitude of 75.0 newtons and points due north. Find the magnitude and direction of the resultant $\mathbf{F_A} + \mathbf{F_B}$ by using the component method. Specify the direction relative to due east.

45. **ssm** A golfer, putting on a green, requires three strokes to "hole the ball." During the first putt, the ball rolls 5.0 m due east. For the second putt, the ball travels 2.1 m at an angle of 20.0° north of east. The third putt is 0.50 m due north. What displacement (magnitude and direction relative to due east) would have been needed to "hole the ball" on the very first putt?

46. On a safari, a team of naturalists sets out toward a research station located 4.8 km away in a direction 42° north of east. After traveling in a straight line for 2.4 km, they stop and discover that they have been traveling 22° north of east, because their guide misread his compass. What are (a) the magnitude and (b) the direction (relative to due east) of the displacement vector now required to bring the team to the research station?

47. A football player runs the pattern given in the drawing by the three displacement vectors **A**, **B**, and **C**. The magnitudes of these vectors are $A = 5.00$ m, $B = 15.0$ m, and $C = 18.0$ m. Using the component method, find the magnitude and direction θ of the resultant vector **A** + **B** + **C**.

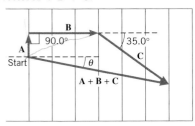

48. A baby elephant is stuck in a mud hole. To help pull it out, game keepers use a rope to apply force $\mathbf{F_A}$, as part a of the drawing shows. By itself, however, force $\mathbf{F_A}$ is insufficient. Therefore, two additional forces $\mathbf{F_B}$ and $\mathbf{F_C}$ are applied, as in part b of the drawing. Each of these additional forces has the same magnitude F. The magnitude of the resultant force acting on the elephant in part b of the drawing is twice that in part a. Find the ratio F/F_A.

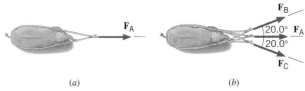

(a)   (b)

*49. **ssm** Vector **A** has a magnitude of 6.00 units and points due east. Vector **B** points due north. (a) What is the magnitude of **B**, if the vector **A** + **B** points 60.0° north of east? (b) Find the magnitude of **A** + **B**.

*50. Two forces act on an object. One has a magnitude of 166 newtons and points at an angle of 60.0° above the $+x$ axis. The second has a magnitude of 284 newtons and points at an angle of 30.0° above the $+x$ axis. A third force is applied and balances to zero the effects of the other two. What are the magnitude and direction of this third force? Specify the direction relative to the negative x axis.

*51. Vector **A** has a magnitude of 188 units and points 30.0° north of west. Vector **B** points 50.0° east of north. Vector **C** points 20.0° west of south. These three vectors add to give a resultant vector that is zero. Using components, find the magnitudes of (a) vector **B** and (b) vector **C**.

*52. A grasshopper makes four jumps. The displacement vectors are (1) 27.0 cm, due west; (2) 23.0 cm, 35.0° south of west; (3) 28.0 cm, 55.0° south of east; and (4) 35.0 cm, 63.0° north of east. Find the magnitude and direction of the resultant displacement. Express the direction with respect to due west.

*53. **ssm** A sailboat race course consists of four legs, defined by the displacement vectors **A**, **B**, **C**, and **D**, as the drawing indicates. The magnitudes of the first three vectors are $A = 3.20$ km, $B = 5.10$ km, and $C = 4.80$ km. The finish line of the course coincides with the starting line. Using the data in the drawing, find the distance of the fourth leg and the angle θ.

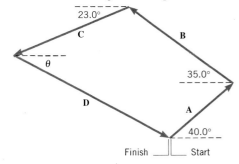

ADDITIONAL PROBLEMS

54. The corners of an equilateral triangle lie on a circle that has a radius of 0.25 m. What is the length of a side of the triangle?

55. A displacement vector **A** has a magnitude of 1.62 km and points due north. Another displacement vector **B** has a magnitude of 2.48 km and points due east. Determine the magnitude and direction of (a) **A** + **B** and (b) **A** − **B**.

56. Consider the equation $v = \frac{1}{3}zxt^2$. The dimensions of the variables x, v, and t are [L], [L]/[T], and [T], respectively. What must be the dimensions of the variable z, such that both sides of the equation have the same dimensions? Show how you determined your answer.

57. **ssm** The speed of an object and the direction in which it moves constitute a vector quantity known as the velocity. An ostrich is running at a speed of 17.0 m/s in a direction of 68.0° north of west. What is the magnitude of the ostrich's velocity component that is directed (a) due north and (b) due west?

58. One acre contains 43 560 ft². How many square meters (m²) are in one acre?

59. Find the resultant of the three displacement vectors in the drawing by means of the component method. The magnitudes of the vectors are $A = 5.00$ m, $B = 5.00$ m, and $C = 4.00$ m.

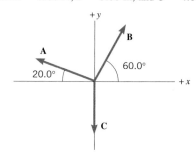

60. A frog hops four times: twice forward, once to the right, and once forward again. Each hop covers a distance of 28 cm. What is the magnitude of the frog's displacement?

61. **ssm** Displacement vector **A** points due east and has a magnitude of 2.00 km. Displacement vector **B** points due north and has a magnitude of 3.75 km. Displacement vector **C** points due west and has a magnitude of 2.50 km. Displacement vector **D** points due south and has a magnitude of 3.00 km. Find the magnitude and direction (relative to due west) of the resultant vector **A** + **B** + **C** + **D**.

*62. A displacement vector **A** has a magnitude of 636 m and points 40.0° above the $-x$ axis. Another displacement vector **B** is added to **A**. The resultant has the same magnitude as **A**, but the opposite direction. Find (a) the x component and (b) the y component of **B**.

*63. Consider the two vectors **A** and **B** in the drawing for problem 59. Determine (a) the vector sum **A** + **B** and (b) the vector difference **A** − **B** using the method of components. In each case, specify both magnitude and direction (relative to the negative x axis).

*64. Three deer, A, B, and C, are grazing in a field. Deer B is located 62 m from deer A at an angle of 51° north of west. Deer C is

located 77° north of east relative to deer A. The distance between deer B and C is 95 m. What is the distance between deer A and C? *(Hint: Consider the law of cosines given in Appendix E.)*

*65. At a picnic, there is a contest in which hoses are used to shoot water at a beach ball from three directions. As a result, three forces act on the ball, $\mathbf{F_1}$, $\mathbf{F_2}$, and $\mathbf{F_3}$ (see the drawing). The magnitudes of $\mathbf{F_1}$ and $\mathbf{F_2}$ are $F_1 = 50.0$ newtons and $F_2 = 90.0$ newtons. Using a scale drawing and the graphical technique, determine (a) the magnitude of $\mathbf{F_3}$ and (b) the angle θ such that the resultant force acting on the ball is zero.

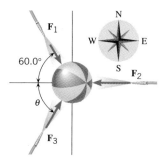

*66. Before starting this problem, review Conceptual Example 6. The force vector $\mathbf{F_A}$ has a magnitude of 90.0 newtons and points due east. The force vector $\mathbf{F_B}$ has a magnitude of 135 newtons and points 75° north of east. Use the graphical method and find the magnitude and direction of (a) $\mathbf{F_A} - \mathbf{F_B}$ (give the direction with respect to due east) and (b) $\mathbf{F_B} - \mathbf{F_A}$ (give the direction with respect to due west).

**67. What are the x and y components of the vector that must be added to the following three vectors, so that the sum of the four vectors is zero? Due east is the $+x$ direction, and due north is the $+y$ direction.

$$\mathbf{A} = 113 \text{ units, } 60.0° \text{ south of west}$$
$$\mathbf{B} = 222 \text{ units, } 35.0° \text{ south of east}$$
$$\mathbf{C} = 177 \text{ units, } 23.0° \text{ north of east}$$

KINEMATICS IN ONE DIMENSION

The motion of these surfers entails three concepts that will be
discussed in this chapter: displacement, velocity, and acceleration.

2.1 DISPLACEMENT

There are always two aspects to any motion. In a purely descriptive sense, there is the movement itself. Is it rapid or slow, for instance? Secondly, there is the issue of what causes the motion or what changes it, which requires that forces be considered. *Mechanics* is the branch of physics that focuses on the motion of objects and the forces that cause the motion to change. There are two parts to mechanics: kinematics and dynamics. *Kinematics* deals with the concepts that are needed to describe motion, without any reference to forces. The present chapter discusses these concepts as they apply to motion in one dimension, while the next chapter treats two-dimensional motion. *Dynamics* deals with the effect that forces have on motion, a topic that we defer until Chapter 4. We turn now to the first of the kinematics concepts to be discussed, which is displacement.

To describe the motion of an object, we must be able to specify the location of the object at all times, and Figure 2.1 shows one way of accomplishing this for one-dimensional motion. In this drawing, the initial position of a car is indicated by the vector labeled $\mathbf{x_0}$. The length of $\mathbf{x_0}$ is the distance of the car from an arbitrarily chosen origin. At a later time the car has moved to a new position, which is indicated by the vector $\mathbf{x}$. The *displacement* of the car $\Delta\mathbf{x}$ (read as "delta x" or "the change in x") is a vector drawn from the initial position to the final position. Displacement is a vector quantity in the sense discussed in Section 1.5, for it conveys both a magnitude (the distance between the initial and final positions) and a direction. The displacement can be related to $\mathbf{x_0}$ and $\mathbf{x}$ by noting from the drawing that

$$\mathbf{x_0} + \Delta\mathbf{x} = \mathbf{x} \quad \text{or} \quad \Delta\mathbf{x} = \mathbf{x} - \mathbf{x_0}$$

Thus, the displacement $\Delta\mathbf{x}$ is the difference between $\mathbf{x}$ and $\mathbf{x_0}$, and the Greek letter delta (Δ) is used to denote this difference. It is important to note that the change in any variable is always the final value minus the initial value.

> ■ **DEFINITION OF DISPLACEMENT**
>
> The displacement is a vector that points from an object's initial position toward its final position and has a magnitude that equals the shortest distance between the two positions.
>
> *SI Unit of Displacement:* meter (m)

The SI unit for displacement is the meter (m), but there are other units as well, such as the centimeter and the inch. When converting between centimeters (cm) and inches (in.), remember that 2.54 cm = 1 in.

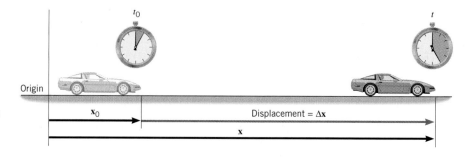

Figure 2.1 The displacement $\Delta\mathbf{x}$ is a vector that points from the initial position to the final position.

Often, we will deal with motion along a straight line. In such a case, a displacement in one direction along the line is assigned a positive value, and a displacement in the opposite direction is assigned a negative value. For instance, assume that a car is moving along an east/west direction and that a positive (+) sign is used to denote a direction due east. Then, $\Delta\mathbf{x} = +500$ m represents a displacement that points to the east and has a magnitude of 500 meters. Conversely, $\Delta\mathbf{x} = -500$ m is a displacement that has the same magnitude but points in the opposite direction, due west.

2.2 SPEED AND VELOCITY

AVERAGE SPEED

One of the most obvious features of an object in motion is how fast it is moving. If a car travels 200 meters in 10 seconds, we say its average speed is 20 meters per second, the *average speed* being the distance traveled divided by the time required to cover the distance:

$$\text{Average speed} = \frac{\text{Distance}}{\text{Elapsed time}} \qquad (2.1)$$

Equation 2.1 indicates that the unit for average speed is the unit for distance divided by the unit for time, or meters per second (m/s) in SI units. The next two examples illustrate how the idea of average speed is used.

EXAMPLE 1 • Distance Run by a Jogger

How far does a jogger run in 1.5 hours (5400 s) if his average speed is 2.22 m/s?

Reasoning and Solution To find the distance run, we rewrite Equation 2.1 as Distance = (Average speed)(Elapsed time):

$$\text{Distance} = (2.22 \text{ m/s})(5400 \text{ s}) = \boxed{12\ 000 \text{ m}}$$

EXAMPLE 2 • A Diving Falcon

With an average speed of 67 m/s, how long does it take a falcon (Figure 2.2) to dive to the ground along a 150-m path?

Reasoning and Solution To find the time of flight, we rewrite Equation 2.1 as Elapsed time = Distance/Average speed:

$$\text{Elapsed time} = \frac{150 \text{ m}}{67 \text{ m/s}} = \boxed{2.2 \text{ s}}$$

Figure 2.2 If we know the average speed of a diving falcon and the distance of the dive, we can determine the time for the falcon to make the dive.

AVERAGE VELOCITY

Speed is a useful idea, because it indicates how fast an object is moving. However, speed does not reveal anything about the direction of the motion. To describe both

Concepts at a Glance

Figure 2.3 *Concepts at a Glance* The concepts of displacement and time are brought together to formulate the concept of velocity. Michael Johnson won the gold medal in the 400 m event in the 1996 Olympics and had a greater displacement per unit time than any of his competitors.

how fast an object moves and the direction of its motion, we need the vector concept of velocity. The concept chart in Figure 2.3 indicates that displacement and the time required for the displacement are brought together to formulate this new idea. In the specific case of the motion illustrated in Figure 2.1, suppose that the car's initial position is x_0 when the time is t_0. A little later the car arrives at the final position x at the time t. The time for the car to travel between these two positions is the *difference* between t and t_0, which is denoted by the symbol Δt:

$$\text{Elapsed time} = t - t_0 = \Delta t$$

Dividing the displacement Δx of the car by the elapsed time Δt gives the ***average velocity*** of the car. It is customary to denote the average value of a quantity by placing a horizontal bar above the symbol representing the quantity. The average velocity, then, is written as $\bar{v}$, as specified in Equation 2.2:

■ **DEFINITION OF AVERAGE VELOCITY**

$$\text{Average velocity} = \frac{\text{Displacement}}{\text{Elapsed time}}$$

$$\bar{v} = \frac{x - x_0}{t - t_0} = \frac{\Delta x}{\Delta t} \tag{2.2}$$

SI Unit of Average Velocity: meter per second (m/s)

Equation 2.2 indicates that the unit for average velocity is the unit for length divided by the unit for time, or meters per second (m/s) in SI units. Velocity can also be expressed in other units, such as kilometers per hour (km/h) or miles per hour (mi/h).

Average velocity is a vector that points in the same direction as the displacement in Equation 2.2. As with displacement, plus and minus signs indicate the two possible directions of the velocity along a straight line. If the displacement points in the positive direction, the average velocity is positive. Conversely, if the displacement points in the negative direction, the average velocity is negative. Example 3 illustrates these features of average velocity.

EXAMPLE 3 • The World's Fastest Jet-Engine Car

A world record of 283 m/s (633 mi/h) for the fastest jet-engine car was set in 1983 by Richard Noble in the car *Thrust 2*. To establish such a record, the driver makes two runs through the course, one in each direction, to nullify wind effects. Figure 2.4*a* shows that the car first travels from left to right and covers a distance of 604 m in 2.12 s. Figure 2.4*b* shows that in the reverse direction, the car covers the same distance in 2.15 s. From these data, determine the average velocity for each run.

Reasoning Average velocity is defined as the displacement divided by the elapsed time. In using this definition we recognize that the displacement is not the same as the distance traveled. Displacement takes the direction of the motion into account, and distance does not. During both runs, the car covers a distance of 604 m. However, for the first run the displacement is $\Delta x = +604$ m, while for the second it is $\Delta x = -604$ m. The plus and minus signs are essential, because the first run is to the right, which is the positive direction, and the second run is in the opposite or negative direction.

In this time-lapse photo of freeway traffic, the velocity of the cars in the left lanes (white headlights) is opposite to that of the cars in the right lanes (red taillights).

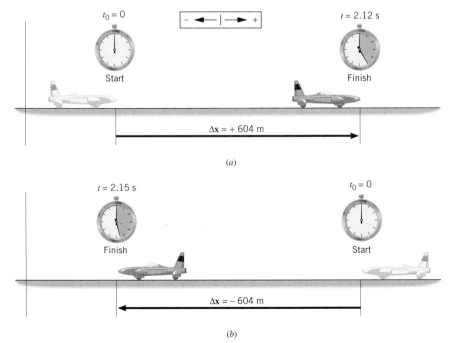

$t_0 = 0$
Start

$t = 2.12$ s
Finish

$\Delta x = +604$ m

(a)

$t = 2.15$ s
Finish

$t_0 = 0$
Start

$\Delta x = -604$ m

(b)

Figure 2.4 The arrows in the box at the top of the drawing indicate the positive and negative directions for the displacements of the car, as explained in Example 3.

Solution According to Equation 2.2, the average velocities are

Run 1
$$\bar{\mathbf{v}} = \frac{\Delta \mathbf{x}}{\Delta t} = \frac{+604 \text{ m}}{2.12 \text{ s}} = \boxed{+285 \text{ m/s}}$$

Run 2
$$\bar{\mathbf{v}} = \frac{\Delta \mathbf{x}}{\Delta t} = \frac{-604 \text{ m}}{2.15 \text{ s}} = \boxed{-281 \text{ m/s}}$$

In these answers the algebraic signs convey the directions of the velocity vectors. In particular, for run 2 the minus sign indicates that the average velocity, like the displacement, points to the left in Figure 2.4*b*. The magnitudes of the velocities are 285 and 281 m/s. The average of these numbers is 283 m/s and is recorded in the record book.

INSTANTANEOUS VELOCITY

Suppose the magnitude of your average velocity for a long trip was 20 m/s. This value, being an average, does not convey any information about how fast you were moving at any instant during the trip. Surely there were times when your car traveled faster than 20 m/s and times when it traveled more slowly. The ***instantaneous velocity*** **v** of the car indicates how fast the car moves and the direction of the motion at each instant of time. The magnitude of the instantaneous velocity is called the ***instantaneous speed,*** and it is the number (with units) indicated by the speedometer.

The instantaneous velocity at any point during a trip can be obtained by measuring the time Δt for the car to travel a *very small* displacement $\Delta \mathbf{x}$. We can then compute the average velocity over this interval. If the time Δt is small enough, the instantaneous velocity does not change much during the measurement. Then, the instantaneous velocity **v** at the point of interest is approximately equal to ($\approx$) the av-

Concepts at a Glance

Change in Velocity (Section 2.2)

Time (Section 1.2)

Acceleration

erage velocity $\overline{\mathbf{v}}$ computed over the interval, or $\mathbf{v} \approx \overline{\mathbf{v}} = \Delta\mathbf{x}/\Delta t$ (for sufficiently small Δt). In fact, in the limit that Δt becomes infinitesimally small, the instantaneous velocity and the average velocity become equal, so that

$$\mathbf{v} = \lim_{\Delta t \to 0} \frac{\Delta\mathbf{x}}{\Delta t} \qquad (2.3)$$

The notation $\lim_{\Delta t \to 0} (\Delta\mathbf{x}/\Delta t)$ means that the ratio $\Delta\mathbf{x}/\Delta t$ is defined by a limiting process in which smaller and smaller values of Δt are used, so small that they approach zero. As smaller values of Δt are used, $\Delta\mathbf{x}$ also becomes smaller. However, the ratio $\Delta\mathbf{x}/\Delta t$ does *not* become zero but, rather, approaches the value of the instantaneous velocity. For brevity, we will use the word *velocity* to mean "instantaneous velocity" and *speed* to mean "instantaneous speed."

In a wide range of motions, the velocity changes from moment to moment. To describe the manner in which it changes, the concept of acceleration is needed, as the next section discusses.

2.3 ACCELERATION

The velocity of a moving object may change in a number of ways. For example, it may increase, as it does when the driver of a car steps on the gas in order to pass the car ahead of him. Or, it may decrease, as it does when the driver applies the brakes to stop at a red light. In either case, the change in velocity may occur over a short or a long time interval. To describe such situations, we introduce the new idea of acceleration. As the concept chart in Figure 2.5 illustrates, this idea emerges when the change in the velocity is combined with the time required for the change to occur.

Figure 2.5 *Concepts at a Glance* To formulate the concept of acceleration, the change in velocity is combined with the time required for the change to occur. In launching himself from the starting block, Michael Johnson changed his velocity from zero to the value at which he ran most of his race. For that brief launch period, his acceleration was substantial. For the remainder of the race, his velocity did not change much and, consequently, he had a relatively small acceleration.

Figure 2.6 During takeoff, the plane accelerates from an initial velocity $\mathbf{v_0}$ to a final velocity $\mathbf{v}$ during the time interval $\Delta t = t - t_0$.

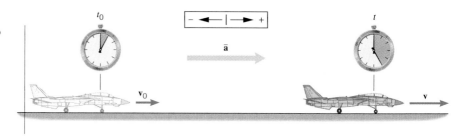

The meaning of *average acceleration* can be illustrated by considering a plane during takeoff. Figure 2.6 focuses attention on how the plane's velocity changes along the runway. During an elapsed time interval $\Delta t = t - t_0$, the velocity changes from an initial value of $\mathbf{v_0}$ to a final value of $\mathbf{v}$, the change in velocity being $\Delta \mathbf{v} = \mathbf{v} - \mathbf{v_0}$. The average acceleration is defined in the following manner, to provide a measure of how much the velocity changes per unit of elapsed time.

■ **DEFINITION OF AVERAGE ACCELERATION**

$$\text{Average acceleration} = \frac{\text{Change in velocity}}{\text{Elapsed time}}$$

$$\overline{\mathbf{a}} = \frac{\mathbf{v} - \mathbf{v_0}}{t - t_0} = \frac{\Delta \mathbf{v}}{\Delta t} \tag{2.4}$$

SI Unit of Average Acceleration: meter per second squared (m/s^2)

The average acceleration $\overline{\mathbf{a}}$ is a vector that points in the same direction as $\Delta \mathbf{v}$, the change in the velocity. Following the usual custom, plus and minus signs indicate the two possible directions for the acceleration vector when the motion is along a straight line.

We are often interested in an object's acceleration at a particular instant of time. The *instantaneous acceleration* $\mathbf{a}$ can be defined by analogy with the procedure used in Section 2.2 for instantaneous velocity:

$$\mathbf{a} = \lim_{\Delta t \to 0} \frac{\Delta \mathbf{v}}{\Delta t} \tag{2.5}$$

Equation 2.5 indicates that the instantaneous acceleration is a limiting case of the average acceleration. When the time interval Δt for measuring the acceleration becomes extremely small (approaching zero in the limit), the average acceleration and the instantaneous acceleration become equal. Moreover, in many situations the acceleration is constant, so the acceleration has the same value at any instant of time. In the future, we will use the word *acceleration* to mean "instantaneous acceleration." Example 4 deals with the acceleration of a plane during takeoff.

A long-distance runner accelerates down a hill.

EXAMPLE 4 • Acceleration and Increasing Velocity

Suppose the plane in Figure 2.6 starts from rest ($\mathbf{v_0} = 0$) when $t_0 = 0$. The plane accelerates down the runway and at $t = 29$ s attains a velocity of $\mathbf{v} = +260$ km/h, where the plus sign indicates the velocity points to the right. Determine the average acceleration of the plane.

Reasoning and Solution The average acceleration of the plane can be found from Equation 2.4 as

$$\overline{\mathbf{a}} = \frac{\mathbf{v} - \mathbf{v_0}}{t - t_0} = \frac{260 \text{ km/h} - 0 \text{ km/h}}{29 \text{ s} - 0 \text{ s}} = \boxed{+9.0 \frac{\text{km/h}}{\text{s}}}$$

The average acceleration calculated in Example 4 is read as "nine kilometers per hour per second." Assuming the acceleration of the plane is constant, a value of $9.0 \dfrac{\text{km/h}}{\text{s}}$ means the velocity changes by 9.0 km/h during each second of the motion. During the first second, the velocity increases from 0 to 9.0 km/h; during the next second, the velocity increases by another 9.0 km/h to 18 km/h, and so on. Figure 2.7 illustrates how the velocity changes during the first three seconds. By the end of the 29th second, the velocity is 260 km/h.

It is customary to express the units for acceleration solely in terms of SI units. One way to obtain SI units for the acceleration in Example 4 is to convert the velocity units from km/h to m/s:

$$260 \frac{\cancel{\text{km}}}{\cancel{\text{h}}} \left(\frac{1000 \text{ m}}{1 \cancel{\text{km}}} \right) \left(\frac{1 \cancel{\text{h}}}{3600 \text{ s}} \right) = 72 \frac{\text{m}}{\text{s}}$$

The average acceleration then becomes

$$\overline{\mathbf{a}} = \frac{72 \text{ m/s} - 0 \text{ m/s}}{29 \text{ s} - 0 \text{ s}} = +2.5 \text{ m/s}^2$$

where we have used $2.5 \dfrac{\text{m/s}}{\text{s}} = 2.5 \dfrac{\text{m}}{\text{s} \cdot \text{s}} = 2.5 \dfrac{\text{m}}{\text{s}^2}$. An acceleration of $2.5 \dfrac{\text{m}}{\text{s}^2}$ is read as "2.5 meters per second per second" (or "2.5 meters per second squared") and means that the velocity changes by 2.5 m/s during each second of the motion.

Example 5 deals with a case where the motion becomes slower as time passes.

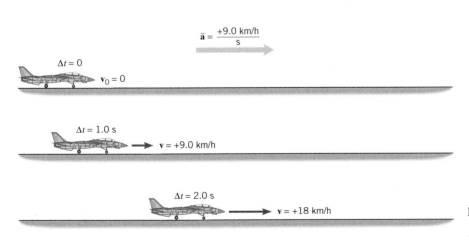

Figure 2.7 An acceleration of $+9.0 \dfrac{\text{km/h}}{\text{s}}$ means that the velocity of the plane changes by +9.0 km/h during each second of the motion. The "+" direction for **a** and **v** is to the right.

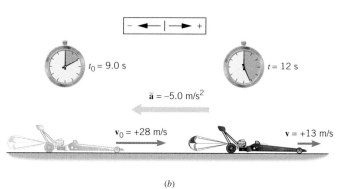

Figure 2.8 (*a*) To slow down, a drag racer deploys a parachute and applies the brakes. (*b*) The velocity of the car is decreasing, giving rise to an average acceleration $\bar{\mathbf{a}}$ that points opposite to the velocity.

EXAMPLE 5 • Acceleration and Decreasing Velocity

A drag racer crosses the finish line, and the driver deploys a parachute and applies the brakes to slow down, as Figure 2.8 illustrates. The driver begins slowing down when $t_0 = 9.0$ s and the car's velocity is $\mathbf{v_0} = +28$ m/s. When $t = 12.0$ s, the velocity has been reduced to $\mathbf{v} = +13$ m/s. What is the average acceleration of the dragster?

Reasoning and Solution According to Equation 2.4, the average acceleration is

$$\bar{\mathbf{a}} = \frac{\mathbf{v} - \mathbf{v_0}}{t - t_0} = \frac{13 \text{ m/s} - 28 \text{ m/s}}{12.0 \text{ s} - 9.0 \text{ s}} = \boxed{-5.0 \text{ m/s}^2}$$

Figure 2.9 shows how the velocity of the dragster changes during the braking, assuming that the acceleration is constant throughout the motion. The acceleration calculated in Example 5 is negative, indicating that the acceleration points to the

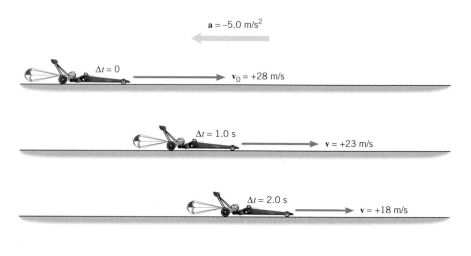

Figure 2.9 Here, an acceleration of -5.0 m/s^2 means the velocity decreases by 5.0 m/s during each second of elapsed time.

left in the drawing. As a result, the acceleration and the velocity point in *opposite* directions. ***Whenever the acceleration and velocity vectors have opposite directions, the object slows down and is said to be "decelerating."*** In contrast, the acceleration and velocity vectors in Figure 2.7 point in the *same* direction, and the object speeds up.

2.4 EQUATIONS OF KINEMATICS FOR CONSTANT ACCELERATION

It is now possible to describe the motion of an object traveling with a constant acceleration along a straight line. To do so, we will use a set of equations known as the equations of kinematics for constant acceleration. These equations entail no new concepts, because they will be obtained by combining the familiar ideas of displacement, velocity, and acceleration, as the concept chart in Figure 2.10 shows. In combining these ideas, it is convenient to assume that the object is located at the origin $\mathbf{x_0} = 0$ when $t_0 = 0$. With this assumption, the displacement $\Delta\mathbf{x} = \mathbf{x} - \mathbf{x_0}$ becomes $\Delta\mathbf{x} = \mathbf{x}$. Furthermore, it is customary to dispense with the use of boldface symbols for the displacement, velocity, and acceleration vectors in the equations that follow. We will, however, continue to convey the directions of these vectors with plus or minus signs.

Consider an object that has an initial velocity of v_0 at time $t_0 = 0$ and moves for a time t with a constant acceleration a. For a complete description of the motion, it is also necessary to know the final velocity and displacement at time t. The final velocity v can be obtained directly from Equation 2.4:

$$\bar{a} = a = \frac{v - v_0}{t} \quad \text{or} \quad v = v_0 + at \tag{2.4}$$

The displacement at time t can be obtained from Equation 2.2, if a value for the average velocity $\bar{v}$ can be obtained. Considering the assumption that $\mathbf{x_0} = 0$ at $t_0 = 0$, we have

$$\bar{v} = \frac{x - x_0}{t - t_0} = \frac{x}{t} \quad \text{or} \quad x = \bar{v}t$$

Because the acceleration is constant, the velocity increases at a constant rate. Thus, the average velocity $\bar{v}$ is midway between the initial and final velocities:

$$\bar{v} = \tfrac{1}{2}(v_0 + v) \quad \text{(constant acceleration)} \tag{2.6}$$

Equation 2.6 applies only if the acceleration is constant and cannot be used when the acceleration is changing. The displacement at time t can now be determined as

$$x = \bar{v}t = \tfrac{1}{2}(v_0 + v)t \quad \text{(constant acceleration)} \tag{2.7}$$

Notice in Equations 2.4 ($v = v_0 + at$) and 2.7 [$x = \tfrac{1}{2}(v_0 + v)t$] that there are five kinematic variables:

1. x = displacement
2. $a = \bar{a}$ = acceleration (constant)
3. v = final velocity at time t
4. v_0 = initial velocity at time $t_0 = 0$
5. t = time elapsed since $t_0 = 0$

Concepts at a Glance

Displacement
(Section 2.1)

Velocity
(Section 2.2)

Equations of Kinematics
for Constant Acceleration

Acceleration
(Section 2.3)

Figure 2.10 *Concepts at a Glance*
The equations of kinematics for constant acceleration are obtained by combining the concepts of displacement, velocity, and acceleration. As long as Michael Johnson's acceleration remains constant, these equations can be used to calculate his displacement and velocity as a function of time.

Each of the two equations contains four of these variables, so if three of them are known, the fourth variable can always be found. Example 6 illustrates how Equations 2.4 and 2.7 are used to describe the motion of an object.

EXAMPLE 6 • The Displacement of a Speedboat

The speedboat in Figure 2.11 has a constant acceleration of $+2.0$ m/s^2. If the initial velocity of the boat is $+6.0$ m/s, find its displacement after 8.0 seconds.

Reasoning The three known variables are listed in the table:

Speedboat Data				
x	a	v	v_0	t
?	$+2.0$ m/s^2		$+6.0$ m/s	8.0 s

We can use $x = \frac{1}{2}(v_0 + v)t$ to find the displacement of the boat if a value for the final velocity v can be found. To find the final velocity, it is necessary to use the value given for the acceleration, because it tells us how the velocity changes, according to $v = v_0 + at$.

Solution The final velocity is

$$v = v_0 + at = 6.0 \text{ m/s} + (2.0 \text{ m/s}^2)(8.0 \text{ s}) = +22 \text{ m/s} \qquad (2.4)$$

The displacement of the boat can now be obtained:

$$x = \tfrac{1}{2}(v_0 + v)t = \tfrac{1}{2}(6.0 \text{ m/s} + 22 \text{ m/s})(8.0 \text{ s}) = \boxed{+110 \text{ m}} \qquad (2.7)$$

A calculator would give the answer as 112 m, but this number must be rounded to 110 m, since the data are accurate to only two significant figures.

The solution to Example 6 involved two steps: finding the final velocity v and then calculating the displacement x. It would be helpful if we could find an equation that allows us to determine the displacement in a single step. Using Example 6 as a guide, we can obtain such an equation by substituting v from Equation 2.4

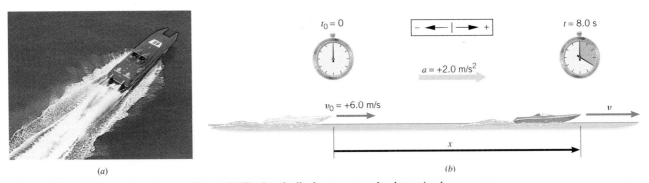

Figure 2.11 (*a*) An accelerating speed boat. (*b*) The boat's displacement can be determined if the boat's acceleration, initial velocity, and time of travel are known.

$(v = v_0 + at)$ for the v that appears in Equation 2.7 $[x = \frac{1}{2}(v_0 + v)t]$:

$$x = \tfrac{1}{2}(v_0 + v)t = \tfrac{1}{2}(v_0 + \boxed{v_0 + at})t = \tfrac{1}{2}(2v_0t + at^2)$$

$$x = v_0t + \tfrac{1}{2}at^2 \qquad \text{(constant acceleration)} \qquad (2.8)$$

You can verify that Equation 2.8 gives the displacement of the speedboat directly without the intermediate step of determining the final velocity. The first term (v_0t) on the right side of this equation represents the displacement that would result if the acceleration were zero, and the velocity remained constant at its initial value of v_0. The second term $(\tfrac{1}{2}at^2)$ gives the additional displacement that arises because the velocity changes (a is not zero) to values that are different from its initial value. We now turn to another illustration of accelerated motion in the next example.

The Physics of...
catapulting a jet from
an aircraft carrier.

• **PROBLEM SOLVING INSIGHT**
"Implied data" is important. For instance, in Example 7 the phrase "starts from rest" means that the initial velocity is zero ($v_0 = 0$).

EXAMPLE 7 • Catapulting a Jet

A jet is taking off from the deck of an aircraft carrier, as Figure 2.12 shows. Starting from rest, the jet is catapulted with a constant acceleration of $+31 \text{ m/s}^2$ along a straight line and reaches a velocity of $+62 \text{ m/s}$. Find the displacement of the jet.

Reasoning The data are as follows:

Jet Data				
x	a	v	v_0	t
?	$+31 \text{ m/s}^2$	$+62 \text{ m/s}$	0	

The initial velocity v_0 is zero, since the jet starts from rest. The displacement x of the aircraft can be obtained from $x = \frac{1}{2}(v_0 + v)t$, if we can determine the time t during which the plane is being accelerated. But t is controlled by the value of the acceleration. With larger accelerations, the jet reaches its final velocity in shorter times, as can be seen by solving $v = v_0 + at$ for t.

Solution Solving Equation 2.4 for t, we find

$$t = \frac{v - v_0}{a} = \frac{62 \text{ m/s} - 0}{31 \text{ m/s}^2} = 2.0 \text{ s}$$

Figure 2.12 (a) A plane is being launched from an aircraft carrier. (b) During the launch, a catapult accelerates the jet down the deck.

(a)

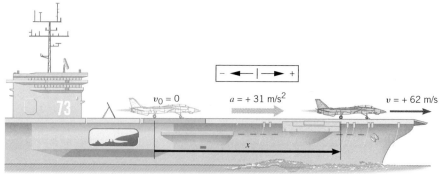

(b)

Table 2.1 Equations of Kinematics for Constant Acceleration

Equation Number	Equation	Variables				
		x	a	v	v_0	t
(2.4)	$v = v_0 + at$		✓	✓	✓	✓
(2.7)	$x = \frac{1}{2}(v_0 + v)t$	✓		✓	✓	✓
(2.8)	$x = v_0 t + \frac{1}{2}at^2$	✓	✓		✓	✓
(2.9)	$v^2 = v_0^2 + 2ax$	✓	✓	✓	✓	

Since the time is now known, the displacement can be found:

$$x = \tfrac{1}{2}(v_0 + v)t = \tfrac{1}{2}(0 + 62 \text{ m/s})(2.0 \text{ s}) = \boxed{+62 \text{ m}} \qquad (2.7)$$

When a, v, and v_0 are known, but the time t is not known, as in Example 7, it is possible to calculate the displacement x in a single step. Solving Equation 2.4 for the time $[t = (v - v_0)/a]$, and then substituting into Equation 2.7 reveals that

$$x = \tfrac{1}{2}(v_0 + v)t = \tfrac{1}{2}(v_0 + v)\boxed{\frac{v - v_0}{a}} = \frac{v^2 - v_0^2}{2a}$$

Solving for v^2 shows that

$$v^2 = v_0^2 + 2ax \qquad \text{(constant acceleration)} \qquad (2.9)$$

It is a straightforward exercise to verify that Equation 2.9 can be used to find the displacement of the jet in Example 7 without having to solve first for the time. Table 2.1 presents a summary of the equations that we have been considering. These equations are called the **equations of kinematics.** Each equation contains four variables, as indicated by the check marks (✓) in the table. The next section shows how to apply the equations of kinematics.

2.5 APPLICATIONS OF THE EQUATIONS OF KINEMATICS

The equations of kinematics can be used for any moving object, as long as the acceleration of the object is constant. However, to avoid errors when using these equations, it helps to follow a few sensible guidelines and to be alert for a few situations that can arise during your calculations.

Decide at the start which directions are to be called positive (+) and negative (−) relative to a conveniently chosen coordinate origin. This decision is arbitrary, but important, for displacement, velocity, and acceleration are vectors, and their directions must always be taken into account. In the examples that follow, the positive and negative directions will be shown in the drawings that accompany the problems. It does not matter which direction is chosen to be positive. However, once the choice has been made, it should not be changed during the course of the calculation. Example 8 illustrates these important issues.

EXAMPLE 8 • Switching Positive and Negative Directions

Repeat Example 6 using the same data, but now assume the negative direction is to the right, rather than to the left (see Figure 2.13).

Reasoning We begin by taking into account the fact that the direction to the right has been chosen to be negative. Since the acceleration a and the initial velocity v_0 point to the right, they are now represented by negative numbers, as indicated in the following data summary. The magnitudes of a and v_0, however, are the same as in Example 6.

	Speedboat Data			
x	a	v	v_0	t
?	-2.0 m/s^2		-6.0 m/s	8.0 s

The displacement of the boat can be found from Equation 2.8 in Table 2.1. This equation is selected because it contains the four variables of interest, x, a, v_0, and t, with x being the only unknown.

Solution According to Equation 2.8, it follows that

$$x = v_0 t + \tfrac{1}{2}at^2 = (-6.0 \text{ m/s})(8.0 \text{ s}) + \tfrac{1}{2}(-2.0 \text{ m/s}^2)(8.0 \text{ s})^2 = \boxed{-110 \text{ m}}$$

The magnitude of the displacement is 110 m, the same as in Example 6. Now, however, the displacement is in the negative direction. Since the negative direction now points to the right, the meaning of this answer is exactly the same as it is in Example 6, that is, the boat moves to the right.

As you reason through a problem before attempting to solve it, be sure to interpret the terms "decelerating" or "deceleration" correctly, should they occur in the problem statement. These terms are the source of frequent confusion, and Conceptual Example 9 offers help in understanding them.

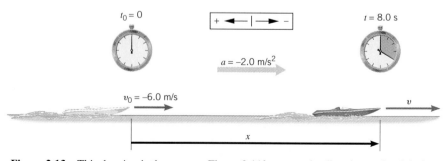

Figure 2.13 This drawing is the same as Figure 2.11*b*, except the direction to the right is now negative.

CONCEPTUAL EXAMPLE 9 • Deceleration Versus Negative Acceleration

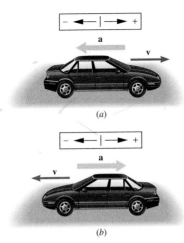

A car is traveling along a straight road and is decelerating. Does the car's acceleration *a* necessarily have a negative value?

Reasoning and Solution We begin with the meaning of the term "decelerating," which has nothing to do with whether the acceleration *a* is positive or negative. The term means only that the acceleration vector points opposite to the velocity vector and indicates that the moving object is slowing down. When a moving object slows down, its instantaneous speed (the magnitude of the instantaneous velocity) decreases. One possibility is that the velocity vector of the car points to the right, in the positive direction, as Figure 2.14a shows. The term "decelerating" implies that the acceleration vector points opposite, or to the left, which is the negative direction. Here, the value of the acceleration *a* would indeed be negative. However, there is another possibility. The car could be traveling to the left, as in Figure 2.14b. Now, since the velocity vector points to the left, the acceleration vector would point opposite or to the right, according to the meaning of the term "decelerating." But right is the positive direction, so the acceleration *a* would have a positive value in Figure 2.14b. We see, then, that *a decelerating object does not necessarily have a negative acceleration.*

Figure 2.14 When a car decelerates along a straight road, the direction of the acceleration vector depends on the direction in which the car is traveling, as Conceptual Example 9 discusses.

Related Homework Material: Problems 22 and 40

Sometimes there are two possible answers to a kinematics problem, each answer corresponding to a different situation. Example 10 discusses one such case.

EXAMPLE 10 • An Accelerating Spacecraft

The spacecraft shown in Figure 2.15a is traveling with a velocity of +3250 m/s. Suddenly the retrorockets are fired, and the spacecraft begins to slow down with an acceleration whose magnitude is 10.0 m/s². What is the velocity of the spacecraft when the displacement of the craft is +215 km, relative to the point where the retrorockets began firing?

The Physics of... the acceleration caused by a retrorocket.

Reasoning Since the spacecraft is slowing down at this stage of the motion, the acceleration must be opposite to the velocity. The velocity points to the right in the drawing, so the acceleration must point to the left, which is the negative direction; thus, $a = -10.0$ m/s². The three known variables are listed as follows:

Spacecraft Data				
x	a	v	v_0	t
+215 000 m	-10.0 m/s²	?	+3250 m/s	

The final velocity v of the spacecraft can be calculated using Equation 2.9, since it contains the four pertinent variables.

Solution From Equation 2.9, we find that

$$v^2 = v_0^2 + 2ax = (3250 \text{ m/s})^2 + 2(-10.0 \text{ m/s}^2)(215\,000 \text{ m}) = 6.3 \times 10^6 \text{ m}^2/\text{s}^2$$

Figure 2.15 (*a*) Because of an acceleration of -10.0 m/s^2, the spacecraft changes its velocity from v_0 to v. (*b*) Continued firing of the retrorockets changes the direction of the craft's motion.

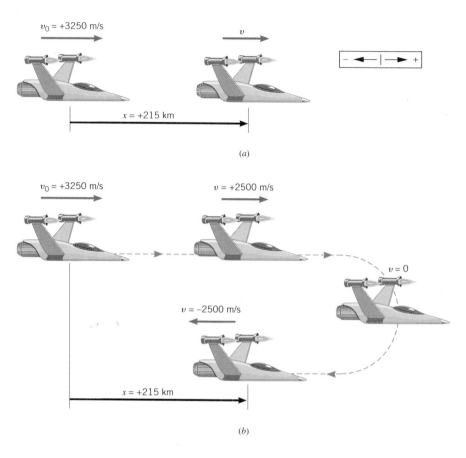

(*a*)

(*b*)

As it comes in for a landing, this eagle has a velocity vector that points to the right. However, since it is decelerating, it has an acceleration vector that points in the opposite direction, or to the left.

The two possible solutions are $v = \pm \sqrt{6.3 \times 10^6 \text{ m}^2/\text{s}^2}$, or

$$v = +2500 \text{ m/s} \quad \text{and} \quad v = -2500 \text{ m/s}$$

Both of these answers correspond to the *same* displacement ($x = +215$ km), but each arises in a different part of the motion. The answer $v = +2500$ m/s corresponds to the situation in Figure 2.15*a*, where the spacecraft has slowed to a speed of 2500 m/s, but is still traveling to the right. The answer $v = -2500$ m/s arises because the retrorockets eventually bring the spacecraft to a momentary halt and cause it to reverse direction, after which it moves toward the left. As the craft moves toward the left, its speed increases due to the continually firing rockets. After a time, the velocity of the craft becomes $v = -2500$ m/s, giving rise to the situation in Figure 2.15*b*. In both parts of the drawing the spacecraft has the same displacement, but a greater travel time is required in part *b* compared to part *a*.

The motion of two objects may be interrelated, so they share a common variable. The fact that the motions are interrelated is an important piece of information. In such cases, data for only two variables need be specified for each object. Example 11 involves the interrelationship between two moving objects, a person and a bus. At first glance it appears that the problem cannot be solved, because only two kinematic variables are specified for each object. However, the time t is the same for the person and the bus. Even though no value is given for the common variable, the problem can be solved. The answer turns out to be interesting.

EXAMPLE 11 • Catching a Bus

In Figure 2.16 a bus has stopped to pick up riders. A woman is running at a constant velocity of +5.0 m/s in an attempt to catch the bus. When she is 11 m from the bus, it pulls away with a constant acceleration of +1.0 m/s². From this point, how much time does it take her to reach the bus if she keeps running with the same velocity?

Reasoning To catch the bus, the woman must run not only 11 m but also the distance traveled by the moving bus, x_{bus}. Figure 2.16 shows the corresponding displacement vectors, and it follows that

$$x_{bus} + 11 \text{ m} = x_{woman}$$

This equation is the starting point for our analysis.

Solution Assuming that the clock is zeroed at the instant when the woman is 11 m from the bus, her displacement is $x_{woman} = (5.0 \text{ m/s})t$, since she runs with a constant velocity. Therefore,

$$x_{bus} + 11 \text{ m} = (5.0 \text{ m/s})t$$

This result reveals that we need a value for x_{bus} to calculate t. The known variables for the bus are summarized below:

Bus Data				
x_{bus}	a	v	v_0	t
?	+1.0 m/s²		0	✓

A check mark has been placed in the box for the time t. Although we do not have an explicit value for t, it is the same for both the bus and the woman. In other words, the time is a "common" variable, and the check mark reminds us of this fact. Equation 2.8 can be used to relate the displacement of the bus and the time:

$$x_{bus} = v_0 t + \tfrac{1}{2}at^2 = \tfrac{1}{2}(1.0 \text{ m/s}^2)t^2$$

This result can be substituted into the expression obtained earlier to give an equation containing only one unknown, the time t:

$$x_{bus} + 11 \text{ m} = (5.0 \text{ m/s})t$$
$$\tfrac{1}{2}(1.0 \text{ m/s}^2)t^2 + 11 \text{ m} = (5.0 \text{ m/s})t$$

Rearranging this result gives

$$(0.50 \text{ m/s}^2)t^2 - (5.0 \text{ m/s})t + 11 \text{ m} = 0$$

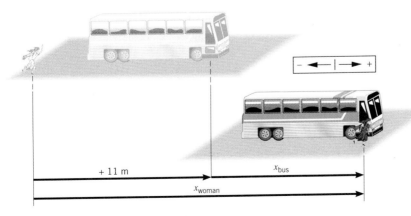

Figure 2.16 A person attempting to catch a bus.

Solving this quadratic equation for t by using the quadratic formula (see Appendix C) reveals that there are two solutions:

$$t = 3.3 \text{ s} \quad \text{and} \quad t = 6.7 \text{ s}$$

Evidently there are two times when the woman can catch the bus. When she catches the bus for the first time at $t = 3.3$ s, she has a greater velocity than the bus. This is because the bus started from rest and at $t = 3.3$ s is still moving slowly. Thus, if she fails to get the driver's attention at this moment, she will run ahead of the bus. Eventually, however, the bus will catch up with her, since it is accelerating while she is maintaining a constant velocity. The two will meet for the second time when $t = 6.7$ s. If the driver does not notice her then, the bus will pull ahead of the woman and leave her behind forever.

Often the motion of an object is divided into segments, each with a different acceleration. When solving such problems, it is important to realize that the final velocity for one segment is the initial velocity for the next segment, as Example 12 illustrates.

EXAMPLE 12 • A Motorcycle Ride

A motorcycle, starting from rest, has an acceleration of $+2.6 \text{ m/s}^2$. After the motorcycle has traveled a distance of 120 m, it slows down with an acceleration of -1.5 m/s^2 until its velocity is $+12$ m/s (see Figure 2.17). What is the total displacement of the motorcycle?

Reasoning The total displacement is the sum of the displacements for the first ("speeding up") and second ("slowing down") segments. The displacement for the first segment is $+120$ m. The displacement for the second segment can be found if the initial velocity for this segment can be determined, since values for two other variables are already known ($a = -1.5 \text{ m/s}^2$ and $v = +12$ m/s). The initial velocity for the second segment can be determined, since it is the final velocity of the first segment.

Solution Recognizing that the motorcycle starts from rest ($v_0 = 0$), we can determine the final velocity v of the first segment from the given data:

Segment 1 Data				
x	a	v	v_0	t
$+120$ m	$+2.6 \text{ m/s}^2$	?	0	

From Equation 2.9 ($v^2 = v_0^2 + 2ax$), it follows that

$$v = \sqrt{v_0^2 + 2ax} = \sqrt{2(2.6 \text{ m/s}^2)(120 \text{ m})} = +25 \text{ m/s}$$

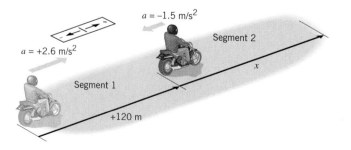

Figure 2.17 This motorcycle ride consists of two segments, each with a different acceleration.

Now we can use $+25$ m/s as the initial velocity for the second segment, along with the remaining data listed below:

Segment 2 Data				
x	a	v	v_0	t
?	-1.5 m/s^2	$+12$ m/s	$+25$ m/s	

The displacement for segment 2 can be obtained by solving $v^2 = v_0^2 + 2ax$ for x:

$$x = \frac{v^2 - v_0^2}{2a} = \frac{(12 \text{ m/s})^2 - (25 \text{ m/s})^2}{2(-1.5 \text{ m/s}^2)} = +160 \text{ m}$$

The total displacement of the motorcycle is $120 \text{ m} + 160 \text{ m} = \boxed{280 \text{ m}}$.

Now that we have seen how the equations of kinematics are applied to various situations, it's a good idea to summarize the reasoning strategy that has been used. This strategy, which is outlined below, will also be used when we consider freely falling bodies in Section 2.6 and two-dimensional motion in Chapter 3.

REASONING STRATEGY

Applying the Equations of Kinematics

1. Make a drawing to represent the situation being studied. When solving kinematics problems, few of us can do without the aid of a drawing to help develop our reasoning and explain it to our coworkers.

2. Decide which directions are to be called positive ($+$) and negative ($-$) relative to a conveniently chosen coordinate origin. Do not change your decision during the course of a calculation.

3. In an organized way, write down the values (with appropriate plus and minus signs) that are given for any of the five kinematic variables (x, a, v, v_0, and t). Be on the alert for "implied data," such as the phrase "starts from rest," which means that the value of the initial velocity is $v_0 = 0$. The data summary boxes used in the examples in the text are a good way of keeping track of this information. In addition, identify the variables that you are being asked to determine.

4. Before attempting to solve a problem, verify that the given information contains values for at least three of the five kinematic variables. Once the three known variables are identified along with the desired unknown variable, the appropriate relation from Table 2.1 can be selected. Remember that the motion of two objects may be interrelated, so they may share a common variable. The fact that the motions are interrelated is an important piece of information. In such cases, data for only two variables need be specified for each object.

5. When the motion of an object is divided into segments, remember that the final velocity of one segment is the initial velocity for the next segment.

6. Keep in mind that there may be two possible answers to a kinematics problem. Try to visualize the different physical situations to which the answers correspond.

2.6 FREELY FALLING BODIES

THE EQUATIONS OF KINEMATICS

Everyone has observed the effect of gravity as it causes objects to fall downward. In the absence of air resistance, it is found that all bodies at the same location above the earth fall vertically with the same acceleration. Furthermore, if the distance of the fall is small compared to the radius of the earth, the acceleration remains essentially constant throughout the fall. This idealized motion, in which air resistance is neglected and the acceleration is nearly constant, is known as *free-fall.* Since the acceleration is constant in free-fall, the equations of kinematics can be used.

The acceleration of a freely falling body is called the *acceleration due to gravity,* and its magnitude is denoted by the symbol g. The acceleration due to gravity is directed downward, toward the center of the earth. Near the earth's surface g is approximately

$$g = 9.80 \text{ m/s}^2 \quad \text{or} \quad 32.2 \text{ ft/s}^2$$

Air-filled
tube

Evacuated
tube

(a)   (b)

Figure 2.18 (*a*) In the presence of air resistance, the acceleration of the rock is greater than that of the paper. (*b*) In the absence of air resistance, both the rock and the paper have the same acceleration.

Unless circumstances warrant otherwise, we will use either of these values for g in subsequent calculations. In reality, however, g decreases with increasing altitude and varies slightly with latitude.

Figure 2.18*a* shows the well-known phenomenon of a rock falling faster than a sheet of paper. The effect of air resistance is responsible for the slower fall of the paper, for when air is removed from the tube, as in Figure 2.18*b*, the rock and the paper have exactly the same acceleration due to gravity. In the absence of air, the rock and the paper both exhibit free-fall motion. Free-fall is closely approximated for objects falling near the surface of the moon, where there is no air to retard the motion. A nice demonstration of lunar free-fall was performed by astronaut David Scott who dropped a hammer and a feather simultaneously from the same height. Both experienced the same acceleration due to lunar gravity and consequently hit the ground at the same time. The acceleration due to gravity near the surface of the moon is approximately one-sixth as large as that on the earth.

When the equations of kinematics are applied to free-fall motion, it is natural to use the symbol y for the displacement that occurs, since the motion occurs in the vertical or y direction. Thus, when using the equations in Table 2.1 for free-fall motion, we will simply replace x with y. There is no significance to this change. The equations have the same algebraic form for either the horizontal or vertical direction, provided that the acceleration remains constant during the motion. We now turn our attention to several examples that illustrate how the equations of kinematics are applied to freely falling bodies.

EXAMPLE 13 • A Falling Stone

A stone is dropped from rest from the top of a tall building, as Figure 2.19 indicates. After 3.00 s of free-fall, what is the displacement y of the stone?

Reasoning The upward direction is chosen as the positive direction. The three known variables are shown in the box below. The initial velocity v_0 of the stone is zero, because the stone is dropped from rest. The acceleration due to gravity is negative, since it points downward in the negative direction.

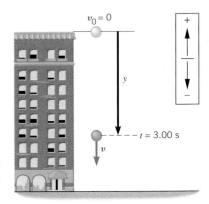

Figure 2.19 The stone, starting with zero velocity at the top of the building, is accelerated downward by gravity.

Stone Data				
y	a	v	v_0	t
?	-9.80 m/s^2		0	3.00 s

Equation 2.8 contains the appropriate variables and offers a direct solution to the problem. Since the stone moves downward and upward is the positive direction, we expect the displacement y to have a negative value.

Solution Using Equation 2.8, we find that

$$y = v_0 t + \tfrac{1}{2}at^2 = \tfrac{1}{2}(-9.80 \text{ m/s}^2)(3.00 \text{ s})^2 = \boxed{-44.1 \text{ m}}$$

The answer for y is negative, as expected.

EXAMPLE 14 • The Velocity of a Falling Stone

After 3.00 s of free-fall, what is the velocity v of the stone in Figure 2.19?

Reasoning Because of the acceleration due to gravity, the magnitude of the stone's downward velocity increases by 9.80 m/s during each second of free-fall. The data for the stone are the same as in Example 13, and Equation 2.4 offers a direct solution for the final velocity. Since the stone is moving downward in the negative direction, the value determined for v should be negative.

Solution Using Equation 2.4, we obtain

$$v = v_0 + at = (-9.80 \text{ m/s}^2)(3.00 \text{ s}) = \boxed{-29.4 \text{ m/s}}$$

The velocity is negative, as expected.

The acceleration due to gravity is always a downward-pointing vector. It describes how the speed increases for an object that is falling freely downward. This same acceleration also describes how the speed decreases for an object moving upward under the influence of gravity alone, in which case the object eventually comes to a momentary halt and then falls back to earth. Examples 15 and 16 show how the equations of kinematics are applied to an object that is moving upward under the influence of gravity.

EXAMPLE 15 • How High Does It Go?

A football game customarily begins with a coin toss to determine who kicks off. The referee tosses the coin up with an initial speed of 8.00 m/s. In the absence of air resistance, how high does the coin go above its point of release?

• **PROBLEM SOLVING INSIGHT**
"Implied data" is important. In Example 15, for instance, the phrase "how high does the coin go" refers to the maximum height, which occurs when the final velocity v in the vertical direction is $v = 0$.

Reasoning The coin is given an upward initial velocity, as in Figure 2.20. But the acceleration due to gravity points downward. Since the velocity and acceleration point in opposite directions, the coin slows down as it moves upward. Eventually, the velocity of the coin becomes $v = 0$ at the highest point. Assuming that the upward direction is positive, the data can be summarized as shown below:

Coin Data				
y	a	v	v_0	t
?	-9.80 m/s^2	0	$+8.00$ m/s	

With these data, we can use Equation 2.9 ($v^2 = v_0^2 + 2ay$) to find the maximum height y.

Solution Rearranging Equation 2.9, we find that the maximum height of the coin above its release point is

$$y = \frac{v^2 - v_0^2}{2a} = \frac{-(8.00 \text{ m/s})^2}{2(-9.80 \text{ m/s}^2)} = \boxed{+3.27 \text{ m}}$$

EXAMPLE 16 • How Long Is It in the Air?

In Figure 2.20, what is the total time the coin is in the air before returning to its release point?

Reasoning During the time the coin travels upward, gravity causes its speed to decrease to zero. On the way down, however, gravity causes the coin to regain the lost speed. Thus, the time for the coin to go up is equal to the time for it to come down. In other words, the total travel time is twice the time for the upward motion. The data for the coin during the upward trip are the same as in Example 15. With these data, we can use Equation 2.4 ($v = v_0 + at$) to find the upward travel time.

Solution Rearranging Equation 2.4, we find that

$$t = \frac{v - v_0}{a} = \frac{-8.00 \text{ m/s}}{-9.80 \text{ m/s}^2} = 0.816 \text{ s}$$

The total up-and-down time is twice this value, or $\boxed{1.63 \text{ s}}$.

It is possible to determine the total time by another method. When the coin is tossed upward and returns to its release point, the displacement for the *entire trip* is $y = 0$. With this value for the displacement, Equation 2.8 ($y = v_0t + \frac{1}{2}at^2$) can be used to find the time for the entire trip directly.

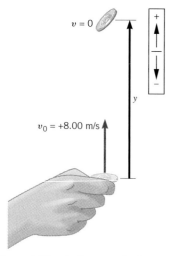

Figure 2.20 At the start of a football game, a referee tosses a coin upward with an initial velocity of $v_0 = +8.00$ m/s. The velocity of the coin is momentarily zero when the coin reaches its maximum height.

Examples 15 and 16 illustrate that the expression "freely falling" does not necessarily mean an object is falling down. A freely falling object is any object moving either upward or downward under the influence of gravity alone. In either case, the object always experiences the same *downward acceleration* due to gravity, a fact that is the focus of the next example.

CONCEPTUAL EXAMPLE 17 • Acceleration Versus Velocity

There are three parts to the motion of the coin in Figure 2.20. On the way up, the coin has a velocity vector that is directed upward and has a decreasing magnitude. At the top of its path, the coin momentarily has a zero velocity. On the way down, the coin has a downward-pointing velocity vector with an increasing magnitude. In the absence of air resistance, does the acceleration of the coin, like the velocity, change from one part of the motion to another?

Reasoning and Solution Since air resistance is absent, the coin is in free-fall motion. Therefore, the acceleration vector is that due to gravity and has the same magnitude and the same direction at all times. It has a magnitude of 9.80 m/s^2 and points downward during both the upward and downward portions of the motion. Furthermore, just because the coin's instantaneous velocity is zero at the top of the motional path, don't think that the acceleration vector is also zero there. Acceleration is the rate at which velocity changes, and the velocity at the top is changing, even though at one instant it is zero. In fact, the acceleration at the top has the same magnitude of 9.80 m/s^2 and the same downward direction as during the rest of the motion. Thus, *the coin's velocity vector changes from moment to moment, but its acceleration vector does not change.*

SYMMETRY IN THE MOTION OF FREELY FALLING BODIES

The motion of an object that is thrown upward and eventually returns to earth contains a symmetry that is useful to keep in mind from the point of view of problem solving. The calculations just completed indicate that a time symmetry exists in free-fall motion, in the sense that the time required for the object to reach maximum height equals the time for it to return to its starting point.

A type of symmetry involving the speed also exists. Figure 2.21 shows the coin considered in Examples 15 and 16. At any displacement y above the point of release, the coin's speed during the upward trip equals the speed at the same point during the downward trip. For instance, when $y = +1.50$ m, Equation 2.9 gives two possible values for the final velocity v, assuming that the initial velocity is $v_0 = +8.00$ m/s:

$$v^2 = v_0{}^2 + 2ay = (8.00 \text{ m/s})^2 + 2(-9.80 \text{ m/s}^2)(1.50 \text{ m}) = 34.6 \text{ m}^2/\text{s}^2$$
$$v = \pm 5.88 \text{ m/s}$$

The value $v = +5.88$ m/s is the velocity of the coin on the upward trip, while $v = -5.88$ m/s is the velocity on the downward trip. The speed in both cases is identical and equals 5.88 m/s. Likewise, the speed just as the coin returns to its point of release is 8.00 m/s, which equals the initial speed. This symmetry involving the speed arises because the coin loses 9.80 m/s in speed each second on the way up and gains back the same amount each second on the way down. In Conceptual Example 18, we use just this kind of symmetry to guide our reasoning as we analyze the motion of a pellet shot from a gun.

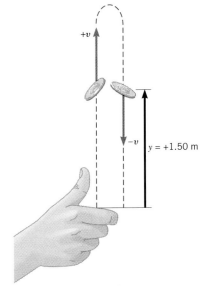

Figure 2.21 For a given displacement along the motional path, the upward speed of the coin is equal to its downward speed, but the two velocities point in opposite directions.

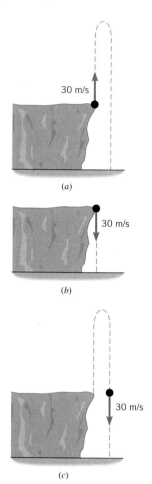

Figure 2.22 (*a*) From the edge of a cliff, a pellet is fired straight upward from a gun. The pellet's initial speed is 30 m/s. (*b*) The pellet is fired straight downward with an initial speed of 30 m/s. (*c*) In Conceptual Example 18 this drawing plays the central role in reasoning that is based on symmetry.

CONCEPTUAL EXAMPLE 18 • Taking Advantage of Symmetry

Figure 2.22*a* shows a pellet, having been fired from a gun, moving straight upward from the edge of a cliff. The initial speed of the pellet is 30 m/s. It goes up and then falls back down, eventually hitting the ground beneath the cliff. In Figure 2.22*b* the pellet has been fired straight downward at the same initial speed. In the absence of air resistance, does the pellet in part *b* strike the ground beneath the cliff with a smaller, a greater, or the same speed as the pellet in part *a*?

Reasoning and Solution Because air resistance is absent, the motion is that of free-fall, and the symmetry inherent in free-fall motion offers an immediate answer to our question. Figure 2.22*c* shows why. This part of the drawing shows the pellet after it has been fired upward and then fallen back down to its starting point. Symmetry indicates that the speed in part *c* is the same as in part *a*, namely, 30 m/s. Thus, part *c* is just like part *b*, where the pellet is actually fired downward with a speed of 30 m/s. Consequently, whether the pellet is fired as in part *a* or part *b*, it moves downward from the cliff edge at a speed of 30 m/s. In either case, there is the same acceleration due to gravity and the same displacement from the cliff edge to the ground below. Under these conditions, ***the pellet reaches the ground with the same speed no matter in which vertical direction it is fired initially.***

Related Homework Material: *Problems 48 and 50*

2.7 GRAPHICAL ANALYSIS OF VELOCITY AND ACCELERATION

Graphical techniques are helpful in understanding the concepts of velocity and acceleration. Suppose a bicyclist is riding with a constant velocity of $v = +4$ m/s. The position x of the bicycle can be plotted along the vertical axis of a graph, while the time t is plotted along the horizontal axis. Since the position of the bike increases by 4 m every second, the graph of x versus t is a straight line. Furthermore, if the bike is assumed to be at $x = 0$ when $t = 0$, the straight line passes through the origin, as Figure 2.23 shows. Each point on this line gives the position of the bike at a particular time. For instance, at $t = 1$ s the position is 4 m, while at $t = 3$ s the position is 12 m.

In constructing the graph in Figure 2.23, we used the fact that the velocity was $+4$ m/s. Suppose, however, that we were given this graph, but did not have any prior knowledge of the velocity. The velocity could be determined by considering what happens to the bike between the times of 1 and 3 s, for instance. The change in time is $\Delta t = 2$ s. During this time interval, the position of the bike changes from $+4$ to $+12$ m, and the change in position is $\Delta x = +8$ m. The ratio $\Delta x/\Delta t$ is called the **slope** of the straight line.

$$\text{Slope} = \frac{\Delta x}{\Delta t} = \frac{+8 \text{ m}}{2 \text{ s}} = +4 \text{ m/s}$$

Notice that the slope is equal to the velocity of the bike. This result is no accident, because $\Delta x/\Delta t$ is the definition of average velocity (see Equation 2.2). Thus, for an object moving with a constant velocity, the slope of the straight line in a position–time graph gives the value of the velocity. Since the position–time graph is a straight line, any time interval Δt can be chosen to calculate the velocity. Choosing a different Δt will yield a different Δx, but the velocity $\Delta x/\Delta t$ will not change. In the real world, objects rarely move with a constant velocity at all times, as the next example illustrates.

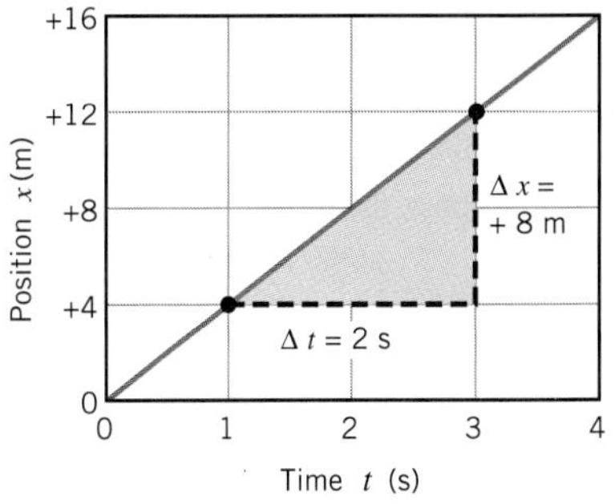

EXAMPLE 19 • A Bicycle Trip

A bicyclist maintains a constant velocity on the outgoing leg of a journey, zero velocity while stopped for lunch, and another constant velocity on the way back. Figure 2.24 shows the position–time graph for such a trip. Using the time and position intervals indicated in the drawing, obtain the velocities for each segment of the trip.

Reasoning and Solution Using Equation 2.2, we find the following velocities:

Segment 1 $\qquad \overline{v} = \dfrac{\Delta x}{\Delta t} = \dfrac{+400 \text{ m}}{200 \text{ s}} = \boxed{+2 \text{ m/s}}$

Segment 2 $\qquad \overline{v} = \dfrac{\Delta x}{\Delta t} = \dfrac{0 \text{ m}}{400 \text{ s}} = \boxed{0 \text{ m/s}}$

Segment 3 $\qquad \overline{v} = \dfrac{\Delta x}{\Delta t} = \dfrac{-400 \text{ m}}{400 \text{ s}} = \boxed{-1 \text{ m/s}}$

In the second segment of the journey the velocity is zero, reflecting the fact that the bike is stationary. Since the position of the bike does not change, segment 2 is a horizontal line that has a zero slope. In the third part of the motion the velocity is negative, because the position of the bike decreases from $x = +800$ m to $x = +400$ m during the 400-s interval shown in the graph. As a result, segment 3 has a negative slope, and the velocity is negative.

If the object is accelerating, its velocity is changing. When the velocity is changing, the x versus t graph is not a straight line, but is a curve, perhaps like that in Fig-

Figure 2.23 A graph of position vs. time for an object moving with a constant velocity of $v = \Delta x/\Delta t = +4$ m/s.

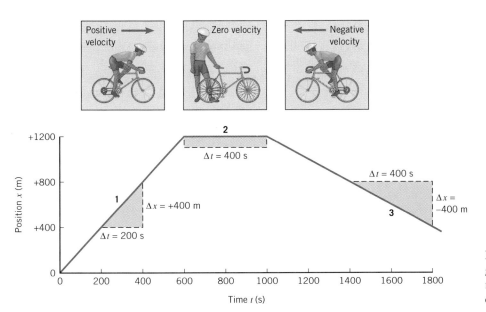

Figure 2.24 This position vs. time graph consists of three straight line segments, each corresponding to a different constant velocity.

Figure 2.25 When the velocity is changing, the position vs. time graph is a curved line. The slope $\Delta x/\Delta t$ of the tangent line drawn to the curve at a given time is the instantaneous velocity at that time.

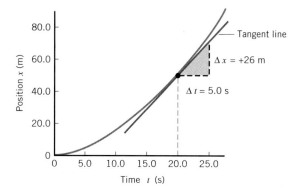

ure 2.25. This curve was drawn using Equation 2.8 ($x = v_0 t + \frac{1}{2}at^2$), assuming an acceleration of $a = 0.26$ m/s^2 and an initial velocity of $v_0 = 0$. The velocity at any instant of time can be determined by measuring the slope of the curve at that instant. The slope at any point along the curve is defined to be the slope of the tangent line drawn to the curve at that point. For instance, in Figure 2.25 a tangent line is drawn at $t = 20.0$ s. To determine the slope of the tangent line, a triangle is constructed using an arbitrarily chosen time interval of $\Delta t = 5.0$ s. The change in x associated with this time interval can be read from the tangent line as $\Delta x = +26$ m. Therefore,

$$\text{Slope of tangent line} = \frac{\Delta x}{\Delta t} = \frac{+26 \text{ m}}{5.0 \text{ s}} = +5.2 \text{ m/s}$$

The slope of the tangent line is the instantaneous velocity, which in this case is $v = +5.2$ m/s. This graphical result can be verified by using Equation 2.4 with $v_0 = 0$: $v = at = (+0.26 \text{ m/s}^2)(20.0 \text{ s}) = +5.2$ m/s.

Insight into the meaning of acceleration can also be gained with the aid of a graphical representation. Consider an object moving with a constant acceleration of $a = +6$ m/s^2. If the object has an initial velocity of $v_0 = +5$ m/s, its velocity at any time is represented by Equation 2.4 as

$$v = v_0 + at = (5 \text{ m/s}) + (6 \text{ m/s}^2)t$$

This relation is plotted as the velocity versus time graph in Figure 2.26. The graph of v versus t is a straight line that intercepts the vertical axis at $v_0 = 5$ m/s. The slope of this straight line can be calculated from the data shown in the drawing:

$$\text{Slope} = \frac{\Delta v}{\Delta t} = \frac{+12 \text{ m/s}}{2 \text{ s}} = +6 \text{ m/s}^2$$

The ratio $\Delta v/\Delta t$ is, by definition, equal to the average acceleration (Equation 2.4), so the slope of the straight line in a velocity–time graph is the average acceleration.

Figure 2.26 A velocity vs. time graph that applies to an object with an acceleration of $\Delta v/\Delta t = +6$ m/s^2. The initial velocity is $v_0 = +5$ m/s when $t = 0$.

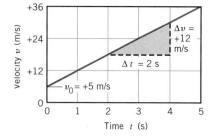

SUMMARY

Displacement is a vector that points from an object's initial position toward its final position. The magnitude of the displacement is the shortest distance between the two positions.

The **average speed** of an object is the distance traveled by the object divided by the time required to cover the distance: Average speed = (Distance)/(Elapsed time).

The **average velocity** $\bar{v}$ of an object is defined as the object's displacement Δx divided by the elapsed time Δt: $\bar{v} = \Delta x/\Delta t$. Average velocity is a vector that has the same direction as the displacement. When the elapsed time Δt is infinitesimally small, the average velocity becomes equal to the **instantaneous velocity v,** the velocity at an instant of time.

Average acceleration $\bar{a}$ is a vector that is equal to the change in the velocity Δv divided by the elapsed time Δt, the "change in velocity" being the final minus the initial velocity: $\bar{a} = \Delta v/\Delta t$. When Δt is infinitesimally small, the average acceleration becomes equal to the **instantaneous ac-**

celeration a. Acceleration is the rate at which the velocity is changing.

When an object moves with a constant acceleration along a straight line, its displacement $x - x_0$, final velocity v, initial velocity v_0, acceleration a, and the elapsed time t are related by the following equations, assuming that $x_0 = 0$ at $t = 0$:

$$x = \bar{v}t = \tfrac{1}{2}(v_0 + v)t \quad \text{and} \quad v = v_0 + at$$

These two equations can be combined algebraically to give two additional equations that are listed in Table 2.1. The equations in this table are known as the **equations of kinematics.**

In **free-fall** motion, an object experiences negligible air resistance and a constant acceleration due to gravity. All objects at the same location above the earth have the same acceleration due to gravity. The acceleration due to gravity is directed toward the center of the earth and has a magnitude of approximately 9.80 m/s² near the earth's surface.

CONCEPTUAL QUESTIONS

1. A honeybee leaves the hive and travels 2 km before returning. Is the displacement for the trip the same as the distance traveled? If not, why not?

2. Two buses depart from Chicago, one going to New York and one to San Francisco. Each bus travels at a speed of 30 m/s. Do they have equal velocities? Explain.

3. Is the average speed of a vehicle a vector or a scalar quantity? Provide a reason for your answer.

4. Often, traffic lights are timed so that if you travel at a certain constant speed, you can avoid all red lights. Discuss how the timing of the lights is determined, considering that the distance between them varies from one light to the next.

5. One of the following statements is incorrect. (a) The car traveled around the track at a constant velocity. (b) The car traveled around the track at a constant speed. Which statement is incorrect and why?

6. Give an example from your own experience in which the velocity of an object is zero for just an instant of time, but its acceleration is not zero.

7. At a given instant of time, a car and a truck are traveling side by side in adjacent lanes of a highway. The car has a greater velocity than the truck. Does the car necessarily have a greater acceleration? Explain.

8. The average velocity for a trip has a positive value. Is it possible for the instantaneous velocity at any point during the trip to have a negative value? Justify your answer.

9. A runner runs half the remaining distance to the finish line every ten seconds. She runs in a straight line and does not ever reverse her direction. Does her acceleration have a constant magnitude? Give a reason for your answer.

10. An object moving with a constant acceleration can certainly slow down. But can an object ever come to a permanent halt if its acceleration truly remains constant? Explain.

11. An experimental vehicle slows down and comes to a halt with an acceleration whose magnitude is 9.80 m/s². After reversing direction in a negligible amount of time, the vehicle speeds up with an acceleration of 9.80 m/s². Other than being horizontal, how is this motion different, if at all, from the motion of a ball that is thrown straight upward, comes to a halt, and falls back to earth?

12. A ball is dropped from rest from the top of a building and strikes the ground with a speed v_f. From ground level, a second ball is thrown straight upward at the same instant that the first ball is dropped. The initial speed of the second ball is $v_0 = v_f$, the same speed with which the first ball will eventually strike the ground. Ignoring air resistance, decide whether the balls cross paths at half the height of the building, above the halfway point, or below the halfway point. Give your reasoning.

13. Two objects are thrown vertically upward, first one, and then, a bit later, the other. Is it possible that both reach the same maximum height at the same instant? Account for your answer.

14. The muzzle velocity of a gun is the velocity of the bullet when it leaves the barrel. The muzzle velocity of one rifle with a short barrel is greater than the muzzle velocity of another rifle that has a longer barrel. In which rifle is the acceleration of the bullet larger? Explain your reasoning.

PROBLEMS

ssm Solution is in the Student Solutions Manual. **www** Solution is available on the World Wide Web at http://www.wiley.com/college/cutnell ⚕ This icon represents a biomedical application.

Section 2.2 Speed and Velocity

1. **ssm** A whale swims due east for a distance of 6.9 km, turns around and goes due west for 1.8 km, and finally turns around again and heads 3.7 km due east. (a) What is the total distance traveled by the whale? (b) What are the magnitude and direction of the displacement of the whale?

2. One afternoon, a couple walks three-fourths of the way around a circular lake, the radius of which is 1.50 km. They start at the west side of the lake and head due south at the beginning of their walk. (a) What is the distance they travel? (b) What are the magnitude and direction (relative to due east) of the couple's displacement?

3. En route to a Hawaiian vacation, a traveler arrives late at the airport at 1:08 pm. His plane is scheduled to depart at 1:22 pm. To catch the flight, he must run 2.1 km to the gate. What must be his minimum average running speed (in m/s)?

4. Sound travels at a constant speed of 343 m/s in air. Approximately how much time (in seconds) does it take for the sound of thunder to travel 1609 m (one mile)?

5. **ssm** A plane is sitting on a runway, awaiting takeoff. On an adjacent parallel runway, another plane lands and passes the stationary plane at a speed of 45 m/s. The arriving plane has a length of 36 m. By looking out of a window (very narrow), a passenger on the stationary plane can see the moving plane. For how long a time is the moving plane visible?

6. The three-toed sloth is the slowest moving land mammal. On the ground, the sloth moves at an average speed of 0.037 m/s, considerably slower than the giant tortoise, which walks at 0.076 m/s. After 12 minutes of walking, how much further would the tortoise have gone relative to the sloth?

7. An 18-year-old runner can complete a 10.0-km course with an average speed of 4.38 m/s. A 50-year-old runner can cover the same distance with an average speed of 4.27 m/s. How much later should the younger runner start in order to finish the course *at the same time* as the older runner?

8. A tourist being chased by an angry bear is running in a straight line toward his car at a speed of 4.0 m/s. The car is a distance *d* away. The bear is 26 m behind the tourist and running at 6.0 m/s. The tourist reaches the car safely. What is the maximum possible value for *d*?

*9. **ssm** **www** A woman and her dog are out for a morning run to the river, which is located 4.0 km away. The woman runs at 2.5 m/s in a straight line. The dog is unleashed and runs back and forth at 4.5 m/s between his owner and the river, until she reaches the river. What is the total distance run by the dog?

*10. A sky diver, with parachute unopened, falls 625 m in 15.0 s. Then she opens her parachute and falls another 356 m in 142 s. What is her average velocity (both magnitude and direction) for the entire fall?

*11. A car makes a 60.0-km trip with an average velocity of 40.0 km/h in a direction due north. The trip consists of three parts. The car moves with a constant velocity of 25 km/h due north for the first 15 km and 62 km/h due north for the next 32 km. With what constant velocity does the car travel for the last 13-km segment of the trip?

**12. You are on a train that is traveling at 3.0 m/s along a level straight track. Very near and parallel to the track is a wall that slopes upward at a 12° angle with the horizontal. As you face the window (0.90 m high, 2.0 m wide) in your compartment, the train is moving to the left, as the drawing indicates. The top edge of the wall first appears at window corner A and eventually disappears at window corner B. How much time passes between appearance and disappearance of the upper edge of the wall?

Section 2.3 Acceleration

13. **ssm** **www** A jogger accelerates from rest to 3.0 m/s in 2.0 s. A car accelerates from 38 to 41 m/s also in 2.0 s. (a) Find the acceleration (magnitude only) of the jogger. (b) Determine the acceleration (magnitude only) of the car. (c) Does the car travel farther than the jogger during the 2.0 s? If so, how much farther?

14. For a standard production car, the highest road-tested acceleration ever reported occurred in 1993, when a Ford RS200 Evolution went from zero to 26.8 m/s (60 mi/h) in 3.275 s. Find the magnitude of the car's acceleration.

15. A sprinter explodes out of the starting block with an acceleration of 2.3 m/s², which she sustains for 2.0 s. Then, her acceleration drops to zero for the rest of the race. What is her speed at (a) *t* = 2.0 s and (b) the end of the race?

16. The velocity of a train is +26.4 m/s. At an average acceleration of −1.50 m/s², how much time is required for the train to decrease its velocity to +9.72 m/s?

17. **ssm** A runner accelerates to a velocity of 5.36 m/s due west in 3.00 s. His average acceleration is 0.640 m/s², also directed due west. What was his velocity when he began accelerating?

18. Starting from rest, a speedboat reaches a speed of 3.2 m/s in 2.0 s. What is the boat's speed after an additional 3.0 s has elapsed, assuming the boat's acceleration remains the same?

*19. A car is traveling along a straight road at a velocity of +31.0 m/s when its engine cuts out. For the next ten seconds, the car slows down, and its average acceleration is $\bar{a}_1$. For the next five seconds, the car slows down further, and its average acceleration is $\bar{a}_2$. The velocity of the car at the end of the fifteen-second period is +24.5 m/s. The ratio of the average acceleration values is $\bar{a}_1/\bar{a}_2 = 1.67$. Find the velocity of the car at the end of the initial ten-second interval.

**20. Two motorcycles are traveling due east with different velocities. However, four seconds later, they have the same velocity. During this four-second interval, motorcycle A has an average acceleration of 2.0 m/s² due east, while motorcycle B has an average acceleration of 4.0 m/s² due east. By how much did the speeds *differ* at the beginning of the four-second interval, and which motorcycle was moving faster?

Section 2.4 Equations of Kinematics for Constant Acceleration, Section 2.5 Applications of the Equations of Kinematics

21. ssm A cheetah, the fastest of all land animals over a short distance, accelerates from rest to 26 m/s. Assuming that the acceleration is constant, find the average speed of the cheetah.

22. Review Conceptual Example 9 as background for this problem. A car is traveling to the left, which is the negative direction. The direction of travel remains the same throughout this problem. The car's initial speed is 27.0 m/s, and during a 5.0-s interval, it changes to a final speed of (a) 29.0 m/s and (b) 23.0 m/s. In each case, find the acceleration (magnitude and algebraic sign) and state whether or not the car is decelerating.

23. (a) What is the magnitude of the average acceleration of a skier who, starting from rest, reaches a speed of 8.0 m/s when going down a slope for 5.0 s? (b) How far does the skier travel in this time?

24. In getting ready to slam-dunk the ball, a basketball player starts from rest and sprints to a speed of 6.0 m/s in 1.5 s. Assuming that the player accelerates uniformly, determine the distance he runs.

25. ssm A jetliner, traveling northward, is landing with a speed of 69 m/s. Once the jet touches down, it has 750 m of runway in which to reduce its speed to 6.1 m/s. Compute the average acceleration (magnitude and direction) of the plane during landing.

26. A truck, traveling at a velocity of 33 m/s due east, comes to a halt by decelerating at 11 m/s². How far does the truck travel in the process of stopping?

27. A speed trap is set up with two pressure-activated strips placed across a highway, 110 m apart. A car is speeding along at 33 m/s, while the speed limit is 21 m/s. At the instant the car activates the first strip, the driver begins slowing down. What minimum deceleration is needed in order that the average speed is within the limit by the time the car crosses the second marker?

28. The length of the barrel of a primitive blowgun is 1.2 m. Upon leaving the barrel, a dart has a speed of 14 m/s. Assuming that the dart is uniformly accelerated, how long does it take for the dart to travel the length of the barrel?

29. ssm www A speed ramp at an airport is basically a large conveyor belt on which you can stand and be moved along. The belt of one speed ramp moves at a constant speed such that a person who stands still on it leaves the ramp 64 s after getting on. Clifford is in a real hurry, however, and skips the speed ramp. Starting from rest with an acceleration of 0.37 m/s², he covers the same distance as the ramp does, but in one-fourth the time. What is the speed at which the belt of the ramp is moving?

*30. A drag racer, starting from rest, speeds up for 402 m with an acceleration of +17.0 m/s². A parachute then opens, slowing the car down with an acceleration of −6.10 m/s². How fast is the racer moving 3.50×10^2 m after the parachute opens?

*31. Suppose a car is traveling at 12.0 m/s, and the driver sees a traffic light turn red. After 0.510 s has elapsed (the reaction time), the driver applies the brakes, and the car decelerates at 6.20 m/s². What is the stopping distance of the car, as measured from the point where the driver first notices the red light?

*32. A speedboat starts from rest and accelerates at +2.01 m/s² for 7.00 s. At the end of this time, the boat continues for an additional 6.00 s with an acceleration of +0.518 m/s². Following this, the boat accelerates at −1.49 m/s² for 8.00 s. (a) What is the velocity of the boat at $t = 21.0$ s? (b) Find the total displacement of the boat.

*33. ssm An object starts from rest at the origin and accelerates in the direction of the $+x$ axis for a time t. The acceleration is constant. The object continues to move with the same acceleration for an additional time of one second. The distance traveled during the time t is one-half of that traveled during the one-second interval. Find the time t.

*34. A cab driver picks up a customer and delivers her 2.00 km away, driving a straight route. The driver accelerates to the speed limit and, upon reaching it, begins to decelerate immediately. The magnitude of the deceleration is three times the magnitude of the acceleration. Find the lengths of the acceleration and deceleration phases of the trip.

*35. A car is traveling at a constant speed of 27 m/s on a highway. At the instant this car passes an entrance ramp, a second car enters the highway from the ramp. The second car starts from rest and has a constant acceleration. What acceleration must it maintain, so that the two cars meet for the first time at the next exit, which is 1.8 km away?

*36. Two soccer players start from rest, 48 m apart. They run directly toward each other, both players accelerating. The first player has an acceleration whose magnitude is 0.50 m/s². The second player's acceleration has a magnitude of 0.30 m/s². (a) How much time passes before they collide? (b) At the instant they collide, how far has the first player run?

**37. ssm A locomotive is accelerating at 1.6 m/s². It passes through a 20.0-m-wide crossing in a time of 2.4 s. After the locomotive leaves the crossing, how much time is required until its speed reaches 32 m/s?

**38. A Boeing 747 "Jumbo Jet" has a length of 59.7 m. The runway on which the plane lands intersects another runway. The width of the intersection is 25.0 m. The plane decelerates through the intersection at a rate of 5.70 m/s² and clears it with a final speed of 45.0 m/s. How much time is needed for the plane to clear the intersection?

**39. In the one-hundred-meter dash a sprinter accelerates from rest to a top speed with an acceleration whose magnitude is 2.68 m/s². After achieving top speed, he runs the remainder of the race without speeding up or slowing down. If the total race is run in 12.0 s, how far does he run during the acceleration phase?

Section 2.6 Freely Falling Bodies

40. In preparation for this problem, review Conceptual Example 9. From the top of a cliff, a person uses a slingshot to fire a pebble straight downward, which is the negative direction. The initial speed of the pebble is 9.0 m/s. (a) What is the acceleration (magnitude and direction) of the pebble during the downward motion? Is the pebble decelerating? Explain. (b) After 0.50 s, how far beneath the cliff-top is the pebble?

41. ssm The greatest height reported for a jump into an airbag is 99.4 m by stunt-man Dan Koko. In 1984 he jumped from rest from the top of the Vegas World Hotel and Casino. He struck the airbag at a speed of 39 m/s (88 mi/h). To assess the effects of air resistance, determine how fast he would have been traveling on impact had air resistance been absent.

42. The drawing shows a device that you can make with a piece of cardboard, which can be used to measure a person's reaction time. Hold the card at the top and suddenly drop it. Ask a friend to try to catch the card between his or her thumb and index finger. Initially, your friend's fingers must be level with the asterisks at the bottom. By noting where your friend catches the card, you can determine his or her reaction time in milliseconds (ms). Calculate the distances d_1, d_2, and d_3.

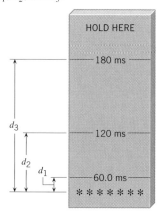

43. An arrow is fired from ground level straight upward with an initial speed of 15 m/s. How long is the arrow in the air before it strikes the ground?

44. A golf ball rebounds from the floor and travels straight upward with a speed of 5.0 m/s. To what maximum height does the ball rise?

45. ssm From her bedroom window a girl drops a water-filled balloon to the ground, 6.0 m below. If the balloon is released from rest, how long is it in the air?

46. Suppose a ball is thrown vertically upward. Eight seconds later it returns to its point of release. What is the initial velocity of the ball?

47. A diver springs upward with an initial speed of 1.8 m/s from a 3.0-m board. (a) Find the velocity with which he strikes the water. [*Hint: When the diver reaches the water, his displacement is $y = -3.0$ m (measured from the board), assuming that the downward direction is chosen as the negative direction.*] (b) What is the highest point he reaches above the water?

48. Review Conceptual Example 18 before attempting this problem. Two identical pellet guns are fired simultaneously from the edge of a cliff. These guns impart an initial speed of 30.0 m/s to each pellet. Gun A is fired straight upward, with the pellet going up and then falling back down, eventually hitting the ground beneath the cliff. Gun B is fired straight downward. In the absence of air resistance, how long after pellet B hits the ground does pellet A hit the ground?

49. ssm A wrecking ball is hanging at rest from a crane when suddenly the cable breaks. The time it takes for the ball to fall halfway to the ground is 1.2 s. Find the time it takes for the ball to fall from rest all the way to the ground.

50. Before working this problem, review Conceptual Example 18. A pellet gun is fired straight downward from the edge of a cliff that is 15 m above the ground. The pellet strikes the ground with a speed of 27 m/s. How far above the cliff edge would the pellet have gone had the gun been fired straight upward?

51. A ball is thrown straight upward and rises to a maximum height of 16 m above its launch point. At what height above its launch point has the speed of the ball decreased to one-half of its initial value?

***52.** Two students, Anne and Joan, are bouncing straight up and down on a trampoline. Anne bounces twice as high as Joan. Assuming both are in free-fall, find the ratio of the time Anne spends between bounces to the time Joan spends.

***53. ssm** A cement block accidentally falls from rest from the ledge of a 53.0-m-high building. When the block is 14.0 m above the ground, a man, 2.00 m tall, looks up and notices that the block is directly above him. How much time, at most, does the man have to get out of the way?

***54.** Two arrows are shot vertically upward. The second arrow is shot after the first one, but while the first is still on its way up. The initial speeds are such that both arrows reach their maximum heights at the same instant, although these heights are different. Suppose that the initial speed of the first arrow is 25.0 m/s and that the second arrow is fired 1.20 s after the first. Determine the initial speed of the second arrow.

***55.** A woman on a bridge 90.0 m high sees a raft floating at a constant speed on the river below. She drops a stone from rest in an attempt to hit the raft. The stone is released when the raft has 6.00 m more to travel before passing under the bridge. The stone hits the water 2.00 m in front of the raft. Find the speed of the raft.

***56.** (a) Just for fun, a person jumps from rest from the top of a tall cliff overlooking a lake. In falling through a distance H, she acquires a certain speed v. Assuming free-fall conditions, how much farther must she fall in order to acquire a speed of $2v$? Express your answer in terms of H. (b) Would the answer to part (a) be different, if this event were to occur on another planet where the acceleration due to gravity had a value other than 9.80 m/s^2? Explain.

***57. ssm www** A spelunker (cave explorer) drops a stone from rest into a hole. The speed of sound is 343 m/s in air, and the sound of the stone striking the bottom is heard 1.50 s after the stone is dropped. How deep is the hole?

*58. A ball is thrown upward from the top of a 25.0-m-tall building. The ball's initial speed is 12.0 m/s. At the same instant, a person is running on the ground at a distance of 31.0 m from the building. What must be the average speed of the person if he is to catch the ball at the bottom of the building?

**59. A roof tile falls from rest from the top of a building. An observer inside the building notices that it takes 0.20 s for the tile to pass her window, whose height is 1.6 m. How far above the top of this window is the roof?

**60. A hot air balloon is ascending straight up at a constant speed of 7.0 m/s. When the balloon is 12.0 m above the ground, a gun fires a pellet straight up from ground level with an initial speed of 30.0 m/s. Along the paths of the balloon and the pellet, there are two places where each of them has the same altitude at the same time. How far above ground level are these places?

Section 2.7 Graphical Analysis of Velocity and Acceleration

61. **ssm** For the first 10.0 km of a marathon, a runner averages a velocity that has a magnitude of 15.0 km/h. For the next 15.0 km, he averages 10.0 km/h, while for the last 15.0 km, he averages 5.0 km/h. Construct, to scale, the position–time graph for the runner.

62. A bus makes a trip according to the position–time graph shown in the drawing. What is the average velocity (magnitude and direction) of the bus during each of the segments labeled A, B, and C? Express your answers in km/h.

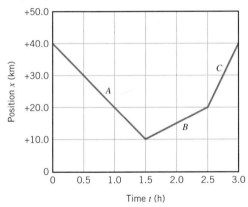

63. A snowmobile moves according to the velocity–time graph

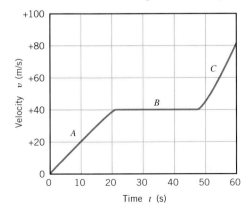

shown in the drawing. What is the snowmobile's average acceleration during each of the segments A, B, and C?

64. A person who walks for exercise produces the position–time graph given with this problem. (a) Without doing any calculations, decide which segments of the graph (A, B, C, or D) indicate positive, negative, and zero average velocities. (b) Calculate the average velocity for each segment to verify your answers to part (a).

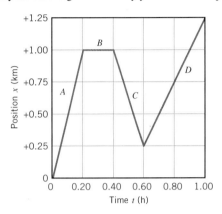

*65. **ssm** A bus makes a trip according to the position–time graph shown in the illustration. What is the average acceleration (in km/h²) of the bus for the entire 3.5-h period shown in the graph?

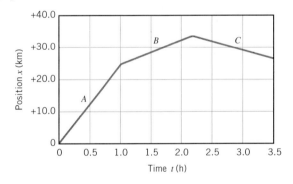

*66. A runner is at the position x = 0 when time t = 0. One hundred meters away is the finish line. Every ten seconds, this runner runs half the remaining distance to the finish line. During each ten-second segment, the runner has a constant velocity. For the first forty seconds of the motion, construct (a) the position–time graph and (b) the velocity–time graph.

**67. Two runners start one hundred meters apart and run toward each other. Each runs ten meters during the first second. During each second thereafter, each runner runs ninety percent of the distance he ran in the previous second. Thus, the velocity of each person changes from second to second. However, during any one second, the velocity remains constant. Make a position–time graph for one of the runners. From this graph, determine (a) how much time passes before the runners collide and (b) the speed with which each is running at the moment of collision.

ADDITIONAL PROBLEMS

68. A bicycle racer is moving at a speed of 14 m/s. To pass another cyclist, the racer speeds up with an acceleration of 1.2 m/s^2. At the end of 6.0 s, how fast is the racer moving?

69. ssm A penny is dropped from rest from the top of the Sears Tower in Chicago. Considering the height of the building is 427 m and ignoring air resistance, find the speed with which the penny strikes the ground.

70. The Concorde jetliner achieves a lift-off speed of 112 m/s in 20.0 s, starting from rest and traveling due east. What is the magnitude and direction of its average acceleration?

71. Two runners in a 1609-m race (one mile) finish with times of 3:52.46 (3 minutes and 52.46 seconds) and 3:52.72. Assuming that both run at their average speeds during the entire time, what distance separates them at the end of the race?

72. A rifle bullet is shot vertically upward. Twenty-three seconds later the bullet has a velocity of 72.0 m/s, downward. What is the velocity of the bullet when the bullet leaves the rifle?

73. ssm A motorcycle has a constant acceleration of 2.5 m/s^2. Both the velocity and acceleration of the motorcycle point in the same direction. How much time is required for the motorcycle to change its speed from (a) 21 to 31 m/s, and (b) 51 to 61 m/s?

74. In 1994, Noureddine Morceli set the men's world record for the 3000-m race in a time of 7 min, 25.11 s. What was his average speed?

75. An automobile starts from rest and accelerates to a final velocity in two stages along a straight road. Each stage occupies the same amount of time. In stage 1, the magnitude of the car's acceleration is 3.0 m/s^2. The magnitude of the car's velocity at the end of stage 2 is 2.5 times greater than it is at the end of stage 1. Find the magnitude of the acceleration in stage 2.

76. An arrow is shot straight up with an initial speed of 50.0 m/s. After reaching its maximum height, the arrow starts down. On the descent, a slight breeze blows the arrow laterally, and it strikes a tree limb that is 30.0 m above the ground. Determine the velocity of the arrow just before the arrow strikes the limb.

*77. ssm A bicyclist makes a trip that consists of three parts, each in the same direction (due north) along a straight road. During the first part, she rides for 22 minutes at an average speed of 7.2 m/s. During the second part, she rides for 36 minutes at an average speed of 5.1 m/s. Finally, during the third part, she rides for 8.0 minutes at an average speed of 13 m/s. (a) How far has the bicyclist traveled during the entire trip? (b) What is the average velocity of the bicyclist for the trip?

*78. Along a straight road through town, there are three speed-limit signs. They occur in the following order: 55, 35, and 25 mi/h, with the 35-mi/h sign being midway between the other two. Obeying these speed limits, the smallest possible time t_A that a driver can spend on this part of the road is to travel between the first and second signs at 55 mi/h and between the second and third signs at 35 mi/h. More realistically, a driver could slow down

from 55 to 35 mi/h with a constant deceleration and then do a similar thing from 35 to 25 mi/h. This alternative requires a time t_B. Find the ratio t_B/t_A.

*79. Suppose that the first one-fourth of the distance between two points is covered with an average velocity of +18 m/s. The average velocity for the remainder of the trip is +51 m/s. What is the average velocity for the entire trip?

*80. A race driver has made a pit stop to refuel. After refueling, he leaves the pit area with an acceleration whose magnitude is 6.0 m/s^2, and after 4.0 s he enters the main speedway. At the same instant, another race car that is on the speedway and traveling at a constant speed of 70.0 m/s overtakes and passes the entering car. If the entering car maintains its acceleration, how much time is required for it to catch the other car?

*81. ssm A log is floating on swiftly moving water. A stone is dropped from rest from a 75-m-high bridge and lands on the log as it passes under the bridge. If the log moves with a constant speed of 5.0 m/s, what is the horizontal distance between the log and the bridge when the stone is released?

*82. In a historical movie, two knights on horseback start from rest 88.0 m apart and ride directly toward each other to do battle. Sir George's acceleration has a magnitude of 0.300 m/s^2, while Sir Alfred's has a magnitude of 0.200 m/s^2. Relative to Sir George's starting point, where do the knights collide?

*83. In reaching her destination, a backpacker walks with an average velocity of 1.34 m/s, due west. This average velocity results, because she hikes for 6.44 km with an average velocity of 2.68 m/s, due west, turns around, and hikes with an average velocity of 0.447 m/s, due east. How far east did she walk?

**84. A ball is dropped from rest from the top of a cliff that is 24 m high. From ground level, a second ball is thrown straight upward at the same instant that the first ball is dropped. The initial speed of the second ball is exactly the same as that with which the first ball eventually hits the ground. In the absence of air resistance, the motions of the balls are just the reverse of each other. Determine how far below the top of the cliff the balls cross paths.

**85. ssm A train has a length of 92 m and starts from rest with a constant acceleration at time $t = 0$. At this instant, a car just reaches the end of the train. The car is moving with a constant velocity. At a time $t = 14$ s, the car just reaches the front of the train. Ultimately, however, the train pulls ahead of the car, and at time $t = 28$ s, the car is again at the rear of the train. Find the magnitudes of (a) the car's velocity and (b) the train's acceleration.

**86. A football player, starting from rest at the line of scrimmage, accelerates along a straight line for a time of 3.0 s. Then, during a negligible amount of time, he changes the magnitude of his acceleration to a value of 1.1 m/s^2. With this acceleration, he continues in the same direction for another 2.0 s, until he reaches a speed of 6.4 m/s. What is the value of his acceleration (assumed to be constant) during the initial 3.0-s period?

KINEMATICS IN TWO DIMENSIONS

Kangaroos are famous for their jumping ability. Once launched,
a kangaroo follows a familiar two-dimensional arc-shaped path that is determined
by the launch velocity and the acceleration due to gravity.

3.1 DISPLACEMENT, VELOCITY, AND ACCELERATION

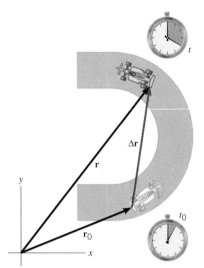

In Chapter 2 the concepts of displacement, velocity, and acceleration are used to describe an object moving along a horizontal or vertical straight line in one dimension. There are also situations in which the motion is along a curved path that lies in a plane. Such two-dimensional motion can also be described using the same concepts. In Grand Prix racing, for example, the course follows a curved road, and Figure 3.1 shows a race car at two different positions along it. These positions are identified by the vectors $\mathbf{r}$ and $\mathbf{r_0}$, which are drawn from an arbitrary coordinate origin. The *displacement* $\Delta\mathbf{r}$ of the car is the vector drawn from the initial position at time t_0 to the final position at time t. The magnitude of $\Delta\mathbf{r}$ is the shortest distance between the two positions. From the drawing it is evident that $\mathbf{r}$ is the vector sum of $\mathbf{r_0}$ and $\Delta\mathbf{r}$, so $\mathbf{r} = \mathbf{r_0} + \Delta\mathbf{r}$, or

$$\text{Displacement} = \Delta\mathbf{r} = \mathbf{r} - \mathbf{r_0}$$

Figure 3.1 The displacement $\Delta\mathbf{r}$ of the car is a vector that points from the initial position of the car at time t_0 to the final position at time t. The magnitude of $\Delta\mathbf{r}$ is the shortest distance between the two positions.

The displacement here is defined as it is in Chapter 2. Now, however, the displacement vector can lie anywhere in a plane, rather than just along a straight line.

The average velocity $\overline{\mathbf{v}}$ of the car between two positions is defined in a manner similar to that in Equation 2.2, as the displacement, $\Delta\mathbf{r} = \mathbf{r} - \mathbf{r_0}$, divided by the elapsed time $\Delta t = t - t_0$:

$$\overline{\mathbf{v}} = \frac{\mathbf{r} - \mathbf{r_0}}{t - t_0} = \frac{\Delta\mathbf{r}}{\Delta t} \qquad (3.1)$$

Since both sides of Equation 3.1 must agree in direction, the average velocity vector has the same direction as the displacement. The velocity of the car at an instant of time is its *instantaneous velocity* $\mathbf{v}$. The average velocity becomes equal to the instantaneous velocity $\mathbf{v}$ in the limit as Δt becomes infinitesimally small:

$$\mathbf{v} = \lim_{\Delta t \to 0} \frac{\Delta\mathbf{r}}{\Delta t}$$

Figure 3.2 illustrates that the instantaneous velocity $\mathbf{v}$ is tangent to the path of the car. The drawing also shows the vector components $\mathbf{v}_x$ and $\mathbf{v}_y$ of the velocity, which are parallel to the x and y axes, respectively. Using the components of a vector is advantageous when describing two-dimensional motion.

The *average acceleration* $\overline{\mathbf{a}}$ is defined just as it is for one-dimensional motion, namely, as the change in velocity, $\Delta\mathbf{v} = \mathbf{v} - \mathbf{v_0}$, divided by the elapsed time Δt:

$$\overline{\mathbf{a}} = \frac{\mathbf{v} - \mathbf{v_0}}{t - t_0} = \frac{\Delta\mathbf{v}}{\Delta t} \qquad (3.2)$$

The average acceleration vector has the same direction as the change in velocity. In the limit that the elapsed time becomes infinitesimally small, the average acceleration becomes equal to the *instantaneous acceleration* $\mathbf{a}$:

$$\mathbf{a} = \lim_{\Delta t \to 0} \frac{\Delta\mathbf{v}}{\Delta t}$$

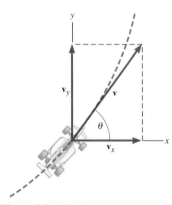

Figure 3.2 The instantaneous velocity $\mathbf{v}$ and its two vector components $\mathbf{v}_x$ and $\mathbf{v}_y$.

The acceleration has a vector component $\mathbf{a}_x$ along the x direction and a vector component $\mathbf{a}_y$ along the y direction.

Table 3.1 Equations of Kinematics for Constant Acceleration in Two-Dimensional Motion

x Component		Variable	y Component
x		Displacement	y
a_x		Acceleration	a_y
v_x		Final velocity	v_y
v_{0x}		Initial velocity	v_{0y}
t		Elapsed time	t
$v_x = v_{0x} + a_x t$	(3.3a)		$v_y = v_{0y} + a_y t$ (3.3b)
$x = \frac{1}{2}(v_{0x} + v_x)t$	(3.4a)		$y = \frac{1}{2}(v_{0y} + v_y)t$ (3.4b)
$x = v_{0x}t + \frac{1}{2}a_x t^2$	(3.5a)		$y = v_{0y}t + \frac{1}{2}a_y t^2$ (3.5b)
$v_x^2 = v_{0x}^2 + 2a_x x$	(3.6a)		$v_y^2 = v_{0y}^2 + 2a_y y$ (3.6b)

3.2 EQUATIONS OF KINEMATICS IN TWO DIMENSIONS

To understand how displacement, velocity, and acceleration are applied to two-dimensional motion, consider a spacecraft equipped with two engines that are mounted perpendicular to each other. The craft is far from other bodies, so the only forces acting on it are those produced by its own engines. As Figure 3.3 indicates, the spacecraft is assumed to be at the coordinate origin when $t_0 = 0$, so that $\mathbf{r_0} = 0$. At a later time t, the displacement is $\Delta\mathbf{r} = \mathbf{r} - \mathbf{r_0} = \mathbf{r}$. Relative to the x and y axes, the displacement $\mathbf{r}$ has vector components of $\mathbf{x}$ and $\mathbf{y}$, respectively.

In Figure 3.4 only the engine oriented along the x direction is firing, and the vehicle accelerates along this direction. It is assumed that the velocity in the y direction is zero, and it remains zero, since the y engine is turned off. The motion of the spacecraft along the x direction is described by the five kinematic variables x, a_x, v_x, v_{0x}, and t. Here the symbol "x" reminds us that we are dealing with the x components of the displacement, velocity, and acceleration vectors. The variables x, a_x, v_x, and v_{0x} are scalar components (or "components," for short). As discussed in Section 1.7, these components are positive or negative numbers (with units), depending on whether the associated vector components point along the $+x$ or the $-x$ axis. If the spacecraft has a constant acceleration along the x direction, the motion is exactly like that described in Chapter 2, and the equations of kinematics can be used. For convenience, these equations are written in the left column of Table 3.1.

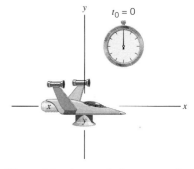

Figure 3.3 At $t_0 = 0$, the spacecraft is assumed to be at the coordinate origin, so $\mathbf{r_0} = 0$.

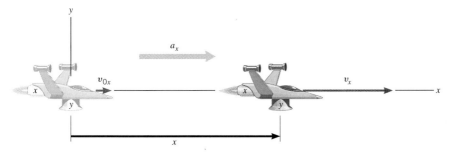

Figure 3.4 The spacecraft is moving with a constant acceleration a_x parallel to the x axis. There is no motion in the y direction, and the y engine is turned off.

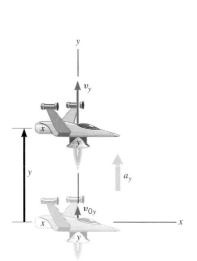

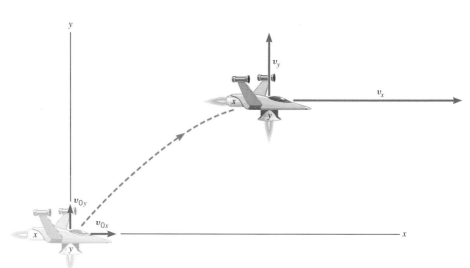

Figure 3.5 The spacecraft is moving with a constant acceleration a_y parallel to the y axis. There is no motion in the x direction, and the x engine is turned off.

Figure 3.6 The two-dimensional motion of the spacecraft can be viewed as the combination of the separate x and y motions.

Figure 3.5 is analogous to Figure 3.4, except that now only the y engine is firing, and the spacecraft accelerates along the y direction. Such a motion can be described in terms of the kinematic variables y, a_y, v_y, v_{0y}, and t. And if the acceleration along the y direction is constant, these variables are related by the equations of kinematics, as written in the right column of Table 3.1. Like their counterparts in the x direction, the components, y, a_y, v_y, and v_{0y}, may be positive ($+$) or negative ($-$) numbers (with units).

If both engines of the spacecraft are firing *at the same time,* the resulting motion takes place in part along the x axis and in part along the y axis, as Figure 3.6 illus-

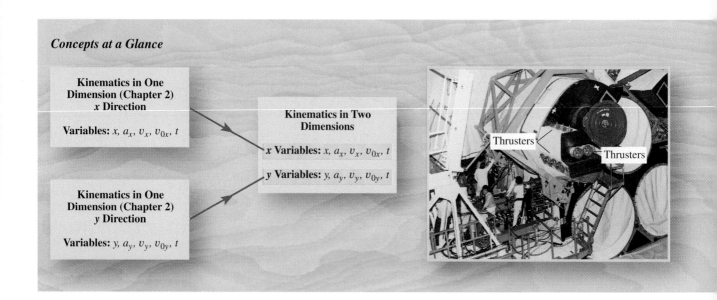

Concepts at a Glance

Kinematics in One Dimension (Chapter 2) x Direction
Variables: x, a_x, v_x, v_{0x}, t

Kinematics in Two Dimensions
x Variables: x, a_x, v_x, v_{0x}, t
y Variables: y, a_y, v_y, v_{0y}, t

Kinematics in One Dimension (Chapter 2) y Direction
Variables: y, a_y, v_y, v_{0y}, t

Thrusters

Thrusters

trates. The thrust of each engine gives the vehicle a corresponding acceleration component. The x engine accelerates the ship in the x direction and causes a change in the x component of the velocity. Likewise, the y engine causes a change in the y component of the velocity. ***It is important to realize that the x part of the motion occurs exactly as it would if the y part did not occur at all. Similarly, the y part of the motion occurs exactly as it would if the x part of the motion did not exist.*** In other words, the x and y motions are independent of each other. Because of this independence, situations that involve two-dimensional kinematics can be viewed as being constructed from two one-dimensional situations, one for the x direction and one for the y direction, as the concept chart in Figure 3.7 emphasizes. In Example 1, this is the approach taken to deal with a moving spacecraft.

EXAMPLE 1 • A Moving Spacecraft

In the x direction, the spacecraft in Figure 3.6 has an initial velocity component of $v_{0x} = +22$ m/s and an acceleration component of $a_x = +24$ m/s². In the y direction, the analogous quantities are $v_{0y} = +14$ m/s and $a_y = +12$ m/s². The directions to the right and upward have been chosen as the positive directions. After a time of 7.0 s, find (a) x and v_x, (b) y and v_y, and (c) the final velocity (magnitude and direction) of the spacecraft.

Figure 3.7 *Concepts at a Glance* In two dimensions, motion along the x direction and motion along the y direction are independent of each other. As a result, each can be analyzed separately according to the procedures for one-dimensional kinematics discussed in Chapter 2. On the space shuttle Challenger, motion in these directions is controlled by thrusters, as shown in the left photograph. The center photograph shows the Challenger in orbit with its sidefiring thruster activated, while the right photograph shows a similar view of the activated upfiring thruster.

Reasoning The motion in the x direction and the motion in the y direction can be treated separately, each as a one-dimensional motion. We will follow this approach in parts (a) and (b) to obtain the location and velocity components of the spacecraft. Then, in part (c) the velocity components will be combined to give the final velocity.

Solution

(a) The data for the motion in the x direction are listed below.

x-Direction Data				
x	a_x	v_x	v_{0x}	t
?	$+24$ m/s^2	?	$+22$ m/s	7.0 s

The x component of the craft's displacement can be found by using Equation 3.5a.

$$x = v_{0x}t + \tfrac{1}{2}a_xt^2 = (22 \text{ m/s})(7.0 \text{ s}) + \tfrac{1}{2}(24 \text{ m/s}^2)(7.0 \text{ s})^2 = \boxed{+740 \text{ m}}$$

The velocity component v_x can be calculated with the aid of Equation 3.3a:

$$v_x = v_{0x} + a_xt = (22 \text{ m/s}) + (24 \text{ m/s}^2)(7.0 \text{ s}) = \boxed{+190 \text{ m/s}}$$

(b) The data for the motion in the y direction are listed below.

y-Direction Data				
y	a_y	v_y	v_{0y}	t
?	$+12$ m/s^2	?	$+14$ m/s	7.0 s

Proceeding in the same manner as in part (a), we find that

$$\boxed{y = +390 \text{ m}} \quad \text{and} \quad \boxed{v_y = +98 \text{ m/s}}$$

(c) Figure 3.8 shows the velocity of the vehicle and its x and y vector components. The magnitude v of the velocity can be found by using the Pythagorean theorem:

$$v = \sqrt{v_x^2 + v_y^2} = \sqrt{(190 \text{ m/s})^2 + (98 \text{ m/s})^2} = \boxed{210 \text{ m/s}}$$

The direction of the velocity vector is given by the angle θ in the drawing:

$$\tan \theta = \frac{v_y}{v_x} \quad \text{or} \quad \theta = \tan^{-1}\left(\frac{v_y}{v_x}\right) = \tan^{-1}\left(\frac{98 \text{ m/s}}{190 \text{ m/s}}\right) = \boxed{27°}$$

After 7.0 s, the spacecraft has a velocity of 210 m/s in a direction of 27° above the positive x axis. At this time the craft is located at a point that is 740 m to the right and 390 m above the origin, as in Figure 3.6.

• **PROBLEM SOLVING INSIGHT**
When the motion is two-dimensional, the time variable t has the same value for both the x and y directions.

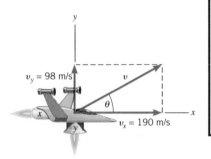

Figure 3.8 The magnitude of the velocity vector gives the speed of the spacecraft, and the angle θ gives the direction of travel relative to the positive x direction.

The following Reasoning Strategy gives an overview of how the equations of kinematics are applied to describe motion in two dimensions, such as that in Example 1. We will use this strategy for solving other types of kinematics problems in the rest of this chapter.

REASONING STRATEGY

Applying the Equations of Kinematics in Two Dimensions

1. Make a drawing to represent the situation being studied.

2. Decide which directions are to be called positive (+) and negative (−) relative to a conveniently chosen coordinate origin. Do not change your decision during the course of a calculation. There are positive and negative directions for both the *x* axis and the *y* axis.

3. Remember that the time variable *t* has the same value for the part of the motion along the *x* axis and the part along the *y* axis.

4. In an organized way, write down the values (with appropriate + and − signs) that are given for any of the five kinematic variables associated with the *x* direction and the *y* direction. Be on the alert for "implied data," such as the phrase "starts from rest," which means that the values of the initial velocity components are zero: $v_{0x} = 0$ and $v_{0y} = 0$. The data summary boxes used in the examples are a good way of keeping track of this information. In addition, identify the variables that you are being asked to determine.

5. Before attempting to solve a problem, verify that the given information contains values for at least three of the kinematic variables. Do this for the *x* and the *y* direction of the motion. Once the three known variables are identified along with the desired unknown variable, the appropriate relations from Table 3.1 can be selected.

6. When the motion of an object is divided into "segments," remember that the final velocity for one segment becomes the initial velocity for the next segment.

7. Keep in mind that there may be two possible answers to a kinematics problem. Try to visualize the different physical situations to which the answers correspond.

3.3 PROJECTILE MOTION

The biggest thrill in baseball is a home run. The motion of the ball on its curving path into the stands is a common type of two-dimensional motion called "projectile motion." A good description of projectile motion can often be obtained with the assumption that air resistance is absent. With this assumption, we follow the procedure outlined earlier in Figure 3.7. In the absence of air resistance, the moving object (the projectile) does not slow down in the horizontal or *x* direction, so that the *x* component of the acceleration is $a_x = 0$, and the *x* component of the velocity remains the same as its initial value ($v_x = v_{0x}$). In the vertical or *y* direction, however, the projectile experiences the effect of gravity, so that a_y is the acceleration due to gravity. If the path or trajectory of the projectile is near the earth's surface, a_y has a magnitude of 9.80 m/s². Thus, the phrase "projectile motion" means that $a_x = 0$

Concepts at a Glance

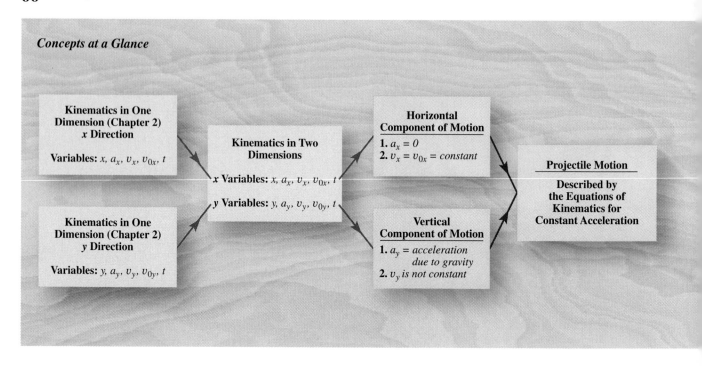

and a_y equals the acceleration due to gravity. The concept chart in Figure 3.9 summarizes these considerations and is an extension of the earlier chart in Figure 3.7. Example 2 and other examples in this section illustrate how the equations of kinematics are applied to projectile motion.

EXAMPLE 2 • A Falling Care Package

Figure 3.10 shows an airplane moving horizontally with a constant velocity of $+115$ m/s at an altitude of 1050 m. The directions to the right and upward have been chosen as the positive directions. The plane releases a "care package" that falls to the ground along a curved trajectory. Ignoring air resistance, determine the time required for the package to hit the ground.

Reasoning The time required for the package to hit the ground is the time it takes for the package to fall through a vertical distance of 1050 m. In falling, however, it moves both downward and to the right. These two parts of the motion occur independently, so we can focus solely on the vertical part. We note that the package is not moving in the y direction at the instant of release, so that $v_{0y} = 0$. (The package is moving initially in the horizontal or x direction, not in the y direction.) Furthermore, when the package hits the ground, the y component of its displacement is $y = -1050$ m, as the drawing shows. The acceleration is that due to gravity, so $a_y = -9.80$ m/s^2. These data are summarized as follows:

• **PROBLEM SOLVING INSIGHT**
The variables y, a_y, v_y, and v_{0y} are scalar components. Therefore, an algebraic sign $(+$ or $-)$ must be included with each one to denote direction.

y-Direction Data				
y	a_y	v_y	v_{0y}	t
-1050 m	-9.80 m/s^2		0	?

With these data, Equation 3.5b ($y = v_{0y}t + \frac{1}{2}a_y t^2$) can be used to find the fall-time.

Figure 3.9 *Concepts at a Glance* In projectile motion, the horizontal component of the acceleration is zero, and the vertical component of the acceleration is the acceleration due to gravity. The diver in this time-lapse photograph exhibits projectile motion, assuming that the effects of air resistance can be ignored.

Solution Since $v_{0y} = 0$, it follows from Equation 3.5b that $y = \frac{1}{2}a_y t^2$ and

$$t = \sqrt{\frac{2y}{a_y}} = \sqrt{\frac{2(-1050\ \text{m})}{-9.80\ \text{m/s}^2}} = \boxed{14.6\ \text{s}}$$

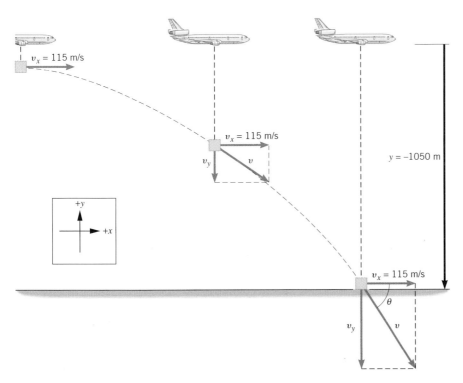

Figure 3.10 The package falling from the plane is an example of projectile motion, as Examples 2 and 3 discuss.

Figure 3.11 Package A and package B are released simultaneously at the same height and strike the ground at the same time, because their y variables (y, a_y, and v_{0y}) are the same.

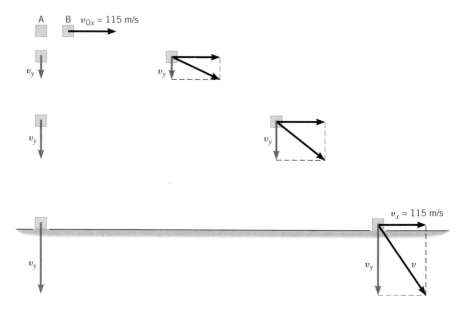

The freely falling package in Example 2 picks up vertical speed on the way down. But the horizontal component of the velocity retains its initial value of $v_{0x} = +115$ m/s throughout the entire descent. Since the plane also travels at a constant horizontal velocity of $+115$ m/s, it remains directly above the falling package. The pilot always sees the package directly beneath the plane, as the dashed vertical lines in Figure 3.10 show. This result is a direct consequence of the fact that the package has no acceleration in the horizontal direction. In reality, air resistance would slow down the package, and it would not remain directly beneath the plane during the descent. To further emphasize this point, Figure 3.11 illustrates what happens to two packages that are released simultaneously from the same height. Package B is given an initial velocity component of $v_{0x} = +115$ m/s in the horizontal direction, as in Example 2, and the package follows the curved path shown in the figure. Package A, on the other hand, is dropped from a stationary balloon and falls straight down toward the ground, since $v_{0x} = 0$. Both packages hit the ground at the same time.

Not only do the packages in Figure 3.11 reach the ground at the same time, but the y components of their velocities are also equal at all points on the way down. However, package B does hit the ground with a greater speed than package A. Remember, speed is the magnitude of the velocity vector, and the velocity of B has an x component, whereas the velocity of A does not. The magnitude and direction of the velocity vector for package B at the instant just before the package hits the ground is computed in Example 3.

EXAMPLE 3 • The Velocity of the Care Package

For the situation shown in Figure 3.10, find the speed of the package and the direction of the velocity vector just before the package hits the ground.

Reasoning Since the speed v of the package is given by $v = \sqrt{v_x^2 + v_y^2}$, it is necessary to know values for v_x and v_y at the instant before impact. The component v_x is constant with a value of 115 m/s. The component v_y can be determined by using Equation 3.3b and the data from Example 2 ($a_y = -9.80$ m/s^2, $v_{0y} = 0$, $t = 14.6$ s). Thus, we expect that the final speed v of the package will be greater than 115 m/s.

Solution From Equation 3.3b, it follows that

$$v_y = v_{0y} + a_y t = (-9.80 \text{ m/s}^2)(14.6 \text{ s}) = -143 \text{ m/s}$$

The speed of the package at the instant before impact is

$$v = \sqrt{(115 \text{ m/s})^2 + (-143 \text{ m/s})^2} = \boxed{184 \text{ m/s}}$$

which is greater than 115 m/s, as expected. The velocity vector makes an angle θ with the horizontal, as Figure 3.10 indicates:

$$\cos \theta = \frac{v_x}{v} \quad \text{or} \quad \theta = \cos^{-1} \left(\frac{v_x}{v} \right) = \cos^{-1} \left(\frac{115 \text{ m/s}}{184 \text{ m/s}} \right) = \boxed{51.3°}$$

• **PROBLEM SOLVING INSIGHT**
The speed of a projectile at any location along its path is the magnitude v of its velocity at that location: $v = \sqrt{v_x{}^2 + v_y{}^2}$. Thus, both the horizontal and vertical velocity components contribute to the speed.

An important feature of projectile motion is that there is no acceleration in the horizontal or x direction. Conceptual Example 4 discusses an interesting implication of this feature.

CONCEPTUAL EXAMPLE 4 • I Shot a Bullet into the Air . . . .

Suppose you are driving in a convertible with the top down. The car is moving to the right at a constant velocity. As Figure 3.12 illustrates, you point a rifle straight upward and fire it. In the absence of air resistance, where would the bullet land—behind you, ahead of you, or in the barrel of the rifle?

Reasoning and Solution If air resistance were present, it would slow down the bullet and cause it to land behind you, toward the rear of the car. However, air resistance is absent, so we must consider the bullet's motion more carefully. Before the rifle is fired, the bullet, rifle, and car are moving together, so the bullet and rifle have the same horizontal velocity as the car. When the rifle is fired, the bullet is given an additional velocity component in the vertical direction; the bullet retains its initial horizontal velocity component since the rifle is pointed straight up. Because there is no air resistance to slow it down, the bullet experiences no acceleration in the horizontal direction. Thus, the horizontal velocity component of the bullet does not change from its initial value, and remains matched to that of the car. As a result, *the bullet remains directly above the rifle at all times during the flight and would fall directly back into the barrel of the rifle,* as the drawing indicates. This situation is analogous to that in Figure 3.10, where the care package, as it falls, remains directly below the plane.

Related Homework Material: *Question 12 and Problem 36*

Often projectiles, like footballs and baseballs, are sent into the air at an angle with respect to the ground. From a knowledge of the projectile's initial velocity, a wealth of information can be obtained about the motion. For instance, Example 5 demonstrates how to calculate the maximum height reached by the projectile.

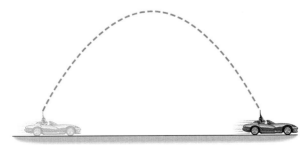

Figure 3.12 The car is moving with a constant velocity to the right, and the rifle is pointed straight up. In the absence of air resistance, a bullet fired from the rifle has no acceleration in the horizontal direction. As a result, the bullet would land back in the barrel of the rifle.

Figure 3.13 A football is kicked with an initial speed of v_0 at an angle of θ above the ground. The ball attains a maximum height H and a range R.

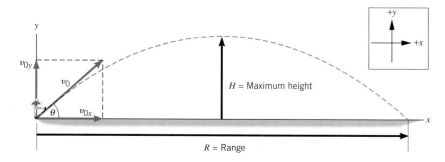

EXAMPLE 5 • The Height of a Kickoff

A placekicker kicks a football at an angle of $\theta = 40.0°$ above the horizontal axis, as Figure 3.13 shows. The initial speed of the ball is $v_0 = 22$ m/s. Ignore air resistance and find the maximum height H that the ball attains.

Reasoning The maximum height reached by the ball is a characteristic of the vertical part of the motion. The vertical part of the motion can be treated separately from the horizontal part. Making use of this fact, we calculate the vertical component of the initial velocity:

$$v_{0y} = v_0 \sin \theta = (22 \text{ m/s}) \sin 40.0° = +14 \text{ m/s}$$

• **PROBLEM SOLVING INSIGHT**
When a projectile reaches maximum height, the vertical component of its velocity is momentarily zero ($v_y = 0$). However, the horizontal component of its velocity is not zero.

The maximum height H can be determined by noting that the y component of the velocity v_y decreases as the ball moves upward. Eventually, $v_y = 0$ at the top of the trajectory. The data below can be used in Equation 3.6b ($v_y^2 = v_{0y}^2 + 2a_y y$) to find the maximum height:

y-Direction Data				
y	a_y	v_y	v_{0y}	t
$H = ?$	-9.80 m/s^2	0	$+14$ m/s	

Solution From Equation 3.6b, we find that

$$y = H = \frac{v_y^2 - v_{0y}^2}{2a_y} = \frac{-(14 \text{ m/s})^2}{2(-9.80 \text{ m/s}^2)} = \boxed{+10 \text{ m}}$$

The height H depends only on the y variables; the same height would have been reached had the ball been thrown *straight up* with an initial velocity of $v_{0y} = +14$ m/s.

It is also possible to find the total time or "hang time" during which the football in Figure 3.13 is in the air. Example 6 shows how to determine this time.

EXAMPLE 6 • The Time of Flight of a Kickoff

The Physics of...
the "hang time" of a football.

For the motion illustrated in Figure 3.13, ignore air resistance and use the data from Example 5 to determine the time of flight between kickoff and landing.

Reasoning Given the initial velocity, it is the acceleration due to gravity that determines how long the ball stays in the air. Thus, to find the time of flight we deal with the vertical part of the motion. Since the ball starts at ground level and returns to ground level, the displacement in the y direction is zero. The initial velocity component in the y direction is the same as that in Example 5, i.e., $v_{0y} = +14$ m/s. Therefore, we have

y-Direction Data				
y	a_y	v_y	v_{0y}	t
0	-9.80 m/s^2		$+14$ m/s	?

The time of flight can be determined from Equation 3.5b.

Solution   Using Equation 3.5b ($y = v_{0y}t + \frac{1}{2}a_y t^2$), we find

$$0 = (14 \text{ m/s})t + \tfrac{1}{2}(-9.80 \text{ m/s}^2)t^2 = [(14 \text{ m/s}) + \tfrac{1}{2}(-9.80 \text{ m/s}^2)t]t$$

There are two solutions to this equation. One is given by

$$(14 \text{ m/s}) + \tfrac{1}{2}(-9.80 \text{ m/s}^2)t = 0 \quad \text{or} \quad t = 2.9 \text{ s}$$

The other is given by $t = 0$. The solution we seek is $\boxed{t = 2.9 \text{ s}}$, because $t = 0$ corresponds to the initial kickoff.

Another important feature of projectile motion is called the "range." The range, as Figure 3.13 shows, is the horizontal distance traveled between launching and landing, assuming the projectile returns to the *same vertical level* at which it was fired. Example 7 shows how to obtain the range.

EXAMPLE 7 • The Range of a Kickoff

For the motion shown in Figure 3.13 and discussed in Examples 5 and 6, ignore air resistance and calculate the range R of the projectile.

Reasoning   The range is a characteristic of the horizontal part of the motion. Thus, our starting point is to determine the horizontal component of the initial velocity:

$$v_{0x} = v_0 \cos\theta = (22 \text{ m/s}) \cos 40.0° = +17 \text{ m/s}$$

Recall from Example 6 that the time of flight is $t = 2.9$ s. Since there is no acceleration in the x direction, v_x remains constant, and the range is simply the product of $v_x = v_{0x}$ and the time.

Solution   The range is

$$x = R = v_{0x}t = (17 \text{ m/s})(2.9 \text{ s}) = \boxed{+49 \text{ m}}$$

The range in the previous example depends on the angle θ at which the projectile is fired above the horizontal. When air resistance is absent, the maximum range results when $\theta = 45°$.

The examples considered thus far have used information about the initial location and velocity of a projectile to determine the final location and velocity. Example 8 deals with the opposite situation and illustrates how the final parameters can be used with the equations of kinematics to determine the initial parameters.

Figure 3.14 The velocity and location of the baseball upon landing can be used to determine its initial velocity, as Example 8 illustrates.

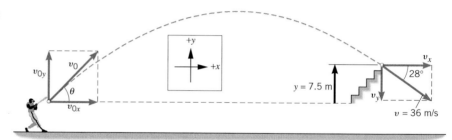

EXAMPLE 8 • A Home Run

A baseball player hits a home run, and the ball lands in the left-field seats, 7.5 m above the point at which the ball was hit. The ball lands with a velocity of 36 m/s at an angle of 28° below the horizontal (see Figure 3.14). Ignoring air resistance, find the initial velocity with which the ball leaves the bat.

Reasoning To find the initial velocity, we must determine the initial speed v_0 and the angle θ in the drawing. These quantities are related to the horizontal and vertical components of the initial velocity (v_{0x} and v_{0y}) by the relations

$$v_0 = \sqrt{v_{0x}^2 + v_{0y}^2} \quad \text{and} \quad \tan \theta = \frac{v_{0y}}{v_{0x}}$$

Therefore, it is necessary to find v_{0x} and v_{0y}, which we will do with the equations of kinematics.

Solution Since air resistance is being ignored, the horizontal component of the velocity v_x remains constant throughout the motion. Thus,

$$v_{0x} = v_x = (36 \text{ m/s}) \cos 28° = 32 \text{ m/s}$$

The value for v_{0y} can be obtained from Equation 3.6b and the data displayed below (see the insert in Figure 3.14 for the positive and negative directions):

y-Direction Data				
y	a_y	v_y	v_{0y}	t
+7.5 m	−9.80 m/s²	(−36 sin 28°) m/s	?	

$$v_y^2 = v_{0y}^2 + 2a_y y \quad \text{or} \quad v_{0y} = \sqrt{v_y^2 - 2a_y y}$$

$$v_{0y} = \sqrt{[(-36 \sin 28°) \text{ m/s}]^2 - 2(-9.80 \text{ m/s}^2)(7.5 \text{ m})} = 21 \text{ m/s}$$

The initial speed v_0 and angle θ of the baseball are

$$v_0 = \sqrt{v_{0x}^2 + v_{0y}^2} = \sqrt{(32 \text{ m/s})^2 + (21 \text{ m/s})^2} = \boxed{38 \text{ m/s}}$$

$$\theta = \tan^{-1}\left(\frac{v_{0y}}{v_{0x}}\right) = \tan^{-1}\left(\frac{21 \text{ m/s}}{32 \text{ m/s}}\right) = \boxed{33°}$$

In projectile motion, the magnitude of the acceleration due to gravity affects the trajectory in a significant way. The next example illustrates the effect by comparing trajectories on the earth and on the moon.

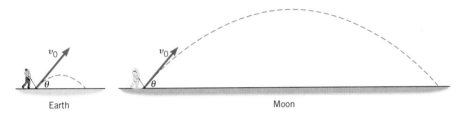

Figure 3.15 Golf balls are launched with identical velocities from the surfaces of the earth and the moon. Because the acceleration due to gravity is smaller on the moon, the maximum height and range of the moon ball are greater than that of the earth ball.

CONCEPTUAL EXAMPLE 9 • Comparing Projectile Motions on the Earth and on the Moon

In 1971 Apollo 14 landed on the moon, and astronaut Alan Shepard walked on its surface. In a moment of whimsy, he hit a golf ball, which was launched upward at an angle and followed the familiar curved trajectory. However, the trajectory differed from what it would have been on earth, because the acceleration due to gravity on the moon is about 6 times smaller than that on earth. Consider the same ball launched upward at the same angle and speed from the surface of the earth. Does the earth ball or the moon ball attain (a) the greater height and (b) the greater range?

Reasoning and Solution

(a) Both balls reach their highest points when the vertical components of their velocities decrease to zero. However, when the moon ball moves upward, the vertical component of its velocity decreases 6 times more slowly than that of the earth ball. The slower rate occurs because the acceleration due to gravity on the moon is 6 times less than that on earth. As a result, it takes the moon ball 6 times longer to reach its maximum height. But we can show that the maximum height is directly proportional to the travel time. The height y of the ball at any instant is given by Equation 3.4b as the product of the average velocity component in the y direction $[\frac{1}{2}(v_{0y} + v_y)]$ and the time t: $y = \frac{1}{2}(v_{0y} + v_y)t$. Since the maximum height H is reached when the final velocity component in the y direction is zero ($v_y = 0$), we find that $H = \frac{1}{2}v_{0y}t$. The initial velocity component in the y direction v_{0y} is the same for both the earth and moon balls, so the maximum height is directly proportional to the time to reach this height. Since the travel time is 6-fold greater on the moon, ***the maximum height of the moon ball is 6-fold greater than that of the earth ball,*** as Figure 3.15 illustrates.

(b) The range of a projectile is also proportional to the time that the projectile is in flight. Thus, ***the moon ball—being in flight 6 times longer than the earth ball— has a 6-fold greater range.***

Related Homework Material: Problems 18 and 19

Section 2.6 points out that certain types of symmetry with respect to time and velocity are present for freely falling bodies. These symmetries are also found in projectile motion, since projectiles are falling freely in the vertical direction. In particular, the time required for a projectile to reach its maximum height H is equal to the time spent returning to the ground. In addition, Figure 3.16 shows that the speed v of the object at any height above the ground on the upward part of the trajectory is equal to the speed v at the same height on the downward part. Although the two speeds are the same, the velocities are different, because they point in different directions. Conceptual Example 10 shows how to use this type of projectile motion symmetry in your reasoning.

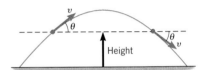

Figure 3.16 The speed v of a projectile at a given height above the ground is the same on the upward and downward parts of the trajectory. The velocities are different, however, since they point in different directions.

Figure 3.17 Two stones are thrown off the cliff with identical initial speeds v_0, but at equal angles θ that are below and above the horizontal. Conceptual Example 10 compares the velocities with which the stones hit the water beneath the cliff.

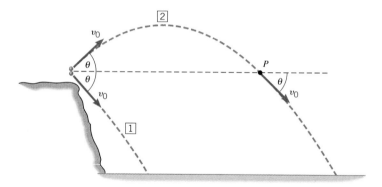

CONCEPTUAL EXAMPLE 10 • Two Ways to Throw a Stone

From the top of a cliff overlooking a lake, a person throws two stones. The stones have identical initial speeds v_0, but stone 1 is thrown downward at an angle θ below the horizontal, while stone 2 is thrown upward at the same angle above the horizontal, as Figure 3.17 shows. Neglect air resistance and decide which stone, if either, strikes the water with the greater velocity.

Reasoning and Solution We might guess that stone 1, being hurled downward, would strike the water with the greater velocity. To show that this is not true, let's follow the path of stone 2 as it rises to its maximum height and falls back to earth. Notice point P in the drawing, where stone 2 returns to its initial height; here the speed of stone 2 is v_0, but its velocity is directed at an angle θ below the horizontal. This is exactly the type of projectile symmetry illustrated in Figure 3.16. At this point, then, stone 2 has a velocity that is identical to the velocity with which stone 1 is thrown downward from the top of the cliff. From this point on, the velocity of stone 2 changes in exactly the same way as that for stone 1, so ***both stones strike the water with the same velocity.***

Related Homework Material: Problems 25 and 33

Another type of symmetry also exists. If you have ever experimented with a stream of water from a hose, you might be aware that there are two possible angles at which to point the nozzle such that the range of the water is about the same. Figure 3.18 illustrates three trajectories that have initial angles of 20°, 45°, and 70°. In the absence of air resistance, the 45° angle gives rise to the maximum range, while the 20° and 70° angles give rise to *identical ranges* that are less than the maximum. There are always two angles giving the same range; one is less than 45° and the

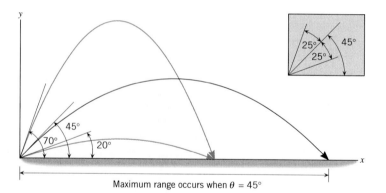

Figure 3.18 In the absence of air resistance, the 20° trajectory and the 70° trajectory have identical ranges. Note from the insert that the two angles are 25° on either side of the 45° line.

other is greater than 45°. The drawing indicates that the two angles are symmetrically placed about the 45° line. In other words, the 20° angle is 25° below the 45° angle, while the 70° angle is 25° above the 45° angle, as the insert in the picture emphasizes.

In all the examples in this section, the projectiles follow a curved trajectory. In general, if the only acceleration is that due to gravity, the shape of the path can be shown to be a *parabola*.

*3.4 RELATIVE VELOCITY

To someone hitchhiking along a highway, two cars speeding by in adjacent lanes seem like a blur. But, if the cars have the same velocity, each driver sees the other remaining in place, one lane away. The hitchhiker observes a velocity of perhaps 30 m/s. But each driver observes the velocity of the other to be zero. Clearly, the velocity of an object is relative to the observer who is making the measurement.

Figure 3.19 illustrates the concept of relative velocity by showing a passenger walking toward the front of a moving train. The people sitting on the train see the passenger walking with a velocity of +2.0 m/s, where the plus sign denotes a direction to the right. Suppose the train is moving with a velocity of +9.0 m/s relative to an observer standing on the ground. Then the ground-based observer would see the passenger moving with a velocity of +11 m/s, due in part to the walking motion and in part to the train's motion. As an aid in describing relative velocity, let us define the following symbols:

Wesley Snipes and Woody Harrelson demonstrate the concept of relative velocity as they make their way along the top of a speeding train in the movie *Money Train*.

$\mathbf{v}_{\boxed{\text{PT}}}$ = velocity of the $\boxed{\text{Passenger}}$ relative to the $\boxed{\text{Train}}$ = +2.0 m/s

$\mathbf{v}_{\boxed{\text{TG}}}$ = velocity of the $\boxed{\text{Train}}$ relative to the $\boxed{\text{Ground}}$ = +9.0 m/s

$\mathbf{v}_{\boxed{\text{PG}}}$ = velocity of the $\boxed{\text{Passenger}}$ relative to the $\boxed{\text{Ground}}$ = +11 m/s

In terms of these symbols, the situation in Figure 3.19 is summarized as follows:

$$\mathbf{v}_{\text{PG}} = \mathbf{v}_{\text{PT}} + \mathbf{v}_{\text{TG}} \tag{3.7}$$

or

$$\mathbf{v}_{\text{PG}} = (2.0 \text{ m/s}) + (9.0 \text{ m/s}) = +11 \text{ m/s}$$

According to Equation 3.7, $\mathbf{v}_{\text{PG}}$ is the vector sum of $\mathbf{v}_{\text{PT}}$ and $\mathbf{v}_{\text{TG}}$, and this sum is shown in the drawing. Had the passenger been walking toward the rear of the train, rather than the front, the velocity relative to the ground-based observer would have been $\mathbf{v}_{\text{PG}} = (-2.0 \text{ m/s}) + (9.0 \text{ m/s}) = +7.0 \text{ m/s}$.

Each velocity symbol in Equation 3.7 contains a two-letter subscript. The first letter in the subscript refers to the body that is moving, while the second letter indi-

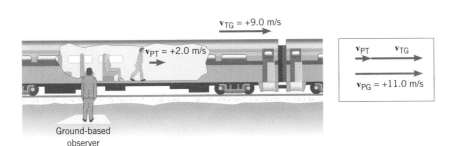

Figure 3.19 The velocity $\mathbf{v}_{\text{PG}}$ of the passenger relative to the ground-based observer is the vector sum of the velocity $\mathbf{v}_{\text{PT}}$ of the passenger relative to the train and the velocity $\mathbf{v}_{\text{TG}}$ of the train relative to the ground: $\mathbf{v}_{\text{PG}} = \mathbf{v}_{\text{PT}} + \mathbf{v}_{\text{TG}}$.

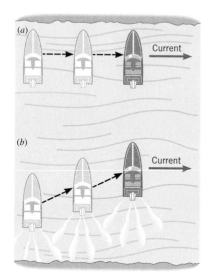

Figure 3.20 (a) A boat with its engine turned off is carried along by the current. (b) With the engine turned on, the boat moves across the river in a diagonal fashion.

cates the object relative to which the velocity is measured. For example, $\mathbf{v}_{TG}$ and $\mathbf{v}_{PG}$ are the velocities of the **T**rain and **P**assenger measured relative to the **G**round. Similarly, $\mathbf{v}_{PT}$ is the velocity of the **P**assenger measured by an observer sitting on the **T**rain.

The ordering of the subscript symbols in Equation 3.7 follows a definite pattern. The first subscript (P) on the left side of the equation is also the first subscript on the right side of the equation. Likewise, the last subscript (G) on the left side of the equation is also the last subscript on the right side of the equation. The third subscript (T) appears only on the right side of the equation as the two "inner" subscripts. The colored boxes below emphasize the pattern of the symbols in the subscripts:

$$\mathbf{v}_{\boxed{PG}} = \mathbf{v}_{\boxed{P}\boxed{T}} + \mathbf{v}_{T\boxed{G}}$$

In other situations, the subscripts will not necessarily be P, G, and T, but will be compatible with the names of the objects involved in the motion.

Equation 3.7 has been presented in connection with one-dimensional motion, but the result is also valid for two-dimensional motion. Figure 3.20 depicts a common situation that deals with relative velocity in two dimensions. Part *a* of the drawing shows a boat being carried downstream by a river; the engine of the boat is turned off. In part *b*, the engine has been turned on, and now the boat moves across the river in a diagonal fashion because of the combined motion produced by the current and the engine. The list below gives the velocities for this type of motion and the objects relative to which they are measured:

$$\mathbf{v}_{\boxed{BW}} = \text{velocity of the } \boxed{\text{Boat}} \text{ relative to the } \boxed{\text{Water}}$$
$$\mathbf{v}_{\boxed{WS}} = \text{velocity of the } \boxed{\text{Water}} \text{ relative to the } \boxed{\text{Shore}}$$
$$\mathbf{v}_{\boxed{BS}} = \text{velocity of the } \boxed{\text{Boat}} \text{ relative to the } \boxed{\text{Shore}}$$

The velocity $\mathbf{v}_{BW}$ of the boat relative to the water is the velocity measured by an observer who, for instance, is floating on an inner tube and drifting downstream with the current. When the engine is turned off, the boat also drifts downstream with the current and $\mathbf{v}_{BW}$ is zero. When the engine is turned on, however, the boat can move relative to the water, and $\mathbf{v}_{BW}$ is no longer zero. The velocity $\mathbf{v}_{WS}$ of the water relative to the shore is the velocity of the current measured by an observer on the shore. The velocity $\mathbf{v}_{BS}$ of the boat relative to the shore is due to the combined motion of the boat relative to the water and the motion of the water relative to the shore. In symbols,

$$\mathbf{v}_{\boxed{BS}} = \mathbf{v}_{\boxed{B}W} + \mathbf{v}_{W\boxed{S}}$$

The ordering of the subscripts in this equation is identical to that in Equation 3.7, although the letters have been changed to reflect a different physical situation. Example 11 illustrates the concept of relative velocity in two dimensions.

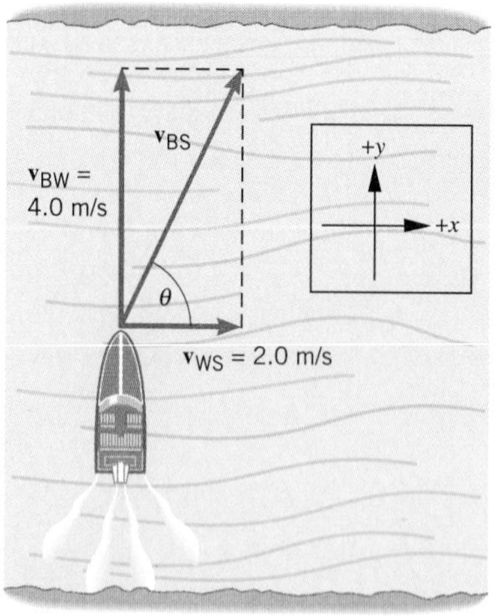

EXAMPLE 11 • Crossing a River

The engine of a boat drives it across a river that is 1800 m wide. The velocity $\mathbf{v}_{BW}$ of the boat relative to the water is 4.0 m/s, directed perpendicular to the current, as in Figure 3.21. The velocity $\mathbf{v}_{WS}$ of the water relative to the shore is 2.0 m/s. (a) What is the velocity $\mathbf{v}_{BS}$ of the boat relative to the shore? (b) How long does it take for the boat to cross the river?

Reasoning

(a) The velocity $\mathbf{v}_{BS}$ of the boat relative to the shore is the vector sum of the velocity

Figure 3.21 The velocity $\mathbf{v}_{BS}$ of the boat relative to the shore is the vector sum of the velocity $\mathbf{v}_{BW}$ of the boat relative to the water and the velocity $\mathbf{v}_{WS}$ of the water relative to the shore: $\mathbf{v}_{BS} = \mathbf{v}_{BW} + \mathbf{v}_{WS}$.

$\mathbf{v_{BW}}$ of the boat relative to the water and the velocity $\mathbf{v_{WS}}$ of the water relative to the shore: $\mathbf{v_{BS}} = \mathbf{v_{BW}} + \mathbf{v_{WS}}$. Since $\mathbf{v_{BW}}$ and $\mathbf{v_{WS}}$ are both known, we can use this relation among the relative velocities, with the aid of trigonometry, to find the magnitude and directional angle of $\mathbf{v_{BS}}$.

(b) The component of $\mathbf{v_{BS}}$ that is parallel to the width of the river (see Figure 3.21) determines how fast the boat is moving across the river; this parallel component is $v_{BS} \sin\theta = v_{BW} = 4.0$ m/s. The time for the boat to cross the river is equal to the width of the river divided by the magnitude of this velocity component.

Solution

(a) Since the vectors $\mathbf{v_{BW}}$ and $\mathbf{v_{WS}}$ are perpendicular (see Figure 3.21), the magnitude of $\mathbf{v_{BS}}$ can be determined from the Pythagorean theorem:

$$v_{BS} = \sqrt{v_{BW}^2 + v_{WS}^2} = \sqrt{(4.0 \text{ m/s})^2 + (2.0 \text{ m/s})^2} = \boxed{4.5 \text{ m/s}}$$

Thus, the boat moves at a speed of 4.5 m/s with respect to an observer on the shore. The direction of the boat relative to the shoreline is given by the angle θ in the drawing:

$$\tan\theta = \frac{v_{BW}}{v_{WS}} \quad \text{or} \quad \theta = \tan^{-1}\left(\frac{v_{BW}}{v_{WS}}\right) = \tan^{-1}\left(\frac{4.0 \text{ m/s}}{2.0 \text{ m/s}}\right) = \boxed{63°}$$

(b) The time t for the boat to cross the river is

$$t = \frac{\text{Width}}{v_{BS} \sin\theta} = \frac{1800 \text{ m}}{4.0 \text{ m/s}} = \boxed{450 \text{ s}}$$

Occasionally, situations arise when two vehicles are in relative motion, and it is useful to know the relative velocity of one with respect to the other. Example 12 considers this type of relative motion.

EXAMPLE 12 • Approaching an Intersection

Figure 3.22*a* shows two cars approaching an intersection along perpendicular roads. The cars have the following velocities:

$\mathbf{v}\boxed{\text{AG}}$ = velocity of $\boxed{\text{car A}}$ relative to the $\boxed{\text{Ground}}$ = 25.0 m/s, eastward

$\mathbf{v}\boxed{\text{BG}}$ = velocity of $\boxed{\text{car B}}$ relative to the $\boxed{\text{Ground}}$ = 15.8 m/s, northward

Find the magnitude and direction of $\mathbf{v_{AB}}$, where

$\mathbf{v}\boxed{\text{AB}}$ = velocity of $\boxed{\text{car A}}$ as measured by a passenger in $\boxed{\text{car B}}$

Reasoning To find $\mathbf{v_{AB}}$, we use an equation whose subscripts follow the order outlined earlier. Thus,

$$\mathbf{v}\boxed{\text{AB}} = \mathbf{v}\boxed{\text{A}}\text{G} + \mathbf{v}_\text{G}\boxed{\text{B}}$$

In this equation, the term $\mathbf{v_{GB}}$ is the velocity of the ground relative to a passenger in car B, rather than $\mathbf{v_{BG}}$, which is given as 15.8 m/s, northward. In other words, the subscripts are reversed. However, $\mathbf{v_{GB}}$ is related to $\mathbf{v_{BG}}$ according to

$$\mathbf{v_{GB}} = -\mathbf{v_{BG}}$$

This relationship reflects the fact that a passenger in car B, moving northward relative to the ground, looks out the car window and sees objects on the ground moving southward, in the opposite direction. Therefore, the equation $\mathbf{v_{AB}} = \mathbf{v_{AG}} + \mathbf{v_{GB}}$ may be used to find $\mathbf{v_{AB}}$, provided we recognize $\mathbf{v_{GB}}$ as a vector that points opposite to the given velocity $\mathbf{v_{BG}}$. With this in mind, Figure 3.22*b* illustrates how $\mathbf{v_{AG}}$ and $\mathbf{v_{GB}}$

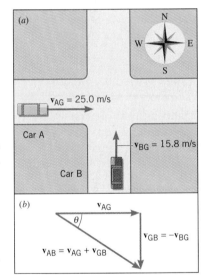

Figure 3.22 Two cars are approaching an intersection along perpendicular roads.

• **PROBLEM SOLVING INSIGHT**
In general, the velocity of object R relative to object S is always the negative of the velocity of object S relative to R: $\mathbf{v_{RS}} = -\mathbf{v_{SR}}$.

are added vectorially to give $\mathbf{v}_{AB}$.

Solution From the vector triangle shown in Figure 3.22b, the magnitude and direction of $\mathbf{v}_{AB}$ can be calculated as

$$v_{AB} = \sqrt{v_{AG}^2 + v_{GB}^2} = \sqrt{(25.0 \text{ m/s})^2 + (-15.8 \text{ m/s})^2} = \boxed{29.6 \text{ m/s}}$$

and

$$\cos \theta = \frac{v_{AG}}{v_{AB}} \quad \text{or} \quad \theta = \cos^{-1}\left(\frac{v_{AG}}{v_{AB}}\right) = \cos^{-1}\left(\frac{25.0 \text{ m/s}}{29.6 \text{ m/s}}\right) = \boxed{32.4°}$$

Thus, passengers in car B measure the velocity of car A to be 29.6 m/s in a direction 32.4° south of east.

SUMMARY

Motion that occurs in two dimensions can be described in terms of the time t and the x and y components of four vectors: the displacement, the acceleration, the final velocity, and the initial velocity. The motion can be analyzed by treating the x and y components of the vectors separately. The directions of these components are conveyed by assigning a plus ($+$) or minus ($-$) sign to each one. When the acceleration vector is constant, the x components of these vectors (x, a_x, v_x, and v_{0x}) are related by the equations of kinematics, as are the y components (y, a_y, v_y, and v_{0y}).

Table 3.1 summarizes the equations of kinematics for two-dimensional motion.

Projectile motion is a kind of two-dimensional motion that occurs when the moving object (the projectile) experiences only the acceleration due to gravity, which acts in the vertical direction. The acceleration of the projectile has no horizontal component ($a_x = 0$), the effects of air resistance being negligible. The vertical component of the acceleration is a_y, and it equals the acceleration due to gravity.

CONCEPTUAL QUESTIONS

1. Suppose an object could move in three dimensions. What additions to the equations of kinematics in Table 3.1 would be necessary to describe three-dimensional motion?

2. An object is thrown upward at an angle θ above the ground, eventually returning to earth. (a) Is there any place along the trajectory where the velocity and acceleration are perpendicular? If so, where? (b) Is there any place where the velocity and acceleration are parallel? If so, where?

3. Is the acceleration of a projectile equal to zero when it reaches the top of its trajectory? If not, why not?

4. In baseball, the pitcher's mound is raised to compensate for the fact that the ball falls downward as it travels from the pitcher toward the batter. If baseball were played on the moon, would the pitcher's mound have to be higher than, lower than, or the same height as it is on earth? Give your reasoning.

5. A tennis ball is hit into the air and moves along an arc. Neglecting air resistance, where along the arc is the speed of the ball (a) a minimum and (b) a maximum? Justify your answers.

6. Suppose there is a wind blowing parallel to the ground and toward the kicker in Figure 3.13. Then the acceleration component in the horizontal direction would not be zero. How would you expect the time of flight of the football to be affected, if at all? Explain.

7. A wrench is accidentally dropped from the top of the mast on a sailboat. Will the wrench hit at the same place on the deck whether the sailboat is at rest or moving with a constant velocity? Justify your answer.

8. A rifle, at a height H above the ground, fires a bullet parallel to the ground. At the same instant and at the same height, a second bullet is dropped from rest. In the absence of air resistance, which bullet strikes the ground first? Explain.

9. Two projectiles are launched from ground level at the same angle above the horizontal, and both return to ground level. Projectile A has a launch speed that is twice that of projectile B. Assuming that air resistance is absent, sketch the trajectories of both projectiles. If your drawings are to be accurate, what should be the ratio of the maximum heights in your drawings and what should be the ratio of the ranges? Justify your answers.

10. In Figure 3.18 the 20° trajectory and the 70° trajectory have the same range. Which of the times of flight for these two trajectories is longer? Give your reasoning.

11. A leopard springs upward at a 45° angle and then falls back to the ground. Does the leopard, at any point on its trajectory, ever have a speed that is one-half its initial value? Give your reasoning.

12. As background for this question, review Conceptual Example 4. A football quarterback throws a pass on the run and then keeps running without changing his velocity. Can he throw the pass and then catch it himself? Give your reasoning.

13. On a riverboat cruise, a plastic bottle is accidentally dropped overboard. A passenger on the boat estimates that the boat pulls ahead of the bottle by 5 meters each second. Is it possible to conclude that the boat is moving at 5 m/s with respect to the shore? Account for your answer.

14. A plane takes off at St. Louis, flies straight to Denver, and then returns the same way. The plane flies at the same speed with respect to the ground during the entire flight, and there are no head winds or tail winds. Since the earth revolves around its axis once a day, you might expect that the times for the outbound trip and the return trip differ, depending on whether the plane flies against the earth's rotation or with it. However, under the conditions given, the two flight times are identical. Explain why.

15. A child is playing on the floor of a recreational vehicle (RV), as it moves along the highway at a constant velocity. He has a toy cannon, which shoots a marble at a fixed angle and speed with respect to the floor. The cannon can be aimed toward the front or the rear of the RV. Is the range toward the front the same as, less than, or greater than the range toward the rear? Answer this question (a) from the child's point of view and (b) from the point of view of an observer standing still on the ground. Justify your answers.

16. Three swimmers can swim equally fast relative to the water. They have a race to see who can swim across a river in the least time. Swimmer A swims perpendicular to the current and lands on the far shore downstream, because the current has swept him in that direction. Swimmer B swims upstream at an angle to the current and lands on the far shore directly opposite the starting point. Swimmer C swims downstream at an angle to the current in an attempt to take advantage of the current. Who crosses the river in the least time? Account for your answer.

PROBLEMS

---

ssm Solution is in the Student Solutions Manual. **www** Solution is available on the World Wide Web at http://www.wiley.com/college/cutnell ⚕ This icon represents a biomedical application.

Section 3.1 Displacement, Velocity, and Acceleration

1. ssm In diving to a depth of 750 m, an elephant seal also moves 460 m due east of his starting point. What is the magnitude of the seal's displacement?

2. A radar antenna is tracking a satellite orbiting the earth. At a certain time, the radar screen shows the satellite to be 162 km away. The radar antenna is pointing upward at an angle of 62.3° from the ground. Find the x and y components (in km) of the position of the satellite.

3. A baseball player hits a triple and ends up on third base. A baseball "diamond" is a square, each side of length 27.4 m, with home plate and the three bases on the four corners. What is the magnitude of his displacement?

4. The altitude of a hang glider is increasing at a rate of 6.80 m/s. At the same time, the shadow of the glider moves along the ground at a speed of 15.5 m/s when the sun is directly overhead. Find the magnitude of the glider's velocity.

5. ssm A jetliner is moving at a speed of 245 m/s. The vertical component of the plane's velocity is 40.6 m/s. Determine the magnitude of the horizontal component of the plane's velocity.

6. A dart is thrown upward at an angle of 25° above the horizontal. The vertical component of the dart's velocity is $v_y = 2.2$ m/s. Determine the x component of the velocity.

7. A wild horse starts from rest and runs in a straight line 29° north of west. After 36 s of running in this direction, the horse has a speed of 12 m/s. (a) What is the magnitude of the horse's average acceleration? Assuming that north and east are the positive directions, find the component of the horse's acceleration that points along (b) the north–south line and (c) the east–west line.

8. An archer draws back an arrow and then releases it from rest. The arrow is propelled forward with an average acceleration of 2400 m/s^2 and remains in contact with the bow string for a time of 0.025 s. What is the speed of the arrow when it just leaves the string?

9. ssm www In a mall, a shopper rides up an escalator between floors. At the top of the escalator, the shopper turns right and walks 9.00 m to a store. The magnitude of the shopper's displacement from the bottom of the escalator is 16.0 m. The vertical distance between the floors is 6.00 m. At what angle is the escalator inclined above the horizontal?

*10. A bird watcher meanders through the woods, walking 0.50 km due east, 0.75 km due south, and 2.15 km in a direction 35.0° north of west. The time required for this trip is 2.50 h. Determine the magnitude and direction (relative to due west) of the bird watcher's (a) displacement and (b) average velocity. Use kilometers and hours for distance and time, respectively.

*11. The earth moves around the sun in a nearly circular orbit of radius 1.50×10^{11} m. During the three summer months (an elapsed time of 7.89×10^6 s), the earth moves one-fourth of the distance around the sun. (a) What is the average speed of the earth? (b) What is the magnitude of the average velocity of the earth during this period?

Section 3.2 Equations of Kinematics in Two Dimensions, Section 3.3 Projectile Motion

12. The initial velocity of a spacecraft is 2650 m/s, directed at an angle of 30.0° above the x axis. Two engines then fire for a time of 475 s. One gives the spacecraft an acceleration in the $+x$ direction of $a_x = 6.30$ m/s^2. The other produces an acceleration in the $+y$ direction of $a_y = 2.85$ m/s^2. What is the speed of the spacecraft when the engines shut off?

13. ssm Suppose that the plane in Example 2 is traveling with twice the horizontal velocity, that is, with a velocity of $+230$ m/s. If all other factors remain the same, determine the time required for the package to hit the ground.

14. A quarterback throws a pass to a receiver, who catches it at the same height as the pass is thrown. The initial velocity of the ball is 15.0 m/s, at an angle of 25.0° above the horizontal. What is the horizontal component of the ball's velocity when the receiver catches it?

15. The punter on a football team tries to kick a football so that it stays in the air for a long "hang time." If the ball is kicked with an initial velocity of 25.0 m/s at an angle of 60.0° above the ground, what is the "hang time"?

16. A rock climber throws a small first aid kit to another climber who is higher up the mountain. The initial velocity of the kit is 11 m/s at an angle of 65° above the horizontal. At the instant when the kit is caught, it is traveling horizontally, so its vertical speed is zero. What is the vertical height between the two climbers?

17. ssm www A diver runs horizontally with a speed of 1.20 m/s off a platform that is 10.0 m above the water. What is his speed just before striking the water?

18. Review Conceptual Example 9 as background for this problem. The acceleration due to gravity on the moon has a magnitude of 1.62 m/s^2. Examples 5–7 deal with a placekicker kicking a football. Assume that the ball is kicked on the moon instead of on the earth. Find (a) the maximum height H and (b) the range that the ball would attain on the moon. Verify that your answers are consistent with the conclusions reached in Conceptual Example 9.

19. Review Conceptual Example 9 in preparation for this problem. On a distant planet, golf is just as popular as it is on earth. A golfer tees off and drives the ball 3.5 times farther than he would have on earth, given the same initial velocities on both planets. The ball is launched at a speed of 45 m/s at an angle of 29° above the horizontal. When the ball lands, it is at the same level as the tee. On the distant planet, what is (a) the maximum height and (b) the range of the ball?

20. During a baseball game a fly ball is hit to center field and is caught 115 m from home plate. Just when the ball is caught, a runner on third base takes off for home, and the center fielder throws the ball to the catcher standing on home plate. The runner takes 3.50 s to reach home, while the baseball is thrown with a velocity whose *horizontal* component is 41 m/s. Which reaches home first, the runner or the ball, and by how much time?

21. ssm A golf ball rolls off a horizontal cliff with an initial speed of 11.4 m/s. The ball falls a vertical distance of 15.5 m into a lake below. (a) How much time does the ball spend in the air? (b) What is the speed v of the ball just before it strikes the water?

22. If a projectile has a launching angle of 52.0° above the horizontal and an initial speed of 18.0 m/s, what is the highest barrier that the projectile can clear?

23. A car drives straight off the edge of a cliff that is 54 m high. The police at the scene of the accident note that the point of impact is 130 m from the base of the cliff. How fast was the car traveling when it went over the cliff?

24. The 1994 Winter Olympics included the aerials competition in skiing. In this event skiers speed down a ramp that slopes sharply upward at the end. The sharp upward slope launches them into the air, where they perform acrobatic maneuvers. In the women's competition, the end of a typical launch ramp is directed 63° above the horizontal. With this launch angle, a skier attains a height of 13 m above the end of the ramp. What is the skier's launch speed?

25. ssm As preparation for this problem, review Conceptual Example 10. The drawing shows an empty fuel tank being dropped by two different planes. At the moment of release each plane has the same speed of 135 m/s and each tank is at the same height of 2.00 km above the ground. While the speeds are the same, the velocities are different at the instant of release, because one plane is flying at an angle of 15.0° above the horizontal and the other is flying at an angle of 15.0° below the horizontal. Find the magnitude and direction of the velocity with which the fuel tank hits the ground if it is from (a) plane A and (b) plane B. In each part, give the directional angles with respect to the horizontal.

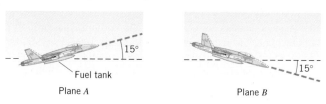

Plane A   Plane B

26. A tennis ball is struck such that it leaves the racket horizontally with a speed of 28.0 m/s. The ball hits the court at a horizontal distance of 19.6 m from the racket. What is the height of the tennis ball when it leaves the racket?

27. A motorcycle daredevil is attempting to jump across as many buses as possible (see the drawing). The takeoff ramp makes an angle of 18.0° above the horizontal, and the landing ramp is identical to the takeoff ramp. The buses are parked side by side, and

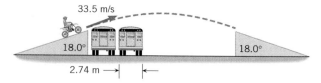

each bus is 2.74 m wide. The cyclist leaves the ramp with a speed of 33.5 m/s. What is the maximum number of buses over which the cyclist can jump?

28. A horizontal rifle is fired at a bull's-eye. The muzzle speed of the bullet is 670 m/s. The barrel is pointed directly at the center of the bull's-eye, but the bullet strikes the target 0.025 m below the center. What is the horizontal distance between the end of the rifle and the bull's-eye?

29. ssm An eagle is flying horizontally at 6.0 m/s with a fish in its claws. It accidentally drops the fish. (a) How much time passes before the fish's speed doubles? (b) How much additional time would be required for the fish's speed to double again?

30. A criminal is escaping across a rooftop and runs off the roof horizontally, landing on the roof of an adjacent building. The horizontal distance between the two buildings is 3.4 m, and the roof of the adjacent building is 2.0 m below the jumping off point. What would be the minimum speed needed by the criminal?

31. Suppose the water at the top of Niagara Falls has a horizontal speed of 2.7 m/s just before it cascades over the edge of the falls. At what vertical distance below the edge does the velocity vector of the water point downward at a 75° angle below the horizontal?

32. An archer is standing inside a building whose ceiling is 11 m high. An arrow is shot from ground level at an initial speed of 62 m/s. Calculate the angle of firing (above the horizontal) that gives the greatest possible range inside the building.

*33. ssm www As preparation for this problem, review Conceptual Example 10. The two stones described there have identical initial speeds of $v_0 = 13.0$ m/s and are thrown at an angle $\theta = 30.0°$, one below the horizontal and one above the horizontal. What is the distance *between* the points where the stones strike the ground?

*34. A rock, thrown horizontally from the top of a lighthouse, strikes the water 2.6 s later. A *straight line* is drawn from the top of the lighthouse to the point where the rock strikes the water. This line makes an angle of 35° with respect to the lighthouse. Calculate the initial speed of the rock.

*35. A soccer player kicks the ball toward a goal that is 29.0 m in front of him. The ball leaves his foot at a speed of 19.0 m/s and an angle of 32.0° above the ground. Find the speed of the ball when the goalie catches it in front of the net. *(Note: The answer is not 19.0 m/s.)*

*36. Review Conceptual Example 4 before beginning this problem. You are traveling in a convertible with the top down. The car is moving at a constant velocity of 25 m/s, due east along flat ground. You throw a tomato straight upward at a speed of 11 m/s. How far has the car moved when you get a chance to catch the tomato?

*37. ssm An airplane is flying with a velocity of 240 m/s at an angle of 30.0° with the horizontal, as the drawing shows. When the altitude of the plane is 2.4 km, a flare is released from the plane. The flare hits the target on the ground. What is the angle θ?

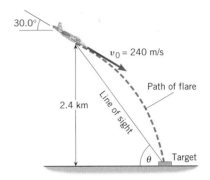

*38. A diver springs upward from a board that is three meters above the water. At the instant she contacts the water her speed is 8.90 m/s and her body makes an angle of 75.0° with respect to the horizontal surface of the water. Determine her initial velocity, both magnitude and direction.

*39. After leaving the end of a ski ramp, a ski jumper lands downhill at a point that is displaced 55 m horizontally from the end of the ramp. His velocity, just before landing, is 25 m/s and points in a direction 38° below the horizontal. Neglecting air resistance and any lift that he experiences while airborne, find his initial velocity (magnitude and direction) when he left the end of the ramp.

*40. Stones are thrown horizontally with the same velocity from the tops of two different buildings. One stone lands twice as far from the base of the building from which it was thrown as does the other stone. Find the ratio of the height of the taller building to the height of the shorter building.

*41. ssm An Olympic long jumper leaves the ground at an angle of 23° and travels through the air for a horizontal distance of 8.7 m before landing. What is the takeoff speed of the jumper?

*42. The lob in tennis is an effective tactic when your opponent is near the net. It consists of lofting the ball over his head, forcing him to move quickly away from the net (see the drawing). Suppose that you loft the ball with an initial speed of 15.0 m/s, at an angle of 50.0° above the horizontal. At this instant your opponent is 10.0 m away from the ball. He begins moving away from you 0.30 s later, hoping to reach the ball and hit it back at the moment that it is 2.10 m above its launch point. With what minimum average speed must he move? (Ignore the fact that he can stretch, so that his racket can reach the ball before he does.)

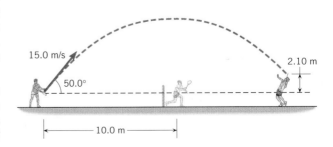

**43. Water from a garden hose that is pointed 25° above the horizontal lands directly on a sunbather lying on the ground 4.4 m away in the horizontal direction. If the hose is held 1.4 m above the ground, at what speed does the water leave the nozzle?

**44. A baseball is hit into the air at an initial speed of 36.6 m/s and an angle of 50.0° above the horizontal. At the same time, the center fielder starts running away from the batter and catches the ball 0.914 m above the level at which it was hit. If the center fielder is initially 1.10×10^2 m from home plate, what must be his average speed?

**45. ssm From the top of a tall building, a gun is fired. The bullet leaves the gun at a speed of 340 m/s, parallel to the ground. As the drawing shows, the bullet puts a hole in a window of another building and hits the wall that faces the window. Using the data in the drawing, determine the distances D and H, which locate the point where the gun was fired. Assume that the bullet does not slow down as it passes through the window.

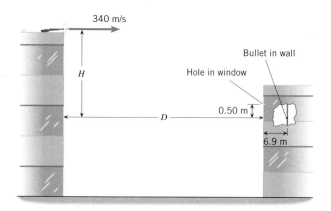

**46. Two cannons are mounted as shown in the drawing and rigged to fire simultaneously. They are used in a circus act in which two clowns serve as human cannon balls. The clowns are fired toward each other and collide at a height of 1.00 m above the muzzles of the cannons. Clown A is launched at a 75.0° angle, with a speed of 9.00 m/s. The horizontal separation between the clowns as they leave the cannons is 6.00 m. Find the launch speed v_{0B} and the launch angle θ_B (>45.0°) for clown B.

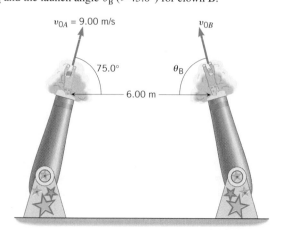

**47. A small can is hanging from the ceiling. A rifle is aimed directly at the can, as the figure illustrates. At the instant the gun is fired, the can is released. Ignore air resistance and show that the bullet will always strike the can, regardless of the initial speed of the bullet. Assume that the bullet strikes the can before the can reaches the ground.

Section 3.4 Relative Velocity

48. A bus has a velocity of 25 m/s, due south. A passenger walks to the back of the bus with a speed of 1.0 m/s relative to the bus. What is the velocity (magnitude and direction) of the passenger relative to a person standing on the ground?

49. ssm Two passenger trains are passing each other on adjacent tracks. Train A is moving east with a speed of 13 m/s, and train B is traveling west with a speed of 28 m/s. (a) What is the velocity (magnitude and direction) of train A as seen by the passengers in train B? (b) What is the velocity (magnitude and direction) of train B as seen by the passengers in train A?

50. At some airports there are speedramps to help passengers get from one place to another. A speedramp is a moving conveyor belt that you can either stand or walk on. Suppose a speedramp has a length of 105 m and is moving at a speed of 2.0 m/s relative to the ground. In addition, suppose you can cover this distance in 75 s when walking on the ground. If you walk at the same rate with respect to the speedramp that you walk with respect to the ground, how long does it take for you to travel the 105 m using the speedramp?

51. Two cars are approaching each other in adjacent lanes. Initially the cars are 1200 m apart. Each car has a speed of 25 m/s relative to the ground. How much time elapses before the cars pass each other?

52. The escalator that leads down into a subway station has a length of 30.0 m and a speed of 1.8 m/s relative to the ground. A student is coming out of the station by running in the wrong direction on this escalator. The local record time for this trick is 11 s. Relative to the escalator, what speed must the student exceed in order to beat the record?

53. ssm A swimmer, capable of swimming at a speed of 1.4 m/s in still water (i.e., the swimmer can swim with a speed of 1.4 m/s relative to the water), starts to swim directly across a 2.8-km-wide river. However, the current is 0.91 m/s, and it carries the swimmer downstream. (a) How long does it take the swimmer to cross the river? (b) How far downstream will the swimmer be upon reaching the other side of the river?

54. A remote-controlled model airplane is flying due east in still air. The airplane travels with a speed of 22.6 m/s relative to the air. A wind suddenly begins to blow from the north toward the south with a speed of 8.70 m/s. Find the velocity (magnitude and direction) of the airplane as seen by the controller standing on the ground. Determine the directional angle relative to due east.

55. You are in a hot-air balloon that, relative to the ground, has a velocity of 6.0 m/s in a direction due east. You see a hawk moving directly away from the balloon in a direction due north. The speed of the hawk relative to you is 2.0 m/s. What are the magnitude and direction of the hawk's velocity relative to the ground? Express the directional angle relative to due east.

56. A person looking out the window of a stationary train notices that raindrops are falling vertically down at a speed of 5.0 m/s relative to the ground. When the train moves at a constant velocity, the raindrops make an angle of 25° when they move past the window, as the drawing shows. How fast is the train moving?

***57.** **ssm** Mario, a hockey player, is skating due south at a speed of 7.0 m/s relative to the ice. A teammate passes the puck to him. The puck has a speed of 11.0 m/s and is moving in a direction of 22° west of south, relative to the ice. What are the magnitude and direction (relative to due south) of the puck's velocity, as observed by Mario?

***58.** A ferry boat is traveling in a direction 35.1° north of east with a speed of 5.12 m/s relative to the water. A passenger is walking with a velocity of 2.71 m/s due east relative to the boat. What is the velocity (magnitude and direction) of the passenger with respect to the water? Determine the directional angle relative to due east.

***59.** An oceanliner is heading due north with a speed of 8.5 m/s relative to the water. A small sailboat is heading 45° east of north with a speed of 1.0 m/s relative to the water. Find the relative velocity (magnitude and direction) of the sailboat as observed by the passengers on the oceanliner. Determine the directional angle relative to due north.

***60.** Two twins set out to row separately across a swiftly moving river. They have identical canoes and can row at the same speed relative to the water. Twin A heads straight across the river but, due to the current, is carried downstream before reaching the opposite bank. Twin B heads upstream at an angle of 55° from the bank so as to arrive at the opposite bank at a point that is directly across from the starting point. Determine the ratio of the time it takes twin A to cross the river to the time it takes twin B to cross.

****61.** **ssm www** A Coast Guard ship is traveling at a constant velocity of 4.20 m/s, due east, relative to the water. On his radar screen the navigator detects an object that is moving at a constant velocity. The object is located at a distance of 2310 m with respect to the ship, in a direction 32.0° south of east. Six minutes later, he notes that the object's position relative to the ship has changed to 1120 m, 57.0° south of west. What are the magnitude and direction of the velocity of the object relative to the water? Express the direction as an angle with respect to due west.

ADDITIONAL PROBLEMS

62. A mountain-climbing expedition establishes two intermediate camps, labeled A and B in the drawing, above the base camp. What is the magnitude Δr of the displacement between camp A and camp B?

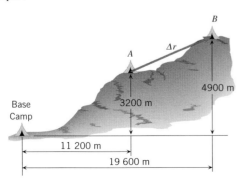

63. Michael Jordan, of the Chicago Bulls basketball team, has some fanatic fans. They claim that he is able to jump and remain in the air for two full seconds from launch to landing. Evaluate this claim by calculating the maximum height that such a jump would attain. For comparison, Jordan's maximum jump height has been estimated at about one meter.

64. A bullet is fired from a rifle that is held 1.6 m above the ground in a horizontal position. The initial speed of the bullet is 1100 m/s. Find (a) the time it takes for the bullet to strike the ground and (b) the horizontal distance traveled by the bullet.

65. **ssm** A fire hose ejects a stream of water at an angle of 35.0° above the horizontal. The water leaves the nozzle with a speed of 25.0 m/s. Assuming that the water behaves like a projectile, how far from a building should the fire hose be located to hit the highest possible fire?

66. Two cars with different velocities are approaching an intersection. Car A, traveling due east, has a speed of 15 m/s. Car B, traveling due north, has a speed of 21 m/s. What is the velocity (magnitude and direction) of B as seen by the passengers in A? Express the directional angle relative to due north.

67. A spacecraft is traveling with a velocity of $v_{0x} = 5480$ m/s along the $+x$ direction. Two engines are turned on for a time of 842 s. One engine gives the spacecraft an acceleration in the $+x$ direction of $a_x = 1.20$ m/s^2, while the other gives it an acceleration in the $+y$ direction of $a_y = 8.40$ m/s^2. At the end of the firing, find (a) v_x and (b) v_y.

68. At a certain point along its trajectory, a golf ball has the following velocity components: $v_x = +14.0$ m/s and $v_y = +18.9$ m/s, with upward and to the right being positive. (a) What is the speed at this point? (b) At what angle does the velocity vector point relative to the horizontal direction?

69. **ssm** A hot-air balloon is rising straight up with a speed of 3.0 m/s. A ballast bag is released from rest relative to the balloon when it is 9.5 m above the ground. How much time elapses before the ballast bag hits the ground?

70. When chasing a hare along a flat stretch of ground, a greyhound leaps into the air at a speed of 10.0 m/s, at an angle of 31° above the horizontal. (a) What is the range of his leap and (b) for how much time is he in the air?

*71. A soccer ball is kicked and launched at an angle of 11.0° above the ground. All other factors remaining the same, at what angle should the ball be launched to double the maximum height that it attains?

*72. During a hot-air balloon rally, two balloons are rising upward. Balloon A is rising straight up with a speed of 0.50 m/s relative to the ground. Due to a breeze, balloon B is moving away from A. Balloon B rises at an angle of 75° above the horizontal at a speed of 0.40 m/s relative to the ground. What is the speed of B relative to A?

*73. **ssm** A small aircraft is headed due south with a speed of 57.8 m/s with respect to still air. Then, for 9.00×10^2 s a wind blows the plane so that it moves in a direction 45.0° west of south, even though the plane continues to point due south. The plane travels 81.0 km with respect to the ground in this time. Determine the velocity (magnitude and direction) of the wind with respect to the ground. Determine the directional angle relative to due south.

*74. In the javelin throw at a track and field event, the javelin is launched at a speed of 29 m/s at an angle of 36° above the horizontal. As the javelin travels upward, its velocity points above the horizontal at an angle that decreases as time passes. How much time is required for the angle to be reduced from 36° at launch to 18°?

*75. A golfer standing on a fairway hits a shot to a green that is elevated 6.0 m above the point where she is standing. If the ball leaves her club with a velocity of 43 m/s at an angle of 40.0° to the ground, find (a) the time for the ball to come down on the green, and (b) the speed of the ball just before impact.

**76. A jetliner can fly 6.00 hours on a full load of fuel. Without any wind it flies at a speed of 2.40×10^2 m/s. The plane is to make a round-trip by heading due west for a certain distance, turning around, and then heading due east for the return trip. During the entire flight, however, the plane encounters a 57.8 m/s wind from the jet stream, which blows from west to east. What is the maximum distance that the plane can travel due west and just be able to return home?

77. **ssm The drawing shows an exaggerated view of a rifle that has been "sighted in" for a 91.4-meter target. If the muzzle speed of the bullet is $v_0 = 427$ m/s, what are the two possible angles θ_1 and θ_2 between the rifle barrel and the horizontal such that the bullet will hit the target? One of these angles is so large that it is never used in target shooting. (*Hint: The following trigonometric identity may be useful:* $2 \sin \theta \cos \theta = \sin 2\theta$.)

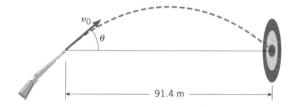

**78. A projectile is launched at an angle that yields the maximum possible range. Show, for this particular angle, that the range is four times greater than the maximum height reached by the projectile.

FORCES AND NEWTON'S LAWS OF MOTION

The cork accelerates out of a bottle of champagne
because the gas within the bottle applies a force to the cork
that exceeds the force that static friction applies.

4.1 THE CONCEPTS OF FORCE AND MASS

In common usage, a ***force*** is a push or a pull, as the examples in Figure 4.1 illustrate. In basketball, a player launches a shot by pushing on the ball. The tow bar attached to a speeding boat pulls a water skier. Forces such as those that launch the basketball or pull the skier are called ***contact forces***, because they arise from the physical contact between two objects. There are circumstances, however, in which two objects exert forces on one another even though they are not touching. One example of such a ***noncontact force*** occurs when a skydiver is pulled toward the earth because of the force of gravity. The earth exerts this force even when it is not in direct contact with the skydiver. In Figure 4.1, arrows are used to represent the forces. It is appropriate to use arrows, because a force is a vector quantity and has both a magnitude and a direction. The direction of the arrow gives the direction of the force, and the length is proportional to the strength, or magnitude, of the force.

The word ***mass*** is just as familiar as the word force in our vocabulary. A massive supertanker, for instance, is one that contains an enormous amount of mass. As we will see in the next section, such massive objects are difficult to get moving and are hard to stop once they are in motion. In comparison, a penny does not contain much mass. The emphasis here is on the amount of mass, and the idea of direction is of no concern. Therefore, mass is a scalar quantity.

During the seventeenth century, Isaac Newton, starting with the work of Galileo, developed three important laws that deal with force and mass. Collectively they are called "Newton's laws of motion" and provide the basis for understanding the effect that forces have on an object. Because of the importance of these laws, a separate section will be devoted to each one.

Figure 4.1 The arrow labeled **F** represents the force that acts on the basketball, the water skier, and the skydiver.

(a)

(b)

(c)

4.2 NEWTON'S FIRST LAW OF MOTION

THE FIRST LAW

To gain some insight into Newton's first law, think about the game of ice hockey (Figure 4.2). If a player does not hit a stationary puck, it will remain at rest on the ice. After the puck is struck, however, it coasts on its own across the ice, slowing down only slightly because of friction. Since ice is very slippery, there is only a relatively small amount of friction to slow down the puck. In fact, if it were possible to remove all friction and wind resistance, and if the rink were infinitely large, the puck would coast forever in a straight line at a constant speed. Left on its own, the puck would lose none of the velocity imparted to it at the time it was struck. This is the essence of Newton's first law of motion:

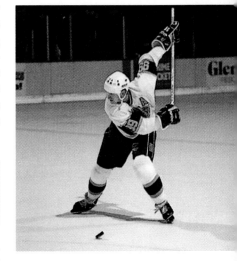

Figure 4.2 The game of ice hockey can give some insight into Newton's laws of motion.

■ **NEWTON'S FIRST LAW OF MOTION**

An object continues in a state of rest or in a state of motion at a constant speed along a straight line, unless compelled to change that state by a net force.

The first law uses the phrase "net force." Often, several forces act simultaneously on a body, and ***the net force is the vector sum of all of them***. Individual forces matter only to the extent that they contribute to the total. For instance, if friction and other opposing forces were absent, a car could travel forever at 30 m/s in a straight line, without using any gas after it has come up to speed. In reality gas is needed, but only so that the engine can produce the necessary force to cancel opposing forces such as friction. This cancellation ensures that there is no net force to change the state of motion of the car.

When an object moves at a constant speed along a straight line, its velocity is constant. Newton's first law indicates that a state of rest (zero velocity) and a state of constant velocity are completely equivalent, in the sense that neither one requires the application of a net force to sustain it. *The purpose served when a net force acts on an object is not to sustain the object's velocity, but, rather, to change it.*

INERTIA AND MASS

A greater net force is required to change the velocity of some objects than of others. For instance, a net force that is just enough to cause a bicycle to pick up speed will cause an imperceptible change in the motion of a freight train. In comparison to the bicycle, the train has a much greater tendency to remain at rest. Accordingly, we say that the train has more ***inertia*** than the bicycle. Quantitatively, the inertia of an object is measured by its ***mass***. The following definition of inertia and mass indicates why Newton's first law is sometimes called the law of inertia:

■ **DEFINITION OF INERTIA AND MASS**

Inertia is the natural tendency of an object to remain at rest or in motion at a constant speed along a straight line. The mass of an object is a quantitative measure of inertia.

SI Unit of Inertia and Mass: kilogram (kg)

The SI unit for mass is the kilogram (kg), whereas the units in the CGS system and the BE system are the gram (g) and the slug (sl), respectively. Conversion factors between these units are given on the page facing the inside of the front cover. Figure 4.3 gives the masses of various objects, ranging from a penny to a super-tanker. The larger the mass, the greater is the inertia. Often the words "mass" and "weight" are used interchangeably, but this is incorrect. Mass and weight are different concepts, and Section 4.7 will discuss the distinction between them.

Figure 4.4 shows a useful application of inertia. Automobile seat belts unwind freely when pulled gently, so they can be buckled. But in an accident, they hold you safely in place. One seat-belt mechanism consists of a ratchet wheel, a locking bar, and a pendulum. The belt is wound around a spool mounted on the ratchet wheel. While the car is at rest or moving at a constant velocity, the pendulum hangs straight down, and the locking bar rests horizontally, as the black-lined part of the drawing shows. Consequently, nothing prevents the ratchet wheel from turning, and the seat belt can be pulled out easily. When the car suddenly slows down in an accident, however, the relatively massive, lower part of the pendulum keeps moving forward because of its inertia. The pendulum swings on its pivot into the position shown in color and causes the locking bar to block the rotation of the ratchet wheel, thus preventing the seat belt from unwinding.

Inertia also lies at the heart of automobile safety air bag systems. When a severe collision occurs, an air bag is inflated with nitrogen gas, which originates from a sodium azide cartridge. Figure 4.5 shows one mechanism used to sense that a collision has occurred, fire the cartridge, and release the nitrogen gas. This mechanism includes a firing pin and a steel ball that fits snugly in a short cylinder. The firing pin is pushed against a spring and held in place by a restraint mounted on a lever. The restraint is shaped like a half moon. A counterbalancing spring applies a force to the lower end of the lever to hold the half moon against the firing pin. When the car decelerates suddenly in a collision, the ball's inertia keeps it moving forward, so that it pushes on the upper end of the lever. This push overcomes the effect of the counterbalancing spring. The lever rotates clockwise about its pivot, and the firing pin is released. The compressed spring behind the firing pin rams it into a detonator cap, which ignites and causes the sodium azide cartridge to fire. Within four-hundredths of a second after the collision occurs, the air bag is fully inflated.

The Physics of...
seat belts.

The Physics of...
air bags.

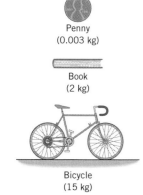

Penny
(0.003 kg)

Book
(2 kg)

Bicycle
(15 kg)

Car
(2000 kg)

Jetliner
(1.2×10^5 kg)

Supertanker
(1.5×10^8 kg)

Figure 4.3 The masses of various objects.

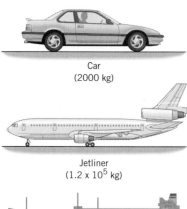

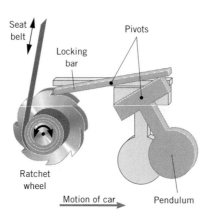

Seat belt

Locking bar

Pivots

Ratchet wheel

Motion of car

Pendulum

Figure 4.4 Inertia plays a central role in one seat belt mechanism. The parts of the drawing not outlined in black show what happens when the car suddenly slows down, as in an accident.

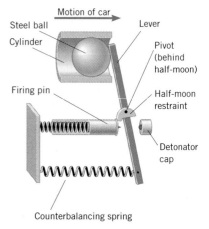

Motion of car

Steel ball

Cylinder

Firing pin

Lever

Pivot (behind half-moon)

Half-moon restraint

Detonator cap

Counterbalancing spring

Figure 4.5 Automobile safety air bag systems use inertia in the mechanism that senses when a collision has occurred and causes an air bag to deploy.

AN INERTIAL REFERENCE FRAME

Newton's first law (and also the second law) can appear to be invalid to certain observers. Suppose, for instance, that you are a passenger riding in a friend's car. While the car moves at a constant speed along a straight line, you do not feel the seat pushing against your back to any unusual extent. This experience is consistent with the first law, which indicates that in the absence of a net force you should move with a constant velocity. Suddenly the driver floors the gas pedal. Immediately you feel the seat pressing against your back as the car accelerates. Therefore, you sense that a force is being applied to you. The first law leads you to believe that your motion should change, and, relative to the ground outside, your motion does change. But *relative to the car*, you can see that your motion does *not* change, because you remain stationary with respect to the car. Clearly, Newton's first law does not hold for observers who use the accelerating car as a frame of reference. As a result, such a reference frame is said to be noninertial. All accelerating reference frames are noninertial. In contrast, observers for whom the law of inertia is valid are said to be using **inertial reference frames** for their observations, as defined below:

> ■ **DEFINITION OF AN INERTIAL REFERENCE FRAME**
>
> An inertial reference frame is one in which Newton's law of inertia is valid.

The acceleration of an inertial reference frame is zero, so it moves with a constant velocity. All of Newton's laws of motion are valid in inertial reference frames, and when we apply these laws, we will be assuming such a reference frame. In particular, the earth itself is a good approximation of an inertial reference frame.

4.3 NEWTON'S SECOND LAW OF MOTION

THE SECOND LAW

Newton's first law indicates that if no net force acts on an object, then the velocity of the object remains unchanged. The second law deals with what happens when a net force does act. Consider a hockey puck once again. When a player strikes a stationary puck, he causes the velocity of the puck to change. In other words, he makes the puck accelerate. The cause of the acceleration is the force that the hockey stick applies. As long as this force acts, the velocity increases, and the puck accelerates. Now, suppose another player strikes the puck and applies twice as much force as the first player does. The greater force produces a greater acceleration. In fact, if the friction between the puck and the ice is negligible, and if there is no wind resistance, the acceleration of the puck is directly proportional to the force. Twice the force produces twice the acceleration. Moreover, the acceleration is a vector quantity, just as the force is, and points in the same direction as the force.

Often, several forces act on an object simultaneously. Friction and wind resistance, for instance, do have some effect on a hockey puck. In such cases, it is the net force, or the vector sum of all the forces acting, that is important. Mathematically, the net force is written as $\Sigma\mathbf{F}$, where the Greek capital letter Σ (sigma) denotes the vector sum. Newton's second law states that the acceleration is proportional to the net force acting on the object.

In Newton's second law, the net force is only one of two factors that determine the acceleration. The other factor is the inertia or mass of the object. After all, the same net force that imparts an appreciable acceleration to a hockey puck (small mass) will impart very little acceleration to a semitrailer truck (large mass). Newton's second law states that for a given net force, the magnitude of the acceleration is inversely proportional to the mass. Twice the mass means one-half the acceleration, if the same net force acts on both objects. Thus, the second law shows how the acceleration depends on both the net force and the mass, as summarized below in Equation 4.1.

> ### ■ NEWTON'S SECOND LAW OF MOTION
>
> When a net external force $\Sigma\mathbf{F}$ acts on an object of mass m, the acceleration $\mathbf{a}$ that results is directly proportional to the net force and has a magnitude that is inversely proportional to the mass. The direction of the acceleration is the same as the direction of the net force.
>
> $$\mathbf{a} = \frac{\Sigma\mathbf{F}}{m} \quad \text{or} \quad \Sigma\mathbf{F} = m\mathbf{a} \tag{4.1}$$
>
> **SI Unit of Force:** $\text{kg} \cdot \text{m/s}^2 = $ newton (N)

It is important to note that the net force in Equation 4.1 includes only the forces that the environment exerts on the object of interest. Such forces are called **external forces,** to distinguish them from **internal forces.** Internal forces are forces that one part of an object exerts on another part of the object and are not included in Equation 4.1.

According to Equation 4.1, the SI unit for force is the unit for mass (kg) times the unit for acceleration (m/s^2), or

$$\text{SI unit for force} = (\text{kg})\left(\frac{\text{m}}{\text{s}^2}\right) = \frac{\text{kg} \cdot \text{m}}{\text{s}^2}$$

The combination of $\text{kg} \cdot \text{m/s}^2$ is called a *newton* (N) and is a derived SI unit, not a base unit; 1 newton = 1 N = 1 $\text{kg} \cdot \text{m/s}^2$.

In the CGS system, the procedure for establishing the unit of force is the same as with SI units, except that mass is expressed in grams (g) and acceleration in cm/s^2. The resulting unit for force is the *dyne*; 1 dyne = 1 $\text{g} \cdot \text{cm/s}^2$.

In the BE system, the unit for force is defined to be the pound (lb),* while the unit for acceleration is ft/s^2. With this procedure, Newton's second law can then be used to obtain the unit for mass:

$$\text{BE unit for force} = \text{lb} = m\left(\frac{\text{ft}}{\text{s}^2}\right)$$

$$m = \frac{\text{lb} \cdot \text{s}^2}{\text{ft}}$$

The combination of $\text{lb} \cdot \text{s}^2/\text{ft}$ is the unit for mass in the BE system and is called the *slug* (sl); 1 slug = 1 sl = 1 $\text{lb} \cdot \text{s}^2/\text{ft}$.

Table 4.1 summarizes the various units for mass, acceleration, and force. Con-

---

* We refer here to the gravitational version of the BE system, in which a force of one pound is defined to be the pull of the earth on a certain standard body at a location where the acceleration due to gravity is 32.174 ft/s^2.

Table 4.1 Units for Mass, Acceleration, and Force

System	Mass	Acceleration	Force
SI	kilogram (kg)	meter/second^2 (m/s^2)	newton (N)
CGS	gram (g)	centimeter/second^2 (cm/s^2)	dyne (dyn)
BE	slug (sl)	foot/second^2 (ft/s^2)	pound (lb)

version factors between force units from different systems are provided on the page facing the inside of the front cover.

FREE-BODY DIAGRAMS AND THE SECOND LAW

When using the second law to calculate the acceleration, it is necessary to determine the net force that acts on the object. In this determination a *free-body diagram* helps enormously. A free-body diagram is a diagram that represents the object and the forces that act on it. Only the forces that *act on the object* appear in a free-body diagram. Forces that the object exerts on its environment are not included. Example 1 illustrates the use of a free-body diagram.

> • **PROBLEM SOLVING INSIGHT**
> A free-body diagram is very helpful when applying Newton's second law. Always start a problem by drawing the free-body diagram.

EXAMPLE 1 • Pushing a Stalled Car

Two people are pushing a stalled car, as Figure 4.6*a* indicates. The mass of the car is 1850 kg. One person applies a force of 275 N to the car, while the other applies a force of 395 N. Both forces act in the same direction. A third force of 560 N also acts on the car, but in a direction opposite to that in which the people are pushing. This force arises because of friction and the extent to which the pavement opposes the motion of the tires. Find the acceleration of the car.

Reasoning According to Newton's second law of motion, the acceleration is the net force divided by the mass of the car. To determine the net force acting on the car, we use the free-body diagram in Figure 4.6*b*. In this diagram, the car is represented as a dot, and the motion of the car is chosen to be along the *x* axis. The diagram makes it clear that the forces all act along one direction. Therefore, they can be added as colinear vectors to obtain the net force.

Solution The net force is

$$\Sigma F = +275 \text{ N} + 395 \text{ N} - 560 \text{ N} = +110 \text{ N}$$

The acceleration can now be obtained:

$$a = \frac{\Sigma F}{m} = \frac{+110 \text{ N}}{1850 \text{ kg}} = \boxed{+0.059 \text{ m/s}^2} \qquad (4.1)$$

The plus sign indicates that the acceleration points along the +*x* axis, in the same direction as the net force.

> • **PROBLEM SOLVING INSIGHT**
> The direction of the acceleration is always the same as the direction of the net force.

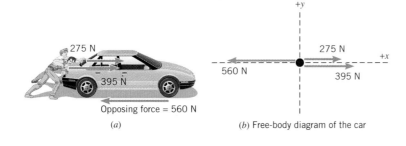

(a)

(b) Free-body diagram of the car

Figure 4.6 (*a*) Two people push a stalled car, in opposition to a force created by friction and the pavement. (*b*) A free-body diagram that shows the horizontal forces acting on the car.

4.4 THE VECTOR NATURE OF NEWTON'S SECOND LAW OF MOTION

When a football player throws a pass, the direction of the force he applies to the ball is important. Both the force and the resulting acceleration of the ball are vector quantities, as are all forces and accelerations. The directions of these vectors can be taken into account in two dimensions by using x and y components. The net force $\Sigma\mathbf{F}$ in Newton's second law has components ΣF_x and ΣF_y, while the acceleration $\mathbf{a}$ has components a_x and a_y. Consequently, Newton's second law, as expressed in Equation 4.1, can be written in an equivalent form as two equations, one for the x components and one for the y components:

$$\Sigma F_x = ma_x \tag{4.2a}$$

$$\Sigma F_y = ma_y \tag{4.2b}$$

This procedure is similar to that employed in Chapter 3 for the equations of two-dimensional kinematics (see Table 3.1). The components themselves in Equations 4.2a and 4.2b will be either positive or negative numbers, depending on whether they point along the positive or negative x or y axis. The remainder of this section deals with examples that show how these equations are used.

EXAMPLE 2 • Applying Newton's Second Law Using Components

A man is stranded on a raft (mass of man and raft = 1300 kg), as shown in Figure 4.7a. By paddling, he causes an average force **P** of 17 N to be applied to the raft in a direction due east (the $+x$ direction). The wind also exerts a force **A** on the raft. This

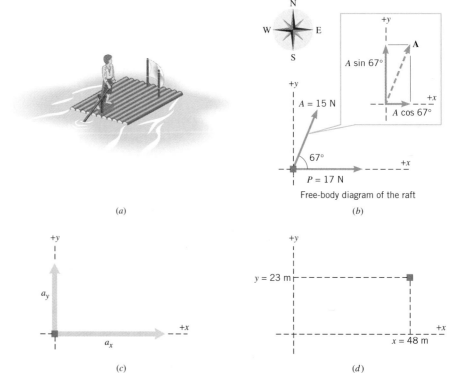

Figure 4.7 (a) A man is paddling a raft, as in Example 2. (b) The free-body diagram shows the forces **P** and **A** that act on the raft. Vertical forces perpendicular to the raft play no role in the example and are omitted from the diagram for clarity. (c) The raft's acceleration components a_x and a_y. (d) In 65 s, the components of the raft's displacement are $x = 48$ m and $y = 23$ m.

force has a magnitude of 15 N and points 67° north of east. Ignoring any resistance from the water, find the x and y components of the raft's acceleration.

Reasoning Since the mass of the man and the raft is known, Newton's second law can be used to determine the acceleration components from the given forces. According to the form of the second law in Equations 4.2a and 4.2b, the acceleration component in a given direction is obtained by dividing the component of the net force in that direction by the mass. As an aid in determining the components ΣF_x and ΣF_y of the net force, we use the free-body diagram in Figure 4.7b. In this diagram, the direction due east is the $+x$ direction.

Solution Figure 4.7b shows the force components:

Force	x Component	y Component
P	$+17$ N	0
A	$+(15\text{ N})\cos 67° = +6$ N	$+(15\text{ N})\sin 67° = +14$ N
	$\Sigma F_x = +17\text{ N} + 6\text{ N} = +23$ N	$\Sigma F_y = +14$ N

The plus signs indicate that ΣF_x points in the direction of the $+x$ axis, and ΣF_y points in the direction of the $+y$ axes. The x and y components of the acceleration point in the directions of ΣF_x and ΣF_y, respectively, and can now be calculated:

$$a_x = \frac{\Sigma F_x}{m} = \frac{+23\text{ N}}{1300\text{ kg}} = \boxed{+0.018\text{ m/s}^2} \qquad (4.2a)$$

$$a_y = \frac{\Sigma F_y}{m} = \frac{+14\text{ N}}{1300\text{ kg}} = \boxed{+0.011\text{ m/s}^2} \qquad (4.2b)$$

These acceleration components are shown in Figure 4.7c.

• **PROBLEM SOLVING INSIGHT**
Applications of Newton's second law always involve the net external force, which is the vector sum of all the external forces that act on an object.

EXAMPLE 3 • The Displacement of a Raft

At the moment the forces **P** and **A** begin acting on the raft in Example 2, the velocity of the raft is 0.15 m/s, in a direction due east (the $+x$ direction). Assuming that the forces are maintained for 65 s, find the x and y components of the raft's displacement during this time interval.

Reasoning Once the net force acting on an object and the object's mass have been used in Newton's second law to determine the acceleration, it becomes possible to use the equations of kinematics to describe the resulting motion. We know from Example 2 that the acceleration components are $a_x = +0.018$ m/s^2 and $a_y = +0.011$ m/s^2, and it is given here that the initial velocity components are $v_{0x} = +0.15$ m/s and $v_{0y} = 0$. Thus, Equation 3.5a ($x = v_{0x}t + \frac{1}{2}a_x t^2$) and Equation 3.5b ($y = v_{0y}t + \frac{1}{2}a_y t^2$) can be used with $t = 65$ s to determine the x and y components of the raft's displacement.

Solution According to Equations 3.5a and 3.5b, the x and y components of the displacement are

$$x = v_{0x}t + \tfrac{1}{2}a_x t^2 = (0.15\text{ m/s})(65\text{ s}) + \tfrac{1}{2}(0.018\text{ m/s}^2)(65\text{ s})^2 = \boxed{48\text{ m}}$$

$$y = v_{0y}t + \tfrac{1}{2}a_y t^2 = (0)(65\text{ s}) + \tfrac{1}{2}(0.011\text{ m/s}^2)(65\text{ s})^2 = \boxed{23\text{ m}}$$

Figure 4.7d shows the final location of the raft.

The two elephants exert action and re-action forces on each other.

4.5 NEWTON'S THIRD LAW OF MOTION

Imagine you are in a football game. You line up facing your opponent, the ball is snapped, and the two of you crash together. No doubt, you feel a force. But think about your opponent. He too feels something, for while he is applying a force to you, you are applying a force to him. In other words, there isn't just one force on the line of scrimmage; there is a pair of forces. Newton was the first to realize that all forces occur in pairs and there is no such thing as an isolated force, existing all by itself. His third law of motion deals with this fundamental characteristic of forces.

■ **NEWTON'S THIRD LAW OF MOTION**

Whenever one body exerts a force on a second body, the second body exerts an oppositely directed force of equal magnitude on the first body.

The third law is often called the "action–reaction" law, for it is sometimes quoted as follows: "For every action (force) there is an equal, but opposite, reaction."

Figure 4.8 illustrates how the third law applies to an astronaut who is drifting just outside a spacecraft and who pushes on the spacecraft with a force $\mathbf{F}$. According to the third law, the spacecraft pushes back on the astronaut with a force $-\mathbf{F}$ that is equal in magnitude, but opposite in direction. In Example 4, we examine the accelerations produced by each of these forces.

Figure 4.8   The astronaut pushes on the spacecraft with a force $+\mathbf{F}$. According to Newton's third law, the spacecraft simultaneously pushes back on the astronaut with a force $-\mathbf{F}$.

EXAMPLE 4 • The Accelerations Produced by Action and Reaction Forces

Suppose that the mass of the spacecraft in Figure 4.8 is $m_S = 11\ 000$ kg and that the mass of the astronaut is $m_A = 92$ kg. In addition, assume that the astronaut exerts a force of $\mathbf{F} = +36$ N on the spacecraft. Find the accelerations of the spacecraft and the astronaut.

Reasoning According to Newton's third law, when the astronaut applies the force $\mathbf{F} = +36$ N to the spacecraft, the spacecraft applies a reaction force $-\mathbf{F} = -36$ N to the astronaut. As a result, the spacecraft and the astronaut accelerate in opposite directions. Although the action and reaction forces have the same magnitude, they do not create accelerations of the same magnitude, because the spacecraft and the astronaut have different masses. According to Newton's second law, the astronaut, having a much smaller mass, will experience a much larger acceleration.

Solution The acceleration of the spacecraft is

$$\mathbf{a}_S = \frac{\mathbf{F}}{m_S} = \frac{+36\ \text{N}}{11\ 000\ \text{kg}} = \boxed{+0.0033\ \text{m/s}^2}$$

The acceleration of the astronaut is

$$\mathbf{a}_A = \frac{-\mathbf{F}}{m_A} = \frac{-36\ \text{N}}{92\ \text{kg}} = \boxed{-0.39\ \text{m/s}^2}$$

• **PROBLEM SOLVING INSIGHT**
Even though the magnitudes of the action and reaction forces are always equal, these forces do not necessarily produce accelerations that have equal magnitudes, since each force acts on a different object that may have a different mass.

There is a clever application of Newton's third law in some rental trailers. As Figure 4.9 illustrates, the tow bar connecting the trailer to the rear bumper of a car contains a mechanism that can automatically actuate brakes on the trailer wheels.

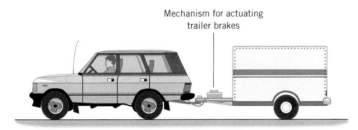

Mechanism for actuating
trailer brakes

Figure 4.9 Some rental trailers include an automatic brake-actuating mechanism.

This mechanism works without the need for electrical connections between the car and the trailer. When the driver applies the car brakes, the car slows down. Because of inertia, however, the trailer continues to roll forward and begins pushing against the bumper. In reaction, the bumper pushes back on the tow bar. The reaction force is used by the mechanism in the tow bar to "push the brake pedal" for the trailer.

The Physics of...
automatic trailer brakes.

4.6 TYPES OF FORCES: AN OVERVIEW

Newton's three laws of motion make it clear that forces play a central role in determining the motion of an object. In the next four sections we discuss some common forces: the gravitational force (Section 4.7), the normal force (Section 4.8), frictional forces (Section 4.9), and the tension force (Section 4.10). No matter which of these forces is present, however, its effect on the motion is to contribute to the net force, which determines the acceleration according to Newton's second law. Thus, from the point of view of the second law, these forces are treated in exactly the same way, as the concept chart in Figure 4.10 illustrates.

In nature there are two general types of forces, fundamental and nonfundamental. Fundamental forces are the ones that are truly unique, in the sense that all other forces can be explained in terms of them. Only three fundamental forces have been discovered:

1. Gravitational force
2. Strong nuclear force
3. Electroweak force

Figure 4.10 *Concepts at a Glance*
When any of the external forces listed here act on an object, they must be included as part of the net force $\Sigma\mathbf{F}$ in any application of Newton's second law. In this tug-of-war, each of the four external forces acts on the team members.

Concepts at a Glance

External Forces

1. Gravitational Force
(Section 4.7)
2. Normal Force
(Section 4.8)
3. Frictional Forces
(Section 4.9)
4. Tension Force
(Section 4.10)

Newton's Second Law

$\Sigma\mathbf{F} = m\mathbf{a}$

The gravitational force is discussed in the next section. The strong nuclear force plays a primary role in the stability of the nucleus of the atom (see Section 31.2). The electroweak force is a single force that manifests itself in two ways (see Section 32.6). One manifestation is the electromagnetic force that electrically charged particles exert on one another (see Sections 18.5, 21.2, and 21.8). The other manifestation is the so-called weak nuclear force that plays a role in the radioactive disintegration of certain nuclei (see Section 31.5).

Except for the gravitational force, all of the forces discussed in this chapter are nonfundamental, because they are related to the electromagnetic force. They arise from the interactions between the electrically charged particles that comprise atoms and molecules. Our understanding of which forces are fundamental is continually evolving. For instance, in the 1860s and 1870s James Clerk Maxwell showed that the electric force and the magnetic force could be explained as manifestations of a single electromagnetic force. Then, in the 1970s, Sheldon Glashow (1932–), Abdus Salam (1926–), and Steven Weinberg (1933–) presented the theory that explains how the electromagnetic force and the weak nuclear force are related to the electroweak force. They received a Nobel prize in 1979 for their achievement. Today, efforts continue that have the goal of reducing further the number of fundamental forces.

4.7 THE GRAVITATIONAL FORCE

NEWTON'S LAW OF UNIVERSAL GRAVITATION

Objects fall downward because of gravity, and Chapters 2 and 3 discuss how to describe the effects of gravity by using a value of $g = 9.80 \text{ m/s}^2$ for the downward acceleration it causes. However, nothing has been said about why g is 9.80 m/s^2. The reason is fascinating, as we will now see.

The acceleration due to gravity is like any other acceleration, and Newton's second law indicates that it must be caused by a net force. In addition to his famous three laws of motion, Newton also provided a coherent understanding of the ***gravitational force***. His "law of universal gravitation" is stated as follows:

■ **NEWTON'S LAW OF UNIVERSAL GRAVITATION**

Every particle in the universe exerts an attractive force on every other particle. A particle is a piece of matter, small enough in size to be regarded as a mathematical point. For two particles, which have masses m_1 and m_2 and are separated by a distance r, the force that each exerts on the other is directed along the line joining the particles (see Figure 4.11) and has a magnitude given by

$$F = G \frac{m_1 m_2}{r^2} \tag{4.3}$$

The symbol G denotes the universal gravitational constant, whose value is found experimentally to be

$$G = 6.672\,59 \times 10^{-11} \text{ N} \cdot \text{m}^2/\text{kg}^2$$

Figure 4.11 The two particles, whose masses are m_1 and m_2, are attracted by gravitational forces $+\mathbf{F}$ and $-\mathbf{F}$.

The constant G that appears in Equation 4.3 is called the ***universal gravitational constant***, because it has the same value for all pairs of particles anywhere in the universe, no matter what their separation. The value for G was first measured in an experiment by the English scientist Henry Cavendish (1731–1810), more than a century after Newton proposed his law of universal gravitation.

To see the main features of Newton's law of universal gravitation, look at the two particles in Figure 4.11. They have masses m_1 and m_2 and are separated by a distance r. In the picture, it is assumed that a force pointing to the right is positive. The gravitational forces point along the line joining the particles and are

$\quad$ $+\mathbf{F}$, the gravitational force exerted on m_1 by m_2
$\quad$ $-\mathbf{F}$, the gravitational force exerted on m_2 by m_1

These two forces have equal magnitudes and opposite directions. They act on different bodies, causing them to be mutually attracted. In fact, these forces are an action–reaction pair, as required by Newton's third law. Example 5 shows that the magnitude of the gravitational force is extremely small for ordinary values of the masses and the distance between them.

EXAMPLE 5 • Gravitational Attraction

What is the magnitude of the gravitational force that acts on each particle in Figure 4.11, assuming $m_1 = 12$ kg (approximately the mass of a bicycle), $m_2 = 25$ kg, and $r = 1.2$ m?

Reasoning and Solution The magnitude of the gravitational force can be found using Equation 4.3:

$$F = G\frac{m_1 m_2}{r^2} = (6.67 \times 10^{-11} \text{ N} \cdot \text{m}^2/\text{kg}^2)\frac{(12 \text{ kg})(25 \text{ kg})}{(1.2 \text{ m})^2} = \boxed{1.4 \times 10^{-8} \text{ N}}$$

For comparison, you exert a force of about 1 N when pushing a doorbell, so that the gravitational force is exceedingly small in circumstances such as those here. This result is due to the fact that G itself is very small. However, if one of the bodies has a large mass, like that of the earth (5.98×10^{24} kg), the gravitational force can be large.

As expressed by Equation 4.3, Newton's law of gravitation applies only to particles. However, most familiar objects are too large to be considered particles. Nevertheless, the law of universal gravitation can be applied to such objects with the aid of calculus. Newton was able to prove that an object of finite size can be considered to be a particle for purposes of using the gravitation law, provided the mass of the object is distributed with spherical symmetry about its center. Thus, Equation 4.3 can be applied when each object is a sphere whose mass is spread uniformly over its entire volume. Figure 4.12 shows this kind of application, assuming that the earth and the moon are such uniform spheres of matter. In this case, r is the distance *between the centers of the spheres* and not the distance between the outer surfaces. The gravitational forces that the spheres exert on each other are the same as if the entire mass of each was concentrated at its center. Even if the objects are not uniform spheres, Equation 4.3 can be used to a good degree of approximation if the sizes of the objects are small relative to the distance of separation r.

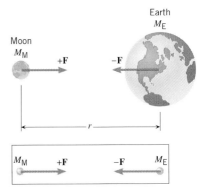

Figure 4.12 The gravitational force that each uniform sphere of matter exerts on the other is the same as if each sphere were a particle with its mass concentrated at its center. The earth (mass M_E) and the moon (mass M_M) approximate such uniform spheres.

WEIGHT

The weight of an object arises because of the gravitational pull of the earth.

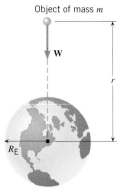

Object of mass m

R_E

Mass of Earth = M_E

Figure 4.13 On or near the earth, the weight **W** of an object is the gravitational force exerted on the object by the earth.

> ### ■ DEFINITION OF WEIGHT
>
> The weight of an object on the earth is the gravitational force that the earth exerts on the object. The weight always acts downward, toward the center of the earth. On another astronomical body, the weight is the gravitational force exerted on the object by that body.
>
> **SI Unit of Weight:** newton (N)

Using W for the magnitude of the weight,* m for the mass of the object, and M_E for the mass of the earth, it follows from Equation 4.3 that

$$W = G\frac{M_E m}{r^2} \tag{4.4}$$

Equation 4.4 and Figure 4.13 both emphasize that an object has weight whether or not it is resting on the earth's surface, because the gravitational force is acting even when the distance r is not equal to the radius R_E of the earth. However, the gravitational force becomes weaker as r increases, since r is in the denominator of Equation 4.4. Example 6 demonstrates how the weight changes, depending on where an object is located with respect to the center of the earth.

EXAMPLE 6 • The Hubble Space Telescope

The mass of the Hubble space telescope is 11 600 kg. Determine the weight of the telescope (a) when it was resting on the earth and (b) as it is in its orbit 596 km above the earth's surface.

Reasoning The weight of the Hubble Space Telescope is the gravitational force exerted on it by the earth. According to Equation 4.4, the weight varies inversely as the square of the radial distance r. Thus, we expect the telescope's weight on the earth's surface (r smaller) to be greater than its weight in orbit (r larger).

Solution

(a) On the earth's surface, the weight is given by Equation 4.4 with $r = 6.38 \times 10^6$ m (the earth's radius):

$$W = G\frac{M_E m}{r^2} = \frac{(6.67 \times 10^{-11}\ \text{N·m}^2/\text{kg}^2)(5.98 \times 10^{24}\ \text{kg})(11\ 600\ \text{kg})}{(6.38 \times 10^6\ \text{m})^2}$$

$$\boxed{W = 1.14 \times 10^5\ \text{N}}$$

(b) When the telescope is 596 km above the surface, its distance from the center of the earth is

$$r = 6.38 \times 10^6\ \text{m} + 596 \times 10^3\ \text{m} = 6.98 \times 10^6\ \text{m}$$

The weight now can be calculated as in part (a), except the new value of r must be used: $\boxed{W = 0.950 \times 10^5\ \text{N}}$. As expected, the weight is less in orbit.

---

* Often, the word "weight" and the phrase "magnitude of the weight" are used interchangeably, even though weight is a vector. Generally, the context makes it clear when the direction of the weight vector must be taken into account.

The space age has forced us to broaden our ideas about weight. For instance, an astronaut weighs only about one-sixth as much on the moon as on the earth. To obtain the weight of the astronaut on the moon from Equation 4.4, it is only necessary to replace M_E by M_M (the mass of the moon) and let $r = R_M$ (the radius of the moon).

RELATION BETWEEN MASS AND WEIGHT

Although massive objects weigh a lot on the earth, mass and weight are not the same quantity. As Section 4.2 discusses, mass is a quantitative measure of inertia. As such, mass is an intrinsic property of matter and does not change as an object is moved from one location to another. Weight, on the other hand, is the gravitational force acting on the object and can vary, depending on how far the object is above the earth's surface or whether the object is located near another body such as the moon.

The relation between weight W and mass m can be written in one of two ways:

$$W = \boxed{G\,\frac{M_E}{r^2}}\,m \qquad (4.4)$$

$$W = m\,\boxed{g} \qquad (4.5)$$

The first of these is Newton's law of universal gravitation, while the second is Newton's second law incorporating the acceleration g due to gravity. These expressions make the distinction between mass and weight stand out. The weight of an object whose mass is m depends on the values for the universal gravitation constant G, the mass M_E of the earth, and the distance r. These three parameters together determine the acceleration g due to gravity. The specific value of $g = 9.80$ m/s^2 applies only when r equals the radius R_E of the earth. For larger values of r, as would be the case on top of a mountain, the effective value of g is less than 9.80 m/s^2. The fact that g decreases as the distance r increases means that the weight likewise decreases. The mass of the object, however, does not depend on these effects and does not change. Conceptual Example 7 further explores the difference between mass and weight.

• **PROBLEM SOLVING INSIGHT**
Mass and weight are different quantities. They cannot be interchanged when solving problems.

CONCEPTUAL EXAMPLE 7 • Mass Versus Weight

A vehicle is being designed for use in exploring the moon's surface and is being tested on earth, where it weighs roughly six times more than it will on the moon. In one test, the acceleration of the vehicle is measured. To achieve the same acceleration on the moon, will the net force acting on the vehicle be greater than, less than, or the same as that required on earth?

Reasoning and Solution The net force $\Sigma\mathbf{F}$ required to accelerate the vehicle is specified by Newton's second law as $\Sigma\mathbf{F} = m\mathbf{a}$, where m is the vehicle's mass and $\mathbf{a}$ is the acceleration. For a given acceleration, the net force depends only on the mass. But the mass is an intrinsic property of the vehicle and is the same on the moon as it is on the earth. Therefore, the same net force would be required for a given acceleration on the moon as on the earth. Do not be misled by the fact that the vehicle weighs more on earth. The greater weight occurs only because the earth's mass and radius are different than the moon's. In any event, *in Newton's second law, the net force is proportional to the vehicle's mass, not its weight.*

Related Homework Material: Problems 21 and 102

The lunar exploration vehicle that this astronaut is driving on the moon has the same mass that it has on the earth. However, its weight is different on the moon than on the earth, as Conceptual Example 7 discusses.

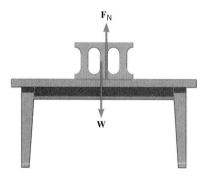

Figure 4.14 Two forces act on the block, its weight **W** and the normal force **F**$_N$ exerted by the surface of the table.

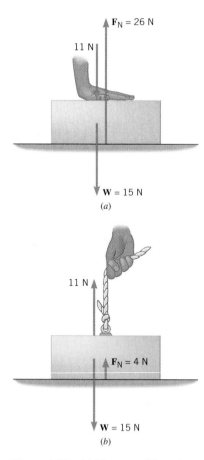

Figure 4.15 (*a*) The normal force is greater than the weight of the box, because the box is being pressed downward with an 11-N force. (*b*) The normal force is smaller than the weight, because the rope supplies an upward force of 11 N that partially supports the box.

4.8 THE NORMAL FORCE

THE NORMAL FORCE AND NEWTON'S THIRD LAW

In many situations, an object is in contact with a surface, such as a tabletop. Because of the contact, there is a force acting on the object. The present section discusses only one component of this force, the component that acts perpendicular to the surface. The next section discusses the component that acts parallel to the surface. The perpendicular component is called the ***normal force,*** where the word "normal" is used as a synonym for the word "perpendicular."

> ■ **DEFINITION OF THE NORMAL FORCE**
>
> The normal force **F**$_N$ is one component of the force that a surface exerts on an object with which it is in contact, namely, the component that is perpendicular to the surface.

Figure 4.14 shows a block resting on a horizontal table and identifies the two forces that act on the block, the weight **W** and the normal force **F**$_N$. To understand how an inanimate object, such as a tabletop, can exert the normal force, think about what happens when you sit on a mattress. Your weight causes the springs in the mattress to compress. As a result, the compressed springs exert an upward force (the normal force) on you. In a similar manner, the weight of the block causes invisible "atomic springs" in the surface of the table to compress, thus producing a normal force on the block.

Newton's third law plays an important role in connection with the normal force. In Figure 4.14, for instance, the block exerts a force on the table by pressing down on it. Consistent with the third law, the table exerts an oppositely directed force of equal magnitude on the block. This reaction force is the normal force. The magnitude of the normal force indicates how hard the two objects are pressing against each other.

If an object is resting on a horizontal surface and there are no vertically acting forces except the object's weight and the normal force, the magnitudes of these two forces are equal, i.e., $F_N = W$. This is the situation in Figure 4.14. The weight must be balanced by the normal force for the object to remain at rest on the table. If the magnitudes of these forces were not equal, there would be a net force acting on the block, and the block would accelerate either upward or downward, in accord with Newton's second law.

If other forces in addition to **W** and **F**$_N$ act in the vertical direction, the magnitudes of the normal force and the weight are no longer equal. In Figure 4.15*a*, for instance, a box whose weight is 15 N is being pushed downward against a table. The pushing force has a magnitude of 11 N. Thus, the total downward force exerted on the box is 26 N, and this must be balanced by the upward-acting normal force if the box is to remain at rest. In this situation, then, the normal force is 26 N, which is considerably larger than the weight of the box.

Figure 4.15*b* illustrates a different situation. Here, the box is being pulled upward by a rope that applies a force of 11 N. The net force acting on the box due to its weight and the rope is only 4 N, downward. To balance this force, the normal force needs to be only 4 N. It is not hard to imagine what would happen if the force applied by the rope were increased to 15 N—exactly equal to the weight of the box. In this situation, the normal force would become zero. In fact, the table could be removed, since the block would be supported entirely by the rope. The situations in Figure 4.15 are consistent with the idea that the magnitude of the normal force indicates how hard two objects are pressing against each other. Clearly, the box and the table are pressing against each other harder in part *a* of the picture than in part *b*.

Like the box and the table in Figure 4.15, various parts of the human body press against one another and exert normal forces. Example 8 illustrates the remarkable ability of the human skeleton to withstand a wide range of normal forces.

EXAMPLE 8 • A Balancing Act

In a circus balancing act, a woman performs a headstand on top of a man's head, as Figure 4.16*a* illustrates. The woman weighs 490 N, and the man's head and neck weigh 50 N. It is primarily the seventh cervical vertebra in the spine that supports all the weight above the shoulders. What is the normal force that this vertebra exerts on the neck and head of the man (a) before the act and (b) during the act?

Reasoning To begin, we draw a free-body diagram for the neck and head of the man. Before the act, there are only two forces, the weight of the man's head and neck, and the normal force. During the act, an additional force is present due to the woman's weight. In both cases, the upward and downward forces must balance for the head and neck to remain at rest. This condition of balance will lead us to values for the normal force.

Solution

(a) Figure 4.16*b* shows the free-body diagram for the man's head and neck before the act. The only forces acting are the normal force $\mathbf{F_N}$ and the 50-N weight. These two forces must balance for the man's head and neck to remain at rest. Therefore, the seventh cervical vertebra exerts a normal force of $\boxed{F_N = 50 \text{ N}}$.

(b) Figure 4.16*c* shows the free-body diagram that applies during the act. Now, the total downward force exerted on the man's head and neck is 50 N + 490 N = 540 N, which must be balanced by the upward normal force, so that $\boxed{F_N = 540 \text{ N}}$.

The Physics of...
a circus balancing act.

Figure 4.16 (*a*) A balancing act and free-body diagrams for the man's body above the shoulders (*b*) before the act and (*c*) during the act. For convenience, the scales used for the vectors in parts *b* and *c* are different.

(*a*)

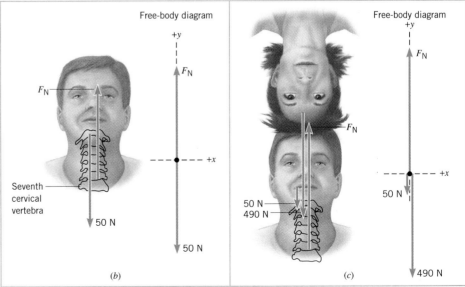

(*b*)

(*c*)

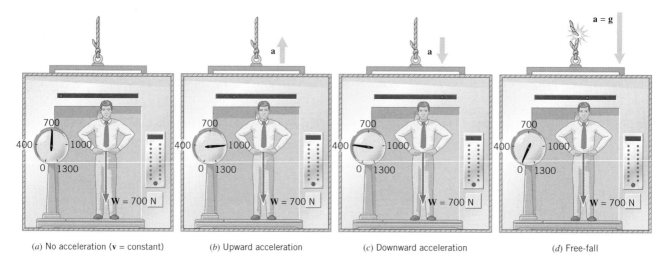

(a) No acceleration (**v** = constant)   (b) Upward acceleration   (c) Downward acceleration   (d) Free-fall

Figure 4.17 (a) When the elevator is not accelerating, the scale registers the true weight (W = 700 N) of the person. (b) When the elevator accelerates upward, the apparent weight (1000 N) exceeds the true weight. (c) When the elevator accelerates downward, the apparent weight (400 N) is less than the true weight. (d) The apparent weight is zero if the elevator falls freely, that is, if it falls with the acceleration of gravity.

In summary, the normal force does not necessarily have the same magnitude as the weight of the object. The value of the normal force depends on what other forces are present. It also depends on whether the objects in contact are accelerating. In one situation that involves accelerating objects, the magnitude of the normal force can be regarded as a kind of "apparent weight," as we will now see.

APPARENT WEIGHT

Usually, the weight of an object can be determined with the aid of a scale. However, even though a scale is working properly, there are situations in which it does not give the correct weight. In such situations, the reading on the scale gives only the "apparent" weight, rather than the gravitational force or "true" weight. The apparent weight is the force that the object exerts on the scale with which it is in contact.

To see the discrepancies that can arise between true weight and apparent weight, consider the scale in the elevator in Figure 4.17. The reasons for the discrepancies will be explained shortly. A person whose true weight is 700 N steps on the scale. If the elevator is at rest or moving with a constant velocity (either upward or downward), the scale registers the true weight, as Figure 4.17a illustrates.

If the elevator is accelerating, the apparent weight and the true weight are not equal. When the elevator accelerates upward, the apparent weight is greater than the true weight, as Figure 4.17b shows. Conversely, if the elevator accelerates downward, as in part c, the apparent weight is less than the true weight. In fact, if the elevator falls freely, so its acceleration is equal to the acceleration due to gravity, the apparent weight becomes zero, as part d indicates. In a situation such as this, where the apparent weight is zero, the person is said to be "weightless." The apparent weight, then, does not equal the true weight if the scale and the person on it are accelerating.

The discrepancies between true weight and apparent weight can be understood with the aid of Newton's second law. Figure 4.18 shows a free-body diagram of the person in the elevator. The two forces that act on him are the true weight **W** = m**g** and the normal force **F**$_N$ exerted by the platform of the scale. Applying Newton's second law in the vertical direction gives

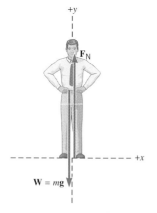

Figure 4.18 A free-body diagram showing the forces acting on the person riding in the elevator of Figure 4.17. **W** is the true weight and **F**$_N$ is the normal force exerted on the person by the platform of the scale.

$$\Sigma F_y = +F_N - mg = ma$$

where a is the acceleration of the elevator and person. In this result, the symbol g stands for the magnitude of the acceleration due to gravity. It can never be a negative quantity. However, the acceleration a may be either positive or negative, depending on whether the elevator is accelerating upward ($+$) or downward ($-$). Solving for the normal force F_N shows that

$$\underbrace{F_N}_{\substack{\text{Apparent} \\ \text{weight}}} = \underbrace{mg}_{\substack{\text{True} \\ \text{weight}}} + ma \qquad (4.6)$$

In Equation 4.6, F_N is the magnitude of the normal force exerted on the person by the platform of the scale. But in accord with Newton's third law, F_N is also the magnitude of the downward force that the person exerts on the scale, namely, the apparent weight.

Equation 4.6 contains all the features shown in Figure 4.17. If the elevator is not accelerating, $a = 0$, and the apparent weight equals the true weight. If the elevator accelerates upward, the acceleration a is positive, and the equation shows that the apparent weight is greater than the true weight. If the elevator accelerates downward, a is negative, and the apparent weight is less than the true weight. If the elevator falls freely, $a = -g$, and the apparent weight is zero. The scale registers an apparent weight of zero, because when both the person and the scale fall freely, they cannot push against one another. In this text, when the weight is given, it is assumed to be the true weight, unless stated otherwise.

Figure 4.19 The tread design on a tire helps to preserve friction between the tire surface and the wet road.

4.9 STATIC AND KINETIC FRICTIONAL FORCES

When an object is in contact with a surface, there is a force acting on the object, and the previous section discusses the component of this force that is perpendicular to the surface. When the object moves or attempts to move along the surface, there is also a component of the force that is parallel to the surface. This parallel force component is called *friction.*

In many situations considerable engineering effort is expended trying to reduce friction. For example, oil is used to reduce the friction that causes wear and tear in the pistons and cylinder walls of an automobile engine. Sometimes, however, friction is absolutely necessary. Without friction, car tires could not provide the traction needed to move the car. In fact, the raised tread on a tire is designed to maintain friction. On a wet road, the spaces in the tread pattern (see Figure 4.19) provide places for water to collect without coming between the tire surface and the road surface, where it would reduce friction and allow the tire to slip.

Surfaces that appear to be highly polished can actually look quite rough when examined under a microscope. Such an examination reveals that two surfaces in contact touch only at relatively few spots, as Figure 4.20 illustrates. The microscopic area of contact for these spots is substantially less than the apparent macroscopic area of contact between the surfaces—perhaps thousands of times less. At these contact points the molecules of the different bodies are close enough together to exert strong attractive intermolecular forces on one another, leading to what are known as "cold welds." Frictional forces are associated with these welded spots, but the exact details of how frictional forces arise are not well understood. However, some empirical relations have been developed that allow us to account for the effects of friction.

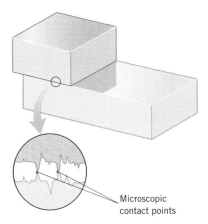

Microscopic
contact points

Figure 4.20 Even when two highly polished surfaces are in contact, they touch only at a relatively few points.

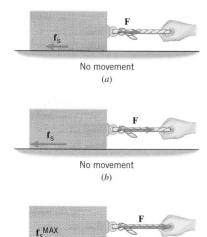

No movement
(a)

No movement
(b)

Just when movement begins
(c)

Figure 4.21 Applying a small force **F** to the block, as in parts *a* and *b*, produces no movement, because the static frictional force **f**$_s$ exactly balances the applied force. (*c*) The block just begins to move when the applied force is slightly greater than the maximum static frictional force **f**$_s^{MAX}$.

Figure 4.21 helps to explain the main features of the type of friction known as *static friction*. The block in this drawing is initially at rest on a table, and as long as there is no attempt to move the block, there is no static frictional force. Then, a horizontal force **F** is applied to the block by means of a rope. If **F** is small, as in part *a,* experience tells us that the block still does not move. Why? It does not move because the static frictional force **f**$_s$ exactly cancels the effect of the applied force. The direction of **f**$_s$ is opposite to that of **F**, and the magnitude of **f**$_s$ equals the magnitude of the applied force, $f_s = F$. Increasing the applied force in Figure 4.21 by a small amount still does not cause the block to move. There is no movement because the static frictional force also increases, by an amount that cancels out the increase in the applied force (see part *b* of the drawing). If the applied force continues to increase, however, there comes a point when the block finally "breaks away" and begins to slide. The force just before breakaway represents the *maximum static frictional force* **f**$_s^{MAX}$ that the table can exert on the block (see part *c* of the drawing). Any applied force that is greater than **f**$_s^{MAX}$ cannot be balanced by static friction, and the resulting net force accelerates the block to the right.

Experimental evidence shows that, to a good degree of approximation, the maximum static frictional force between a pair of dry, unlubricated surfaces has two main characteristics. It is independent of the apparent macroscopic area of contact between the objects, provided that the surfaces are hard or nondeformable. For instance, in Figure 4.22 the maximum static frictional force that the surface of the table can exert on the block is the same, whether the block is resting on its largest side or its smallest side. The other main characteristic of **f**$_s^{MAX}$ is that its magnitude is proportional to the magnitude of the normal force **F**$_N$. As Section 4.8 points out, the magnitude of the normal force indicates how hard the two surfaces are being pressed together. The harder the surfaces are pressed together, the larger is f_s^{MAX}, presumably because the number of "cold-welded," microscopic contact points is increased. Equation 4.7 expresses the proportionality between f_s^{MAX} and F_N with the aid of a proportionality constant μ_s, which is called the *coefficient of static friction.*

■ STATIC FRICTIONAL FORCE

The magnitude f_s of the static frictional force can have any value from zero up to a maximum value of f_s^{MAX}, depending on the applied force. In other words, $f_s \leq f_s^{MAX}$, where the symbol "≤" is read as "less than or equal to." The equality holds only when f_s attains its maximum value, which is

$$f_s^{MAX} = \mu_s F_N \tag{4.7}$$

μ_s is the coefficient of static friction, and F_N is the magnitude of the normal force.

It should be emphasized that Equation 4.7 relates only the magnitudes of **f**$_s^{MAX}$ and **F**$_N$, *not the vectors themselves.* This equation does not imply that the directions of the vectors are the same. In fact, **f**$_s^{MAX}$ is parallel to the surface, while **F**$_N$ is perpendicular to the surface.

The coefficient of static friction, being the ratio of the magnitudes of two forces $(\mu_s = f_s^{MAX}/F_N)$, is a unitless number. It depends on the type of material from which each surface is made (steel on wood, rubber on concrete, etc.), the condition of the surfaces (polished, rough, lubricated, etc.), and other variables such as tem-

Figure 4.22 The maximum static frictional force **f**$_s^{MAX}$ would be the same, no matter which side of the block is in contact with the table.

perature. Typical values for μ_s range from about 0.01 for smooth surfaces to about 1.5 for rough surfaces. Example 9 illustrates the use of Equation 4.7 for the maximum static frictional force.

EXAMPLE 9 • The Force Needed to Start a Sled Moving

A sled is resting on a horizontal patch of snow, and the coefficient of static friction is $\mu_s = 0.350$. The sled and its rider have a total mass of 38.0 kg. Determine the horizontal force needed to start the sled barely moving.

Reasoning and Solution To set the sled into motion, sufficient force is needed to overcome the maximum force of static friction, the magnitude of which is $f_s^{MAX} = \mu_s F_N$, according to Equation 4.7. Since the sled does not accelerate in the vertical direction, there can be no net force acting vertically on the sled. As a result, the normal force and the weight of the sled (and its rider) must balance, so that $F_N = mg$. Thus, the horizontal force must just barely exceed

$$f_s^{MAX} = \mu_s F_N = \mu_s mg = (0.350)(38.0 \text{ kg})(9.80 \text{ m/s}^2) = \boxed{130 \text{ N}}$$

To maneuver his way up Devil's Tower in Wyoming, this rock climber uses the static frictional force between his hands and feet and the vertical rock walls.

Once two surfaces begin sliding over one another, the static frictional force is no longer of any concern. Instead, a type of friction known as ***kinetic* friction*** comes into play. The kinetic frictional force opposes the relative sliding motion. If you have ever pushed an object across a floor, you may have noticed that it takes less force to keep the object sliding than it takes to get it going in the first place. In other words, the kinetic frictional force is usually less than the static frictional force.

Experimental evidence indicates that the kinetic frictional force $\mathbf{f}_k$ has three main characteristics, to a good degree of approximation. It is independent of the apparent area of contact between the surfaces (see Figure 4.22). It is independent of the speed of the sliding motion, if the speed is small. And lastly, the magnitude of the kinetic frictional force is proportional to the magnitude of the normal force. Equation 4.8 expresses this proportionality with the aid of a proportionality constant μ_k, which is called the ***coefficient of kinetic friction.***

■ KINETIC FRICTIONAL FORCE

The magnitude f_k of the kinetic frictional force is given by

$$f_k = \mu_k F_N \qquad (4.8)$$

In Equation 4.8, μ_k is the coefficient of kinetic friction, and F_N is the magnitude of the normal force.

Equation 4.8, like Equation 4.7, is a relationship between only the magnitudes of the frictional and normal forces. The directions of these forces are perpendicular. Moreover, like the coefficient of static friction, the coefficient of kinetic friction is a unitless number and depends on the type and condition of the two surfaces that are in contact. Values for μ_k are typically less than those for μ_s, reflecting the fact that kinetic friction is generally less than static friction. The next example illustrates the effect of kinetic friction.

---

* The word kinetic is derived from the Greek word *kinetikos*, meaning "of motion."

The kinetic frictional force is just what the doctor ordered for this itchy back.

EXAMPLE 10 • Sled Riding

A sled is traveling at 4.00 m/s along a horizontal stretch of snow, as Figure 4.23a illustrates. The coefficient of kinetic friction is $\mu_k = 0.0500$. How far does the sled go before stopping?

Reasoning The sled comes to a halt because the kinetic frictional force opposes the motion and causes the sled to slow down. Therefore, we will determine the kinetic frictional force and use it in Newton's second law to find the acceleration of the sled. Knowing the acceleration, we can determine the stopping distance by employing the appropriate equation of kinematics, as discussed in Chapter 3.

Solution To determine the magnitude f_k of the kinetic frictional force, it is necessary to know the magnitude F_N of the normal force, since $f_k = \mu_k F_N$. Part b of Figure 4.23 shows the free-body diagram for the sled. Since the sled does not accelerate in the vertical direction, there can be no net force acting vertically on the sled. As a result, the normal force and the weight **W** must balance, so the magnitude of the normal force is $F_N = mg$. The magnitude of the kinetic frictional force is

$$f_k = \mu_k F_N = \mu_k mg \qquad (4.8)$$

The kinetic frictional force is the only force acting on the sled in the x direction, so it is the net force. Newton's second law then gives the acceleration of the sled and rider as

$$a_x = \frac{-f_k}{m} = \frac{-\mu_k mg}{m} = -\mu_k g = -(0.0500)(9.80 \text{ m/s}^2) = -0.490 \text{ m/s}^2 \quad (4.2a)$$

The minus sign before f_k arises because the kinetic frictional force opposes the sliding motion of the sled and is directed to the left, or along the $-x$ axis, in Figure 4.23b. Therefore, the acceleration also points along the $-x$ axis. Notice that the acceleration does not depend on the mass of the sled and rider, since it cancels algebraically. The stopping distance x can be obtained with the aid of Equation 3.6a from the equations of kinematics ($v_x^2 = v_{0x}^2 + 2a_x x$), where the initial velocity is $v_{0x} = 4.00$ m/s and the final velocity is $v_x = 0$:

$$x = \frac{v_x^2 - v_{0x}^2}{2a_x} = \frac{-(4.00 \text{ m/s})^2}{2(-0.490 \text{ m/s}^2)} = \boxed{16.3 \text{ m}}$$

Static friction opposes the impending relative motion between two objects, while kinetic friction opposes the relative sliding motion that actually does occur. In either case, *relative motion* is opposed. However, this opposition to relative motion does not mean that friction prevents or works against the motion of *all* objects. For instance, the foot of a person walking exerts a force on the earth, and the earth exerts a reaction force on the foot. This reaction force is a static frictional force, and it

Figure 4.23 (a) The moving sled decelerates because of the kinetic frictional force. (b) Three forces act on the moving sled, its weight **W**, the normal force F_N, and the kinetic frictional force, f_k. The free-body diagram for the sled shows these forces.

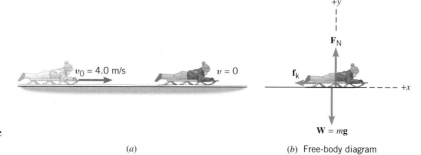

opposes the impending backward motion of the foot, propelling the person forward in the process. Kinetic friction can also cause an object to move, all the while opposing relative motion, as it does in Example 10. In this example the kinetic frictional force acts on the sled and opposes the relative motion of the sled and the earth. Newton's third law indicates, however, that since the earth exerts the kinetic frictional force on the sled, the sled must exert a reaction force on the earth. In response, the earth accelerates, but because of the earth's huge mass, the motion is too slight to be noticed.

4.10 THE TENSION FORCE

Forces are often applied by means of cables or ropes that are used to pull on an object. For instance, Figure 4.24*a* shows a force **T** being applied to the right end of a rope that is attached to a box. Each particle in the rope, in turn, applies a force to its neighbor. As a result of this process, the force is applied to the box at the other end, as part *b* of the drawing shows.

In situations such as that in Figure 4.24, we say that "the force **T** is applied to the box because of the tension in the rope," meaning that the tension in the rope and the force applied to the box have the same magnitude. However, the word "tension" is more commonly used to mean the tendency of the rope to be pulled apart. To see the relationship between these two uses of the word "tension," consider the left end of the rope, which applies the force **T** to the box. In accordance with Newton's third law, the box applies a reaction force to the rope. The reaction force has the same magnitude as **T** but is oppositely directed. In other words, a force $-\mathbf{T}$ acts on the left end of the rope. Thus, forces of equal magnitude act on opposite ends of the rope, as in Figure 4.24*c*, and tend to pull the rope apart.

In the previous discussion, we have used the concept of a "massless" rope ($m = 0$) without saying so. In reality, a massless rope does not exist, but it is useful as an idealization when applying Newton's second law. According to the second law, a net force is required to accelerate an object that has mass. In contrast, no net force is needed to accelerate a massless rope, since $\Sigma\mathbf{F} = m\mathbf{a}$ and $m = 0$. Thus, when a force **T** is applied to one end of a massless rope, none of the force is needed to accelerate the rope. As a result, the force **T** is also applied undiminished to the object attached at the other end, as we assumed in Figure 4.24.* If the rope had mass, however, some of the force **T** would have to be used to accelerate the rope. The force applied to the box would, then, be less than **T**, and the tension would be different at different locations along the rope. In this text we will assume that a rope connecting one object to another is massless, unless stated otherwise. The ability of a massless rope to transmit tension undiminished from one end to the other is not affected when the rope passes around objects such as the pulley in Figure 4.25 (provided the pulley itself is massless and frictionless).

4.11 EQUILIBRIUM APPLICATIONS OF NEWTON'S LAWS OF MOTION

Have you ever been so upset that it took days to recover your "equilibrium?" In this context, the word "equilibrium" refers to a balanced state of mind, one that is not

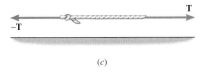

(a)

(b)

(c)

Figure 4.24 (*a*) A force **T** is being applied to the right end of the rope. (*b*) The force is transmitted to the box. (*c*) Forces are applied to both ends of the rope. These forces have equal magnitudes and opposite directions.

Figure 4.25 The force **T** applied at one end of a massless rope is transmitted undiminished to the other end, even when the rope bends around a pulley, provided the pulley is also massless and there is no friction.

---

* If a rope is not accelerating, **a** is zero in the second law, and $\Sigma\mathbf{F} = m\mathbf{a} = 0$, regardless of the mass of the rope. Then, the rope can be ignored, no matter what mass it has.

Concepts at a Glance

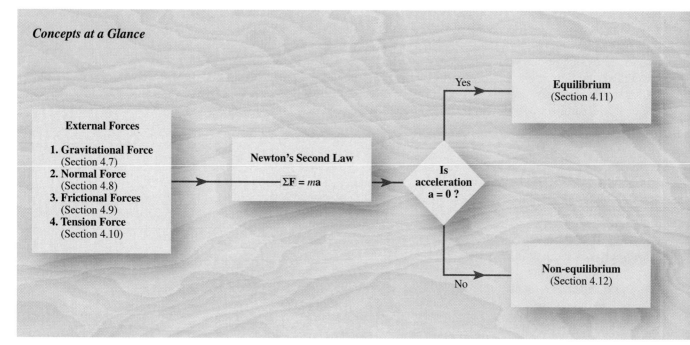

Figure 4.26 *Concepts at a Glance*
Both equilibrium and nonequilibrium problems can be solved with the aid of Newton's second law. For equilibrium situations, such as the tight-rope walker in the left photograph, the acceleration **a** is zero, while for nonequilibrium situations, such as the freely falling performer in the right photograph, it is not zero.

changing wildly. In physics, the word "equilibrium" also refers to a lack of change, but in the sense that the velocity of an object isn't changing. If its velocity doesn't change, an object is not accelerating. Our definition of equilibrium, then, is as follows:

■ DEFINITION OF EQUILIBRIUM

An object is in equilibrium when it has zero acceleration.

The concept chart in Figure 4.26, which is an expanded version of the chart in Figure 4.10, illustrates how we will use this definition when applying Newton's second law to solve problems in this section. When the acceleration vector is zero, we have an equilibrium situation, as the upper right-hand portion of the drawing indicates. Since the acceleration is zero, all of its components are also zero; in two dimensions, this means that $a_x = 0$ and $a_y = 0$. Substituting these values into the second law ($\Sigma F_x = ma_x$ and $\Sigma F_y = ma_y$) shows that the x component and the y component of the net force must each be zero. Thus, in two dimensions, the equilibrium condition is expressed by two equations:

$$\Sigma F_x = 0 \qquad (4.9\text{a})$$
$$\Sigma F_y = 0 \qquad (4.9\text{b})$$

In other words, the forces acting on an object in equilibrium must balance. When the acceleration is not zero, the forces acting on an object do not balance. Then, as the lower right-hand portion of Figure 4.26 indicates, we have a nonequilibrium situation. Section 4.12 deals with nonequilibrium applications of Newton's second law.

In using Equations 4.9a and 4.9b to solve equilibrium problems, we will use the following five-step reasoning strategy:

Equilibrium   Non-equilibrium

REASONING STRATEGY

Analyzing Equilibrium Situations

1. Select the object (often called the "system") to which Equations 4.9a and 4.9b are to be applied. Generally, this will be the object about which the most information is known. It may be that two or more objects are connected by means of a rope or a cable. Then, it may be necessary to treat each object separately according to the following steps.

2. Draw a free-body diagram for each object chosen above. As Section 4.3 discusses, a free-body diagram is a drawing that represents the object and shows the forces that act on it, each force with its proper direction. Be sure to include only forces that act on the object. *Do not include forces that the object exerts on its environment.*

3. Choose a set of x, y axes for each object and resolve all forces in the free-body diagram into components that point along these axes. Select the axes so that as many forces as possible point directly along the x axis or the y axis. Such a choice minimizes the number of calculations needed to determine the force components.

4. Apply Equations 4.9a and 4.9b by setting the sum of the x components of the forces equal to zero and the sum of the y components of the forces equal to zero.

5. Solve the two equations obtained in Step 4 for the desired unknown quantities, remembering that two equations can yield answers for only two unknowns at most.

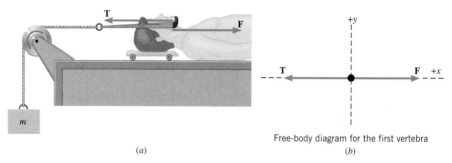

Free-body diagram for the first vertebra

(a)   (b)

Figure 4.27   (a) A traction device for the neck. (b) The free-body diagram for the first vertebra. Only the horizontal forces are shown for the sake of clarity.

Example 11 is a straightforward illustration of how these steps are followed, because only two forces act together to establish the equilibrium.

EXAMPLE 11 • Traction for the Neck

During recuperation from a neck injury, the cervical vertebrae are kept under tension by means of a traction device, as Figure 4.27a illustrates. The device creates tension in the vertebrae by pulling to the left on the head with a force **T**, which, in effect, is applied to the first vertebra at the top of the spine. This vertebra remains in equilibrium, because it is simultaneously pulled to the right by a force **F** that is supplied by the next vertebra in line. The force **F** comes about in reaction to the pulling effect of force **T**, in accord with Newton's third law. If it is desired that **F** have a magnitude of 34 N, how much mass m should be suspended from the rope?

Reasoning Since the forces **T** and **F** act on the first cervical vertebra, we choose it as the object for analysis. In Figure 4.27b, the free-body diagram for this vertebra shows only the two forces **T** and **F**. Friction between the head and the table is assumed to be negligible, since the head rests on a small rolling platform. Since the first cervical vertebra is in equilibrium, we can set the net force equal to zero and find the force **T**. With a knowledge of **T**, we can then determine the mass that is suspended from the rope.

Solution The condition for equilibrium is that the net force in the x direction is zero:

$$\Sigma F_x = F - T = 0$$

so that $F = T$. But $T = mg$, so that the necessary mass is

$$m = \frac{F}{g} = \frac{34\text{ N}}{9.80\text{ m/s}^2} = \boxed{3.5\text{ kg}}$$

The next example also deals with a traction device, but now three forces act together to bring about the equilibrium.

EXAMPLE 12 • Traction for the Foot

Figure 4.28a shows a traction device used with a foot injury. The weight of the 2.2-kg object creates a tension in the rope that passes around the pulleys. Therefore, tension forces T_1 and T_2 are applied to the pulley on the foot. It may seem surprising

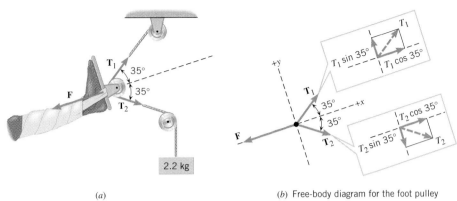

Figure 4.28 (*a*) A traction device for the foot. (*b*) The free-body diagram for the pulley on the foot.

(*a*)

(*b*) Free-body diagram for the foot pulley

that the rope applies a force to either side of the foot pulley. A similar effect occurs when you place a finger inside a rubber band and push downward. You can feel each side of the rubber band pulling upward on the finger. The foot pulley is kept in equilibrium, because the foot also applies a force **F** to it. This force arises in reaction (Newton's third law) to the pulling effect of the forces **T₁** and **T₂**. Ignoring the weight of the foot, find the magnitude of **F**.

Reasoning The forces **T₁**, **T₂**, and **F** keep the pulley on the foot at rest. The pulley, therefore, has no acceleration and is in equilibrium. As a result, the sum of the x components and the sum of the y components of the three forces must each be zero. Figure 4.28*b* shows the free-body diagram of the pulley on the foot. The x axis is chosen to be along the direction of force **F**, and the components of the forces are indicated in the drawing. (See Section 1.7 for a review of vector components.)

Solution Since the sum of the y components of the forces must be equal to zero, it follows that

$$\Sigma F_y = +T_1 \sin 35° - T_2 \sin 35° = 0$$

or $T_1 = T_2$. In other words, the magnitudes of the tension forces are equal. In addition, the sum of the x components of the forces is zero, so we have that

$$\Sigma F_x = +T_1 \cos 35° + T_2 \cos 35° - F = 0$$

Solving for F and letting $T_1 = T_2 = T$, we find that $F = 2T \cos 35°$. However, the tension T in the rope is determined by the weight of the 2.2-kg object: $T = mg = (2.2 \text{ kg})(9.8 \text{ m/s}^2) = 22$ N. Therefore,

$$F = 2(22 \text{ N}) \cos 35° = \boxed{36 \text{ N}}$$

• **PROBLEM SOLVING INSIGHT**
Choose the orientation of the x, y axes for convenience. Here, the axes have been rotated so the x axis points along the force F. Since F does not have a component along the y axis, the analysis is simplified.

Example 13 presents another situation in which three forces are responsible for the equilibrium of an object. However, in this example all the forces have different magnitudes.

EXAMPLE 13 • Replacing an Engine

An automobile engine has a weight **W**, whose magnitude is 3150 N. This engine is being positioned above an engine compartment, as Figure 4.29*a* illustrates. To position the engine, a worker is using a rope. Find the tension $\mathbf{T_1}$ in the supporting cable and the tension $\mathbf{T_2}$ in the positioning rope.

Reasoning Under the influence of the forces $\mathbf{W}$, $\mathbf{T_1}$, and $\mathbf{T_2}$ the ring is at rest and therefore in equilibrium. Consequently, the sum of the x components and the sum of the y components of these forces must each be zero, $\Sigma F_x = 0$ and $\Sigma F_y = 0$. By using these relations, we can find T_1 and T_2. Figure 4.29*b* shows the free-body diagram of the ring and the force components for a suitable x, y axes system.

Solution The free-body diagram shows the components for each of the three forces, and the components are listed in the following table:

Force	x Component	y Component
$\mathbf{T_1}$	$-T_1 \sin 10.0°$	$+T_1 \cos 10.0°$
$\mathbf{T_2}$	$+T_2 \sin 80.0°$	$-T_2 \cos 80.0°$
$\mathbf{W}$	0	-3150 N

• **PROBLEM SOLVING INSIGHT**
When an object is in equilibrium, as here in Example 13, the net force is zero, $\Sigma F = 0$. This does not mean that each individual force is zero. It means that the vector sum of all the forces is zero.

The plus signs in the table denote components that point along the positive axes, while the minus signs denote components that point along the negative axes. Setting the sum of the x components and the sum of the y components equal to zero leads to the following two equations:

$$\Sigma F_x = -T_1 \sin 10.0° + T_2 \sin 80.0° = 0$$
$$\Sigma F_y = +T_1 \cos 10.0° - T_2 \cos 80.0° - 3150 \text{ N} = 0$$

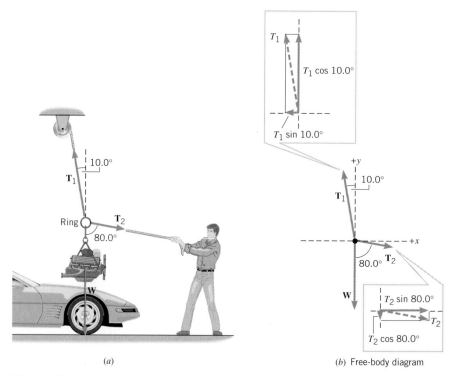

(a)   (b) Free-body diagram

Figure 4.29 (*a*) The ring is in equilibrium because of the three forces $\mathbf{T_1}$ (the tension force in the supporting cable), $\mathbf{T_2}$ (the tension force in the positioning rope), and $\mathbf{W}$ (the weight of the engine). (*b*) The free-body diagram for the ring.

Solving the first of these equations for T_1 shows that

$$T_1 = \left(\frac{\sin 80.0°}{\sin 10.0°} \right) T_2 = 5.67 T_2$$

Substituting this expression for T_1 into the second equation gives

$$(5.67T_2) \cos 10.0° - T_2 \cos 80.0° - 3150 \text{ N} = 0$$

which can be solved to show that $\boxed{T_2 = 582 \text{ N}}$. Since $T_1 = 5.67T_2$, it follows that $\boxed{T_1 = 3.30 \times 10^3 \text{ N}}$.

An object can be moving and still be in equilibrium, provided there is no acceleration. Example 14 illustrates such a case, and the solution is again obtained using the five-step reasoning strategy summarized at the beginning of the section.

EXAMPLE 14 • Equilibrium at Constant Velocity

A jet plane is flying with a constant speed along a straight line, at an angle of 30.0° above the horizontal, as Figure 4.30*a* indicates. The plane has a weight **W** whose magnitude is 86 500 N, and its engines provide a forward thrust **T** of 103 000 N. In addition, the lift force **L** (directed perpendicular to the wings) and the force **R** of air resistance (directed opposite to the motion) act on the plane. Find **L** and **R**.

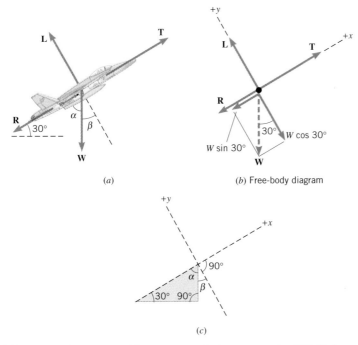

(a)

(b) Free-body diagram

(c)

Figure 4.30 (*a*) A plane moves with a constant velocity at an angle of 30.0° above the horizontal due to the action of four forces, the weight **W**, the lift **L**, the engine thrust **T**, and the air resistance **R**. (*b*) The free-body diagram for the plane. (*c*) This geometry occurs often in physics.

Reasoning Figure 4.30*b* shows the free-body diagram of the plane, including the forces **W**, **L**, **T**, and **R**. Since the plane is not accelerating, it is in equilibrium, and the sum of the *x* components and the sum of the *y* components of these forces must be zero. The lift force **L** and the force **R** of air resistance can be obtained from these equilibrium conditions. To calculate the components, we have chosen axes in the free-body diagram that are rotated by 30.0° from their usual horizontal-vertical positions. This has been done purely for convenience, since the weight **W** is then the only force that does not lie along either axis.

Solution When determining the components of the weight, it is necessary to realize that the angle β in Figure 4.30*a* is equal to 30.0°. Part *c* of the drawing focuses attention on the geometry that is responsible for this fact. There it can be seen that $\alpha + \beta = 90°$ and $\alpha + 30° = 90°$, with the result that $\beta = 30°$. Geometry similar to that in Figure 4.30*c* occurs often in physics. The table below lists the components of the forces that act on the jet.

Force	*x* Component	*y* Component
W	$-(86\ 500\ \text{N}) \sin 30.0°$	$-(86\ 500\ \text{N}) \cos 30.0°$
L	0	$+L$
T	$+103\ 000\ \text{N}$	0
R	$-R$	0

Setting the sum of the *x* components and the sum of the *y* components of the forces equal to zero yields

$$\Sigma F_x = -(86\ 500\ \text{N}) \sin 30.0° + 103\ 000\ \text{N} - R = 0 \qquad (4.9\text{a})$$

$$\Sigma F_y = -(86\ 500\ \text{N}) \cos 30.0° + L = 0 \qquad (4.9\text{b})$$

These equations can be solved to show that $\boxed{R = 59\ 800\ \text{N}}$ and $\boxed{L = 74\ 900\ \text{N}}$.

4.12 NONEQUILIBRIUM APPLICATIONS OF NEWTON'S LAWS OF MOTION

When an object is accelerating, it is not in equilibrium, as indicated in Figure 4.26. The forces acting on it are not balanced, so the net force is not zero in Newton's second law. However, with one exception, the reasoning strategy followed in solving nonequilibrium problems is identical to that used in equilibrium situations. The exception occurs in step 4 of the five steps outlined at the beginning of the last section. Since the object is now accelerating, the representation of Newton's second law in Equations 4.2a and 4.2b applies instead of Equations 4.9a and 4.9b:

$$\Sigma F_x = ma_x \quad (4.2\text{a}) \qquad \text{and} \qquad \Sigma F_y = ma_y \quad (4.2\text{b})$$

Example 15 uses these equations in a situation where the forces are applied in directions similar to those in Example 12, except that now an acceleration is present.

EXAMPLE 15 • Towing a Supertanker

A supertanker (mass $= 1.50 \times 10^8$ kg) is being towed by two tugboats, as in Figure 4.31*a*. The tensions in the towing cables apply the forces **T**₁ and **T**₂ at equal angles of 30.0° with respect to the tanker's axis. In addition, the tanker's engines produce a forward drive force **D**, whose magnitude is 75.0×10^3 N. Moreover, the water applies an opposing force **R**, whose magnitude is 40.0×10^3 N. The tanker moves forward with an acceleration that points along the tanker's axis and has a magnitude of 2.00×10^{-3} m/s². Find the magnitudes of the tensions T_1 and T_2.

Reasoning The unknown forces T_1 and T_2 contribute to the net force that accelerates the tanker. An essential step in determining T_1 and T_2, therefore, is to analyze the net force, which we will do using components. The various force components can be found readily by referring to the free-body diagram for the tanker in Figure 4.31*b*, where the ship's axis is chosen as the *x* axis. We will then use Newton's second law in its component form, $\Sigma F_x = ma_x$ and $\Sigma F_y = ma_y$, to obtain the magnitudes of T_1 and T_2.

Solution The individual force components are summarized as follows:

Force	*x* Component	*y* Component
T_1	$+T_1 \cos 30.0°$	$+T_1 \sin 30.0°$
T_2	$+T_2 \cos 30.0°$	$-T_2 \sin 30.0°$
D	$+D$	0
R	$-R$	0

Since the acceleration points along the *x* axis, there is no *y* component of the acceleration. Consequently, the sum of the *y* components of the forces must be zero:

$$\Sigma F_y = +T_1 \sin 30.0° - T_2 \sin 30.0° = 0$$

This result shows that the magnitudes of the tensions in the cables are equal, $T_1 = T_2$. Since the ship accelerates along the *x* direction, the sum of the *x* components of the forces is not zero. The second law indicates that

$$\Sigma F_x = T_1 \cos 30.0° + T_2 \cos 30.0° + D - R = ma_x$$

Using the fact that $T_1 = T_2 = T$ and the values given for D, R, m, and a_x, we find that

$$2T \cos 30.0° + 75.0 \times 10^3 \text{ N} - 40.0 \times 10^3 \text{ N} = (1.50 \times 10^8 \text{ kg})(2.00 \times 10^{-3} \text{ m/s}^2)$$

$$\boxed{T = 1.53 \times 10^5 \text{ N}}$$

Figure 4.31 (*a*) Four forces act on a supertanker: T_1 and T_2 are the tension forces due to the towing cables, **D** is the forward drive force produced by the tanker's engines, and **R** is the force with which the water opposes the tanker's motion. (*b*) The free-body diagram for the tanker.

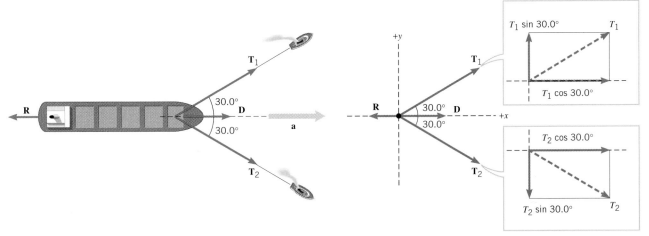

(*a*)

(*b*) Free-body diagram for the tanker

It often happens that two objects are connected somehow, perhaps by a drawbar like that used when a truck pulls a trailer. If the tension in the connecting device is of no interest, the objects can be treated as a single composite object when applying Newton's second law. However, if it is necessary to find the tension, as in the next example, then the second law must be applied separately to at least one of the objects.

EXAMPLE 16 • Hauling a Trailer

An 8500-kg truck is hauling a 27 000-kg trailer along a level road, as Figure 4.32*a* illustrates. The acceleration is 0.78 m/s². Ignoring the retarding forces of friction and air resistance, determine (a) the magnitude of the tension in the horizontal drawbar between the trailer and the truck and (b) the force **D** that propels the truck forward.

Reasoning Since the truck and the trailer accelerate along the horizontal direction and friction is being ignored, only forces that have components in the horizontal direction are of interest here. Therefore, Figure 4.32 omits the weight and the normal force, since they act vertically. Note, however, that these vertical forces balance, since there is no vertical component to the acceleration. To determine the tension force **T** in the drawbar, we draw the free-body diagram for the trailer and then apply Newton's second law, $\Sigma F_x = ma_x$. Similarly, we can determine the force **D** that propels the truck forward by drawing the free-body diagram for the truck and then applying Newton's second law to relate the net force to the acceleration.

Solution

(a) The free-body diagram for the trailer is shown in Figure 4.32*b*. There is only one horizontal force acting on the trailer, the tension force **T** due to the drawbar. Therefore, it is straightforward to obtain the tension from $\Sigma F_x = m_2 a_x$, since the mass of the trailer and the acceleration are known:

$$T = (27\ 000 \text{ kg})(0.78 \text{ m/s}^2) = \boxed{21\ 000 \text{ N}}$$

(b) Two horizontal forces act on the truck, as the free-body diagram in Figure 4.32*b* shows. One is the desired force **D**. The other is the force **T'**. According to Newton's third law, **T'** is the force with which the trailer pulls back on the truck, in reaction to the truck pulling forward. If the drawbar has negligible mass, the magnitude of **T'** is equal to the magnitude of **T**, namely, 21 000 N. Since the magnitude of **T'**, the mass

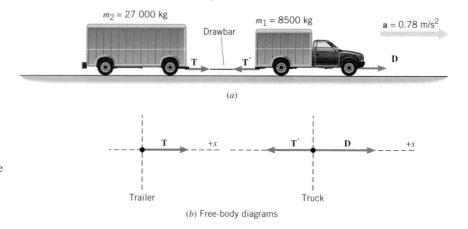

(a)

Figure 4.32 (*a*) The force **D** acts on the truck and propels it forward. The drawbar exerts the tension force **T'** on the truck and the tension force **T** on the trailer. (*b*) The free-body diagrams for the truck and the trailer, ignoring the vertical forces.

(*b*) Free-body diagrams

of the truck, and the acceleration are known, $\Sigma F_x = m_1 a_x$ can be used to determine the drive force:

$$D - (21\,000\text{ N}) = (8500\text{ kg})(0.78\text{ m/s}^2) \quad \text{or} \quad \boxed{D = 28\,000\text{ N}}$$

In Section 4.11 we examined situations where the net force acting on an object is zero, and in this section we have considered two examples where the net force is not zero. Conceptual Example 17 illustrates a common situation where the net force is zero at certain times but is not zero at other times.

CONCEPTUAL EXAMPLE 17 • The Motion of a Water Skier

Figure 4.33 shows a water skier at four different times:

(a) The skier is floating motionless in the water (part *a* of the figure).

(b) The skier is being pulled out of the water and up onto the skis (part *b*).

(c) The skier is moving at a constant speed along a straight line (part *c*).

(d) The skier has let go of the tow rope and is slowing down (part *d*).

For each phase of the motion, explain whether the net force acting on the skier is zero.

Reasoning and Solution According to Newton's second law, if an object has zero acceleration, the net force acting on it is zero. In such a case, the object is in equilibrium. Conversely, if the object has an acceleration, the net force acting on it is not zero. Such an object is not in equilibrium. We will apply this criterion to each of the four phases of the motion to decide if the net force is zero.

(a) The skier is floating motionless in the water, so her velocity and acceleration are both zero. Therefore, the net force acting on her is zero, and she is in equilibrium.

(b) As the skier is being pulled up and out of the water, her velocity is increasing. Thus, she is accelerating, and the net force acting on her is not zero. The skier is not in equilibrium. The net force is shown in Figure 4.33*b*.

(c) The skier is now moving at a constant speed along a straight line, so her velocity is constant. Since her velocity is constant, her acceleration is zero. Thus, the net force acting on her is zero, and she is again in equilibrium.

(d) After the skier lets go of the tow rope, her speed decreases, so she is decelerating. Thus, the net force acting on her is not zero, and she is not in equilibrium. The net force is shown in Figure 4.33*d*.

Notice in (a) and (c) that the net force acting on the skier is zero, even though she is at rest in (a) and moving with a constant velocity in (c). In contrast, in (b) and (d) there is a net force acting on her, and it has a different direction in each case.

Related Homework Material: Problem 68

Figure 4.33 A water skier (*a*) floating in water, (*b*) being pulled up by the boat, (*c*) moving at a constant velocity, and (*d*) slowing down.

(*a*)   (*b*)   (*c*)   (*d*)

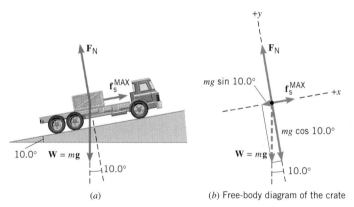

Figure 4.34 (*a*) A crate on a truck is kept from slipping by the static frictional force $\mathbf{f}_s^{\text{MAX}}$. The other forces that act on the crate are its weight $\mathbf{W}$ and the normal force $\mathbf{F}_N$. (*b*) The free-body diagram for the crate.

The force of gravity is often present among the forces that affect the acceleration of an object. Examples 18–20 deal with typical situations.

EXAMPLE 18 • Hauling a Crate

A flatbed truck is carrying a crate up a 10.0° hill, as Figure 4.34*a* illustrates. The coefficient of static friction between the truck bed and the crate is $\mu_s = 0.350$. Find the maximum acceleration that the truck can attain before the crate begins to slip backward relative to the truck.

Reasoning The crate will not slip as long as it has the same acceleration as the truck. Therefore, a net force must act on the crate to accelerate it, and the static frictional force $\mathbf{f}_s$ contributes in a major way to this net force. Since it is necessary to prevent the crate from slipping backward, the static frictional force must be directed forward, up the hill. As the acceleration of the truck increases, $\mathbf{f}_s$ must also increase to produce a corresponding increase in the acceleration of the crate. However, the static frictional force can increase only until its maximum magnitude $f_s^{\text{MAX}} = \mu_s F_N$ is reached, at which point the crate and the truck have the maximum acceleration $\mathbf{a}^{\text{MAX}}$. If the acceleration of the truck increases even more, the crate will slip. To find $\mathbf{a}^{\text{MAX}}$, we focus our attention on the crate, and part *b* of the drawing shows its free-body diagram. The three forces acting on the crate at the instant slipping begins are its weight $\mathbf{W} = m\mathbf{g}$, the normal force $\mathbf{F}_N$ exerted by the truck bed, and the maximum static frictional force $\mathbf{f}_s^{\text{MAX}}$.

Solution Using the *x* components of the forces (see the free-body diagram for the magnitudes of the components) and Newton's second law ($\Sigma F_x = ma_x$), we have

$$\Sigma F_x = -mg \sin 10.0° + \mu_s F_N = ma^{\text{MAX}}$$

Before this equation can be solved for a^{MAX}, however, a value is needed for F_N. This value can be obtained by considering the force components along the *y* axis. Since the crate does not accelerate along this axis, the sum of the *y* components of the forces must be zero according to Newton's second law ($\Sigma F_y = ma_y = 0$):

$$\Sigma F_y = -mg \cos 10.0° + F_N = 0 \quad \text{or} \quad F_N = mg \cos 10.0°$$

• **PROBLEM SOLVING INSIGHT**
The magnitude F_N of the normal force is not necessarily equal to the weight of the object.

Substituting this result for F_N into the previous equation and eliminating the mass m algebraically reveals that

$$-g \sin 10.0° + \mu_s g \cos 10.0° = a^{MAX}$$

Letting $g = 9.80 \text{ m/s}^2$ and $\mu_s = 0.350$, we find that $\boxed{a^{MAX} = 1.68 \text{ m/s}^2}$.

EXAMPLE 19 • Accelerating Blocks

Block 1 ($m_1 = 8.00$ kg) is moving on a frictionless 30.0° incline. This block is connected to block 2 ($m_2 = 22.0$ kg) by a cord that passes over a massless and frictionless pulley (see Figure 4.35*a*). Find the acceleration of each block and the tension in the cord.

Reasoning Since both blocks accelerate, there must be a net force acting on each one. The key to solving this problem is to realize that Newton's second law can be used separately for each block to relate the net force and the acceleration. It is also important to realize that both blocks have accelerations of the same magnitude *a*, since they move as a unit.

Solution We assume that block 1 accelerates up the incline and choose this direction to be the +*x* axis. If block 1 in reality accelerates down the incline, then the value obtained for the acceleration will be a negative number. Three forces act on block 1: (1) $\mathbf{W}_1$ is its weight [$W_1 = m_1 g = (8.00 \text{ kg})(9.80 \text{ m/s}^2) = 78.4$ N], (2) $\mathbf{T}$ is the force applied because of the tension in the cord, and (3) $\mathbf{F}_N$ is the normal force that the incline exerts. Figure 4.35*b* shows the free-body diagram for block 1. The weight is the only force that does not point along the *x*, *y* axes, and its *x* and *y* components are given in the diagram. Applying the second law to the motion of block 1 shows that

$$\Sigma F_x = -W_1 \sin 30.0° + T = m_1 a_x$$
$$-(78.4 \text{ N}) \sin 30.0° + T = (8.00 \text{ kg})a$$

where we have set $a_x = a$. This equation cannot be solved as it stands, since both T and a are unknown quantities. To complete the solution, we next consider block 2.

Two forces act on block 2, as the free-body diagram in Figure 4.35*b* indicates: (1) $\mathbf{W}_2$ is its weight [$W_2 = m_2 g = (22.0 \text{ kg})(9.80 \text{ m/s}^2) = 216$ N] and (2) $\mathbf{T}'$ is exerted as a result of block 1 pulling back on the connecting cord. If the masses of the cord

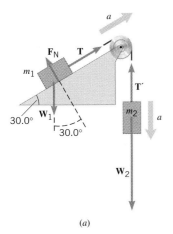

(a)

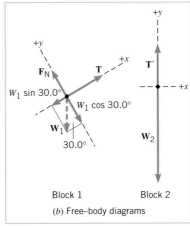

(b) Free–body diagrams

Figure 4.35 (*a*) Three forces act on block 1: its weight $\mathbf{W}_1$, the normal force $\mathbf{F}_N$, and the force $\mathbf{T}$ due to the tension in the cord. Two forces act on block 2: its weight $\mathbf{W}_2$ and the force $\mathbf{T}'$ due to the tension. The acceleration is labeled *a*. (*b*) The free-body diagrams for the two blocks.

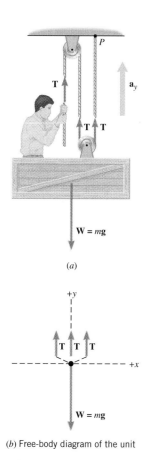

(a)

+y
|
|
T   T   T
- - - - - - ● - - - - - - +x
|
|
W = mg

(b) Free-body diagram of the unit

Figure 4.36 (a) A window washer pulls down on the rope to hoist the scaffold up the side of a building. The force **T** results from the effort of the window washer and acts on him and the scaffold in three places, as discussed in Example 20. (b) The free-body diagram of the unit comprising the man and the scaffold.

and the frictionless pulley are negligible, the magnitudes of **T'** and **T** are the same: $T' = T$. Applying the second law to block 2 reveals that

$$\Sigma F_y = T' - W_2 = m_2 a_y$$
$$+T - 216\ \text{N} = (22.0\ \text{kg})(-a)$$

The acceleration a_y has been set equal to $-a$ since block 2 moves downward along the $-y$ axis in the free-body diagram, consistent with the original assumption that block 1 moves up the incline. Now there are two equations in two unknowns, and following the procedure discussed in Appendix C, they may be solved simultaneously to give the values of the tension T and the acceleration a:

$$\boxed{T = 86.3\ \text{N}} \quad \text{and} \quad \boxed{a = 5.89\ \text{m/s}^2}$$

EXAMPLE 20 • Hoisting a Scaffold

A window washer on a scaffold is hoisting the scaffold up the side of a building by pulling downward on a rope, as in Figure 4.36a. The magnitude of the pulling force is 540 N, and the combined mass of the person and the scaffold is 155 kg. Find the upward acceleration of the unit.

Reasoning The person and the scaffold form a single unit, on which the rope exerts a force in three places. The left end of the rope exerts an upward force **T** on the person's hands. This force arises because the person pulls downward with a 540-N force, and the rope exerts an oppositely directed force of equal magnitude on the person, in accord with Newton's third law. Thus, the magnitude of the upward force **T** is 540 N and is the magnitude of the tension in the rope. If the masses of the rope and each pulley are negligible, and if the pulleys are friction-free, the tension is transmitted undiminished along the rope. Then, a 540-N tension force **T** acts upward on the left side of the scaffold pulley (see part *a* of the drawing). A tension force is also applied to the point *P*, where the rope attaches to the roof. The roof pulls back on the rope in accord with the third law, and this pull leads to the 540-N tension force **T** that acts on the right side of the scaffold pulley. In addition to the three upward forces, the weight of the unit must be taken into account [$W = mg = (155\ \text{kg}) \times (9.80\ \text{m/s}^2) = 1520\ \text{N}$]. Part *b* of the drawing shows the free-body diagram.

Solution Newton's second law ($\Sigma F_y = ma_y$) can be applied to calculate the acceleration a_y:

$$\Sigma F_y = +T + T + T - W = ma_y$$
$$3(540\ \text{N}) - 1520\ \text{N} = (155\ \text{kg})a_y \quad \text{or} \quad a_y = \boxed{0.65\ \text{m/s}^2}$$

SUMMARY

Newton's first law of motion or **law of inertia** states that an object continues in a state of rest or in a state of motion at a constant speed along a straight line unless compelled to change that state by a net force. **Inertia** is the natural tendency of an object to remain at rest or in motion at a constant speed along a straight line. The **mass** of a body is a quantitative measure of inertia and is measured in an SI unit called the **kilogram** (kg). An **inertial reference frame** is one in which Newton's law of inertia is valid.

Newton's second law of motion states that when a net force $\Sigma \mathbf{F}$ acts on an object of mass m, the acceleration **a** of the object is given by $\Sigma \mathbf{F} = m\mathbf{a}$. The SI unit of force is the **newton** (N). When determining the net force, a **free-body diagram** is helpful. A free-body diagram is a diagram that represents the object and the forces acting on it.

Newton's third law of motion, often called the **action–reaction law,** states that whenever one object exerts a force on a second object, the second object exerts an oppositely directed force of equal magnitude on the first object.

Newton's law of universal gravitation states that every particle in the universe exerts an attractive force on every other particle. For two particles that are separated by a distance r and have masses m_1 and m_2, the law states that the magnitude of this attractive force is $F = Gm_1m_2/r^2$, while its direction lies along the line between the particles. The constant G has a value of $G = 6.673 \times 10^{-11} \, \text{N} \cdot \text{m}^2/\text{kg}^2$ and is called the universal gravitational constant. The gravitational force is one of nature's three fundamental forces. The **weight** of an object on earth is the gravitational force that the earth exerts on the object.

The **apparent weight** is the force that an object exerts on the platform of a scale and may be larger or smaller than the true weight, if the object and the scale are accelerating.

A surface exerts a force on an object with which it is in contact. The component of the force perpendicular to the surface is called the **normal force**. The component parallel to the surface is called friction. The **force of static friction** between two surfaces opposes any impending relative motion of the surfaces. The magnitude of the static friction force depends on the magnitude of the applied force and can assume any value up to a maximum of $f_s^{\text{MAX}} = \mu_s F_N$,

where μ_s is the **coefficient of static friction** and F_N is the magnitude of the normal force. The **force of kinetic friction** between two surfaces sliding against one another opposes the relative motion of the surfaces. This force has a magnitude given by $f_k = \mu_k F_N$, where μ_k is the **coefficient of kinetic friction**.

The word **tension** is commonly used to mean the tendency of a rope to be pulled apart due to forces that are applied at each end. Because of tension, a rope transmits a force from one end to the other. When a rope is accelerating, the force is transmitted undiminished only if the rope is massless. Neither the tension force, the normal force, nor the friction force is one of nature's fundamental forces.

An object is in **equilibrium** when the object moves at a constant velocity (which may be zero), or, in other words, when it is not accelerating. The sum of the forces that act on an object in equilibrium is zero. Under equilibrium conditions in two dimensions, the separate sums of the force components in the x direction and in the y direction must each be zero: $\Sigma F_x = 0$ and $\Sigma F_y = 0$. If an object is not in equilibrium, then Newton's second law must be used to account for the acceleration: $\Sigma F_x = ma_x$ and $\Sigma F_y = ma_y$. Sections 4.11 and 4.12 discuss a five-step reasoning strategy that facilitates the use of Newton's laws of motion.

CONCEPTUAL QUESTIONS

1. The instructions for mounting a phono cartridge on the tone arm of a stereo turntable say to "adjust the tracking force of the cartridge so it is less than three grams." From the point of view of correct physics, is anything wrong with this statement? Explain.

2. Why do you lunge forward when your car suddenly comes to a halt? Why are you pressed backward against the seat when your car rapidly accelerates? In your explanation, refer to the most appropriate one of Newton's three laws of motion.

3. A bird feeder of large mass is hung from a tree limb, as the drawing shows. A cord attached to the bottom of the feeder has been left dangling free. Curiosity gets the best of a child, who pulls on the dangling cord in an attempt to see what's in the

feeder. The dangling cord is cut from the same source as the cord attached to the limb. Is the cord between the feeder and the limb more likely to snap with a slow continuous pull or a sudden downward pull? Give your reasoning.

4. Is a net force being applied to an object when the object is moving downward (a) with a constant acceleration of $9.80 \, \text{m/s}^2$ and (b) with a constant velocity of $9.80 \, \text{m/s}$? Explain.

5. Newton's second law indicates that when a net force acts on an object, it must accelerate. Does this mean that when two or more forces are applied to an object simultaneously, it must accelerate? Explain.

6. A father and his seven-year-old daughter are facing each other on ice skates. With their hands, they push off against one another. (a) Compare the magnitudes of the pushing forces that they experience. (b) Which one, if either, experiences the larger acceleration? Account for your answers.

7. A gymnast is bouncing on a trampoline. After a high bounce the gymnast comes down and hits the elastic surface of the trampoline. In so doing the gymnast applies a force to the trampoline. (a) Describe the effect this force has on the elastic surface. (b) The surface applies a reaction force to the gymnast. Describe the effect that this reaction force has on the gymnast.

8. According to Newton's third law, when you push on an object, the object pushes back on you with an oppositely directed force of equal magnitude. If the object is a massive crate resting on the floor, it will probably not move. Some people think that the reason the crate does not move is that the two oppositely directed pushing forces cancel. Explain why this logic is faulty and why the crate does not move.

9. Three particles have identical masses. Each particle experiences only the gravitational forces due to the other two particles. How should the particles be arranged so each one experiences a net gravitational force that has the same magnitude? Give your reasoning.

10. When a body is moved from sea level to the top of a mountain, what changes—the body's mass, its weight, or both? Explain.

11. The force of air resistance acts to oppose the motion of an object moving through the air. A ball is thrown upward and eventually returns to the ground. (a) As the ball moves upward, is the net force that acts on the ball greater than, less than, or equal to its weight? Justify your answer. (b) Repeat part (a) for the downward motion of the ball.

12. Object A weighs twice as much as object B at the same spot on the earth. Would the same be true at a given spot on Mars? Account for your answer.

13. Does the acceleration of a freely falling object depend to any extent on the location, i.e., whether the object is on top of Mt. Everest or in Death Valley, California? Explain.

14. A "bottle rocket" is a type of fireworks that has a long thin tail that you insert into an empty bottle, to provide a launch platform. One of these rockets is fired with the bottle pointing vertically upward. An identical rocket is fired with the bottle lying on its side, pointing horizontally. In which case does the rocket leave the bottle with the greater acceleration? Explain, ignoring air resistance and friction.

15. A 10-kg suitcase is placed on a scale that is in an elevator. Is the elevator accelerating up or down when the scale reads (a) 75 N and (b) 120 N? Justify your answers.

16. A stack of books whose true weight is 165 N is placed on a scale in an elevator. The scale reads 165 N. Can you tell from this information whether the elevator is moving with a constant velocity of 2 m/s upward or 2 m/s downward or whether the elevator is at rest? Explain.

17. Suppose you are in an elevator that is moving upward with a constant velocity. A scale inside the elevator shows your weight to be 600 N. (a) Does the scale register a value that is greater than, less than, or equal to 600 N during the time when the elevator slows down as it comes to a stop? (b) What is the reading when the elevator is stopped? (c) How does the value registered on the scale compare to 600 N during the time when the elevator picks up speed again on its way back down? Give your reasoning in each case.

18. A person has a choice of either pushing or pulling a sled at a constant velocity, as the drawing illustrates. Friction is present. If the angle θ is the same in both cases, does it require less force to push or to pull? Account for your answer.

19. Suppose that the coefficients of static and kinetic friction have values such that $\mu_s = 2.0\mu_k$ for a crate in contact with a cement floor. Does this mean that the magnitude of the static frictional force acting on the crate at rest would always be twice the magnitude of the kinetic frictional force acting on the moving crate? Give your reasoning.

20. A box rests on the floor of an elevator. Because of static friction, a force is required to start the box sliding across the floor when the elevator is (a) stationary, (b) accelerating upward, and (c) accelerating downward. Rank the forces required in these three situations in ascending order, i.e., smallest first. Explain.

21. A rope is used in a tug-of-war between two teams of five people each. Both teams are equally strong, so neither team wins. An identical rope is tied to a tree, and the same ten people pull just as hard on the loose end as they did in the contest. In both cases, the people pull steadily with no jerking. Which rope, if either, is more likely to break? Justify your answer.

22. A stone is thrown from the top of a cliff. As the stone falls, is it in equilibrium? Explain, ignoring air resistance.

23. Can an object ever be in equilibrium if the object is acted on by only (a) a single nonzero force, (b) two forces that point in mutually perpendicular directions, and (c) two forces that point in directions that are not perpendicular? Account for your answers.

24. A circus performer hangs stationary from a rope. She then begins to climb upward by pulling herself up, hand-over-hand. When she starts climbing, is the tension in the rope less than, equal to, or greater than it is when she hangs stationary? Explain.

25. During the final stages of descent, a sky diver with an open parachute approaches the ground with a constant velocity. The wind does not blow him from side to side. Is the sky diver in equilibrium and, if so, what forces are responsible for the equilibrium?

26. A weight hangs from a ring at the middle of a rope, as the drawing illustrates. Can the person who is pulling on the right end of the rope ever make the rope perfectly horizontal? Explain your answer in terms of the forces that act on the ring.

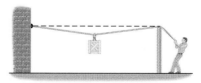

27. A freight train is accelerating on a level track. Other things being equal, would the tension in the coupling between the engine and the first car change if some of the cargo in the last car were transferred to any one of the other cars? Account for your answer.

PROBLEMS

ssm Solution is in the Student Solutions Manual. **www** Solution is available on the World Wide Web at http://www.wiley.com/college/cutnell ⚕ This icon represents a biomedical application.

Section 4.3 Newton's Second Law of Motion

1. ssm A person with a blackbelt in karate has a fist that has a mass of 0.70 kg. Starting from rest, this fist attains a velocity of 8.0 m/s in 0.15 s. What is the magnitude of the average net force applied to the fist to achieve this level of performance?

2. A bicycle has a mass of 13.1 kg, and its rider has a mass of 81.7 kg. The rider is pumping hard, so that a horizontal net force of 9.78 N accelerates them. What is the acceleration?

3. An airplane has a mass of 31 000 kg and takes off under the influence of a constant net force of 37 000 N. What is the net force that acts on the plane's 78-kg pilot, assuming that he has the same acceleration as the plane?

4. Scientists are experimenting with a kind of gun that may eventually be used to fire payloads directly into orbit. In one test, this gun accelerates a 5.0-kg projectile from rest to a speed of 4.0×10^3 m/s. The net force accelerating the projectile is 4.9×10^5 N. How much time is required for the projectile to come up to speed?

5. ssm When a 0.058-kg tennis ball is served, it accelerates from rest to a speed of 45 m/s. The impact with the racket gives the ball a constant acceleration over a distance of 0.44 m. What is the magnitude of the net force acting on the ball?

6. During a circus performance, a 72-kg human cannonball is shot out of an 18-m-long cannon. If the human cannonball spends 0.95 s in the cannon, determine the average net force exerted on him in the barrel of the cannon.

7. A catapult on an aircraft carrier is capable of accelerating a plane from 0 to 56.0 m/s in a distance of 80.0 m. Find the average net force that the catapult exerts on a 13 300-kg jet.

*8. An arrow, starting from rest, leaves the bow with a speed of 25.0 m/s. If the average force exerted on the arrow by the bow were doubled, all else remaining the same, with what speed would the arrow leave the bow?

*9. ssm www Two forces $\mathbf{F_A}$ and $\mathbf{F_B}$ are applied to an object. The larger force is $\mathbf{F_A}$. When both forces point due east, the object's acceleration has a magnitude of 0.50 m/s². However, when $\mathbf{F_A}$ points due east and $\mathbf{F_B}$ points due west, the acceleration has a magnitude of 0.40 m/s². Find the ratio F_A/F_B of the magnitudes of the forces.

Section 4.4 The Vector Nature of Newton's Second Law of Motion, Section 4.5 Newton's Third Law of Motion

10. A force vector has a magnitude of 720 N and a direction of 38° north of east. Determine the magnitude and direction of the components of the force that point along the north-south line and along the east-west line.

11. Two forces, $\mathbf{F_1}$ and $\mathbf{F_2}$, act on the 5.00-kg block shown in the drawing. The magnitudes of the forces are $F_1 = 45.0$ N and $F_2 = 25.0$ N. What is the horizontal acceleration (magnitude and direction) of the block?

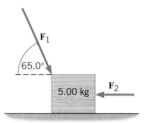

12. Two forces act on an object (mass = 4.00 kg), as in the drawing. Find the magnitude and direction (relative to the x axis) of the acceleration of the object.

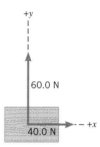

13. ssm Two forces act on an object (mass = 3.00 kg), as in the drawing. Find the magnitude and direction (relative to the x axis) of the acceleration of the object.

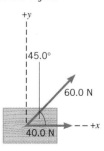

14. Two skaters, an 82-kg man and a 48-kg woman, are standing on ice. Neglect any friction between the skate blades and the ice. The woman pushes on the man with a force of 45 N due east. Determine the accelerations (magnitude and direction) of the man and the woman.

15. A duck has a mass of 2.5 kg. As the duck paddles, a force of 0.10 N acts on it in a direction due east. In addition, the current of the water exerts a force of 0.20 N in a direction of 52° south of east. When these forces begin to act, the velocity of the duck is 0.11 m/s in a direction due east. Find the magnitude and direction (relative to due east) of the displacement that the duck undergoes in 3.0 s while the forces are acting.

*16. A space probe has two engines. Each generates the same amount of force when fired, and the directions of these forces can be independently adjusted. When the engines are fired simultaneously and each applies its force in the same direction, the probe, starting from rest, takes 28 s to travel a certain distance. How long does it take to travel the same distance, again starting from rest, if the engines are fired simultaneously and the forces that they apply to the probe are perpendicular?

*17. **ssm www** A 325-kg boat is sailing 15.0° north of east at a speed of 2.00 m/s. Thirty seconds later, it is sailing 35.0° north of east at a speed of 4.00 m/s. During this time, three forces act on the boat, a 31.0-N force directed 15.0° north of east (due to an auxiliary engine), a 23.0-N force directed 15.0° south of west (resistance due to the water), and $\mathbf{F_W}$ (due to the wind). Find the magnitude and direction of the force $\mathbf{F_W}$. Express the direction as an angle with respect to due east.

**18. At a time when mining asteroids has become feasible, astronauts have connected a line between their 3500-kg space tug and a 6200-kg asteroid. Using their ship's engine, they pull on the asteroid with a force of 490 N. Initially the tug and the asteroid are at rest, 450 m apart. How much time does it take for the ship and the asteroid to meet?

Section 4.7 The Gravitational Force

19. The mass of one small ball is 0.001 50 kg, and the mass of another is 0.870 kg. If the center-to-center distance between these two balls is 0.100 m, find the magnitude of the gravitational force that each exerts on the other.

20. A rock of mass 45 kg accidentally breaks loose from the edge of a cliff and falls straight down. The magnitude of the air resistance that opposes its downward motion is 250 N. What is the magnitude of the acceleration of the rock?

21. **ssm** In preparation for this problem, review Conceptual Example 7. A space traveler whose mass is 115 kg leaves earth. What are his weight and mass (a) on earth and (b) in interplanetary space where there are no nearby planetary objects?

22. On earth, two parts of a space probe weigh 11 000 N and 3400 N. These parts are separated by a center-to-center distance of 12 m and may be treated as uniform spherical objects. Find the magnitude of the gravitational force that each part exerts on the other out in space, far from any other objects.

23. The drawing shows three particles far away from any other objects and located on a straight line. The masses of these particles are $m_A = 363$ kg, $m_B = 517$ kg, and $m_C = 154$ kg. Find the

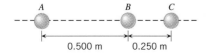

magnitude and direction of the net gravitational force acting on (a) particle A, (b) particle B, and (c) particle C.

24. The drawing (not to scale) shows one alignment of the sun, earth, and moon. The gravitational force $\mathbf{F_{SM}}$ that the sun exerts on the moon is perpendicular to the force $\mathbf{F_{EM}}$ that the earth exerts on the moon. The masses are: mass of sun = 1.99×10^{30} kg, mass of earth = 5.98×10^{24} kg, mass of moon = 7.35×10^{22} kg. The distances shown in the drawing are $r_{SM} = 1.50 \times 10^{11}$ m and $r_{EM} = 3.85 \times 10^8$ m. Determine the magnitude of the net gravitational force on the moon.

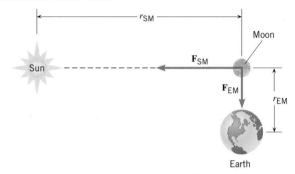

25. **ssm** Mars has a mass of 6.46×10^{23} kg and a radius of 3.39×10^6 m. (a) What is the acceleration due to gravity on Mars? (b) How much would a 65-kg person weigh on this planet?

26. As pointed out in the text, your weight W_M on the moon is approximately one-sixth of its value W_E on the earth. The masses and radii of the moon and the earth are given on the inside of the front cover. Use these data and calculate a more exact value (to 3 significant digits) for the ratio W_M/W_E.

27. The mass of a robot is 5450 kg. This robot weighs 3620 N more on planet A than it does on planet B. Both planets have the same radius of 1.33×10^7 m. What is the difference $M_A - M_B$ in the masses of these planets?

28. A space traveler weighs 580 N on earth. What will the traveler weigh on another planet whose radius is three times that of the earth and whose mass is twice that of the earth?

29. **ssm** Synchronous communications satellites are placed in a circular orbit that is 3.59×10^7 m above the surface of the earth. What is the magnitude of the acceleration due to gravity at this distance?

*30. Three uniform spheres are located at the corners of an equilateral triangle. Each side of the triangle has a length of 1.20 m. Two of the spheres have a mass of 2.80 kg each. The third sphere (mass unknown) is released from rest. Considering only the gravitational forces that the spheres exert on each other, what is the magnitude of the initial acceleration of the third sphere?

*31. At a distance H above the surface of a planet, the true weight of a remote probe is one percent less than its true weight on the surface. The radius of the planet is R. Find the ratio H/R.

*32. A spacecraft is on a journey to the moon. The masses of the earth and moon are, respectively, 5.98×10^{24} kg and $7.35 \times$

10^{22} kg. The distance between the centers of the earth and the moon is 3.85×10^8 m. At what point, as measured from the center of the earth, does the gravitational force exerted on the craft by the earth balance the gravitational force exerted by the moon? This point lies on a line between the centers of the earth and the moon.

*33. **ssm www** Several people are riding in a hot-air balloon. The combined mass of the people and balloon is 310 kg. The balloon is motionless in the air, because the downward-acting weight of the people and balloon is balanced by an upward-acting "buoyant" force. If the buoyant force remains constant, how much mass should be dropped overboard so the balloon acquires an upward acceleration of 0.15 m/s²?

*34. Jupiter is the largest planet in our solar system, having a mass and radius that are, respectively, 318 and 11.2 times that of earth. Suppose that an object falls from rest near the surface of each planet, and that each object falls the same distance before striking the ground. Determine the ratio of the time of fall on Jupiter to that on earth.

**35. Two particles are located on the x axis. Particle 1 has a mass m and is at the origin. Particle 2 has a mass $2m$ and is at $x = +L$. Where on the x axis should a third particle be located so that the magnitude of the gravitational force on *both* particle 1 and particle 2 doubles? Express your answer in terms of L. Note that there are two answers.

Section 4.8 The Normal Force, Section 4.9 Static and Kinetic Frictional Forces

36. A 95.0-kg person stands on a scale in an elevator. What is the apparent weight when the elevator is (a) accelerating upward with an acceleration of 1.80 m/s², (b) moving upward at a constant speed, and (c) accelerating downward with an acceleration of 1.30 m/s²?

37. **ssm** A rocket blasts off from rest and attains a speed of 45 m/s in 15 s. An astronaut has a mass of 57 kg. What is the astronaut's apparent weight during takeoff?

38. A 55-kg person crouches on a scale and jumps straight up. As the person springs up, the reading on the scale suddenly rises to 622 N. What is the acceleration of the person at this instant?

39. A person exerts a horizontal force of 267 N in attempting to push a freezer across a room, but the freezer does not move. What is the static frictional force that the floor exerts on the freezer?

40. A block whose weight is 45.0 N rests on a horizontal table. A horizontal force of 36.0 N is applied to the block. The coefficients of static and kinetic friction are 0.650 and 0.420, respectively. Will the block move under the influence of the force, and, if so, what will be the block's acceleration? Explain your reasoning.

41. **ssm** A 60.0-kg crate rests on a level floor at a shipping dock. The coefficients of static and kinetic friction are 0.760 and 0.410, respectively. What horizontal pushing force is required to (a) just start the crate moving and (b) slide the crate across the dock at a constant speed?

42. A cup of coffee is sitting on a table in an airplane that is flying at a constant altitude and a constant velocity. The coefficient of static friction between the cup and the table is 0.30. Suddenly, the plane accelerates, its altitude remaining constant. What is the maximum acceleration that the plane can have without the cup sliding backward on the table?

43. A 92-kg baseball player slides into second base. The coefficient of kinetic friction between the player and the ground is $\mu_k = 0.61$. (a) What is the magnitude of the frictional force? (b) If the player comes to rest after 1.2 s, what is his initial speed?

44. A 6.00-kg box is sliding across the horizontal floor of an elevator. The coefficient of kinetic friction between the box and the floor is 0.360. Determine the kinetic frictional force that acts on the box when the elevator is (a) stationary, (b) accelerating upward with an acceleration whose magnitude is 1.20 m/s², and (c) accelerating downward with an acceleration whose magnitude is 1.20 m/s².

*45. **ssm** A skater with an initial speed of 7.60 m/s is gliding across the ice. Air resistance is negligible. (a) The coefficient of kinetic friction between the ice and the skate blades is 0.100. Find the deceleration caused by kinetic friction. (b) How far will the skater travel before coming to rest?

*46. During a shuffleboard game (played on a horizontal surface) a disk is given an initial speed of 6.80 m/s. The coefficient of kinetic friction between the disk and surface is 0.290. How much time passes before the disk comes to rest?

*47. The drawing shows a large cube (mass = 25 kg) being accelerated across a horizontal frictionless surface by a horizontal force **P**. A small cube (mass = 4.0 kg) is in contact with the front surface of the large cube and will slide downward unless **P** is sufficiently large. The coefficient of static friction between the cubes is 0.71. What is the smallest magnitude that **P** can have in order to keep the small cube from sliding downward?

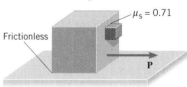

**48. While moving in, a new homeowner is pushing a box across the floor at a constant velocity. The coefficient of kinetic friction between the box and the floor is 0.41. The pushing force is directed downward at an angle θ below the horizontal. When θ is greater than a certain value, it is not possible to move the box, no matter how large the pushing force is. Find that value of θ.

Section 4.11 Equilibrium Applications of Newton's Laws of Motion

49. **ssm** A 12.0-kg lantern is suspended from the ceiling by two vertical wires. What is the tension in each wire?

50. A supertanker (mass = 1.70×10^8 kg) is moving with a constant velocity. Its engines generate a forward thrust of 7.40×10^5 N. Determine (a) the magnitude of the resistive force exerted

on the tanker by the water and (b) the magnitude of the upward buoyant force exerted on the tanker by the water.

51. As preparation for this problem, review Example 14. Suppose that the pilot suddenly jettisons 2800 N of fuel. If the plane is to continue moving with the same velocity under the influence of the same air resistance **R**, by how much does the pilot have to reduce (a) the thrust and (b) the lift?

52. A wire is stretched between the tops of two identical buildings. When a tightrope walker is at the middle of the wire, the tension in the wire is 2220 N. Each half of the wire makes an angle of 8.00° with respect to the horizontal. Find the weight of the performer.

53. ssm A 1.40-kg bottle of vintage wine is lying horizontally in the rack shown in the drawing. The two surfaces on which the bottle rests are 90.0° apart, and the right surface makes an angle of 45.0° with respect to the ground. Each surface exerts a force on the bottle that is perpendicular to the surface. What is the magnitude of each of these forces?

54. A student presses a book between his hands, as the drawing indicates. The forces that he exerts on the front and back covers of the book are perpendicular to the book and are horizontal. The book weighs 31 N. The coefficient of static friction between his hands and the book is 0.40. To keep the book from falling, what is the magnitude of the minimum pressing force that each hand must exert?

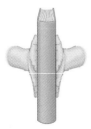

55. The helicopter in the drawing is moving horizontally to the right at a constant velocity. The weight of the helicopter is $W =$ 53 800 N. The lift force **L** generated by the rotating blade makes an angle of 21.0° with respect to the vertical. (a) What is the magnitude of the lift force? (b) Determine the magnitude of the air resistance **R** that opposes the motion.

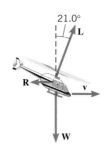

56. A Mercedes–Benz 300SL ($m = 1700$ kg) is parked on a road that rises 15° above the horizontal. What is the magnitude of the static frictional force exerted on the tires by the road?

57. ssm A 20.0-kg sled is being pulled across a horizontal surface at a constant velocity. The pulling force has a magnitude of 80.0 N and is directed at an angle of 30.0° above the horizontal. Determine the coefficient of kinetic friction.

58. The drawing shows a circus clown who weighs 890 N. The coefficient of static friction between the clown's feet and the ground is 0.53. He pulls vertically downward on a rope that passes around three pulleys and is tied around his feet. What is the minimum pulling force that the clown must exert to yank his feet out from under himself?

59. A bicyclist coasts at a constant velocity along a road that slopes downward at an angle of 20.0° with respect to the horizontal. The combined mass of the bicycle and rider is 75.0 kg. Find the magnitude of the resistive force that opposes the motion.

****60.** A 43.8-kg sign is suspended by two wires, as the drawing shows. Find the tension in wire 1 and in wire 2.

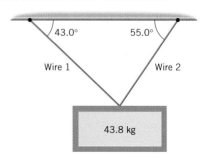

****61. ssm** A 44-kg chandelier is suspended 1.5 m below a ceiling

by three wires, each of which has the same tension and the same length of 2.0 m (see the drawing). Find the tension in any wire.

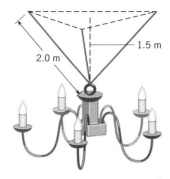

1.5 m

2.0 m

*62. A skier is pulled up a slope at a constant velocity by a tow bar. The slope is inclined at 25.0° with respect to the horizontal. The force applied to the skier by the tow bar is parallel to the slope. The skier's mass is 55.0 kg, and the coefficient of kinetic friction between the skis and the snow is 0.120. Find the magnitude of the force that the tow bar exerts on the skier.

*63. A 0.600-kg kite is being flown at the end of a string. The string is straight and makes an angle of 55.0° above the horizontal. The kite is stationary and the tension in the string is 35.0 N. Determine the force (both magnitude and direction) that the wind exerts on the kite. Specify the angle relative to the horizontal.

64. The weight of the block in the drawing is 88.9 N. The coefficient of static friction between the block and the vertical wall is 0.560. What minimum force **F is required to (a) prevent the block from sliding down the wall (*Hint: The static frictional force exerted on the block is directed upward, parallel to the wall.*) and (b) start the block moving up the wall? (*Hint: The static frictional force is now directed down the wall.*)

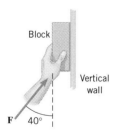

Block

Vertical wall

F 40°

65. ssm A bicyclist is coasting straight down a hill at a constant speed. The mass of the rider and bicycle is 80.0 kg, and the hill is inclined at 15.0° with respect to the horizontal. Air resistance opposes the motion of the cyclist. Later, the bicyclist climbs the same hill at the same constant speed. How much force (directed parallel to the hill) must be applied to the bicycle in order for the bicyclist to climb the hill?

**66. A damp washcloth is hung over the edge of a table to dry. Thus, part (mass = m_{on}) of the washcloth rests on the table and

part (mass = m_{off}) does not. The coefficient of static friction between the table and the washcloth is 0.40. Determine the maximum fraction $[m_{off}/(m_{on} + m_{off})]$ that can hang over the edge without causing the whole washcloth to slide off the table.

Section 4.12 Nonequilibrium Applications of Newton's Laws of Motion

67. A 350-kg sailboat has an acceleration of 0.62 m/s² at an angle of 64° north of east. Find the magnitude and direction of the net force that acts on the sailboat.

68. Review Conceptual Example 17 as background for this problem. The water skier there has a mass of 73 kg. Find the magnitude of the net force acting on the skier when (a) she is accelerated from rest to a speed of 11 m/s in 8.0 s and (b) she lets go of the tow rope and glides to a halt in 21 s.

69. **ssm** A 1380-kg car is moving due east with an initial speed of 27.0 m/s. After 8.00 s the car has slowed down to 17.0 m/s. Find the magnitude and direction of the net force that produces the deceleration.

70. A parachutist (mass = 72.0 kg) is falling straight down with an acceleration of 1.30 m/s². What is the magnitude of the resistive force of the air that opposes the downward motion?

71. A fisherman is fishing from a bridge and is using a "45-N test line." In other words, the line will sustain a maximum force of 45 N without breaking. (a) What is the heaviest fish that can be pulled up vertically, when the line is reeled in at a constant speed? (b) Repeat part (a), assuming that the line is given an upward acceleration of 2.0 m/s².

72. In the drawing, the weight of the block on the table is 422 N and that of the hanging block is 185 N. Ignoring all frictional effects and assuming the pulley to be massless, find (a) the acceleration of the two blocks and (b) the tension in the cord.

422 N

185 N

73. **ssm www** A student is skateboarding down a ramp that is 6.0 m long and inclined at 18° with respect to the horizontal. The initial speed of the skateboarder at the top of the ramp is 2.6 m/s. Neglect friction and find the speed at the bottom of the ramp.

74. A car is towing a boat on a trailer. The driver starts from rest and accelerates to a speed of 11 m/s in a time of 28 s. The combined mass of the boat and trailer is 410 kg. What is the tension in the hitch that connects the trailer to the car?

75. A cable is lifting a construction worker and a crate, as the drawing shows. The weights of the worker and crate are 965 and 1510 N, respectively. The acceleration of the cable is 0.620 m/s², upward. What is the tension in the cable (a) below the worker and (b) above the worker?

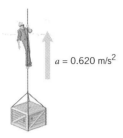

$a = 0.620 \ \text{m/s}^2$

76. A passenger is pulling on the strap of a 15.0-kg suitcase with a force of 70.0 N. The strap makes an angle of 35.0° above the horizontal. A 37.8-N friction force opposes the motion (horizontal) of the suitcase. Determine the acceleration of the suitcase.

77. **ssm** In a supermarket parking lot, an employee is pushing ten empty shopping carts, lined up in a straight line. The acceleration of the carts is 0.050 m/s². The ground is level, and each cart has a mass of 26 kg. (a) What is the net force acting on any one of the carts? (b) Assuming friction is negligible, what is the force exerted by the fifth cart on the sixth cart?

78. A lunar landing craft (mass = 11 400 kg) is about to touch down on the surface of the moon, where the acceleration due to gravity is 1.60 m/s². At an altitude of 165 m the craft's downward velocity is 18.0 m/s. To slow down the craft, a retrorocket is firing to provide an upward thrust. Assuming the descent is vertical, find the magnitude of the thrust needed to reduce the velocity to zero at the instant when the craft touches the lunar surface.

*79. The drawing shows Robin Hood (mass = 82 kg) about to escape from a dangerous situation. With one hand, he is gripping the rope that holds up a chandelier (mass = 220 kg). When he cuts the rope where it is tied to the floor, the chandelier will fall,

and he will be pulled up toward a balcony above. Ignore the friction between the rope and the beams over which it slides, and find (a) the acceleration with which Robin is pulled upward and (b) the tension in the rope while Robin escapes.

*80. A 205-kg log is pulled up a ramp by means of a rope that is parallel to the surface of the ramp. The ramp is inclined at 30.0° with respect to the horizontal. The coefficient of kinetic friction between the log and the ramp is 0.900, and the log has an acceleration of 0.800 m/s². Find the tension in the rope.

*81. **ssm** To hoist himself into a tree, a 72.0-kg man ties one end of a nylon rope around his waist and throws the other end over a branch of the tree. He then pulls downward on the free end of the rope with a force of 358 N. Neglect any friction between the rope and the branch, and determine the man's upward acceleration.

*82. At an airport, luggage is unloaded from a plane into the three cars of a luggage carrier, as the drawing shows. The acceleration of the carrier is 0.12 m/s², and friction is negligible. The coupling bars have negligible mass. By how much would the tension in *each* of the coupling bars A, B, and C change if 39 kg of luggage has been removed from car 2 and placed in (a) car 1 and (b) car 3? If the tension changes, specify whether it increases or decreases.

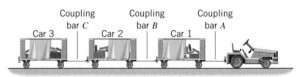

*83. A book is resting on a piece of paper. The paper is resting on a flat table. The coefficient of static friction between the book and the paper is 0.72. If a person pulls on the paper, what is the maximum acceleration that the book can have, before the book begins to slip relative to the paper?

*84. A girl is sledding down a slope that is inclined at 30.0° with respect to the horizontal. A moderate wind is aiding the motion by providing a steady force of 105 N that is parallel to the motion of the sled. The combined mass of the girl and sled is 65.0 kg, and the coefficient of kinetic friction between the runners of the sled and the snow is 0.150. How much time is required for the sled to travel down a 175-m slope, starting from rest?

*85. **ssm** A box is sliding up an incline that makes an angle of 15.0° with respect to the horizontal. The coefficient of kinetic friction between the box and the surface of the incline is 0.180. The initial speed of the box at the bottom of the incline is 1.50 m/s. How far does the box travel along the incline before coming to rest?

*86. A sports car is accelerating up a hill that rises 18° above the horizontal. The coefficient of static friction between the wheels and the road is $\mu_s = 0.88$. It is the static frictional force that propels the car forward. (a) What is the magnitude of the maximum acceleration that the car can have? (b) What is the magnitude of the maximum acceleration if the car is being driven down the hill?

*87. A truck is traveling at a speed of 25.0 m/s along a level road. A crate is resting on the bed of the truck, and th[...] of static friction between the crate and the truck bed is [...] termine the shortest distance in which the truck can com[...] without causing the crate to slip forward relative to the tru[...]

**88. As part *a* of the drawing shows, two blocks are connec[...] a rope that passes over a set of pulleys. One block has a weig[...] 412 N, and the other has a weight of 908 N. The rope and the [...] leys are massless and there is no friction. (a) What is the accelera[...] tion of the lighter block? (b) Suppose that the heavier block is re-moved, and a downward force of 908 N is provided by someone pulling on the rope, as part *b* of the drawing shows. Find the acceleration of the remaining block. (c) Explain why the answers in (a) and (b) are different.

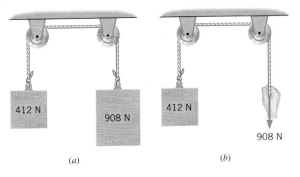

(a)   (b)

**89. ssm A penguin slides at a constant velocity of 1.4 m/s down an icy incline. The incline slopes above the horizontal at an

[...]dle oujhe incline, the penguin slides onto celeration of the [...]cient of kinetic friction be-two strings. [...]e for the incline as for the [...]ired for the penguin to [...]ch of ice? [...]are connected by [...]ulleys. The ob-[...]een the mid-[...] is the ac-[...]h of the

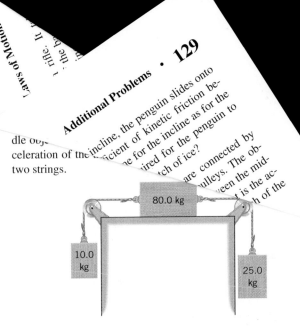

**91. A 5.00-kg block is placed on top of a 12.0-kg block that rests on a frictionless table. The coefficient of static friction between the two blocks is 0.600. What is the maximum horizontal force that can be applied before the 5.00-kg block begins to slip relative to the 12.0-kg block, if the force is applied to (a) the more massive block and (b) the less massive block?

ADDITIONAL PROBLEMS

92. A water skier (mass = 89 kg) is being pulled at a constant velocity. The horizontal pulling force is 350 N. Find (a) the magnitude of the total resistive force exerted on the skier by the water and air and (b) the magnitude of the upward force exerted on the skier by the water.

93. ssm Saturn has an equatorial radius of 6.00×10^7 m and a mass of 5.67×10^{26} kg. (a) Compute the acceleration of gravity at the equator of Saturn. (b) How many times greater is a person's weight on Saturn compared to that on earth?

94. A net force of 525 N gives an object an acceleration of 4.20 m/s². (a) What net force is needed to give the object an acceleration of 13.7 m/s²? (b) What net force is required to keep the object moving at a constant velocity?

95. A stuntman is being pulled along a rough road at a constant velocity, by a cable attached to a moving truck. The cable is parallel to the ground. The mass of the stuntman is 109 kg, and the coefficient of kinetic friction between the road and him is 0.870. Find the tension in the cable.

96. Our solar system is in the Milky Way galaxy. The nearest galaxy is Andromeda, a distance of 2×10^{22} m away. The masses of the Milky Way and Andromeda galaxies are 7×10^{41} and 6×10^{41} kg, respectively. Treat the galaxies as particles and find

the magnitude of the gravitational force exerted on the Milky Way by the Andromeda galaxy.

97. ssm Three forces act on a moving object. One force has a magnitude of 80.0 N and is directed due north. Another has a magnitude of 60.0 N and is directed due west. What must be the magnitude and direction of the third force, such that the object continues to move with a constant velocity?

98. When a parachute opens, it develops a large drag force with the air. This upward force is initially greater than the weight of the sky diver and, thus, slows him down. Suppose the weight of the sky diver is 915 N and the drag force has a magnitude of 1027 N. What is the magnitude and direction of the acceleration?

99. A rescue helicopter is lifting a man (weight = 822 N) from a capsized boat by means of a cable and harness. (a) What is the tension in the cable when the man is given an initial upward acceleration of 1.10 m/s²? (b) What is the tension during the remainder of the rescue when he is pulled upward at a constant velocity?

100. A woman stands on a scale in a moving elevator. Her mass is 60.0 kg, and the combined mass of the elevator and scale is an additional 815 kg. Starting from rest, the elevator accelerates upward. During the acceleration, there is a tension of 9410 N in the hoisting cable. What does the scale read during the acceleration?

101. **ssm** ... in preparation for this prob-
2.50 × 10⁻³ ... ace exploration vehicle (mass =
it exits the ... forward acceleration of 0.220 m/s². To
celeration ... orward ... ration on the moon, the vehicle's engines
erted on ... ve force of 1.43 × 10³ N. What is the magni-

102. ...
lem. ... tional force that acts on the vehicle on the moon?
5.90 ...
act ...

... spacecraft has a mass of 3.50 × 10⁴ kg and is drifting
... ng a straight line through deep space. Its speed is 1820 m/s. An
engine is suddenly turned on and provides a thrust of 2240 N in
the direction of the spacecraft's velocity. (a) Find the acceleration
of the spacecraft. (b) What is the distance (in km) traveled while
the spacecraft increases its speed to 2310 m/s?

104. A 1580-kg car is traveling with a speed of 15.0 m/s. What
is the magnitude of the horizontal net force that is required to
bring the car to a halt in a distance of 50.0 m?

*105. **ssm** Two objects (45.0 and 21.0 kg) are connected by a
massless string that passes over a massless, frictionless pulley.
The pulley hangs from the ceiling. Find (a) the acceleration of the
objects and (b) the tension in the string.

*106. Traveling at a speed of 16.1 m/s, the driver of an automobile suddenly locks the wheels by slamming on the brakes. The
coefficient of kinetic friction between the tires and the road is
0.720. How far does the car skid before coming to a halt? Ignore
the effects of air resistance.

*107. A toboggan slides down a hill and has a constant velocity.
The angle of the hill is 11.3° with respect to the horizontal. What
is the coefficient of kinetic friction between the surface of the hill
and the toboggan?

*108. During a storm, a limb falls from a tree. It comes to rest
across a barbed wire fence, one-fifth of the way between two
fence posts, which are 4.00 m apart. Thus, the wire between the
posts is divided into a shorter section and a longer section. The
limb exerts a downward force of 151 N on the wire, depressing it
0.200 m below the horizontal. Find the tension in the section of
wire that is (a) shorter and (b) longer.

*109. **ssm** A person whose weight is 5.20 × 10² N is being
pulled up vertically by a rope from the bottom of a cave that is
35.1 m deep. The maximum tension that the rope can withstand
without breaking is 569 N. What is the shortest time, starting from
rest, in which the person can be brought out of the cave?

*110. A billiard ball strikes the cushion of a pool table perpendicularly. The mass of the ball is 0.38 kg. The ball approaches the
cushion with a velocity of +2.1 m/s and rebounds with a velocity
of −2.0 m/s. The ball remains in contact with the cushion for a
time of 3.3 × 10⁻³ s. What is the average force (magnitude and
direction) exerted on the ball by the cushion?

*111. A crate is resting on the bed of a moving truck. The coefficient of static friction between the crate and the truck bed is 0.40.
The driver hits the brakes. Assuming the truck is traveling on level
ground, determine the maximum deceleration that the truck can
have without the crate slipping forward relative to the truck.

*112. A neutron star has a mass of 2.0 × 10³⁰ kg (about the mass
of our sun) and a radius of 5.0 × 10³ m (about the height of a
good-sized mountain). Suppose an object falls from rest near the
surface of such a star. How fast would it be moving after it had
fallen a distance of 0.010 m? (Assume that the gravitational force
is constant over the distance of the fall, and that the star is not rotating.)

*113. **ssm** A person is trying to judge if a picture (mass =
1.10 kg) is properly positioned by temporarily pressing it against
a wall. The pressing force is perpendicular to the wall. The coefficient of static friction between the picture and the wall is 0.660.
What is the minimum amount of pressing force that must be used?

**114. A small sphere is hung by a string from the ceiling of a
van. When the van is stationary, the sphere hangs vertically. However, when the van accelerates, the sphere swings backward so
that the string makes an angle of θ with respect to the vertical.
(a) Derive an expression for the magnitude a of the acceleration of
the van in terms of the angle θ and the magnitude g of the acceleration due to gravity. (b) Find the acceleration of the van when θ =
10.0°. (c) What is the angle θ when the van moves with a constant
velocity?

**115. A sofa rests on the horizontal bed of a moving van. The coefficient of static friction between the sofa and van bed is 0.30.
The van starts from rest and accelerates for a time of 5.1 s. What
is the maximum distance that the van can travel in this time period
without having the sofa slide?

**116. In the drawing, the rope and the pulleys are massless, and
there is no friction. Find (a) the tension in the rope and (b) the acceleration of the 10.0-kg block. (*Hint: The larger mass moves
twice as far as the smaller mass.*)

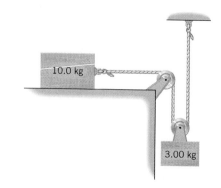

10.0 kg

3.00 kg

117. **ssm A 225-kg crate rests on a surface that is inclined
above the horizontal at an angle of 20.0°. A horizontal force (magnitude = 535 N and parallel to the ground, not the incline) is required to start the crate moving down the incline. What is the coefficient of static friction between the crate and the incline?

DYNAMICS OF UNIFORM CIRCULAR MOTION

In traveling on a vertical circular arc, these planes experience a net force and an acceleration that point toward the center of the circle.

5.1 UNIFORM CIRCULAR MOTION

There are many examples of motion on a circular path. Of the many possibilities, we want to single out those that satisfy the following definition:

> ■ **DEFINITION OF UNIFORM CIRCULAR MOTION**
>
> Uniform circular motion is the motion of an object traveling at a constant (uniform) speed on a circular path.

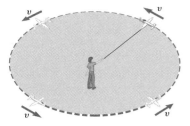

Figure 5.1 The motion of an airplane flying at a constant speed on a horizontal circular path is an example of uniform circular motion.

As an example of uniform circular motion, Figure 5.1 shows a model airplane on a guideline. The speed of the plane is the magnitude of the velocity vector **v**, and since the speed is constant, the vectors in the drawing have the same magnitude at various points on the circle.

Sometimes it is more convenient to describe uniform circular motion by specifying the period of the motion, rather than the speed. The **period T** is the time required to travel once around the circle, that is, to make one complete revolution. There is a relationship between period and speed, since speed v is the distance traveled (circumference of the circle $= 2\pi r$) divided by the time T:

$$v = \frac{2\pi r}{T} \tag{5.1}$$

If the radius is known, as in Example 1, the speed can be calculated from the period or vice versa.

EXAMPLE 1 • A Tire-Balancing Machine

The wheel of a car has a radius of 0.29 m and is being rotated at 830 revolutions per minute (rpm) on a tire-balancing machine. Determine the speed (in m/s) at which the outer edge of the wheel is moving.

Reasoning The speed v can be obtained directly from $v = 2\pi r/T$, but first the period T is needed. The period is the time for one revolution. We need to express the period in seconds, because the problem asks for the speed in meters per second.

Solution Since the tire makes 830 revolutions in one minute, the number of minutes required for a single revolution is

$$\frac{1}{830 \text{ revolutions/min}} = 1.2 \times 10^{-3} \text{ min/revolution}$$

Therefore, the period is $T = 1.2 \times 10^{-3}$ min, which corresponds to 0.072 s. Equation 5.1 can now be used to find the speed:

$$v = \frac{2\pi r}{T} = \frac{2\pi(0.29 \text{ m})}{0.072 \text{ s}} = \boxed{25 \text{ m/s}}$$

The definition of uniform circular motion emphasizes that the magnitude of the velocity vector is constant. It is equally significant that the direction of the vector is *not constant*. In Figure 5.1, for instance, the velocity vector changes direction as the

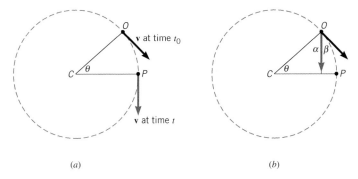

(a)   (b)

Figure 5.2 (*a*) For an object in uniform circular motion, the velocity vector **v** has different directions at different points on the circle. (*b*) The velocity vector has been removed from point *P*, shifted parallel to itself, and redrawn with its tail at point *O*.

plane moves around the circle. Any change in the velocity vector, even if it is only a change in direction, means that an acceleration is occurring. This particular acceleration is called "centripetal acceleration," because it points toward the center of the circle, as the next section explains.

5.2 CENTRIPETAL ACCELERATION

In this section we determine how the magnitude a_c of the centripetal acceleration depends on the speed v of the object and the radius r of the circular path. We will see that $a_c = v^2/r$.

In Figure 5.2*a* an object (symbolized by a dot ●) is in uniform circular motion. At time t_0 the velocity is tangent to the circle at point *O*, while at a later time t the velocity is tangent at point *P*. As the object moves from *O* to *P*, the radius traces out the angle θ, and the velocity vector changes direction. To emphasize the change, part *b* of the picture shows the velocity vector removed from point *P*, shifted parallel to itself, and redrawn with its tail at point *O*. The angle β between the two vectors indicates the change in direction. Since the radii *CO* and *CP* are perpendicular to the tangents at points *O* and *P*, respectively, it follows that $\alpha + \beta = 90°$ and $\alpha + \theta = 90°$. Therefore, angle β and angle θ are equal.

As always, acceleration is the change $\Delta\mathbf{v}$ in velocity divided by the elapsed time Δt, or $\mathbf{a} = \Delta\mathbf{v}/\Delta t$. Figure 5.3*a* shows the two velocity vectors oriented at the angle θ with respect to one another, together with the vector $\Delta\mathbf{v}$ that represents the change in velocity. The change $\Delta\mathbf{v}$ is the increment that must be added to the velocity at time t_0, so that the resultant velocity has the new direction after an elapsed time $\Delta t = t - t_0$. Figure 5.3*b* shows the sector of the circle *COP*. In the limit of a very small elapsed time Δt, the arc length *OP* can be approximated as a straight line whose length is the distance $v\,\Delta t$ traveled by the object. In this limit, *COP* is an isosceles triangle, as is the triangle in part *a* of the drawing. Since both triangles have equal apex angles θ, they are similar, so that

$$\frac{\Delta v}{v} = \frac{v\,\Delta t}{r}$$

This equation can be solved for $\Delta v/\Delta t$, to show that the magnitude a_c of the centripetal acceleration is given by $a_c = v^2/r$.

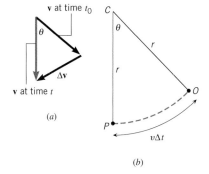

(a)

(b)

Figure 5.3 (*a*) The direction of the velocity vector at time *t* differs from the direction at time t_0 by the angle θ. (*b*) When the object moves along the circle from *O* to *P*, the radius *r* traces out the same angle θ. To facilitate a comparison to part *a*, the sector *COP* has been rotated clockwise by 90° relative to its orientation in Figure 5.2.

Centripetal acceleration is a vector quantity and, therefore, has a direction as well as a magnitude. The direction is toward the center of the circle, and Conceptual Example 2 helps us to set the stage for explaining this important fact.

CONCEPTUAL EXAMPLE 2 • Which Way Will the Object Go?

In Figure 5.4 an object, such as a model airplane on a guideline, is in uniform circular motion. The object is symbolized by a dot (●), and at point O it is suddenly released from its circular path. For instance, the guideline for a model plane is suddenly cut. Does the object move along the straight tangent line between points O and A or along the circular arc between points O and P?

Reasoning and Solution Newton's first law of motion guides our reasoning. An object continues in a state of rest or in a state of motion at a constant speed along a straight line unless compelled to change that state by a net force. When the object is suddenly released from its circular path, there is no longer a net force being applied to the object. In the case of a model airplane, the guideline cannot apply a force, since it is cut. Gravity certainly acts on the plane, but the wings provide a lift force that balances the weight of the plane. In the absence of a net force, then, the plane or any object would continue to move at a constant speed along a straight line in the direction it had at the time of release. This speed and direction are given in Figure 5.4 by the velocity vector **v**. As a result, *the object would move along the straight line between points O and A, not on the circular arc between points O and P.*

Related Homework Material: Problems 4 and 6

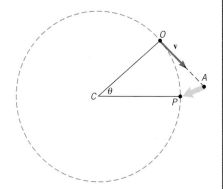

Figure 5.4 If an object (●) moving on a circular path were released from its path at point O, it would move along the straight tangent line OA in the absence of a net force. To remain on the circular arc OP, the object must accelerate toward the center of the circle.

As Example 2 discusses, the object in Figure 5.4 would travel on a tangent line if it were suddenly released from its circular path at point O. It would then move in a straight line to point A in the time it would have taken to travel on the circle to point P. It is as if the object drops through the distance AP in the process of remaining on the circle, and AP is directed toward the center of the circle in the limit that the angle θ is small. Thus, the object accelerates toward the center of the circle at every moment. The acceleration is called *centripetal acceleration,* because the word "centripetal" means "center-seeking."

■ **CENTRIPETAL ACCELERATION**

MAGNITUDE: The centripetal acceleration of an object moving with a speed v on a circular path of radius r has a magnitude a_c given by

$$a_\text{c} = \frac{v^2}{r} \tag{5.2}$$

DIRECTION: The centripetal acceleration vector always points toward the center of the circle and continually changes direction as the object moves.

The following example illustrates the effect of the radius r on the centripetal acceleration.

EXAMPLE 3 • The Effect of Radius on Centripetal Acceleration

The bobsled track at the 1994 Olympics in Lillehammer, Norway contained turns with radii of 33 m and 24 m, as Figure 5.5 illustrates. Find the centripetal acceleration at each turn for a speed of 34 m/s, a speed that was achieved in the 2-man event. Express the answers as multiples of $g = 9.8$ m/s^2.

Reasoning In each case, the magnitude of the centripetal acceleration can be obtained from $a_c = v^2/r$. Since the radius r is in the denominator on the right side of this expression, we expect the acceleration to be smaller when r is larger.

Solution From $a_c = v^2/r$ it follows that

Radius = 33 m $a_c = \dfrac{(34 \text{ m/s})^2}{33 \text{ m}} = 35 \text{ m/s}^2 = \boxed{3.6\,g}$

Radius = 24 m $a_c = \dfrac{(34 \text{ m/s})^2}{24 \text{ m}} = 48 \text{ m/s}^2 = \boxed{4.9\,g}$

The centripetal acceleration is indeed smaller when the radius is larger. In fact, with r in the denominator on the right of $a_c = v^2/r$, the acceleration approaches zero when the radius becomes very large. Uniform circular motion along the arc of an infinitely large circle entails no acceleration, because it is just like motion at a constant speed along a straight line.

In Section 4.11 we learned that an object is in equilibrium when it has zero acceleration. Conceptual Example 4 discusses whether an object undergoing uniform circular motion can ever be at equilibrium.

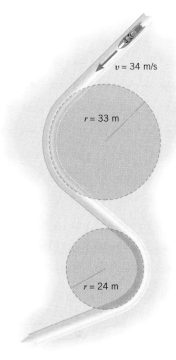

Figure 5.5 This bobsled travels at the same speed around two curves with different radii. For the turn with the larger radius, the sled has a smaller centripetal acceleration.

CONCEPTUAL EXAMPLE 4 • Uniform Circular Motion and Equilibrium

A car moves at a constant speed, and there are three parts to the motion. It moves along a straight line toward a circular turn, goes around the turn, and then moves away along a straight line. In each of these parts, is the car in equilibrium?

Reasoning and Solution An object in equilibrium has no acceleration, according to the definition given in Section 4.11. As the car approaches the turn, both the speed and direction of the motion are constant. Thus, the velocity vector does not change, and there is no acceleration. The same is true as the car moves away from the turn. For these parts of the motion, then, the car is in equilibrium. As the car goes around the turn, however, the direction of travel changes, and the car has a centripetal acceleration that is characteristic of uniform circular motion. Because of this acceleration, the car is not in equilibrium during the turn. In general, ***an object that is in uniform circular motion can never be in equilibrium.***

Related Homework Material: *Problem 13*

Because of the different centripetal accelerations, going around each turn in Figure 5.5 would feel different, as most drivers know from experiences with tight turns (smaller r) and gentle turns (larger r). The feeling is associated with the force that is present in uniform circular motion, and we now turn to this topic.

5.3 CENTRIPETAL FORCE

Newton's second law indicates that whenever an object accelerates, there must be a net force to create the acceleration. Thus, in uniform circular motion there must be a net force to produce the centripetal acceleration. According to the second law, the net force F_c is equal to the product of the object's mass and its acceleration (see the concept chart in Figure 5.6, which is an expanded version of the chart in Figure 4.10). This net force is called the *centripetal force* and points in the same direction as the centripetal acceleration, that is, toward the center of the circle.

■ CENTRIPETAL FORCE

MAGNITUDE: The centripetal force is the name given to the net force required to keep an object of mass m, moving at a speed v, on a circular path of radius r and has a magnitude of

$$F_c = \frac{mv^2}{r} \qquad (5.3)$$

DIRECTION: The centripetal force always points toward the center of the circle and continually changes direction as the object moves.

Figure 5.6 *Concepts at a Glance*
Uniform circular motion entails a centripetal acceleration $a_c = v^2/r$. The net force required to produce this acceleration is called the centripetal force and is given by Newton's second law of motion as $F_c = mv^2/r$. In the death spiral in pairs figure skating, the woman is spun around in a circle by her partner, and the centripetal force is largely provided by a component of the force applied to her by his hand.

The phrase "centripetal force" does not denote a new and separate force created by nature. The phrase merely labels the net force pointing toward the center of the circular path, and this net force is the vector sum of all the force components that point along the radial direction.

In some cases, it is easy to identify the source of the centripetal force, as when a model airplane on a guideline flies in a horizontal circle. The only force pulling the plane inward is the tension in the line, so this force alone (or a component of it) is the centripetal force. Example 5 illustrates the fact that higher speeds require greater tensions.

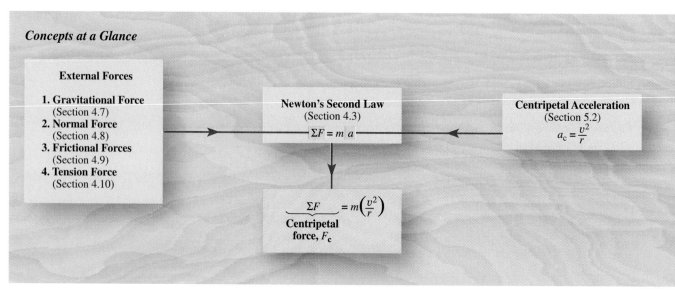

Concepts at a Glance

External Forces	Newton's Second Law (Section 4.3)	Centripetal Acceleration (Section 5.2)
1. **Gravitational Force** (Section 4.7) 2. **Normal Force** (Section 4.8) 3. **Frictional Forces** (Section 4.9) 4. **Tension Force** (Section 4.10)	$\Sigma F = m\,a$	$a_c = \dfrac{v^2}{r}$

$$\underbrace{\Sigma F}_{\text{Centripetal force, } F_c} = m\left(\frac{v^2}{r}\right)$$

EXAMPLE 5 • The Effect of Speed on Centripetal Force

The model airplane in Figure 5.7 has a mass of 0.90 kg and moves at a constant speed on a circle that is parallel to the ground. Find the tension T in the guideline (length = 17 m) for speeds of 19 and 38 m/s.

Reasoning Because gravity pulls downward on the plane, the guideline slopes downward at an angle with respect to the ground. As a result, the radius of the circle is slightly less than the 17-m length of the guideline, and only the horizontal component of the tension points toward the center of the circle to provide the centripetal force. For simplicity, however, we ignore the effect of gravity and assume that the guideline is parallel to the ground. With this assumption, the radius of the circle is 17 m, and the entire tension, rather than just a component of it, provides the centripetal force.

Solution Equation 5.3 gives the tension directly: $F_c = T = mv^2/r$.

Speed = 19 m/s $T = \dfrac{(0.90 \text{ kg})(19 \text{ m/s})^2}{17 \text{ m}} = \boxed{19 \text{ N}}$

Speed = 38 m/s $T = \dfrac{(0.90 \text{ kg})(38 \text{ m/s})^2}{17 \text{ m}} = \boxed{76 \text{ N}}$

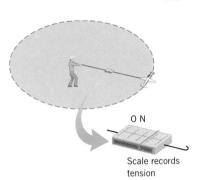

Figure 5.7 The scale records the tension in the guideline. See Example 5.

Conceptual Example 6 deals with another case where it is easy to identify the source of the centripetal force.

CONCEPTUAL EXAMPLE 6 • A Trapeze Act

In a circus, a man hangs upside down from a trapeze, legs bent over the bar and arms downward, holding his partner (see Figure 5.8). Is it harder for the man to hold his partner when the partner hangs straight downward and is stationary or when the partner is swinging through the straight-down position?

Figure 5.8 The man hanging upside down from the trapeze has a harder job holding his partner when the team is swinging through the straight-down position than when they are hanging straight down and stationary.

• PROBLEM SOLVING INSIGHT
When using an equation to obtain a numerical answer, algebraically solve for the unknown variable in terms of the known variables. Then substitute in the numbers for the known variables, as this example shows.

Reasoning and Solution When the man and his partner are stationary, the man's arms must support his partner's weight. When the two are swinging, however, the man's arms must do an additional job. Then, the partner is moving on a circular arc and has a centripetal acceleration. The man's arms must exert an additional pull so that there will be sufficient centripetal force to produce this acceleration. Because of the additional pull, *it is harder for the man to hold his partner while swinging than while stationary.*

Related Homework Material: *Problem 47*

When a car moves at a steady speed around an unbanked curve, the centripetal force keeping the car on the curve comes from the static friction between the road and the tires, as Figure 5.9 indicates. It is static, rather than kinetic friction, because the tires are not slipping with respect to the radial direction. If the static frictional force is insufficient, given the speed and the radius of the turn, the car will skid off the road. Example 7 shows how an icy road can limit safe driving.

EXAMPLE 7 • Road Conditions and Safe Driving

Compare the maximum speeds at which a car can safely negotiate an unbanked turn (radius = 50.0 m) in dry weather (coefficient of static friction = 0.900) and icy weather (coefficient of static friction = 0.100).

Reasoning At the maximum speed, the maximum centripetal force acts on the tires, and static friction must provide it. The magnitude of the maximum force of static friction is specified by Equation 4.7 as $f_s^{MAX} = \mu_s F_N$, where μ_s is the coefficient of static friction and F_N is the magnitude of the normal force. Our strategy, then, is to find the normal force, substitute it in the expression for the maximum force of static friction, and then equate the result to mv^2/r. Experience indicates that the maximum speed for the dry road will be greater than for the icy road.

Solution Since the car does not accelerate in the vertical direction, the weight mg of the car is balanced by the normal force, so $F_N = mg$. From Equation 5.3 it follows, then, that

$$F_c = \mu_s F_N = \mu_s mg = \frac{mv^2}{r}$$

Consequently, $\mu_s g = v^2/r$, and

$$v = \sqrt{\mu_s g r}$$

The mass m of the car has been eliminated algebraically from this result. All cars, heavy or light, have the same maximum speed. The maximum speeds can now be calculated:

Dry road ($\mu_s = 0.900$)　$v = \sqrt{(0.900)(9.80 \text{ m/s}^2)(50.0 \text{ m})} = \boxed{21.0 \text{ m/s}}$

Icy road ($\mu_s = 0.100$)　$v = \sqrt{(0.100)(9.80 \text{ m/s}^2)(50.0 \text{ m})} = \boxed{7.00 \text{ m/s}}$

As expected, the dry road allows the greater maximum speed.

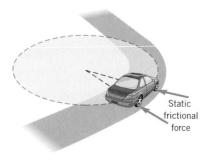

Static frictional force

Figure 5.9 When the car moves without skidding around a curve, static friction between the road and the tires provides the centripetal force to keep the car on the road.

The passenger in Figure 5.9 must also experience a centripetal force to remain on the circular path. However, if the upholstery is very slippery, there may not be enough static friction to keep him in place as the car makes a tight turn at high speed. Then, when viewed from inside the car, he appears to be thrown toward the outside of the curve. What really happens is that the passenger slides off on a tan-

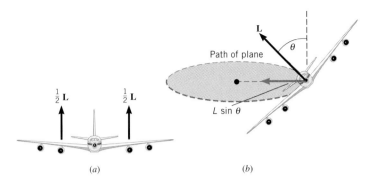

Figure 5.10 (a) The air exerts an upward lifting force $\frac{1}{2}$L on each wing. (b) When a plane executes a circular turn, the plane banks at an angle θ. The lift component $L \sin \theta$ is directed toward the center of the circle and provides the centripetal force.

gent to the circle, until he encounters a source of centripetal force to keep him in place while the car turns. This occurs when the passenger bumps into the side of the car, which pushes on him with the necessary force. Clearly, the centripetal force can come from different sources. It can come from the static friction between the passenger and the upholstery, from a seat belt, or from the side of the car when the passenger bumps against it.

Sometimes the source of the centripetal force is not obvious. A pilot making a turn, for instance, banks or tilts the plane at an angle to create the centripetal force. As a plane flies, the air pushes upward on the wing surfaces with a net lifting force **L** that is perpendicular to the wing surfaces, as Figure 5.10a shows. Part b of the drawing illustrates that when the plane is banked at an angle θ, a component $L \sin \theta$ of the lifting force is directed toward the center of the turn. It is this component of the lift that provides the centripetal force. Greater speeds and/or tighter turns require greater centripetal forces. In such situations, the pilot must bank the plane at a larger angle, so that a larger component of the lift points toward the center of the turn. The technique of banking into a turn also has an application in the construction of high-speed roadways, where the road itself is banked to achieve a similar effect, as the next section discusses.

The Physics of...
flying an airplane in a banked turn.

5.4 BANKED CURVES

When a car travels without skidding around an unbanked curve, the static frictional force between the tires and the road provides the centripetal force. From day to day, however, the friction may vary, depending on the condition of the pavement (ice, oil slicks, etc.). The reliance on friction can be eliminated completely for a given speed if the curve is banked at an angle relative to the horizontal, much in the same way that a plane is banked while making a turn.

Figure 5.11a shows a car going around a friction-free banked curve. The radius of the curve is r, where r is measured parallel to the horizontal and not to the slanted road surface. Part b shows the normal force $\mathbf{F_N}$ that the road applies to the car, the normal force being perpendicular to the road surface. Because the roadbed makes an angle θ with respect to the horizontal, the normal force has a component $F_N \sin \theta$ pointing toward the center C of the circle, and this component provides the centripetal force:

$$F_c = F_N \sin \theta = \frac{mv^2}{r}$$

The vertical component of the normal force is $F_N \cos \theta$, and since the car does not accelerate in the vertical direction, this component must balance the weight mg of

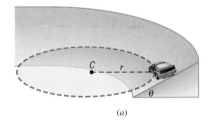

(a)

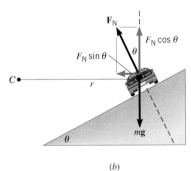

(b)

Figure 5.11 (a) A car travels on a circle of radius r on a frictionless banked road. The banking angle is θ, and the center of the circle is at C. (b) The forces acting on the car are its weight $m\mathbf{g}$ and the normal force $\mathbf{F_N}$. A component $F_N \sin \theta$ of the normal force provides the centripetal force.

the car. Therefore, $F_N \cos \theta = mg$. Dividing this equation into the previous one shows that

$$\frac{F_N \sin \theta}{F_N \cos \theta} = \frac{mv^2/r}{mg}$$

$$\tan \theta = \frac{v^2}{rg} \qquad (5.4)$$

Equation 5.4 indicates that, for a given speed v, the centripetal force needed for a turn of radius r can be obtained from the normal force by banking the turn at an angle θ, independent of the mass of the vehicle. Greater speeds and smaller radii require more steeply banked curves, that is, larger values of θ. At a speed that is too small for a given θ, a car would slide down a frictionless banked curve; at a speed that is too large, a car would slide off the top. The next example deals with a famous banked curve.

The Physics of...
the Daytona International Speedway.

EXAMPLE 8 • The Daytona 500

The Daytona 500 is the major event of the NASCAR (National Association for Stock Car Auto Racing) season. It is held at the Daytona International Speedway in Daytona, Florida. The turns in this oval track have a maximum radius (at the top) of $r = 316$ m and are banked steeply, with $\theta = 31°$ (see Figure 5.11). Suppose these turns were frictionless. At what speed would the race cars have to travel around them?

Reasoning and Solution From Equation 5.4, it follows that

$$v = \sqrt{rg \tan \theta} = \sqrt{(316 \text{ m})(9.80 \text{ m/s}^2)\tan 31°} = \boxed{43 \text{ m/s}}$$

This speed corresponds to 96 mph. Drivers negotiate the turns at the Daytona 500 at speeds up to 195 mph, however. These higher speeds necessitate a greater centripetal force than that implied by Equation 5.4 for a frictionless turn. In reality, the turns are not frictionless, and static friction provides the additional centripetal force.

5.5 SATELLITES IN CIRCULAR ORBITS

THE RELATION BETWEEN ORBITAL RADIUS AND ORBITAL SPEED

Today there are many satellites in orbit about the earth. The ones in circular orbits are examples of uniform circular motion. Like a model airplane on a guideline, each satellite is kept on its circular path by a centripetal force. The gravitational pull of the earth provides the centripetal force and acts like an invisible guideline for the satellite.

There is only one speed that a satellite can have if the satellite is to remain in an orbit with a fixed radius. To see how this fundamental characteristic arises, consider the gravitational force acting on the satellite of mass m in Figure 5.12. Since the gravitational force is the only force acting on the satellite in the radial direction, it alone provides the centripetal force. Therefore, using Newton's law of gravitation (Equation 4.3), we have

$$F_c = G \frac{mM_E}{r^2} = \frac{mv^2}{r}$$

where G is the universal gravitational constant, M_E is the mass of the earth, and r is

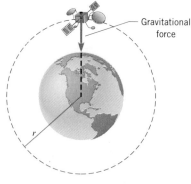

Figure 5.12 For a satellite in circular orbit around the earth, the gravitational force provides the required centripetal force.

Gravitational force

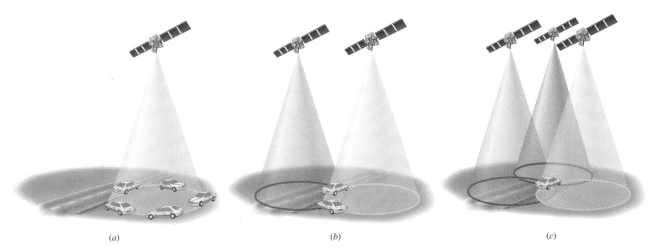

(a)   *(b)*   *(c)*

Figure 5.13 The Navstar Global Positioning System (GPS) of satellites can be used with a GPS receiver to locate an object, such as a car, on the earth. (*a*) One satellite identifies the car as being somewhere on a circle. (*b*) A second satellite places it on another circle, which identifies two possibilities for the exact spot. (*c*) A third satellite provides the means for deciding where the car is.

the distance from the center of the earth to the satellite. Solving for the speed v of the satellite gives

$$v = \sqrt{\frac{GM_{\mathrm{E}}}{r}} \qquad (5.5)$$

If the satellite is to remain in an orbit of radius r, the speed must have precisely this value. Note that the radius r of the orbit is in the denominator in Equation 5.5. This means that the closer the satellite is to the earth, the smaller is the value for r and the greater the orbital speed must be. Once in orbit at the correct speed, the satellite continues in uniform circular motion forever, assuming that effects such as friction due to residual atmosphere do not reduce the speed.

Many applications of satellite technology affect our lives. One of the most recent is a network of satellites called the Global Positioning System (GPS), which can be used to determine the position of an object on the earth's surface to within 15 m or less. Figure 5.13 illustrates how the system works. Each GPS satellite carries a highly accurate atomic clock, whose time is transmitted to the ground continually by means of radio waves. In the drawing, a car carries a computerized GPS receiver that can detect the waves and is synchronized to the satellite clock. The receiver can, therefore, determine the distance between the car and a satellite from a knowledge of the travel time of the waves and the speed at which they move. This speed, as we will see in Chapter 24, is the speed of light and is known with great precision. A measurement using a single satellite locates the car to be somewhere on a circle, as Figure 5.13*a* shows, while a measurement using a second GPS satellite locates the car on another circle. The intersection of the circles reveals two possible positions for the car, as in Figure 5.13*b*. With the aid of a third satellite, a third circle can be established, which intersects the other two and identifies the car's exact position, as in Figure 5.13*c*. The orbits of the GPS satellites place three or more of them in view simultaneously at any instant from any spot on earth. The GPS technique has the potential for a major impact on our lives in many ways, including navigational systems for automobiles, boats, and planes, as well as portable systems that tell hikers and people with visual impairments where they are located.

The mass m of the satellite does not appear in Equation 5.5, having been eliminated algebraically. ***Consequently, for a given orbit, a satellite with a large mass has exactly the same orbital speed as a satellite with a small mass.*** However, more effort is certainly required to lift the larger-mass satellite into orbit. The orbital speed of one famous artificial satellite is determined in the following example.

▶ **The Physics of...**
the Global Positioning System.

EXAMPLE 9 • Orbital Speed of the Hubble Space Telescope

Determine the speed of the Hubble space telescope (see Figure 5.14) orbiting at a height of 596 km above the earth's surface.

Reasoning Before Equation 5.5 can be applied, the orbital radius r must be determined *relative to the center of the earth*. Since the radius of the earth is approximately 6.38×10^6 m, and the height of the telescope above the earth's surface is 0.596×10^6 m, the orbital radius is $r = 6.98 \times 10^6$ m.

Solution The orbital speed is

$$v = \sqrt{\frac{GM_E}{r}} = \sqrt{\frac{(6.67 \times 10^{-11}\ \text{N} \cdot \text{m}^2/\text{kg}^2)(5.98 \times 10^{24}\ \text{kg})}{6.98 \times 10^6\ \text{m}}}$$

$$\boxed{v = 7.56 \times 10^3\ \text{m/s}\ (16\ 900\ \text{mi/h})}$$

Equation 5.5 applies to man-made earth satellites or to the moon, which is a natural satellite of the earth. It also applies for circular orbits about any astronomical object, provided M_E is replaced by the mass of the object on which the orbit is centered. Example 10, for instance, shows how scientists have applied this equation to conclude that a supermassive black hole is probably located at the center of the galaxy known as M87. This galaxy is located at a distance of about 50 million light-years away from the earth. (One light-year is the distance that light travels in a year, or 9.5×10^{15} m.)

EXAMPLE 10 • A Supermassive Black Hole

The Hubble telescope has detected the light being emitted from different regions of galaxy M87, which is shown in Figure 5.15. The black circle identifies the center of the galaxy. From the characteristics of this light, astronomers have determined an orbiting speed of 7.5×10^5 m/s for matter located at a distance of 5.7×10^{17} m from the center. Find the mass M of the object located at the galactic center.

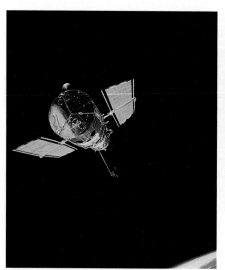

Figure 5.14 The Hubble space telescope.

Figure 5.15 This image of the ionized gas (yellow) at the heart of galaxy M87 was obtained by the Hubble space telescope. The circle identifies the center of the galaxy, at which a black hole is thought to exist.

Reasoning and Solution Replacing M_E in Equation 5.5 with M gives $v = \sqrt{GM/r}$, which can be solved to show that

$$M = \frac{v^2 r}{G} = \frac{(7.5 \times 10^5 \text{ m/s})^2 (5.7 \times 10^{17} \text{ m})}{6.67 \times 10^{-11} \text{ N} \cdot \text{m}^2/\text{kg}^2} = \boxed{4.8 \times 10^{39} \text{ kg}}$$

This is an incredibly large mass. The ratio of this mass to the mass of our sun is $(4.8 \times 10^{39} \text{ kg})/(2.0 \times 10^{30} \text{ kg}) = 2.4 \times 10^9$, so that matter equivalent to 2.4 billion of our suns is located at the center of galaxy M87. Considering that the volume of space in which this matter is located contains relatively few visible stars, researchers have concluded that the data provide strong evidence for the existence of a supermassive black hole. The object located at the center of galaxy M87 is called a "black hole," because its tremendous mass prevents even light from escaping from it. The light that forms the image in Figure 5.15 comes not from the black hole itself, but from matter that surrounds it.

THE PERIOD OF THE SATELLITE

The period T of a satellite is the time required for one orbital revolution. As in any uniform circular motion, the period is related to the speed of the motion by $v = 2\pi r/T$. Substituting v from Equation 5.5 shows that

$$\sqrt{\frac{GM_E}{r}} = \frac{2\pi r}{T}$$

Solving this expression for the period T gives

$$T = \frac{2\pi r^{3/2}}{\sqrt{GM_E}} \tag{5.6}$$

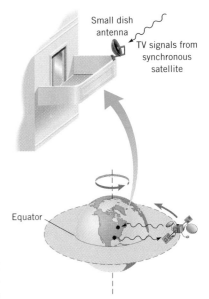

Although derived for earth orbits, Equation 5.6 can also be used for calculating the periods of those planets that have nearly circular orbits about the sun, if M_E is replaced by the mass M_S of the sun and r is interpreted as the distance between the center of the planet and the center of the sun. The fact that the period is proportional to the three-halves power of the orbital radius is known as Kepler's third law, for it is one of the laws discovered by Johannes Kepler (1571–1630) during his studies of the motions of the planets. Kepler's third law also holds for elliptical orbits, which will be discussed in Chapter 9.

An important application of Equation 5.6 occurs in the field of communications, where "synchronous satellites" are put into a circular orbit that is in the plane of the equator, as Figure 5.16 shows. The period of such a satellite is chosen to be one day, which is also the time it takes for the earth to turn once about its axis. Therefore, these satellites move around their orbits in a way that is synchronized with the rotation of the earth. For earth-based observers, synchronous satellites have the useful characteristic of appearing in fixed positions in the sky and can serve as "stationary" relay stations for communication signals sent up from the earth's surface. In fact, this is exactly what is done in the digital satellite systems that have become popular as an alternative to cable TV. As the blowup in Figure 5.16 indicates, a small "dish" antenna mounted on your house picks up the digital TV signals relayed to earth by the synchronous satellite. After being decoded, these signals are delivered directly to your TV set. All synchronous satellites are placed in orbit at the same height above the earth's surface, as Example 11 shows.

Figure 5.16 A synchronous satellite orbits the earth once per day on a circular path that lies in the plane of the equator. Digital satellite system television uses synchronous satellites as relay stations for TV signals that are sent up from the earth's surface and then rebroadcast down toward your own small dish antenna.

The Physics of...
digital satellite system TV.

EXAMPLE 11 • The Orbital Radius for Synchronous Satellites

What is the height above the earth's surface at which all synchronous satellites (regardless of mass) must be placed in orbit?

Reasoning Knowing the period T of a synchronous satellite is one day, we can use Equation 5.6 to find the distance r from the center of the earth. However, the problem asks for the height H of the satellite above the earth's surface. Therefore, to find H, we will have to take into account the fact that the earth itself has a radius of 6.38×10^6 m.

Solution A period of one day* corresponds to $T = 8.64 \times 10^4$ s. In using this value it is convenient to rearrange the equation $T = 2\pi r^{3/2}/\sqrt{GM_E}$ as follows:

$$r^{3/2} = \frac{T\sqrt{GM_E}}{2\pi} = \frac{(8.64 \times 10^4 \text{ s})\sqrt{(6.67 \times 10^{-11} \text{ N} \cdot \text{m}^2/\text{kg}^2)(5.98 \times 10^{24} \text{ kg})}}{2\pi}$$

By squaring and then taking the cube root, we find that $r = 4.23 \times 10^7$ m. Since the radius of the earth is approximately 6.38×10^6 m, the height of the satellite above the earth's surface is

$$H = 4.23 \times 10^7 \text{ m} - 0.64 \times 10^7 \text{ m} = \boxed{3.59 \times 10^7 \text{ m (22 300 mi)}}$$

Figure 5.17 As they orbit the earth, these NASA astronauts float around in a state of apparent weightlessness.

5.6 APPARENT WEIGHTLESSNESS AND ARTIFICIAL GRAVITY

The idea of life on board an orbiting satellite conjures up visions of astronauts floating around in a state of "weightlessness," as in Figure 5.17. Actually, this state should be called "apparent weightlessness," because it is similar to the condition of zero apparent weight that occurs in an elevator during free-fall. Conceptual Example 12 explores this similarity.

CONCEPTUAL EXAMPLE 12 • Apparent Weightlessness and Free-fall

Figure 5.18 shows a person on a scale in a freely falling elevator and in a satellite that is in a circular orbit. In each case, what is the apparent weight recorded by the scale?

Reasoning and Solution As Section 4.8 discusses, apparent weight is the force that an object exerts on the platform of a scale. In the freely falling elevator in Figure 5.18a, the apparent weight is zero. The reason is that both the scale and the person on it fall together and, therefore, cannot push against one another. In the orbiting satellite in Figure 5.18b, both the person and the scale are in uniform circular motion. Objects in uniform circular motion continually accelerate or "fall" toward the center of the circle, in order to remain on the circular path. Consequently, at all times, the scale in Figure 5.18b and the person on it both "fall" with the same acceleration toward the center of the orbit and cannot push against one another. Thus, ***the apparent weight in the satellite is zero, just as it is in the freely falling elevator.*** The

The Physics of...
apparent weightlessness.

---

* Successive appearances of the sun define the solar day of 24 h or 8.64×10^4 s. The sun moves against the background of the stars, however, and the time required for the earth to turn once on its axis relative to the fixed stars is 23 h 56 min, which is called the sidereal day. The sidereal day should be used in Example 11, but the neglect of this effect introduces an error of less than 0.4 percent in the answer.

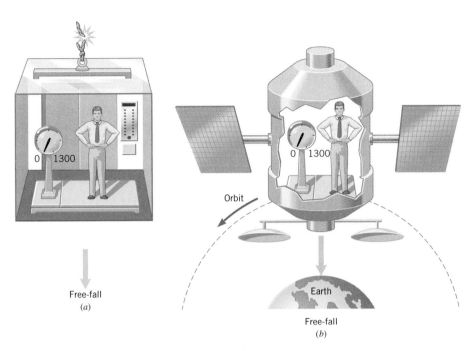

Figure 5.18 (a) During free-fall, the elevator accelerates downward with the acceleration of gravity, and the apparent weight of the person is zero.
(b) The orbiting space station is also in free-fall toward the center of the earth.

Orbit

Earth

Free-fall
(a)

Free-fall
(b)

only difference between the satellite and the elevator is that the satellite moves in a circle, so that its "falling" does not bring it closer to the earth. In contrast to the apparent weight, the true weight is the gravitational force ($F = GmM_E/r^2$) that the earth exerts on an object, and this force is not zero in a freely falling elevator or aboard an orbiting satellite.

The physiological effects of prolonged apparent weightlessness are only partially known. To minimize such effects, it is likely that artificial gravity will be provided in large space stations of the future. To help explain artificial gravity, Figure 5.19 shows a space station rotating about an axis. Because of the rotational motion, any object located at a point P on the interior surface of the station experiences a centripetal force directed toward the axis. The surface of the station provides this force by pushing on the feet of an astronaut. The centripetal force can be adjusted to match the astronaut's earth weight by properly selecting the rotational speed of the space station, as Examples 13 and 14 illustrate.

Figure 5.19 The surface of the rotating space station pushes on an object with which it is in contact and thereby provides the centripetal force needed to keep the object moving on a circular path.

EXAMPLE 13 • Artificial Gravity

At what speed must the surface of the space station ($r = 1700$ m) move in Figure 5.19, so that the astronaut at point P experiences a push on his feet that equals his earth weight?

Reasoning and Solution The earth weight of the astronaut (mass $= m$) is mg, and with this substitution, Equation 5.3 can be used to determine the required speed: $F_c = mg = mv^2/r$. Solving this equation for the speed, we find that

$$v = \sqrt{rg} = \sqrt{(1700 \text{ m})(9.80 \text{ m/s}^2)} = \boxed{130 \text{ m/s}}$$

The Physics of...
artificial gravity.

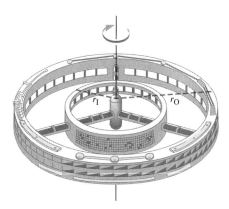

Figure 5.20 The outer ring (radius = r_O) of this rotating space laboratory simulates the acceleration of gravity on earth, while the inner ring (radius = r_I) simulates the acceleration of gravity on Mars.

Roller coasters are often the most popular rides at amusement parks. Designing modern roller coasters, with their corkscrew vertical loops, requires a thorough understanding of centripetal force.

EXAMPLE 14 • A Rotating Space Laboratory

A space laboratory is rotating to create artificial gravity, as Figure 5.20 indicates. Its period of rotation is chosen so the outer ring (r_O = 2150 m) simulates the acceleration of gravity on earth (9.80 m/s²). What should be the radius r_I of the inner ring, so it simulates the acceleration of gravity on the surface of Mars (3.72 m/s²)?

Reasoning It is important to realize that the value given for either acceleration corresponds to the centripetal acceleration $a_c = v^2/r$ in the corresponding ring. Moreover, the speed v and radius r are related according to $v = 2\pi r/T$ (Equation 5.1), where T is the period of the motion. Substituting this relation for the speed into the expression for the centripetal acceleration gives

$$a_c = \frac{v^2}{r} = \frac{(2\pi r/T)^2}{r} = \frac{4\pi^2 r}{T^2}$$

To solve this problem, we apply this expression to the outer and inner rings, both of which have the same period T. The periods of the two rings are the same, because the laboratory is rigid. All points on a rigid object require the same time to make one revolution.

Solution Applying the result that $a_c = 4\pi^2 r/T^2$, we find that

$$\underbrace{9.80 \text{ m/s}^2 = \frac{4\pi^2(2150 \text{ m})}{T^2}}_{\text{Outer ring}} \quad \text{and} \quad \underbrace{3.72 \text{ m/s}^2 = \frac{4\pi^2 r_I}{T^2}}_{\text{Inner ring}}$$

Dividing the inner-ring expression by the outer-ring expression and eliminating the common algebraic factors of $4\pi^2$ and T^2 gives

$$\frac{3.72 \text{ m/s}^2}{9.80 \text{ m/s}^2} = \frac{r_I}{2150 \text{ m}} \quad \text{or} \quad r_I = \boxed{816 \text{ m}}$$

*5.7 VERTICAL CIRCULAR MOTION

Motorcycle stunt drivers often perform a feat in which they drive their cycles around a vertical circular track, as in Figure 5.21a. Usually, the speed varies in this stunt, decreasing as the cycle moves upward and increasing as the cycle comes downward. When the speed of travel on a circular path changes from moment to moment, the motion is said to be "nonuniform." Nonetheless, we can use the con-

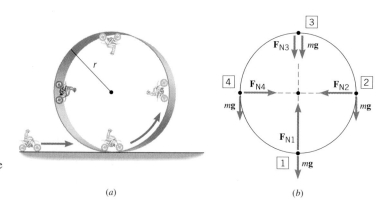

Figure 5.21 (a) A vertical loop-the-loop motorcycle stunt. (b) The normal force $\mathbf{F}_N$ and the weight $m\mathbf{g}$ of the cycle and the rider are shown here at four locations.

cepts that apply to uniform circular motion in order to gain considerable insight into the motion that occurs on a vertical circle.

As the speed changes in the motorcycle trick, the magnitude of the centripetal force also changes. There are four points on the circle, however, where the centripetal force can be identified easily, as Figure 5.21b indicates. As you look at Figure 5.21b, keep in mind that the centripetal force is not a new and separate force of nature. Instead, at each point the centripetal force is the net sum of all the force components oriented along the radial direction and points toward the center of the circle. The drawing shows only the weight of the cycle plus rider (magnitude = mg) and the normal force pushing on the cycle (magnitude = F_N). The propulsion and braking forces are omitted for simplicity and, in any event, do not act in the radial direction. The magnitude of the centripetal force at each of the four points is given as follows in terms of mg and F_N:

$$(1) \quad F_{c1} = \frac{mv_1^2}{r} = F_{N1} - mg \qquad (3) \quad F_{c3} = \frac{mv_3^2}{r} = F_{N3} + mg$$

$$(2) \quad F_{c2} = \frac{mv_2^2}{r} = F_{N2} \qquad (4) \quad F_{c4} = \frac{mv_4^2}{r} = F_{N4}$$

• **PROBLEM SOLVING INSIGHT**
The centripetal force is the name given to the net force that points toward the center of a circular path. As shown here, there may be several forces that contribute to this net force.

As the cycle goes around, the magnitude of the normal force changes. It changes because the speed changes and because the weight does not have the same effect at every point on the circle. At the bottom, the normal force and the weight oppose one another, giving a centripetal force of magnitude $F_{N1} - mg$. At the top, the normal force and the weight reinforce each other to provide a centripetal force whose magnitude is $F_{N3} + mg$. At points 2 and 4 on either side, only F_{N2} and F_{N4} provide the centripetal force. The weight is tangent to the circle at points 2 and 4 and has no component pointing toward the center. If the speed at each of the four points is known, along with the mass and the radius, the normal forces can be determined.

Riders who perform the loop-the-loop trick know that they must have at least a minimum speed at the top of the circle to remain on the track. This speed can be determined by considering the centripetal force at point 3. The speed v_3 in the equation $mv_3^2/r = F_{N3} + mg$ is a minimum when F_{N3} is zero. Then, the speed is given by $v_3 = \sqrt{rg}$. At this speed, the track does not exert a normal force to keep the cycle on the circle, because the weight mg provides all the centripetal force. Under these conditions, the rider experiences an apparent weightlessness like that discussed in Section 5.6, because for an instant the rider and the cycle are falling freely toward the center of the circle.

The Physics of...
the loop-the-loop motorcycle stunt.

SUMMARY

In **uniform circular motion,** an object of mass m travels at a constant speed v on a circular path of radius r. The **period** T of the motion is the time required to make one revolution. The speed, the period, and the radius are related according to $v = 2\pi r/T$. The velocity vector in such motion is always changing direction, and, therefore, an acceleration exists. This acceleration is called **centripetal acceleration,** and its magnitude is $a_c = v^2/r$, while its direction is toward the center of the circle. To create this acceleration, a net force pointing toward the center of the circle is needed. This net force is called the **centripetal force,** and its magnitude is $F_c = mv^2/r$.

There are many examples of uniform circular motion, including the **banking of curves** and the **orbiting of satellites.** The angle θ at which a friction-free curve is banked depends on the radius r of the curve and the speed v at which the curve is to be negotiated, according to $\tan \theta = v^2/(rg)$. The speed and the period of a satellite in a circular orbit about the earth depend on the radius of the orbit, according to $v = \sqrt{GM_E/r}$ and $T = 2\pi r^{3/2}/\sqrt{GM_E}$, where G is the universal gravitational constant and M_E is the mass of the earth. **Apparent weightlessness** and **artificial gravity** in satellites can be explained in terms of uniform circular motion.

CONCEPTUAL QUESTIONS

1. The speedometer of your car shows that you are traveling at a constant speed of 35 m/s. Is it possible that your car is accelerating? If so, explain how this could happen.

2. Consider two people, one on the earth's surface at the equator and the other at the north pole. Which has the larger centripetal acceleration? Explain.

3. The equations of kinematics (Equations 3.3–3.6) describe the motion of an object that has a constant acceleration. These equations cannot be applied to uniform circular motion. Why not?

4. Is it possible for an object to have an acceleration when the velocity of the object is constant? When the speed of the object is constant? In each case, give your reasoning.

5. A car is traveling at a constant speed along the road *ABCDE* shown in the drawing. Sections *AB* and *DE* are straight. Rank the accelerations in each of the four sections according to magnitude, listing the smallest first. Justify your ranking.

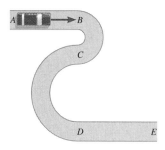

6. Other things being equal, would it be easier to drive at high speed around an unbanked horizontal curve on the moon as compared to driving around the same curve on the earth? Explain.

7. A bug lands on a windshield wiper. Explain why the bug is more likely to be dislodged when the wipers are turned on at the high rather than the low setting.

8. What is the chance of a light car safely rounding an unbanked curve on an icy road as compared to that of a heavy car: worse, the same as, or better? Assume that both cars have the same speed and are equipped with identical tires. Account for your answer.

9. The drawing shows two identical balls connected to two identical strings. The arrangement is held by the free end of string *A*

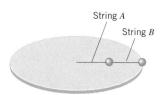

and whirled around in a horizontal circle, with the strings nearly parallel to the floor. As the balls are whirled faster and faster, which string is likely to break first? Justify your answer.

10. A propeller, operating under test conditions, is being made to rotate at ever faster speeds. Explain what is likely to happen, and why, when the maximum rated speed of the propeller is exceeded.

11. A penny is placed on a rotating turntable. Where on the platter does the penny require the largest centripetal force to remain in place? Give your reasoning.

12. Explain why a real airplane must bank as it flies in a circle, while a model airplane on a guideline can fly in a circle without banking.

13. Would a change in the earth's mass affect (a) the banking of airplanes as they turn, (b) the banking of roadbeds, (c) the speeds with which satellites are put into circular orbits, and (d) the performance of the loop-the-loop motorcycle stunt? In each case, give your reasoning.

14. A stone is tied to a string and whirled around in a circle at a constant speed. Is the string more likely to break when the circle is horizontal or when it is vertical? Account for your answer, assuming the constant speed is the same in each case.

15. 🜂 A fighter plane is diving toward the earth. At the last instant, the pilot pulls out of the dive on a vertical circle and begins climbing upward. In such maneuvers, the pilot can black out if too much blood drains from his head. Why does blood drain from his head?

16. Two cars are identical, except for the type of tread design on their tires. The cars are driven at the same speed and enter the same unbanked horizontal turn. Car A cannot negotiate the turn, but car B can. Which tread design yields a larger coefficient of static friction between the tires and the road? Give your reasoning in terms of centripetal force.

PROBLEMS

ssm Solution is in the Student Solutions Manual. **www** Solution is available on the World Wide Web at http://www.wiley.com/college/cutnell 🜂 This icon represents a biomedical application.

Section 5.1 Uniform Circular Motion, Section 5.2 Centripetal Acceleration

1. **ssm** How long does it take a plane, traveling at a speed of 110 m/s, to fly once around a circle whose radius is 2850 m?

2. The tips of the blades in a food blender are moving with a

speed of 21 m/s in a circle that has a radius of 0.053 m. How much time does it take for the blades to make one revolution?

3. A horse races once around a circular track in a time of 118 s, with a speed of 17 m/s. What is the radius of the track?

4. Review Conceptual Example 2 in preparation for this problem. In Figure 5.4, an object, after being released from its circular path, travels the distance OA in the same time it would have moved from O to P on the circle. The speed of the object on and off the circle remains constant at the same value. Suppose that the radius of the circle in Figure 5.4 is 3.6 m and the angle θ is 25°. What is the distance OA?

5. **ssm** Computer-controlled display screens provide drivers in the Indianapolis 500 with a variety of information about how their cars are performing. For instance, as a car is going through a turn, a speed of 221 mi/h (98.8 m/s) and a centripetal acceleration of 3.00 g (three times the acceleration due to gravity) are displayed. Determine the radius of the turn (in meters).

6. Review Conceptual Example 2 as background for this problem. One kind of slingshot consists of a pocket that holds a pebble and is whirled on a circle of radius r. The pebble is released from the circle at the angle θ, so that it will hit the target. The distance to the target from the center of the circle is d. (See the drawing, which is not to scale.) The circular path is parallel to the ground, and the target lies in the plane of the circle. The distance d is ten times the radius r. Ignore the effect of gravity in pulling the stone downward after it is released, and find the angle θ.

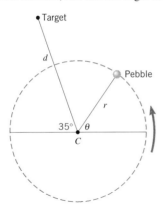

7. The blade of a windshield wiper moves through an angle of 90.0° in 0.28 s. The tip of the blade moves on the arc of a circle that has a radius of 0.76 m. What is the magnitude of the centripetal acceleration of the tip of the blade?

*8. The earth rotates once per day about an axis passing through the north and south poles, an axis that is perpendicular to the plane of the equator. Assuming the earth is a sphere with a radius of 6.38×10^6 m, determine the speed and centripetal acceleration of a person situated (a) at the equator and (b) at a latitude of 30.0° north of the equator.

*9. **ssm** The large blade of a helicopter is rotating in a horizontal circle. The length of the blade is 6.7 m, measured from its tip to the center of the circle. Find the ratio of the centripetal acceleration at the end of the blade to that which exists at a point located 3.0 m from the center of the circle.

*10. A centrifuge is a device in which a small container of material is rotated at a high speed on a circular path. Such a device is used in medical laboratories, for instance, to cause the

more dense red blood cells to settle through the less dense blood serum and collect at the bottom of the container. Suppose the centripetal acceleration of the sample is 6.25×10^3 times larger than the acceleration due to gravity. How many revolutions per minute is the sample making, if it is located at a radius of 5.00 cm from the axis of rotation?

Section 5.3 Centripetal Force

11. A 125-kg crate rests on the flatbed of a truck that moves at a speed of 15.0 m/s around an unbanked curve whose radius is 66.0 m. The crate does not slip relative to the truck. Obtain the magnitude of the static frictional force that the truck bed exerts on the crate.

12. A car is safely negotiating an unbanked circular turn at a speed of 21 m/s. The maximum static frictional force acts on the tires. Suddenly a wet patch in the road reduces the maximum static frictional force by a factor of three. If the car is to continue safely around the curve, to what speed must the driver slow the car?

13. **ssm** Review Example 3, which deals with the bobsled in Figure 5.5. Also review Conceptual Example 4. The mass of the sled and its two riders in Figure 5.5 is 350 kg. Find the magnitude of the centripetal force that acts on the sled during the turn with a radius of (a) 33 m and (b) 24 m.

14. A child is twirling a 0.0120-kg ball on a string in a horizontal circle whose radius is 0.100 m. The ball travels once around the circle in 0.500 s. (a) Determine the centripetal force acting on the ball. (b) If the speed is doubled, does the centripetal force double? If not, by what factor does the centripetal force increase?

15. A model airplane is being flown on a guideline that can sustain at most 180 N of tension. The mass of the plane is 0.75 kg, and its speed is 28 m/s. Assuming that the guideline is nearly parallel to the ground, what is the radius of the smallest horizontal circle in which the plane can be flown?

16. A stone has a mass of 6.0×10^{-3} kg and is wedged into the tread of an automobile tire, as the drawing shows. The coefficient of static friction between the stone and each side of the tread channel is 0.90. When the tire surface is rotating at 13 m/s, the stone flies out of the tread. The magnitude F_N of the normal force that each side of the tread channel exerts on the stone is 1.8 N. Assume that only static friction supplies the centripetal force, and determine the radius r of the tire.

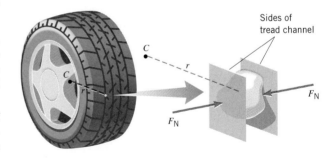

*17. **ssm www** The hammer throw is a track and field event in which a 7.3-kg ball (the "hammer") is whirled around in a circle several times and released. It then moves upward on the familiar curving path of projectile motion and eventually returns to earth, some distance away. The world record for this distance is 86.75 m, achieved in 1986 by Yuriy Sedykh. Ignore air resistance and the fact that the ball is released above the ground rather than at ground level. Furthermore, assume that the ball is whirled on a circle that has a radius of 1.8 m and that its velocity at the instant of release is directed 41° above the horizontal. Find the magnitude of the centripetal force acting on the ball just prior to the moment of release.

*18. What is the *minimum* coefficient of static friction necessary to allow a penny to rotate along with a $33\frac{1}{3}$-rpm record (diameter = 0.300 m), when the penny is placed at the outer edge of the record?

*19. A "swing" ride at a carnival consists of chairs that are swung in a circle by 12.0-m cables attached to a vertical rotating pole, as the drawing shows. Suppose the total mass of a chair and its occupant is 220 kg. (a) Determine the tension in the cable attached to the chair. (b) Find the speed of the chair.

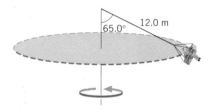

65.0°   12.0 m

**20. A rigid massless rod is rotated about one end in a horizontal circle. There is a mass m_1 attached to the center of the rod and a mass m_2 attached to the outer end of the rod. The inner section of the rod sustains three times as much tension as the outer section. Find the ratio m_2/m_1.

Section 5.4 Banked Curves

21. **ssm** A curve of radius 120 m is banked at an angle of 18°. At what speed can it be negotiated under icy conditions where friction is negligible?

22. At what angle should a curve of radius 150 m be banked, so cars can travel safely at 25 m/s without relying on friction?

23. On a banked race track, the smallest circular path on which cars can move has a radius of 112 m, while the largest has a radius of 165 m, as the drawing illustrates. The height of the outer wall is 18 m. Find (a) the smallest and (b) the largest speed at which cars can move on this track without relying on friction.

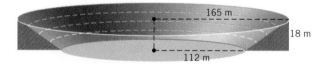

165 m

18 m

112 m

*24. There is a similarity between a plane banking into a turn and a car going around a banked curve. The lifting force **L** in Figure 5.10 plays the same role as the normal force $\mathbf{F_N}$ in Figure 5.11. (a) Derive an expression that relates the banking angle to the speed of the plane, the radius of the turn, and the acceleration due to gravity. (b) At what angle with respect to the horizontal should a plane be banked when traveling at 195 m/s around a turn whose radius is 8250 m?

*25. **ssm www** Refer to problem 24 before attempting to solve this problem. A jet ($m = 2.00 \times 10^5$ kg), flying at 123 m/s, banks to make a horizontal circular turn. The radius of the turn is 3810 m. Calculate the necessary lifting force.

**26. The drawing shows a baggage carousel at an airport. Your suitcase has not slid all the way down the slope and is going around at a constant speed on a circle ($r = 11.0$ m) as the carousel turns. The coefficient of static friction between the suitcase and the carousel is 0.760, and the angle θ in the drawing is 36.0°. How much time is required for your suitcase to go around once?

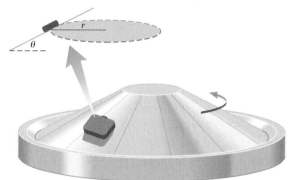

r

θ

Section 5.5 Satellites in Circular Orbits, Section 5.6 Apparent Weightlessness and Artificial Gravity

27. A satellite is placed in orbit 6.00×10^5 m above the surface of Jupiter. Jupiter has a mass of 1.90×10^{27} kg and a radius of 7.14×10^7 m. Find the orbital speed of the satellite.

28. Suppose the surface (radius = r) of the space station in Figure 5.19 is rotating at 35.8 m/s. What must be the value of r for the astronauts to weigh one-half of their earth weight?

29. **ssm www** A satellite is in a circular orbit around an unknown planet. The satellite has a speed of 1.70×10^4 m/s, and the radius of the orbit is 5.25×10^6 m. A second satellite also has a circular orbit around this same planet. The orbit of this second satellite has a radius of 8.60×10^6 m. What is the orbital speed of the second satellite?

30. Venus rotates slowly about its axis, the period being 243 days. The mass of Venus is 4.87×10^{24} kg. Determine the radius for a synchronous satellite in orbit about Venus.

31. The moon orbits the earth at a distance of 3.85×10^8 m. Assume that this distance is between the centers of the earth and the moon and that the mass of the earth is 5.98×10^{24} kg. Find the

period for the moon's motion around the earth. Express the answer in days and compare it to the length of a month.

32. The earth travels around the sun once a year in an approximately circular orbit whose radius is 1.50×10^{11} m. From these data determine (a) the orbital speed of the earth and (b) the mass of the sun.

***33.** **ssm** A satellite has a mass of 5850 kg and is in a circular orbit 4.1×10^5 m above the surface of a planet. The period of the orbit is two hours. The radius of the planet is 4.15×10^6 m. What is the true weight of the satellite when it is at rest on the planet's surface?

***34.** The earth orbits the sun once a year at a distance of 1.50×10^{11} m. Venus orbits the sun at a distance of 1.08×10^{11} m. These distances are between the centers of the planets and the sun. How long (in earth days) does it take for Venus to make one orbit around the sun?

***35.** Two satellites, A and B, are in different circular orbits about the earth. The orbital speed of satellite A is twice that of satellite B. Find the ratio (T_A/T_B) of the periods of the satellites.

****36.** To create artificial gravity, the space station shown in the drawing is rotating at a rate of 1.00 rpm. The radii of the cylindrically shaped chambers have the ratio $r_A/r_B = 4.00$. Each chamber A simulates an acceleration due to gravity of 10.0 m/s². Find values for (a) r_A, (b) r_B, and (c) the acceleration due to gravity that is simulated in chamber B.

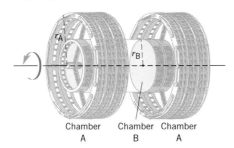

Chamber A   Chamber B   Chamber A

Section 5.7 Vertical Circular Motion

37. **ssm** The condition of apparent weightlessness for the passengers can be created for a brief instant when a plane flies over the top of a vertical circle. At a speed of 215 m/s, what is the radius of the vertical circle that the pilot must use?

38. A fighter pilot dives his plane toward the ground at 230 m/s. He pulls out of the dive on a vertical circle. What is the minimum radius of the circle, so that the normal force exerted on the pilot by his seat never exceeds three times his weight?

39. The maximum tension that a 0.50-m string can tolerate is 14 N. A 0.25-kg ball attached to this string is being whirled in a vertical circle. What is the maximum speed that the ball can have (a) at the top of the circle and (b) at the bottom of the circle? *(Hint: The tension serves the same purpose as the normal force in Figure 5.21.)*

40. A downhill skier, whose mass is 50.0 kg, attains a speed of

21.0 m/s just as she reaches the point where a jump is necessary (point A in the drawing). When she leaves the ground, her velocity is horizontal. In other words, point A is at the bottom of the circular arc AB (radius = 27.0 m). Determine the normal force acting on the skis at point A.

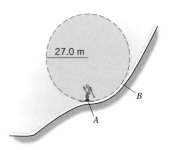

41. **ssm** A 2100-kg demolition ball swings at the end of a 15-m cable on the arc of a vertical circle. At the lowest point of the swing, the ball is moving at a speed of 7.6 m/s. Determine the tension in the cable. *(Hint: The tension serves the same purpose as the normal force at point 1 in Figure 5.21.)*

***42.** A roller-blader is traveling up one side of a hill and down the other side, trying to go fast enough to lose contact with the ground as she goes over the top. The crest is a circular arc with a radius of 20.0 m. Determine the minimum speed that she needs to achieve at the top of the hill.

***43.** A stone is tied to a string (length = 0.950 m) and whirled in a circle at the same constant speed in two different ways. First, the circle is horizontal and the string is nearly parallel to the ground. Next, the circle is vertical. In the vertical case the maximum tension in the string is 10.0% larger than the tension that exists when the circle is horizontal. Determine the speed of the stone.

****44.** In an automatic clothes drier, a hollow cylinder moves the clothes on a vertical circle (radius $r = 0.32$ m), as the drawing shows. The appliance is designed so that the clothes tumble gently as they dry. This means that when a piece of clothing reaches an angle of θ above the horizontal, it loses contact with the wall of the cylinder and falls onto the clothes below. How many revolutions per second should the cylinder make in order that the clothes lose contact with the wall when $\theta = 70.0°$?

ADDITIONAL PROBLEMS

45. ssm A 0.015-kg ball is shot from the plunger of a pinball machine. Because of a centripetal force of 0.028 N, the ball follows a circular arc whose radius is 0.25 m. What is the speed of the ball?

46. A bicycle chain is wrapped around a rear sprocket ($r = 0.039$ m) and a front sprocket ($r = 0.10$ m). The chain moves with a speed of 1.4 m/s around the sprockets, while the bike moves at a constant velocity. Find the magnitude of the acceleration of a chain link that is in contact with (a) the rear sprocket, (b) neither sprocket, and (c) the front sprocket.

47. For background pertinent to this problem, review Conceptual Example 6. In Figure 5.8 the man hanging upside down is holding a partner who weighs 475 N. Assume that the partner moves on a circle that has a radius of 6.50 m. At a swinging speed of 4.00 m/s, what force must the man apply to his partner?

48. A rocket is used to place a synchronous satellite in orbit about the earth. What is the speed of the satellite in orbit?

49. ssm Speedboat A negotiates a curve whose radius is 120 m. Speedboat B negotiates a curve whose radius is 240 m. Each boat experiences the same centripetal acceleration. What is the ratio v_A/v_B of the speeds of the boats?

50. A satellite circles the earth in an orbit whose radius is twice the earth's radius. The earth's mass is 5.98×10^{24} kg, and its radius is 6.38×10^6 m. What is the period of the satellite?

51. A dust particle of mass 1.5×10^{-7} kg is located on the outer edge of a compact disc recording (radius = 0.060 m). The disc is rotating at 3.5 rev/s. Find the magnitude of the centripetal force that acts on the dust particle.

52. The coefficient of static friction between the tires of a motorcycle and the road is $\mu_s = 0.56$. At a speed of 19 m/s, what is the radius of the tightest unbanked turn that the driver can handle?

53. ssm A motorcycle has a constant speed of 25.0 m/s as it passes over the top of a hill whose radius of curvature is 126 m. The mass of the motorcycle and driver is 342 kg. Find the magnitude of (a) the centripetal force and (b) the normal force that acts on the cycle.

*54. A computer is reading data from a rotating CD-ROM. At a point that is 0.030 m from the center of the disc, the centripetal acceleration is 120 m/s². What is the centripetal acceleration at a point that is 0.050 m from the center of the disc?

*55. A racetrack has the shape of an inverted cone, as the drawing shows. On this surface the cars race in circles that are parallel to the ground. For a speed of 27.0 m/s, at what value of the distance d should a driver locate his car, if he wishes to stay on a circular path without depending on friction?

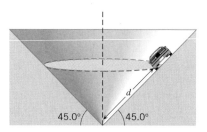

*56. Each of the space shuttle's main engines is fed liquid hydrogen by a high-pressure pump. Turbine blades inside the pump rotate at 617 rev/s. A point on one of the blades traces out a circle with a radius of 0.020 m as the blade rotates. (a) What is the magnitude of the centripetal acceleration that the blade must sustain at this point? (b) Express this acceleration as a multiple of $g = 9.80$ m/s².

**57. ssm www Redo Example 5 without ignoring the effect of gravity. That is, the guideline now slopes downward due to the weight of the plane. For purposes of significant figures, use 0.900 kg for the mass of the plane, 17.0 m for the length of the guideline, and 19.0 and 38.0 m/s for the speeds.

**58. At amusement parks, there is a popular ride where the floor of a rotating cylindrical room falls away, leaving the backs of the riders "plastered" against the wall. Suppose the radius of the room is 3.30 m and the speed of the wall is 10.0 m/s when the floor falls away. (a) What is the source of the centripetal force acting on the riders? (b) How much centripetal force acts on a 55.0-kg rider? (c) What is the minimum coefficient of static friction that must exist between a rider's back and the wall, if the rider is to remain in place when the floor drops away?

**59. A block is hung by a string from the inside roof of a van. When the van goes straight ahead at a speed of 28 m/s, the block hangs vertically down. But when the van maintains this same speed around an unbanked curve (radius = 150 m), the block swings toward the outside of the curve. Then the string makes an angle θ with the vertical. Find θ.

WORK AND ENERGY

For their size, leafcutter ants can carry surprisingly heavy loads uphill and do relatively large amounts of work.

Figure 6.1 Work is done when a force **F** pushes a car through a displacement **s**.

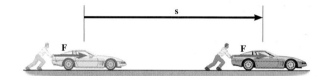

6.1 WORK DONE BY A CONSTANT FORCE

DEFINITION OF WORK

Work is a familiar concept. For example, it takes work to push a heavy object such as a stalled car. In fact, more work is done when the pushing force is greater or when the displacement of the car is greater. Force and displacement, then, are the two essential elements of work, as Figure 6.1 illustrates. The drawing shows a constant pushing force **F** that points in the same direction as the resulting displacement **s** of the car. In such a case, the work W is defined as the magnitude F of the force times the magnitude s of the displacement: $W = Fs$. The work done to push a car is the same whether the car is moved north to south or east to west, provided that the amount of force used and the distance moved are the same. Work does not convey directional information and, therefore, is a scalar quantity.

The equation $W = Fs$ indicates that the unit of work is the unit of force times the unit of distance, or the newton · meter in SI units. One newton · meter is referred to as a *joule* (J) (rhymes with "cool"), in honor of James Joule (1818–1889) and his research into the nature of work, energy, and heat. Table 6.1 summarizes the units for work in several systems of measurement.

The definition of work as $W = Fs$ does have one surprising feature. If the distance s is zero, the work is zero, even if a force is applied. Pushing on an immovable object, such as a brick wall, may tire your muscles, but there is no work done of the type we are discussing. In physics, the idea of work is intimately tied up with the idea of motion. If there is no movement of the object, the work done by the force acting on the object is zero.

Often, the force and displacement do not point in the same direction. For instance, Figure 6.2a shows a luggage carrier being pulled to the right by a force that is applied along the handle. The force is directed at an angle θ relative to the displacement. In such a case, only the component of the force along the displacement is used in defining work. As part b of the drawing illustrates, this component is $F \cos \theta$, and it appears in the general definition of work as follows:

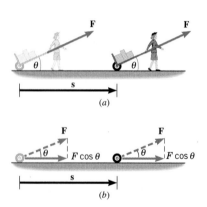

Figure 6.2 (a) Work can be done by a force **F** that points at an angle θ relative to the displacement **s**. (b) The force component that points along the displacement is $F \cos \theta$.

■ **DEFINITION OF WORK DONE BY A CONSTANT FORCE**

The work done on an object by a constant* force **F** is

$$W = (F \cos \theta)s \qquad (6.1)$$

where F is the magnitude of the force, s is the magnitude of the displacement, and θ is the angle between the force and the displacement.

SI Unit of Work: newton · meter = joule (J)

---

* Section 6.9 discusses the work done by a variable force.

Table 6.1   Units of Measurement for Work

System	Force	×	Distance	=	Work
SI	newton (N)		meter (m)		joule (J)
CGS	dyne (dyn)		centimeter (cm)		erg
BE	pound (lb)		foot (ft)		foot · pound (ft · lb)

When the force points in the same direction as the displacement, then $\theta = 0°$, and Equation 6.1 reduces to $W = Fs$. The following example illustrates how Equation 6.1 is used to calculate work.

EXAMPLE 1 • Pulling a Luggage Carrier

Find the work done by a 45.0-N force in pulling the luggage carrier in Figure 6.2*a* at an angle $\theta = 50.0°$ for a distance $s = 75.0$ m.

Reasoning and Solution   According to Equation 6.1, the work done on the luggage carrier by the 45.0-N force is

$$W = (F \cos \theta)s = [(45.0 \text{ N}) \cos 50.0°](75.0 \text{ m}) = \boxed{2170 \text{ J}}$$

The answer is expressed in newton · meters or joules (J).

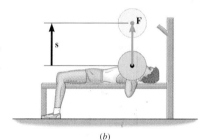
(*a*)

The definition of work in Equation 6.1 takes into account only the component of the force in the direction of the displacement. The force component perpendicular to the displacement does no work. To do work, there must be a force *and* a displacement, and since there is no displacement in the perpendicular direction, there is no work done by the perpendicular component of the force.

(*b*)

POSITIVE AND NEGATIVE WORK

Work can be either positive or negative, depending on whether a component of the force points in the same direction as the displacement or in the opposite direction. Example 2 illustrates how positive and negative work arise.

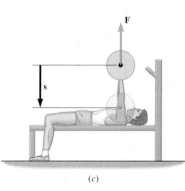
(*c*)

EXAMPLE 2 • Bench-pressing

The weight lifter in Figure 6.3*a* is bench-pressing a barbell whose weight is 710 N. In part *b* of the figure, he raises the barbell a distance of 0.65 m above his chest, and in part *c* he lowers it the same distance. The weight is raised and lowered at a constant velocity. Determine the work done on the barbell by the weight lifter during (a) the lifting phase and (b) the lowering phase.

Reasoning   To calculate the work, it is necessary to know the force exerted by the weight lifter. The barbell is raised and lowered at a constant velocity and, therefore, is in equilibrium. Consequently, the force **F** exerted by the weight lifter must balance the weight of the barbell, so $F = 710$ N. During the lifting phase, the force **F** and displacement **s** are in the same direction, as Figure 6.3*b* shows. The angle between

Figure 6.3   (*a*) In the bench press, work is done during both the lifting and lowering phases of the barbell's motion. (*b*) During the lifting phase, the force **F** does positive work on the barbell. (*c*) During the lowering phase, the force does negative work.

them is $\theta = 0°$. When the barbell is lowered, however, the force and displacement are in opposite directions, as part c of the drawing indicates. The angle between the force and the displacement is now $\theta = 180°$. With these observations, we may find the work.

Solution

(a) During the lifting phase, the work done by the force **F** is given by Equation 6.1 as

$$W = (F \cos \theta)s = [(710 \text{ N}) \cos 0°](0.65 \text{ m}) = \boxed{460 \text{ J}}$$

(b) The work done during the lowering phase is

$$W = (F \cos \theta)s = [(710 \text{ N}) \cos 180°](0.65 \text{ m}) = \boxed{-460 \text{ J}}$$

since $\cos 180° = -1$. The work is negative, because the force is opposite to the displacement. Weight lifters call each complete up-and-down movement of the barbell a repetition, or "rep." The lifting of the weight is referred to as the positive part of the rep, and the lowering is known as the negative part of the rep.

The Physics of...
positive and negative "reps" in
weight lifting.

As Example 2 illustrates, work can be positive or negative. If the force has a component in the *same* direction as the displacement of the object, the work done by the force is *positive*. On the other hand, if a force component points in the direction *opposite* to the displacement, the work is *negative*. If the force is perpendicular to the displacement, the force has no component in the direction of the displacement, and the work is zero ($\theta = 90°$ in Equation 6.1). The fact that work can be positive, negative, or zero is explored further in Conceptual Example 3.

CONCEPTUAL EXAMPLE 3 · Bicycling and Work

Figure 6.4 shows a cyclist coasting down one hill and up another, under the influence of two forces: the weight **W** of the bicycle and rider and the normal force **F$_N$** exerted on the tires by the ground. For the downward and upward parts of the motion, determine whether each of these forces does positive or negative work, or no work at all.

Reasoning and Solution Along with the forces, Figure 6.4 shows the displacement **s** of the bicycle for each part of the motion. Notice that the normal force **F$_N$** is always perpendicular to the displacement. Thus, the normal force does no work ($\theta = 90°$ in Equation 6.1), regardless of whether the cyclist is going downhill or uphill. In contrast, the weight does work in both parts of the motion. When the bicycle is heading downhill in Figure 6.4a, a component of the weight points in the direction of the displacement (θ is less than $90°$ and $\cos \theta$ is a positive number in Equation 6.1). As a result, the weight does positive work during the downhill ride. In part b of the drawing, the cyclist is headed uphill, and once again the weight has a component along the displacement. Now, however, this component points opposite to the displacement (θ is greater than $90°$ and $\cos \theta$ is a negative number in Equation 6.1). Therefore, the weight does negative work on the uphill ride. In summary, ***the weight of an object can do positive or negative work, depending on how the weight is oriented relative to the displacement.***

Related Homework Material: *Problems 7 and 8*

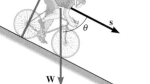

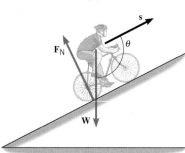

Figure 6.4 The drawing shows the weight **W**, the normal force **F$_N$**, and the displacement **s** when the bicyclist travels (*a*) downhill and (*b*) uphill.

The next example deals with the work done by a static frictional force when it acts on a crate that is resting on the bed of an accelerating truck.

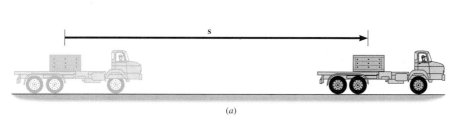

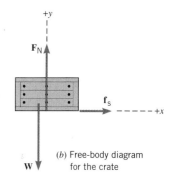

Figure 6.5 (*a*) The truck and crate are accelerating to the right for a distance of $s = 65$ m.
(*b*) The free-body diagram for the crate.

(*b*) Free-body diagram
for the crate

EXAMPLE 4 • Accelerating a Crate

Figure 6.5*a* shows a 120-kg crate on the flatbed of a truck that is moving with an acceleration of $a = +1.5$ m/s² along the positive *x* axis. The crate does not slip with respect to the truck, as the truck undergoes a displacement of $s = 65$ m. What is the total work done on the crate by all of the forces acting on it?

Reasoning The free-body diagram in Figure 6.5*b* shows the forces that act on the crate. These are (1) the weight **W** of the crate, (2) the normal force $\mathbf{F}_N$ exerted by the flatbed, and (3) the static frictional force $\mathbf{f}_s$, which is exerted by the flatbed in the forward direction and keeps the crate from slipping backward. The weight and the normal force are directed perpendicular to the displacement, so they do no work. Only the static frictional force does work, since it acts on the crate in the *x* direction. To determine the frictional force, we note that the crate does not slip relative to the truck, and, therefore, must have the same acceleration of $a = +1.5$ m/s² as does the truck. The force creating this acceleration is the static frictional force, and knowing the mass of the crate and its acceleration, we can use Newton's second law to obtain its magnitude. Then, with values for the frictional force and the displacement, we can determine the total work done on the crate.

Solution From Newton's second law, we find that the magnitude f_s of the static frictional force is

$$f_s = ma = (120 \text{ kg})(1.5 \text{ m/s}^2) = 180 \text{ N}$$

The total work is that done by the static frictional force and is

$$W = (f_s \cos \theta)s = (180 \text{ N})(\cos 0°)(65 \text{ m}) = \boxed{1.2 \times 10^4 \text{ J}} \qquad (6.1)$$

The work is positive, because the frictional force is in the same direction as the displacement ($\theta = 0°$).

6.2 THE WORK–ENERGY THEOREM AND KINETIC ENERGY

Most people expect that if you do work, you should get something as a result. In physics, when a net force performs work on an object, there is always a result from the effort. The result is a change in the ***kinetic energy*** of the object. The important relationship that relates work to the change in kinetic energy is known as the ***work–energy theorem.*** As the concept chart in Figure 6.6 outlines, we will obtain

Concepts at a Glance

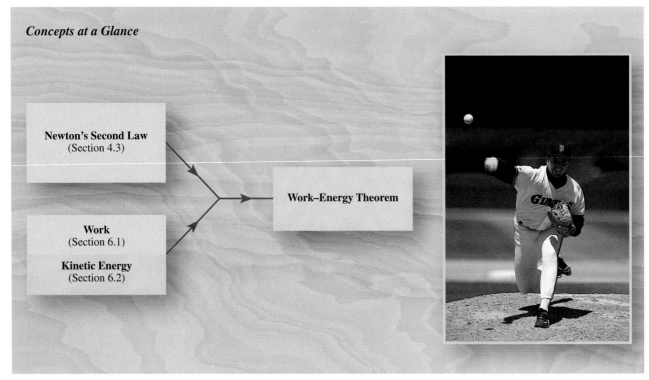

Figure 6.6 *Concepts at a Glance*
In this section, the familiar concept of work and the new concept of kinetic energy will be integrated with Newton's second law of motion to produce the work–energy theorem. In throwing a baseball, pitcher John Burkett does work on it. This work causes the ball's kinetic energy to increase in accordance with the work–energy theorem.

this theorem by combining Newton's second law of motion with the now-familiar concept of work and a new idea called kinetic energy.

To gain some insight into the idea of kinetic energy and the work–energy theorem, look at Figure 6.7, where a constant net external force F is acting on an airplane of mass m. This net force is the vector sum of all the external forces acting on the plane, and, for simplicity, its direction is assumed to be the same as the direction of the displacement **s**. According to Newton's second law, the net force produces an acceleration a, given by $a = F/m$. Consequently, the speed of the plane changes from an initial value of v_0 to a final value of v_f. Multiplying both sides of $F = ma$ by the distance s gives

$$Fs = mas$$

The term as on the right side can be related to v_0 and v_f by using Equation 2.9

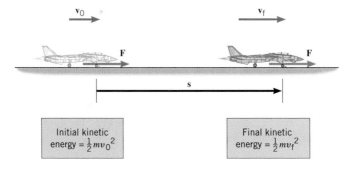

Figure 6.7 A constant net external force **F** acts over a displacement **s** and does work on the plane. As a result of the work done, the plane's kinetic energy changes.

$(v_f^2 = v_0^2 + 2as)$, which is one of the equations of kinematics. Solving this equation to give $as = \frac{1}{2}(v_f^2 - v_0^2)$ and substituting this result into $Fs = mas$ shows that

$$\underbrace{Fs}_{\text{Work}} = \underbrace{\tfrac{1}{2}mv_f^2}_{\substack{\text{Final} \\ \text{kinetic} \\ \text{energy}}} - \underbrace{\tfrac{1}{2}mv_0^2}_{\substack{\text{Initial} \\ \text{kinetic} \\ \text{energy}}}$$

This expression is the work–energy theorem. Its left side is the work W done by the net external force, while its right side involves the difference between two terms, each of which has the form $\frac{1}{2}(\text{mass})(\text{speed})^2$. The quantity $\frac{1}{2}(\text{mass})(\text{speed})^2$ is called kinetic energy and plays a significant role in physics, as we will soon see.

■ **DEFINITION OF KINETIC ENERGY**

The kinetic energy KE of an object with mass m and speed v is given by

$$\text{KE} = \tfrac{1}{2}mv^2 \tag{6.2}$$

SI Unit of Kinetic Energy: joule (J)

The SI unit of kinetic energy is the same as the unit for work, the joule. Kinetic energy, like work, is a scalar quantity. These are not surprising observations, for work and kinetic energy are closely related, as is clear from the following statement of the work–energy theorem.

■ **THE WORK–ENERGY THEOREM**

When a net external force does work W on an object, the kinetic energy of the object changes from its initial value of KE_0 to a final value of KE_f, the difference between the two values being equal to the work:

$$W = \text{KE}_f - \text{KE}_0 = \tfrac{1}{2}mv_f^2 - \tfrac{1}{2}mv_0^2 \tag{6.3}$$

The work–energy theorem may be derived for any direction of the force relative to the displacement, not just the situation in Figure 6.7. In fact, the force may even vary from point to point along a path that is curved rather than straight, and the theorem remains valid. According to the work–energy theorem, a moving object has kinetic energy, because work was done to accelerate the object from rest to a speed v.* Conversely, an object with kinetic energy can perform work, if it is allowed to push or pull on another object.

Example 5 illustrates the work–energy theorem and considers a single force that does work to change the kinetic energy of a space probe.

---

* Strictly speaking, the work–energy theorem, as given by Equation 6.3, applies only to a single particle, which occupies a mathematical point in space. A macroscopic object, however, is a collection or system of particles and is spread out over a region of space. Therefore, when a force is applied to a macroscopic object, the point of application of the force may be anywhere on the object. To take into account this and other factors, a discussion of work and energy is required that is beyond the scope of this text. The interested reader may refer to A. B. Arons, *The Physics Teacher,* October 1989, p. 506.

EXAMPLE 5 • A Space Probe

A space probe of mass $m = 5.00 \times 10^4$ kg is traveling at a speed of $v_0 = 1.10 \times 10^4$ m/s through deep space. No forces act on the probe except that generated by its own engine. The engine exerts a constant external force $\mathbf{F}$ of 4.00×10^5 N, directed parallel to the displacement (Figure 6.8). The engine fires continually while the probe moves in a straight line for a displacement $\mathbf{s}$ of 2.50×10^6 m. Determine the final speed of the probe.

Reasoning Since the force $\mathbf{F}$ is the only force acting on the probe, it is the net external force, and the work it does causes the kinetic energy of the spacecraft to change. The work can be calculated directly from Equation 6.1 and is positive, because $\mathbf{F}$ points in the same direction as the displacement $\mathbf{s}$. According to the work–energy theorem ($W = \mathrm{KE}_f - \mathrm{KE}_0$), a positive value for W means that the kinetic energy increases. Thus, we can use the work–energy theorem to find the final kinetic energy and, hence, the final speed of the spacecraft. Since the kinetic energy increases, we expect the final speed to be greater than the initial speed.

Solution From the definition of work (Equation 6.1), we have

$$W = (F \cos \theta)s = [(4.00 \times 10^5 \text{ N}) \cos 0°](2.50 \times 10^6 \text{ m}) = 1.00 \times 10^{12} \text{ J}$$

where $\theta = 0°$, because the force and displacement are in the same direction (see the drawing). Since $W = \mathrm{KE}_f - \mathrm{KE}_0$ according to the work–energy theorem, the final kinetic energy of the probe is

$$\mathrm{KE}_f = W + \mathrm{KE}_0$$
$$= (1.00 \times 10^{12} \text{ J}) + \tfrac{1}{2}(5.00 \times 10^4 \text{ kg})(1.10 \times 10^4 \text{ m/s})^2 = 4.03 \times 10^{12} \text{ J}$$

The final kinetic energy is $\mathrm{KE}_f = \tfrac{1}{2}mv_f^2$, so the final speed is

$$v_f = \sqrt{\frac{2(\mathrm{KE}_f)}{m}} = \sqrt{\frac{2(4.03 \times 10^{12} \text{ J})}{(5.00 \times 10^4 \text{ kg})}} = \boxed{1.27 \times 10^4 \text{ m/s}}$$

As expected, the final speed of the probe is greater than the initial speed.

In Example 5 only the force of the engine does work on the spacecraft. If several external forces act on an object, they must be added together vectorially to give the net force. The work done by the net external force can then be related to the change in the object's kinetic energy by using the work–energy theorem, as in the next example.

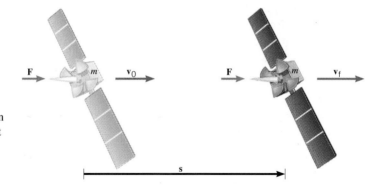

Figure 6.8 The engine of the space probe generates a force $\mathbf{F}$ that points in the same direction as the displacement $\mathbf{s}$. The force performs positive work, causing the probe to gain kinetic energy.

EXAMPLE 6 • Downhill Skiing

A 58-kg skier is coasting down a 25° slope, as Figure 6.9*a* shows. A kinetic frictional force of magnitude $f_k = 70$ N opposes her motion. Near the top of the slope, the skier's speed is $v_0 = 3.6$ m/s. Ignoring air resistance, determine the speed v_f at a point that is displaced 57 m downhill.

Reasoning As in Example 5, we will use the work–energy theorem to find the final speed. In using this theorem, we note that the work done by the net external force is needed, not just the work done by any single force. The net force is the vector sum of the forces shown in the free-body diagram in Figure 6.9*b*. In this diagram, the normal force $\mathbf{F_N}$ is balanced by the component of the skier's weight ($mg \cos 25°$) perpendicular to the slope, because no acceleration occurs along that direction. Therefore, the net force is in the direction of the displacement and does positive work. According to the work–energy theorem, the kinetic energy increases when the net force does positive work, so we expect the final speed to be greater than the initial speed.

Solution The net external force in Figure 6.9*b* points along the $+x$ axis and is

$$F = mg \sin 25° - f_k = (58 \text{ kg})(9.80 \text{ m/s}^2) \sin 25° - 70 \text{ N} = +170 \text{ N}$$

The work done by the net force is

$$W = (F \cos \theta)s = [(170 \text{ N}) \cos 0°](57 \text{ m}) = 9700 \text{ J} \qquad (6.1)$$

where $\theta = 0°$ because the net force and the displacement point in the same direction. From the work–energy theorem ($W = \text{KE}_f - \text{KE}_0$), it follows that the final kinetic energy of the skier is

$$\text{KE}_f = W + \text{KE}_0 = 9700 \text{ J} + \tfrac{1}{2}(58 \text{ kg})(3.6 \text{ m/s})^2 = 10\,100 \text{ J}$$

Since the final kinetic energy is $\text{KE}_f = \tfrac{1}{2}mv_f^2$, the final speed of the skier is

$$v_f = \sqrt{\frac{2(\text{KE}_f)}{m}} = \sqrt{\frac{2(10\,100 \text{ J})}{58 \text{ kg}}} = \boxed{19 \text{ m/s}}$$

which is greater than the initial speed, as expected.

Example 6 emphasizes that ***the work–energy theorem deals with the work done by the net external force. The work–energy theorem does not apply to the work done by an individual force,*** unless that force happens to be the only one present, in which case it is the net force. If the work done by the net force is *positive,* as in

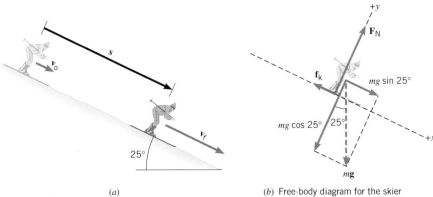

(*a*)

(*b*) Free-body diagram for the skier

Figure 6.9 (*a*) A skier coasting downhill. (*b*) The free-body diagram for the skier.

Example 6, the kinetic energy of the object *increases*. If the work done is *negative*, the kinetic energy *decreases*. If the work is zero, the kinetic energy, and hence the speed, of the object remain the same. Conceptual Example 7 explores these ideas further.

CONCEPTUAL EXAMPLE 7 • Work and Kinetic Energy

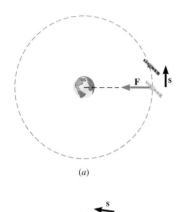

(a)

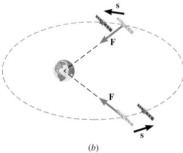

(b)

Figure 6.10 (*a*) In a circular orbit, the gravitational force **F** is always perpendicular to the displacement **s** of the satellite and does no work. (*b*) In an elliptical orbit, there can be a component of the force along the displacement, and, consequently, work can be done.

Figure 6.10 illustrates a satellite moving about the earth in a circular orbit and in an elliptical orbit. The only external force that acts on the satellite is the gravitational force. For these two orbits, determine whether the kinetic energy of the satellite changes during the motion.

Reasoning and Solution For the circular orbit in Figure 6.10*a* the gravitational force **F** does no work on the satellite, since the force is perpendicular to the instantaneous displacement **s** at all times. The gravitational force is the only external force acting on the satellite, so it is the net force. Thus, the work done by the net force is zero, and according to the work–energy theorem (Equation 6.3), the kinetic energy of the satellite (and, hence its speed) remains the same everywhere on the orbit.

In contrast to the case of a circular orbit, the gravitational force does do work on a satellite in an elliptical orbit. For example, as the satellite moves toward the earth in the top part of Figure 6.10*b*, there is a component of the gravitational force that points in the same direction as the displacement. (This situation is like that in Figure 6.4*a* where the bicycle is coasting downhill.) Consequently, the gravitational force does positive work on the satellite during this part of the orbit, and the kinetic energy of the satellite increases. When the satellite moves away from the earth, as in the lower part of Figure 6.10*b*, the gravitational force has a component that points opposite to the displacement. (This situation is similar to the bicycle coasting uphill in Figure 6.4*b*.) Now, the gravitational force does negative work, and the kinetic energy of the satellite decreases.

Can you identify the two places on an elliptical orbit where the gravitational force does no work?

Related Homework Material: *Problem 16*

6.3 GRAVITATIONAL POTENTIAL ENERGY

WORK DONE BY THE FORCE OF GRAVITY

The gravitational force is a well-known force that can do positive or negative work, and Figure 6.11 helps to show how the work can be determined. This drawing depicts a basketball of mass m moving vertically downward, the force of gravity mg being the only force acting on the ball. The initial height of the ball is h_0, and the final height is h_f, both distances measured from the earth's surface. The displacement **s** is downward and has a magnitude of $s = h_0 - h_f$. To calculate the work W_{gravity} done on the ball by the force of gravity, we use $W = (F \cos \theta)s$ with $F = mg$ and $\theta = 0°$, since the force and displacement are in the same direction:

$$W_{\text{gravity}} = (mg \cos 0°)(h_0 - h_f) = mg(h_0 - h_f) \qquad (6.4)$$

Equation 6.4 is valid for *any path* taken between the initial and final heights, and

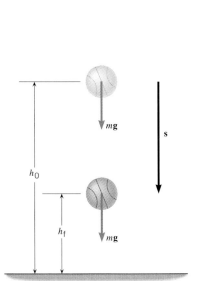

Figure 6.11 Gravity exerts a force $m\mathbf{g}$ on the basketball. Work is done by the gravitational force as the basketball falls from a height of h_0 to a height of h_f.

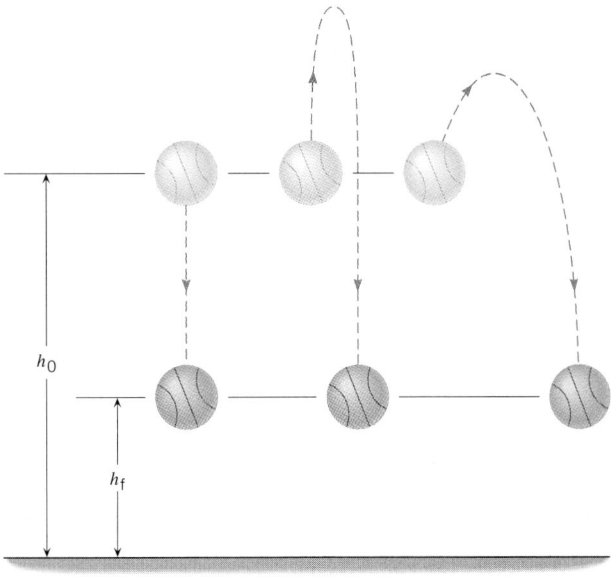

Figure 6.12 An object can move along different paths in going from an initial height of h_0 to a final height of h_f. In each case, the work done by the gravitational force is the same $[W_{\text{gravity}} = mg(h_0 - h_f)]$, since the change in vertical distance $(h_0 - h_f)$ is the same.

not just for the straight-down path shown in Figure 6.11. For example, the same expression can be derived for the three paths shown in Figure 6.12. Thus, only the *difference in vertical distances* $(h_0 - h_f)$ need be considered when calculating the work done by gravity. Since the difference in the vertical distances is the same for each path in the drawing, the work done by gravity is the same in each case. We are assuming here that the difference in heights is small compared to the radius of the earth, so that the magnitude g of the acceleration due to gravity is the same at every height. Moreover, for positions close to the earth's surface, we can use the value of $g = 9.80 \text{ m/s}^2$.

Since only the difference between h_0 and h_f appears in Equation 6.4, the vertical distances themselves need not be measured from the earth. For instance, they could be measured relative to a zero level that is one meter above the ground, and $h_0 - h_f$ would still have the same value. Example 8 illustrates how the work done by gravity is used in conjunction with the work–energy theorem.

EXAMPLE 8 • A Gymnast on a Trampoline

A gymnast springs vertically upward from a trampoline as in Figure 6.13*a*. The gymnast leaves the trampoline at a height of 1.20 m and reaches a maximum height of 4.80 m before falling back down. All heights are measured with respect to the ground. Ignoring air resistance, determine (a) the initial speed v_0 with which the gymnast leaves the trampoline and (b) the speed of the gymnast after falling back to a height of 3.50 m.

(a)

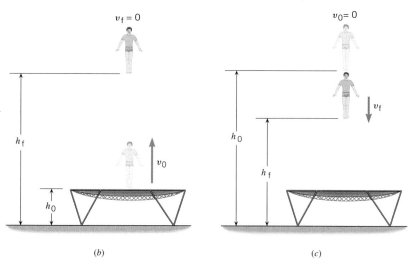

(b)   (c)

Figure 6.13 (a) A gymnast bounces on a trampoline. (b) The gymnast moves upward with an initial speed v_0 and reaches maximum height with a final speed of zero. (c) Starting at the top with an initial speed of zero, the gymnast begins to fall. In parts b and c, the values for h_0 are different, and so are the values for h_f.

Reasoning We can find the speed of the gymnast (mass = m) by using the work–energy theorem, provided the work done by the net external force can be determined. Since only the gravitational force acts on the gymnast in the air, the gravitational force is the net force, and we can evaluate the work by using the relation $W_{\text{gravity}} = mg(h_0 - h_f)$.

Solution

(a) Figure 6.13b shows the gymnast moving upward. The initial and final heights are $h_0 = 1.20$ m and $h_f = 4.80$ m, respectively. The initial speed v_0 follows from the work–energy theorem ($W = W_{\text{gravity}} = KE_f - KE_0$). Since the final speed and kinetic energy are zero at the highest point ($KE_f = 0$), this theorem reduces to $W_{\text{gravity}} = -KE_0 = -\frac{1}{2}mv_0^2$. Solving for the initial speed and substituting $W_{\text{gravity}} = mg(h_0 - h_f)$ gives

$$v_0 = \sqrt{\frac{-2W_{\text{gravity}}}{m}} = \sqrt{\frac{-2mg(h_0 - h_f)}{m}} = \sqrt{-2g(h_0 - h_f)}$$

$$v_0 = \sqrt{-2(9.80 \text{ m/s}^2)(1.20 \text{ m} - 4.80 \text{ m})} = \boxed{8.40 \text{ m/s}}$$

(b) Part c of the drawing shows the gymnast on the way down. The calculation here proceeds in the same fashion as in part (a), except now the initial position is at the top of the flight path, so $h_0 = 4.80$ m and $h_f = 3.50$ m. Since the initial speed and kinetic energy are zero at the top of the flight path ($KE_0 = 0$), the work–energy theorem indicates that $W_{\text{gravity}} = KE_f - KE_0 = KE_f = \frac{1}{2}mv_f^2$. Thus, the final speed at a height of 3.50 m is

$$v_f = \sqrt{\frac{2W_{\text{gravity}}}{m}} = \sqrt{\frac{2mg(h_0 - h_f)}{m}} = \sqrt{2g(h_0 - h_f)}$$

$$v_f = \sqrt{2(9.80 \text{ m/s}^2)(4.80 \text{ m} - 3.50 \text{ m})} = \boxed{5.05 \text{ m/s}}$$

GRAVITATIONAL POTENTIAL ENERGY

We have seen that an object in motion has kinetic energy. Energy also occurs in other forms. For example, an object may possess energy by virtue of its position relative to the earth; such an object is said to have gravitational potential energy. A pile driver, for instance, is used by construction workers to pound "piles" or structural support beams into the ground. The pile driver contains a massive hammer that is raised to a height h above the ground and then dropped (see Figure 6.14). As a result, the hammer has the potential to do the work of driving the pile into the ground. The greater the height of the hammer, the greater is the potential for doing work, and the greater is the gravitational potential energy.

Now, let's obtain an expression for the gravitational potential energy of an object at a given height above the ground. Our starting point is Equation 6.4 for the work done by the gravitational force as an object moves from an initial height h_0 to a final height h_f:

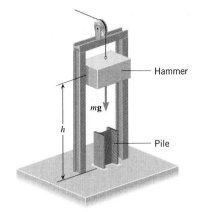

Figure 6.14 In a pile driver, the gravitational potential energy of the hammer relative to the ground is PE = mgh.

$$W_{gravity} = \underbrace{mgh_0}_{\substack{\text{Initial} \\ \text{gravitational} \\ \text{potential energy,} \\ \text{PE}_0}} - \underbrace{mgh_f}_{\substack{\text{Final} \\ \text{gravitational} \\ \text{potential energy,} \\ \text{PE}_f}} \qquad (6.4)$$

This equation indicates that the work done by the gravitational force is equal to the difference between the initial and final values of the quantity mgh. The value of mgh is larger when the height is larger and smaller when the height is smaller. We are led, then, to identify the quantity mgh as the ***gravitational potential energy.*** The concept of potential energy is associated only with a type of force known as a "conservative" force. We will discuss this type of force in Section 6.4.

■ DEFINITION OF GRAVITATIONAL POTENTIAL ENERGY

The gravitational potential energy PE is the energy that an object of mass m has by virtue of its position relative to the surface of the earth. That position is measured by the height h of the object relative to an arbitrary zero level:

$$PE = mgh \qquad (6.5)$$

SI Unit of Gravitational Potential Energy: joule (J)

Gravitational potential energy, like work and kinetic energy, is a scalar quantity, and has the same SI unit as they do, the joule. It is the *difference* between two potential energies that is related by Equation 6.4 to the work done by the force of gravity. Therefore, the zero level for the heights can be taken anywhere, as long as both h_0 and h_f are measured relative to the same zero level. The gravitational potential energy depends on both the object and the earth (m and g, respectively), as well as the height h. Therefore, the gravitational potential energy belongs to the object and the earth as a system, although one often speaks of the object alone as possessing the gravitational potential energy.

6.4 CONSERVATIVE FORCES AND NONCONSERVATIVE FORCES

CONSERVATIVE FORCES

The gravitational force has an interesting property that when an object is moved from one place to another, the work done by the gravitational force does not depend on the choice of path. In Figure 6.12, for instance, an object moves from an initial height h_0 to a final height h_f along several different paths. As Section 6.3 discusses, the work done by gravity depends only on the initial and final heights, and not on the path between these heights. For this reason, the gravitational force is called a *conservative force,* according to version 1 of the following definition:

> ■ **DEFINITION OF A CONSERVATIVE FORCE**
>
> **Version 1**   A force is conservative when the work it does on a moving object is independent of the path of the motion between the object's initial and final positions.
>
> **Version 2**   A force is conservative when it does no net work on an object moving around a closed path, starting and finishing at the same point.

The gravitational force is our first example of a conservative force. Later, we will encounter other examples, such as the elastic force of a spring and the electrical force of electrically charged particles. With each conservative force we will associate a potential energy, as we have already done in the gravitational case (see Equation 6.5). For other conservative forces, however, the algebraic form of the potential energy will differ from that in Equation 6.5.

Version 1 of the definition given above is not the only way to describe a conservative force. Figure 6.15 helps us to illustrate another way, which leads to version 2 of the definition. The picture shows a roller coaster car racing through dips and double dips, ultimately returning to its starting point. This kind of path, which begins and ends at the same place, is called a *closed* path. Gravity provides the only force that does work on the car, assuming that there is no friction or air resistance. Of course, the track exerts a normal force, but this force is always directed perpendicular to the motion and, hence, does no work. On the downward parts of the trip, the gravitational force does positive work, increasing the car's kinetic energy. Conversely, on the upward parts of the motion, the gravitational force does negative work, decreasing the car's kinetic energy. Over the entire trip, the gravitational force does as much positive work as negative work, so the net work is zero, and the car returns to its starting point with the same kinetic energy it had at the start. Therefore, consistent with version 2 of the definition of a conservative force, $W_{\text{gravity}} = 0$ for a closed path.

Start

Figure 6.15   A roller coaster track is an example of a closed path.

NONCONSERVATIVE FORCES

Not all forces are conservative forces. A force is nonconservative if the work it does on an object moving between two points depends on the path of the motion between the points. The kinetic frictional force is one example of a nonconservative force. When an object slides over a surface, the kinetic frictional force points opposite to the sliding motion and does negative work. Between any two points, greater amounts of work are done over longer paths between the points, so that the work depends on the choice of path. Thus, the kinetic frictional force is nonconservative. Air resistance is another nonconservative force. The concept of potential energy is not defined for a nonconservative force.

For a closed path, the total work done by a nonconservative force is not zero as it is for a conservative force. In Figure 6.15, for instance, a frictional force would oppose the motion and slow down the car. Unlike gravity, friction would do negative work on the car throughout the entire trip, on *both* the up and down parts of the motion. Assuming that the car makes it back to the starting point, the car would have *less* kinetic energy than it had originally. Table 6.2 gives some examples of conservative and nonconservative forces.

Table 6.2 Examples of Conservative and Nonconservative Forces
Conservative Forces
Gravitational force (Ch. 4)
Elastic spring force (Ch. 10)
Electric force (Ch. 18, 19)
Nonconservative Forces
Static and kinetic frictional forces
Air resistance
Tension
Normal force
Propulsion force of a rocket or a motor

THE WORK–ENERGY THEOREM

In most everyday situations, both conservative forces (such as gravity) and nonconservative forces (such as friction and air resistance) act simultaneously on an object. Therefore, we write the work W done by the net external force as $W = W_c + W_{nc}$, where W_c is the work done by the conservative forces and W_{nc} is the work done by the nonconservative forces. According to the work–energy theorem, the work done by the net external force is equal to the change in the object's kinetic energy, or $W_c + W_{nc} = \frac{1}{2}mv_f^2 - \frac{1}{2}mv_0^2$. If the only conservative force acting is the gravitational force, then $W_c = W_{gravity} = mg(h_0 - h_f)$, and the work–energy theorem becomes

$$mg(h_0 - h_f) + W_{nc} = \tfrac{1}{2}mv_f^2 - \tfrac{1}{2}mv_0^2$$

The work done by the gravitational force can be moved to the right side of this equation, with the result that

$$W_{nc} = (\tfrac{1}{2}mv_f^2 - \tfrac{1}{2}mv_0^2) + (mgh_f - mgh_0) \qquad (6.6)$$

In terms of kinetic and potential energies, we find that

$$W_{nc} = \underbrace{(KE_f - KE_0)}_{\substack{\text{Change in} \\ \text{kinetic energy}}} + \underbrace{(PE_f - PE_0)}_{\substack{\text{Change in} \\ \text{gravitational} \\ \text{potential energy}}} \qquad (6.7a)$$

Equation 6.7a states that the net work done by all the external nonconservative forces equals the change in the object's kinetic energy plus the change in its gravitational potential energy. It is customary to use the delta symbol (Δ) to denote such changes; thus, $\Delta KE = (KE_f - KE_0)$ and $\Delta PE = (PE_f - PE_0)$. With the delta notation, the work–energy theorem takes the form

$$W_{nc} = \Delta KE + \Delta PE \qquad (6.7b)$$

In the next two sections, we will show why the form of the work–energy theorem expressed by Equations 6.7a and 6.7b is useful.

• **PROBLEM SOLVING INSIGHT**
The change in both the kinetic and the potential energy is always the final value minus the initial value: $\Delta KE = KE_f - KE_0$ and $\Delta PE = PE_f - PE_0$.

Concepts at a Glance

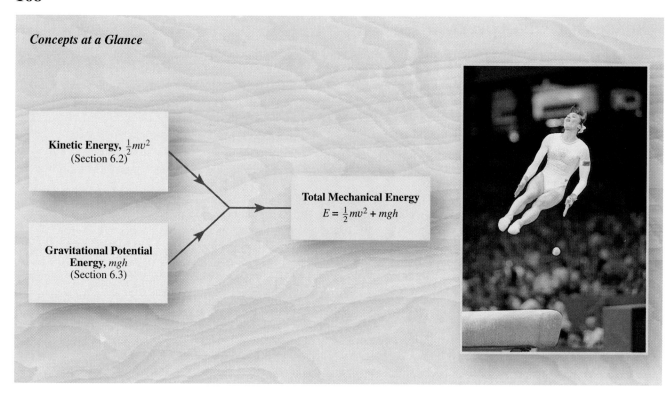

Figure 6.16 *Concepts at a Glance* The total mechanical energy E is formed by combining the concepts of kinetic energy and potential energy. Gymnast Svetlana Boguinskaia has both types of energy as she vaults through the air.

Concepts at a Glance

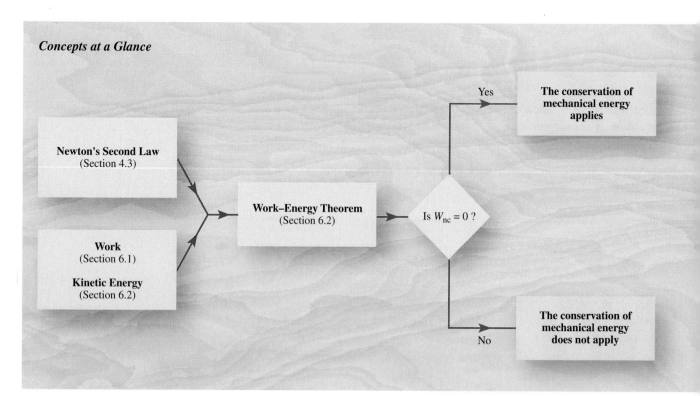

6.5 THE CONSERVATION OF MECHANICAL ENERGY

The work–energy theorem has led us to consider two types of energy: kinetic energy and potential energy. As the concept chart in Figure 6.16 illustrates, the sum of these two kinds of energy is called the ***total mechanical energy*** E, so that $E = KE + PE$. The idea of total mechanical energy will prove to be very useful in describing the motion of objects in this and other chapters.

The work–energy theorem of Equation 6.7a can be expressed in terms of the total mechanical energy as

$$W_{nc} = (KE_f - KE_0) + (PE_f - PE_0)$$
$$= \underbrace{(KE_f + PE_f)}_{E_f} - \underbrace{(KE_0 + PE_0)}_{E_0}$$

or

$$W_{nc} = E_f - E_0 \qquad (6.8)$$

Equation 6.8 states that W_{nc}, the net work done by external nonconservative forces, changes the total mechanical energy from an initial value of E_0 to a final value of E_f.

The conciseness of the work–energy theorem in the form $W_{nc} = E_f - E_0$ allows an important basic principle of physics to stand out. This principle is known as the conservation of mechanical energy, and the concept chart in Figure 6.17 outlines the logic that we will follow in presenting the principle. This chart is an expanded version of the chart in Figure 6.6. We will begin with the work–energy theorem as

Figure 6.17 *Concepts at a Glance*
The work–energy theorem leads to the principle of conservation of mechanical energy under circumstances in which $W_{nc} = 0$, where W_{nc} is the net work done by external nonconservative forces. The force of air resistance that acts on a parachute is a nonconservative force. Thus, the conservation of mechanical energy does not apply during the fall.

given in Equation 6.8 and ask "is the net work W_{nc} done by external nonconservative forces equal to zero?" If the answer to this question is no, then the conservation of mechanical energy does not apply, as the lower right-hand side of Figure 6.17 indicates. If the answer is yes, however, we will see that the conservation principle does apply, as the upper right-hand side of Figure 6.17 shows. In such a case, Equation 6.8 reduces to

$$E_f = E_0 \tag{6.9}$$

This result indicates that the final mechanical energy is equal to the initial mechanical energy. Consequently, the total mechanical energy *remains constant all along the path between the initial and final points,* never varying from the initial value of E_0. A quantity that stays constant throughout the motion is said to be "conserved." The fact that the total mechanical energy is conserved when $W_{nc} = 0$ is called the ***principle of conservation of mechanical energy.***

■ THE PRINCIPLE OF CONSERVATION OF MECHANICAL ENERGY

The total mechanical energy (E = KE + PE) of an object remains constant as the object moves, provided that the net work done by external nonconservative forces is zero.

The principle of conservation of mechanical energy offers keen insight into the way in which the physical universe operates. While the sum of the kinetic and potential energies at any point is conserved, the two forms may be interconverted or transformed into one another. Kinetic energy of motion is converted into potential energy of position, for instance, when a moving object coasts up a hill. Conversely, potential energy is converted into kinetic energy when an object above the earth's surface is allowed to fall. Figure 6.18 illustrates such transformations of energy for a bobsled run, assuming that nonconservative forces, such as friction and wind resistance, can be ignored. The normal force, being directed perpendicular to the path, does no work. Only the force of gravity does work, so the total mechanical energy E remains constant at all points along the run. The conservation principle is well known for the ease with which it can be applied, as in the following examples.

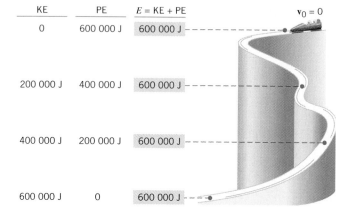

Figure 6.18 If friction and wind resistance are ignored, a bobsled run illustrates how kinetic and potential energy can be interconverted, while the total mechanical energy remains constant. The total mechanical energy is 600 000 J, being all potential energy at the top and all kinetic energy at the bottom.

KE	PE	E = KE + PE
0	600 000 J	600 000 J
200 000 J	400 000 J	600 000 J
400 000 J	200 000 J	600 000 J
600 000 J	0	600 000 J

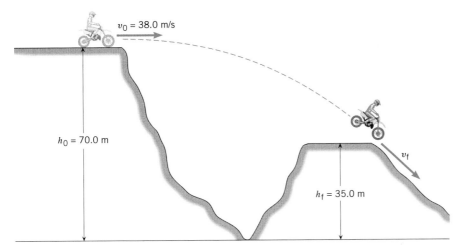

Figure 6.19 A daredevil jumping a canyon.

EXAMPLE 9 • A Daredevil Motorcyclist

A motorcyclist is trying to leap across the canyon shown in Figure 6.19 by driving horizontally off the cliff. When it leaves the cliff, the cycle has a speed of 38.0 m/s. Ignoring air resistance, find the speed with which the cycle strikes the ground on the other side.

Reasoning Once the cycle leaves the cliff, no forces other than gravity act on the cycle, since air resistance is being ignored. Thus, the work done by external nonconservative forces is zero, $W_{nc} = 0$. Accordingly, the principle of conservation of mechanical energy holds, so the total mechanical energy is the same at the final and initial positions of the motorcycle. We will use this important observation to determine the final speed of the cyclist.

Solution The principle of conservation of mechanical energy is written as

$$\underbrace{\tfrac{1}{2}mv_f^2 + mgh_f}_{E_f} = \underbrace{\tfrac{1}{2}mv_0^2 + mgh_0}_{E_0} \qquad (6.9)$$

The mass m of the rider and cycle can be eliminated algebraically from this equation, since m appears as a factor in every term. Solving for v_f gives

$$v_f = \sqrt{v_0^2 + 2g(h_0 - h_f)}$$

$$v_f = \sqrt{(38.0 \text{ m/s})^2 + 2(9.80 \text{ m/s}^2)(70.0 \text{ m} - 35.0 \text{ m})} = \boxed{46.2 \text{ m/s}}$$

• **PROBLEM SOLVING INSIGHT**
Be on the alert for factors, such as the mass m here in Example 9, that sometimes can be eliminated algebraically when using the conservation of mechanical energy.

Examples 10–12 emphasize that the principle of conservation of mechanical energy can be applied even when forces act perpendicular to the path of a moving object.

EXAMPLE 10 • The Favorite Swimming Hole

A 6.00-m rope is tied to a tree limb and used as a swing. A person starts from rest with the rope held in a horizontal orientation, as in Figure 6.20. Ignoring friction and air resistance, determine how fast the person is moving at the lowest point on the circular arc of the swing.

Figure 6.20 During the downward swing, the tension **T** in the rope acts perpendicular to the circular arc and, hence, does no work on the person. Therefore, the principle of conservation of mechanical energy applies, and the speed v_f can be determined.

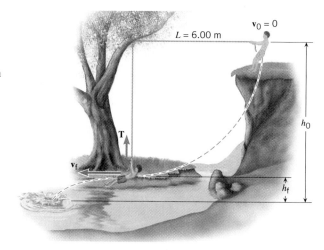

• **PROBLEM SOLVING INSIGHT**
When nonconservative forces are perpendicular to the motion, we can still use the principle of conservation of mechanical energy, because such "perpendicular" forces do no work.

Reasoning The principle of conservation of mechanical energy provides a straightforward way to find the speed of the person at the lowest point on the arc. But does the principle really apply here? After all, a nonconservative force does act on the person, namely, the tension **T** in the rope. The tension, however, does no work, because it points perpendicular to the circular path of the motion. Thus, $W_{nc} = 0$, and the conservation principle is applicable.

Solution From Equation 6.9 we have

$$\underbrace{\tfrac{1}{2}mv_f^2 + mgh_f}_{E_f} = \underbrace{\tfrac{1}{2}mv_0^2 + mgh_0}_{E_0} \qquad (6.9)$$

The mass m can be eliminated algebraically, and since $v_0 = 0$, it follows that $\tfrac{1}{2}v_f^2 + gh_f = gh_0$. Therefore,

$$v_f = \sqrt{2g(h_0 - h_f)} = \sqrt{2(9.80 \text{ m/s}^2)(6.00 \text{ m})} = \boxed{10.8 \text{ m/s}}$$

where we have used the fact that $h_0 - h_f = 6.00$ m (see the drawing).

Example 10 illustrates that the tension **T** in the rope plays no role in determining the final speed of the person, because the tension does no work. Conceptual Example 11 illustrates another interesting consequence of this fact.

CONCEPTUAL EXAMPLE 11 • The Favorite Swimming Hole, Revisited

Consider two possible paths by which the person in Figure 6.21 can reach the water. On path 1, the person lets go of the rope on the downward part of the swing. On path 2, the person swings upward before letting go. Ignoring air resistance, for which path is the speed upon entering the water greater?

Reasoning and Solution One could argue that the person strikes the water with the greater speed for path 1, since he is moving downward at the release point. Actually, however, the speed is the same for both paths. Here's why. In each case, the motion starts from rest at the top of the cliff, so the total mechanical energy is the same at the beginning of either path. There is no air resistance, and the work done by the tension in the rope is zero, as discussed in Example 10. Thus, the net work done by external nonconservative forces is zero, $W_{nc} = 0$, and the total mechanical energy is

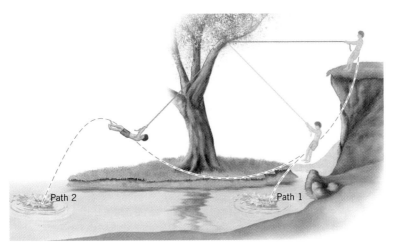

Figure 6.21 A person can follow path 1 or path 2 into the water, depending on where he lets go of the rope. Conceptual Example 11 discusses his speed upon entering the water via these two paths.

Path 2

Path 1

conserved, regardless of the path taken to the water. Since the total mechanical energy at the start is the same for each path, it must be the same at the end, when the person enters the water with only kinetic energy. Therefore, *the speed upon entering the water is the same no matter what path is followed.*

Related Homework Material: Problems 38 and 40

The next example illustrates how the conservation of mechanical energy is applied to the breathtaking drop of a roller coaster.

▶ **The Physics of...**
a giant roller coaster.

EXAMPLE 12 • The Magnum XL-200

One of the fastest roller coasters in the world is the Magnum XL-200 at Cedar Point Park in Sandusky, Ohio (Figure 6.22). The ride includes a vertical drop of 59.4 m. Assume that the coaster has a speed of nearly zero as it crests the top of the hill. Neglect friction and find the speed of the riders at the bottom of the hill.

Reasoning Since we are neglecting friction, we may set the work done by the frictional force equal to zero. A normal force acts on each rider, but this force is perpendicular to the motion, so it does not do any work. Thus, the work done by external nonconservative forces is zero, and we may use the principle of conservation of mechanical energy to find the speed of the riders at the bottom of the hill.

Solution The principle of conservation of mechanical energy states that

$$\underbrace{\tfrac{1}{2}mv_f^2 + mgh_f}_{E_f} = \underbrace{\tfrac{1}{2}mv_0^2 + mgh_0}_{E_0} \qquad (6.9)$$

The mass m of the rider appears as a factor in every term in this equation and can be eliminated algebraically. Solving for the final speed gives

$$v_f = \sqrt{v_0^2 + 2g(h_0 - h_f)} = \sqrt{2(9.80 \text{ m/s}^2)(59.4 \text{ m})} = \boxed{34.1 \text{ m/s (about 76 mph)}}$$

where the initial speed of the coaster is assumed to be zero, $v_0 = 0$, and the vertical height of the hill is $h_0 - h_f = 59.4$ m.

When applying the principle of conservation of mechanical energy to solving problems, we have been using the following reasoning strategy:

Figure 6.22 The Magnum XL-200 roller coaster; it is one of the fastest roller coasters in the world and includes a vertical drop of 59.4 m.

REASONING STRATEGY

Applying the Principle of Conservation of Mechanical Energy

1. Identify the external conservative and nonconservative forces that act on the object. In order for this principle to apply, any nonconservative forces that are present must act perpendicular to the displacement of the object, so they do no work; $W_{nc} = 0$.

2. Choose the location where the gravitational potential energy is taken to be zero. This location is arbitrary, but must not be changed during the course of solving a problem.

3. Set the final total mechanical energy of the object equal to the initial total mechanical energy, as in Equation 6.9. The total mechanical energy is the sum of the kinetic and potential energies.

6.6 NONCONSERVATIVE FORCES AND THE WORK–ENERGY THEOREM

Most moving objects experience one or more nonconservative forces, such as friction, air resistance, and propulsive forces. The work W_{nc} done by the net external nonconservative force is not zero. In these situations, the difference between the final and initial total mechanical energies is equal to W_{nc}, according to $W_{nc} = E_f - E_0$ (Equation 6.8). Consequently, the total mechanical energy is not conserved, as Figure 6.17 indicates. The next two examples illustrate how Equation 6.8 is used when nonconservative forces are present and do work.

EXAMPLE 13 • The Magnum XL-200, Revisited

In Example 12, we ignored friction. In reality, however, friction is present when the roller coaster descends the hill. The actual speed of the riders at the bottom is 32.2 m/s, which is less than that determined in Example 12. Assuming again that the coaster has a speed of nearly zero at the top of the hill, find the work done by the nonconservative frictional force on a 55.0-kg rider during the descent from a height h_0 to a height h_f, where $h_0 - h_f = 59.4$ m.

Reasoning Since the speed at the top of the hill is nearly zero and the final speed and the vertical drop are given, we can determine the initial and final total mechanical energies of the rider. The work–energy theorem, $W_{nc} = E_f - E_0$, can then be used to determine the work W_{nc} done by the nonconservative frictional force. We expect this work to be negative, because the frictional force acts opposite to the displacement of the rider.

Solution The work–energy theorem is

$$W_{nc} = \underbrace{(\tfrac{1}{2}mv_f^2 + mgh_f)}_{E_f} - \underbrace{(\tfrac{1}{2}mv_0^2 + mgh_0)}_{E_0} \tag{6.8}$$

Since $v_0 = 0$ at the top of the hill, this equation can be written as

$$W_{nc} = \tfrac{1}{2}mv_f^2 - mg(h_0 - h_f)$$

$$W_{nc} = \tfrac{1}{2}(55.0 \text{ kg})(32.2 \text{ m/s})^2 - (55.0 \text{ kg})(9.80 \text{ m/s}^2)(59.4 \text{ m}) = \boxed{-3500 \text{ J}}$$

The work done by friction is negative, as expected.

• **PROBLEM SOLVING INSIGHT**
As illustrated here and in Example 4, a frictional force can do negative or positive work. It does negative work when it has a component opposite to the displacement and slows down the object. It does positive work when it has a component in the direction of the displacement and speeds up the object.

EXAMPLE 14 • Fireworks

A fireworks rocket (0.20 kg) is launched from rest and follows an erratic flight path to reach the point P, as Figure 6.23 shows. Point P is 29 m above the starting point. In the process, 425 J of work is done on the rocket by the nonconservative force generated by the burning propellant. Ignoring air resistance and the mass lost due to the burning propellant, find the speed v_f of the rocket at the point P.

Reasoning The only nonconservative force acting on the rocket is the force generated by the burning propellant, and the work done by this force is $W_{nc} = 425$ J. Because work is done by a nonconservative force, we use the work–energy theorem in the form $W_{nc} = E_f - E_0$ to find the final speed v_f of the rocket.

Solution From the work–energy theorem we have

$$W_{nc} = (\tfrac{1}{2}mv_f^2 + mgh_f) - (\tfrac{1}{2}mv_0^2 + mgh_0) \tag{6.8}$$

Setting $v_0 = 0$ and solving for the final speed of the rocket, we get

$$v_f = \sqrt{\frac{2[W_{nc} - mg(h_f - h_0)]}{m}}$$

$$v_f = \sqrt{\frac{2[425\ \text{J} - (0.20\ \text{kg})(9.80\ \text{m/s}^2)(29\ \text{m})]}{0.20\ \text{kg}}} = \boxed{61\ \text{m/s}}$$

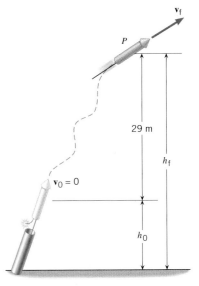

Figure 6.23 A fireworks rocket, moving along an erratic flight path, reaches a point P that is 29 m above the launch point.

6.7 POWER

In many situations, the time it takes to do work is just as important as the amount of work that is done. Consider two automobiles that are identical in all respects (e.g., same mass), except that one has a "souped-up" engine. The car with the "souped-up" engine can go from 0 to 27 m/s (60 mph) in 4 seconds, while the other car requires 8 seconds to achieve the same speed. Each engine does work in accelerating its car, but one does it more quickly. Where cars are concerned, we have come to associate the quicker performance with an engine that has a larger horsepower rating. A large horsepower rating means that the engine can do a large amount of work in a short time. In physics, the horsepower rating is just one way to measure an engine's ability to generate power. The idea of ***power*** incorporates both the concepts of work and time, for power is work done per unit time.

> ■ **DEFINITION OF AVERAGE POWER**
>
> Average power $\overline{P}$ is the average rate at which work W is done, and it is obtained by dividing W by the time t required to perform the work:
>
> $$\overline{P} = \frac{\text{Work}}{\text{Time}} = \frac{W}{t} \tag{6.10}$$
>
> ***SI Unit of Power:*** joule/s = watt (W)

Since both work and time are scalar quantities, power is also a scalar quantity. The unit in which power is expressed is that of work divided by time, or a joule per second in SI units. One joule per second is called a watt (W), in honor of James Watt (1736–1819), the developer of the steam engine. The unit of power in the BE

Table 6.3   Units of Measurement for Power

System	Work	÷	Time	=	Power
SI	joule (J)		second (s)		watt (W)
CGS	erg		second (s)		erg per second (erg/s)
BE	foot · pound (ft · lb)		second (s)		foot · pound per second (ft · lb/s)

system is the foot-pound per second (ft · lb/s), although the familiar horsepower (hp) unit is frequently used for specifying the power generated by electric motors and internal combustion engines:

$$1 \text{ horsepower} = 550 \text{ foot} \cdot \text{pounds/second} = 746 \text{ watts}$$

Table 6.3 summarizes the units for power in the various systems of measurement.

An alternative expression for power can be obtained from Equation 6.1, which indicates that the work W done when a constant net force of magnitude F points in the same direction as the displacement is $W = (F \cos 0°)s = Fs$. Dividing both sides of this equation by the time t it takes for the force to move the object through the distance s, we obtain

$$\frac{W}{t} = \frac{Fs}{t}$$

But W/t is the average power $\overline{P}$, and s/t is the average speed $\overline{v}$, so that

$$\overline{P} = F\overline{v} \qquad (6.11)$$

The next example illustrates the use of Equation 6.11.

EXAMPLE 15 • The Power to Accelerate a Car

A 1.10×10^3-kg car, starting from rest, accelerates for 5.00 s. The magnitude of the acceleration is $a = 4.60 \text{ m/s}^2$. Determine the average power generated by the net force that accelerates the vehicle.

Reasoning   We can find the average power by using the relation $\overline{P} = F\overline{v}$, provided the magnitude F of the net force and the average speed $\overline{v}$ of the car can be determined. The net force can be obtained from Newton's second law, and the average speed can be calculated from the equations of kinematics.

Solution   According to Newton's second law, the magnitude of the net force that acts on the car is

$$F = ma = (1.10 \times 10^3 \text{ kg})(4.60 \text{ m/s}^2) = 5060 \text{ N}$$

Since the car starts from rest ($v_0 = 0$) and has a constant acceleration, the average speed $\overline{v}$ of the car is one-half of its final speed v:

$$\overline{v} = \tfrac{1}{2}(v_0 + v) = \tfrac{1}{2}v \qquad (2.6)$$

Because the initial speed of the car is zero, the final speed of the car after 5.00 s is the product of its acceleration and time:

$$v = v_0 + at = (4.60 \text{ m/s}^2)(5.00 \text{ s}) = 23.0 \text{ m/s} \qquad (2.4)$$

Thus, the average speed is $\bar{v} = 11.5$ m/s, and the average power is

$$\bar{P} = F\,\bar{v} = (5060\ \text{N})(11.5\ \text{m/s}) = \boxed{5.82 \times 10^4\ \text{W}\ (78.0\ \text{hp})}$$

6.8 OTHER FORMS OF ENERGY AND THE CONSERVATION OF ENERGY

Up to now, we have considered only two types of energy, kinetic energy and gravitational potential energy. There are many other types, however. Electrical energy is used to run electrical appliances. Energy in the form of heat is utilized in cooking food. Moreover, the work done by the kinetic frictional force often appears as heat, as you can experience by rubbing your hands back and forth. Chemical energy is the energy stored in the molecules of fuels and food. When gasoline is burned, some of the stored chemical energy is released and does the work of moving cars, airplanes, and boats. The chemical energy stored in food provides the energy needed for metabolic processes.

One of the most controversial forms of energy is nuclear energy. The research of many scientists, most notably Albert Einstein, led to the discovery that mass itself is one manifestation of energy. Einstein's famous equation, $E_0 = mc^2$, describes how mass m and energy E_0 are related, where c is the speed of light in a vacuum and has a value of 3.00×10^8 m/s. Because the speed of light is so large, this equation implies that very small masses are equivalent to large amounts of energy. The relationship between mass and energy will be discussed further in Chapter 28.

We have seen that kinetic energy can be converted into gravitational potential energy and vice versa. In general, energy of all types can be converted from one form to another. Part of the chemical energy stored in food is transformed into the kinetic energy of walking and into the thermal energy needed to keep our bodies at a temperature near 98.6 °F. Similarly, in a moving car the chemical energy of gasoline is converted into kinetic energy, as well as electrical energy (to operate the radio, headlights, and air conditioner), and heat (to warm the car during the winter). Whenever energy is transformed from one form to another, it is found that no energy is gained or lost in the process; the total of all the energies before the process is equal to the total of the energies after the process. This observation leads to the following important principle:

■ THE PRINCIPLE OF CONSERVATION OF ENERGY

Energy can neither be created nor destroyed, but can only be converted from one form to another.

Learning how to convert energy from one form to another more efficiently is one of the main goals of modern science and technology.

6.9 WORK DONE BY A VARIABLE FORCE

Up to now we have considered the work done by a constant force (constant in both magnitude and direction). This work W is given by Equation 6.1 as $W = (F\cos\theta)s$. Quite often, situations arise in which the magnitude of the force is not

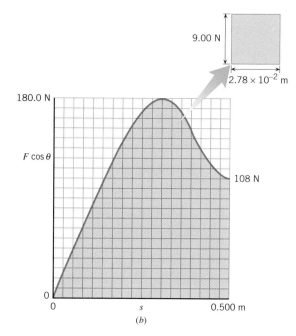

Figure 6.24 (*a*) A compound bow. (*b*) A plot of $F \cos \theta$ versus *s* as the bowstring is drawn back.

The Physics of...
the compound bow.

constant, but changes with the displacement of the object. For instance, Figure 6.24*a* shows an archer using a high-tech compound bow. This type of bow consists of a series of pulleys and strings that produce a force-versus-displacement graph like that in Figure 6.24*b*. One of the key features of the compound bow is that the force rises to a maximum as the string is being drawn back, and then falls to 60% of this maximum value when the string is fully drawn. The reduced force at $s = 0.500$ m makes it much easier for the archer to hold the fully drawn bow while aiming the arrow.

When the force varies with the displacement, as in Figure 6.24*b*, we cannot use Equation 6.1 to find the work, because this equation is valid only when the force is constant. However, we can use a graphical method to determine the work done by a variable force. In this method we divide the total displacement into very small segments, Δs_1, Δs_2, and so on (see Figure 6.25*a*). For each segment, the *average value* of the force component in that segment is indicated by a short horizontal line. For example, the short horizontal line for segment Δs_1 is labeled $(F \cos \theta)_1$ in Figure 6.25*a*. We can then use this average value as the constant force component in Equation 6.1 and determine an approximate value for the work ΔW_1 done during the first segment: $\Delta W_1 = (F \cos \theta)_1 \Delta s_1$. But this work is just the area of the colored rectangle in the drawing. The word "area" here refers to the area of a rectangle that has a width of Δs_1 and a height of $(F \cos \theta)_1$; it does not mean an area in square meters, such as the area of a parcel of land. In a like manner, we can calculate an approximate value for the work for each segment. Then we add the results for the segments to get, approximately, the work W done by the variable force:

$$W \approx (F \cos \theta)_1 \Delta s_1 + (F \cos \theta)_2 \Delta s_2 + \cdots$$

The symbol "$\approx$" means "approximately equal to." The right side of this equation is the sum of all the rectangular areas in Figure 6.25*a* and is an approximate value for

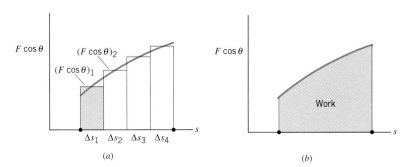

the area shaded in color under the graph in Figure 6.25*b*. If the rectangles are made narrower and narrower by decreasing each Δs, the right side of this equation eventually becomes equal to the area under the graph. Thus, we define the work done by a variable force as follows: ***The work done by a variable force in moving an object is equal to the area under the graph of F cos θ versus s.***

Example 16 illustrates how to use this graphical method to determine the approximate work done when a high-tech compound bow is drawn.

EXAMPLE 16 • Work and the Compound Bow

Find the work that the archer must do in drawing back the string of the compound bow in Figure 6.24 from 0 to 0.500 m.

Reasoning The work done is equal to the colored area under the curved line in Figure 6.24*b*. For convenience, this area is divided into a number of small squares, each square having an area of $(9.00 \text{ N})(2.78 \times 10^{-2} \text{ m}) = 0.250$ J. The area can then be found by counting the number of squares under the curve and multiplying by the area per square.

Solution We estimate that there are 242 colored squares in the drawing. Since each square represents 0.250 J of work, the total work done is

$$W = (242 \text{ squares})\left(0.250 \, \frac{\text{J}}{\text{square}}\right) = \boxed{60.5 \text{ J}}$$

When the arrow is fired, part of this work is imparted to it as kinetic energy.

SUMMARY

The **work** W done by a constant force acting on an object is $W = (F \cos \theta)s$, where F is the magnitude of the force, s is the magnitude of the displacement, and θ is the angle between the force and the displacement vectors. Work is a scalar quantity and can be positive or negative, depending on whether the force has a component that points in the same direction as the displacement or in the opposite direction. The work is zero if the force is perpendicular ($\theta = 90°$) to the displacement.

The **kinetic energy** KE of an object of mass m and speed v is KE $= \frac{1}{2}mv^2$. The **work–energy theorem** states that the work W done by the net external force acting on an object equals the difference between the object's final kinetic energy KE_f and initial kinetic energy KE_0: $W = \text{KE}_f - \text{KE}_0$. If the net force does positive work, the kinetic energy increases; if the net force does negative work, the kinetic energy decreases.

The **work done by the force of gravity** on an object of

mass m is $W_{\text{gravity}} = mg(h_0 - h_f)$, where h_0 and h_f are the initial and final heights of the object, respectively.

Gravitational potential energy PE is the energy that an object has by virtue of its position. For an object near the surface of the earth, the gravitational potential energy is given by PE $= mgh$, where h is the height of the object relative to an arbitrary zero level.

A **conservative force** is one that, in moving an object between two points, does the same work, independent of the path taken between the points. Alternatively, a force is conservative if the work it does in moving an object around any closed path is zero. A force is a **nonconservative force** if the work it does on an object moving between two points depends on the path of the motion between the points.

The **total mechanical energy** E is the sum of the kinetic energy and the potential energy: $E = \text{KE} + \text{PE}$. The work–energy theorem can be expressed in an alternative form as $W_{\text{nc}} = E_f - E_0$, where W_{nc} is the net work done by the external nonconservative forces, and E_f and E_0 are the final and initial total mechanical energies, respectively.

The **principle of conservation of mechanical energy** states that the total mechanical energy E remains constant along the path of an object, provided that the net work done by external nonconservative forces is zero. While E is constant, however, KE and PE may be transformed into one another.

Average power $\overline{P}$ is the work done per unit time, $\overline{P} = $ Work/Time, or the rate at which work is done.

The **principle of conservation of energy** states that energy can neither be created nor destroyed, but can only be transformed from one form to another.

CONCEPTUAL QUESTIONS

1. Two forces $\mathbf{F}_1$ and $\mathbf{F}_2$ are acting on the box shown in the drawing, causing the box to move across the floor. The two force vectors are drawn to scale. Which force does more work? Justify your answer.

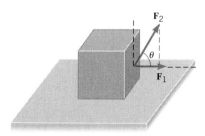

2. A box is being moved with a velocity $\mathbf{v}$ by a force $\mathbf{P}$ (parallel to $\mathbf{v}$) along a level horizontal floor. The normal force is $\mathbf{F}_N$, the kinetic frictional force is $\mathbf{f}_k$, and the weight of the box is $m\mathbf{g}$. Decide which forces do positive, zero, or negative work. Provide a reason for each of your answers.

3. A force does positive work on a particle that has a displacement pointing in the $+x$ direction. This same force does negative work on a particle that has a displacement pointing in the $+y$ direction. In what quadrant does the force lie? Account for your answer.

4. A sailboat is moving at a constant velocity. (a) Is work being done by a net external force acting on the boat? Explain. (b) Recognizing that the wind propels the boat forward and the water resists the boat's motion, what does your answer in part (a) imply about the work done by the wind's force compared to the work done by the water's resistive force?

5. A ball has a speed of 15 m/s. Only one external force acts on the ball. After this force acts, the speed of the ball is 7 m/s. Has the force done positive or negative work? Explain.

6. Is it possible for the same object to have different velocities at two different times and yet have the same kinetic energy at each time? Explain.

7. A slow-moving car may have more kinetic energy than a fast-moving motorcycle. How is this possible?

8. A net external force acts on a particle. This net force is not zero. Is this sufficient information to conclude that (a) the velocity of the particle changes, (b) the kinetic energy of the particle changes, and (c) the speed of the particle changes? Give your reasoning in each case.

9. The speed of a particle doubles and then doubles again, because a net external force acts on it. Does the net force do more work during the first or the second doubling? Justify your answer.

10. A shopping bag is hanging straight down from your hand as you walk across a horizontal floor at a constant velocity. (a) Does the force that your hand exerts on the bag's handle do any work? Explain. (b) Does this force do any work while you are riding up an escalator at a constant velocity? Give a reason for your answer.

11. In a simulation on earth, an astronaut in his space suit climbs up a vertical ladder. On the moon, the same astronaut makes the same climb. In which case does the gravitational potential energy of the astronaut change by a greater amount? Account for your answer.

12. A net external nonconservative force does positive work on a particle, and both its kinetic and potential energies change. What, if anything, can you conclude about (a) the change in the particle's total mechanical energy and (b) the individual changes in the kinetic and potential energies? Justify your answers.

13. Suppose the total mechanical energy of an object is conserved. (a) If the kinetic energy decreases, what must be true about the gravitational potential energy? (b) If the potential energy decreases, what must be true about the kinetic energy? (c) If the kinetic energy does not change, what must be true about the potential energy?

14. A person is riding on a Ferris wheel. When the wheel makes one complete turn, is the net work done by the gravitational force positive, negative, or zero? Justify your answer.

15. Consider the following two situations in which the retarding effects of friction and air resistance are negligible. Car A approaches a hill. The driver turns off the engine at the bottom of the hill, and the car coasts up the hill. Car B, its engine running, is driven up the hill at a constant speed. Which situation is an example of the principle of conservation of mechanical energy? Provide a reason for your answer.

16. A trapeze artist, starting from rest, swings downward on the bar, lets go at the bottom of the swing, and falls freely to the net. An assistant, standing on the same platform as the trapeze artist, jumps from rest straight downward. Friction and air resistance are negligible. (a) On which person, if either, does gravity do the greatest amount of work? Explain. (b) Who, if either, strikes the net with a greater speed? Why?

17. The drawing shows an empty fuel tank being released by three different jet planes. At the moment of release, each plane has the same speed and each tank is at the same height above the ground. However, the directions of travel are different. In the absence of air resistance, do the tanks have different speeds when they hit the ground? If so, which tank has the largest speed and which has the smallest speed? Explain.

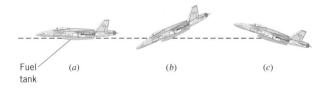

Fuel tank   (a)   (b)   (c)

18. Is it correct to conclude that one engine is doing twice the work of another just because it is generating twice the power? Explain, neglecting friction and taking into account the time of operation of the engines.

PROBLEMS

Section 6.1 Work

1. ssm During a tug-of-war, team A pulls on team B by applying a force of 1100 N to the rope between them. How much work does team A do if it pulls team B toward them a distance of 2.0 m?

2. A tow rope, oriented parallel to the water, pulls a water-skier for a distance of 86 m. The work done by the tension in the rope is 1.9×10^4 J. Find the magnitude of the tension.

3. You are moving into an apartment. Your weight is 685 N and that of your belongings is 915 N. (a) How much work does the elevator do in lifting you and your belongings up five stories (15.2 m) at a constant velocity? (b) How much work does the elevator do on you (without belongings) on the downward trip, which is also made at a constant velocity? Be sure you include the correct sign for the work.

4. A person pulls a toboggan for a distance of 35.0 m along the snow with a rope directed 25.0° above the snow. The tension in the rope is 94.0 N. (a) How much work is done on the toboggan by the tension force? (b) How much work is done if the same tension is directed parallel to the snow?

5. ssm Suppose in Figure 6.2 that $+1.10 \times 10^3$ J of work are done by the force **F** (magnitude $= 30.0$ N) in moving the luggage carrier a distance of 50.0 m. At what angle θ is the force oriented with respect to the ground?

6. The drawing shows a boat being pulled by two locomotives through a canal of length 2.00 km. The tension in each cable is

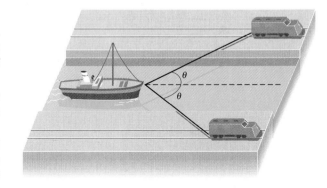

5.00×10^3 N, and $\theta = 20.0°$. What is the net work done on the boat by the two locomotives?

7. As preparation for this problem, review Conceptual Example 3. Suppose that both the hills in this example slope 21° above the horizontal and that the mass of the bicycle and rider is 86 kg. What is the work done by the combined weight of the bicycle and its rider when the cyclist (a) coasts 180 m down one hill and (b) pumps 75 m up the other? Pay particular attention to determining the angle θ in parts a and b of Figure 6.4.

8. Review Conceptual Example 3 as background for this problem. The drawing shows a plane diving toward the ground and then climbing back upward. During each of these motions, the lift force **L** acts perpendicular to the displacement **s**, which has the

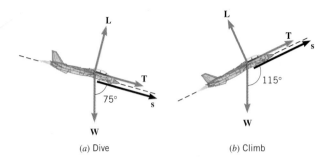

(a) Dive

(b) Climb

same magnitude of 1.7×10^3 m in each case. The engines of the plane exert a thrust **T**, which points in the direction of the displacement and has the same magnitude during the dive and the climb. The weight **W** of the plane has a magnitude of 5.9×10^4 N. In both motions, net work is performed due to the combined action of the forces **L**, **T**, and **W**. (a) Is more net work done during the dive or the climb? Explain. (b) Find the difference between the net work done during the dive and the climb.

*9. **ssm** A husband and wife take turns pulling their child in a wagon along a horizontal sidewalk. Each exerts a constant force and pulls the wagon through the same displacement. They do the same amount of work, but the husband's pulling force F_H is directed 58° above the horizontal, while the wife's pulling force F_W is directed 38° above the horizontal. What is the ratio F_H/F_W of the magnitudes of the pulling forces?

*10. A 2.40×10^2-N force is pulling an 85.0-kg refrigerator across a horizontal surface. The force acts at an angle of 20.0° above the surface. The coefficient of kinetic friction is 0.200, and the refrigerator moves a distance of 8.00 m. Find (a) the work done by the pulling force, and (b) the work done by the kinetic frictional force.

*11. A 1.00×10^2-kg crate is being pulled across a horizontal floor by a force **P** that makes an angle of 30.0° above the horizontal. The coefficient of kinetic friction is 0.200. What should be the magnitude of **P**, so that the net work done by it and the kinetic frictional force is zero?

12. A 1200-kg car is being driven up a 5.0° hill, as the drawing illustrates. The frictional force is directed opposite to the motion of the car and has a magnitude of $f = 5.0 \times 10^2$ N. The force **F is applied to the car by the road and propels the car forward. In addition to these two forces, two other forces act on the car: its weight **W**, and the normal force F_N directed perpendicular to the road surface. The length of the road up the hill is 3.0×10^2 m. What

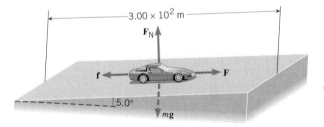

should be the magnitude of **F**, so that the net work done by all the forces acting on the car is $+150\ 000$ J?

Section 6.2 The Work–Energy Theorem and Kinetic Energy

13. **ssm** The hammer throw is a track and field event in which a 7.3-kg ball (the "hammer"), starting from rest, is whirled around in a circle several times and released. It then moves upward on the familiar curving path of projectile motion. In one throw, the hammer is given a speed of 29 m/s. For comparison, a .22 caliber bullet has a mass of 0.0026 kg and, starting from rest, exits the barrel of a gun with a speed of 410 m/s. Determine the work done to launch the motion of (a) the hammer and (b) the bullet.

14. A 0.075-kg arrow is fired horizontally. The bowstring exerts an average force of 65 N on the arrow over a distance of 0.90 m. With what speed does the arrow leave the bow?

15. A 65.0-kg jogger is running at a speed of 5.30 m/s. (a) What is the kinetic energy of the jogger? (b) How much work is done by the net external force that accelerates the jogger to 5.30 m/s from rest?

16. As background for this problem, review Conceptual Example 7. A 7420-kg satellite has an elliptical orbit, as in Figure 6.10b. The point on the orbit that is farthest from the earth is called the *apogee* and is at the far right side of the drawing. The point on the orbit that is closest to the earth is called the *perigee* and is at the far left side of the drawing. Suppose that the speed of the satellite is 2820 m/s at the apogee and 8450 m/s at the perigee. Find the work done by the gravitational force when the satellite moves from (a) the apogee to the perigee and (b) the perigee to the apogee.

17. **ssm www** Two cars, A and B, are traveling with the same speed of 40.0 m/s, each having started from rest. Car A has a mass of 1.20×10^3 kg, and car B has a mass of 2.00×10^3 kg. Compared to the work required to bring car A up to speed, how much *additional* work is required to bring car B up to speed?

18. The work done by the net external force acting on a tennis ball (mass = 0.058 kg) is the same as that on a billiard ball (mass = 0.38 kg). The final speed of the tennis ball is v_T, while that of the billiard ball is v_B, both balls being at rest initially. Find the ratio v_T/v_B of the final speeds.

19. A 5.0×10^4-kg space probe is traveling at a speed of 11 000 m/s through deep space. Retrorockets are fired along the line of motion to reduce the probe's speed. The retrorockets generate a force of 4.0×10^5 N over a distance of 2500 km. What is the final speed of the probe?

20. When a 0.045-kg golf ball takes off after being hit, its speed is 41 m/s. (a) How much work is done on the ball by the club? (b) Assume that the force of the golf club acts parallel to the motion of the ball and that the club is in contact with the ball for a distance of 0.010 m. Ignore the weight of the ball and determine the average force applied to the ball by the club.

*21. **ssm** A sled is being pulled across a horizontal patch of snow. Friction is negligible. The pulling force points in the same direction as the sled's displacement, which is along the $+x$ axis.

As a result, the kinetic energy of the sled increases by 38%. By what percentage would the sled's kinetic energy have increased if this force had pointed 62° above the +x axis?

*22. The speed of a hockey puck decreases from 45.00 to 44.67 m/s in coasting 16 m across the ice. Find the coefficient of kinetic friction between the puck and the ice.

*23. A power boat of mass 480 kg is cruising at a constant speed of 8.9 m/s. The propeller provides a drive force of 760 N. The driver of the boat shuts off the engine, and the boat coasts to a halt. Assume—contrary to fact—that the resistive force due to the water is constant, independent of the boat's speed. (a) How far does the boat coast? (b) How much time does it take for the boat to come to rest after the engine is turned off?

*24. In screeching to a halt, a car leaves skid marks that are 65 m long. The coefficient of kinetic friction between the tires and the road is $\mu_k = 0.71$. How fast was the car going before the driver applied the brakes?

*25. **ssm www** A 6200-kg satellite is in a circular earth orbit that has a radius of 7.6×10^6 m. A net external force must act on the satellite to make it change to a circular orbit that has a radius of 5.3×10^6 m. What work must the net external force do?

**26. The model airplane in Figure 5.7 is flying at a speed of 22 m/s on a horizontal circle of radius 16 m. The mass of the plane is 0.90 kg. The person holding the guideline pulls it in until the radius of the circle becomes 14 m. The plane speeds up, and the tension in the guideline becomes four times greater. What is the net work done on the plane?

Section 6.3 Gravitational Potential Energy,
Section 6.4 Conservative Forces
and Nonconservative Forces

27. Relative to the ground, what is the gravitational potential energy of a 55.0-kg person who is at the top of the Sears Tower, a height of 443 m above the ground?

28. A 0.15-kg ball is thrown 9.0 m straight up. (a) Find the work done by the gravitational force. Be sure to include the correct sign. (b) What is the change ($\Delta PE = PE_f - PE_0$) in the gravitational potential energy?

29. **ssm** A bicyclist rides 5.0 km due east, while the resistive force from the air has a magnitude of 3.0 N and points due west. The rider then turns around and rides 5.0 km due west, back to her starting point. The resistive force from the air on the return trip has a magnitude of 3.0 N and points due east. (a) Find the work done by the resistive force during the round trip. (b) Based on your answer to part (a), is the resistive force a conservative force? Explain.

30. A shot-putter puts a shot (weight = 71.1 N) that leaves his hand at a distance of 1.52 m above the ground. (a) Find the work done by the gravitational force when the shot has risen to a height of 2.13 m above the ground. Include the correct sign for the work. (b) Determine the change ($\Delta PE = PE_f - PE_0$) in the gravitational potential energy of the shot.

31. A pole-vaulter just clears the bar at 5.80 m and falls back to the ground. The change in the vaulter's potential energy is -3.70×10^3 J. What is his weight?

32. A 75.0-kg skier rides a 2830-m-long lift to the top of a mountain. The lift makes an angle of 14.6° with the horizontal. What is the change in the skier's gravitational potential energy?

33. **ssm** "Rocket man" has a propulsion unit strapped to his back. He starts from rest on the ground, fires the unit, and is propelled straight upward. At a height of 16 m, his speed is 5.0 m/s. His mass, including the propulsion unit, has the approximately constant value of 136 kg. Find the work done by the force generated by the propulsion unit.

Section 6.5 The Conservation of Mechanical Energy

34. A gymnast is swinging on a high bar. The distance between his waist and the bar is 1.1 m, as the drawing shows. At the top of the swing his speed is momentarily zero. Ignoring friction and treating the gymnast as if all of his mass is located at his waist, find his speed at the bottom of the swing.

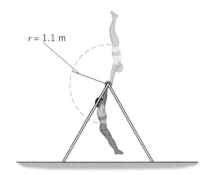

$r = 1.1$ m

35. A slingshot fires a pebble from the top of a building at a speed of 10.0 m/s. The building is 20.0 m tall. Ignoring air resistance, find the speed with which the pebble strikes the ground when the pebble is fired (a) horizontally, (b) vertically straight up, and (c) vertically straight down.

36. A water-skier lets go of the tow rope upon leaving the end of a jump ramp at a speed of 14.0 m/s. As the drawing indicates, the skier has a speed of 13.0 m/s at the highest point of the jump. Ignoring air resistance, determine the skier's height H above the *top of the ramp* at the highest point.

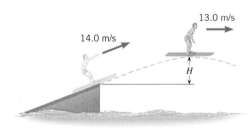

13.0 m/s

14.0 m/s

H

37. **ssm** A 2.00-kg rock is released from rest at a height of 20.0 m. Ignore air resistance and determine the kinetic energy,

gravitational potential energy, and total mechanical energy at each of the following heights: 20.0, 10.0, and 0 m.

38. Before starting this problem, review Conceptual Example 11, which discusses the situations depicted in Figure 6.21. This figure shows two possible paths by which a person, starting from rest at the top of a cliff, can enter the water below. Suppose that he enters the water at a speed of 13.0 m/s via path 1. How fast is he moving on path 2 when he releases the rope at a height of 5.20 m above the water?

39. A cyclist approaches the bottom of a gradual hill at a speed of 11 m/s. The hill is 5.0 m high, and the cyclist estimates that she is going fast enough to coast up and over it without peddling. Ignoring air resistance and friction, find the speed at which the cyclist crests the hill.

40. Conceptual Example 11 provides background for this problem. For both paths into the water in Figure 6.21, the person starts at rest at the top of the cliff. On path 2, the person releases the rope 0.70 m higher above the water than on path 1. His speed at the point of release on path 1 is 7.0 m/s. What is his speed at the point of release on path 2?

41. ssm www Two pole-vaulters just clear the bar at the same height. The first lands at a speed of 8.90 m/s, while the second lands at a speed of 9.00 m/s. The first vaulter clears the bar at a speed of 1.00 m/s. Ignore air resistance and friction and determine the speed at which the second vaulter clears the bar.

*42. A grappling hook, attached to a 1.5-m rope, is whirled in a circle that lies in the vertical plane. The lowest point on this circle is at ground level. The hook is whirled at a constant rate of three revolutions per second. In the absence of air resistance, to what maximum height can the hook be cast?

*43. A wrecking ball swings at the end of a 10.0-m cable on a vertical circular arc. The crane operator manages to give the ball a speed of 6.00 m/s as the ball passes through the lowest point of its swing and then gives the ball no further assistance. Friction and air resistance are negligible. What speed v_f does the ball have when the cable makes an angle of 30.0° with respect to the vertical?

*44. A skier starts from rest at the top of a hill. The skier coasts down the hill and up a second hill, as the drawing illustrates. The crest of the second hill is circular, with a radius of $r = 36$ m. Neglect friction and air resistance. What must be the height h of the first hill so that the skier just loses contact with the snow at the crest of the second hill?

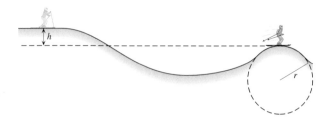

*45. ssm A water slide is constructed so that swimmers, starting

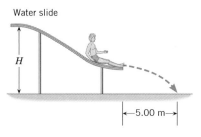

Water slide

from rest at the top of the slide, leave the end of the slide traveling horizontally. As the drawing shows, one person is observed to hit the water 5.00 m from the end of the slide in a time of 0.500 s after leaving the slide. Ignoring friction and air resistance, find the height H in the drawing.

**46. A swing is made from a rope that will tolerate a maximum tension of 8.00×10^2 N without breaking. Initially, the swing hangs vertically. The swing is then pulled back at an angle of 60.0° with respect to the vertical and released from rest. What is the mass of the heaviest person who can ride the swing?

**47. The drawing shows a version of the loop-the-loop trick for a small car. If the car is given an initial speed of 4.0 m/s, what is the largest value that the radius r can have if the car is to remain in contact with the circular track at all times?

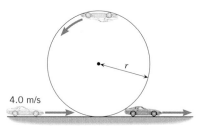

Section 6.6 Nonconservative Forces and the Work–Energy Theorem

48. A basketball player makes a jump shot. The 0.600-kg ball is released at a height of 2.00 m above the floor with a speed of 7.20 m/s. The ball goes through the net 3.00 m above the floor at a speed of 4.20 m/s. What is the work done on the ball by air resistance, a nonconservative force?

49. ssm A roller coaster (375 kg) moves from A (5.00 m above the ground) to B (20.0 m above the ground). Two nonconservative forces are present: friction does -2.00×10^4 J of work on the car, and a chain mechanism does $+3.00 \times 10^4$ J of work to help the car up a long climb. What is the change in the car's kinetic energy, $\Delta KE = KE_f - KE_0$, from A to B?

50. An 83.0-kg student, starting from rest, slides down a 11.8-m-high water slide. On the way down, friction (a nonconservative force) does -6.50×10^3 J of work on him. How fast is he going at the bottom of the slide?

51. A 55.0-kg skateboarder starts out with a speed of 1.80 m/s. He does $+80.0$ J of work on himself by pushing with his feet against the ground. In addition, friction does -265 J of work on him. In both cases, the forces doing the work are nonconservative. The final speed of the skateboarder is 6.00 m/s. (a) Calculate the change ($\Delta PE = PE_f - PE_0$) in the gravitational potential energy. (b) How much has the vertical height of the skater changed, and is the skater above or below the starting point?

52. A pitcher throws a 0.140-kg baseball, and it approaches the bat at a speed of 40.0 m/s. The bat does $W_{nc} = 70.0$ J of work on the ball in hitting it. Ignoring air resistance, determine the speed of the ball after the ball leaves the bat and is 25.0 m above the point of impact.

53. ssm In a pinball machine, a flipper is a small paddle that the player can activate to hit the ball. Suppose that the 0.10-kg ball is sliding toward the flipper at a speed of 0.80 m/s when it is struck. The flipper acts on the ball over a displacement that has a magnitude of 0.010 m, and the speed of the ball increases to 1.2 m/s. (a) Assuming that the force acts in the direction of the displacement and that the displacement is parallel to the ground, find the magnitude of the average force that the flipper exerts on the ball. (b) Repeat (a) when the displacement is directed at 15° above the ground.

*54. A gymnast is bouncing on a trampoline. On the upward part of the motion, the mat of the trampoline pushes on the gymnast with a nonconservative force over a distance of 0.300 m. After leaving the mat, the 55.0-kg gymnast rises into the air for an additional 2.00 m before falling back down. What average force does the mat exert on the gymnast?

*55. A basketball of mass 0.60 kg is dropped from rest from a height of 1.22 m. It rebounds to a height of 0.69 m. (a) How much mechanical energy was lost during the collision with the floor? (b) A basketball player dribbles the ball from a height of 1.22 m by exerting a constant downward force on it for a distance of 0.13 m. In dribbling, the player compensates for the mechanical energy lost during each bounce. If the ball now returns to a height of 1.22 m, what is the magnitude of the force?

*56. At the end of the semester, you are loading your gear into a van to go home. In loading the van, you can either lift a 45.0-kg crate straight up (method 1) or push it up a 3.00-m-long ramp that is inclined at an angle θ above the horizontal (method 2). In both methods, the crate is moved at a constant speed, and the vertical distance from the ground to the floor of the van has the same value h (see the drawing). In method 1, no friction is present. In

method 2, however, a frictional force does −485 J of work on the crate. In applying your nonconservative pushing force to the crate, you do twice as much work in method 2 as in method 1. Find the angle θ.

*57. ssm At a carnival, you can try to ring a bell by striking a target with a 9.00-kg hammer. In response, a 0.400-kg metal piece is sent upward toward the bell, which is 5.00 m above. Suppose that 25.0% of the hammer's kinetic energy is used to do the (nonconservative) work of sending the metal piece upward. How fast must the hammer be moving when it strikes the target, so that the bell just barely rings?

**58. A 3.00-kg model rocket is launched vertically straight up with sufficient initial speed to reach a maximum height of 1.00×10^2 m, even though air resistance (a nonconservative force) performs -8.00×10^2 J of work on the rocket. How high would the rocket have gone without air resistance?

Section 6.7 Power

59. A person is making homemade ice cream. She exerts a force of magnitude 22 N on the free end of the crank handle, and this end moves in a circular path of radius 0.28 m. The force is always applied parallel to the motion of the handle. If the handle is turned once every 1.3 s, what is the average power being expended?

60. In an action-adventure movie, the hero lifts the 91-kg villain straight upward through a distance of 1.2 m in 0.51 s at a constant speed. What power does the hero produce while doing this?

61. ssm One kilowatt·hour (kWh) is the amount of work or energy generated when one kilowatt of power is supplied for a time of one hour. A kilowatt·hour is the unit of energy used by power companies when figuring your electric bill. Determine the number of joules of energy in one kilowatt·hour.

62. A 3.00×10^2-kg piano is being lifted at a steady speed from ground level straight up to an apartment 10.0 m above the ground. The crane that is doing the lifting produces a steady power of 4.00×10^2 W. How much time does it take to lift the piano?

*63. A car accelerates uniformly from rest to 27 m/s in 7.0 s along a level stretch of road. Ignoring friction, determine the average power required to accelerate the car if (a) the weight of the car is 1.2×10^4 N, and (b) the weight of the car is 1.6×10^4 N.

*64. A cheetah is after its prey, which has a mass that is one-half that of the cheetah. When both start from rest, the cheetah sustains 2.50 times the acceleration as does its prey in the same time period. Find the ratio of the average power generated by the cheetah to that generated by its prey.

**65. ssm The motor of a ski boat generates an average power of 7.50×10^4 W when the boat is moving at a constant speed of 12 m/s. When the boat is pulling a skier at the same speed, the engine must generate an average power of 8.30×10^4 W. What is the tension in the tow rope that is pulling the skier?

**66. A motorcycle (mass of cycle plus rider = 2.50×10^2 kg) is traveling at a steady speed of 20.0 m/s. The force of air resistance acting on the cycle and rider is 2.00×10^2 N. Find the power nec-

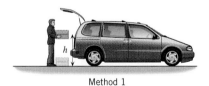

Method 1

Method 2

essary to sustain this speed if (a) the road is level and (b) the road is sloped upward at 37.0° with respect to the horizontal.

**67. A 1900-kg car experiences a combined force of air resistance and friction that has the same magnitude whether the car goes up or down a hill at 27 m/s. Going up a hill, the car's engine needs to produce 47 hp more power to sustain the constant velocity than it does going down the same hill. At what angle is the hill inclined above the horizontal?

Section *6.9 Work Done by a Variable Force

68. The force component along the displacement varies with the magnitude of the displacement, as shown in the graph. Find the work done by the force in the interval from (a) 0 to 1.0 m, (b) 1.0 to 2.0 m, and (c) 2.0 to 4.0 m. *(Note: In the last interval the force component is negative, so the work is negative.)*

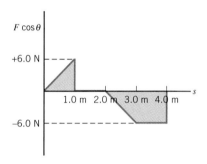

69. **ssm** The drawing shows the force-versus-displacement graph for two different bows. These graphs give the force that

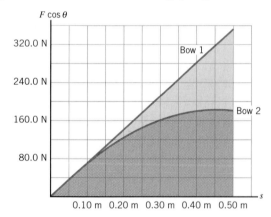

an archer must apply to draw the bow string. (a) For which bow is more work required to draw the bow fully from $s = 0$ to $s = 0.50$ m? Give your reasoning. (b) Estimate the additional work required for the bow identified in part (a) compared to the other bow.

70. Review Example 16, in which the work done in drawing the bowstring in Figure 6.24 from $s = 0$ to $s = 0.500$ m is determined. In part *b* of the figure, the force component $F \cos \theta$ reaches a maximum at $s = 0.306$ m. Find the percentage of the total work that is done when the bowstring is moved (a) from $s = 0$ to 0.306 m and (b) from $s = 0.306$ to 0.500 m.

71. The graph shows the net external force component $F \cos \theta$ along the displacement as a function of the magnitude s of the displacement. The graph applies to a 65-kg ice skater. How much work does the net force component do on the skater from (a) 0 to 3.0 m and (b) 3.0 m to 6.0 m? (c) If the initial speed of the skater is 1.5 m/s when $s = 0$, what is the speed when $s = 6.0$ m?

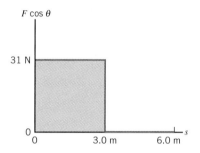

72. A net external force is applied to a 6.00-kg object that is initially at rest. The net force component along the displacement of the object varies with the magnitude of the displacement as shown in the drawing. (a) How much work is done by the net force? (b) What is the speed of the object at $s = 20.0$ m?

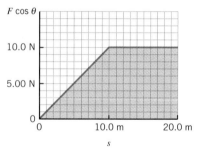

ADDITIONAL PROBLEMS

73. **ssm** A pole-vaulter approaches the takeoff point at a speed of 9.00 m/s. Assuming that only this speed determines the height to which he can rise, find the maximum height at which the vaulter can clear the bar.

74. The cable of a large crane applies a force of 2.2×10^4 N to a demolition ball as it lifts the ball vertically upward a distance of

7.6 m. (a) How much work does this force do on the ball? (b) Is the work positive or negative? Explain.

75. A softball pitcher has a "windmill" windup in which a 0.25-kg ball moves on a vertical arc of radius $r = 0.51$ m. To accelerate the ball, she exerts a 28-N force parallel to the ball's motion along the circular arc. The speed of the ball is 12 m/s at the top of the arc. With what speed is the ball released one-half a revolution later at the bottom of the arc?

76. A water-skier whose mass is 70.3 kg has an initial speed of 6.10 m/s. Later, the speed of the skier is 11.3 m/s. Determine the work done by the net external force acting on the skier.

77. ssm The brakes of a truck cause it to slow down by applying a retarding force of 3.0×10^3 N to the truck over a distance of 850 m. What is the work done by this force on the truck? Is the work positive or negative? Why?

78. A 5.00×10^2-kg hot-air balloon takes off from rest at the surface of the earth. The nonconservative wind and lift forces take the balloon up, doing $+9.70 \times 10^4$ J of work on the balloon in the process. At what height above the surface of the earth does the balloon have a speed of 8.00 m/s?

***79.** A 73-kg sprinter, starting from rest, reaches a speed of 7.0 m/s in 1.8 s, with a negligible effect due to air resistance. The sprinter then runs the remainder of the race at a steady speed of 7.0 m/s under the influence of a 34-N force due to air resistance. What is the average power needed (a) to accelerate the runner and (b) to sustain the steady speed at which most of the race is run?

***80.** A 55-kg box is being pushed a distance of 7.0 m across the floor by a force **P** whose magnitude is 150 N. The force **P** is parallel to the displacement of the box. The coefficient of kinetic friction is 0.25. Determine the work done on the box by each of the *four* forces that act on the box. Be sure to include the proper plus or minus sign for the work done by each force.

***81. ssm** The (nonconservative) force propelling a 1.50×10^3-kg car up a mountain road does 4.70×10^6 J of work on the car. The car starts from rest at sea level and has a speed of 27.0 m/s at an altitude of 2.00×10^2 m above sea level. Obtain the work done on the car by the combined forces of friction and air resistance, both of which are nonconservative forces.

***82.** A particle, starting from point A in the drawing, is projected down the curved runway. Upon leaving the runway at point B, the particle is traveling straight upward and reaches a height of 4.00 m above the floor before falling back down. Ignoring friction and air resistance, find the speed of the particle at point A.

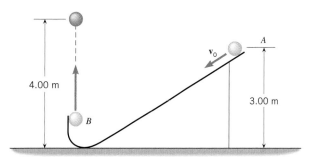

***83.** A 55.0-kg person jumps from rest off a 10.0-m-high tower straight down into the water. Neglect air resistance during the descent. She comes to rest 3.00 m under the surface of the water. Determine the average force that the water exerts on the diver. This force is nonconservative.

****84.** A person starts from rest at the top of a large frictionless spherical surface, and slides into the water below (see the drawing). At what angle θ does the person leave the surface? (*Hint: When the person leaves the surface the normal force is zero.*)

****85. ssm www** A truck is traveling at 11.1 m/s down a hill when the brakes on all four wheels lock. The hill makes an angle of 15.0° with respect to the horizontal. The coefficient of kinetic friction between the tires and the road is 0.750. How far does the truck skid before coming to a stop?

IMPULSE AND MOMENTUM

During this fireworks display, the exploding rockets produce a shower of burning fragments. The concepts of impulse and momentum discussed in this chapter aid us in understanding such explosions.

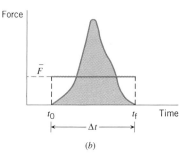

(a)

(b)

Figure 7.1 (*a*) When a bat strikes a ball, (*b*) the magnitude of the force exerted on the ball rises to a maximum value and then returns to zero when the ball leaves the bat. The time interval during which the force acts is Δt, and the magnitude of the average force is $\overline{F}$.

7.1 THE IMPULSE–MOMENTUM THEOREM

There are many situations in which the force acting on an object is not constant, but varies with time. For instance, Figure 7.1*a* shows a baseball being hit, and part *b* of the figure illustrates how the force applied to the ball by the bat might change during the time of contact. The magnitude of the force is zero at the instant t_0 just before the bat touches the ball. During contact, the force rises to a maximum and then returns to zero at the time t_f when the ball leaves the bat. The time interval $\Delta t = t_f - t_0$ during which the bat and ball are in contact is quite short, being only a few-thousandths of a second, although the maximum force can be very large, often exceeding thousands of newtons. For comparison, the graph also shows the magnitude $\overline{F}$ of the average force exerted on the ball during the time of contact. Figure 7.2 depicts other situations in which a time-varying force is applied to a ball.

Figure 7.2 In each of these situations the force applied to the ball varies with time. The time of contact is usually very small, but the maximum force can be large.

To describe how a time-varying force affects the motion of an object, we will introduce two new ideas, the impulse of a force and the linear momentum of an object. As the concept chart in Figure 7.3 illustrates, these two ideas will be integrated with Newton's second law of motion from Chapter 4 to produce an important result known as the impulse–momentum theorem. This theorem will play a central role in how we describe a collision between objects, such as that between a ball and a bat. Notice the close analogy between this concept chart and that in Figure 6.6, where the notions of work and kinetic energy are integrated with Newton's second law to produce the work–energy theorem.

If a baseball is to be hit well, both the size of the force and the time of contact are important. When a large average force acts on the ball for a long enough time, the ball is hit solidly. Therefore, we bring together the average force and the time of contact, calling the product of the two the **impulse** of the force.

■ **DEFINITION OF IMPULSE**

The impulse of a force is the product of the average force $\overline{\mathbf{F}}$ and the time interval Δt during which the force acts:

$$\text{Impulse} = \overline{\mathbf{F}}\,\Delta t \qquad (7.1)$$

Impulse is a vector quantity and has the same direction as the average force.

SI Unit of Impulse: newton · second (N · s)

When a ball is hit, it responds to the value of the impulse. A large impulse produces a large response, that is, the ball departs from the bat with a large velocity. However, we know from experience that the more massive the ball, the less velocity it has after leaving the bat. Therefore, both mass and velocity play a role in how an object responds to a given impulse. The effect of mass and velocity is included in the concept of **linear momentum,** which is defined as follows:

Figure 7.3 *Concepts at a Glance* In this section, two new concepts, impulse and linear momentum, will be integrated with Newton's second law of motion to produce the impulse–momentum theorem, a theorem used for describing collisions between objects. When the cue in this photograph strikes the ball during a pool game, an impulse is imparted to the ball. In response, the momentum of the ball changes, in accord with the impulse–momentum theorem.

Concepts at a Glance

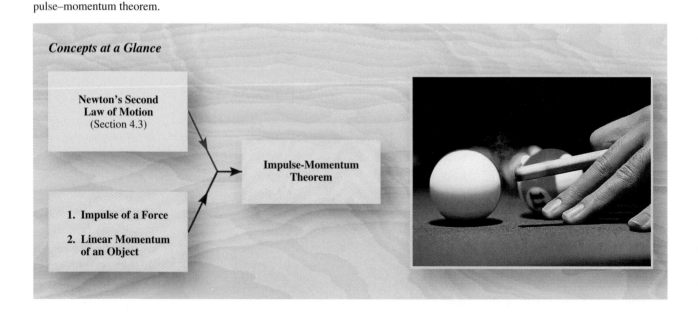

■ DEFINITION OF LINEAR MOMENTUM

The linear momentum **p** of an object is the product of the object's mass m and velocity **v**:

$$\mathbf{p} = m\mathbf{v} \qquad (7.2)$$

Linear momentum is a vector quantity that points in the same direction as the velocity.

SI Unit of Momentum: kilogram · meter/second (kg · m/s)

Newton's second law of motion can now be used to reveal a relationship between impulse and momentum. Figure 7.4 shows a ball approaching a bat with an initial velocity of $\mathbf{v}_0$, being struck by the bat, and then leaving with a final velocity of $\mathbf{v}_f$. When the velocity of an object changes from $\mathbf{v}_0$ to $\mathbf{v}_f$ during a time interval Δt, the average acceleration $\overline{\mathbf{a}}$ is given by Equation 2.4 as

$$\overline{\mathbf{a}} = \frac{\mathbf{v}_f - \mathbf{v}_0}{\Delta t}$$

According to Newton's second law, $\overline{\mathbf{F}} = m\overline{\mathbf{a}}$, the cause of the acceleration is an average net force $\overline{\mathbf{F}}$. Thus,

$$\overline{\mathbf{F}} = m\left(\frac{\mathbf{v}_f - \mathbf{v}_0}{\Delta t}\right) = \frac{m\mathbf{v}_f - m\mathbf{v}_0}{\Delta t} \qquad (7.3)$$

In this result, the numerator on the far right is the final momentum minus the initial momentum, which is the change in momentum. Thus, the average net force is given by the change in momentum per unit of time.* Multiplying both sides of Equation 7.3 by Δt yields Equation 7.4, which is known as the *impulse–momentum theorem.*

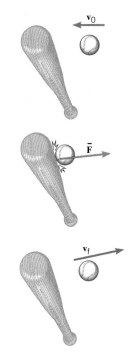

Figure 7.4 When a bat hits a ball, an average force $\overline{\mathbf{F}}$ is applied to the ball. As a result, the ball's velocity changes from an initial value of $\mathbf{v}_0$ (top drawing) to a final value of $\mathbf{v}_f$ (bottom drawing).

■ IMPULSE–MOMENTUM THEOREM

When a net force **F** acts on an object, the impulse of the net force is equal to the change in momentum of the object:

$$\underbrace{\overline{\mathbf{F}}\,\Delta t}_{\text{Impulse}} = \underbrace{m\mathbf{v}_f}_{\substack{\text{Final} \\ \text{momentum}}} - \underbrace{m\mathbf{v}_0}_{\substack{\text{Initial} \\ \text{momentum}}} \qquad (7.4)$$

Impulse = Change in momentum

During a collision, it is often difficult to measure the average force $\overline{\mathbf{F}}$, so it is not easy to determine the impulse, $\overline{\mathbf{F}}\,\Delta t$, directly. On the other hand, it is usually straightforward to measure the mass and velocity of an object, so that its momentum just after and just before the collision, $m\mathbf{v}_f$ and $m\mathbf{v}_0$, can be found. Thus, the impulse–momentum theorem allows us to gain information about the impulse indirectly by simply measuring the change in momentum that the impulse causes. Then,

---

* The equality between force and the time rate of change of momentum is the version of the second law of motion presented originally by Newton.

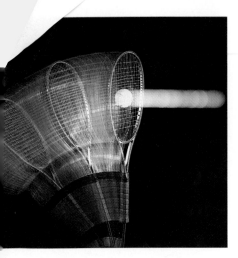

The tennis racquet applies an impulse to the ball. As a result, the outgoing ball has a momentum that is different than that of the incoming ball.

• **PROBLEM SOLVING INSIGHT**
Momentum is a vector quantity and, as such, has a magnitude and a direction. For motion in one dimension, be sure to indicate the direction by assigning a plus or minus sign to it, as this example indicates.

armed with a knowledge of the contact time Δt, we can evaluate the average force. Examples 1 and 2 illustrate how the theorem is used in this way.

EXAMPLE 1 • A Well-Hit Ball

A baseball ($m = 0.14$ kg) has an initial velocity of $\mathbf{v_0} = -38$ m/s as it approaches a bat. We have chosen the direction of approach as the negative direction. The bat applies a force that is much larger than the weight of the ball, and the ball departs from the bat with a final velocity of $\mathbf{v_f} = +58$ m/s. (a) Determine the impulse applied to the ball by the bat. (b) Assuming that the time of contact is $\Delta t = 1.6 \times 10^{-3}$ s, find the average net force exerted on the ball by the bat.

Reasoning In hitting the ball, the bat imparts an impulse to it. We cannot use Equation 7.1 (**Impulse** $= \overline{\mathbf{F}} \, \Delta t$) to determine the impulse, since the average net force $\overline{\mathbf{F}}$ is not known. We can find the impulse, however, by turning to the impulse–momentum theorem, which states that the impulse is equal to the ball's final momentum minus its initial momentum. With values for the impulse and the time of contact, Equation 7.1 can be used to determine the average net force applied to the ball. But in the present case, the net force is just the force applied by the bat, assuming that the ball's weight is negligible in comparison.

Solution

(a) We begin by finding the initial and final momenta of the ball:

Initial momentum $\mathbf{p_0} = m\mathbf{v_0} = (0.14 \text{ kg})(-38 \text{ m/s}) = -5.3 \text{ kg} \cdot \text{m/s}$ (7.2)

Final momentum $\mathbf{p_f} = m\mathbf{v_f} = (0.14 \text{ kg})(+58 \text{ m/s}) = +8.1 \text{ kg} \cdot \text{m/s}$ (7.2)

According to the impulse-momentum theorem, the impulse applied to the ball is:

$$\text{Impulse} = m\mathbf{v_f} - m\mathbf{v_0} \qquad (7.4)$$
$$= (8.1 \text{ kg} \cdot \text{m/s}) - (-5.3 \text{ kg} \cdot \text{m/s}) = \boxed{+13.4 \text{ kg} \cdot \text{m/s}}$$

(b) Now that the impulse is known, the contact time can be used in Equation 7.1 to find the average force:

$$\overline{\mathbf{F}} = \frac{\text{Impulse}}{\Delta t} = \frac{+13.4 \text{ kg} \cdot \text{m/s}}{1.6 \times 10^{-3} \text{ s}} = \boxed{+8400 \text{ N}}$$

The force is positive, indicating that it points opposite to the velocity of the approaching ball. A force of 8400 N corresponds to 1900 lb, such a large value being necessary to change the ball's momentum during the brief contact time.

EXAMPLE 2 • A Rain Storm

During a storm, rain comes straight down with a velocity of $\mathbf{v_0} = -15$ m/s and hits the roof of a car perpendicularly (see Figure 7.5). The mass of rain per second that strikes the car roof is 0.060 kg/s. Assuming that the rain comes to rest upon striking the car roof ($\mathbf{v_f} = 0$), find the average force exerted by the rain on the car roof.

Reasoning This example differs from Example 1 in an important way. Example 1 gives information about the ball and asks for the force applied to the ball. In contrast, the present example gives information about the rain, but doesn't ask for the force applied to the rain. Instead, it asks for the force acting on the car roof. However, the force exerted on the roof by the rain and the force exerted on the rain by the roof have equal magnitudes and opposite directions, according to Newton's law of action and reaction (see Section 4.5). Thus, we will find the force exerted on the rain and then apply the law of action and reaction to obtain the force on the roof.

Solution The average force $\overline{\mathbf{F}}$ needed to reduce the rain's velocity from $\mathbf{v}_0 = -15$ m/s to $\mathbf{v}_f = 0$ is given by Equation 7.3 as

$$\overline{\mathbf{F}} = \frac{m\mathbf{v}_f - m\mathbf{v}_0}{\Delta t} = -\left(\frac{m}{\Delta t}\right)\mathbf{v}_0$$

The term $m/\Delta t$ is the mass of rain per second that strikes the car roof, so that $m/\Delta t = 0.060$ kg/s. Thus, the average force acting on the rain is

$$\overline{\mathbf{F}} = -(0.060 \text{ kg/s})(-15 \text{ m/s}) = +0.90 \text{ N}$$

This force is in the positive or upward direction. This is a reasonable result, since the roof must exert an upward force on each downward-moving raindrop in order to bring it to rest. According to the action–reaction law, the force exerted on the roof by the rain also has a magnitude of 0.90 N but points downward:

$$\text{Force on roof} = \boxed{-0.90 \text{ N}}$$

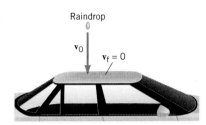

Figure 7.5 A raindrop falling on a car roof has an initial velocity of $\mathbf{v}_0$ just before striking the roof. The final velocity of the raindrop is $\mathbf{v}_f = 0$, because it comes to rest on the roof.

As you reason through problems such as those in Examples 1 and 2, take advantage of the impulse–momentum theorem. It is a powerful statement that can lead to significant insights. The following Conceptual Example further illustrates its use.

CONCEPTUAL EXAMPLE 3 • Hailstones Versus Raindrops

In Example 2 rain is falling on the roof of a car and exerts a force on it. Instead of rain, suppose hail is falling. The hail comes straight down at a mass rate of $m/\Delta t = 0.060$ kg/s and an initial velocity of $\mathbf{v}_0 = -15$ m/s. The hail strikes the roof perpendicularly, just as the rain does in Example 2. However, unlike rain, hail usually does not come to rest after striking a surface. Instead, the hailstones bounce off the roof of the car. If hail fell instead of rain, would the force on the roof be smaller than, equal to, or greater than that calculated in Example 2?

Reasoning and Solution The raindrops and the hailstones fall in exactly the same way. That is, they both fall with the same initial velocity and mass rate, and they both strike the roof perpendicularly. However, there is a difference, and it is an important one: The raindrops come to rest (see Figure 7.5), while the hailstones bounce upward after striking the roof (see Figure 7.6). According to the impulse–momentum theorem, the impulse that acts on an object is given by the change in the momentum of the object. This change is $m\mathbf{v}_f - m\mathbf{v}_0 = m\Delta\mathbf{v}$ and is proportional to the change in velocity $\Delta\mathbf{v}$. For a hailstone, the change in velocity is from $\mathbf{v}_0$ (downward) to $\mathbf{v}_f$ (upward). This is a larger change than for the raindrop, for which the change is only from $\mathbf{v}_0$ (downward) to zero. Therefore, the change in momentum is greater for the hailstone, and, correspondingly, a greater impulse must act on the hailstone. But an impulse is the product of the average force and the time interval Δt. Since the same amount of mass falls in the same time interval in either case, Δt is the same for hailstones as for raindrops. The greater impulse acting on the hailstones, then, means that the car roof must exert a greater force on the hailstones than on the raindrops. Conversely, according to Newton's action–reaction law, **the car roof experiences a greater force from the hailstones than from the raindrops.** This is one reason why a hailstorm often damages cars more than a rainstorm does.

Related Homework Material: Problems 8, 50

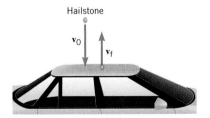

Figure 7.6 Hailstones falling on this car roof have a downward velocity of $\mathbf{v}_0$ just before striking the roof. They rebound from the roof with an upward velocity of $\mathbf{v}_f$.

In the next section we discuss a principle that is based on the impulse–momentum theorem, namely, the principle of conservation of linear momentum.

7.2 THE PRINCIPLE OF CONSERVATION OF LINEAR MOMENTUM

It is worthwhile comparing the impulse–momentum theorem to the work–energy theorem discussed in Chapter 6. The impulse–momentum theorem states that the impulse produced by a net force is equal to the change in the object's momentum, while the work–energy theorem states that the work done by a net force is equal to the change in the object's kinetic energy. The work–energy theorem leads directly to the principle of conservation of mechanical energy (see Figure 6.17), and the impulse–momentum theorem also leads to a conservation theorem, known as the conservation of linear momentum. The concept chart in Figure 7.7, which is an extension of the chart in Figure 7.3, presents an overview of how this new conservation principle arises. Our first task will be to apply the impulse–momentum theorem to each object in a system (or collection) of objects. Then, we will examine the external forces that act on the system. If the sum of these forces is zero, we will see that the impulse–momentum theorem leads to the principle of conservation of linear momentum (upper right box in the drawing). In this and subsequent sections we will apply this conservation principle to a variety of real-life examples.

We begin by applying the impulse–momentum theorem to a midair collision between two objects. The two objects (masses m_1 and m_2) are approaching each other with initial velocities $\mathbf{v}_{01}$ and $\mathbf{v}_{02}$, as Figure 7.8a shows. The collection of objects being studied is referred to as the "system." The objects interact during the collision

Figure 7.7 *Concepts at a Glance*
The impulse–momentum theorem leads to the principle of conservation of linear momentum when the sum of the external forces acting on an object is zero. Note the analogy between this drawing and Figure 6.17, which illustrates how the work–energy theorem leads to the principle of conservation of mechanical energy. In the collision illustrated in the photograph, the conservation of linear momentum means that the total linear momentum of the cue ball and the other balls before the collision is equal to the total linear momentum after the collision.

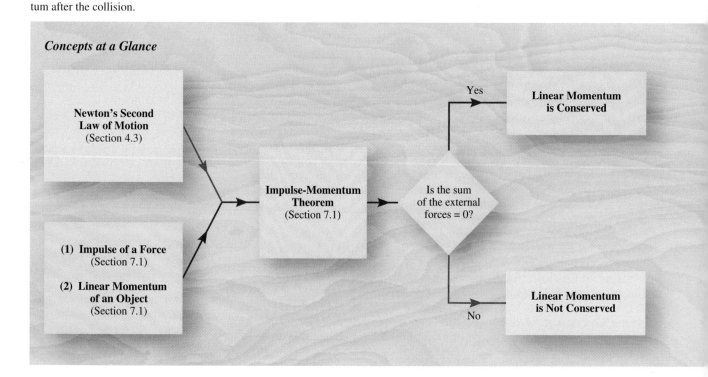

Concepts at a Glance

Newton's Second Law of Motion (Section 4.3)

(1) Impulse of a Force (Section 7.1)

(2) Linear Momentum of an Object (Section 7.1)

Impulse-Momentum Theorem (Section 7.1)

Is the sum of the external forces = 0?

Yes — Linear Momentum is Conserved

No — Linear Momentum is Not Conserved

in part *b* of the drawing and then depart with the final velocities $\mathbf{v_{f1}}$ and $\mathbf{v_{f2}}$ shown in part *c*. Because of the collision, the initial and final velocities are not the same.

There are two types of forces acting on the system:

1. **Internal forces** — Forces that the objects within the system exert on each other.
2. **External forces** — Forces exerted on the objects by agents that are external to the system.

During the collision shown in Figure 7.8*b*, $\mathbf{F_{12}}$ is the force exerted on object 1 by object 2, while $\mathbf{F_{21}}$ is the force exerted on object 2 by object 1. These two forces are action–reaction forces that are equal in magnitude but opposite in direction, so $\mathbf{F_{12}} = -\mathbf{F_{21}}$. They are also classified as internal forces, since they are the forces that the two objects within the system exert on each other. The force of gravity also acts on the objects, their weights being $\mathbf{W_1}$ and $\mathbf{W_2}$. These weights are external forces, because they are applied by the earth, which is outside the system. Friction and air resistance would also be considered external forces, although these forces are ignored here for the sake of simplicity. The impulse–momentum theorem, as applied to each object, gives the following results:

Object 1
$$(\underbrace{\mathbf{W_1}}_{\substack{\text{External} \\ \text{force}}} + \underbrace{\overline{\mathbf{F}}_{\mathbf{12}}}_{\substack{\text{Internal} \\ \text{force}}}) \Delta t = m_1 \mathbf{v_{f1}} - m_1 \mathbf{v_{01}}$$

Object 2
$$(\underbrace{\mathbf{W_2}}_{\substack{\text{External} \\ \text{force}}} + \underbrace{\overline{\mathbf{F}}_{\mathbf{21}}}_{\substack{\text{Internal} \\ \text{force}}}) \Delta t = m_2 \mathbf{v_{f2}} - m_2 \mathbf{v_{02}}$$

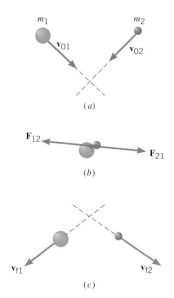

Figure 7.8 (*a*) The velocities of the two objects before the collision are $\mathbf{v_{01}}$ and $\mathbf{v_{02}}$. (*b*) During the collision, each object exerts a force on the other. These forces are labeled $\mathbf{F_{12}}$ and $\mathbf{F_{21}}$. (*c*) The velocities after the collision are $\mathbf{v_{f1}}$ and $\mathbf{v_{f2}}$.

Adding these equations produces a single result for the system as a whole:

$$\underbrace{(\mathbf{W}_1 + \mathbf{W}_2 + \overline{\mathbf{F}}_{12} + \overline{\mathbf{F}}_{21})}_{\substack{\text{External} \\ \text{forces} \quad \text{Internal} \\ \text{forces}}} \Delta t = \underbrace{(m_1 \mathbf{v}_{f1} + m_2 \mathbf{v}_{f2})}_{\substack{\text{Total final} \\ \text{momentum } \mathbf{P}_f}} - \underbrace{(m_1 \mathbf{v}_{01} + m_2 \mathbf{v}_{02})}_{\substack{\text{Total initial} \\ \text{momentum } \mathbf{P}_0}}$$

On the right side of this equation, the quantity $m_1 \mathbf{v}_{f1} + m_2 \mathbf{v}_{f2}$ is the vector sum of the final momenta for each object, or the total final momentum $\mathbf{P}_f$ of the system. Likewise, $m_1 \mathbf{v}_{01} + m_2 \mathbf{v}_{02}$ is the total initial momentum $\mathbf{P}_0$. Therefore, the result above can be rewritten as

$$\left(\begin{matrix} \textbf{Sum of average} \\ \textbf{external forces} \end{matrix} + \begin{matrix} \textbf{Sum of average} \\ \textbf{internal forces} \end{matrix} \right) \Delta t = \mathbf{P}_f - \mathbf{P}_0 \tag{7.5}$$

The advantage of the internal/external force classification is that the internal forces always add together to give zero, as a consequence of Newton's law of action–reaction; $\overline{\mathbf{F}}_{12} = -\overline{\mathbf{F}}_{21}$ so that $\overline{\mathbf{F}}_{12} + \overline{\mathbf{F}}_{21} = 0$. Cancellation of the internal forces occurs no matter how many parts there are to the system and allows us to ignore the internal forces, as Equation 7.6 indicates:

$$(\textbf{Sum of average external forces}) \, \Delta t = \mathbf{P}_f - \mathbf{P}_0 \tag{7.6}$$

We developed this result with gravity as the only external force. But, in general, the sum of the external forces on the left includes *all* external forces.

With the aid of Equation 7.6, it is possible to see how the conservation of linear momentum arises. Suppose that the sum of the external forces is zero. A system for which this is true is called an ***isolated system.*** Then Equation 7.6 indicates that

$$0 = \mathbf{P}_f - \mathbf{P}_0 \quad \text{or} \quad \mathbf{P}_f = \mathbf{P}_0 \tag{7.7}$$

In other words, the final total momentum of the isolated system after the objects in Figure 7.8 collide is the same as the initial total momentum. This is an example of a general principle known as the ***principle of conservation of linear momentum.***

■ PRINCIPLE OF CONSERVATION OF LINEAR MOMENTUM

The total linear momentum of an isolated system remains constant (is conserved). An isolated system is one for which the vector sum of the external forces acting on the system is zero.

This principle applies to a system containing any number of objects, regardless of the internal forces, provided the system is isolated. Whether the system is isolated depends on whether the vector sum of the external forces is zero. Judging whether a force is internal or external depends on which objects are included in the system, as Conceptual Example 4 illustrates.

CONCEPTUAL EXAMPLE 4 • Is the Total Momentum Conserved?

Imagine two balls on a billiard table that is friction-free. Use the momentum conservation principle in answering the following questions. (a) Just before the balls collide, suppose that a hole opens in the table beneath them. In this hypothetical situation, is the total momentum of the two-ball system the same before and after the collision? (b) In a normal situation, when there is no hole in the table, is the total momentum of the two-ball system the same before and after the collision? (c) Answer part (b) for a system that contains only one of the balls.

Reasoning and Solution

(a) Figure 7.9*a* shows this hypothetical situation. The dashed outline around the two balls identifies them as the system. The only external forces that act on this system are $\mathbf{W}_1$ and $\mathbf{W}_2$, which are the weights of the balls. The forces that one ball exerts on the other during the collision are not included in the drawing, because they are internal forces and cannot cause the total momentum of the system to change. Clearly, the vector sum of the external forces is not zero and points downward. This external force causes the balls to accelerate downward and gives the system a downward-pointing momentum that increases during the collision. Thus, the total momentum of the two-ball system is not the same before and after the collision, and ***the total momentum is not conserved.***

(b) Figure 7.9*b* shows the balls colliding on a normal table. Again, the dashed outline emphasizes that both balls are part of the system. As in the first part of this example, the collision forces are not shown, because they are internal forces. The external forces include the weights $\mathbf{W}_1$ and $\mathbf{W}_2$ of the balls. There are now two additional external forces, however: the upward-pointing normal force that the table exerts on each ball. These normal forces are $\mathbf{F}_{N1}$ and $\mathbf{F}_{N2}$. Since the balls do not accelerate in the vertical direction, the normal forces must balance the weights, so that the vector sum of the four external forces in Figure 7.9*b* is zero. Thus, there is no net external force to change the total momentum of the two-ball system, and it is the same before and after the collision. ***The total momentum of this two-ball system is conserved.***

(c) In Figure 7.9*c* only one ball is included in the system, as the dashed outline indicates. The forces that act on this system are all external and include the weight $\mathbf{W}_1$ of the ball and the normal force $\mathbf{F}_{N1}$. As in part *b* of the drawing, these two forces balance. However, there is a third external force to consider. Ball 2 is outside the system, so the force $\mathbf{F}_{12}$ that it applies to the system during the collision is now an external force. As a result, the vector sum of the three external forces is not zero, and the net external force causes the total momentum of the one-ball system to be different after the collision than it is before the collision. ***The total momentum of this one-ball system is not conserved.***

Next, we apply the principle of conservation of linear momentum to the problem of assembling a freight train.

EXAMPLE 5 • Assembling a Freight Train

A freight train is being assembled in a switching yard, and Figure 7.10 shows two boxcars being coupled together. Car 1 has a mass of $m_1 = 65 \times 10^3$ kg and moves at a velocity of $v_{01} = +0.80$ m/s. Car 2, with a mass of $m_2 = 92 \times 10^3$ kg and a velocity of $v_{02} = +1.2$ m/s, overtakes car 1 and couples to it. Neglecting friction, find the common velocity v_f of the two cars after they become coupled.

Reasoning The two boxcars constitute the system. The sum of the external forces acting on the system is zero, because the weight of each car is balanced by a corresponding normal force, and friction is being neglected. Thus, the system is isolated, and the principle of conservation of linear momentum applies. The coupling forces

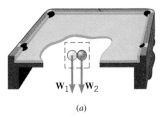

(a)

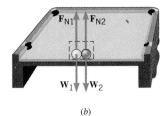

(b)

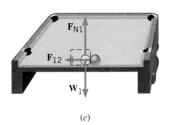

(c)

Figure 7.9 (*a*) In this hypothetical situation, a hole opens up beneath two colliding billiard balls, which have weights $\mathbf{W}_1$ and $\mathbf{W}_2$. The dashed box emphasizes that both balls are included in the system. (*b*) The balls collide on a normal billiard table, where $\mathbf{F}_{N1}$ and $\mathbf{F}_{N2}$ are the normal forces that the table applies to the balls. The dashed box emphasizes that both balls are again included in the system. (*c*) The balls collide on a normal billiard table, but now only ball 1 is included in the system, as the dashed box indicates.

(a)   *(b)*

Figure 7.10 (*a*) The boxcar on the left eventually catches up with the other boxcar and couples to it. (*b*) The coupled cars move together with a common velocity after the collision.

that each car exerts on the other are internal forces and do not affect the applicability of this principle.

Solution Momentum conservation indicates that

$$\underbrace{(m_1 + m_2)v_{\mathrm{f}}}_{\substack{\text{Total momentum}\\\text{after collision}}} = \underbrace{m_1v_{01} + m_2v_{02}}_{\substack{\text{Total momentum}\\\text{before collision}}}$$

This equation can be solved for v_{f}, the common velocity of the two cars after the collision:

$$v_{\mathrm{f}} = \frac{m_1v_{01} + m_2v_{02}}{m_1 + m_2}$$

$$= \frac{(65 \times 10^3 \text{ kg})(0.80 \text{ m/s}) + (92 \times 10^3 \text{ kg})(1.2 \text{ m/s})}{(65 \times 10^3 \text{ kg} + 92 \times 10^3 \text{ kg})} = \boxed{+1.0 \text{ m/s}}$$

In the previous example it can be seen that the velocity of car 1 increases, while the velocity of car 2 decreases as a result of the collision. The acceleration and deceleration arise at the moment the cars become coupled, because the cars exert internal forces on each other. These forces are equal in magnitude and opposite in direction, in accord with Newton's third law. The powerful feature of the momentum conservation principle is that it allows us to determine the changes in velocity without knowing what the internal forces are. Example 6 further illustrates this feature.

EXAMPLE 6 • Ice Skaters

Starting from rest, two skaters "push off" against each other on smooth level ice, where friction is negligible. As Figure 7.11*a* shows, one is a woman ($m_1 = 54$ kg), and one is a man ($m_2 = 88$ kg). Part *b* of the drawing shows that the woman moves away with a velocity of $v_{\mathrm{f}1} = +2.5$ m/s. Find the "recoil" velocity $v_{\mathrm{f}2}$ of the man.

Reasoning For a system consisting of the two skaters on level ice, the sum of the external forces is zero. This is because the weight of each skater is balanced by a corresponding normal force and the ice is assumed to be frictionless. The skaters, then, constitute an isolated system, and the principle of conservation of linear momentum applies. We expect that the man should have a smaller recoil speed for the following reason. The internal forces that the man and woman exert on each other during pushoff have equal magnitudes, but opposite directions, according to Newton's action–reaction law. The man, having the larger mass, experiences a smaller acceleration and, hence, a smaller recoil speed.

Solution The total momentum of the skaters before they push on each other is zero, since they are at rest. Momentum conservation requires that the total momentum remains zero after the skaters have separated, as in part *b* of the drawing:

$$\underbrace{m_1v_{\mathrm{f}1} + m_2v_{\mathrm{f}2}}_{\substack{\text{Total momentum}\\\text{after pushing}}} = \underbrace{0}_{\substack{\text{Total momentum}\\\text{before pushing}}}$$

Solving for the recoil velocity of the man gives

$$v_{\mathrm{f}2} = \frac{-m_1v_{\mathrm{f}1}}{m_2} = \frac{-(54 \text{ kg})(2.5 \text{ m/s})}{88 \text{ kg}} = \boxed{-1.5 \text{ m/s}}$$

(a)

$v_{\mathrm{f}2}$ $v_{\mathrm{f}1} = +2.5$ m/s

(b)

Figure 7.11 (*a*) In the absence of friction, two skaters pushing on each other constitute an isolated system. (*b*) As the skaters move away, the total linear momentum of the system remains zero, which is what it was initially.

The minus sign indicates that the man moves to the left in the drawing. After the skaters separate, the total momentum of the system remains zero, because momentum is a vector quantity, and the momenta of the man and the woman have equal magnitudes but opposite directions.

In Example 6 one skater recoils from the other when they push off. Conceptual Example 7 also deals with recoil, this time in a situation often seen in movies. With the momentum conservation principle to guide our reasoning, we will find that you don't always see correct physics in movies.

CONCEPTUAL EXAMPLE 7 • Firing Blanks

The Physics of...
firing a blank cartridge.

Movies often show someone firing a gun loaded with blanks, as in Figure 7.12. In a blank cartridge the lead bullet is removed, and the end of the shell casing is crimped shut to prevent the gunpowder from spilling out. When a gun fires a blank, is the recoil greater than, the same as, or less than when the gun fires a standard bullet?

Reasoning and Solution   Consider the bullet and gunpowder in a standard cartridge as one part of the system and the rest of the cartridge and the gun, together with the person firing it, as the other part. These two parts are like the two skaters in Example 6, who use their hands to push off. Here, the exploding gunpowder enables the lead bullet and the rest of the system to push off against one another. As in Example 6, we apply the momentum conservation principle. Initially, the two-part system including the gun is at rest and, therefore, has zero total momentum. Momentum conservation dictates that the total momentum is also zero after the gun is fired. However, the lead bullet and the gas from the burning gunpowder are sent off with a large forward momentum. The only way that the total momentum of the system can remain zero is for the rest of the system to recoil with an equally large backward momentum. Now, consider a blank cartridge. Without a lead bullet, a much smaller amount of mass is shot forward when the gun fires. Correspondingly, the forward momentum created upon firing is smaller than with the standard bullet, and the size of the backward momentum required by momentum conservation is also smaller. ***Thus, the recoil is less when a blank is fired than when a standard bullet is fired.***

Related Homework Material: *Problems 20, 46*

Figure 7.12   When a blank cartridge is fired in a movie, is the recoil the same as when a standard bullet is fired? See Conceptual Example 7. (From the movie *Thelma and Louise.*)

It is important to realize that the total linear momentum may be conserved even when the kinetic energies of the individual parts of a system do change. In Example 6, for instance, the initial kinetic energy is zero since the skaters are stationary. But after they push off, the skaters are moving, so each has kinetic energy. The kinetic energy changes, because work is done by the internal force that each skater exerts on the other. This work causes the kinetic energy to increase, consistent with the work–energy theorem (see Section 6.2). However, internal forces cannot change the total linear momentum of a system, since the total linear momentum of an isolated system is conserved in the presence of such forces.

When applying the principle of conservation of linear momentum, we have been following a definite reasoning strategy that is summarized below.

REASONING STRATEGY

Applying the Principle of Conservation of Linear Momentum

1. Decide which objects are included in the system.
2. Relative to the system that you have chosen, identify the internal forces and the external forces.
3. Verify that the system is isolated. In other words, verify that the sum of the external forces applied to the system is zero. Only if the sum of the external forces is zero can the conservation principle be applied. If the sum of the external forces is not zero, consider a different system for analysis.
4. Set the total final momentum of the isolated system equal to the total initial momentum. Remember that linear momentum is a vector. If necessary, apply the conservation principle separately to the various vector components.

7.3 COLLISIONS IN ONE DIMENSION

As discussed in the last section, the total linear momentum is conserved when two objects collide, provided they constitute an isolated system. When the objects are atoms or subatomic particles, it is often found, in addition, that the total kinetic energy of the system is conserved. In other words, the total kinetic energy of the particles before the collision equals the total kinetic energy of the particles after the collision. In such a case, whatever kinetic energy is gained by one particle is lost by the other.

In contrast, when two macroscopic objects collide, such as two cars, the total kinetic energy after the collision is generally less than that before the collision. During such a collision, kinetic energy is lost mainly in two ways. First, it can be converted into heat because of friction. Second, kinetic energy is lost whenever an object suffers permanent distortion and does not return to its original shape. In this case, energy is spent in creating the damage, as in an automobile collision. With very hard objects, such as a solid steel ball and a marble floor, the permanent distortion suffered upon collision is much smaller than with softer objects and, consequently, less kinetic energy is lost during the collision.

Collisions are often classified according to whether the total kinetic energy changes during the collision:

1. **Elastic collision**—One in which the total kinetic energy of the system after the collision is equal to the total kinetic energy before the collision.

2. **Inelastic collision**—One in which the total kinetic energy of the system is *not* the same before and after the collision; if the objects stick together after colliding, the collision is said to be completely inelastic.

The boxcars coupling together in Figure 7.10 is an example of a completely inelastic collision. When a collision is completely inelastic, the greatest amount of kinetic energy is lost. Example 8 shows how one particular elastic collision is described using the conservation of linear momentum and the fact that no kinetic energy is lost during the collision.

EXAMPLE 8 • A Collision in One Dimension

As Figure 7.13 illustrates, a ball of mass $m_1 = 0.250$ kg and velocity $v_{01} = +5.00$ m/s collides head-on with a ball of mass $m_2 = 0.800$ kg that is initially at rest. No external forces act on the balls. If the collision is elastic, what are the velocities of the balls after the collision?

Reasoning The total linear momentum of the two-ball system is conserved, because no external forces act on the system. Momentum conservation applies whether or not the collision is elastic:

$$\underbrace{m_1 v_{f1} + m_2 v_{f2}}_{\substack{\text{Total momentum} \\ \text{after collision}}} = \underbrace{m_1 v_{01} + 0}_{\substack{\text{Total momentum} \\ \text{before collision}}}$$

For an elastic collision, the total kinetic energy is the same before and after the collision

$$\underbrace{\tfrac{1}{2} m_1 v_{f1}^2 + \tfrac{1}{2} m_2 v_{f2}^2}_{\substack{\text{Total kinetic energy} \\ \text{after collision}}} = \underbrace{\tfrac{1}{2} m_1 v_{01}^2 + 0}_{\substack{\text{Total kinetic energy} \\ \text{before collision}}}$$

We expect that ball 1, having the smaller mass, will rebound to the left after striking ball 2, which is more massive. Ball 2 will be driven to the right in the process. One solution to the equations above, then, should give a value for v_{f1} that is negative (the left direction in Figure 7.13) and a value for v_{f2} that is positive.

Solution The equations above are simultaneous equations containing the two unknown quantities v_{f1} and v_{f2}. To solve them, we begin by rearranging the equation expressing momentum conservation to show that $v_{f2} = m_1(v_{01} - v_{f1})/m_2$. Substituting this result into the equation expressing the conservation of kinetic energy leads to the following equation for v_{f1}:

$$v_{f1} = \left(\frac{m_1 - m_2}{m_1 + m_2} \right) v_{01} \qquad (7.8a)$$

The expression for v_{f1} in Equation 7.8a can be substituted into either of the equations obtained in the Reasoning to show that

$$v_{f2} = \left(\frac{2m_1}{m_1 + m_2} \right) v_{01} \qquad (7.8b)$$

• **PROBLEM SOLVING INSIGHT**
As long as the net external force is zero, the conservation of linear momentum applies to any type of collision. This is true whether the collision is elastic or inelastic.

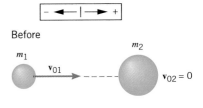

Figure 7.13 A 0.250-kg ball, traveling with an initial velocity of $v_{01} = +5.00$ m/s, undergoes an elastic collision with a 0.800-kg ball that is initially at rest.

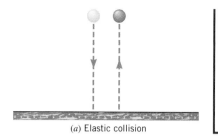

(a) Elastic collision

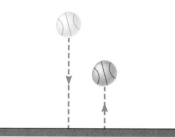

(b) Inelastic collision

(c) Completely inelastic collision

Figure 7.14 (a) A hard steel ball would rebound to its original height after striking a hard marble surface if the collision were elastic. (b) A partially deflated basketball has little bounce on a soft asphalt surface. (c) A deflated basketball has no bounce at all.

With the given values for m_1, m_2, and v_{01}, Equations 7.8 yield the following values for v_{f1} and v_{f2}:

$$\boxed{v_{f1} = -2.62 \text{ m/s}} \quad \text{and} \quad \boxed{v_{f2} = +2.38 \text{ m/s}}$$

The negative value for v_{f1} indicates that ball 1 rebounds to the left after the collision in Figure 7.13, while the positive value for v_{f2} indicates that ball 2 moves to the right, as expected.

We can get a feel for an elastic collision by dropping a steel ball onto a hard surface, such as a marble floor. If the collision is elastic, the ball will rebound to its original height, as Figure 7.14a illustrates. In contrast, a partially deflated basketball exhibits little rebound from a relatively soft asphalt surface, as in part b, indicating that a fraction of the ball's kinetic energy is dissipated during the inelastic collision. The completely deflated basketball in part c has no bounce at all, and a maximum amount of kinetic energy is lost during the completely inelastic collision.

The next example illustrates a completely inelastic collision in a device called a "ballistic pendulum." This device can be used to measure the speed of a bullet.

EXAMPLE 9 • A Ballistic Pendulum

A ballistic pendulum is sometimes used in laboratories to measure the speed of a projectile, such as a bullet. The ballistic pendulum shown in Figure 7.15a consists of a block of wood (mass $m_2 = 2.50$ kg) suspended by a wire of negligible mass. A bullet (mass $m_1 = 0.0100$ kg) is fired with a speed v_{01}. Just after the bullet collides with it, the block (with the bullet in it) has a speed v_f and then swings to a maximum height of 0.650 m above the initial position (see part b of the drawing). Find the speed of the bullet.

Reasoning The physics of the ballistic pendulum can be divided into two parts. First, there is the completely inelastic collision between the bullet and the block. Second, there is the resulting motion of the block and bullet as they swing upward. The total momentum of the system (block plus bullet) is conserved during the collision, because the suspension wire supports the system's weight, which means that the sum of the external forces acting on the system is nearly zero. Furthermore, as the system swings upward, the principle of conservation of mechanical energy applies, because nonconservative forces do no work. The tension force in the wire does no work because it acts perpendicular to the motion. And air resistance is negligible during the swing.

Solution Applying the momentum conservation principle, we find that

$$\underbrace{(m_1 + m_2)v_f}_{\substack{\text{Total momentum} \\ \text{after collision}}} = \underbrace{m_1 v_{01}}_{\substack{\text{Total momentum} \\ \text{before collision}}}$$

This equation can be solved for the initial speed v_{01} of the bullet:

$$v_{01} = \frac{m_1 + m_2}{m_1} v_f$$

▶ **The Physics of...**
measuring the speed of a bullet.

To determine v_{01}, a value is needed for the speed v_f immediately after the collision. This value can be obtained from the maximum height to which the system swings, by using the principle of conservation of mechanical energy:

$$\underbrace{(m_1 + m_2)gh_f}_{\substack{\text{Total mechanical energy} \\ \text{at the top of the swing,} \\ \text{all potential}}} = \underbrace{\tfrac{1}{2}(m_1 + m_2)v_f^2}_{\substack{\text{Total mechanical energy} \\ \text{at the bottom of the} \\ \text{swing, all kinetic}}}$$

It follows that

$$v_f = \sqrt{2gh_f} = \sqrt{2(9.80 \text{ m/s}^2)(0.650 \text{ m})} = 3.57 \text{ m/s}$$

With this value for v_f, it is now possible to determine the speed of the bullet:

$$v_{01} = \frac{m_1 + m_2}{m_1}v_f = \frac{0.0100 \text{ kg} + 2.50 \text{ kg}}{0.0100 \text{ kg}}(3.57 \text{ m/s}) = \boxed{+896 \text{ m/s}}$$

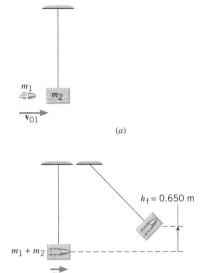

Figure 7.15 (*a*) A bullet approaches a ballistic pendulum. (*b*) The block and bullet swing upward after the collision.

7.4 COLLISIONS IN TWO DIMENSIONS

The collisions discussed so far have been "head-on" or one-dimensional collisions, in the sense that the velocities of the objects all point along a single line before and after contact is made. Collisions often occur, however, in two or three dimensions. Figure 7.16 shows a two-dimensional case in which two balls collide on a horizontal frictionless table.

For the system consisting of the two balls, the external forces include the weights of the balls and the corresponding normal forces produced by the table. Since each weight is balanced by a normal force, the sum of the external forces is

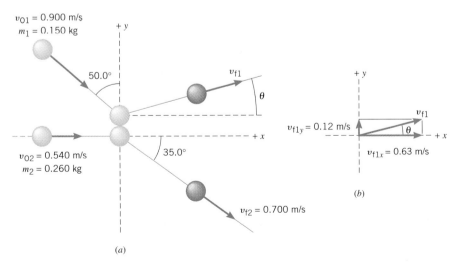

Figure 7.16 (*a*) Top view of two balls colliding on a horizontal frictionless table. (*b*) This part of the drawing shows the *x* and *y* components of the velocity of ball 1 after the collision.

zero, and the total momentum of the system is conserved, as Equation 7.7 indicates ($\mathbf{P_f} = \mathbf{P_0}$). Momentum is a vector quantity, however, and in two dimensions the x and y components of the total momentum are conserved separately. In other words, Equation 7.7 is equivalent to the following two equations:

x Component $\qquad\qquad\qquad\qquad\qquad P_{fx} = P_{0x}$ $\qquad\qquad\qquad$ (7.9a)

y Component $\qquad\qquad\qquad\qquad\qquad P_{fy} = P_{0y}$ $\qquad\qquad\qquad$ (7.9b)

Example 10 shows how to deal with a two-dimensional collision when the total linear momentum of the system is conserved.

EXAMPLE 10 • A Collision in Two Dimensions

For the data given in Figure 7.16, use momentum conservation to determine the magnitude and direction of the velocity of ball 1 after the collision.

Reasoning The magnitude and direction of the velocity of ball 1 can be obtained once the components v_{f1x} and v_{f1y} of this velocity are known. Thus, we will begin by determining these components with the aid of the momentum conservation principle.

Solution Applying momentum conservation (Equation 7.9a) to the x component of the total momentum, we find that

x Component

$$\underbrace{(0.150\text{ kg})(v_{f1x})}_{\text{Ball 1, after}} + \underbrace{(0.260\text{ kg})(0.700\text{ m/s})(\cos 35.0°)}_{\text{Ball 2, after}}$$

$$= \underbrace{(0.150\text{ kg})(0.900\text{ m/s})(\sin 50.0°)}_{\text{Ball 1, before}} + \underbrace{(0.260\text{ kg})(0.540\text{ m/s})}_{\text{Ball 2, before}}$$

• **PROBLEM SOLVING INSIGHT**
Momentum, being a vector quantity, has a magnitude and a direction. In two dimensions, take into account the direction by using vector components and assigning a plus or minus sign to each component, as illustrated in this example.

This equation can be solved to show that $v_{f1x} = +0.63$ m/s. Applying momentum conservation (Equation 7.9b) to the y component of the total momentum, we find that

y Component

$$\underbrace{(0.150\text{ kg})(v_{f1y})}_{\text{Ball 1, after}} + \underbrace{(0.260\text{ kg})[-(0.700\text{ m/s})(\sin 35.0°)]}_{\text{Ball 2, after}}$$

$$= \underbrace{(0.150\text{ kg})[-(0.900\text{ m/s})(\cos 50.0°)]}_{\text{Ball 1, before}} + \underbrace{0}_{\text{Ball 2, before}}$$

The solution to this equation reveals that $v_{f1y} = +0.12$ m/s.

Figure 7.16*b* shows the x and y components of the final velocity of ball 1. The magnitude of the velocity is

$$v_{f1} = \sqrt{(0.63\text{ m/s})^2 + (0.12\text{ m/s})^2} = \boxed{0.64\text{ m/s}}$$

The direction of the velocity is given by the angle θ:

$$\theta = \tan^{-1}\left(\frac{0.12\text{ m/s}}{0.63\text{ m/s}}\right) = \boxed{11°}$$

7.5 CENTER OF MASS

In previous sections, we have encountered situations in which objects interact with one another, such as the two colliding billiard balls in Conceptual Example 4 and the two skaters pushing off in Example 6. In these kinds of situations, the mass of the system is located in several places and the various contributions to the total move about during the interaction. It is possible, however, to speak of a kind of average location for the total mass by introducing a concept known as the *center of mass* (abbreviated as "cm"). With the aid of this concept, we will be able to gain additional insight into the principle of conservation of linear momentum.

The center of mass is a point that represents the average location for the total mass of a system. Figure 7.17, for example, shows two particles of mass m_1 and m_2 that are located on the x axis at the positions x_1 and x_2, respectively. The distance x_{cm} of the center-of-mass point from the origin is defined to be

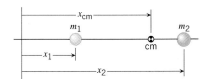

Figure 7.17 The center of mass "cm" of the two particles is located on a line between them and lies closer to the more massive one.

Center of mass
$$x_{cm} = \frac{m_1 x_1 + m_2 x_2}{m_1 + m_2} \qquad (7.10)$$

Each term in the numerator of this equation is the product of a particle's mass and position, while the denominator is the total mass of the system. If the two masses are equal, we expect the average location of the total mass to be midway between the particles. With $m_1 = m_2 = m$, Equation 7.10 becomes $x_{cm} = (mx_1 + mx_2)/(m + m) = \frac{1}{2}(x_1 + x_2)$, which indeed corresponds to the point midway between the particles in Figure 7.17. Alternatively, suppose that $m_1 = 5.0$ kg and $x_1 = 2.0$ m, while $m_2 = 12$ kg and $x_2 = 6.0$ m. Then we expect the average location of the total mass to be located closer to particle 2, since it is more massive. Equation 7.10 is also consistent with this expectation, for it gives

$$x_{cm} = \frac{(5.0 \text{ kg})(2.0 \text{ m}) + (12 \text{ kg})(6.0 \text{ m})}{5.0 \text{ kg} + 12 \text{ kg}} = 4.8 \text{ m}$$

If a system contains more than two particles, the center-of-mass point can be determined by appropriately generalizing Equation 7.10. For three particles, for instance, the numerator in Equation 7.10 would contain a third term $m_3 x_3$, and the total mass in the denominator would be $m_1 + m_2 + m_3$. For a macroscopic object, which contains many, many particles, the center-of-mass point is located at the geometric center of the object, provided that the mass is distributed symmetrically. Such would be the case for a billiard ball. For objects such as a golf club, the mass is not distributed symmetrically and the center-of-mass point is not located at the geometric center of the club. The driver used to launch a golf ball from the tee, for instance, has more mass in the club head than in the handle, so the center-of-mass point is closer to the head than to the handle.

To see how the center of mass concept is related to momentum conservation, suppose that the two particles in a system are moving, as they would be during a collision. With the aid of Equation 7.10, we can readily determine the velocity v_{cm} of the center-of-mass point. During a time interval Δt, the particles experience displacements of Δx_1 and Δx_2, as Figure 7.18 shows. They have different displacements during this time, because they have different velocities. Equation 7.10 can be used to find the displacement Δx_{cm} of the center of mass by replacing x_{cm} by Δx_{cm},

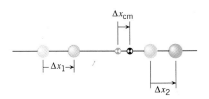

Figure 7.18 During a time interval Δt, the displacements of the particles are Δx_1 and Δx_2, while the displacement of the center of mass is Δx_{cm}.

x_1 by Δx_1, and x_2 by Δx_2:

$$\Delta x_{cm} = \frac{m_1 \Delta x_1 + m_2 \Delta x_2}{m_1 + m_2}$$

Now we divide both sides of this equation by the time interval Δt. In the limit as Δt becomes infinitesimally small, the ratio $\Delta x_{cm}/\Delta t$ becomes equal to the instantaneous velocity v_{cm} of the center of mass. (See Section 2.2 for a review of instantaneous velocity.) Likewise, the ratios $\Delta x_1/\Delta t$ and $\Delta x_2/\Delta t$ become equal to the instantaneous velocities v_1 and v_2, respectively. Thus, we have

Velocity of center of mass $\qquad v_{cm} = \dfrac{m_1 v_1 + m_2 v_2}{m_1 + m_2}$ (7.11)

The numerator $(m_1 v_1 + m_2 v_2)$ on the right-hand side in Equation 7.11 is the momentum of particle 1 $(m_1 v_1)$ plus the momentum of particle 2 $(m_2 v_2)$, which is the total linear momentum of the system. In an isolated system, the total linear momentum does not change because of an interaction such as a collision. Therefore, Equation 7.11 indicates that the velocity v_{cm} of the center of mass does not change either. To emphasize this important point, consider the collision discussed in Example 8. With the data from that example, we can apply Equation 7.11 to determine the velocity of the center of mass before and after the collision:

Before collision $\qquad v_{cm} = \dfrac{(0.250 \text{ kg})(+5.00 \text{ m/s}) + (0.800 \text{ kg})(0)}{0.250 \text{ kg} + 0.800 \text{ kg}} = +1.19 \text{ m/s}$

After collision $\qquad v_{cm} = \dfrac{(0.250 \text{ kg})(-2.62 \text{ m/s}) + (0.800 \text{ kg})(+2.38 \text{ m/s})}{0.250 \text{ kg} + 0.800 \text{ kg}} = +1.19 \text{ m/s}$

Thus, the velocity of the center of mass is the same before and after the objects interact during a collision in which the total linear momentum is conserved.

SUMMARY

The **impulse** of a force is the product of the average force $\overline{F}$ and the time interval Δt during which the force acts: **Impulse** $= \overline{F} \Delta t$. Impulse is a vector that points in the same direction as the average force.

The **linear momentum p** of an object is the product of the object's mass m and velocity $\mathbf{v}$: $\mathbf{p} = m\mathbf{v}$. Linear momentum is a vector that points in the direction of the velocity. The total momentum of a system of objects is the vector sum of the momenta of the individual objects.

The **impulse–momentum theorem** states that an impulse produces a change in an object's momentum, according to $\overline{F} \Delta t = m\mathbf{v}_f - m\mathbf{v}_0$, where $m\mathbf{v}_f$ is the final momentum and $m\mathbf{v}_0$ is the initial momentum.

The **principle of conservation of linear momentum** states that the total linear momentum of an isolated system remains constant. An isolated system is one for which the sum of the external forces acting on the system is zero. External forces are those that agents external to the system exert on objects within the system. In contrast, internal forces

are those that the objects within the system exert on each other.

An **elastic collision** is one in which the total kinetic energy is the same before and after the collision. An **inelastic collision** is one in which the total kinetic energy is not the same before and after the collision. If the objects stick together after colliding, the collision is said to be **completely inelastic.**

The location x_{cm} of the **center of mass** of two objects lying on the x axis is given by $x_{cm} = \dfrac{m_1 x_1 + m_2 x_2}{m_1 + m_2}$, where m_1 and m_2 are the masses of the objects, and x_1 and x_2 are their distances from the coordinate origin. If the objects are moving with velocities v_1 and v_2, the velocity v_{cm} of the center of mass is given by $v_{cm} = \dfrac{m_1 v_1 + m_2 v_2}{m_1 + m_2}$. If the total linear momentum of a system of objects remains constant during an interaction such as a collision, the velocity of the center of mass also remains constant.

CONCEPTUAL QUESTIONS

1. Two identical automobiles have the same speed, one traveling east and one traveling west. Do these cars have the same momentum? Explain.

2. In Times Square in New York City, people celebrate on New Year's Eve. Some just stand around, but many move about randomly. Consider a system comprised of all of these people. Approximately, what is the total linear momentum of this system at any given instant? Justify your answer.

3. If two different objects have the same momentum, do they necessarily have the same kinetic energy? Give your reason.

4. (a) Can a single object have kinetic energy but no momentum? (b) Can a system of two or more objects have a total kinetic energy that is not zero but a total momentum that is zero? Account for your answers.

5. An airplane is flying horizontally with a constant momentum during a time interval Δt. (a) With the aid of Equation 7.4, decide whether a net impulse is acting on the plane during this time interval. (b) In the horizontal direction, the thrust generated by the engines and air resistance both act on the plane. What does the answer in part (a) imply about the impulse of the thrust compared to the impulse of the resistive force?

6. You have a choice. You may get hit head-on either by an adult moving slowly on a bicycle or by a child that is moving twice as fast on a bicycle. The mass of the child and bicycle is one-half that of the adult and bicycle. Considering only the issues of mass and velocity, which collision do you prefer? Or doesn't it matter? Account for your answer.

7. An object slides along the surface of the earth and slows down because of kinetic friction. If the object itself is considered as the system, the kinetic frictional force must be identified as an external force that, according to Equation 7.4, decreases the momentum of the system. (a) If *both* the object and the earth are considered to be part of the system, is the force of kinetic friction still an external force? (b) Can the friction force change the total linear momentum of the two-body system? Give your reasoning for both answers.

8. When driving a golf ball, a good "follow-through" helps to increase the distance of the drive. A good follow-through means that the club head is kept in contact with the ball as long as possible. Using the impulse–momentum theorem, explain why this technique allows you to hit the ball farther.

9. The drawing shows a garden sprinkler that whirls around a

vertical axis. From each of the four arms of the sprinkler, water exits through a tapered nozzle. Because of this nozzle, the water leaves each nozzle with a speed that is greater than the speed inside the arm. (a) Apply the impulse–momentum theorem to deduce the direction of the force applied to the water. (b) Then, with the aid of Newton's third law, explain how the water causes the whirling motion.

10. In movies, Superman hovers stationary in midair, grabs a villain by the neck, and throws him forward. Superman, however, remains stationary. Using the conservation of linear momentum, explain what is wrong with this scene.

11. A satellite explodes in outer space, far from any other body, sending thousands of pieces in all directions. How does the linear momentum of the satellite before the explosion compare with the total linear momentum of all the pieces after the explosion? Account for your answer.

12. You are a passenger on a jetliner that is flying at a constant velocity. You get up from your seat and walk toward the front of the plane. Because of this action, your forward momentum increases. What, if anything, happens to the forward momentum of the plane? Give your reasoning.

13. In a movie, the hero jumps straight down from a bridge onto a small boat that is moving at constant velocity. The horizontal velocity component of the boat does not change. Has trick photography been used to produce this scene? Explain.

14. On a distant asteroid, a large catapult is used to "throw" chunks of stone into space. Could such a device be used as a propulsion system to move the asteroid closer to the earth? Explain.

15. A collision occurs between three moving billiard balls such that no net external force acts on the three-ball system. Is the momentum of *each* ball conserved during the collision? If so, explain why. If not, what quantity is conserved?

16. In an elastic collision, is the kinetic energy of *each* object the same before and after the collision? Explain.

17. Review Example 8. Now, suppose both objects have the same mass, $m_1 = m_2$. Describe what happens to the velocities of both objects as a result of the collision, using Equations 7.8a and 7.8b to justify your answers.

18. Where would you expect the center of mass of a doughnut to be located? Why?

19. Would you expect the center of mass of a baseball bat to be located halfway between the ends of the bat, nearer the lighter end or nearer the heavier end? Provide a reason for your answer.

20. A sunbather is lying on a floating raft that is stationary. She then gets up and walks to one end of the raft. Consider the sunbather and raft as an isolated system. (a) What is the velocity of the center of mass of this system while she is walking? Why? (b) Does the raft itself move while she is walking? If so, what is the direction of the raft's velocity relative to that of the sunbather? Provide a reason for your answer.

PROBLEMS

Section 7.1 The Impulse–Momentum Theorem

1. ssm A soccer player kicks a ball with an average force of $+1400$ N, and her foot remains in contact with the ball for a time of 7.9×10^{-3} s. What is the impulse (magnitude and direction) of this force?

2. A freight train moves due north with a speed of 1.4 m/s. The mass of the train is 4.5×10^5 kg. How fast would an 1800-kg automobile have to be moving due north to have the same momentum as the train?

3. A 62.0-kg person, standing on a diving board, dives straight down into the water. Just before striking the water, her speed is 5.50 m/s. At a time of 1.65 s after entering the water, her speed is reduced to 1.10 m/s. What is the average net force (magnitude and direction) that acts on her when she is in the water?

4. (a) The earth travels with an approximate speed of 29.9 km/s (66 900 mi/h) in its journey around the sun. The mass of the earth is 5.98×10^{24} kg. Find the magnitude of its linear momentum. (b) Is the direction of the earth's linear momentum constant? If not, describe how it changes and specify the force that causes it to change.

5. ssm A volleyball is spiked so that its incoming velocity of $+4.0$ m/s is changed to an outgoing velocity of -21 m/s. The mass of the volleyball is 0.35 kg. What impulse does the player apply to the ball?

6. A woman, driving a golf ball off a tee, gives the ball a velocity of $+28$ m/s. The mass of the ball is 0.045 kg, and the duration of the impact with the golf club is 6.0×10^{-3} s. (a) What is the change in momentum of the ball? (b) Determine the average force applied to the ball by the club.

7. A 46-kg skater is standing still in front of a wall. By pushing against the wall she propels herself backward with a velocity of -1.2 m/s. Her hands are in contact with the wall for 0.80 s. Ignore friction and wind resistance. Find the magnitude and direction of the average force she exerts on the wall (which has the same magnitude, but opposite direction, as the force that the wall applies to her).

8. Before starting this problem, review Conceptual Example 3. Suppose that the hail described there bounces off the roof of the car with a velocity of $+15$ m/s. Calculate the force exerted by the hail on the roof. Compare your answer to that obtained in Example 2 for the rain, and verify that your answer is consistent with the conclusion reached in Conceptual Example 3.

*9. ssm www A golf ball strikes a hard, smooth floor at an angle of 30.0° and, as the drawing shows, rebounds at the same angle. The mass of the ball is 0.047 kg, and its speed is 45 m/s just before and after striking the floor. What is the magnitude of the

impulse applied to the golf ball by the floor? *(Hint: Note that only the vertical component of the ball's momentum changes during impact with the floor.)*

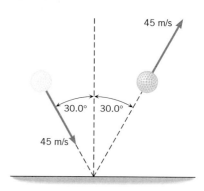

*10. A basketball ($m = 0.60$ kg) is dropped from rest. Just before striking the floor, the magnitude of the basketball's momentum is 3.1 kg·m/s. At what height was the basketball dropped?

*11. A stream of water strikes a stationary turbine blade, as the drawing illustrates. The incident water stream has a velocity of $+18.0$ m/s, while the exiting water stream has a velocity of -18.0 m/s. The mass of water per second that strikes the blade is 25.0 kg/s. Find the magnitude of the average force exerted on the water by the blade.

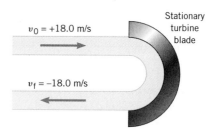

*12. A 0.500-kg ball is dropped from rest at a point 1.20 m above the floor. The ball rebounds straight upward to a height of 0.700 m. What is the magnitude and direction of the impulse applied to the ball by the floor?

**13. ssm Starting with an initial speed of 5.00 m/s at a height of 0.300 m, a 1.50-kg ball swings downward and strikes a 4.60-kg ball that is at rest, as the drawing shows. (a) Using the principle of conservation of mechanical energy, find the speed of the 1.50-kg ball just before impact. (b) Assuming that the collision is elastic, find the velocities (magnitude and direction) of both balls just after the collision. (c) How high does each ball swing after the collision, ignoring air resistance?

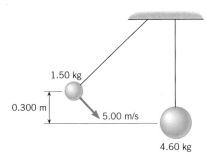

1.50 kg

0.300 m

5.00 m/s

4.60 kg

**14. A dump truck is being filled with sand. The sand falls straight downward from rest from a height of 2.00 m above the truck bed, and the mass of sand that hits the truck per second is 55.0 kg/s. The truck is parked on the platform of a weight scale. By how much does the scale reading exceed the weight of the truck and sand?

Section 7.2 The Principle of Conservation of Linear Momentum

15. With the engines off, a spaceship is coasting at a velocity of +230 m/s through outer space. It fires a rocket straight ahead at an enemy vessel. The mass of the rocket is 1300 kg, and the mass of the spaceship (not including the rocket) is 4.0×10^6 kg. The firing of the rocket brings the spaceship to a halt. What is the velocity of the rocket?

16. ⚕ For tests using a *ballistocardiograph,* a patient lies on a horizontal platform that is supported on jets of air. Because of the air jets, the friction impeding the horizontal motion of the platform is negligible. Each time the heart beats, blood is pushed out of the heart in a direction that is nearly parallel to the platform. Since momentum must be conserved, the body and the platform recoil, and this recoil can be detected to provide information about the heart. For each beat, suppose that 0.050 kg of blood is pushed out of the heart with a velocity of +0.25 m/s and that the mass of the patient and platform is 85 kg. Assuming that the patient does not slip with respect to the platform, and that the patient and platform start from rest, determine the recoil velocity.

17. **ssm www** A lumberjack (mass = 98 kg) is standing at rest on one end of a floating log (mass = 230 kg) that is also at rest. The lumberjack runs to the other end of the log, attaining a velocity of +3.6 m/s relative to the shore, and then hops onto an identical floating log that is initially at rest. Neglect any friction and resistance between the logs and the water. (a) What is the velocity of the first log just before the lumberjack jumps off? (b) Determine the velocity of the second log if the lumberjack comes to rest on it.

18. An astronaut is motionless in outer space. Upon command, the propulsion unit strapped to his back ejects some gas with a velocity of +32 m/s, and the astronaut recoils with a velocity of −0.30 m/s. After the gas is ejected, the mass of the astronaut is 160 kg. What is the mass of the ejected gas?

19. A 1550-kg car, traveling with a velocity of +12.0 m/s, plows into a 1220-kg stationary car. During the collision, the two cars lock bumpers and then move together as a unit. (a) What is their common velocity just after the impact? (b) What fraction of the initial kinetic energy remains after the collision?

20. Review Conceptual Example 7 as background for this problem. The lead female character in the movie *Diamonds Are Forever* is standing at the edge of an offshore oil rig. As she fires the gun, she is driven back over the edge and into the sea. Suppose the mass of a bullet is 0.010 kg, and its velocity is +720 m/s. Her mass (including the gun) is 51 kg. (a) What recoil velocity does she acquire in response to a single shot from a stationary position, assuming that no external force keeps her in place? (b) Under the same assumption, what would be her recoil velocity if, instead, she shoots a blank cartridge that ejects a mass of 5.0×10^{-4} kg at a velocity of +720 m/s?

*21. ssm During July 1995 the comet Shoemaker–Levy 9 smashed into Jupiter in a spectacular fashion. The comet actually consisted of 21 distinct pieces, the largest of which had a mass of about 4.0×10^{12} kg and a speed of 6.0×10^4 m/s. Jupiter, the largest planet in the solar system, has a mass of 1.9×10^{27} kg and an orbital speed of 1.3×10^4 m/s. If this piece of the comet had hit Jupiter head on, what would be the *change* in Jupiter's orbital speed (not its final speed)?

*22. Adolf and Ed are wearing harnesses and are hanging from the ceiling by means of ropes attached to them. They are face to face and push off against one another. Adolf has a mass of 120 kg, and Ed has a mass of 78 kg. Following the push, Adolf swings upward to a height of 0.65 m above his starting point. To what height above his starting point does Ed rise?

*23. An automobile has a mass of 2300 kg and a velocity of +16 m/s. It makes a rear-end collision with a stationary car whose mass is 1800 kg. The cars lock bumpers and skid off together with the wheels locked. (a) What is the velocity of the two cars just after the collision? (b) Find the impulse (magnitude and direction) that acts on the skidding cars from just after the collision until they come to a halt. (c) If the coefficient of kinetic friction between the wheels of the cars and the pavement is $\mu_k = 0.80$, determine how far the cars skid before coming to rest.

**24. A wagon is coasting at a speed v_A along a straight and level road. When ten percent of the wagon's mass is thrown off the wagon, parallel to the ground and in the forward direction, the wagon is brought to a halt. If the direction in which this mass is thrown is exactly reversed, but the speed of this mass relative to the wagon remains the same, the wagon accelerates to a new speed v_B. Calculate the ratio v_B/v_A.

**25. ssm Two people are standing on a 2.0-m-long platform, one at each end. The platform floats parallel to the ground on a cushion of air, like a hovercraft. One person throws a 6.0-kg ball to the other, who catches it. The ball travels nearly horizontally. Excluding the ball, the total mass of the platform and people is 118 kg. Because of the throw, this 118-kg mass recoils. How far does it move before coming to rest again?

Section 7.3 and Section 7.4 Collisions in One and Two Dimensions

26. A 45-kg swimmer runs with a horizontal velocity of +5.1 m/s off a boat dock into a stationary 12-kg rubber raft. Find the velocity that the swimmer and raft would have after the impact, if there were no friction and resistance due to the water.

27. In a football game, a receiver is standing still, having just caught a pass. Before he can move, a tackler, running at a velocity of +4.5 m/s, grabs him. The tackler holds onto the receiver, and the two move off together with a velocity of +2.6 m/s. The mass of the tackler is 115 kg. Assuming that momentum is conserved, find the mass of the receiver.

28. A boy is at rest on a skateboard. The total mass of the boy and the skateboard is 30.0 kg. He is at the foot of a hill when he catches a 5.00-kg ball and, as a result, moves up the hill. Just before he catches the ball, it is moving nearly parallel to the ground and has a speed of 10.0 m/s. Ignore the effects of friction and gravity during the short time the ball is being caught. Through what maximum vertical height will the boy coast up the hill?

29. ssm A golf ball bounces down a flight of steel stairs, striking each stair once on the way down. The ball starts at the top step with a vertical velocity component of zero. If all the collisions with the stairs are elastic, and if the vertical height of the staircase is 3.00 m, determine the bounce height when the ball reaches the bottom of the stairs. Neglect air resistance.

30. A cue ball (mass = 0.165 kg) is at rest on a frictionless pool table. The ball is hit dead center by a pool stick which applies an impulse of +1.50 N·s to the ball. The ball then slides along the table and makes an elastic head-on collision with a second ball of equal mass that is initially at rest. Find the velocity of the second ball just after it is struck.

31. A 0.150-kg projectile is fired with a velocity of +715 m/s at a 2.00-kg wooden block that rests on a frictionless table. The velocity of the block, immediately after the projectile passes through it, is +40.0 m/s. Find the velocity with which the projectile exits from the block.

32. The drawing shows a collision between two pucks on an air-hockey table. Puck A has a mass of 0.025 kg and is moving along the x axis with a velocity of +5.5 m/s. It makes a collision with puck B, which has a mass of 0.050 kg and is initially at rest. After the collision, the two pucks fly apart with the angles shown in the drawing. Find the final speed of (a) puck A and (b) puck B.

33. ssm www A 5.00-kg ball, moving to the right at a velocity of +2.00 m/s on a frictionless table, collides head-on with a stationary 7.50-kg ball. Find the final velocities of the balls if the collision is (a) elastic and (b) completely inelastic.

*__**34.**__ A 60.0-kg person, running horizontally with a velocity of +3.80 m/s, jumps onto a 12.0-kg sled that is initially at rest. (a) Ignoring the effects of friction during the collision, find the velocity of the sled and person as they move away. (b) The sled and person coast 30.0 m on level snow before coming to rest. What is the coefficient of kinetic friction between the sled and the snow?

*__**35.**__ A 0.010-kg bullet is fired straight up at a falling wooden block that has a mass of 2.0 kg. The bullet has a speed of 750 m/s when it strikes the block. The block originally was dropped from rest from the top of a building and had been falling for a time t when the collision with the bullet occurs. As a result of the collision, the block (with the bullet in it) reverses direction, rises, and comes to a momentary halt at the top of the building. Find the time t.

*__**36.**__ Three guns are mounted on a circle, 120.0° apart. They are aimed at the center of the circle, and each fires a bullet simultaneously. One bullet has an unknown mass and a speed of 575 m/s. The other two bullets have the same mass of 4.50×10^{-3} kg and the same speed of 324 m/s. The bullets collide at the center and mash into a *stationary* lump. What is the unknown mass?

*__**37.** **ssm**__ A 50.0-kg skater is traveling due east at a speed of 3.00 m/s. A 70.0-kg skater is moving due south at a speed of 7.00 m/s. They collide and hold on to each other after the collision, managing to move off at an angle θ south of east, with a speed of v_f. Find (a) the angle θ and (b) the speed v_f, assuming that friction can be ignored.

*__**38.**__ A mine car, whose mass is 440 kg, rolls at a speed of 0.50 m/s on a horizontal track, as the drawing shows. A 150-kg chunk of coal has a speed of 0.80 m/s when it leaves the chute. Determine the velocity of the car/coal system after the coal has come to rest in the car.

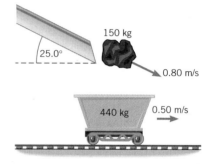

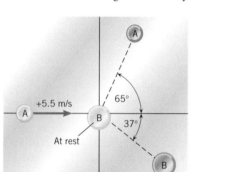

Before collision   After collision

*__**39.**__ A ball is dropped from rest at the top of a 6.10-m-tall building, falls straight downward, collides inelastically with the ground, and bounces back. The ball loses 10.0% of its kinetic energy every time it collides with the ground. How many bounces can the ball make and still reach a window sill that is 2.44 m above the ground?

**40. Two identical balls are traveling toward each other with velocities of -4.0 and $+7.0$ m/s, and they experience an elastic head-on collision. Obtain the velocities (magnitude and direction) of each ball after the collision.

Section 7.5 Center of Mass

41.   ssm The earth and moon are separated by a center-to-center distance of 3.85×10^8 m. The mass of the earth is 5.98×10^{24} kg and that of the moon is 7.35×10^{22} kg. How far does the center of mass lie from the center of the earth?

42. Consider the two moving boxcars in Example 5. Determine the velocity of their center of mass (a) before and (b) after the col-

lision. (c) Should your answer in part (b) be less than, greater than, or equal to the common velocity v_f of the two coupled cars after the collision? Justify your answer.

43. The carbon monoxide molecule (CO) consists of a carbon atom and an oxygen atom separated by a distance of 1.13×10^{-10} m. The mass m_C of the carbon atom is 0.750 times the mass m_O of the oxygen atom: $m_C = 0.750 \, m_O$. Determine the location of the center of mass of this molecule relative to the carbon atom.

*44. Ball 1 and ball 2 are each moving in the $+x$ direction. The ratio of their velocities is $v_2/v_1 = 4.0$. A force is then applied to ball 2, causing its velocity to triple. The velocity of ball 1 remains unchanged. As a result, the velocity of the center of mass of the system doubles. Find the ratio m_2/m_1 of the masses of the balls.

ADDITIONAL PROBLEMS

45.   ssm Batman (mass = 91 kg) jumps straight down from a bridge into a boat (mass = 510 kg) in which a criminal is fleeing. The velocity of the boat is initially $+11$ m/s. What is the velocity of the boat after Batman lands in it?

46. As background for this problem, review Conceptual Example 7. A person is standing on a sheet of ice that is so slippery that friction may be ignored. This individual fires a gun parallel to the ground. When a standard cartridge is used, a 14-g bullet is shot forward with a speed of 280 m/s, and the person recoils with a speed of v_C. When a blank cartridge is used, a mass of 0.10 g is shot forward with a speed of 55 m/s, and the recoil speed is v_B. Find the ratio v_B/v_C and verify that it is consistent with the conclusion reached in Conceptual Example 7.

47. Two arrows are fired horizontally with the same speed of 30.0 m/s. Each arrow has a mass of 0.100 kg. One is fired due east and the other due south. Find the magnitude and direction of the total momentum of this two-arrow system. Specify the direction with respect to due east.

48. Two joggers are running with the same velocity of 4.00 m/s due south. Their masses are 90.0 and 55.0 kg. (a) Find the magnitude of the total linear momentum and the total kinetic energy of the two-jogger system. (b) Repeat part (a), assuming that one of the joggers is running due north at 4.00 m/s.

49.   ssm In a science fiction novel two enemies, Bonzo and Ender, are fighting in outer space. From stationary positions they push against each other. Bonzo flies off with a velocity of $+1.5$ m/s, while Ender recoils with a velocity of -2.5 m/s. (a) Without doing any calculations, decide which person has the greater mass. Give your reasoning. (b) Determine the ratio of the masses (m_{Bonzo}/m_{Ender}) of these two people.

50. Refer to Conceptual Example 3 as an aid in understanding this problem. A hockey goalie is standing on ice. Another player fires a puck ($m = 0.17$ kg) at the goalie with a velocity of $+65$ m/s. (a) If the goalie catches the puck with his glove in a time of 5.0×10^{-3} s, what is the average force (magnitude and di-

rection) exerted on the goalie by the puck? (b) Instead of catching the puck, the goalie slaps it with his stick and returns the puck straight back to the player with a velocity of -65 m/s. The puck and stick are in contact for a time of 5.0×10^{-3} s. Now, what is the average force exerted on the goalie by the puck? Verify that your answers to parts (a) and (b) are consistent with the conclusion of Conceptual Example 3.

51. Two men pushing a stalled car generate a net force of $+680$ N for 7.2 s. What is the final momentum of the car?

52. Two balls are approaching each other head-on. Their velocities are $+9.70$ and -11.8 m/s. Determine the velocity of the center of mass of the two balls if (a) they have the same mass and (b) if the mass of one ball ($v = 9.70$ m/s) is twice the mass of the other ball ($v = -11.8$ m/s).

53.   ssm A two-stage rocket moves in space at a constant velocity of 4900 m/s. The two stages are then separated by a small explosive charge placed between them. Immediately after the explosion the velocity of the 1200-kg upper stage is 5700 m/s in the same direction as before the explosion. What is the velocity (magnitude and direction) of the 2400-kg lower stage after the explosion?

54. A 1200-kg car has an initial velocity of 13 m/s, eastward. The driver applies the brakes and brings the car to rest in 6.0 s. Determine the average net force (magnitude and direction) exerted on the car.

*55. A person stands in a stationary canoe and throws a 5.00-kg stone with a velocity of 8.00 m/s at an angle of 30.0° above the horizontal. The person and canoe have a combined mass of 105 kg. Ignoring air resistance and effects of the water, find the horizontal recoil velocity (magnitude and direction) of the canoe.

*56. A 3.00-kg block of wood rests on the muzzle opening of a vertically oriented rifle, the stock of the rifle being firmly planted on the ground. When the rifle is fired, an 8.00-g bullet (velocity = 8.00×10^2 m/s, straight upward) becomes completely embedded in the block. (a) Using the conservation of linear momentum, find the velocity of the block/bullet system immediately after the collision. (b) Ignoring air resistance, determine how high the block/bullet system rises above the muzzle opening of the rifle.

*57. **ssm** By accident, a large plate is dropped and breaks into three pieces. The pieces fly apart parallel to the floor. As the plate falls, its momentum has only a vertical component, and no component parallel to the floor. After the collision, the component of the total momentum parallel to the floor must remain zero, since the net external force acting on the plate has no component parallel to the floor. Using the data shown in the drawing, find the masses of pieces 1 and 2.

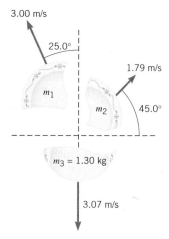

*58. The drawing shows the bond lengths and angles in the nitric acid (HNO_3) molecule, which is planar. The masses of the atoms

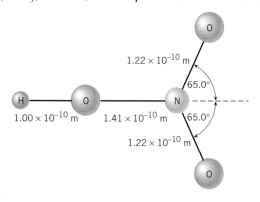

are $m_H = 1.67 \times 10^{-27}$ kg, $m_N = 23.3 \times 10^{-27}$ kg, and $m_O = 26.6 \times 10^{-27}$ kg. Locate the center of mass of this molecule relative to the hydrogen atom. (*Hint: The oxygen atoms are located symmetrically on either side of the H−O−N line.*)

*59. A fireworks rocket is moving at a speed of 50.0 m/s. The rocket suddenly breaks into two pieces of equal mass, and they fly off with velocities $\mathbf{v}_1$ and $\mathbf{v}_2$, as shown in the drawing. What is the magnitude of (a) $\mathbf{v}_1$ and (b) $\mathbf{v}_2$?

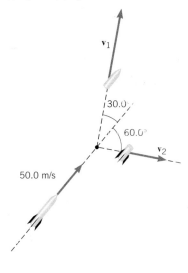

**60. A 1080-kg car moves with a speed of 28.0 m/s and is headed 30.0° north of east. A second car has a mass of 1630 kg and is headed due south. A third car has a mass of 1350 kg and is headed due west. What are the speeds of the second and third cars if the total linear momentum of the three-car system is zero?

61. **ssm **www** A cannon of mass 5.80×10^3 kg is rigidly bolted to the earth so it can recoil only by a negligible amount. The cannon fires an 85.0-kg shell horizontally with an initial velocity of +551 m/s. Suppose the cannon is then unbolted from the earth, and no external force hinders its recoil. What would be the velocity of a shell fired by this loose cannon? (*Hint: In both cases assume that the burning gunpowder imparts the same kinetic energy to the system.*)

ROTATIONAL KINEMATICS

These whirling windmills illustrate the central theme of this chapter, rotational kinematics.
Rotational motion is described by using the concepts of angular displacement, angular
velocity, and angular acceleration, concepts that are analogous to those used in
Chapters 2 and 3 for describing linear motion.

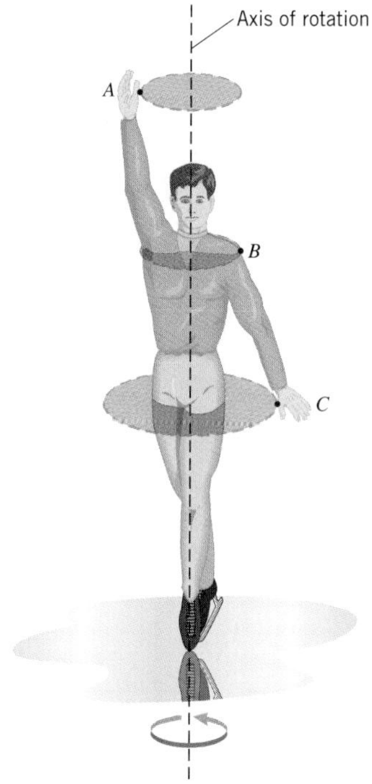

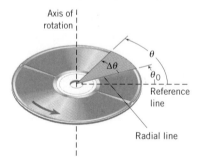

Figure 8.1 When an object rotates, points on the object, such as *A*, *B*, or *C*, move on circular paths. The centers of the circles form a line that is the axis of rotation.

8.1 ROTATIONAL MOTION AND ANGULAR DISPLACEMENT

In the simplest kind of rotation, points on a rigid object move on circular paths. In Figure 8.1, for example, we see the circular paths traversed by points *A*, *B*, and *C* on a spinning skater. The centers of all such circular paths define a line, called the *axis of rotation.*

The angle through which a rigid object rotates about a fixed axis is called the *angular displacement.* Figure 8.2 shows how the angular displacement is measured for a rotating compact disc (CD). Here, the axis of rotation passes through the center of the disc and is perpendicular to its surface. On the surface of the CD we draw a radial line, which is a line that intersects the axis of rotation perpendicularly. As the CD turns, we observe the angle through which this line moves relative to a convenient reference line. The radial line moves from its initial orientation at angle θ_0 to a final orientation at angle θ (Greek letter theta). In the process, the line sweeps out the angle $\Delta\theta = \theta - \theta_0$. The angle $\Delta\theta$ is the angular displacement. A rotating object may rotate either counterclockwise or clockwise, and standard convention calls a counterclockwise displacement positive and a clockwise displacement negative.

■ **DEFINITION OF ANGULAR DISPLACEMENT**

When a rigid body rotates about a fixed axis, the angular displacement is the angle $\Delta\theta$ swept out by a line passing through any point on the body and intersecting the axis of rotation perpendicularly. By convention, the angular displacement is positive if it is counterclockwise and negative if it is clockwise.

SI Unit of Angular Displacement: radian (rad)*

Angular displacement is often expressed in one of three units. The first is the familiar *degree,* and it is well known that there are 360 degrees in a circle. The second unit is the *revolution (rev),* one revolution representing one complete turn of 360°. The most useful unit from a scientific viewpoint is the SI unit called the *radian (rad).* Figure 8.3 shows how the radian is defined, again using a CD as an

Figure 8.2 The angular displacement of a CD is the angle $\Delta\theta$ swept out by a radial line as the disc turns about its axis of rotation.

Figure 8.3 In radian measure, the angle θ is defined to be the arc length s divided by the radius r.

---

* The radian is neither a base nor a derived SI unit. It is regarded as a supplementary SI unit.

example. The picture focuses attention on a point P on the disc. This point starts out on the reference line, so that $\theta_0 = 0$, and the angular displacement is $\Delta\theta = \theta - \theta_0 = \theta$. As the disc rotates, the point traces out an arc of length s, which is measured along a circle of radius r. Equation 8.1 defines the angle θ in radians:

$$\theta \text{ (in radians)} = \frac{\text{Arc length}}{\text{Radius}} = \frac{s}{r} \tag{8.1}$$

According to this definition, an angle in radians is the ratio of two lengths, for example, meters/meters. In calculations, therefore, the radian is treated as a unitless number and has no effect on other units that it multiplies or divides.

To convert between degrees and radians, it is only necessary to remember that the arc length of an entire circle of radius r is the circumference $2\pi r$. Therefore, according to Equation 8.1, *the number of radians that corresponds to 360° or one revolution is*

$$\theta = \frac{2\pi r}{r} = 2\pi \text{ rad}$$

Thus, 2π rad corresponds to 360°, so the number of degrees in one radian is

$$1 \text{ rad} = \frac{360°}{2\pi} = 57.3°$$

It is useful to express an angle θ in radians, because then the arc length s subtended at any radius r can be calculated by multiplying θ by r. Example 1 illustrates this point and also shows how to convert between degrees and radians.

EXAMPLE 1 • Adjacent Synchronous Satellites

Synchronous or "stationary" communications satellites are put into an orbit whose radius is $r = 4.23 \times 10^7$ m, as Figure 8.4 illustrates. The orbit is in the plane of the equator, and two adjacent satellites have an angular separation of $\theta = 2.00°$. Find the arc length s (see the drawing) that separates the satellites.

Reasoning Since the radius r and the angle θ are known, we may find the arc length s by using the relation θ (in radians) $= s/r$. But first, the angle must be converted to radians from degrees.

Solution To convert 2.00° into radians, we use the fact that 2π radians is equivalent to 360°:

$$2.00° = (2.00 \text{ degrees})\left(\frac{2\pi \text{ radians}}{360 \text{ degrees}}\right) = 0.0349 \text{ radians}$$

From Equation 8.1, it follows that

$$s = r\theta = (4.23 \times 10^7 \text{ m})(0.0349 \text{ rad}) = \boxed{1.48 \times 10^6 \text{ m (920 miles)}}$$

The radian, being a unitless quantity, is dropped from the final result, leaving the answer expressed in meters.

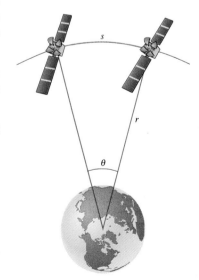

Figure 8.4 Two adjacent synchronous satellites have an angular separation of $\theta = 2.00°$. The distances and angles have been exaggerated for clarity.

Conceptual Example 2 takes advantage of the radian as a unit for measuring angles and explains the spectacular phenomenon of a total solar eclipse.

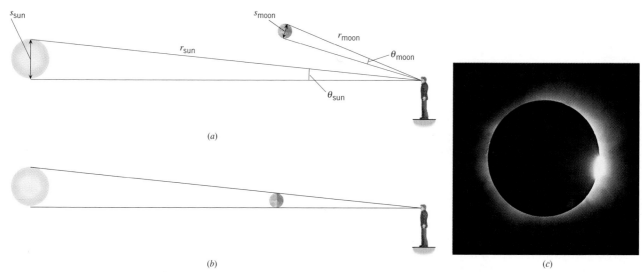

(a)

(b)

(c)

Figure 8.5 (a) The angles subtended by the moon and sun at the eyes of the observer are θ_{moon} and θ_{sun}. (The distances and angles are exaggerated for the sake of clarity.) (b) Since the moon and sun subtend approximately the same angle, the moon blocks nearly all the sun's light from reaching the observer's eyes. (c) The result is a total solar eclipse.

The Physics of...
a total solar eclipse.

CONCEPTUAL EXAMPLE 2 • A Total Eclipse of the Sun

The diameter of the sun is about 400 times greater than that of the moon. By coincidence, the sun is also about 400 times farther from the earth than is the moon. For an observer on earth, compare the angle subtended by the moon to the angle subtended by the sun, and show why this result leads to a total solar eclipse.

Reasoning and Solution Figure 8.5a shows a person on earth viewing the moon and the sun. The angle θ_{moon} subtended by the moon is given by Equation 8.1 as the arc length s_{moon} divided by the distance r_{moon} from the earth to the moon: $\theta_{\text{moon}} = s_{\text{moon}}/r_{\text{moon}}$. The earth and moon are sufficiently far apart that the arc length s_{moon} is very nearly equal to the moon's diameter. Similar reasoning applies for the sun, so that the angle subtended by the sun is $\theta_{\text{sun}} = s_{\text{sun}}/r_{\text{sun}}$. But since $s_{\text{sun}} \approx 400\ s_{\text{moon}}$ (where the symbol "$\approx$" means "approximately equal to") and $r_{\text{sun}} \approx 400\ r_{\text{moon}}$, the two angles are approximately equal: $\theta_{\text{moon}} \approx \theta_{\text{sun}}$. Figure 8.5b shows what happens when the moon comes between the sun and the earth. *Since the angle subtended by the moon is nearly equal to the angle subtended by the sun, the moon blocks most of the sun's light from reaching the observer's eyes,* and a total solar eclipse like that in Figure 8.5c occurs.

Related Homework Material: Problem 16

8.2 ANGULAR VELOCITY AND ANGULAR ACCELERATION

ANGULAR VELOCITY

In Section 2.2 we introduced the idea of linear velocity to describe how fast an object moves and the direction of its motion. A rotating object also exhibits motion. However, it is an angular type of motion that occurs about an axis of rotation. To describe the angular motion we now introduce the idea of angular velocity, which, as the concept chart in Figure 8.6 illustrates, is obtained by combining the angular displacement and the time required for the displacement to occur. A comparison of this drawing with the concept chart in Figure 2.3 suggests that angular displacement and time are combined to produce the notion of angular velocity in the same

Concepts at a Glance

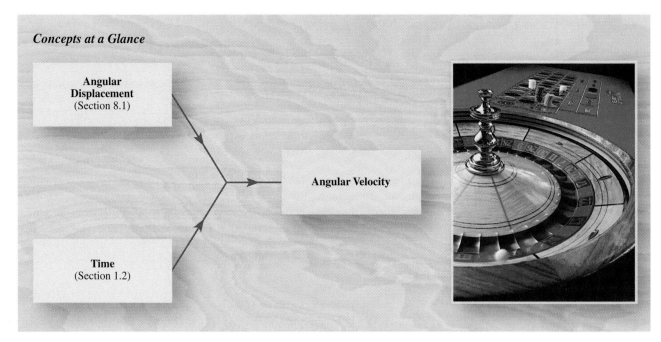

Figure 8.6 *Concepts at a Glance* Angular displacement and time are combined to produce the concept of angular velocity. This spinning roulette wheel has an angular velocity that depends on how fast the wheel is spinning as well as the direction of its spin.

manner as linear displacement and time are combined to produce the notion of linear velocity.

According to Equation 2.2 ($\overline{\mathbf{v}} = \Delta\mathbf{x}/\Delta t$), the average linear velocity is the linear displacement of the object divided by the time required for the displacement to occur. For rotational motion about a fixed axis, the ***average angular velocity*** $\overline{\omega}$ (Greek letter omega) is obtained in an analogous way, as the angular displacement divided by the elapsed time during which the displacement occurs.

■ **DEFINITION OF AVERAGE ANGULAR VELOCITY**

$$\frac{\text{Average angular}}{\text{velocity}} = \frac{\text{Angular displacement}}{\text{Elapsed time}}$$

$$\overline{\omega} = \frac{\theta - \theta_0}{t - t_0} = \frac{\Delta\theta}{\Delta t} \tag{8.2}$$

SI Unit of Angular Velocity: radian per second (rad/s)

The SI unit for angular velocity is the radian per second (rad/s), although other units such as revolutions per minute (rev/min) are encountered. In agreement with the sign convention adopted for angular displacement, angular velocity is positive when the rotation is counterclockwise and negative when it is clockwise. Example 3 shows how the concept of average angular velocity is applied to a gymnast.

EXAMPLE 3 • Gymnast on a High Bar

A gymnast on a high bar swings through two revolutions in a time of 1.90 s, as Figure 8.7 suggests. Find the average angular velocity (in rad/s) of the gymnast.

Reasoning The average angular velocity of the gymnast in rad/s is the angular displacement in radians divided by the elapsed time. However, the angular displacement is given as two revolutions, so we begin by converting this value into radian measure.

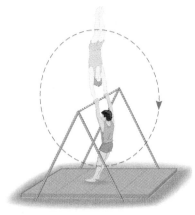

Figure 8.7 Swinging on a high bar.

Solution The angular displacement (in radians) of the gymnast is

$$\Delta\theta = -2.00 \text{ revolutions} \left(\frac{2\pi \text{ radians}}{1 \text{ revolution}} \right) = -12.6 \text{ radians}$$

where the minus sign denotes that the gymnast rotates clockwise. The average angular velocity is

$$\overline{\omega} = \frac{\Delta\theta}{\Delta t} = \frac{-12.6 \text{ rad}}{1.90 \text{ s}} = \boxed{-6.63 \text{ rad/s}} \tag{8.2}$$

The *instantaneous angular velocity* ω is the angular velocity that exists at any given instant. To measure it, we follow the same procedure used in Chapter 2 for the instantaneous linear velocity. In this procedure, a small angular displacement $\Delta\theta$ occurs during a small time interval Δt. The time interval is so small that it approaches zero ($\Delta t \rightarrow 0$), and in this limit, the measured average angular velocity, $\overline{\omega} = \Delta\theta/\Delta t$, becomes the instantaneous angular velocity ω:

$$\omega = \lim_{\Delta t \rightarrow 0} \overline{\omega} = \lim_{\Delta t \rightarrow 0} \frac{\Delta\theta}{\Delta t} \tag{8.3}$$

The magnitude of the instantaneous angular velocity, without reference to whether it is a positive or negative quantity, is called the *instantaneous angular speed.* If a rotating object has a constant angular velocity, the instantaneous value and the average value are the same.

ANGULAR ACCELERATION

In linear motion, a changing velocity means that an acceleration is occurring. Such is also the case in rotational motion; a changing angular velocity means that an *angular acceleration* is occurring. There are many examples of angular acceleration. For instance, as a compact disc recording is played, the disc turns with an angular velocity that is continually decreasing. And when the push buttons of an electric blender are changed from a lower setting to a higher setting, the angular velocity of the blades increases. The concept chart in Figure 8.8 illustrates that the notion of angular acceleration is formulated by combining the change in the angular velocity with the time required for the change to occur. A comparison of this drawing with the concept chart in Figure 2.5 reveals the close analogy between angular acceleration and linear acceleration. When the linear velocity of an object changes, Equation 2.4 ($\overline{\mathbf{a}} = \Delta\mathbf{v}/\Delta t$) defines the average linear acceleration as the change in velocity per unit time. When the angular velocity changes from an initial value of ω_0 at time t_0 to a final value of ω at time t, the average angular acceleration $\overline{\alpha}$ (Greek letter alpha) is defined similarly:

■ **DEFINITION OF AVERAGE ANGULAR ACCELERATION**

$$\begin{array}{c} \text{Average angular} \\ \text{acceleration} \end{array} = \frac{\begin{array}{c} \text{Change in} \\ \text{angular velocity} \end{array}}{\text{Elapsed time}}$$

$$\overline{\alpha} = \frac{\omega - \omega_0}{t - t_0} = \frac{\Delta\omega}{\Delta t} \tag{8.4}$$

SI Unit of Average Angular Acceleration: radian per second squared (rad/s²)

Concepts at a Glance

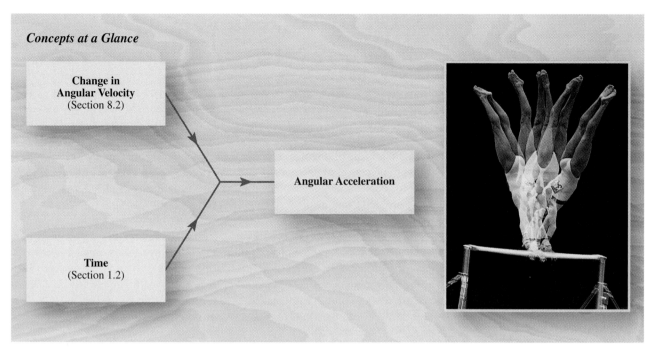

Figure 8.8 *Concepts at a Glance*
To define the concept of angular acceleration, we bring together the change in angular velocity and the time required for the change to occur. Gymnast Dominique Dawes has an angular acceleration because her angular velocity changes as she rotates around the bar at the 1996 Olympics.

The SI unit for average angular acceleration is the unit for angular velocity divided by the unit for time, or (rad/s)/s = rad/s^2. An angular acceleration of $+5$ rad/s^2, for example, means that the angular velocity of the rotating object increases by 5 radians per second during each second of acceleration.

The ***instantaneous angular acceleration*** α is the angular acceleration at a given instant. Previously, in discussing linear motion, a condition of constant acceleration was assumed, so that the average and instantaneous accelerations were identical ($\overline{\mathbf{a}} = \mathbf{a}$). Similarly, we will assume here that the angular acceleration is constant, so that the instantaneous angular acceleration α and the average angular acceleration $\overline{\alpha}$ are the same ($\overline{\alpha} = \alpha$). The next example illustrates the concept of angular acceleration.

EXAMPLE 4 • A Jet Revving Its Engines

A jet awaiting clearance for takeoff is momentarily stopped on the runway. As the engines idle, the fan blades are rotating with an angular velocity of -110 rad/s, where the negative sign indicates a clockwise rotation (see Figure 8.9). As the plane takes off, the angular velocity of the blades reaches -330 rad/s in a time of 14 s. Find the angular acceleration, assuming it to be constant.

Reasoning and Solution The angular acceleration is constant, so it is equal to the average acceleration. Applying Equation 8.4, we find

$$\overline{\alpha} = \frac{\omega - \omega_0}{t - t_0} = \frac{(-330 \text{ rad/s}) - (-110 \text{ rad/s})}{14 \text{ s}} = \boxed{-16 \text{ rad/s}^2}$$

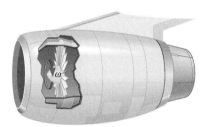

Thus, the magnitude of the angular velocity increases by 16 rad/s during each second that the blades are accelerating. The negative sign in the answer indicates that the direction of the angular acceleration is also in the clockwise direction.

Figure 8.9 The fan blades of a jet engine are accelerating in a clockwise direction.

8.3 THE EQUATIONS OF ROTATIONAL KINEMATICS

In Chapters 2 and 3 the concepts of displacement, velocity, and acceleration were introduced. By combining these concepts, we developed a set of equations called the equations of kinematics for constant acceleration (see Tables 2.1 and 3.1). These equations were a great aid in solving problems involving linear motion in one and two dimensions. As the concept chart in Figure 8.10 shows, we will now integrate the ideas of angular displacement, angular velocity, and angular acceleration to produce a set of analogous equations called the equations of rotational kinematics for constant angular acceleration. These equations will prove very useful in solving problems that involve rotational motion.

A complete description of rotational motion requires values for the angular displacement $\Delta\theta$, the angular acceleration α, the final angular velocity ω, the initial angular velocity ω_0, and the elapsed time Δt. In Example 4, for instance, only the angular displacement of the fan blades during the 14-s interval is missing. Such missing information can be calculated, however. For convenience in the calculations, we assume that the orientation of the rotating object is given by $\theta_0 = 0$ at time $t_0 = 0$. Then, the angular displacement becomes $\Delta\theta = \theta - \theta_0 = \theta$, and the time interval becomes $\Delta t = t - t_0 = t$.

In Example 4, the angular velocity of the fan blades changes at a constant rate from an initial value of $\omega_0 = -110$ rad/s to a final value of $\omega = -330$ rad/s. Therefore, the average angular velocity is midway between the initial and final values:

$$\overline{\omega} = \tfrac{1}{2}[(-110 \text{ rad/s}) + (-330 \text{ rad/s})] = -220 \text{ rad/s}$$

Figure 8.10 *Concepts at a Glance* The equations of rotational kinematics for constant angular acceleration are obtained by integrating the notions of angular displacement, angular velocity, and angular acceleration. The rotational motion of this Ferris wheel can be described by using the equations of rotational kinematics, provided that the angular acceleration is constant.

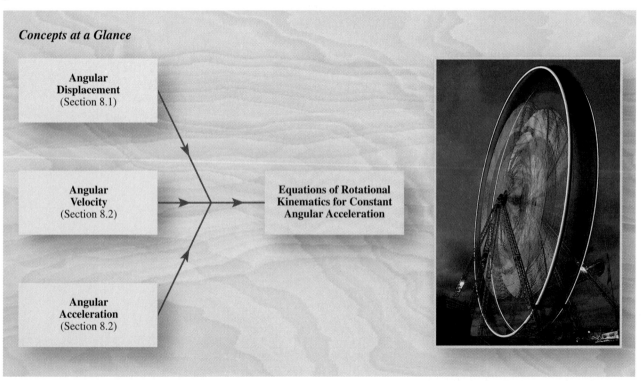

Concepts at a Glance

Angular Displacement (Section 8.1)

Angular Velocity (Section 8.2)

Angular Acceleration (Section 8.2)

Equations of Rotational Kinematics for Constant Angular Acceleration

Table 8.1 **The Equations of Kinematics for Rotational and Linear Motion**

Rotational Motion (α = constant)		Linear Motion (a = constant)	
$\omega = \omega_0 + \alpha t$	(8.4)	$v = v_0 + at$	(2.4)
$\theta = \frac{1}{2}(\omega_0 + \omega)t$	(8.6)	$x = \frac{1}{2}(v_0 + v)t$	(2.7)
$\theta = \omega_0 t + \frac{1}{2}\alpha t^2$	(8.7)	$x = v_0 t + \frac{1}{2}at^2$	(2.8)
$\omega^2 = \omega_0^2 + 2\alpha\theta$	(8.8)	$v^2 = v_0^2 + 2ax$	(2.9)

In other words, when the angular acceleration is constant, the average angular velocity is given by

$$\overline{\omega} = \tfrac{1}{2}(\omega_0 + \omega) \tag{8.5}$$

With a value for the average angular velocity, Equation 8.2 can be used to obtain the angular displacement of the fan blades:

$$\theta = \overline{\omega}t = (-220 \text{ rad/s})(14 \text{ s}) = -3100 \text{ rad}$$

In general, when the angular acceleration is constant. the angular displacement can be obtained from

$$\theta = \overline{\omega}t = \tfrac{1}{2}(\omega_0 + \omega)t \tag{8.6}$$

This equation and Equation 8.4 provide a complete description of rotational motion under the condition of constant angular acceleration. Equation 8.4 (with $t_0 = 0$) and Equation 8.6 are compared with the analogous results for linear motion in the first two lines of Table 8.1. The purpose of this comparison is to emphasize that the mathematical forms of Equations 8.4 and 2.4 are identical, as are the forms of Equations 8.6 and 2.7. Of course, the symbols used for the rotational variables are different from those used for the linear variables, as Table 8.2 indicates.

In Chapter 2, Equations 2.4 and 2.7 are used to derive the remaining two equations of kinematics (Equations 2.8 and 2.9). These additional equations convey no new information but are convenient to have when solving problems. Similar derivations can be carried out here. The results are listed as Equations 8.7 and 8.8 in Table 8.1 and can be inferred directly from their counterparts in linear motion by making the substitution of symbols indicated in Table 8.2. The four equations in the left column of Table 8.1 are called the ***equations of rotational kinematics for constant angular acceleration.*** The following example illustrates that they are used in the same fashion as the equations of linear kinematics.

Table 8.2 **Symbols Used in Rotational and Linear Kinematics**

Rotational Motion	Quantity	Linear Motion
θ	Displacement	x
ω_0	Initial velocity	v_0
ω	Final velocity	v
α	Acceleration	a
t	Time	t

Figure 8.11 The angular velocity of the blades in an electric blender changes each time a different push button is chosen.

EXAMPLE 5 • Blending with a Blender

The blades of an electric blender are whirling with an angular velocity of +375 rad/s while the "puree" button is pushed in, as Figure 8.11 shows. When the "blend" button is pressed, the blades accelerate and reach a greater angular velocity after the blades have rotated through an angular displacement of +44.0 rad (seven revolutions). The angular acceleration has a constant value of +1740 rad/s². Find the final angular velocity of the blades.

Reasoning   The three known variables are listed in the table below, along with a question mark indicating that a value for the final angular velocity ω is being sought.

θ	α	ω	ω_0	t
+44.0 rad	+1740 rad/s²	?	+375 rad/s	

We can use Equation 8.8, because it relates the angular variables θ, α, ω, and ω_0.

Solution   From Equation 8.8 we find that

$$\omega^2 = \omega_0{}^2 + 2\alpha\theta$$
$$= (375 \text{ rad/s})^2 + 2(1740 \text{ rad/s}^2)(44.0 \text{ rad}) = 2.94 \times 10^5 \text{ rad}^2/\text{s}^2$$
$$\boxed{\omega = +542 \text{ rad/s}}$$

The negative root is disregarded, since the blades do not reverse their direction of rotation.

The equations of rotational kinematics can be used with any self-consistent set of units for θ, α, ω, ω_0, and t. Radians are used in Example 5 only because data are given in terms of radians. Had the data for θ, α, and ω_0 been provided in rev, rev/s², and rev/s, respectively, then Equation 8.8 could have been used to determine the answer for ω directly in rev/s. In any case, the reasoning strategy for applying the kinematics equations is summarized below.

REASONING STRATEGY

Applying the Equations of Rotational Kinematics

1. Make a drawing to represent the situation being studied, showing the direction of rotation.

2. Decide which direction of rotation is to be called positive (+) and which direction is to be called negative (−). In this text we choose the counter-clockwise direction to be positive and the clockwise direction to be nega-tive, but this is arbitrary. However, do not change your decision during the course of a calculation.

3. In an organized way, write down the values (with appropriate + and − signs) that are given for any of the five rotational kinematic variables (θ, α, ω, ω_0, and t). Be on the alert for implied data, such as the phrase "starts from rest," which means that the value of the initial angular veloc-ity is $\omega_0 = 0$. The data box in Example 5 is a good way to keep track of this information. In addition, identify the variable(s) that you are being asked to determine.

4. Before attempting to solve a problem, verify that the given information contains values for at least three of the five kinematic variables. Once the three variables are identified for which values are known, the appropriate relation from Table 8.1 can be selected.

5. When the rotational motion is divided into segments, the final angular velocity of one segment becomes the initial angular velocity for the next segment.

6. Keep in mind that there may be two possible answers to a kinematics problem. Try to visualize the different physical situations to which the answers correspond.

8.4 ANGULAR VARIABLES AND TANGENTIAL VARIABLES

In the familiar ice-skating stunt known as "crack-the-whip," a number of skaters attempt to maintain a straight line as they skate around the one person (the pivot) who remains in place. Figure 8.12 shows each skater moving on a circular arc and includes the corresponding velocity vector at the instant portrayed in the picture. For every individual skater, the vector is drawn tangent to the appropriate circle and therefore is called the ***tangential velocity*** $\mathbf{v_T}$. The magnitude of the tangential velocity is referred to as the ***tangential speed.***

Of all the skaters involved in the stunt, the one farthest from the pivot has the hardest job. Why? Because, in keeping the line straight, this skater covers more distance than anyone else. To accomplish this, he must skate faster than anyone else and, thus, must have the largest tangential speed. In fact, the line remains straight only if each person skates with the correct tangential speed. Those skaters closer to the pivot must move with smaller tangential speeds than those farther out, as indicated by the magnitudes of the tangential velocity vectors drawn in Figure 8.12.

With the aid of Figure 8.13, it is possible to show that the tangential speed of any skater is directly proportional to his distance r from the pivot, assuming a given angular speed for the rotating line. When the line rotates as a rigid unit for a time t, it sweeps out the angle θ shown in the drawing. The distance s through which a skater moves along a circular arc can be calculated from the relation $s = r\theta$, provided θ is measured in radians. Dividing both sides of this equation by t gives $s/t = r(\theta/t)$. The term s/t is the tangential speed v_T (e.g., in meters/second) of the skater, while the term θ/t is the angular speed ω (in radians/second) at which the line rotates:

$$v_T = r\omega \qquad (\omega \text{ in rad/s}) \qquad (8.9)$$

Thus, for a given angular speed ω, the tangential speed v_T is directly proportional to the radius r. In this expression, the terms v_T and ω refer to the magnitudes of the tangential and angular velocities, respectively, and are numbers without algebraic signs.

It is important to emphasize that the angular speed ω in $v_T = r\omega$ must be expressed in radian measure (e.g., in rad/s); no other units, such as revolutions per second, are acceptable. This restriction arises because the equation was derived by using the definition of radian measure, $s = r\theta$.

The real challenge for the "crack-the-whip" skaters is to keep the line straight while making it pick up angular speed, that is, while giving it an angular acceleration. To make the angular speed of the line increase, each skater must increase his tangential speed, since the two speeds are related according to $v_T = r\omega$. Of course, the fact that a skater must skate faster and faster means that he must accelerate in the tangential direction, and his tangential acceleration a_T can be related to the angular acceleration α of the straight line. If time is measured relative to $t_0 = 0$, the definition of linear acceleration is given by Equation 2.4 as $a_T = (v_T - v_{T0})/t$, where v_T and v_{T0} are the final and initial tangential speeds, respectively. Substituting $v_T = r\omega$ for the tangential speed shows that

$$a_T = \frac{v_T - v_{T0}}{t} = \frac{(r\omega) - (r\omega_0)}{t} = r\left(\frac{\omega - \omega_0}{t}\right)$$

Since $\alpha = (\omega - \omega_0)/t$ according to Equation 8.4, it follows that

$$a_T = r\alpha \qquad (\alpha \text{ in rad/s}^2) \qquad (8.10)$$

The Physics of...
"crack-the-whip."

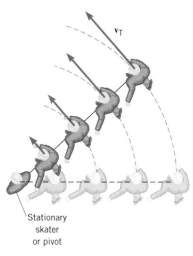

Stationary skater or pivot

Figure 8.12 When doing a stunt known as "crack-the-whip," each skater along the radial line moves on a circular arc. The tangential velocity $\mathbf{v_T}$ of each skater is represented by an arrow that is tangent to each arc.

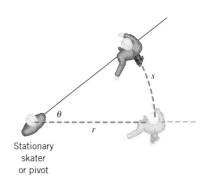

Stationary skater or pivot

Figure 8.13 During a time t, the line of skaters sweeps through an angle θ. An individual skater, located at a distance r from the stationary skater, moves through a distance s on a circular arc.

This result shows that for a given value of α the tangential acceleration a_T is proportional to the radius r, so the skater farthest from the pivot must have the largest tangential acceleration. In this expression, the terms a_T and α refer to the magnitudes of the numbers involved, without reference to any algebraic sign. Moreover, as is the case for ω in $v_T = r\omega$, only radian measure can be used for α in $a_T = r\alpha$.

There is an advantage to using the angular velocity ω and the angular acceleration α to describe the rotational motion of a rigid object, for these angular quantities describe the motion of the *entire object*. In contrast, the tangential quantities v_T and a_T describe only the motion of a single point on the object, and Equations 8.9 and 8.10 indicate that different points located at different distances r have different tangential velocities and accelerations. Example 6 stresses this advantage.

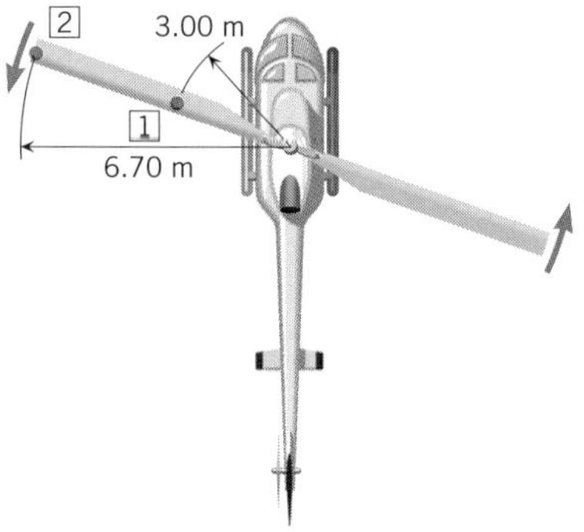

Figure 8.14 Points 1 and 2 on the rotating blade of the helicopter have the same angular speed and acceleration, but they have *different* tangential speeds and accelerations.

• **PROBLEM SOLVING INSIGHT**
When using Equations 8.9 and 8.10 to relate tangential and angular quantities, remember that the angular quantity is always expressed in radian measure. These equations are not valid if angles are expressed in degrees or revolutions.

EXAMPLE 6 • A Helicopter Blade

A helicopter blade has an angular speed of $\omega = 6.50$ rev/s and an angular acceleration of $\alpha = 1.30$ rev/s^2. For points 1 and 2 on the blade in Figure 8.14, find the magnitudes of (a) the tangential speeds and (b) the tangential accelerations.

Reasoning Since the radius r of each point and the angular speed ω of the helicopter blade are known, we can find the tangential speed v_T of each point by using the relation $v_T = r\omega$. However, since this equation can be used only with radian measure, the angular speed ω must be converted to rad/s from rev/s. In a similar manner, the tangential acceleration a_T of points 1 and 2 can be found using the relation $a_T = r\alpha$, provided the angular acceleration α is expressed in rad/s^2 rather than in rev/s^2.

Solution

(a) Converting the angular speed ω to rad/s from rev/s, we obtain

$$\omega = \left(6.50\ \frac{\text{rev}}{\text{s}}\right)\left(\frac{2\pi\ \text{rad}}{1\ \text{rev}}\right) = 40.8\ \frac{\text{rad}}{\text{s}}$$

The tangential speed of each point is

Point 1 $v_T = r\omega = (3.00\ \text{m})(40.8\ \text{rad/s}) = \boxed{122\ \text{m/s (273 mph)}}$ (8.9)

Point 2 $v_T = r\omega = (6.70\ \text{m})(40.8\ \text{rad/s}) = \boxed{273\ \text{m/s (611 mph)}}$ (8.9)

The rad unit, being dimensionless, does not appear in the final answers.

(b) Converting the angular acceleration α to rad/s^2 from rev/s^2, we find

$$\alpha = \left(1.30\ \frac{\text{rev}}{\text{s}^2}\right)\left(\frac{2\pi\ \text{rad}}{1\ \text{rev}}\right) = 8.17\ \frac{\text{rad}}{\text{s}^2}$$

The tangential accelerations can now be determined:

Point 1 $a_T = r\alpha = (3.00\ \text{m})(8.17\ \text{rad/s}^2) = \boxed{24.5\ \text{m/s}^2}$ (8.10)

Point 2 $a_T = r\alpha = (6.70\ \text{m})(8.17\ \text{rad/s}^2) = \boxed{54.7\ \text{m/s}^2}$ (8.10)

8.5 CENTRIPETAL ACCELERATION AND TANGENTIAL ACCELERATION

When an object picks up speed as it moves around a circle, it has a tangential acceleration, as discussed in the last section. In addition, the object also has a centripetal acceleration, as emphasized in Chapter 5. That chapter deals with ***uniform circular motion,*** in which a particle moves at a constant tangential speed on a circular path. The tangential speed v_T is the magnitude of the tangential velocity vector. Even though the magnitude of the tangential velocity is constant, an acceleration is present, since the direction of the velocity changes continually. Because the resulting acceleration points toward the center of the circle, it is called the centripetal acceleration. Figure 8.15*a* shows the centripetal acceleration $\mathbf{a}_c$ for a model airplane flying on a guide wire, the magnitude of $\mathbf{a}_c$ being

$$a_c = \frac{v_T^2}{r} \tag{5.2}$$

The subscript "T" has now been included in this equation as a reminder that it is the tangential speed that appears in the numerator.

The centripetal acceleration can be expressed in terms of the angular speed ω by using $v_T = r\omega$ (Equation 8.9):

$$a_c = \frac{v_T^2}{r} = \frac{(r\omega)^2}{r} = r\omega^2 \qquad (\omega \text{ in rad/s}) \tag{8.11}$$

Only radian measure, such as rad/s, can be used for ω in this result, since the derivation depends on the relation $v_T = r\omega$, which presumes radian measure.

While considering uniform circular motion in Chapter 5, we ignored the details of how the motion is established in the first place. In Figure 8.15*b*, for instance, the engine of the plane produces a thrust in the tangential direction, and this force leads to a tangential acceleration. In response, the tangential speed of the plane increases from moment to moment, until the situation shown in the drawing results. While the tangential speed is changing, the motion is called ***nonuniform circular motion.***

Figure 8.15*b* illustrates an important feature of nonuniform circular motion. Since both the direction and the magnitude of the tangential velocity are changing, the airplane experiences two acceleration components simultaneously. The changing direction means that there is a centripetal acceleration $\mathbf{a}_c$. The magnitude of $\mathbf{a}_c$ at any moment can be calculated using the value of the instantaneous angular speed and the radius: $a_c = r\omega^2$. The fact that the magnitude of the tangential velocity is changing means that there is also a tangential acceleration $\mathbf{a}_T$. The magnitude of $\mathbf{a}_T$ can be determined from the angular acceleration α according to $a_T = r\alpha$, as the previous section explains. Alternatively, a_T can be calculated using Newton's second law, $F_T = ma_T$, if the magnitude F_T of the net tangential force and the mass m are known. Figure 8.15*b* shows the two acceleration components. The total acceleration is given by the vector sum of $\mathbf{a}_c$ and $\mathbf{a}_T$. Since $\mathbf{a}_c$ and $\mathbf{a}_T$ are perpendicular, the magnitude of the total acceleration $\mathbf{a}$ can be obtained from the Pythagorean theorem as $a = \sqrt{a_c^2 + a_T^2}$, while the angle ϕ in the drawing can be determined from $\tan \phi = a_T/a_c$. The next example applies these concepts to a discus thrower.

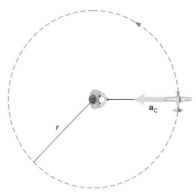

(a) Uniform circular motion

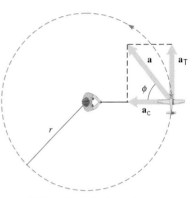

(b) Nonuniform circular motion

Figure 8.15 (*a*) If a model airplane flying on a guide wire has a constant tangential speed, the motion is uniform circular motion, and the plane experiences only a centripetal acceleration $\mathbf{a}_c$. (*b*) Nonuniform circular motion occurs when the tangential speed changes. Then there is a tangential acceleration $\mathbf{a}_T$ in addition to the centripetal acceleration.

EXAMPLE 7 • A Discus Thrower

Discus throwers often warm up by standing with both feet flat on the ground and throwing the discus with a twisting motion of their bodies. Figure 8.16*a* illustrates a top view of such a warm-up throw. Starting from rest, the thrower accelerates the discus to a final angular velocity of $+15.0$ rad/s in a time of 0.270 s before releasing it. During the acceleration, the discus moves on a circular arc of radius 0.810 m. Find (a) the magnitude a of the total acceleration of the discus just before it is released and (b) the angle ϕ that the total acceleration makes with the radius at this moment.

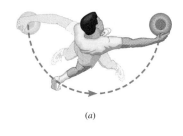

(b)

Figure 8.16 (a) A discus thrower. (b) The total acceleration **a** of the discus just before the discus is released. The total acceleration is the vector sum of the centripetal acceleration **a**$_c$ and the tangential acceleration **a**$_T$.

Reasoning Since the tangential speed of the discus is increasing, the discus will simultaneously experience a tangential acceleration **a**$_T$ and a centripetal acceleration **a**$_c$ that are oriented at right angles to each other. The magnitude of the total acceleration is $a = \sqrt{a_c^2 + a_T^2}$, where a_c and a_T are the magnitudes of the centripetal and tangential accelerations. The angle ϕ in Figure 8.16b is given by $\phi = \tan^{-1}(a_T/a_c)$. The centripetal acceleration can be evaluated from $a_c = r\omega^2$. The tangential acceleration follows from $a_T = r\alpha$, where α can be found from the definition of angular acceleration in Equation 8.4.

Solution

(a) The magnitude of the centripetal acceleration is

$$a_c = r\omega^2 = (0.810 \text{ m})(15.0 \text{ rad/s})^2 = 182 \text{ m/s}^2 \qquad (8.11)$$

The magnitude of the tangential acceleration is $a_T = r\alpha$. But $\alpha = (\omega - \omega_0)/t$ according to Equation 8.4, so

$$a_T = r\alpha = r\left(\frac{\omega - \omega_0}{t}\right)$$

$$a_T = (0.810 \text{ m})\left(\frac{15.0 \text{ rad/s} - 0}{0.270 \text{ s}}\right) = 45.0 \text{ m/s}^2$$

Just before the moment of release, the total acceleration of the discus is

$$a = \sqrt{a_c^2 + a_T^2} = \sqrt{(182 \text{ m/s}^2)^2 + (45.0 \text{ m/s}^2)^2} = \boxed{187 \text{ m/s}^2}$$

(b) The angle ϕ in Figure 8.16b is given by

$$\phi = \tan^{-1}\left(\frac{a_T}{a_c}\right) = \tan^{-1}\left(\frac{45.0 \text{ m/s}^2}{182 \text{ m/s}^2}\right) = \boxed{13.9°}$$

8.6 ROLLING MOTION

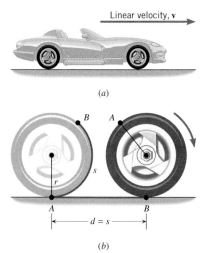

Linear velocity, **v**

(a)

(b)

Figure 8.17 (a) An automobile moves with a linear speed v. (b) If the tires roll and do not slip, the distance d, through which an axle moves, equals the circular arc length s along the outer edge of a tire.

Rolling motion is a familiar situation that involves rotation, as Figure 8.17 illustrates for the case of an automobile tire. The essence of rolling motion is that there is *no slipping* at the point of contact where the tire touches the ground. To a good approximation, the tires on a normally moving automobile roll and do not slip. On the other hand, the squealing tires that accompany the start of a drag race are rotating, but they are not rolling as they rapidly spin and slip against the ground.

When the tires in Figure 8.17 roll, there is a relationship between the angular speed at which the tires rotate and the linear speed (assumed constant) at which the car moves forward. In part b of the drawing, consider the points labeled A and B on the left tire. Between these points we apply a coat of red paint to the tread of the tire; the length of this circular arc of paint is s. The tire then rolls to the right until point B comes in contact with the ground. As the tire rolls, all the paint comes off the tire and sticks to the ground, leaving behind the horizontal red line shown in the drawing. The axle of the wheel moves through a linear distance d, which is equal to the length of the horizontal strip of paint. Since the tire does not slip, the distance d must be equal to the circular arc length s, measured along the outer edge of the tire: $d = s$. Dividing both sides of this equation by the elapsed time t shows that $d/t = s/t$. The term d/t is the speed at which the axle moves parallel to the ground, namely, the linear speed v of the car. The term s/t is the tangential speed v_T at which a point on the outer edge of the tire moves relative to the axle. In addition, v_T

is related to the angular speed ω about the axle according to $v_T = r\omega$ (Equation 8.9). Therefore, it follows that

$$v = r\omega \qquad (\omega \text{ in rad/s}) \qquad (8.12)$$

Linear speed

Tangential speed, v_T

If the car in Figure 8.17 has a linear acceleration **a** parallel to the ground, a point on the tire's outer edge experiences a tangential acceleration $\mathbf{a}_T$ relative to the axle. The same kind of reasoning used in the last paragraph reveals that the magnitudes of these accelerations are the same and that they are related to the angular acceleration α of the wheel relative to the axle:

$$a = r\alpha \qquad (\alpha \text{ in rad/s}^2) \qquad (8.13)$$

Linear acceleration

Tangential acceleration, a_T

Equations 8.12 and 8.13 may be applied to any rolling motion, because the object does not slip against the surface on which it is rolling. Example 8 illustrates the basic features of rolling motion.

EXAMPLE 8 • An Accelerating Car

An automobile starts from rest and for 20.0 s has a constant linear acceleration of 0.800 m/s² to the right. During this period, the tires do not slip. The radius of the tires is 0.330 m. At the end of the 20.0-s interval what is the angle through which each wheel has rotated?

Reasoning As the car accelerates, the tires rotate faster and faster. Thus, each tire has an angular acceleration, which must be taken into account when we determine the angular displacement of each wheel. Since the tires roll without slipping, the magnitude α of the angular acceleration is related to the magnitude a of the linear acceleration of the car by $a = r\alpha$ (Equation 8.13). Therefore, we find that

$$\alpha = \frac{a}{r} = \frac{0.800 \text{ m/s}^2}{0.330 \text{ m}} = 2.42 \text{ rad/s}^2$$

The angular data, then, are as follows:

θ	α	ω	ω_0	t
?	-2.42 rad/s²		0	20.0 s

The angular acceleration is negative, because the tires rotate increasingly rapidly in the clockwise direction as the car moves to the right (see Figure 8.17). The angular displacement θ is given in terms of α, ω_0, and t by Equation 8.7.

Solution From Equation 8.7, we find that

$$\theta = \omega_0 t + \tfrac{1}{2}\alpha t^2 = \tfrac{1}{2}(-2.42 \text{ rad/s}^2)(20.0 \text{ s})^2 = \boxed{-484 \text{ rad}}$$

The angular displacement θ is negative, because the wheel rotates in the clockwise direction.

The rolling motion of an automobile tire has some surprising features, as Conceptual Example 9 discusses.

CONCEPTUAL EXAMPLE 9 • A Rolling Tire

A car is moving with a linear velocity **v** relative to the ground. The tires are rolling and not slipping. With respect to the ground, what is the velocity of a point on the tire that (a) touches the ground and (b) is on the very top of the tire, directly above the ground?

Reasoning and Solution

(a) To help guide our reasoning, Figure 8.18*a* shows the tread on a moving tank. An intriguing aspect of this motion is that the segment of the tread in contact with the ground is stationary, for it does not move with respect to the ground. In fact, tank treads are often designed with metal projections that dig into the ground and prevent the tread from slipping against the ground. A rolling automobile tire is like the tread on a moving tank, although a perfectly round tire would have only one point in contact with the ground at any instant, as in Figure 8.18*b*. According to our discussion of relative velocity in Section 3.4, the velocity $\mathbf{v}_{PG}$ of the contact point with respect to the ground is $\mathbf{v}_{PG} = \mathbf{v}_{PA} + \mathbf{v}_{AG}$, where the subscript "A" stands for "axle." In this equation, $\mathbf{v}_{PA}$ is the velocity of the contact point with respect to the axle, or in other words, it is the tangential velocity $\mathbf{v}_T$. The term $\mathbf{v}_{AG}$ is the velocity of the axle with respect to the ground, or, equivalently, it is the linear velocity **v** of the car. We find, then, that $\mathbf{v}_{PG} = \mathbf{v}_T + \mathbf{v}$. However, the velocities $\mathbf{v}_T$ and **v** have opposite directions where the tire is in contact with the ground, as Figure 8.18*b* indicates. Since the tire rolls without slipping, the magnitudes of $\mathbf{v}_T$ and **v** are the same. Because $\mathbf{v}_T$ and **v** have opposite directions and equal magnitudes, we conclude that $\mathbf{v}_{PG} = \mathbf{v}_T + \mathbf{v} = 0$. Thus, *the point where the wheel touches the ground has zero velocity.*

(b) While the segment of the tank tread in contact with the ground is stationary, the top of the tread moves in the same direction as the tank itself. The top of the tire behaves in a similar way. As Figure 8.18*b* shows, the velocities $\mathbf{v}_T$ and **v** have the same directions at the top of the tire. Consequently, *the velocity of the top point with respect to the ground is $\mathbf{v}_{PG} = \mathbf{v}_T + \mathbf{v} = 2\mathbf{v}$, or twice the linear velocity of the car.* Likewise, the top segment of the tank tread also moves with a velocity of $2\mathbf{v}$.

Related Homework Material: Problem 74

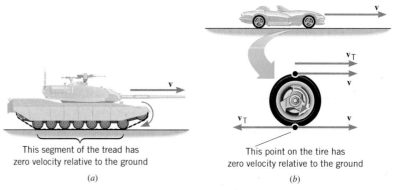

This segment of the tread has zero velocity relative to the ground

(a)

This point on the tire has zero velocity relative to the ground

(b)

Figure 8.18 (*a*) A tank moves with a linear velocity **v**. The segment of the tread touching the ground does not move relative to the ground, while the top of the tread moves in the same direction as the tank. (*b*) The point where the tire touches the ground has zero velocity, while the top of the tire has a velocity of $2\mathbf{v}$.

*8.7 THE VECTOR NATURE OF ANGULAR VARIABLES

We have presented angular velocity and angular acceleration by making the analogy between angular variables and linear variables. Like the linear velocity and the linear acceleration, the angular quantities are also vectors and have a direction as well as a magnitude. As yet, however, we have not discussed the directions of these vectors.

When a rigid object rotates about a fixed axis, it is the axis that identifies the motion, and the angular velocity vector points along this axis. Figure 8.19 shows how to determine the direction using a *right-hand rule:*

> ***Right-Hand Rule.*** Grasp the axis of rotation with your right hand, so that your fingers circle the axis in the same sense as the rotation. Your extended thumb points along the axis in the direction of the angular velocity vector.

Note that no part of the rotating object moves in the direction of the angular velocity vector.

Angular acceleration arises when the angular velocity changes, and the acceleration vector also points along the axis of rotation. The acceleration vector has the same direction as the *change* in the angular velocity. That is, when the angular velocity is increasing, the angular acceleration vector points in the same direction as the angular velocity. Conversely, when the angular velocity is decreasing, the angular acceleration vector points in the direction opposite to the angular velocity.

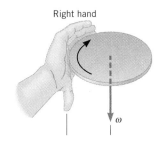

Figure 8.19 The angular velocity vector ω of a rotating object points along the axis of rotation. The direction along the axis depends on the sense of the rotation and can be determined with the aid of a right-hand rule (see text).

SUMMARY

When a rigid body rotates about a fixed axis, the **angular displacement** is the angle swept out by a line passing through any point on the body and intersecting the axis of rotation perpendicularly. The **radian (rad)** is the SI unit of angular displacement. In radians, the angle θ is defined as the circular arc length s traveled by a point on the rotating body divided by the radial distance r of the point from the axis: $\theta = s/r$.

The **average angular velocity** $\overline{\omega}$ is the angular displacement $\Delta\theta$ divided by the elapsed time Δt: $\overline{\omega} = \Delta\theta/\Delta t$. As Δt approaches zero, the average angular velocity becomes equal to the **instantaneous angular velocity** ω. The magnitude of the instantaneous angular velocity is called the **instantaneous angular speed.**

The **average angular acceleration** $\overline{\alpha}$ is the change $\Delta\omega$ in the angular velocity divided by the elapsed time Δt: $\overline{\alpha} = \Delta\omega/\Delta t$. As Δt approaches zero, the average angular acceleration becomes equal to the **instantaneous angular acceleration** α.

When a rigid body rotates with constant angular acceleration about a fixed axis, the angular displacement θ, the final angular velocity ω, the initial angular velocity ω_0, the angular acceleration α, and the elapsed time t are related as follows, assuming that $\theta_0 = 0$ at $t_0 = 0$:

$$\omega = \omega_0 + \alpha t \quad \text{and} \quad \theta = \tfrac{1}{2}(\omega_0 + \omega)t$$

These two equations can be combined algebraically to give two additional equations, Equations 8.7 and 8.8. Together, the four equations are known as the **equations of rotational kinematics for constant angular acceleration.** These equations may be used with any self-consistent set of units and are not restricted to radian measure.

When a rigid body rotates through an angle θ about a fixed axis, any point on the body moves on a circular arc of length s and radius r. Such a point has a tangential velocity $\mathbf{v}_T$ and, possibly, a tangential acceleration $\mathbf{a}_T$. The **angular and tangential variables are related** by the equations $s = r\theta$, $v_T = r\omega$, and $a_T = r\alpha$, where ω and α are, respectively, the magnitudes of the angular velocity and the angular acceleration. These equations refer to the magnitudes of the variables involved, without reference to positive or negative signs, and only radian measure can be used when applying them.

A point on an object rotating with **nonuniform circular**

motion experiences a total acceleration that is the vector sum of two perpendicular acceleration components, the centripetal acceleration $\mathbf{a}_c$ and the tangential acceleration $\mathbf{a}_T$; $\mathbf{a} = \mathbf{a}_c + \mathbf{a}_T$.

The essence of **rolling motion** is that there is no slipping at the point where the object touches the surface upon which it is rolling. As a result, the tangential speed v_T of a point on the outer edge of a rolling object, measured rela- tive to the axis through the center of the object, is equal to the linear speed v with which the object moves parallel to the surface. In other words, $v = v_T$ ($= r\omega$), where r is the radial distance of the point from the axis. The magnitudes of the tangential acceleration a_T and the linear acceleration a of a rolling object are similarly related: $a = a_T$ ($= r\alpha$).

CONCEPTUAL QUESTIONS

1. In the drawing, the flat triangular sheet *ABC* is lying in the plane of the paper. This sheet is going to rotate about an axis that also lies in the plane of the paper and passes through point *A*. Draw two such axes that are oriented so that points *B* and *C* will move on circular paths having the same radii.

2. A pair of scissors is being used to cut a string. Does each blade of the scissors have the same angular velocity (both magnitude and direction) at a given instant? Give your reasoning.

3. An electric clock is hanging on the wall in the living room. The clock is unplugged, and the second hand comes to a halt over a brief period of time. During this period, what is the direction of the angular acceleration of the second hand? Why?

4. The earth rotates once per day about its axis. Where on the earth's surface should you stand in order to have the smallest possible tangential speed? Justify your answer.

5. A thin rod rotates at a constant angular speed. Consider the tangential speed of each point on the rod for the case when the axis of rotation is perpendicular to the rod (a) at its center and (b) at one end. Explain for each case whether there are any points on the rod that have the same tangential speeds.

6. A car is up on a hydraulic lift at a garage. The wheels are free to rotate, and the drive wheels are rotating with a constant angular velocity. Does a point on the rim of a wheel have (a) a tangential acceleration and (b) a centripetal acceleration? In each case, give your reasoning.

7. Two points are located on a rigid wheel that is rotating with an increasing angular velocity about a fixed axis. The axis is perpendicular to the wheel at its center. Point 1 is located on the rim, and point 2 is halfway between the rim and the axis. At any given instant, which point (if either) has the greater (a) angular velocity, (b) angular acceleration, (c) tangential speed, (d) tangential acceleration, and (e) centripetal acceleration? Provide a reason for each of your answers.

8. A building is located on the earth's equator. Which has the greatest tangential speed due to the earth's rotation, the top floor, the bottom floor, or neither? Justify your answer.

9. Section 5.6 discusses how the uniform circular motion of a space station can be used to create "artificial" gravity for the astro- nauts. This can be done by adjusting the angular speed of the space station, so the centripetal acceleration at the astronaut's feet equals *g*, the magnitude of the acceleration due to gravity (see Figure 5.19). If such an adjustment is made, will the acceleration due to the "artificial" gravity be greater than, equal to, or less than *g* at the astronaut's head? Account for your answer.

10. It is possible, but not very practical, to build a clock in which the tips of the second hand, the minute hand, and the hour hand move with the same tangential speed. Explain why not.

11. Explain why a point on the rim of a tire has an accelera- tion when the tire is on a car that is moving at a constant linear velocity.

12. A bicycle is turned upside down, the front wheel is spinning (see the drawing), and there is an angular acceleration. At the in- stant shown, there are six points on the wheel that have arrows as- sociated with them. Which of the following quantities could the arrows represent: (a) tangential velocity, (b) tangential accelera- tion, (c) centripetal acceleration? In each case, answer why the ar- rows do or do not represent the quantity.

13. Suppose that the speedometer of a truck is set to read the lin- ear speed of the truck, but uses a device that actually measures the angular speed of the tires. If larger diameter tires are mounted on the truck, will the reading on the speedometer be correct? If not, will the reading be greater than or less than the true linear speed of the truck? Why?

14. The blades of a fan rotate more and more slowly after the fan is shut off. Eventually they stop rotating altogether. In such a situation, we sometimes assume that the angular acceleration of the blades is constant and apply the equations of rotational kine-

matics as an approximation. Explain why the angular acceleration can never really be constant in this kind of situation.

15. Rolling motion is one example that involves rotation about an axis that is not fixed. Give three other examples. In each case, identify the axis of rotation and explain why it is not fixed.

PROBLEMS

Section 8.1 Rotational Motion and Angular Displacement, Section 8.2 Angular Velocity and Angular Acceleration

1. ssm In Europe, surveyors often measure angles in *grads*. There are 100 grads in one-quarter of a circle. How many grads are there in one radian?

2. For each of the following angles, give its equivalent in radians: (a) 75°, (b) 1.5 revolutions.

3. A gymnast on a trampoline does a double backward somersault. What is the magnitude of the angular displacement (in radians) of the gymnast?

4. A pulsar is a rapidly rotating neutron star that emits a pulse of radio waves each time it makes one revolution. For one particular pulsar, the time between two successive pulses is 0.033 s. Determine the average angular speed (in rad/s) of this pulsar.

5. ssm A diver completes $3\frac{1}{2}$ somersaults in 1.7 s. What is the average angular speed (in rad/s) of the diver?

6. A certain CD has a playing time of 74 minutes. When the music starts, the CD is rotating at an angular speed of 4.8×10^2 revolutions per minute (rpm). At the end of the music, the CD is rotating at 2.1×10^2 rpm. Find the magnitude of the average angular acceleration of the CD. Express your answer in rad/s².

7. A string trimmer is a grass and weed cutting tool that utilizes a 0.21-m length of nylon "string," rotating at 650 rad/s about an axis perpendicular to one end of the string. (a) What is the time needed for the string to sweep out an angle of 0.61 rad? (b) Assuming that the length of the string does not change, find the distance through which the tip of the string moves during this time.

8. An automatic drier spins wet clothes at an angular speed of 5.2 rad/s. Starting from rest, the drier reaches its operating speed with an average angular acceleration of 4.0 rad/s². How long does it take the drier to come up to speed?

9. ssm An electric fan is running on HIGH. After the LOW button is pressed, the angular speed of the fan decreases to 83.8 rad/s in 1.75 s. The deceleration is 42.0 rad/s². Determine the initial angular speed of the fan.

*10. A space station consists of two donut-shaped living chambers, A and B, that have the radii shown in the drawing. As the station rotates, an astronaut in chamber A is moved 2.40×10^2 m along a circular arc. How far along a circular arc is an astronaut in chamber B moved during the same time?

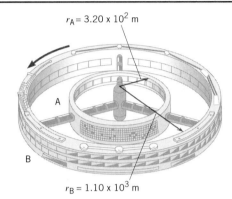

$r_A = 3.20 \times 10^2$ m

A

B

$r_B = 1.10 \times 10^3$ m

*11. Suppose that wedge-shaped pieces are cut from a pie, so that the arc length along the outer crust of each piece exactly equals the radius of the piece. After all such pieces are eaten, what is the apex angle (in radians) of the remaining portion?

*12. Two people start at the same place and walk around a circular lake in opposite directions. One has an angular speed of 1.7×10^{-3} rad/s, while the other has an angular speed of 3.4×10^{-3} rad/s. How long will it be before they meet?

*13. ssm The drawing shows a device that can be used to measure the speed of a bullet. The device consists of two rotating disks, separated by a distance of $d = 0.850$ m, and rotating with an angular speed of 95.0 rad/s. The bullet first passes through the left disk and then through the right disk. It is found that the angular displacement between the two bullet holes is $\theta = 0.240$ rad. From these data, determine the speed of the bullet.

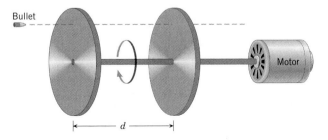

Bullet

Motor

d

*14. A stroboscope is a light that flashes on and off at a constant rate. It can be used to illuminate a rotating object, and if the flashing rate is adjusted properly, the object can be made to appear stationary. What is the shortest time between flashes of light that will make a three-bladed propeller appear stationary when it is rotating with an angular speed of 16.7 rev/s?

** 15. A quarterback throws a pass that is a perfect spiral. In other words, the football does not wobble, but spins smoothly about an axis passing through each end of the ball. Suppose the ball spins at 7.7 rev/s. In addition, the ball is thrown with a linear speed of 19 m/s at an angle of 55° with respect to the ground. If the ball is caught at the same height at which it left the quarterback's hand, how many revolutions has the ball made while in the air?

** 16. Review Conceptual Example 2 before attempting this problem. The moon has a diameter of 3.48×10^6 m and is a distance of 3.85×10^8 m from the earth. The sun has a diameter of 1.39×10^9 m and is 1.50×10^{11} m from the earth. (a) Determine (in radians) the angles subtended by the moon and the sun, as measured by a person standing on the earth. (b) Based on your answers to part (a), decide whether a total eclipse of the sun is really "total." Give your reasoning. (c) If the eclipse is not really total, determine the ratio, expressed as a percentage, of the apparent circular area of the moon to the apparent circular area of the sun.

Section 8.3 The Equations of Rotational Kinematics

17. **ssm** The drill bit of a variable-speed electric drill has a constant angular acceleration of 2.50 rad/s². The initial angular speed of the bit is 5.00 rad/s. After 4.00 s, (a) what angle has the bit turned through and (b) what is the bit's angular speed?

18. A figure skater is spinning with an angular velocity of +15 rad/s. She then comes to a stop over a brief period of time. During this time, her angular displacement is +5.1 rad. Determine (a) her average angular acceleration and (b) the time during which she comes to rest.

19. An airliner arrives at the terminal, and the pilot shuts off the engines. The initial angular velocity of the fan blades is 1800 rad/s, and it takes 120 s for them to come to rest. What is the angular displacement of the blades?

20. A basketball player is balancing a spinning basketball on the tip of his finger. The angular velocity of the ball slows down from 18.5 to 14.1 rad/s, during which the angular displacement is 85.1 rad. Determine the time it takes for the ball to slow down.

21. **ssm www** A flywheel has a constant angular deceleration of 2.0 rad/s². (a) Find the angle through which the flywheel turns as it comes to rest from an angular speed of 220 rad/s. (b) Find the time required for the flywheel to come to rest.

22. The angular velocity of a fan in an air conditioning unit increases to +180 rad/s from +130 rad/s in 14 revolutions. What is the angular acceleration in rad/s²?

23. The angular speed of the rotor in a centrifuge increases from 420 to 1420 rad/s in a time of 5.00 s. (a) Obtain the angle through which the rotor turns. (b) What is the angular acceleration?

* 24. After 10.0 s, a spinning roulette wheel has slowed down to an angular speed of 1.88 rad/s. During this time, the wheel rotates

through an angle of 44.0 rad. Determine the angular acceleration of the wheel.

* 25. **ssm** A spinning wheel on a fireworks display is initially rotating in a counterclockwise direction. The wheel has an angular acceleration of −4.00 rad/s². Because of this acceleration, the angular velocity of the wheel changes from its initial value to a final value of −25.0 rad/s. While this change occurs, the angular displacement of the wheel is zero. (Note the similarity to that of a ball being thrown vertically upward, coming to a momentary halt, and then falling downward to its initial position.) Find the time required for the change in the angular velocity to occur.

* 26. A dentist causes the bit of a high-speed drill to accelerate from an angular speed of 1.05×10^4 rad/s to an angular speed of 3.14×10^4 rad/s. In the process, the bit turns through 1.88×10^4 rad. Assuming a constant angular acceleration, how long would it take the bit to reach its maximum speed of 7.85×10^4 rad/s, starting from rest?

* 27. The drive propeller of a ship starts from rest and accelerates at 3.50×10^{-3} rad/s² for 1.80×10^3 s. For the next 3.60×10^3 s, the propeller rotates at a constant angular speed. Then it decelerates at 2.00×10^{-3} rad/s², until it slows (without reversing direction) to an angular speed of 5.00 rad/s. Find the total angular displacement of the propeller.

* 28. At the local swimming hole, a favorite trick is to run horizontally off a cliff that is 8.3 m above the water. One diver runs off the edge of the cliff, tucks into a "ball," and rotates on the way down with an average angular speed of 1.6 rev/s. Ignore air resistance and determine the number of revolutions she makes while on the way down.

** 29. **ssm www** A child, hunting for his favorite wooden horse, is running on the ground around the edge of a stationary merry-go-round. The angular speed of the child has a constant value of 0.250 rad/s. At the instant the child spots the horse, one-quarter of a turn ($\pi/2$ rad) away, the merry-go-round begins to move (in the direction the child is running) with a constant angular acceleration of 0.0100 rad/s². What is the shortest time it takes for the child to catch up with the horse?

Section 8.4 Angular Variables and Tangential Variables

30. A fisherman is hauling in a fish at a speed of 0.14 m/s. The fishing line is being spooled onto a reel whose radius is 0.030 m. What is the angular speed of the reel?

31. An auto race is held on a circular track. A car completes one lap in a time of 18.9 s, with an average tangential speed of 42.6 m/s. Find (a) the average angular speed and (b) the radius of the track.

32. Our sun rotates in a circular orbit about the center of the Milky Way galaxy. The radius of the orbit is 2.2×10^{20} m, and the angular speed of the sun is 1.2×10^{-15} rad/s. (a) What is the tangential speed of the sun? (b) How long (in years) does it take for the sun to make one revolution around the center?

33. **ssm** A disk (radius = 2.00 mm) is attached to a high-speed drill at a dentist's office and is turning at 7.85×10^4 rad/s. Determine the tangential speed of a point on the outer edge of this disk.

34. The earth has a radius of 6.38×10^6 m and turns on its axis once every 23.9 h. (a) What is the tangential speed (in m/s) of a person living in Ecuador, a country that lies on the equator? (b) At what latitude (i.e., the angle θ in the drawing) is the tangential speed one-third that of a person living in Ecuador?

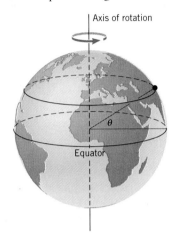

35. The take-up reel of a cassette tape has a radius of 0.025 m. Find the length of tape that passes around the reel in 7.1 s when the reel rotates at an average angular speed of 1.9 rad/s.

36. A meter stick is rotating with a constant angular acceleration of 12.0 rad/s² about an axis that passes through one end of the stick and is perpendicular to it. What point on the stick has a tangential acceleration whose magnitude equals that of the acceleration due to gravity?

*37. **ssm** A compact disc (CD) contains music on a spiral track. Music is put onto a CD with the assumption that, during playback, the music will be detected at a *constant tangential speed* at any point. Since $v_T = r\omega$, a CD rotates at a smaller angular speed for music near the outer edge and a larger angular speed for music near the inner part of the disc. For music at the outer edge ($r = 0.0568$ m), the angular speed is 3.50 rev/s. Find (a) the constant tangential speed at which music is detected and (b) the angular speed (in rev/s) for music at a distance of 0.0249 m from the center of a CD.

*38. A thin rod (length = 1.50 m) is oriented vertically, with its bottom end attached to the floor by means of a frictionless hinge. The mass of the rod may be ignored, compared to the mass of the object fixed to the top of the rod. The rod, starting from rest, tips over and rotates downward. (a) What is the angular speed of the rod just before it strikes the floor? (*Hint: Consider using the principle of conservation of mechanical energy.*) (b) What is the magnitude of the angular acceleration of the rod just before it strikes the floor?

*39. A thin straight rod is rotating with an angular speed of 3.14 rad/s about a vertical axis, as the drawing shows. In one sec-

ond, the end of the rod moves through a circular arc whose length equals the length of the rod. Find the angle θ.

**40. The drawing shows a golf ball passing through a windmill at a miniature golf course. The windmill has 8 blades and rotates at an angular speed of 1.25 rad/s. The opening between successive blades is equal to the width of a blade. A golf ball of diameter 4.50×10^{-2} m is just passing by one of the rotating blades. What must be the *minimum* speed of the ball so that it will not be hit by the next blade?

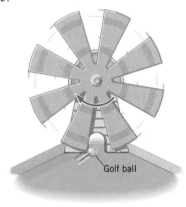

Golf ball

Section 8.5 Centripetal Acceleration and Tangential Acceleration

41. **ssm** During a tennis serve, a racket is given an angular acceleration of magnitude 160 rad/s². At the top of the serve, the racket has an angular speed of 14 rad/s. If the distance between the top of the racket and the shoulder is 1.5 m, find the magnitude of the total acceleration of the top of the racket.

42. Two Formula-one racing cars are negotiating a circular turn with the same *angular* speed. However, the path of car A has a radius of 24 m, while that of car B has a radius of 18 m. Determine the ratio of the centripetal acceleration of car A to that of car B.

43. A ceiling fan has two different angular speed settings: $\omega_1 = 440$ rev/min and $\omega_2 = 110$ rev/min. What is the ratio a_1/a_2 of the centripetal accelerations of a given point on a fan blade?

44. A floppy disk for a personal computer rotates with a constant angular speed of 31.4 rad/s about an axis perpendicular to the disk at its center. (a) Find the tangential speed of a point that is 0.0410 m from the center of the disk. (b) What is the magnitude of the centripetal acceleration at this point?

45. ssm A 220-kg speedboat is negotiating a circular turn (radius = 32 m) around a buoy. During the turn, the engine causes a net tangential force of magnitude 550 N to be applied to the boat. The initial tangential speed of the boat going into the turn is 5.0 m/s. (a) Find the tangential acceleration. (b) After the boat is 2.0 s into the turn, find the centripetal acceleration.

*46. The front and rear sprockets on a bicycle have radii of 9.00 and 5.10 cm, respectively. The angular speed of the front sprocket is 9.40 rad/s. Determine (a) the linear speed (in cm/s) of the chain as it moves between the sprockets and (b) the centripetal acceleration (in cm/s^2) of the chain as it passes around the rear sprocket.

*47. A thin rigid rod is rotating with a constant angular acceleration about an axis that passes perpendicularly through one of its ends. At one instant, the total acceleration vector (centripetal plus tangential) at the other end of the rod makes a 60.0° angle with respect to the rod and has a magnitude of 15.0 m/s^2. The rod has an angular speed of 2.00 rad/s at this instant. What is the rod's length?

**48. The blades of a windmill start from rest and rotate with an angular acceleration of 22.0 rad/s^2. At any point on a blade, how much time passes before the magnitude of the tangential acceleration equals the magnitude of the centripetal acceleration?

**49. ssm An electric drill starts from rest and rotates with a constant angular acceleration. After the drill has rotated through a certain angle, the magnitude of the centripetal acceleration of a point on the drill is twice the magnitude of the tangential acceleration. What is the angle?

Section 8.6 Rolling Motion

Note: All problems in this section assume that there is no slipping of the surfaces in contact during the rolling motion.

50. Suppose you are riding a stationary exercise bicycle, and the electronic meter indicates that the wheel is rotating at 9.1 rad/s. The wheel has a radius of 0.45 m. If you ride the bike for 2100 s, how far would you have gone if the bike could move?

51. A bicycle travels a linear distance of 515 m. Through what angle (in radians) does each tire ($r = 0.356$ m) rotate?

52. An automobile tire has a radius of 0.330 m, and its center moves forward with a linear speed of $v = 15.0$ m/s. (a) Determine the angular speed of the wheel. (b) Relative to the axle, what is the tangential speed of a point located 0.175 m from the axle?

53. ssm www A motorcycle accelerates uniformly from rest and reaches a linear speed of 22.0 m/s in a time of 9.00 s. The radius of each tire is 0.280 m. What is the magnitude of the angular acceleration of each tire?

54. On an open-reel tape deck, the tape is being pulled past the playback head at a constant linear speed of 0.381 m/s. (a) Using the data in part *a* of the drawing, find the angular speed of the take-up reel. (b) After 2.40 × 10^3 s, the take-up reel is almost full, as part *b* of the drawing indicates. Find the average angular accel-

eration of the reel and specify whether the acceleration indicates an increasing or decreasing angular velocity.

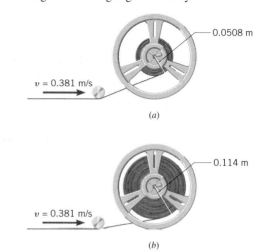

(a)

(b)

55. A car is traveling with a speed of 15.0 m/s along a straight horizontal road. The wheels have a radius of 0.480 m. If the car speeds up with a linear acceleration of 2.00 m/s^2 for 5.00 s, find the angular displacement of each wheel during this period.

*56. A ball of radius 0.200 m rolls along a horizontal table top with a constant linear speed of 3.60 m/s. The ball rolls off the edge and falls a vertical distance of 2.10 m before hitting the floor. What is the angular displacement of the ball while it is in the air?

*57. ssm A person lowers a bucket into a well by turning the hand crank, as the drawing illustrates. The crank handle moves with a constant tangential speed of 1.20 m/s on its circular path. Find the linear speed with which the bucket moves down the well.

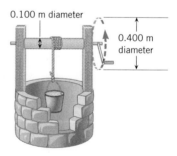

*58. The two-gear combination shown in the drawing is being used to hoist the load L with a constant upward speed of 2.50 m/s.

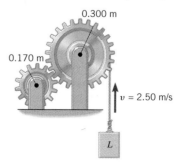

The rope attached to the load is being wound onto a cylinder behind the big gear. The depth of the teeth of the gears is negligible compared to the radii. Determine the angular velocity (magnitude and direction) of (a) the larger gear and (b) the smaller gear.

59. The chain on a bicycle connects the front sprocket (radius = 12.5 cm) to the rear sprocket (radius = 3.50 cm). The wheels have a radius of 40.0 cm. The rear wheel and the rear sprocket turn together as a single unit. The rider is peddling so the front sprocket makes 1.50 revolutions each second. What is the linear speed (in m/s) of the bicycle?

60. A bicycle is rolling down a circular hill whose radius is 9.00 m. As the drawing illustrates, the angular displacement of the bicycle is 0.960 rad. The radius of each wheel is 0.400 m. What is the angle (in radians) through which each tire rotates?

ADDITIONAL PROBLEMS

61. ssm A race car travels with a constant tangential speed of 75.0 m/s around a circular track of radius 625 m. Find (a) the magnitude of the car's total acceleration and (b) the direction of its total acceleration relative to the radial direction.

62. On a wristwatch, what is the angular velocity (magnitude and direction) of (a) the second hand and (b) the hour hand? Express your answers in rad/s.

63. The drawing shows the blade of a chain saw. The rotating sprocket tip at the end of the guide bar has a radius of 3.5×10^{-2} m. The linear speed of a chain link at point A is 7.7 m/s. Find the angular speed of the sprocket tip in rev/s.

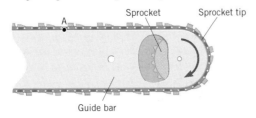

64. The shaft of a pump starts from rest and has an angular acceleration of 3.00 rad/s^2 for 18.0 s. At the end of this interval, what is (a) the shaft's angular speed and (b) the angle through which the shaft has turned?

65. ssm An electric circular saw is designed to reach its final angular speed, starting from rest, in 1.50 s. Its average angular acceleration is 328 rad/s^2. Obtain its final angular speed.

66. A train is rounding a circular curve whose radius is 2.00×10^2 m. At one instant, the train has an angular acceleration of 1.50×10^{-3} rad/s^2 and an angular speed of 0.0500 rad/s. (a) Find the magnitude of the total acceleration (centripetal plus tangential) of the train. (b) Determine the angle of the total acceleration relative to the radial direction.

67. The planet Venus orbits the sun in a nearly perfect circle of radius 1.08×10^{11} m. The angular speed of Venus is 1.63 rev/yr. (There are 3.16×10^7 s in one year.) Determine the tangential speed of Venus.

68. A baton twirler throws a spinning baton directly upward. As it goes up and returns to the twirler's hand, the baton turns through four revolutions. Ignoring air resistance and assuming that the average angular speed of the baton is 1.80 rev/s, determine the height to which the center of the baton travels above the point of release.

69. ssm www The drawing shows a view (from beneath) of the platter on a belt-drive turntable. The platter has an angular speed of 3.49 rad/s. The pulley on the motor shaft has a radius of 0.0127 m. Assuming that the belt does not slip, determine the angular speed of the motor shaft.

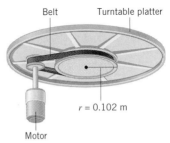

70. A fan blade, whose angular acceleration is a constant 2.00 rad/s^2, rotates through an angle of 285 radians in 11.0 s. How long did it take the blade, starting from rest, to reach the *beginning* of the 11.0-s interval?

71. A rectangular plate is rotating with a constant angular acceleration about an axis that passes perpendicularly through one corner, as the drawing shows. The tangential acceleration measured at

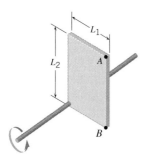

(Hint: Review the paragraph just before Equation 8.12 that discusses how the distance traveled by the axle of a wheel is related to the circular arc length along the outer edge of the wheel.)

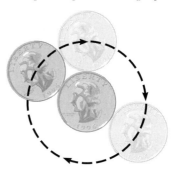

corner A has twice the magnitude of that measured at corner B. What is the ratio L_1/L_2 of the lengths of the sides of the rectangle?

**72. One type of slingshot can be made from a length of rope and a leather pocket for holding the stone. The stone can be thrown by whirling it rapidly in a horizontal circle and releasing it at the right moment. Such a slingshot is used to throw a stone from the edge of a cliff, the point of release being 20.0 m above the base of the cliff. The stone lands on the ground below the cliff at a point X. The horizontal distance of point X from the base of the cliff (directly beneath the point of release) is thirty times the radius of the circle on which the stone is whirled. Determine the angular speed of the stone at the moment of release.

73. **ssm Take two quarters and lay them on a table. Press down on one quarter so it cannot move. Then, starting at the 12:00 position, roll the other quarter along the edge of the stationary quarter, as the drawing suggests. How many revolutions does the rolling quarter make when it travels once around the circumference of the stationary quarter? Surprisingly, the answer is *not* one revolution.

**74. Review Conceptual Example 9 before attempting this problem. The drawing shows a tire of radius R on a moving car and identifies a point P that is at a distance H above the ground. The velocity of this point with respect to the ground has the same magnitude v as the linear velocity of the car. Find the ratio H/R.

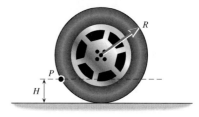

ROTATIONAL DYNAMICS

When the climber grabs onto an overhang, a number of forces act on him. As this chapter discusses, the location of each force is just as important as the force itself, both of which are taken into account by the notion of a torque.

9.1 THE EFFECTS OF FORCES AND TORQUES ON THE MOTION OF RIGID OBJECTS

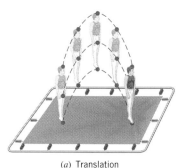

(a) Translation

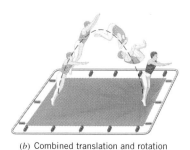

(b) Combined translation and rotation

Figure 9.1 An example of (a) translational motion and (b) combined translational and rotational motion.

Most rigid objects, like a propeller or a wheel, have their mass spread out and not concentrated at a single point. As a result, objects can move in a number of ways. Figure 9.1a illustrates one possibility called translational motion, in which all points on the body travel on parallel paths (not necessarily straight lines). In pure translation there is no rotation of any line in the body. Because translational motion can occur along a curved line, it is often called curvilinear motion or linear motion. Another possibility is rotational motion, which may occur in combination with translational motion, as is the case for the somersaulting gymnast in Figure 9.1b.

We have seen many examples of how a net force affects the linear motion of an object by causing it to accelerate. When there is no net force, there is no acceleration, and the object is in equilibrium. Since rigid objects can also rotate, however, we need to expand our concept of equilibrium and take into account the possibility of rotation. A net external force causes linear motion to change, but what causes rotational motion to change? For example, something causes the propeller on a speed boat to change its rotational velocity while the boat is speeding up. Is it simply a net force? As it turns out, it is a net external torque, not a net force, that causes the rotational velocity to change. Just as greater net forces cause greater linear accelerations, greater net torques cause greater rotational or angular accelerations.

Figure 9.2 helps to explain the idea of torque. When you push on a door with a force **F** as in part a, the door opens more quickly when the force is larger. Other things being equal, a larger force generates a larger torque. However, the door does not open as quickly if you apply the same force at a point closer to the hinge, as in part b, because the force now produces less torque. Furthermore, if your push is directed nearly at the hinge, as in part c, you will have a hard time opening the door at all, because the torque is nearly zero. In summary, the torque depends on the magnitude of the force, on the point where the force is applied relative to the axis of rotation (the hinge in Figure 9.2), and on the direction of the force.

For simplicity, we deal with situations in which the force lies in a plane that is perpendicular to the axis of rotation. In Figure 9.3, for instance, the axis is perpendicular to the page and the force lies in the plane of the paper. The drawing shows the line of action and the lever arm of the force, two concepts that are important in the definition of torque. The **line of action** is an extended line drawn colinear with the force. The **lever arm** is the distance ℓ between the line of action and the axis of rotation, measured on a line that is perpendicular to both. The torque is represented by the symbol τ (Greek letter *tau*), and its magnitude is defined as the magnitude of the force times the lever arm:

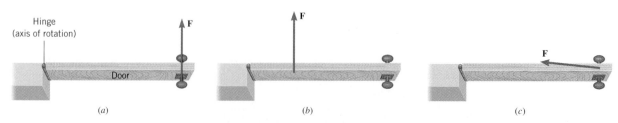

Figure 9.2 It is easier to open a door with a force of a given magnitude by (a) pushing at the door's outer edge than by (b) pushing closer to the axis of rotation (the hinge). (c) Pushing nearly into the hinge makes it very difficult to open the door.

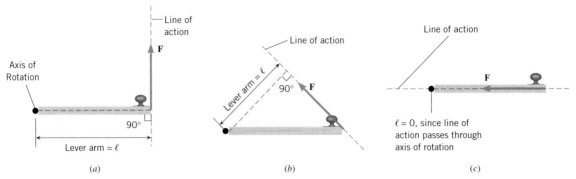

(a)   (b)   (c)

Figure 9.3 In this top view, the hinges of a door appear as a block dot (●) and define the axis of rotation. The line of action and lever arm are illustrated for a force applied to the door (a) perpendicularly and (b) at an angle. (c) The lever arm is zero whenever the line of action passes through the axis of rotation.

■ **DEFINITION OF TORQUE**

Magnitude:

$$\text{Torque} = (\text{Magnitude of the force}) \times (\text{Lever arm})$$
$$\tau = F\ell \qquad (9.1)$$

Direction: The torque is positive if the force tends to produce a counterclockwise rotation about the axis, and negative if the force tends to produce a clockwise rotation.

SI Unit of Torque: newton·meter (N·m)

Equation 9.1 indicates that forces of the same magnitude can produce *different* torques, depending on the value of the lever arm, and Example 1 illustrates this important feature.

EXAMPLE 1 • Different Lever Arms, Different Torques

In Figure 9.3 a force whose magnitude is 55 N is applied to a door. However, the lever arms are different in the three parts of the drawing: (a) $\ell = 0.80$ m, (b) $\ell = 0.60$ m, and (c) $\ell = 0$. Find the magnitude of the torque in each case.

Reasoning In each case the lever arm is the perpendicular distance between the axis of rotation and the line of action of the force. Note that in part *a* this perpendicular distance is equal to the width of the door. In parts *b* and *c*, however, the lever arm is less than the width of the door. Because the lever arm is different in each case, the torque is different, even though the magnitude of the applied force is the same.

Solution Equation 9.1 gives the following values for the torques:

(a) $\tau = F\ell = (55\text{ N})(0.80\text{ m}) = \boxed{44\text{ N}\cdot\text{m}}$

(b) $\tau = F\ell = (55\text{ N})(0.60\text{ m}) = \boxed{33\text{ N}\cdot\text{m}}$

(c) $\tau = F\ell = (55\text{ N})(0) = \boxed{0}$

In part *c* the line of action of *F* passes through the axis of rotation (the hinge). Hence, the lever arm is zero, and the torque is zero.

Muscles and tendons in our bodies produce torques about various joints. Example 2 illustrates how the Achilles tendon produces a torque about the ankle joint.

EXAMPLE 2 • The Achilles Tendon

Figure 9.4*a* shows the ankle joint and the Achilles tendon attached to the heel at point *P*. The tendon exerts a force of magnitude $F = 720$ N, as Figure 9.4*b* indicates.

The Physics of...
the Achilles tendon.

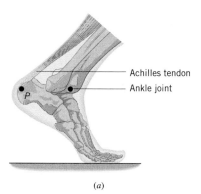

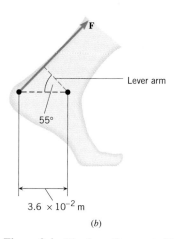

(a)

(b)

Figure 9.4 The force **F** generated by the Achilles tendon produces a clockwise (negative) torque about the ankle joint.

Determine the torque (magnitude and direction) of this force about the ankle joint, which is located 3.6×10^{-2} m away from point P.

Reasoning To calculate the magnitude of the torque, it is necessary to have a value for the lever arm ℓ. It is important to realize, however, that the lever arm is not the given distance of 3.6×10^{-2} m. Instead, the lever arm is the perpendicular distance between the axis of rotation at the ankle joint and the line of action of the force **F**. In Figure 9.4b this distance is indicated by the dashed red line.

Solution From the drawing, it can be seen that the lever arm is $\ell = (3.6 \times 10^{-2} \text{ m}) \cos 55° = 2.1 \times 10^{-2}$ m . The magnitude of the torque is

$$\tau = F\ell = (720 \text{ N})(2.1 \times 10^{-2} \text{ m}) = 15 \text{ N} \cdot \text{m} \qquad (9.1)$$

The force **F** tends to produce a clockwise rotation about the ankle joint, so the torque is negative: $\boxed{\tau = -15 \text{ N} \cdot \text{m}}$

Now that we have introduced the idea of torque, the next step is to incorporate it into the concept of equilibrium. The following section shows that torques, as well as forces, must be balanced in order for a rigid object to be in equilibrium.

9.2 RIGID OBJECTS IN EQUILIBRIUM

If a rigid body is in equilibrium, its motion does not change. By "motion" we mean both linear and rotational motion. An object whose motion is not changing has no acceleration of any kind. Therefore, the net force $\Sigma\mathbf{F}$ applied to the object must be zero, since $\Sigma\mathbf{F} = m\mathbf{a}$ and $\mathbf{a} = 0$. For two-dimensional motion, the condition $\Sigma\mathbf{F} = 0$ means that the x and y components of the net force are separately zero: $\Sigma F_x = 0$ and $\Sigma F_y = 0$ (Equations 4.9a and 4.9b). When calculating the net force, we include only *external forces,* that is, those forces applied to the object by external agents. We ignore the forces that one internal part of an object exerts on another internal part. Internal forces can be ignored because they occur in action–reaction pairs, which have no effect as far as the motion of the entire object is concerned. Each action–reaction pair consists of oppositely directed forces of equal magnitude, and the effect of one force cancels the effect of the other.

A net external torque causes rotational motion to change. But there is no change in the motion of a rigid body in equilibrium, so there can be no net torque acting under equilibrium conditions. The sum of the positive torques must be balanced by the sum of the negative torques. Using the symbol $\Sigma\tau$ to represent the net external torque (the sum of all positive and negative torques), we write this condition as

$$\Sigma\tau = 0 \qquad (9.2)$$

and expand our notion of equilibrium by taking it into account, as illustrated by the concept chart in Figure 9.5 and summarized below.

■ **EQUILIBRIUM OF A RIGID BODY**

A rigid body is in equilibrium if it has zero translational acceleration and zero angular acceleration. In equilibrium, the sum of the externally applied forces is zero, and the sum of the externally applied torques is zero:

$$\Sigma F_x = 0 \quad \text{and} \quad \Sigma F_y = 0 \qquad (4.9a \text{ and } 4.9b)$$

$$\Sigma\tau = 0 \qquad (9.2)$$

Concepts at a Glance

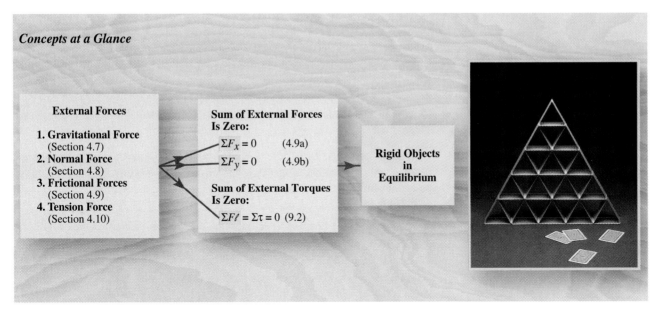

Figure 9.5 *Concepts at a Glance* A rigid object is in equilibrium when the sum of the external forces is zero and when the sum of the external torques is zero. This pyramid of stacked cards is in equilibrium.

The reasoning strategy for analyzing the forces and torques acting on a body in equilibrium is given below. The first four steps of the strategy are essentially the same as those outlined in Section 4.11, when only forces were being considered. Steps 5 and 6 have been added to account for any external torques that may be present. Example 3 illustrates how this reasoning strategy is applied to a diving board.

REASONING STRATEGY

Applying the Conditions of Equilibrium to a Rigid Body

1. Select the object to which the conditions for equilibrium are to be applied.

2. Draw a free-body diagram that shows all the external forces acting on the object, each force with its proper direction.

3. Choose a convenient set of x, y axes and resolve all forces into components that lie along these axes.

4. Apply the conditions that specify the balance of forces at equilibrium: $\Sigma F_x = 0$ and $\Sigma F_y = 0$.

5. Select a convenient axis of rotation. Identify the point where each external force acts on the object, and calculate the torque produced by each force about the axis of rotation. Set the sum of the torques about this axis equal to zero: $\Sigma \tau = 0$.

6. Solve the equations in steps 4 and 5 for the desired unknown quantities.

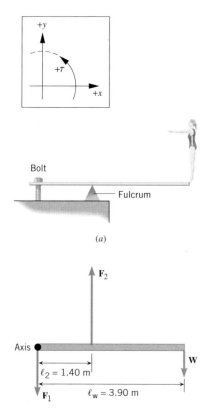

Bolt

Fulcrum

(a)

F_2

Axis

$\ell_2 = 1.40$ m

$\ell_w = 3.90$ m

F_1

W

(b) Free-body diagram of the diving board

Figure 9.6 (a) A diver stands at the end of a diving board. (b) The free-body diagram for the diving board. The box at the upper left shows the positive x and y directions for the forces, as well as the positive (counterclockwise) direction for the torques.

EXAMPLE 3 • A Diving Board

A woman whose weight is 530 N is poised at the right end of a diving board, whose length is 3.90 m. The board has negligible weight and is bolted down at the left end, while being supported 1.40 m away by a fulcrum, as Figure 9.6a shows. Find the forces F_1 and F_2 that the bolt and the fulcrum, respectively, exert on the board.

Reasoning Part b of the figure shows the free-body diagram of the diving board. Three forces act on the board: F_1, F_2, and the force due to the diver's weight W. In choosing the directions of F_1 and F_2 we have used our intuition: F_1 points downward, because the bolt must pull in that direction to counteract the tendency of the board to rotate clockwise about the fulcrum; F_2 points upward because the board pushes downward against the fulcrum and, in reaction, the fulcrum pushes upward on the board.

Solution Since the board is in equilibrium, the sum of the vertical forces must be zero:

$$\Sigma F_y = -F_1 + F_2 - 530\text{ N} = 0 \qquad (4.9b)$$

Similarly, the sum of the torques must be zero, $\Sigma \tau = 0$. For calculating torques, we select an axis that passes through the left end of the board and is perpendicular to the page. (We will see shortly that this choice is arbitrary.) The force F_1 produces no torque since it passes through the axis and has a zero lever arm, while F_2 creates a counterclockwise (positive) torque, and W produces a clockwise (negative) torque. The free-body diagram shows the lever arms for the torques:

$$\Sigma \tau = +F_2\ell_2 - W\ell_w = +F_2(1.40\text{ m}) - (530\text{ N})(3.90\text{ m}) = 0 \qquad (9.2)$$

$$\boxed{F_2 = 1480\text{ N}}$$

This value for F_2 can be substituted into Equation 4.9b above to show that $\boxed{F_1 = 950\text{ N}}$.

In Example 3 the sum of the external torques is calculated using an axis that passes through the left end of the diving board. *However, the location of the axis is completely arbitrary, because if an object is in equilibrium, it is in equilibrium with respect to any axis whatsoever.* Thus, the sum of the external torques is zero, no matter where the axis is placed. One usually chooses the location of the rotational axis so that the lines of action of one or more of the unknown forces pass through the axis. Such a choice simplifies the torque equation, because the torques produced by these forces are zero. For instance, in Example 3 the torque due to the force F_1 does not appear in the equation representing the balance of torques (Equation 9.2), because the lever arm of this force is zero.

In a calculation of torque, the lever arm of the force must be determined relative to the axis of rotation. In Example 3 the lever arms are obvious, but sometimes a little care is needed in determining them, as in the next example.

EXAMPLE 4 • Fighting a Fire

In Figure 9.7a an 8.00-m ladder of weight $W_L = 355$ N leans against a smooth vertical wall. The term "smooth" means that the wall can exert only a normal force directed perpendicular to the wall and cannot exert a frictional force directed parallel to it. A firefighter, whose weight is $W_F = 875$ N, stands 6.30 m from the bottom of the ladder. Assume that the weight of the ladder acts at the ladder's center and neglect the weight of the hose. Find the forces that the wall and the ground exert on the ladder.

Reasoning Part *b* of the figure shows the free-body diagram of the ladder. The following forces are exerted on the ladder:

1. Its weight $\mathbf{W_L}$
2. A force due to the weight $\mathbf{W_F}$ of the firefighter
3. The force $\mathbf{P}$ applied to the top of the ladder by the wall and directed perpendicular to the wall
4. The forces $\mathbf{G}_x$ and $\mathbf{G}_y$, which are the horizontal and vertical components of the force exerted by the ground on the bottom of the ladder.

The ground, unlike the wall, is not smooth, so that the force $\mathbf{G}_x$ is produced by static friction and prevents the ladder from slipping. The force $\mathbf{G}_y$ is the normal force applied to the ladder by the ground. The ladder is in equilibrium, so the sum of these forces must be zero and the sum of the torques produced by these forces must be zero.

Solution Since the ladder is in equilibrium, the net force acting on it is zero:

$$\Sigma F_x = G_x - P = 0 \tag{4.9a}$$
$$\Sigma F_y = G_y - W_L - W_F$$
$$\qquad = G_y - 355\text{ N} - 875\text{ N} = 0 \quad \text{or} \quad \boxed{G_y = 1230\text{ N}}$$

Equation 4.9a cannot be solved as it stands, because it contains two unknown variables. However, another equation can be obtained from the fact that the net torque acting on an object in equilibrium is zero. In calculating torques, it is convenient to use an axis at the left end of the ladder, directed perpendicular to the page, as Figure 9.7c indicates. This axis is convenient, because $\mathbf{G}_x$ and $\mathbf{G}_y$ produce no torques about it, their lever arms being zero. Consequently, these forces will not appear in the equation representing the balance of torques. The lever arms for the remaining forces are shown in part *c* as red, dashed lines. The following list summarizes these forces, their lever arms, and the torques:

Force	Lever arm	Torque
$W_L = 355\text{ N}$	$\ell_L = (4.00\text{ m})\cos 50.0°$	$-W_L\ell_L$
$W_F = 875\text{ N}$	$\ell_F = (6.30\text{ m})\cos 50.0°$	$-W_F\ell_F$
P	$\ell_P = (8.00\text{ m})\sin 50.0°$	$+P\ell_P$

Setting the sum of the torques equal to zero gives

$$\Sigma\tau = -W_L\ell_L - W_F\ell_F + P\ell_P \tag{9.2}$$
$$\qquad = -(355\text{ N})(4.00\cos 50.0°\text{ m}) - (875\text{ N})(6.30\cos 50.0°\text{ m})$$
$$\qquad\quad + P(8.00\sin 50.0°\text{ m}) = 0$$

This equation can be solved to give $\boxed{P = 727\text{ N}}$. Equation 4.9a indicates that $G_x = P$, so $\boxed{G_x = 727\text{ N}}$.

To a large extent the directions of the forces acting on an object in equilibrium can be deduced using intuition. Sometimes, however, the direction of an unknown force is not obvious, and it is inadvertently drawn reversed in the free-body diagram. This kind of mistake causes no difficulty. ***Choosing the direction of an unknown force backward in the free-body diagram simply means that the value determined for the force will be a negative number,*** as the next example illustrates.

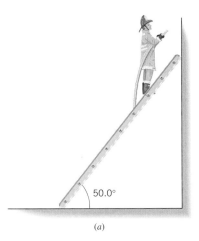

(a)

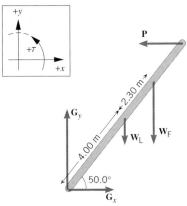

(b) Free-body diagram of the ladder

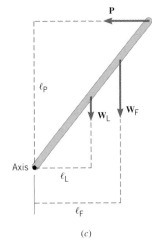

(c)

Figure 9.7 (*a*) A ladder leaning against a smooth wall. (*b*) The free-body diagram for the ladder. (*c*) Three of the forces that act on the ladder and their lever arms. The axis of rotation is at the lower end of the ladder and is perpendicular to the page.

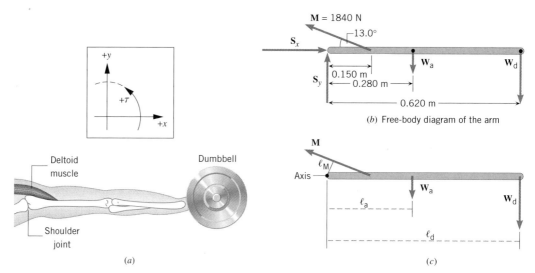

(b) Free-body diagram of the arm

(c)

Figure 9.8 (a) The fully extended, horizontal arm of a bodybuilder supports a dumbbell. (b) The free-body diagram for the arm. (c) Three of the forces that act on the arm and their lever arms. The axis of rotation is at the left end of the arm and is perpendicular to the page. In parts b and c the force vectors are not drawn to scale.

EXAMPLE 5 • Bodybuilding

A bodybuilder, strengthening his shoulder muscles, holds a dumbbell of weight W_d as in Figure 9.8a. His arm is extended horizontally and weighs $W_a = 31.0$ N. The deltoid muscle is assumed to be the only muscle acting and is attached to the arm as shown. The maximum force **M** that the deltoid muscle can supply to keep the arm horizontal has a magnitude of 1840 N. Figure 9.8b shows the distances that locate where the various forces act on the arm. What is the weight of the heaviest dumbbell that can be held, and what are the horizontal and vertical force components, S_x and S_y, that the shoulder joint applies to the left end of the arm?

Reasoning Figure 9.8b is the free-body diagram for the arm. Note that S_x is directed to the right, because the deltoid muscle pulls the arm in toward the shoulder joint, and the joint pushes back in accordance with Newton's third law. The direction of the force S_y, however, is less obvious, and we are alert for the possibility that the direction chosen in the free-body diagram might be backward. If so, the value obtained for S_y will turn out negative.

Solution The arm is in equilibrium, so the net force acting on it is zero:

$$\Sigma F_x = S_x - (1840 \text{ N}) \cos 13.0° = 0 \quad \text{or} \quad \boxed{S_x = 1790 \text{ N}} \qquad (4.9a)$$

$$\Sigma F_y = S_y + (1840 \text{ N}) \sin 13.0° - 31.0 \text{ N} - W_d = 0 \qquad (4.9b)$$

Equation 4.9b cannot be solved at this point, because it contains two unknowns. However, since the arm is in equilibrium, the torques acting on the arm must balance, and this fact provides another equation. To calculate torques, we choose an axis through the left end of the arm and perpendicular to the page. With this axis, the torques due to S_x and S_y are zero, because the line of action of each force passes through the axis and the lever arm of each force is zero. The list below summarizes the remaining forces, their lever arms (see Figure 9.8c), and the torques.

Force	Lever arm	Torque
$W_a = 31.0$ N	$\ell_a = 0.280$ m	$-W_a\ell_a$
W_d	$\ell_d = 0.620$ m	$-W_d\ell_d$
$M = 1840$ N	$\ell_M = (0.150 \text{ m}) \sin 13.0°$	$+M\ell_M$

The condition specifying a zero net torque is

$$\Sigma\tau = -W_a\ell_a - W_d\ell_d + M\ell_M \qquad (9.2)$$
$$= -(31.0\text{ N})(0.280\text{ m}) - W_d(0.620\text{ m}) + (1840\text{ N})(0.150\text{ m})\sin 13.0° = 0$$

Equation 9.2 can be solved to show that the maximum dumbbell weight is $\boxed{W_d = 86.1\text{ N}}$. This value for W_d can be substituted in Equation 4.9b to show that the value for S_y is $\boxed{S_y = -297\text{ N}}$. The minus sign indicates that the choice of direction for S_y in the free-body diagram is wrong. In reality, S_y has a magnitude of 297 N but is directed downward, not upward.

• **PROBLEM SOLVING INSIGHT**
When a force is negative, such as $S_y = -297$ N in this example, it means that the direction of the force is opposite to that chosen originally.

Sometimes it is necessary to choose different objects for analysis at different stages in a problem dealing with the equilibrium of a rigid object. Example 6 focuses on this important point.

EXAMPLE 6 • The Iron Cross

In gymnastics the "iron cross" is performed on the still rings (see Figure 9.9). A gymnast who performs this feat weighs 655 N, and each of his arms is 0.600 m long, as measured from the shoulder joint. Holding his arms parallel to the ground, he uses his latissimus dorsi muscles and pushes with each hand against the rings to support himself, as in part a of the drawing. The latissimus dorsi muscle attaches to the arm

Figure 9.9 (*a*) Three external forces act on the gymnast: his weight (655 N) and the tensions T_1 and T_2 in the ropes. (*b*) The free-body diagram for the arm. (*c*) Two of the forces that act on the arm and their lever arms. In parts b and c the force vectors and the distances are not drawn to scale.

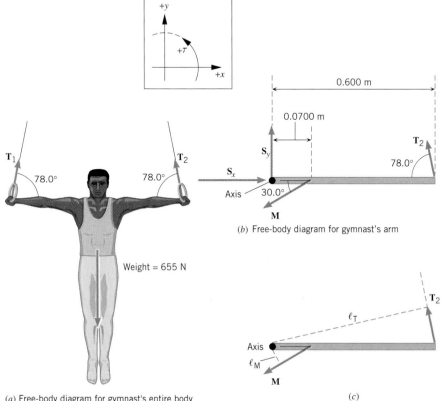

(*b*) Free-body diagram for gymnast's arm

(*a*) Free-body diagram for gymnast's entire body

(*c*)

Gymnast Chainey Umphrey.

The Physics of...
the "iron cross" gymnastics
routine.

at a distance of 0.0700 m from the shoulder joint and, as an approximation, is assumed to be the only muscle acting. It pulls on the arm with a force **M** that is directed downward and toward the body, as part *b* of the drawing shows. Ignoring the weight of each arm, find (a) the tension in each supporting rope and (b) the forces that are applied to each arm by the shoulder joint and the latissimus dorsi muscle.

Reasoning

(a) Three forces act on the gymnast's body: his weight of 655 N and the tensions T_1 and T_2 in each supporting rope. These forces are shown in Figure 9.9*a*, which is the free-body diagram for the gymnast. The weight of the gymnast is assumed to act at his center of mass, which is located midway between the two rings. Since the gymnast is in equilibrium, the sum of the *x* components of these forces is zero, and the sum of the *y* components of these forces is zero. These two conditions will be sufficient for us to find the magnitude of the tension in each rope.
(b) Since we wish to find the forces that the shoulder and latissimus dorsi muscle exert on the gymnast's arm, we select it as the object for our analysis. Figure 9.9*b* shows the free-body diagram for the arm, and the forces that act on it are as follows:

1. The horizontal and vertical force components, S_x and S_y, exerted by the shoulder joint
2. The force **M** exerted by the latissimus dorsi muscle
3. The force T_2 due to the tension in the rope.

Since the arm is in equilibrium, the sum of the *x* components of the forces is zero and the sum of the *y* components is also zero. In addition, the sum of the torques produced by the forces must also be zero. These three conditions will permit us to find values for the magnitudes of S_x, S_y, and **M**.

Solution

(a) Since the gymnast is in equilibrium, the sum of the horizontal components of the forces acting on his body (see Figure 9.9*a*) is equal to zero:

$$\Sigma F_x = T_1 \cos 78.0° - T_2 \cos 78.0° = 0 \qquad (4.9a)$$

This result indicates that $T_1 = T_2$; that is, the tension in each rope has the same magnitude. The sum of the vertical force components is also equal to zero:

$$\Sigma F_y = T_1 \sin 78.0° + T_2 \sin 78.0° - 655 \text{ N} = 0 \qquad (4.9b)$$

Since $T_1 = T_2$, we can solve this equation for either tension. The result is

$$T_1 = T_2 = \frac{655 \text{ N}}{2 \sin 78.0°} = \boxed{335 \text{ N}}$$

(b) The arm is in equilibrium, so we set the sum of the horizontal force components equal to zero and the sum of the vertical force components equal to zero (see Figure 9.9*b*):

$$\Sigma F_x = S_x - M \cos 30.0° - T_2 \cos 78.0° = 0 \qquad (4.9a)$$
$$\Sigma F_y = S_y - M \sin 30.0° + T_2 \sin 78.0° = 0 \qquad (4.9b)$$

When calculating torques, we select an axis parallel to the ground and perpendicular to the arm at the shoulder joint. For this axis, only the forces **M** and T_2 contribute torques, since the forces S_x and S_y have zero lever arms. The table below summarizes the lever arms and torques for the forces **M** and T_2 (see Figure 9.9*b* and *c*):

Force	Lever arm	Torque
M	$\ell_M = (0.0700 \text{ m}) \sin 30.0°$	$-M\ell_M$
$T_2 = 335 \text{ N}$	$\ell_T = (0.600 \text{ m}) \sin 78.0°$	$+T_2\ell_T$

Setting the sum of the torques equal to zero yields

$$\Sigma\tau = -M(0.0700\text{ m})\sin 30.0° + (335\text{ N})(0.600\text{ m})\sin 78.0° = 0 \quad (9.2)$$

Equation 9.2 can be solved directly to show that $\boxed{M = 5620\text{ N}}$. After substituting the values for M and T_2 into Equations 4.9a and 4.9b, these two equations can be solved to yield

$$\boxed{S_x = 4940\text{ N} \quad\text{and}\quad S_y = 2480\text{ N}}$$

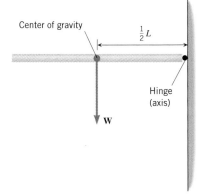

Center of gravity

$\frac{1}{2}L$

Hinge (axis)

W

Figure 9.10 A thin, uniform, horizontal rod of length L is attached to a vertical wall by a hinge. The center of gravity of the rod is at its geometrical center.

9.3 CENTER OF GRAVITY

Often, it is important to know the torque produced by the weight of an *extended* body. In Examples 4 and 5, for instance, it is necessary to determine the torques caused by the weight of the ladder and the arm, respectively. In both cases the weight is considered to act at a definite point for the purpose of calculating the torque. This point is called the ***center of gravity*** (abbreviated "cg").

■ **DEFINITION OF CENTER OF GRAVITY**
The center of gravity of a rigid body is the point at which its weight can be considered to act when calculating the torque due to the weight.

When an object has a symmetric shape and its weight is distributed uniformly, the center of gravity lies at its geometrical center. For instance, consider Figure 9.10, which shows a thin, uniform, horizontal rod of length L attached to a vertical wall by a hinge. The center of gravity of the rod is located at the geometrical center. The lever arm for the weight **W** is $L/2$, and the magnitude of the torque is $\tau = W(L/2)$. In a similar fashion, the center of gravity of any symmetrically shaped and uniform object, such as a sphere, disk, cube, or cylinder, is located at its geometrical center. However, this does not mean that the center of gravity must lie within the object itself. The center of gravity of a compact disc recording, for instance, lies at the center of the hole in the disc and is, therefore, "outside" the object.

Suppose we have a group of objects, with known weights and centers of gravity, and it is necessary to know the center of gravity for the group as a whole. As an example, Figure 9.11*a* shows a group composed of two parts: a horizontal uniform board (weight $\mathbf{W_1}$) and a uniform box (weight $\mathbf{W_2}$) near the left end of the board. The center of gravity of the group can be determined by calculating the net torque created by the board and box about an axis that is picked arbitrarily to lie at the right end of the board. Part *a* of the figure shows the weights $\mathbf{W_1}$ and $\mathbf{W_2}$ and their corresponding lever arms x_1 and x_2. The net torque is $\Sigma\tau = W_1 x_1 + W_2 x_2$. It is also possible to calculate the net torque by treating the total weight $\mathbf{W_1} + \mathbf{W_2}$ as if it were located at the center of gravity and had the lever arm x_{cg}, as part *b* of the drawing indicates: $\Sigma\tau = (W_1 + W_2)x_{cg}$. The two values for the net torque must be the same, so that

$$W_1 x_1 + W_2 x_2 = (W_1 + W_2)x_{cg}$$

This expression can be solved for x_{cg}, which locates the center of gravity relative to the axis:

Center of gravity
$$x_{cg} = \frac{W_1 x_1 + W_2 x_2 + \cdots}{W_1 + W_2 + \cdots} \quad (9.3)$$

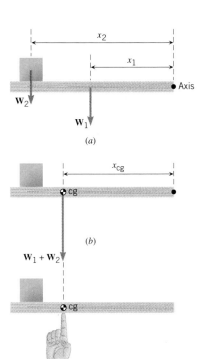

x_2

x_1

Axis

W_2

W_1

(*a*)

x_{cg}

cg

(*b*)

$W_1 + W_2$

cg

(*c*)

Figure 9.11 (*a*) A box rests near the left end of a horizontal board. (*b*) The total weight ($\mathbf{W_1} + \mathbf{W_2}$) acts at the center of gravity of the group. (*c*) The group can be balanced by applying an external force (due to the index finger) at the center of gravity.

The notation "+ · · ·" indicates that the reasoning above can be extended to account for any number of weights distributed along a horizontal line. Figure 9.11c illustrates that the group can be balanced by a single external force (due to the index finger), if the line of action of the force passes through the center of gravity, and if the force is equal in magnitude, but opposite in direction, to the weight of the group. Example 7 demonstrates how to calculate the center of gravity for the human arm.

EXAMPLE 7 • The Center of Gravity of an Arm

The horizontal arm in Figure 9.12 is composed of three parts: the upper arm (weight $W_1 = 17$ N), the lower arm ($W_2 = 11$ N), and the hand ($W_3 = 4.2$ N). The drawing shows the center of gravity of each part, measured with respect to the shoulder joint. Find the center of gravity of the entire arm, relative to the shoulder joint.

Reasoning and Solution The coordinate x_{cg} of the center of gravity is given by

$$x_{cg} = \frac{W_1 x_1 + W_2 x_2 + W_3 x_3}{W_1 + W_2 + W_3} \quad (9.3)$$

$$= \frac{(17 \text{ N})(0.13 \text{ m}) + (11 \text{ N})(0.38 \text{ m}) + (4.2 \text{ N})(0.61 \text{ m})}{17 \text{ N} + 11 \text{ N} + 4.2 \text{ N}} = \boxed{0.28 \text{ m}}$$

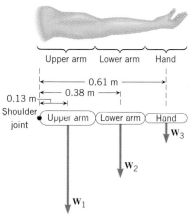

Figure 9.12 The three parts of the human arm, and the weight and center of gravity for each.

The center of gravity plays an important role in determining whether a group of objects remains in equilibrium as the weight distribution within the group changes. A change in the weight distribution causes a change in the position of the center of gravity, and if the change is too great, the group will not remain in equilibrium. Conceptual Example 8 discusses a shift in the center of gravity that led to an embarrassing result.

CONCEPTUAL EXAMPLE 8 • Overloading a Cargo Plane

Figure 9.13a shows a stationary cargo plane with its front landing gear 9 meters off the ground. This accident occurred because the plane was overloaded toward the rear. How did a shift in the center of gravity of the loaded plane cause the accident?

Reasoning and Solution Figure 9.13b shows a drawing of a correctly loaded plane, with the center of gravity located between the front and the main landing gears. The weight **W** of the plane and cargo acts downward at the center of gravity, and the normal forces $\mathbf{F}_{N1}$ and $\mathbf{F}_{N2}$ act upward at the front and at the main landing gear, respectively. With respect to an axis at the main landing gear, the clockwise torque due to **W** balances the counterclockwise torque due to $\mathbf{F}_{N1}$, and the plane remains in equilibrium. Figure 9.13c shows the plane with too much of the cargo loaded toward the rear, just after the plane has begun to rotate counterclockwise. Because of the overloading, the center of gravity has shifted behind the main landing gear. The torque due to **W** is now counterclockwise and is not balanced by any clockwise torque. Due to the unbalanced counterclockwise torque, the plane rotates until its tail hits the ground, which applies an upward force to the tail. The clockwise torque due to this upward force balances the counterclockwise torque due to **W**, and the plane comes again into an equilibrium state. Unfortunately, however, the front landing gear is now nine meters up in the air.

Related Homework Material: *Problems 14, 20*

(a)

Figure 9.13 (a) This stationary cargo plane is sitting on its tail at Los Angeles International Airport, after being overloaded toward the rear. (b) In a correctly loaded plane, the center of gravity is between the front and the main landing gear. (c) When the plane is overloaded toward the rear, the center of gravity shifts behind the main landing gear, and the accident in part a occurs.

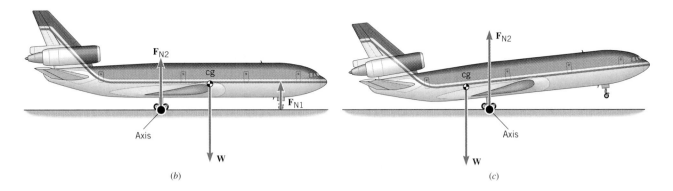

(b)   (c)

The center of gravity of an object with an irregular shape and a nonuniform weight distribution can be found by suspending the object from two different points P_1 and P_2, one at a time. Figure 9.14a shows the object at the moment of release, when its weight **W**, acting at the center of gravity, has a nonzero lever arm ℓ. At this instant the weight produces a torque about the axis. The tension force **T** applied to the object by the suspension cord produces no torque, because its line of action passes through the axis. Hence, in part a there is a net torque applied to the object, and the object begins to rotate. Friction eventually brings the object to rest as in part b, where the center of gravity lies directly below the point of suspension. In

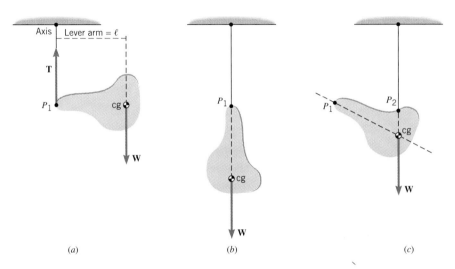

(a)   (b)   (c)

Figure 9.14 The center of gravity (cg) of an object can be located experimentally by suspending the object from two different points, P_1 and P_2, one at a time.

such an orientation, the line of action of the weight passes through the axis, so there is no longer any net torque. In the absence of a net torque the object remains at rest. By suspending the object from a second point (see Figure 9.14c), a second line through the object can be established, along which the center of gravity must also lie. The center of gravity, then, must be at the intersection of the two lines.

The center of gravity is closely related to the center-of-mass concept discussed in Section 7.5. To see why they are related, let's replace each occurrence of the weight in Equation 9.3 by $W = mg$, where m is the mass of a given object and g is the acceleration due to gravity at the location of the object. Suppose that g does not vary over the physical extent of the objects, so that it has the same value at each one. Then g can be algebraically canceled from each term on the right side of Equation 9.3. The resulting equation, which contains only masses and distances, is the same as Equation 7.10, which defines the location of the center of mass. Thus, the two points are identical. For ordinary-sized objects, like cars and boats, the center of gravity coincides with the center of mass.

9.4 NEWTON'S SECOND LAW FOR ROTATIONAL MOTION ABOUT A FIXED AXIS

NET TORQUE, MOMENT OF INERTIA, AND ANGULAR ACCELERATION

The goal of this section is to put Newton's second law into a form that is suitable for describing the rotational motion of a rigid object about a fixed axis. We begin by considering a particle moving on a circular path. Figure 9.15 presents a good approximation of this situation by using a small model plane on a guideline of negligible mass. The plane's engine produces a net external tangential force F_T that gives the plane a tangential acceleration a_T. In accord with Newton's second law, it follows that $F_T = ma_T$. The torque τ produced by this force is $\tau = F_T r$, where the radius r of the circular path is also the lever arm. As a result, the torque is $\tau = ma_T r$. But the tangential acceleration is related to the angular acceleration α according to $a_T = r\alpha$ (Equation 8.10), where α must be expressed in rad/s². With this substitution for a_T, the torque becomes

$$\tau = \underbrace{(mr^2)}_{\substack{\text{Moment} \\ \text{of inertia } I}}\alpha \qquad (9.4)$$

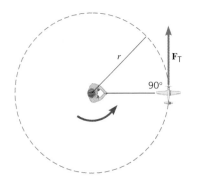

Figure 9.15 A model airplane of mass m is flying on a guideline of length r (top view). A net tangential force F_T acts on the plane.

Equation 9.4 is the form of Newton's second law we have been seeking. It indicates that the net external torque τ is directly proportional to the angular acceleration α. The constant of proportionality is $I = mr^2$, which is called the ***moment of inertia of the particle.*** The SI unit for moment of inertia is kg·m².

If all objects were single particles, it would be just as convenient to use the second law in the form $F_T = ma_T$, as in the form $\tau = I\alpha$. The advantage in using $\tau = I\alpha$ is that it can be applied to any rigid body rotating about a fixed axis, and not just to a particle. To illustrate how this advantage arises, Figure 9.16a shows a flat sheet of material that rotates about an axis perpendicular to the sheet. The sheet is composed of a number of mass particles, $m_1, m_2, \ldots, m_N$, where N is very large. Only four particles are shown for the sake of clarity. Each particle behaves in the same way as the model airplane in Figure 9.15 and obeys the relation $\tau = (mr^2)\alpha$:

$$\tau_1 = (m_1 r_1^2)\alpha$$
$$\tau_2 = (m_2 r_2^2)\alpha$$
$$\vdots$$
$$\tau_N = (m_N r_N^2)\alpha$$

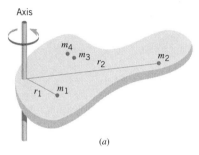

(a)

In these equations each particle has the same angular acceleration α, since the rotating object is assumed to be rigid. Adding together the N equations and factoring out the common value of α, we find that

$$\underbrace{\Sigma\tau}_{\substack{\text{Net} \\ \text{external torque}}} = \underbrace{(\Sigma mr^2)}_{\substack{\text{Moment} \\ \text{of inertia}}}\alpha \qquad (9.5)$$

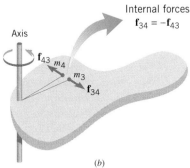

(b)

Figure 9.16 (a) A rigid body consists of a large number of particles, four of which are shown here. (b) The internal forces that particles 3 and 4 exert on each other obey Newton's law of action and reaction.

where $\Sigma\tau = \tau_1 + \tau_2 + \cdots + \tau_N$ is the sum of the torques, and $\Sigma mr^2 = m_1 r_1^2 + m_2 r_2^2 + \cdots + m_N r_N^2$ represents the sum of the individual moments of inertia. The latter quantity is the ***moment of inertia of the body*** and is represented by the symbol I:

Moment of inertia of a body
$$I = \Sigma mr^2 \qquad (9.6)$$

In this equation, r is the perpendicular radial distance of each particle from the axis of rotation. Combining Equation 9.6 with Equation 9.5 gives the following result:

■ **ROTATIONAL ANALOG OF NEWTON'S SECOND LAW FOR A RIGID BODY ROTATING ABOUT A FIXED AXIS**

Net external torque = $\begin{pmatrix}\text{Moment of} \\ \text{inertia}\end{pmatrix} \times \begin{pmatrix}\text{Angular} \\ \text{acceleration}\end{pmatrix}$

$$\Sigma\tau = I\alpha \qquad (9.7)$$

Requirement: α must be expressed in rad/s^2.

The form of the second law for rotational motion, $\Sigma\tau = I\alpha$, is similar to that for translational (linear) motion, $\Sigma F = ma$, and is valid only in an inertial frame. The moment of inertia I plays the same role for rotational motion that the mass m does for translational motion. Thus, I is a measure of the rotational inertia of a body. When using Equation 9.7, α must be expressed in rad/s^2, because the relation $a_T = r\alpha$ (which requires radian measure) was used in the derivation.

When calculating the sum of torques in Equation 9.7, it is necessary to include only the *external torques,* those applied by agents outside the body. The torques produced by internal forces need not be considered, because they always combine to produce a net torque of zero. Internal forces are those that one particle within the body exerts on another particle. They always occur in pairs of oppositely directed forces of equal magnitude, in accord with Newton's third law (see m_3 and m_4 in Figure 9.16b). The forces in such a pair have the same line of action, so they have identical lever arms and produce torques of equal magnitudes. One torque is counterclockwise, while the other is clockwise, the net torque from the pair being zero.

It can be seen from Equation 9.6 that the moment of inertia depends on both the

Figure 9.17 Two particles, masses m_1 and m_2, are attached to the ends of a massless rigid rod. The moment of inertia of this object is different, depending on whether the rod rotates about an axis passing through (a) the end or (b) the center of the rod.

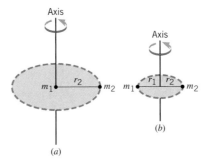

• **PROBLEM SOLVING INSIGHT**
When solving problems involving the moment of inertia, be sure to identify the axis of rotation. The moment of inertia depends on where the axis is. If the location of the axis changes, the moment of inertia may also change.

mass of each particle and its distance from the axis of rotation. The farther a particle is from the axis, the greater is its contribution to the moment of inertia. Therefore, although a rigid object possesses a unique total mass, it does not have a unique moment of inertia, for *the moment of inertia depends on the location and orientation of the axis relative to the particles that make up the object.* Example 9 shows how the moment of inertia can change when the axis of rotation changes.

EXAMPLE 9 • The Moment of Inertia Depends on Where the Axis Is

Figure 9.18 *Concepts at a Glance* An object is in equilibrium when its translational acceleration components, a_x and a_y, and its angular acceleration α are zero. If a_x or a_y or α is not zero, the object has an acceleration and is not in equilibrium. As the cards come tumbling down, they are no longer in equilibrium.

Two particles each have a mass m and are fixed to the ends of a thin rigid rod, whose mass can be ignored. The length of the rod is L. Find the moment of inertia when this object rotates relative to an axis that is perpendicular to the rod at (a) one end and (b) the center. (See Figure 9.17.)

Reasoning When the axis of rotation changes, the distance r between the axis and each particle changes. In determining the moment of inertia using $I = \Sigma mr^2$, we must be careful to use the distances that apply for each axis.

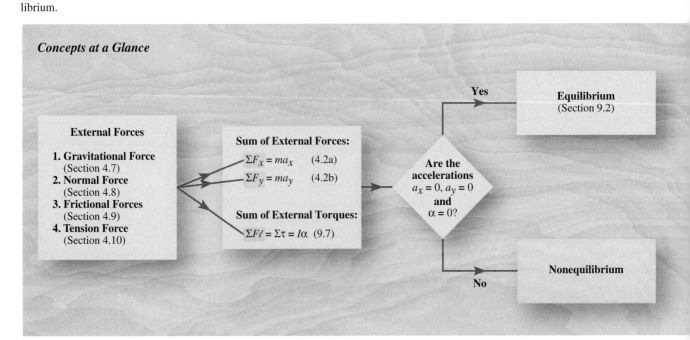

Concepts at a Glance

External Forces

1. **Gravitational Force** (Section 4.7)
2. **Normal Force** (Section 4.8)
3. **Frictional Forces** (Section 4.9)
4. **Tension Force** (Section 4.10)

Sum of External Forces:

$$\Sigma F_x = ma_x \quad (4.2a)$$
$$\Sigma F_y = ma_y \quad (4.2b)$$

Sum of External Torques:

$$\Sigma F\ell = \Sigma\tau = I\alpha \quad (9.7)$$

Are the accelerations $a_x = 0,\ a_y = 0$ **and** $\alpha = 0$?

Yes → **Equilibrium** (Section 9.2)

No → **Nonequilibrium**

Solution

(a) Particle 1 lies on the axis, as part *a* of the drawing shows, and has a zero radial distance: $r_1 = 0$. In contrast, particle 2 moves on a circle whose radius is $r_2 = L$. The moment of inertia is

$$I = \Sigma mr^2 = m_1 r_1^2 + m_2 r_2^2 = \boxed{mL^2} \tag{9.6}$$

(b) Part *b* of the drawing shows that particle 1 no longer lies on the axis but now moves on a circle of radius $r_1 = L/2$. Particle 2 moves on a circle with the same radius, $r_2 = L/2$. Therefore,

$$I = \Sigma mr^2 = m_1 r_1^2 + m_2 r_2^2 = m(L/2)^2 + m(L/2)^2 = \boxed{\tfrac{1}{2}\, mL^2}$$

This value differs from that in part (a), because the axis of rotation is different.

The procedure illustrated in Example 9 can be extended using integral calculus to evaluate the moment of inertia of a rigid object with a continuous mass distribution, and Table 9.1 gives some typical results. These results depend on the total mass of the object, its shape, and the location and orientation of the axis.

APPLICATIONS

When an object is acted upon by forces that may produce torques, the object may possess a translational acceleration *a* as well as an angular acceleration α. In general, we can deal with this type of motion by using Newton's second law for translational motion, $\Sigma F = ma$, together with that for rotational motion, $\Sigma \tau = I\alpha$. The concept chart in Figure 9.18 illustrates the essence of this joint usage. When both *a* and α are zero there is no acceleration of any kind. The object is in equilibrium, and we have the situation already discussed in Section 9.2. On the other hand, if either *a* or α is nonzero we have accelerated motion, and the object is not in equilibrium. Examples 10, 11, and 12 illustrate this type of situation.

Table 9.1 Moments of Inertia for Various Rigid Objects of Mass *M*

Thin-walled hollow cylinder or hoop

$$I = MR^2$$

Solid cylinder or disk

$$I = \tfrac{1}{2}MR^2$$

Thin rod, axis perpendicular to rod and passing through center

$$I = \tfrac{1}{12}ML^2$$

Thin rod, axis perpendicular to rod and passing through one end

$$I = \tfrac{1}{3}ML^2$$

Solid sphere, axis through center

$$I = \tfrac{2}{5}MR^2$$

Solid sphere, axis tangent to surface

$$I = \tfrac{7}{5}MR^2$$

Thin-walled spherical shell, axis through center

$$I = \tfrac{2}{3}MR^2$$

Thin rectangular sheet, axis parallel to one edge and passing through center of other edge

$$I = \tfrac{1}{12}ML^2$$

Thin rectangular sheet, axis along one edge

$$I = \tfrac{1}{3}ML^2$$

EXAMPLE 10 • The Torque of a Turntable Motor

Most turntables can bring a record from rest up to the rated angular speed of $33\frac{1}{3}$ rev/min in one-half a revolution. The platter of one turntable has a moment of inertia of 0.0500 kg·m² (including the effect of the record). Neglecting frictional effects, what net torque (assumed constant) must the turntable motor apply to the platter to achieve this performance?

Reasoning Newton's second law for rotational motion ($\Sigma\tau = I\alpha$) can be used to find the torque, once the angular acceleration is determined. The angular acceleration can be calculated from the data in the table below and the appropriate equation of rotational kinematics.

θ	α	ω	ω_0	t
$-\pi$ rad ($-\frac{1}{2}$ rev)	?	-3.49 rad/s ($-33\frac{1}{3}$ rev/min)	0	

The rotational variables θ and ω are negative, because the rotation of a turntable platter is clockwise when viewed from above. The data for θ and ω have been converted to radian measure, because $\Sigma\tau = I\alpha$ requires that α be expressed in radian measure.

Solution From $\omega^2 = \omega_0^2 + 2\alpha\theta$ (Equation 8.8) it follows that

$$\alpha = \frac{\omega^2 - \omega_0^2}{2\theta} = \frac{(-3.49 \text{ rad/s})^2}{2(-\pi \text{ rad})} = -1.94 \text{ rad/s}^2$$

Newton's second law for rotational motion can now be used to obtain the net torque:

$$\Sigma\tau = I\alpha = (0.0500 \text{ kg·m}^2)(-1.94 \text{ rad/s}^2) = \boxed{-0.0970 \text{ N·m}} \qquad (9.7)$$

Example 10 shows how Newton's second law for rotational motion is used when design considerations demand an adequately large angular acceleration. There are also situations when it is desirable to have as little angular acceleration as possible, and Conceptual Example 11 deals with one of them.

CONCEPTUAL EXAMPLE 11 • Modern Archery and Bow Stabilizers

Archers can shoot with amazing accuracy, especially using modern bows such as the one in Figure 9.19. Notice the bow stabilizer, a long, thin rod that extends from the front of the bow and has a relatively massive cylinder at the tip. Advertisements claim that the stabilizer helps to steady the archer's aim. Could there be any truth to this claim? Explain.

Reasoning and Solution To help explain why the stabilizer works, we have added to the photograph an axis for rotation that passes through the archer's shoulder and is perpendicular to the plane of the paper. Any angular acceleration α about this axis will lead to a rotation of the bow that will degrade the archer's aim. The acceleration will be created by any unbalanced torques that occur while the archer's tensed muscles try to hold the drawn bow. Newton's second law for rotation indicates, however, that the angular acceleration is $\alpha = (\Sigma\tau)/I$. The moment of inertia I is in the denominator on the right side of this equation. Therefore, to the extent that I is larger, a given net torque $\Sigma\tau$ will create a smaller angular acceleration and less disturbance of the aim. It is to increase the moment of inertia of the bow that the stabilizer has been added. The relatively massive cylinder, which is placed far from the axis of rotation (a large value of r in the equation $I = \Sigma mr^2$), is particularly effective in increasing the moment of inertia.

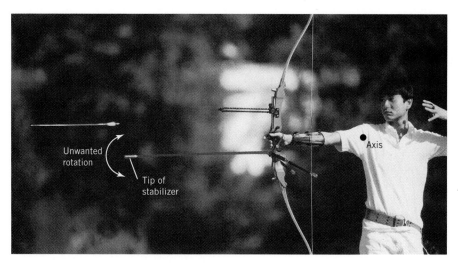

Figure 9.19 The stabilizer helps to steady the archer's aim, as Conceptual Example 11 discusses. Relative to an axis through the archer's shoulder, the moment of inertia of the bow is larger with the stabilizer than without it.

Rotational motion and translation motion sometimes occur together. The next example deals with an interesting situation in which both angular acceleration and translational acceleration must be considered.

EXAMPLE 12 • Hoisting a Crate

A crate that weighs 4420 N is being lifted by the mechanism shown in Figure 9.20a. The two cables are wrapped around their respective pulleys, which have radii of 0.600 and 0.200 m. The pulleys are fastened together to form a "dual" pulley and turn as a single unit about the center axle, relative to which the combined moment of inertia is $I = 50.0 \text{ kg} \cdot \text{m}^2$. If a tension of magnitude $T_1 = 2150 \text{ N}$ is maintained in the cable attached to the motor, find the angular acceleration of the "dual" pulley and the tension in the cable connected to the crate.

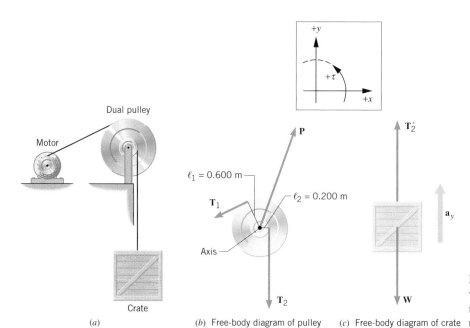

(a)

(b) Free-body diagram of pulley

(c) Free-body diagram of crate

Figure 9.20 (a) The crate is lifted upward by the motor and pulley arrangement. The free-body diagram for (b) the dual pulley and (c) the crate.

Reasoning To determine the angular acceleration of the dual pulley and the tension in the cable attached to the crate, we will apply Newton's second law to the pulley and the crate separately. Three external forces act on the dual pulley, as its free-body diagram shows (Figure 9.20b). These forces are (1) the tension $\mathbf{T_1}$ in the cable connected to the motor, (2) the tension $\mathbf{T_2}$ in the cable attached to the crate, and (3) the reaction force $\mathbf{P}$ exerted on the dual pulley by the axle. The force $\mathbf{P}$ arises because the two cables pull the pulley down and to the left into the axle, and the axle pushes back, thus keeping the pulley in place. The net torque that results from these forces obeys Newton's second law for rotational motion (Equation 9.7). Two external forces act on the crate, as its free-body diagram indicates (Figure 9.20c). These forces are (1) the cable tension $\mathbf{T_2'}$ and (2) the weight $\mathbf{W}$ of crate. The net force that results from these two forces obeys Newton's second law for translational motion (Equation 4.2b).

Solution Using the lever arms ℓ_1 and ℓ_2 shown in part b of the figure, we can apply the second law to the rotational motion of the pulley. We note that the force $\mathbf{P}$ has a zero lever arm, since the line of action of $\mathbf{P}$ passes directly through the axle:

$$\Sigma \tau = T_1 \ell_1 - T_2 \ell_2 = I\alpha \tag{9.7}$$
$$(2150 \text{ N})(0.600 \text{ m}) - T_2(0.200 \text{ m}) = (50.0 \text{ kg} \cdot \text{m}^2)\alpha$$

This equation contains two unknown quantities, so a second equation is needed and may be obtained by applying Newton's second law to the upward translational motion of the crate. In doing so, we note that the magnitude of the tension in the cable attached to the crate is $T_2' = T_2$ and that the mass of the crate is $m = (4420 \text{ N})/(9.80 \text{ m/s}^2) = 451 \text{ kg}$:

$$\Sigma F_y = ma_y \tag{4.2b}$$
$$T_2 - (4420 \text{ N}) = (451 \text{ kg}) \, a_y$$

Because the cable attached to the crate rolls on the pulley without slipping, the linear acceleration a_y of the crate is related to the angular acceleration α of the pulley via Equation 8.13: $a_y = r\alpha = (0.200 \text{ m})\alpha$. With this substitution for a_y, Equation 4.2b becomes

$$T_2 - (4420 \text{ N}) = (451 \text{ kg})(0.200 \text{ m})\alpha$$

This result and Equation 9.7 can be solved simultaneously to yield

$$\boxed{T_2 = 4960 \text{ N}} \quad \text{and} \quad \boxed{\alpha = 6.0 \text{ rad/s}^2}$$

We have seen that Newton's second law for rotational motion, $\Sigma \tau = I\alpha$, has the same form as that for translational motion, $\Sigma F = ma$, so each rotational variable

Table 9.2   Analogies Between Rotational and Translational Concepts

Physical Concept	Rotational	Translational
Displacement	θ	s
Velocity	ω	v
Acceleration	α	a
The cause of acceleration	Torque τ	Force F
Inertia	Moment of inertia I	Mass m
Newton's second law	$\Sigma \tau = I\alpha$	$\Sigma F = ma$
Work	$\tau\theta$	Fs
Kinetic energy	$\frac{1}{2}I\omega^2$	$\frac{1}{2}mv^2$
Momentum	$L = I\omega$	$p = mv$

has a translational analog: torque τ and force F are analogous quantities, as are moment of inertia I and mass m, and angular acceleration α and linear acceleration a. The other physical concepts developed for studying translational motion, such as kinetic energy and momentum, also have rotational analogs. For future reference, Table 9.2 itemizes these concepts and their rotational analogs.

9.5 ROTATIONAL WORK AND ENERGY

ROTATIONAL WORK

Work and energy are among the most fundamental and useful concepts in physics. Chapter 6 discusses their application to translational motion. These concepts are equally useful for rotational motion, provided they are expressed in terms of angular variables.

The work W done by a constant force that points in the same direction as the displacement is $W = Fs$ (Equation 6.1), where F and s are the magnitudes of the force and displacement, respectively. To see how this expression for work can be rewritten using angular variables, consider Figure 9.21. Here a rope is wrapped around a wheel and is under a constant tension F. If the rope is pulled out a distance s, the wheel rotates through an angle $\theta = s/r$ (Equation 8.1), where r is the radius of the wheel and θ is in radians. The work done by the tension force in turning the wheel is $W = Fs = Fr\theta$. But Fr is the torque τ applied to the wheel by the tension, so the rotational work can be written as follows:

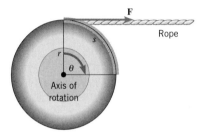

Figure 9.21 The force **F** does work in rotating the wheel through the angle θ.

> ### ■ DEFINITION OF ROTATIONAL WORK
>
> The rotational work W_R done by a constant torque τ in turning an object through an angle θ is
>
> $$W_R = \tau\theta \qquad (9.8)$$
>
> *Requirement:* θ must be expressed in radians.
>
> *SI Unit of Rotational Work:* joule (J)

ROTATIONAL KINETIC ENERGY

In Section 6.2 (The Work–Energy Theorem and Kinetic Energy) we saw that when work is done on an object by a net external force, the translational kinetic energy ($\frac{1}{2}mv^2$) of the object changes. In an analogous manner, when rotational work is done by a net external torque, the rotational kinetic energy of the object changes. A rotating body possesses kinetic energy, because its constituent particles are in motion. If the body is rotating with an angular speed ω, the tangential speed v_T of a particle at a distance r from the axis is $v_T = r\omega$ (Equation 8.9). Figure 9.22 shows two such particles. If a particle's mass is m, its kinetic energy is $\frac{1}{2}mv_T^2 = \frac{1}{2}mr^2\omega^2$. The kinetic energy of the entire rotating body, then, is the sum of the kinetic energies of the particles:

$$\text{Rotational KE} = \Sigma(\tfrac{1}{2}mr^2\omega^2) = \tfrac{1}{2}\underbrace{(\Sigma mr^2)}_{\substack{\text{Moment of}\\\text{inertia, } I}}\omega^2$$

In this result, the angular speed ω is the same for all particles in a rigid body and, therefore, has been factored outside the summation. The term in parentheses is the

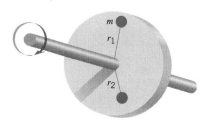

Figure 9.22 The rotating wheel is composed of many particles, two of which are shown.

Concepts at a Glance

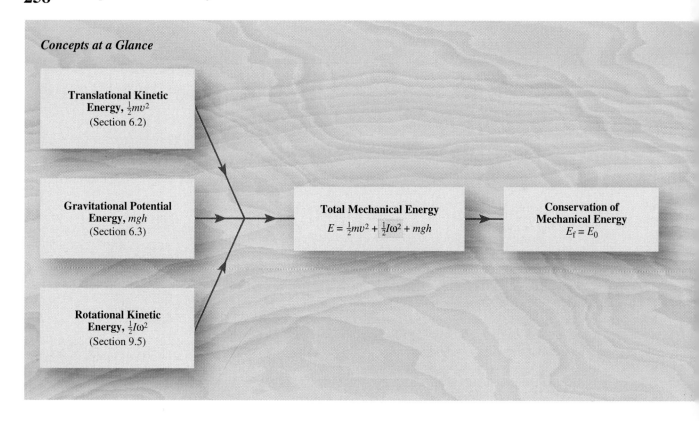

moment of inertia, $I = \Sigma mr^2$, so the rotational kinetic energy takes the form given below:

■ **DEFINITION OF ROTATIONAL KINETIC ENERGY**

The rotational kinetic energy KE_R of a rigid object rotating with an angular speed ω about a fixed axis and having a moment of inertia I is

$$\text{KE}_R = \tfrac{1}{2} I \omega^2 \tag{9.9}$$

Requirement: ω must be expressed in rad/s.

SI Unit of Rotational Kinetic Energy: joule (J)

The concept of kinetic energy is important, for it is one part of an object's total mechanical energy E. The total mechanical energy is the sum of the kinetic energy and potential energy, and it obeys the principle of conservation of mechanical energy (see Section 6.5). Previously, we have used this principle in situations where there is only translational motion, as when a sled slides down a frictionless hill. Now, however, we have seen that translational and rotational motion can occur simultaneously. When a bicycle coasts down a hill, for instance, its tires are both translating and rotating. To use the conservation principle for a rolling bicycle tire, we need only recognize that the total kinetic energy of such an object is the sum of its translational and rotational kinetic energies:

$$\text{KE}_{\text{Total}} = \tfrac{1}{2} m v^2 + \tfrac{1}{2} I \omega^2$$

In this expression, m is the mass, v is the translational speed of the object's center of mass, I is the moment of inertia about an axis through the center of mass, and ω is the angular speed. Incorporating this kinetic energy into the definition of the total

Figure 9.23 *Concepts at a Glance* The conservation of mechanical energy can be applied to an object that has both translational and rotational motion, provided the rotational kinetic energy is included in the total mechanical energy. While this skater is airborne, her translational motion and rotational motion (assumed constant) can be described by the conservation of mechanical energy, to the extent that air resistance is negligible.

mechanical energy, we have

$$E = \underbrace{\tfrac{1}{2}mv^2}_{\substack{\text{Translational} \\ \text{kinetic energy}}} + \underbrace{\tfrac{1}{2}I\omega^2}_{\substack{\text{Rotational} \\ \text{kinetic energy}}} + \underbrace{mgh}_{\substack{\text{Gravitational} \\ \text{potential energy}}}$$

where h is the height of the object relative to an arbitrary zero level. If the total mechanical energy is conserved as an object moves, its final total mechanical energy E_f equals its initial total mechanical energy E_0: $E_f = E_0$. The concept diagram in Figure 9.23, which is an expanded version of that in Figure 6.16, shows how rotational kinetic energy is incorporated into the conservation of mechanical energy. Example 13 illustrates the effect of combined translational and rotational motion.

EXAMPLE 13 • Rolling Cylinders

A thin-walled hollow cylinder (mass $= m_h$, radius $= r_h$) and a solid cylinder (mass $= m_s$, radius $= r_s$) start from rest at the top of an incline (Figure 9.24). Both cylinders start at the same vertical height h_0. All heights are measured relative to an arbitrarily chosen zero level that passes through the center of mass of a cylinder when it is at the bottom of the incline (see the drawing). Ignoring energy losses due to retarding forces, determine which cylinder has the greatest translational speed upon reaching the bottom of the incline.

Reasoning Only the conservative force of gravity does work on the cylinders, so the total mechanical energy is conserved as they roll down the incline. The total mechanical energy E at any height h above the zero level is the sum of the translational kinetic energy ($\tfrac{1}{2}mv^2$), the rotational kinetic energy ($\tfrac{1}{2}I\omega^2$), and the gravitational po-

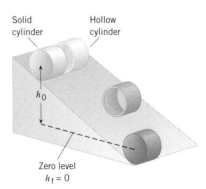

Solid cylinder   Hollow cylinder

h_0

Zero level
$h_f = 0$

Figure 9.24 A hollow cylinder and a solid cylinder start from rest and roll down the incline plane. The conservation of mechanical energy can be used to show that the solid cylinder, having the greatest translational speed, reaches the bottom first.

tential energy (mgh):

$$E = \tfrac{1}{2}mv^2 + \tfrac{1}{2}I\omega^2 + mgh$$

As the cylinders roll down, potential energy is converted into kinetic energy. But the kinetic energy is shared between the translational form ($\tfrac{1}{2}mv^2$) and the rotational form ($\tfrac{1}{2}I\omega^2$). The object with more of its kinetic energy in the translational form will have the greater translational speed at the bottom of the incline. As we shall see, the solid cylinder has the greater translational speed, for more of its mass is located near the rotational axis and, thus, possesses less rotational kinetic energy.

Solution The total mechanical energy E_f at the bottom ($h_f = 0$) is the same as the total mechanical energy E_0 at the top ($h = h_0$, $v_0 = 0$, $\omega_0 = 0$):

$$\tfrac{1}{2}mv_f^2 + \tfrac{1}{2}I\omega_f^2 + mgh_f = \tfrac{1}{2}mv_0^2 + \tfrac{1}{2}I\omega_0^2 + mgh_0$$

$$\tfrac{1}{2}mv_f^2 + \tfrac{1}{2}I\omega_f^2 = mgh_0$$

Since each cylinder rolls without slipping, the final rotational speed ω_f and the final translational speed v_f of its center of mass are related according to Equation 8.12, $\omega_f = v_f/r$, where r is the radius of the cylinder. Substituting this expression for ω_f into the equation above and solving for v_f yields

$$v_f = \sqrt{\frac{2mgh_0}{m + I/r^2}}$$

Setting $m = m_h$, $r = r_h$ and $I = mr_h^2$ for the hollow cylinder and then setting $m = m_s$, $r = r_s$ and $I = \tfrac{1}{2}mr_s^2$ for the solid cylinder (see Table 9.1), we find that the two cylinders have the following translational speeds at the bottom of the incline:

Hollow cylinder $\qquad v_f = \sqrt{gh_0}$

Solid cylinder $\qquad v_f = \sqrt{\dfrac{4gh_0}{3}} = 1.15\sqrt{gh_0}$

The solid cylinder, having the greatest translational speed, arrives at the bottom first.

Concepts at a Glance

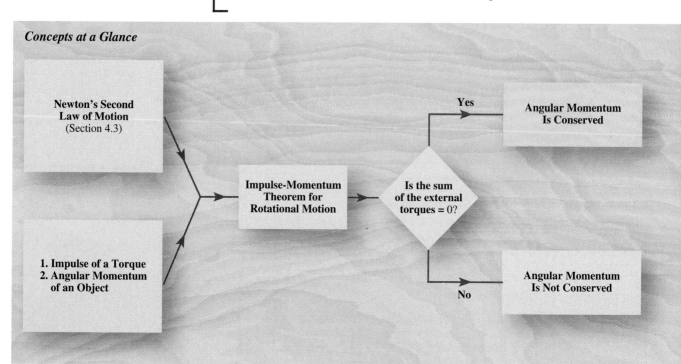

9.6 ANGULAR MOMENTUM

In Chapter 7 the linear momentum p of an object is defined as the product of its mass m and linear velocity v, that is $p = mv$. For rotational motion the analogous concept is called the ***angular momentum L***. The mathematical form of angular momentum is analogous to that of linear momentum, with the mass m and the linear velocity v being replaced with their rotational counterparts, the moment of inertia I and the angular velocity ω.

> ■ **DEFINITION OF ANGULAR MOMENTUM**
>
> The angular momentum L of a body rotating about a fixed axis is the product of the body's moment of inertia I and its angular velocity ω:
>
> $$L = I\omega \qquad (9.10)$$
>
> ***Requirement:*** ω must be expressed in rad/s.
>
> ***SI Unit of Angular Momentum:*** $\text{kg} \cdot \text{m}^2/\text{s}$

Linear momentum is an important concept in physics, because the total momentum P of a system is conserved when the net external force acting on the system is zero. Then, the final linear momentum P_f and the initial linear momentum P_0 of the system are the same: $P_f = P_0$. The conceptual development of the conservation of linear momentum is shown in Figure 7.7. As Table 9.2 indicates, each translational variable is analogous to a corresponding rotational variable. Therefore, if we replace "force" by "torque" and "linear momentum" by "angular momentum" in Figure 7.7, we arrive at the concept chart in Figure 9.25, which displays the development of the conservation of angular momentum. This chart indicates that when the sum of the external torques is zero, the final and initial angular momenta are the same: $L_f = L_0$. This is the ***principle of conservation of angular momentum.***

Figure 9.25 *Concepts at a Glance* The impulse–momentum theorem for rotational motion leads to the principle of conservation of angular momentum when the sum of the external torques is zero. This concept diagram is similar to that in Figure 7.7, which shows how the conservation of linear momentum arises from the impulse–momentum theorem. The angular momentum of the spinning diver is conserved while she is in the air, assuming that air resistance is negligible.

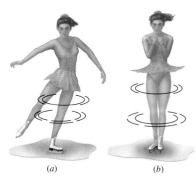

Figure 9.26 (a) A skater spins slowly on one skate, with both arms and one leg outstretched. (b) As she pulls her arms and leg in toward the rotational axis, her moment of inertia I decreases, and the angular velocity ω increases.

The Physics of...
a spinning ice skater.

■ PRINCIPLE OF CONSERVATION OF ANGULAR MOMENTUM

The total angular momentum of a system remains constant (is conserved) if the net external torque acting on the system is zero.

Example 14 illustrates an interesting consequence of the conservation of angular momentum.

CONCEPTUAL EXAMPLE 14 • A Spinning Skater

In Figure 9.26a an ice skater is spinning with both arms and a leg outstretched. In Figure 9.26b she pulls her arms and leg inward. As a result of this maneuver, her spinning motion changes dramatically. Using the principle of conservation of angular momentum, explain how and why it changes.

Reasoning and Solution Choosing the skater as the system, we can apply the conservation principle provided that the net external torque produced by air resistance and by friction between the skates and the ice is negligibly small. We assume that it is. Then the skater in Figure 9.26a would spin forever at the same angular velocity, since her angular momentum is conserved in the absence of a net external torque. In Figure 9.26b the inward movement of her arms and leg involves internal, not external, torques and, therefore, does not change her angular momentum. But angular momentum is the product of the moment of inertia I and angular velocity ω. By moving the mass of her arms and leg inward, the skater decreases the distance r of the mass from the axis of rotation and, consequently, decreases her moment of inertia I ($I = \Sigma mr^2$). If the product of I and ω is to remain constant, then ω must increase. Thus, the consequence of pulling her arms and leg inward is that ***she spins with a larger angular velocity.***

Related Homework Material: *Questions 20, 21, 22 and Problem 56*

The last example in this chapter involves a satellite and illustrates another application of the principle of conservation of angular momentum.

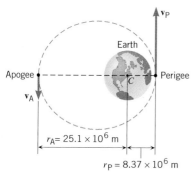

Figure 9.27 A satellite is moving in an elliptical orbit about the earth. The gravitational force exerts no torque on the satellite, so the angular momentum of the satellite is conserved.

EXAMPLE 15 • A Satellite in an Elliptical Orbit

An artificial satellite is placed into an elliptical orbit about the earth, as in Figure 9.27. Telemetry data indicate that its point of closest approach (called the *perigee*) is $r_P = 8.37 \times 10^6$ m from the center of the earth, while its point of greatest distance (called the *apogee*) is $r_A = 25.1 \times 10^6$ m from the center of the earth. The speed of the satellite at the perigee is $v_P = 8450$ m/s. Find its speed v_A at the apogee.

Reasoning The only force of any significance that acts on the satellite is the gravitational force of the earth. However, at any instant, this force is directed toward the center of the earth and passes through the axis about which the satellite instantaneously rotates. Therefore, the gravitational force exerts *no torque* on the satellite (the lever arm is zero). Consequently, the angular momentum of the satellite remains constant at all times.

Solution Since the angular momentum is the same at the apogee (A) and the perigee (P), it follows that $I_A \omega_A = I_P \omega_P$. Furthermore, the orbiting satellite can be considered a point mass, so its moment of inertia is $I = mr^2$. In addition, the angular speed ω of the satellite is related to its tangential speed v_T by $\omega = v_T/r$. If these relations are used at the apogee and perigee, the conservation of angular momentum gives the following result:

$$I_A\omega_A = I_P\omega_P$$

$$(mr_A^2)\left(\frac{v_A}{r_A}\right) = (mr_P^2)\left(\frac{v_P}{r_P}\right)$$

$$v_A = \frac{r_P v_P}{r_A} = \frac{(8.37 \times 10^6 \text{ m})(8450 \text{ m/s})}{25.1 \times 10^6 \text{ m}} = \boxed{2820 \text{ m/s}}$$

The answer is independent of the mass of the satellite. The satellite behaves just like the skater in Figure 9.26, because its speed is greater at the perigee, where the moment of inertia is smaller.

The result in Example 15 indicates that a satellite does not have a constant speed in an elliptical orbit. The speed changes from a maximum at the perigee to a minimum at the apogee; the closer the satellite comes to the earth, the faster it travels. Planets moving around the sun in elliptical orbits exhibit the same kind of behavior, and Johannes Kepler (1571–1630) formulated his famous second law based on observations of such characteristics of planetary motion. Kepler's second law states that, in a given amount of time, a line joining any planet to the sun sweeps out the same amount of area no matter where the planet is on its elliptical orbit, as Figure 9.28 illustrates. The conservation of angular momentum can be used to show why the law is valid, by means of a calculation similar to that in Example 15.

> ▶ **The Physics of...**
> a satellite in orbit about the earth.

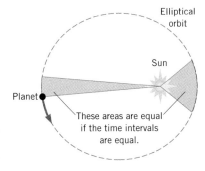

Figure 9.28 Kepler's second law of planetary motion states that a line joining a planet to the sun sweeps out equal areas in equal time intervals.

SUMMARY

The **torque** τ of a force has a magnitude that is given by the magnitude F of the force times the lever arm ℓ: $\tau = F\ell$. The lever arm is the perpendicular distance between the line of action of the force and the axis of rotation.

A rigid body is in **equilibrium** if it has zero translational acceleration and zero angular acceleration, so the net external force and the net external torque acting on the body are zero. For forces acting only in the x, y plane, the conditions for equilibrium are $\Sigma F_x = 0$, $\Sigma F_y = 0$, and $\Sigma \tau = 0$.

The **center of gravity** of a rigid object is the point where its entire weight can be considered to act when calculating the torque due to the weight of the object. For a symmetrical body with uniformly distributed weight, the center of gravity is located at the geometrical center of the body. When a number of objects whose weights W_1, W_2, . . . are distributed along the x axis at locations x_1, x_2, . . . , the center of gravity is located at $x_{cg} = (W_1 x_1 + W_2 x_2 + \cdots)/(W_1 + W_2 + \cdots)$. The **center of mass** is identical to the center of gravity, provided the acceleration due to gravity does not vary over the physical extent of the objects.

For rigid bodies rotating about fixed axes, **Newton's second law for rotational motion** is $\Sigma \tau = I\alpha$. $\Sigma \tau$ is the net external torque applied to a body, I is the moment of inertia of the body, and α is the angular acceleration (in rad/s²). The **moment of inertia** I of an object composed of N particles is $I = (m_1 r_1^2 + m_2 r_2^2 + \cdots + m_N r_N^2)$, where m_1, m_2, . . . , m_N are the masses of the particles and r_1, r_2, . . . , r_N are the perpendicular distances of the particles from the axis of rotation.

The **rotational work** W_R done by a constant torque τ in turning a rigid body through an angular displacement θ (in radians) is expressed by $W_R = \tau\theta$. The **rotational kinetic energy** of an object with angular speed ω (in rad/s) and moment of inertia I is $\text{KE}_R = \frac{1}{2}I\omega^2$. The **total mechanical energy** E of a rigid object is the sum of its translational kinetic energy, its rotational kinetic energy, and its potential energy. The **total mechanical energy is conserved**, provided that the net work done by external nonconservative forces and torques is zero.

The **angular momentum** L of a body rotating with angular velocity ω (in rad/s) about a fixed axis and having a moment of inertia I is $L = I\omega$. The **principle of conservation of angular momentum** states that the total angular momentum of a system remains constant if the net external torque acting on the system is zero.

CONCEPTUAL QUESTIONS

1. Sometimes, even with a wrench, one cannot loosen a nut that is frozen tightly to a bolt. In such a situation, it is often possible to loosen the nut by slipping a long pipe over the wrench handle. The purpose of the pipe is to extend the length of the wrench handle, so that the applied force can be located farther away from the nut. Explain why this trick works.

2. Explain (a) how it is possible for a large force to produce only a small, or even zero, torque, and (b) how it is possible for a small force to produce a large torque.

3. A magnetic tape is being played on a cassette deck. The tension in the tape applies a torque to the supply reel. Assuming the tension remains constant during playback, discuss how this torque varies as the reel becomes empty.

4. A flat rectangular sheet of plywood can rotate about an axis perpendicular to the sheet through one corner. How should a force (acting in the plane of the sheet) be applied to the plywood so as to create the largest possible torque? Give your reasoning.

5. A torque is the product of a force and a distance (lever arm). Work is also the product of a force and a distance. Yet, torque and work *are different*. What is it about the distances that makes torque and work different?

6. Suppose you are standing on a train, both feet together, facing a window. The front of the train is to your left. The train starts moving forward. To keep from falling, you slide your right foot out toward the rear of the train. Explain in terms of torque how this action keeps you from falling over.

7. Starting in the spring, fruit begins to grow on the outer end of a branch on a pear tree. Explain how the center of gravity of the pear-growing branch shifts during the course of the summer.

8. The free-body diagram in the drawing shows the forces that act on a thin rod. The three forces are drawn to scale and lie in the plane of the paper. Are these forces sufficient to keep the rod in equilibrium, or are additional forces necessary? Explain.

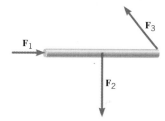

9. An A-shaped step ladder is standing on frictionless ground. The ladder consists of two sections joined at the top and kept from spreading apart by a horizontal crossbar. Draw a free-body diagram showing the forces that keep *one* section of the ladder in equilibrium.

10. The drawing shows a wine rack for a single bottle of wine that seems to defy common sense, as it balances on a table top. Treat the rack and wine bottle as a rigid body and draw the external forces that keep it in equilibrium. In particular, where must the center of gravity of the rigid body be located? Give your reasoning. (*Hint: The physics here is the same as in Figure 9.11c.*)

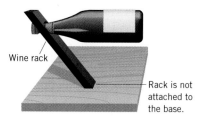

11. A flat triangular sheet of uniform material is shown in the drawing. There are three possible axes of rotation, each perpendicular to the sheet and passing through one corner, A, B, or C. For which axis is the greatest net external torque required to bring the triangle up to an angular speed of 10.0 rad/s in 10.0 s, starting from rest? Explain, assuming that the net torque is kept constant while it is being applied.

12. An object has an angular velocity. It also has an angular acceleration due to torques that are present. Therefore, the angular velocity is changing. What happens to the angular velocity if (a) additional torques are applied so as to make the net torque suddenly equal to zero and (b) all the torques are suddenly removed?

13. The satellite shown in the drawing is initially moving with a constant translational velocity and zero angular velocity through outer space. (a) When the two engines are fired, each generating a thrust of magnitude *T*, will the translational velocity increase, decrease, or remain the same? Why? (b) Explain what will happen to the angular velocity.

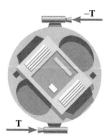

14. Sit-ups are more difficult to do with your hands placed behind your head instead of on your stomach. Why?

15. For purposes of computing the translational kinetic energy of a rigid body, its mass can be considered as concentrated at the center of mass. If one wishes to compute the body's moment of inertia, can the mass be considered as concentrated at the center of mass? If not, why not?

16. A thin sheet of plastic is uniform and has the shape of an equilateral triangle. Consider two axes for rotation. Both are perpendicular to the plane of the triangle, axis *A* passing through the center of the triangle and axis *B* passing through one corner. If the angular speed *ω* about each axis is the same, for which axis does the triangle have the greater rotational kinetic energy? Explain.

17. Suppose the solid cylinder in Example 13 (see Figure 9.24) had ten times more mass than the hollow cylinder. Would the solid cylinder still have the greater translational speed at the bottom? Explain your reasoning.

18. A hoop, a solid cylinder, a spherical shell, and a solid sphere are placed at rest at the top of an incline. All the objects have the same radius. They are then released at the same time. What is the order in which they reach the bottom? Justify your answer.

19. A woman is sitting on the spinning seat of a piano stool with her arms folded. What happens to her (a) angular velocity and (b) angular momentum when she extends her arms outward? Justify your answers.

20. Review Conceptual Example 14 as an aid in answering this question. Suppose the ice cap at the South Pole melted and the water was distributed uniformly over the earth's oceans. Would the earth's angular velocity increase, decrease, or remain the same? Explain.

21. The concept behind this question is discussed in Conceptual Example 14. Many rivers, like the Mississippi River, flow from north to south toward the equator. These rivers often carry a large amount of sediment that they deposit when entering the ocean. What effect does this redistribution of the earth's soil have on the angular velocity of the earth? Why?

22. Conceptual Example 14 provides background for this ques-

tion. A cloud of interstellar gas is rotating. Because the gravitational force pulls the gas particles together, the cloud shrinks, and, under the right conditions, a star may ultimately be formed. Would the angular velocity of the star be less than, equal to, or greater than, the angular velocity of the rotating gas? Justify your answer.

23. A person is hanging motionless from a vertical rope over a swimming pool. She lets go of the rope and drops straight down. After letting go, is it possible for her to curl into a ball and start spinning? Justify your answer.

24. The photograph shows a workman struggling to keep a stack of boxes balanced on a dolly. The man's right foot is on the axle of the dolly. Assuming that the boxes are identical, which one creates the greatest torque with respect to the axle? Why?

PROBLEMS

ssm Solution is in the Student Solutions Manual. **www** Solution is available on the World Wide Web at http://www.wiley.com/college/cutnell ⚕ This icon represents a biomedical application.

Section 9.1 The Effects of Forces and Torques on the Motion of Rigid Objects

1. ssm A post is driven perpendicularly into the ground and serves as the axis about which a gate rotates. A force of 12 N is applied perpendicular to the gate and acts parallel to the ground. How far from the post should the force be applied to produce a torque of 3.0 N·m?

2. You are installing a new spark plug in your car, and the man-

ual specifies that it be tightened to a torque of 45 N·m. Using the data in the drawing, determine the magnitude *F* of the force that you must exert on the wrench.

3. A force of 110 N is applied perpendicularly to the left edge of the rectangle shown in the drawing. (a) Find the torque (magnitude and direction) produced by this force with respect to an axis perpendicular to the plane of the rectangle at corner *A* and (b) with respect to a similar axis at corner *B*.

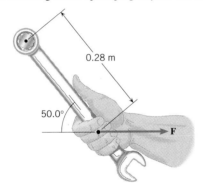

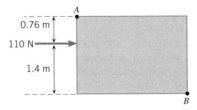

4. In San Francisco a very simple technique is used to turn around a cable car when it reaches the end of its route. The car rolls onto a turntable, which can rotate about a vertical axis through its center. Then, two people push perpendicularly on the car, one at each end, as in the drawing. The turntable is rotated

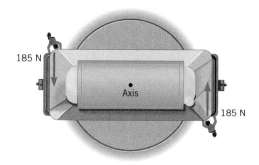

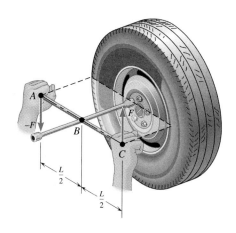

one-half of a revolution to turn the car around. If the length of the car is 9.20 m and each person pushes with a 185-N force, what is the net torque applied to the car?

5. ssm www A square, 0.40 m on a side, is mounted so that it can rotate about an axis that passes through the center of the square. The axis is perpendicular to the plane of the square. A force of 15 N lies in this plane and is applied to the square. What is the magnitude of the maximum torque that such a force could produce?

6. Find the net torque (magnitude and direction) produced by the forces $\mathbf{F}_1$ and $\mathbf{F}_2$ about the rotational axis shown in the drawing. The forces are acting on a thin rigid rod, and the axis is perpendicular to the page.

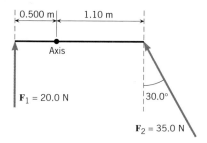

7. The wheel of a car has a radius of 0.330 m. The engine of the car applies a torque of 272 N · m to this wheel, which does not slip against the road surface. Since the wheel does not slip, the road must be applying a force of static friction to the wheel that produces a countertorque. Moreover, the car has a constant velocity, so this countertorque balances the applied torque. What is the magnitude of the static frictional force?

***8.** A pair of forces with equal magnitudes, opposite directions, and different lines of action is called a "couple." When a couple acts on a rigid object, the couple produces a torque that does *not* depend on the location of the axis. The drawing shows a couple acting on a tire wrench, each force being perpendicular to the wrench. Determine an expression for the torque produced by the couple when the axis is perpendicular to the tire and passes through (a) point A, (b) point B, and (c) point C. Express your answers in terms of the magnitude F of the force and the length L of the wrench.

***9. ssm** One end of a meter stick is pinned to a table, so the stick can rotate freely in a plane parallel to the tabletop. Two forces, both parallel to the tabletop, are applied to the stick in such a way that the net torque is zero. One force has a magnitude of 2.00 N and is applied perpendicular to the length of the stick at the free end. The other force has a magnitude of 6.00 N and acts at a 30.0° angle with respect to the length of the stick. Where along the stick is the 6.00-N force applied? Express this distance with respect to the end that is pinned.

****10.** A rotational axis is directed perpendicular to the plane of a square and is located as shown in the drawing. Two forces, $\mathbf{F}_1$ and $\mathbf{F}_2$, are applied to diagonally opposite corners, and act along the sides of the square, first as shown in part a and then as shown in part b of the drawing. In each case the net torque produced by the forces is zero. The square is one meter on a side, and the magnitude of $\mathbf{F}_2$ is three times that of $\mathbf{F}_1$. Find the distances a and b that locate the axis.

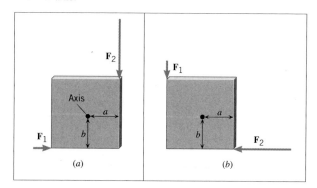

Section 9.2 Rigid Objects in Equilibrium, Section 9.3 Center of Gravity

11. A 71.0-kg boulder is placed on a 2.00-m-long board, at a point that is 1.40 m from one end. Cliff and Will support the board at each end so that it is horizontal. Cliff is nearest the boulder. If the weight of the board is negligible, what force (magnitude only) does (a) Cliff and (b) Will apply to the board?

12. Suppose you need to measure the mass of an object, but have only a 10.0-kg standard mass and a meter stick. Place a knife edge under the 50-cm mark of the meter stick. Then put the 10.0-kg standard mass at the 15-cm mark and place the object so as to balance the standard mass. If the system is balanced when the object is at the 72-cm mark, find the unknown mass.

13. ssm Review Example 4 before attempting this problem. What is the minimum value for the coefficient of static friction between the ladder and the ground, so that the ladder does not slip?

14. Review Conceptual Example 8 before starting this problem. A uniform plank of length 5.0 m and weight 225 N rests horizontally on two supports, with 1.1 m of the plank hanging over the right support (see the drawing). To what distance x can a person who weighs 450 N walk on the overhanging part of the plank before it just begins to tip?

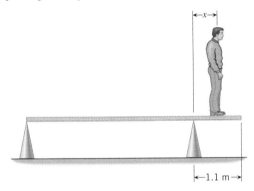

15. The wheels, axle, and handles of a wheelbarrow weigh 60.0 N. The load chamber and its contents weigh 525 N. It is well known that the wheelbarrow is much easier to use if the center of gravity of the load is placed directly over the axle, as in the right wheelbarrow in the drawing. Verify this fact by calculating the vertical lifting force **F** required to support (a) the left wheelbarrow and (b) the right wheelbarrow.

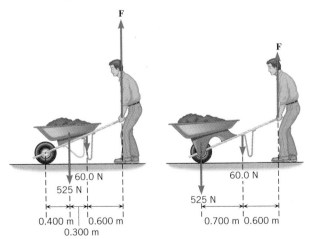

16. In an isometric exercise a person places a hand on a scale and pushes vertically downward, keeping the forearm horizontal. This is possible because the triceps muscle applies an upward

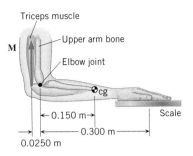

force **M** perpendicular to the arm, as the drawing indicates. The forearm weighs 22.0 N and has a center of gravity as indicated. The scale registers 111 N. Determine the magnitude of **M**.

17. ssm A uniform door (0.81 m wide and 2.1 m high) weighs 140 N and is hung on two hinges that are bolted to a vertical wall. The hinges are 2.1 m apart. Assume that the lower hinge bears all the weight of the door. Find the magnitude and direction of the horizontal component of the force applied **to the door** by (a) the upper hinge and (b) the lower hinge. Determine the magnitude and direction of the force applied **by the door** to (c) the upper hinge and (d) the lower hinge.

18. Three objects are situated on the x axis. Their weights and positions are: 43.0 N at $x = +2.20$ m, 18.0 N at $x = -0.600$ m, and 26.7 N at $x = -1.40$ m. Where on the x axis is (a) the center of gravity and (b) the center of mass of this group of objects?

19. A person exerts a horizontal force of 170 N in the test apparatus shown in the drawing. Find the horizontal force **M** (magnitude and direction) that his flexor muscle exerts on his forearm.

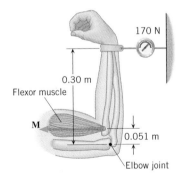

*20. Review Conceptual Example 8 as background material for this problem. A jet transport has a weight of 1.00×10^6 N and is at rest on the runway. The two rear wheels are 15.0 m behind the front wheel, and the plane's center of gravity is 12.6 m behind the front wheel. Determine the normal force exerted by the ground on (a) the front wheel and on (b) each of the two rear wheels.

*21. ssm www A massless, rigid board is placed across two bathroom scales that are separated by a distance of 2.00 m. A person lies on the board. The scale under his head reads 425 N, and the scale under his feet reads 315 N. (a) Find the weight of the person. (b) Locate the center of gravity of the person relative to the scale beneath his head.

*22. A woman who weighs 5.00×10^2 N is leaning against a smooth vertical wall, as the drawing shows. Find (a) the force $\mathbf{F_N}$ (directed perpendicular to the wall) exerted on her shoulder by the wall and the (b) horizontal and (c) vertical components of the force exerted on her shoes by the ground.

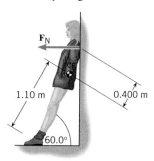

*23. 🔬 A person is sitting with one leg outstretched, so that it makes an angle of 30.0° with the horizontal, as the drawing indicates. The weight of the leg below the knee is 44.5 N with the center of gravity located below the knee joint. The leg is being held in this position because of the force $\mathbf{M}$ applied by the quadriceps muscle, which is attached 0.100 m below the knee joint (see the drawing). Obtain the magnitude of $\mathbf{M}$.

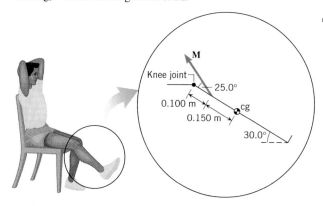

*24. A wrecking ball (weight = 4800 N) is supported by a boom, which may be assumed to be uniform and has a weight of 3600 N. As the drawing shows, a support cable runs from the top of the boom to the tractor. The angle between the support cable and the horizontal is 32°, and the angle between the boom and the horizontal is 48°. Find (a) the tension in the support cable and (b) the

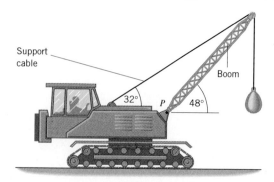

magnitude of the force exerted on the lower end of the boom by the hinge at point P.

*25. **ssm** A uniform board is leaning against a smooth vertical wall. The board is at an angle θ above the horizontal ground. The coefficient of static friction between the ground and the lower end of the board is 0.650. Find the smallest value for the angle θ, such that the lower end of the board does not slide along the ground.

**26. The drawing shows an inverted "A" that is suspended from the ceiling by two vertical ropes. Each leg of the "A" has a length of $2L$ and a weight of 120 N. The horizontal crossbar has a negligible weight. Find the force that the crossbar applies to each leg.

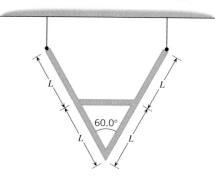

**27. The drawing shows an A-shaped ladder. Both sides of the ladder are equal in length. This ladder is standing on a frictionless horizontal surface, and only the crossbar (which has a negligible mass) of the "A" keeps the ladder from collapsing. The ladder is uniform and has a mass of 20.0 kg. Determine the tension in the crossbar of the ladder.

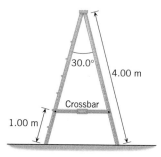

**28. Two vertical walls are separated by a distance of 1.5 m, as the drawing shows. Wall 1 is smooth, while wall 2 is not smooth. A uniform board is propped between them. The coefficient of static friction between the board and wall 2 is 0.98. What is the length of the longest board that can be propped between the walls?

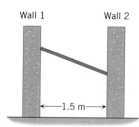

Section 9.4 Newton's Second Law for Rotational Motion About a Fixed Axis

29. **ssm** A clay vase on a potter's wheel experiences an angular acceleration of 8.00 rad/s² due to the application of a 10.0-N·m net torque. Find the total moment of inertia of the vase and potter's wheel.

30. Three particles are located on the *x* axis as follows: (1) 4.00 kg at *x* = 0.300 m, (2) 10.0 kg at *x* = 0.700 m, and (3) 1.50 kg at *x* = 0.960 m. (a) Calculate the moment of inertia of *each* particle with respect to the *y* axis. (b) Find the total moment of inertia. (c) Based on the results above, decide whether it is true that the particle with the smallest mass necessarily contributes the smallest amount to the total moment of inertia. Explain.

31. The blades of a ceiling fan have a total moment of inertia of 0.16 kg·m² and an angular acceleration of 7.0 rad/s². What net torque is being applied to the blades?

32. A uniform solid disk with a mass of 24.3 kg and a radius of 0.314 m is free to rotate about a frictionless axle. Forces of 90.0 and 125 N are applied to the disk, as the drawing illustrates. What is (a) the net torque produced by the two forces and (b) the angular acceleration of the disk?

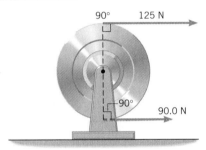

33. **ssm** A bicycle wheel has a radius of 0.330 m and a rim whose mass is 1.20 kg. The wheel has 50 spokes, each with a mass of 0.010 kg. (a) Calculate the moment of inertia of the rim about the axle. (b) Determine the moment of inertia of *any one spoke*, assuming it to be a long, thin rod that can rotate about one end. (c) Find the *total* moment of inertia of the wheel, including the rim and all 50 spokes.

34. A rotating door is made from four rectangular glass panes, as shown in the drawing. The mass of each pane is 85 kg. A person pushes on the outer edge of one pane with a force of *F* =

68 N that is directed perpendicular to the pane. Determine the magnitude of the door's angular acceleration.

35. A cylinder is rotating about an axis that passes through the center of each circular end piece. The cylinder has a radius of 0.0750 m, an angular speed of 88.0 rad/s, and a moment of inertia of 0.850 kg·m². A brake shoe presses against the surface of the cylinder and applies a tangential frictional force to it. The frictional force reduces the angular speed of the cylinder by a factor of two during a time of 5.00 s. (a) Find the magnitude of the angular deceleration of the cylinder. (b) Find the magnitude of the force of friction applied by the brake shoe.

36. A turntable platter has a radius of 0.150 m and is rotating at 3.49 rad/s. When the power is shut off, the platter slows down and comes to rest in 15.0 s, due to a net retarding torque of magnitude 6.20 × 10⁻³ N·m. Assume the platter to be a uniform solid disk. Determine the mass of the platter.

*37. **ssm** A thin, rigid, uniform rod has a mass of 2.00 kg and a length of 2.00 m. (a) Find the moment of inertia of the rod relative to an axis that is perpendicular to the rod at one end. (b) Suppose all the mass of the rod were located at a single point. Determine the perpendicular distance of this point from the axis in part (a), such that this point particle has the same moment of inertia as the rod. This distance is called the ***radius of gyration*** of the rod.

*38. Refer to the drawing for question 13 (not problem 13) and assume that the radius of the satellite is 2.3 m. Suppose that the magnitude of the thrust generated by each of the two engines is *T* = 630 N and that the moment of inertia of the satellite is 660 kg·m². If the satellite is rotating counterclockwise at an angular velocity of 1.2 rad/s at the time the engines begin to fire, how many revolutions will the satellite make during a 5.0-s burst from the engines?

*39. One part of a fireworks display uses a 0.10-m-square platform, on which tubes of gunpowder are mounted along each outer edge (see the drawing). As the powder burns, the square spins about an axis perpendicular to its center. The unit reaches an angular speed of 31 rad/s in one-half second, starting from rest, because each of three tubes generates a force of 0.30 N and the fourth tube generates an unknown force. Assume that the moment of inertia of the entire unit has a constant value of 1.0 × 10⁻³ kg·m² and that the forces are constant. What is the magnitude of the force generated by the fourth tube?

*40. The drawing shows a model for the motion of the human forearm in throwing a dart. Because of the force **M** applied by the triceps muscle, the forearm can rotate about an axis at the elbow joint. Assume that the forearm has the dimensions shown in the drawing and a moment of inertia of 0.065 kg·m² (including the effect of the dart) relative to the axis at the elbow. Assume also

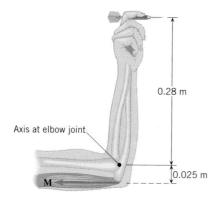

0.28 m

Axis at elbow joint

0.025 m

M

that the force **M** acts perpendicular to the forearm. Ignoring the effect of gravity and any frictional forces, determine the magnitude of the force **M** needed to give the dart a tangential speed of 5.0 m/s in 0.10 s, starting from rest.

*41. **ssm** A stationary bicycle is raised off the ground, and its front wheel ($m = 1.3$ kg) is rotating at an angular velocity of 13.1 rad/s (see the drawing). The front brake is then applied for 3.0 s, and the wheel slows down to 3.7 rad/s. Assume that all the mass of the wheel is concentrated in the rim, the radius of which is 0.33 m. The coefficient of kinetic friction between each brake pad and the rim is $\mu_k = 0.85$. What is the magnitude of the normal force that *each* brake pad applies to the rim?

Brake pads

0.33 m

*42. The *parallel axis theorem* provides a useful way to calculate the moment of inertia I about an arbitrary axis. The theorem states that $I = I_{cm} + Mh^2$, where I_{cm} is the moment of inertia of the object relative to an axis that passes through the center of mass and is parallel to the axis of interest, M is the total mass of the object, and h is the perpendicular distance between the two axes. Use this theorem and information to determine an expression for the moment of inertia of a solid cylinder of radius R relative to an axis that lies on the surface of the cylinder and is perpendicular to the circular ends.

**43. By means of a rope whose mass is negligible, two blocks are suspended over a pulley, as the drawing shows. The pulley can be treated as a uniform solid cylindrical disk. The downward acceleration of the 44.0-kg block is observed to be exactly one-half the acceleration due to gravity. Noting that the tension in the rope is not the same on each side of the pulley, find the mass of the pulley.

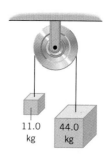

11.0 kg   44.0 kg

**44. The crane shown in the drawing is lifting a 180-kg crate upward with an acceleration of 1.2 m/s². The cable from the crate passes over a solid cylindrical pulley at the top of the boom. The pulley has a mass of 130 kg. The cable is then wound onto a hollow cylindrical drum that is mounted on the deck of the crane. The mass of the drum is 150 kg, and its radius is 0.76 m. The engine applies a counterclockwise torque to the drum in order to wind up the cable. What is the magnitude of this torque? Ignore the mass of the cable.

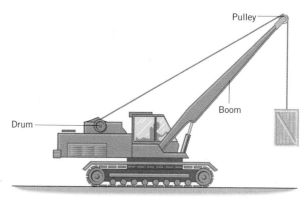

Pulley

Boom

Drum

Section 9.5 Rotational Work and Energy

45. ssm A flywheel is a solid disk that rotates about an axis that is perpendicular to the disk at its center. Rotating flywheels provide a means for storing energy in the form of rotational kinetic energy and are being considered as a possible alternative to batteries in electric cars. The gasoline burned in a 300-mile trip in a typical midsize car produces about 1.2×10^9 J of energy. How fast would a 13-kg flywheel with a radius of 0.30 m have to rotate in order to store this much energy? Give your answer in rev/min.

46. A baseball is thrown such that the translational speed of its center of mass is 31 m/s, and its angular speed about the center of mass is 180 rad/s. Treat the baseball as if it were a uniform solid sphere of radius 3.7 cm. What fraction of the ball's total kinetic energy is rotational kinetic energy?

47. A four-bladed ceiling fan rotates with an angular speed of 30.0 rad/s. The length of each blade is 0.600 m, and the mass of each blade is 2.00 kg. Each blade can be approximated as a uniform thin rod that rotates about one end. Determine the total rotational kinetic energy of the four blades.

48. A car is moving with a speed of 27.0 m/s. Each wheel has a radius of 0.300 m and a moment of inertia of 0.850 kg·m². The car has a total mass (including the wheels) of 1.20×10^3 kg. Find (a) the translational kinetic energy of the entire car, (b) the total rotational kinetic energy of the four wheels, and (c) the total kinetic energy of the car.

49. ssm Three objects lie in the x, y plane. Each rotates about the z axis with an angular speed of 6.00 rad/s. The mass m of each object and its perpendicular distance r from the z axis are as follows: (1) $m_1 = 6.00$ kg and $r_1 = 2.00$ m, (2) $m_2 = 4.00$ kg and $r_2 = 1.50$ m, (3) $m_3 = 3.00$ kg and $r_3 = 3.00$ m. (a) Find the tangential speed of each object. (b) Determine the total kinetic energy of this system using the expression KE $= \frac{1}{2}m_1v_1^2 + \frac{1}{2}m_2v_2^2 + \frac{1}{2}m_3v_3^2$. (c) Obtain the moment of inertia of the system. (d) Find the rotational kinetic energy of the system using the relation $\frac{1}{2}I\omega^2$ to verify that the answer is the same as that in (b).

*50. Review Example 13 before attempting this problem. A marble and a cube are placed at the top of a ramp. Starting from rest at the same height, the marble rolls and the cube slides (no kinetic friction) down the ramp. Determine the ratio of the center-of-mass speed of the cube to the center-of-mass speed of the marble at the bottom of the ramp.

*51. A thin-walled spherical shell is rolling on a surface. What fraction of its total kinetic energy is in the form of rotational kinetic energy about the center of mass?

*52. A thin uniform rod is initially positioned in the vertical direction, with its lower end attached to a frictionless axis that is mounted on the floor. The rod has a length of 2.00 m and is allowed to fall, starting from rest. Find the tangential speed of the free end of the rod, just before the rod hits the floor after rotating through 90°.

*53. ssm Review Example 13 before attempting this problem. A solid cylinder and a thin-walled hollow cylinder (see Table 9.1) have the same mass and radius. They are rolling horizontally toward the bottom of an incline. The center of mass of each has the same translational speed. The cylinders roll up the incline and reach their highest points. Calculate the ratio of the distances ($s_{\text{solid}}/s_{\text{hollow}}$) along the incline through which each center of mass moves.

**54. A tennis ball, starting from rest, rolls down the hill in the drawing. At the end of the hill the ball becomes airborne, leaving at an angle of 35° with respect to the ground. Treat the ball as a thin-walled spherical shell, and determine the range x.

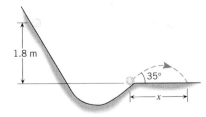

Section 9.6 Angular Momentum

55. For a certain satellite with an apogee distance of $r_A = 1.30 \times 10^7$ m, the ratio of the orbital speed at perigee to the orbital speed at apogee is 1.20. Find the perigee distance r_P.

56. Before starting this problem, review Conceptual Example 14. A woman stands at the center of a platform. The woman and the platform rotate with an angular speed of 5.00 rad/s. Friction is negligible. Her arms are outstretched, and she is holding a dumbbell in each hand. In this position the total moment of inertia of the rotating system (platform, woman, and dumbbells) is 5.40 kg·m². By pulling in her arms, the moment of inertia is reduced to 3.80 kg·m². Find her new angular speed.

57. ssm www Two disks are rotating about the same axis. Disk A has a moment of inertia of 3.4 kg·m² and an angular velocity of $+7.2$ rad/s. Disk B is rotating with an angular velocity of -9.8 rad/s. The two disks are then linked together without the aid of any external torques, so that they rotate as a single unit with an angular velocity of -2.4 rad/s. The axis of rotation for this unit is the same as that for the separate disks. What is the moment of inertia of disk B?

58. A playground carousel is free to rotate about its center on frictionless bearings. The carousel has an angular speed of 3.14 rad/s, a moment of inertia of 125 kg·m², and a radius of 1.50 m. A 40.0-kg person, standing still next to the carousel, jumps onto it very close to the outer edge. Find the resulting angular speed of the carousel and person.

59. A thin rod has a length of 0.20 m and rotates in a circle on a frictionless tabletop. The axis is perpendicular to the length of the rod at one of its ends. The rod has an angular velocity of 0.40 rad/s and a moment of inertia of 1.2×10^{-3} kg·m². A bug standing on the axis decides to crawl out to the other end of the rod. When the bug (mass $= 5.0 \times 10^{-3}$ kg) gets where it's going, what is the angular velocity of the rod?

*60. A flat uniform circular disk (radius $= 2.00$ m, mass $= 1.00 \times 10^2$ kg) is initially stationary. The disk is free to rotate in the horizontal plane about a frictionless axis perpendicular to the center of the disk. A 40.0-kg person, standing 1.25 m from the axis, begins to run on the disk in a circular path and has a tangential speed of 2.00 m/s relative to the ground. (a) Find the resulting angular speed of the disk (in rad/s) and describe the direction of the rotation. (b) Determine the time it takes for a spot marking the starting point to pass again beneath the runner's feet.

*61. ssm A cylindrically shaped space station is rotating about the axis of the cylinder to create artificial gravity. The radius of the cylinder is 82.5 m. The moment of inertia of the station without people is 3.00×10^9 kg·m². Suppose 500 people, with an average mass of 70.0 kg each, live on this station. As they move radially from the outer surface of the cylinder toward the axis, the angular speed of the station changes. What is the maximum possible percentage change in the station's angular speed due to the radial movement of the people?

**62. A small 0.500-kg object moves on a frictionless horizontal table in a circular path of radius 1.00 m. The angular speed is 6.28 rad/s. The object is attached to a string of negligible mass that passes through a small hole in the table at the center of the circle. Someone under the table begins to pull the string downward to make the circle smaller. If the string will tolerate a tension of no more than 105 N, what is the radius of the smallest possible circle on which the object can move?

**63. A platform is rotating at an angular speed of 2.2 rad/s. A block is resting on this platform at a distance of 0.30 m from the axis. The coefficient of static friction between the block and the platform is 0.75. Without any external torque acting on the system, the block is moved toward the axis. Ignore the moment of inertia of the platform and determine the smallest distance from the axis at which the block can be relocated and still remain in place as the platform rotates.

ADDITIONAL PROBLEMS

64. The circular blade on a radial arm saw is turning at 262 rad/s at the instant the motor is turned off. In 18.0 s the speed of the blade is reduced to 85 rad/s. Assume the blade to be a uniform solid disk of radius 0.130 m and mass 0.400 kg. Find the net torque applied to the blade.

65. **ssm** The steering wheel of a car has a radius of 0.19 m, while the steering wheel of a truck has a radius of 0.25 m. The same force is applied in the same direction to each. What is the ratio of the torque produced by this force in the truck to the torque produced in the car?

66. The mass of an automobile is 1160 kg, and the distance between its front and rear axles is 2.54 m. The center of gravity of the car is between the front and rear tires, and the horizontal distance between the center of gravity and the front axle is 1.02 m. Determine the normal force that the ground applies to (a) *each* of the two front wheels and to (b) *each* of the two rear wheels.

67. A baggage carousel at an airport is rotating with an angular speed of 0.20 rad/s when the baggage begins to be loaded onto it. The moment of inertia of the carousel is 1500 kg·m². Ten pieces of baggage with an average mass of 15 kg each are dropped vertically onto the carousel and come to rest at a perpendicular distance of 2.0 m from the axis of rotation. (a) Assuming that no net external torque acts on the system of carousel and baggage, find the final angular speed. (b) In reality, the angular speed of a baggage carousel does not change. Therefore, speaking qualitatively, what can you say about the external torque acting on the system?

68. A lunch tray is being held in one hand, as the drawing illustrates. The mass of the tray itself is 0.200 kg, and its center of gravity is located at its geometrical center. On the tray is a 1.00-kg plate of food and a 0.250-kg cup of coffee. Obtain the force **T** exerted by the thumb and the force **F** exerted by the four fingers. Both forces act perpendicular to the tray, which is being held parallel to the ground.

69. **ssm** A particle is located at each corner of an imaginary cube. Each edge of the cube is 0.25 m long, and each particle has a mass of 0.12 kg. What is the moment of inertia of these particles with respect to an axis that lies along one edge of the cube?

70. One side of a rotating rectangle is three times the length of the other side. A force F_A is applied to one corner of the rectangle and is parallel to the short side. A second force F_B is applied to the diagonally opposite corner and is parallel to the long side. Find the ratio of the magnitudes of the forces F_B/F_A so that the rectangle rotates at a constant angular velocity about an axis perpendicular to the rectangle at its center.

71. A solid disk has a mass of 162 kg and a radius of 1.30 m. This disk rotates about an axis through its center, like a compact disc, and has an angular speed of 18.0 rad/s. If all the kinetic energy of the disk were used to lift a 3.00-kg block, how high could the block be lifted?

*72. A uniform steel beam of length 5.00 m has a weight of 4.50×10^3 N. One end of the beam is bolted to a vertical wall. The beam is held in a horizontal position by a cable attached between the other end of the beam and a point on the wall. The cable makes an angle of 25.0° above the horizontal. A load whose weight is 12.0×10^3 N is hung from the beam at a point that is 3.50 m from the wall. Find (a) the magnitude of the tension in the supporting cable and (b) the magnitude of the force exerted on the end of the beam by the bolt that attaches the beam to the wall.

*73. **ssm** ✍ A man holds a 178-N ball in his hand, with the forearm horizontal (see the drawing). He can support the ball in this position because of the flexor muscle force **M**, which is applied perpendicular to the forearm. The forearm weighs 22.0 N and has a center of gravity as indicated. Find (a) the magnitude of **M** and (b) the magnitude and direction of the force applied by the upper arm bone to the forearm at the elbow joint.

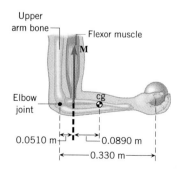

*74. A thin uniform rod is rotating at an angular velocity of 7.0 rad/s about an axis that is perpendicular to the rod at its center. As the drawing indicates, the rod is hinged at two places, one-quarter of the length from each end. Without the aid of external torques, the rod suddenly assumes a "u" shape, with the arms of the "u" parallel to the rotation axis. What is the angular velocity of the rotating "u"?

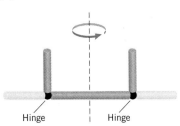

Hinge   Hinge

*75. Review Example 13 before attempting this problem. A bowling ball encounters a 0.760-m vertical rise on the way back to the ball rack, as the drawing illustrates. Ignore frictional losses and assume that the mass of the ball is distributed uniformly. If the translational speed of the ball is 3.50 m/s at the bottom of the rise, find the translational speed at the top.

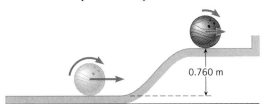

0.760 m

76. The drawing shows the top view of two doors. The doors are uniform and identical. Door A rotates about an axis through its left edge, while door B rotates about an axis through the center. The same force **F** is applied perpendicular to each door at its right edge, and the force remains perpendicular as the door turns. Starting from rest, door A rotates through a certain angle in 3.00 s. How long does it take door B to rotate through the same angle?

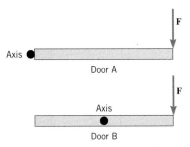

Axis

F

Door A

F

Axis

Door B

77. **ssm www** An inverted "vee" is made of uniform boards and weighs 356 N. Each side has the same length and makes a 30.0° angle with the vertical, as the drawing shows. Find the magnitude of the static frictional force that acts on the lower end of each leg of the "vee."

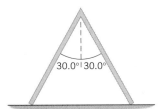

30.0° | 30.0°

ELASTICITY AND SIMPLE HARMONIC MOTION

The web spun by the orb weaver spider is both strong and elastic. As this chapter discusses, the elastic behavior of materials leads to an important kind of vibratory motion, known as simple harmonic motion.

10.1 ELASTIC DEFORMATION

STRETCHING, COMPRESSION, AND YOUNG'S MODULUS

All materials become distorted when they are squeezed or stretched, and many of them, such as rubber, return to their original shape when the squeezing or stretching is removed. Such materials are said to be "elastic." From an atomic viewpoint, elastic behavior has its origin in the forces that atoms exert on each other, and Figure 10.1 symbolizes these forces with the aid of springs. It is well known that a spring can be compressed or stretched when a force is applied to it, and that it returns to its original shape when the force is removed. Thus, because of atomic level "springs," a material tends to return to its initial shape once the forces that cause the deformation are removed.

The interatomic forces that hold the atoms of a solid together are particularly strong, so considerable force must be applied to stretch a solid object. The force needed depends on several factors, as Figure 10.2 illustrates. For two identical rods, part *a* of the drawing indicates that more force is required to produce a greater amount of stretch than a smaller amount. Part *b* indicates another fact: To stretch two equally long rods by the same amount, more force is needed for the rod with the larger cross-sectional area, assuming the rods are made from the same material. Finally, part *c* shows that, for a given amount of stretch, more force is required for a shorter rod than for a longer rod, provided the rods are made from the same material and have the same cross-sectional areas. Simply take a long rubber band, for example, and stretch it an arbitrary amount. Then stretch a much shorter rubber band with the same cross-sectional area by the same amount. You will be easily convinced that more force is required to stretch the shorter rubber band.

Experiments have shown that the elastic behavior described above can be expressed by the following relation, provided that the amount of stretching is small compared to the original length of the object:

$$F = Y\left(\frac{\Delta L}{L_0}\right)A \qquad (10.1)$$

As Figure 10.3 shows, F denotes the magnitude of the stretching force applied perpendicularly to the surface at the end, A is the cross-sectional area of the rod, ΔL is the increase in length, and L_0 is the original length. The term Y is a proportionality constant called *Young's modulus,* after Thomas Young (1773–1829), and its value

Figure 10.1 The forces between atoms act like springs. The atoms are represented by gold spheres, and the springs between some atoms have been omitted for clarity.

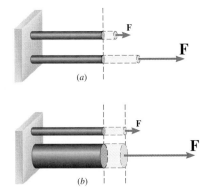

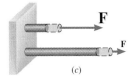

Figure 10.2 The amount of force F needed to stretch a solid rod depends on (*a*) the amount by which the rod is stretched, (*b*) the cross-sectional area of the rod, and (*c*) the unstretched length of the rod.

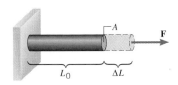

Figure 10.3 In this diagram, **F** denotes the stretching force, A the cross-sectional area, L_0 the original length of the rod, and ΔL the amount of stretch.

Table 10.1 Values for the Young's Modulus of Solid Materials

Material	Young's Modulus Y (N/m^2)
Aluminum	6.9×10^{10}
Bone	
Compression	9.4×10^{9}
Tension	1.6×10^{10}
Brass	9.0×10^{10}
Brick	1.4×10^{10}
Copper	1.1×10^{11}
Mohair	2.9×10^{9}
Nylon	3.7×10^{9}
Pyrex glass	6.2×10^{10}
Steel	2.0×10^{11}
Teflon	3.7×10^{8}
Tungsten	3.6×10^{11}

The Physics of...
bone compression.

Figure 10.4 The entire weight of the balanced group is supported by the legs of the performer who is lying on his back. From the Ringling Brothers and Barnum & Bailey circus.

depends on the nature of the material, as Table 10.1 reveals. Solving Equation 10.1 for Y shows that Young's modulus has units of force per unit area (N/m^2).

It should be noted that the magnitude of the force in Equation 10.1 is proportional to the fractional increase in length $\Delta L/L_0$, rather than the absolute increase ΔL. The magnitude of the force is also proportional to the cross-sectional area A, which need not be circular, but can have any shape (e.g., rectangular).

Forces that are applied as in Figure 10.3 and cause stretching are called "tensile" forces because they create a tension in the material, much like the tension in a rope. Equation 10.1 also applies when the force compresses the material along its length. In this situation, the force is applied in a direction opposite to that shown in Figure 10.3, and ΔL stands for the amount by which the original length L_0 decreases.

Most solids have Young's moduli that are rather large, reflecting the fact that a large force is needed to change the length of a solid object by even a small amount, as Example 1 illustrates.

EXAMPLE 1 • Bone Compression

In a circus act, a performer supports the combined weight (1640 N) of a number of colleagues (see Figure 10.4). Each thighbone (femur) of this performer has a length of 0.55 m and an effective cross-sectional area of 7.7×10^{-4} m^2. Determine the amount by which each thighbone compresses under the extra weight.

Reasoning The additional weight supported by each thighbone is $F = \frac{1}{2}(1640 \text{ N}) = 820$ N, and Table 10.1 indicates that Young's modulus for bone compression is 9.4×10^{9} N/m^2. Since the length and cross-sectional area of the thighbone are also known, we may use Equation 10.1 to find the amount by which the additional weight compresses the thighbone.

Solution The amount of compression ΔL of each thighbone is

$$\Delta L = \frac{FL_0}{YA} = \frac{(820 \text{ N})(0.55 \text{ m})}{(9.4 \times 10^9 \text{ N/m}^2)(7.7 \times 10^{-4} \text{ m}^2)} = \boxed{6.2 \times 10^{-5} \text{ m}}$$

This is a very small change, the fractional decrease being $\Delta L/L_0 = 0.000\,11$.

SHEAR DEFORMATION AND THE SHEAR MODULUS

It is possible to deform a solid object in a way other than stretching or compressing it. For instance, place a book on a rough table and push on the top cover, as in Figure 10.5a. Notice that the top cover, and the pages below it, become shifted relative to the stationary bottom cover. The resulting deformation is called a ***shear deformation*** and occurs because of the combined effect of the force **F** applied (by the hand) to the top of the book and the force **−F** applied (by the table) to the bottom of the book. The directions of the forces are parallel to the covers of the book, each of which has an area A, as illustrated in part b of the drawing. These two forces have equal magnitudes, but opposite directions, and ensure that the book remains in equilibrium. Equation 10.2 gives the magnitude F of the force needed to produce an amount of shear ΔX for an object with thickness L_0:

$$F = S\left(\frac{\Delta X}{L_0}\right)A \qquad (10.2)$$

This equation is very similar to Equation 10.1. The constant of proportionality S is

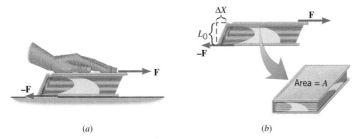

Figure 10.5 (*a*) An example of a shear deformation. The shearing forces **F** and −**F** are applied parallel to the top and bottom covers of the book. In general, shearing forces cause a solid object to change its shape. (*b*) The shear deformation is ΔX. The area of each cover is A, and the thickness of the book is L_0.

called the ***shear modulus*** and, like Young's modulus, has units of force per unit area (N/m^2). The value of S depends on the nature of the material, and Table 10.2 gives some representative values. Example 2 illustrates how to determine the shear modulus of a favorite dessert.

Table 10.2 Values for the Shear Modulus of Solid Materials

Material	Shear Modulus S (N/m^2)
Aluminum	2.4×10^{10}
Bone	8.0×10^{10}
Brass	3.5×10^{10}
Copper	4.2×10^{10}
Lead	5.4×10^{9}
Nickel	7.3×10^{10}
Steel	8.1×10^{10}
Tungsten	1.5×10^{11}

EXAMPLE 2 • J-E-L-L-O

A block of Jell-O is resting on a plate. Figure 10.6*a* gives the dimensions of the block. You are bored, impatiently waiting for dinner, and push tangentially across the top surface with a force of $F = 0.45$ N, as in part *b* of the drawing. The top surface moves a distance $\Delta X = 6.0 \times 10^{-3}$ m relative to the bottom surface. Use this idle gesture to measure the shear modulus of Jell-O.

Reasoning and Solution The shear modulus is given by Equation 10.2 as $S = FL_0/(A\Delta X)$, where $A = (0.070$ m$)(0.070$ m$)$ is the area of the top surface, and $L_0 = 0.030$ m is the thickness of the block:

$$S = \frac{FL_0}{A\,\Delta X} = \frac{(0.45\text{ N})(0.030\text{ m})}{(0.070\text{ m})(0.070\text{ m})(6.0 \times 10^{-3}\text{ m})} = \boxed{460\text{ N/m}^2}$$

Jell-O can be deformed easily, so its shear modulus is significantly less than that of a more rigid material like steel (see Table 10.2).

Although Equations 10.1 and 10.2 are similar, they refer to different kinds of deformations. The tensile force in Figure 10.3 is perpendicular to the surface whose area is A, whereas the shearing force in Figure 10.5 is parallel to that surface. Furthermore, the ratio $\Delta L/L_0$ in Equation 10.1 is different than the ratio $\Delta X/L_0$ in Equation 10.2. The distances ΔL and L_0 are parallel, whereas ΔX and L_0 are perpendicular. Young's modulus refers to a *change in length* of one dimension of a solid object as a result of tensile or compressive forces. In contrast, the shear modulus refers to a *change in shape* of a solid object as a result of shearing forces.

VOLUME DEFORMATION AND THE BULK MODULUS

When a compressive force is applied along one dimension of a solid, the length of that dimension decreases. It is also possible to apply compressive forces so that the size of every dimension (length, width, and depth) decreases, leading to a decrease

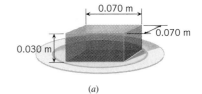

(*a*)

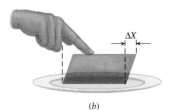

(*b*)

Figure 10.6 (*a*) A block of Jell-O and (*b*) a shearing force applied to it.

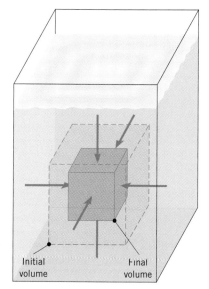

Figure 10.7 The arrows denote the forces that push perpendicularly on every surface of an object immersed in a liquid. The force per unit area is the pressure. When the pressure increases, the volume of the object decreases from an initial value of V_0 to a final value of $V_0 - \Delta V$.

in volume, as Figure 10.7 illustrates. This kind of overall compression occurs, for example, when an object is submerged in a liquid, and the liquid presses inward everywhere on the object. The forces acting in such situations are applied perpendicular to every surface, and it is more convenient to speak of the perpendicular force per unit area, rather than the amount of any one force in particular. The magnitude of the perpendicular force per unit area is called the **pressure P.**

■ **DEFINITION OF PRESSURE**

The pressure P is the magnitude F of the force acting perpendicular to a surface divided by the area A over which the force acts:

$$P = \frac{F}{A} \qquad (10.3)$$

SI Unit of Pressure: $N/m^2 = $ pascal (Pa)

Equation 10.3 indicates that the SI unit for pressure is the unit of force divided by the unit of area, or newton/meter^2 (N/m^2). This unit of pressure is often referred to as a *pascal* (Pa), named after the French scientist Blaise Pascal (1623–1662).

The result of increasing the pressure on an object by an amount ΔP is that the volume of the object decreases by an amount ΔV (see Figure 10.7). Such a pressure increase occurs, for example, when a swimmer dives deeper into the water. Experiment reveals that the amount of pressure increase needed to decrease the volume is directly proportional to the fractional volume change $\Delta V/V_0$, where V_0 is the initial volume:

$$\Delta P = -B\left(\frac{\Delta V}{V_0}\right) \qquad (10.4)$$

This relation is analogous to Equations 10.1 and 10.2, except that the area A in those equations does not appear here explicitly; the area is already taken into account by the concept of pressure (force per unit area). The proportionality constant B is known as the **bulk modulus.** The minus sign occurs because an increase in pressure (ΔP positive) always creates a decrease in volume (ΔV negative), and B is given as a positive quantity. Like Young's modulus and the shear modulus, the bulk modulus has units of force per unit area (N/m^2), and its value depends on the nature of the material. Table 10.3 gives representative values of the bulk modulus and includes liquids as well as solids.

Example 3 illustrates that an extremely large change in pressure causes only a small change in the volume of a steel ball.

Table 10.3 Values for the Bulk Modulus of Solid and Liquid Materials

Material	Bulk Modulus B (N/m^2)
Solids	
Aluminum	7.1×10^{10}
Brass	6.7×10^{10}
Copper	1.3×10^{11}
Lead	4.2×10^{10}
Nylon	6.1×10^{9}
Pyrex glass	2.6×10^{10}
Steel	1.4×10^{11}
Liquids	
Ethanol	8.9×10^{8}
Oil	1.7×10^{9}
Water	2.2×10^{9}

EXAMPLE 3 • **The Mariana Trench**

The Mariana trench is located in the Pacific Ocean, and at one place it is nearly seven miles beneath the surface of the water. The water pressure at the bottom of the trench is enormous, being about $\Delta P = 1.1 \times 10^8$ Pa greater than the pressure at the surface of the ocean. A solid steel ball of volume $V_0 = 0.20$ m^3 is dropped into the ocean and falls to the bottom of the trench. What is the change ΔV in the volume of the ball when it reaches the bottom?

Reasoning As the ball falls, the water pressure increases, and the volume of the ball becomes smaller. The change in volume ΔV can be determined directly from Equation 10.4 as $\Delta V = -(V_0 \, \Delta P/B)$, where B is the bulk modulus of steel.

Solution From Equation 10.4 we have

$$\Delta V = \frac{-V_0\,\Delta P}{B} = \frac{-(0.20\text{ m}^3)(1.1\times10^8\text{ Pa})}{1.4\times10^{11}\text{ N/m}^2} = \boxed{-1.6\times10^{-4}\text{ m}^3}$$

It is evident that the steel ball is compressed very little, even under the large pressure at the bottom of the Mariana trench.

10.2 STRESS, STRAIN, AND HOOKE'S LAW

Equations 10.1, 10.2, and 10.4 specify the amount of force needed for a given amount of elastic deformation, and they are repeated below to emphasize their common features:

$$\frac{F}{A} = Y\left(\frac{\Delta L}{L_0}\right) \tag{10.1}$$

$$\frac{F}{A} = S\left(\frac{\Delta X}{L_0}\right) \tag{10.2}$$

$$\Delta P = -B\left(\frac{\Delta V}{V_0}\right) \tag{10.4}$$

$$\boxed{\text{Stress}} \underset{\text{to}}{\overset{\text{is}}{\text{proportional}}} \boxed{\text{Strain}}$$

The left side of each equation is the magnitude of the force per unit area required to cause an elastic deformation. In general, the ratio of the force to the area is called the **stress.** The right side of each equation involves the change in a quantity (ΔL, ΔX, or ΔV) divided by a quantity (L_0 or V_0) relative to which the change is compared. The terms $\Delta L/L_0$, $\Delta X/L_0$, and $\Delta V/V_0$ are unitless ratios, and each is referred to as the **strain** that results from the applied stress. In the case of stretch and compression, the strain is the fractional change in length, whereas in volume deformation it is the fractional change in volume. In shear deformation the strain refers to a change in shape of the object. Experiments show that these three equations, with constant values for Young's modulus, the shear modulus, and the bulk modulus, apply to a wide range of materials. Therefore, stress and strain are directly proportional to one another, a relationship first discovered by Robert Hooke (1635–1703) and now referred to as **Hooke's law.**

■ **HOOKE'S LAW FOR STRESS AND STRAIN**
Stress is directly proportional to strain.

SI Unit of Stress: newton per square meter = pascal (Pa)

SI Unit of Strain: Strain is a unitless quantity.

In reality, materials obey Hooke's law only up to a certain limit, as Figure 10.8 shows. As long as stress remains proportional to strain, a plot of stress versus strain is a straight line. The point on the graph where the material begins to deviate from straight-line behavior is called the "proportionality limit." Beyond the proportional-

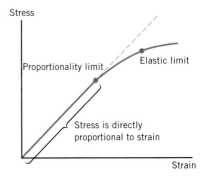

Figure 10.8 Hooke's law (stress is directly proportional to strain) is valid only up to the proportionality limit of a material. Beyond this limit, Hooke's law no longer applies. Beyond the elastic limit, the material remains deformed even when the stress is removed.

ity limit stress and strain are no longer directly proportional. However, if the stress does not exceed the "elastic limit" of the material, the object will return to its original size and shape once the stress is removed. The "elastic limit" is the point beyond which the object no longer returns to its original size and shape when the stress is removed; the object remains permanently deformed.

10.3 THE IDEAL SPRING AND SIMPLE HARMONIC MOTION

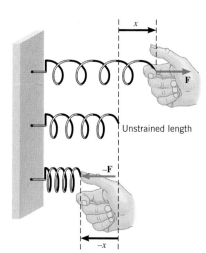

Figure 10.9 An ideal spring obeys the equation, $F = kx$, where **F** is the force applied to the spring, x is the amount of stretch or compression, and k is the spring constant.

Springs are familiar objects that exhibit elastic behavior. They have a great variety of applications, from pogo sticks to automobile suspension systems. Provided the applied force is not large enough to cause permanent deformation, the spring in Figure 10.9 will return to its original length after being stretched or compressed. For relatively small deformations, the force F required to stretch or compress a spring obeys the following equation:

$$F = kx \qquad (10.5)$$

In this expression x denotes the displacement of the spring from its unstrained length. The term k is a proportionality constant called the **spring constant** and has dimensions of force per unit length (N/m).

Equation 10.5 is a Hooke's law type of relationship, as can be seen by comparing it to Equation 10.1 for the stretching or compressing of a solid rod:

$$F = \frac{YA}{L_0} \Delta L$$

$$F = \underbrace{k}\;\underbrace{x}$$

This comparison shows that x is analogous to ΔL, whereas k is analogous to the term YA/L_0, which is a constant for a given object. A spring that behaves according to the Hooke's law relationship $F = kx$ is said to be an **ideal spring.** Example 4 illustrates one application of such a spring.

The Physics of...
a tire pressure gauge.

EXAMPLE 4 • A Tire Pressure Gauge

In a tire pressure gauge, the air in the tire pushes against a plunger attached to a spring when the gauge is pressed against the tire valve, as in Figure 10.10. Suppose the spring constant of the spring is $k = 320$ N/m and the bar indicator of the gauge extends 2.0 cm when the gauge is pressed against the air valve. What force does the air in the tire apply to the spring?

Reasoning and Solution   Since the spring constant is known, the force applied to the spring can be obtained from Equation 10.5:

$$F = kx = (320 \text{ N/m})(0.020 \text{ m}) = \boxed{6.4 \text{ N}}$$

Thus, the exposed length of the bar indicator gives a measure of the force that the air pressure in the tire exerts on the spring. Since pressure is force per unit area and the area of the plunger surface is fixed, the bar indicator can be marked in units of pressure.

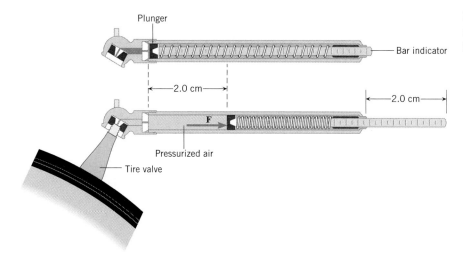

Figure 10.10 In a tire pressure gauge, the pressurized air from the tire exerts a force **F** that compresses a spring.

Sometimes the spring constant k is referred to as the *stiffness* of the spring, because a large value for k means the spring is "stiff," in the sense that a large force is required to stretch or compress it. Conceptual Example 5 examines what happens to the stiffness of a spring when it is cut into two shorter pieces.

CONCEPTUAL EXAMPLE 5 • Are Shorter Springs Stiffer Springs?

Figure 10.11*a* shows a 10-coil spring that has a spring constant k. If this spring is cut in half, so there are two 5-coil springs, what is the spring constant of each of the smaller springs?

Reasoning and Solution It is tempting to say that each 5-coil spring has a spring constant of $\frac{1}{2}k$, or one-half the value for the 10-coil spring. However, the spring constant of each 5-coil spring is $2k$, twice that of the 10-coil spring (see part *b* of the drawing). Here's the logic. Suppose we apply a force that compresses the 10-coil spring by 1 cm, each coil being compressed by 0.1 cm. Let us also compress a 5-coil spring by 1 cm, each coil now being compressed by 0.2 cm. According to Equation 10.5 ($F = kx$), the necessary force is directly proportional to the displacement of the coil. The force needed to compress a single coil by 0.2 cm must be twice as large as the force needed to compress it by 0.1 cm. Thus, the spring constant of the 5-coil spring must be twice that of the 10-coil spring. In general, the spring constant is inversely proportional to the number of coils in the spring, so ***shorter springs are stiffer springs,*** all other things being equal.

Related Homework Material: *Problems 32 and 39*

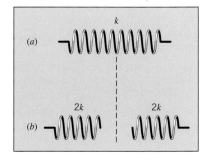

Figure 10.11 (*a*) The 10-coil spring has a spring constant k. (*b*) Each 5-coil spring has a spring constant of $2k$, twice that of the 10-coil spring.

To stretch or compress a spring, a force must be applied to it. In accord with Newton's third law, the spring exerts an oppositely directed force of equal magnitude. This reaction force is applied by the spring to the agent that does the pulling or pushing. In other words, the reaction force is applied to the object attached to the spring. The reaction force is also called a "restoring force," for a reason that will be clarified shortly. The restoring force of an ideal spring is obtained from the relation $F = kx$ by including the minus sign required by Newton's action–reaction law, as indicated in Equation 10.6.

■ HOOKE'S LAW RESTORING FORCE OF AN IDEAL SPRING

The restoring force of an ideal spring is

$$F = -kx \qquad (10.6)$$

where k is the spring constant and x is the displacement of the spring from its unstrained length. The minus sign indicates that the restoring force always points in a direction opposite to the displacement of the spring.

The restoring force of an ideal spring, like the other forces that we have encountered, may contribute to the net external force that causes an object to accelerate. The inclusion of the restoring force of a spring into Newton's second law is illustrated by the concept chart in Figure 10.12, which, except for this force, is identical to the chart in Figure 4.10. Once again, we see the unifying theme of Newton's second law, in that individual forces contribute to the net force, which, in turn, is responsible for the acceleration. However, the restoring force is not constant, but varies with the displacement x of the spring. This implies that the resulting acceleration is also not constant, as we will see in the next section.

Figure 10.13 helps to explain why the phrase "restoring force" is used. In the picture, an object of mass m is attached to a spring on a frictionless table. In part A, the object is at rest. The spring is undeformed and, hence, applies no force to the object, which is in equilibrium. In part B, the spring has been stretched to the right, so it exerts the leftward-pointing force $-\mathbf{F}$. When the object is released, this force pulls it to the left, restoring it toward its equilibrium position. However, consistent with Newton's first law, the moving object has inertia and coasts beyond the equilibrium position, compressing the spring as in part C. The force exerted by the spring now points to the right and, after bringing the object to a momentary halt, acts to restore the object to its equilibrium position. But the object's inertia again carries it beyond the equilibrium position, this time stretching the spring and leading to the restoring force $-\mathbf{F}$ shown in part D. The back-and-forth motion illus-

Figure 10.12 *Concepts at a Glance* The restoring force of a spring may contribute to the net external force $\Sigma\mathbf{F}$ that acts on an object. According to Newton's second law, the resulting acceleration $\mathbf{a}$ is directly proportional to the net force. The spring on this motorcycle shock absorber provides a force that counteracts the forces generated by a bumpy road, thus reducing the acceleration of a jarring ride.

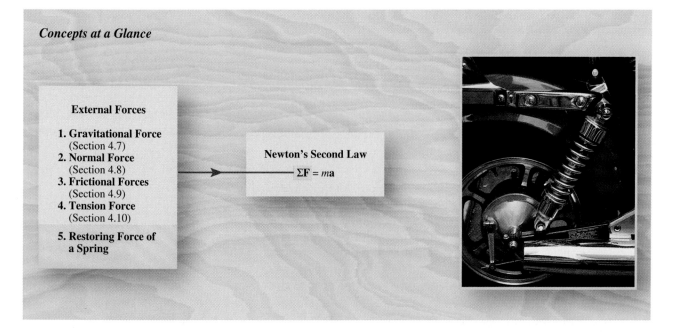

Concepts at a Glance

External Forces

1. **Gravitational Force** (Section 4.7)
2. **Normal Force** (Section 4.8)
3. **Frictional Forces** (Section 4.9)
4. **Tension Force** (Section 4.10)
5. **Restoring Force of a Spring**

Newton's Second Law
$$\Sigma\mathbf{F} = m\mathbf{a}$$

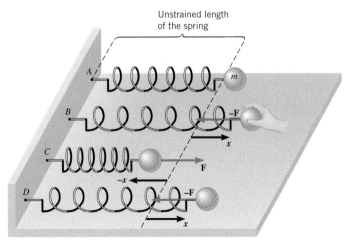

Figure 10.13 The restoring force (see blue arrows) produced by an ideal spring always points opposite to the displacement (see black arrows) of the spring and leads to a back-and-forth motion of the object.

trated in the drawing then repeats itself, continuing forever, since there is no friction acting on the object or the spring.

When the restoring force has the mathematical form given by $F = -kx$, the type of friction-free motion illustrated in Figure 10.13 is designated as "simple harmonic motion." By attaching a pen to the object and moving a strip of paper past it at a steady rate, we can record the position of the vibrating object as time passes. Figure 10.14 illustrates the resulting graphical record of simple harmonic motion. The maximum excursion from equilibrium is the ***amplitude A*** of the motion. The shape of this graph is characteristic of simple harmonic motion and is called "sinusoidal," because it has the shape of a trigonometric sine or cosine function.

When an object attached to a horizontal spring is moved from its equilibrium position and released, the restoring force $F = -kx$ leads to simple harmonic motion. The restoring force also leads to simple harmonic motion when the object is attached to a vertical spring. When the spring is vertical, however, the weight of the object causes the spring to stretch, and the motion occurs with respect to the equi-

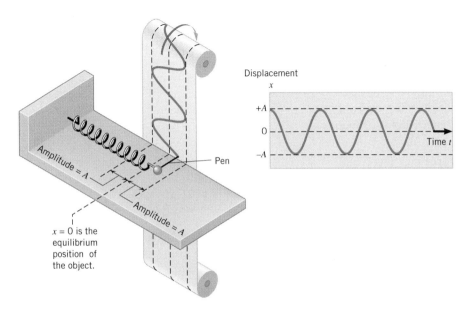

Figure 10.14 When an object moves in simple harmonic motion, a graph of its position as a function of time has a sinusoidal shape with an amplitude A. A pen attached to the object records the graph.

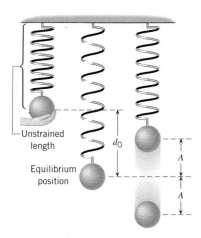

Figure 10.15 The weight of an object on a vertical spring stretches the spring by an amount d_0. Simple harmonic motion of amplitude A occurs with respect to the equilibrium position of the object on the stretched spring.

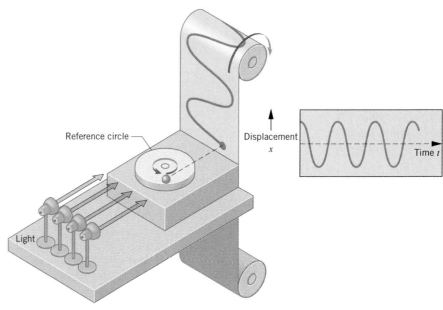

Figure 10.16 The ball mounted on the turntable moves in uniform circular motion, and its shadow, projected on a moving strip of film, executes simple harmonic motion.

librium position of the object on the stretched spring, as Figure 10.15 indicates. The amount of initial stretching d_0 due to the weight can be calculated by equating the weight to the magnitude of the restoring force that supports it; thus, $mg = kd_0$, which gives $d_0 = mg/k$.

10.4 SIMPLE HARMONIC MOTION AND THE REFERENCE CIRCLE

Simple harmonic motion, like any motion, can be described in terms of displacement, velocity, and acceleration, and the model in Figure 10.16 is helpful in explaining these characteristics. The model consists of a small ball attached to the top of a rotating turntable. The ball is moving in uniform circular motion (see Chapter 5) on a circle known as the *reference circle.* As the ball moves, its shadow falls on a strip of film, which is moving upward at a steady rate and records where the shadow is. A comparison of the film with the paper in Figure 10.14 reveals the same kind of patterns, suggesting that the shadow of the ball is a good model for simple harmonic motion.

DISPLACEMENT

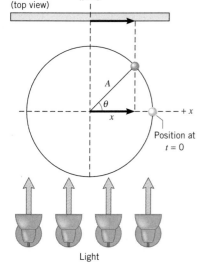

Figure 10.17 The ball's shadow on the film has a displacement x that depends on the angle θ through which the ball has moved on the reference circle.

Figure 10.17 takes a closer look at the reference circle (radius = A) and indicates how to determine the displacement of the shadow on the film. The ball starts on the x axis at $x = +A$ and moves through the angle θ in a time t. Since the circular motion is uniform, the ball moves with a constant angular speed ω (in rad/s). Therefore, the angle has a value (in rad) of $\theta = \omega t$. The displacement x of the shadow is just the projection of the radius A onto the x axis:

$$x = A \cos \theta = A \cos \omega t \qquad (10.7)$$

Figure 10.18 shows a graph of this equation. As time passes, the shadow of the ball oscillates between the values of $x = +A$ and $x = -A$, corresponding to the limiting values of $+1$ and -1 for the cosine of an angle. The radius A of the reference circle, then, is the amplitude of the simple harmonic motion.

As the ball moves one revolution or cycle around the reference circle, its shadow executes one cycle of back-and-forth motion. For any object in simple harmonic motion, the time required to complete one cycle is the *period T*, as Figure 10.18 indicates. The value of T depends on the angular speed ω of the ball, because the greater the angular speed, the shorter the time it takes to complete one revolution. We can obtain the relationship between ω and T by recalling that $\omega = \theta/t$ (Equation 8.6), where θ is the angular displacement of the ball, and t is the time. For one cycle, $\theta = 2\pi$ rad and $t = T$, so that

$$\omega = \frac{2\pi}{T} \tag{10.8}$$

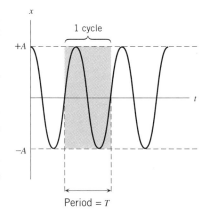

Figure 10.18 For simple harmonic motion, the graph of displacement x versus time t is a sinusoidal curve. The period T is the time required for one complete motional cycle.

Often, instead of the period, it is more convenient to speak of the *frequency f* of the motion, the frequency being just the number of cycles of the motion per second. For example, if an object on a spring completes 10 cycles in one second, the frequency is $f = 10$ cycles/s. The period T, or the time for one cycle, would be $\frac{1}{10}$ s. Thus, frequency and period are related according to

$$f = \frac{1}{T} \tag{10.9}$$

Usually one cycle per second is referred to as one hertz (Hz), the unit being named after Heinrich Hertz (1857–1894). One thousand cycles per second is denoted as one kilohertz (kHz). Thus, five thousand cycles per second, for instance, can be written as 5 kHz.

Using the relationships $\omega = 2\pi/T$ and $f = 1/T$, we can relate the angular speed ω (in rad/s) to the frequency f (in cycles/s or Hz):

$$\omega = \frac{2\pi}{T} = 2\pi f \tag{10.10}$$

Because ω is directly proportional to the frequency f, ω is often called the *angular frequency*.

VELOCITY

The reference circle model can also be used to determine the velocity of an object in simple harmonic motion. Figure 10.19 shows the tangential velocity $\mathbf{v}_T$ of the ball on the reference circle. The drawing indicates that the velocity $\mathbf{v}$ of the shadow is just the x component of the vector $\mathbf{v}_T$, that is, $v = -v_T \sin \theta$, where $\theta = \omega t$. The minus sign is necessary, since $\mathbf{v}$ points to the left, in the direction of the negative x axis. Since the tangential speed v_T is related to the angular speed ω by $v_T = r\omega$ (Equation 8.9) and since $r = A$, it follows that $v_T = A\omega$. Therefore, the velocity in simple harmonic motion is given by

$$v = -A\omega \sin \theta = -A\omega \sin \omega t \tag{10.11}$$

This velocity is *not* constant, but varies between maximum and minimum values as time passes. When the shadow changes direction at either end of the oscillatory motion, the velocity is momentarily zero. When the shadow passes through the $x = 0$

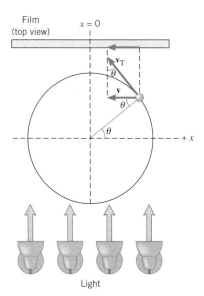

Figure 10.19 The velocity $\mathbf{v}$ of the ball's shadow is the x component of the tangential velocity $\mathbf{v}_T$ of the ball on the reference circle.

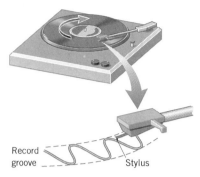

Figure 10.20 A phono stylus (monaural) generates electrical signals by vibrating in simple harmonic motion within a record groove.

▶ **The Physics of...**
a phonograph stylus.

position, the velocity has a maximum magnitude of $A\omega$, since the sine of an angle is between $+1$ and -1:

$$v_{max} = A\omega \qquad (\omega \text{ in rad/s}) \qquad (10.12)$$

Both the amplitude A and the angular frequency ω determine the maximum velocity, as Example 6 emphasizes.

EXAMPLE 6 • The Maximum Speed of a Phono Stylus

In response to a record groove, a phono stylus (monaural) generates electrical signals by vibrating back and forth in simple harmonic motion. (See Figure 10.20.) In a standard test, the frequency of the motion is $f = 1.0$ kHz and the amplitude is $A = 8.0 \times 10^{-6}$ m. What is the maximum speed of the stylus?

Reasoning and Solution The maximum speed v_{max} of the vibrating stylus can be obtained from Equation 10.12 and the fact that $\omega = 2\pi f$:

$$v_{max} = A\omega = (8.0 \times 10^{-6} \text{ m})[2\pi(1.0 \times 10^3 \text{ Hz})] = \boxed{0.050 \text{ m/s}}$$

Simple harmonic motion is not just any kind of vibratory motion. It is a very specific kind and, among other things, must have the velocity given by Equation 10.11. For instance, advertising signs often use a "moving light" display to grab your attention. Conceptual Example 7 examines the back-and-forth motion in one such display, to see if it is simple harmonic motion.

CONCEPTUAL EXAMPLE 7 • Moving Lights

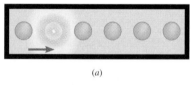

(a)

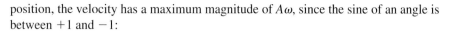

(b)

Figure 10.21 The motion of a lighted bulb is from (a) left to right and then from (b) right to left.

Over the entrance to a restaurant is mounted a strip of equally spaced light bulbs, as Figure 10.21a illustrates. Starting at the left end, each bulb turns on in sequence for one-half second. Thus, a lighted bulb appears to move from left to right. Once the apparent motion of a lighted bulb reaches the right side of the sign, the motion reverses. The lighted bulb then appears to move to the left, as part b of the drawing indicates. Thus, the lighted bulb appears to oscillate back and forth. Is the apparent motion simple harmonic motion?

Reasoning and Solution Since the bulbs are equally spaced and each bulb remains lit for the same amount of time, the apparent motion of a lighted bulb across the sign occurs at a constant speed. If the motion were simple harmonic motion, however, it would not have a constant speed (see Equation 10.11), for it would have zero speed at each end of the sign and increase to a maximum speed at the center of the sign. Therefore, although the apparent motion of a lighted bulb is oscillatory, *it is not simple harmonic motion.*

ACCELERATION

In simple harmonic motion, the velocity is not constant; consequently, there must be an acceleration. This acceleration can also be determined with the aid of the reference circle model. As Figure 10.22 shows, the ball on the reference circle moves in uniform circular motion, and, therefore, has a centripetal acceleration $\mathbf{a}_c$ that points toward the center of the circle. The acceleration $\mathbf{a}$ of the shadow is the x

component of the centripetal acceleration; $a = -a_c \cos \theta$. The minus sign is needed because the acceleration of the shadow points to the left. Recalling that the centripetal acceleration is related to the angular speed ω by $a_c = r\omega^2$ (Equation 8.11) and using $r = A$, we find that $a_c = A\omega^2$. With this substitution, the acceleration in simple harmonic motion becomes

$$a = -A\omega^2 \cos \theta = -A\omega^2 \cos \omega t \qquad (10.13)$$

The acceleration, like the velocity, does *not* have a constant value as time passes. The maximum magnitude of the acceleration is

$$a_{max} = A\omega^2 \qquad (\omega \text{ in rad/s}) \qquad (10.14)$$

Although both the amplitude A and the angular frequency ω determine the maximum value, the frequency has a particularly strong effect, for it is squared. Example 8 shows that the acceleration can be remarkably large in a practical situation.

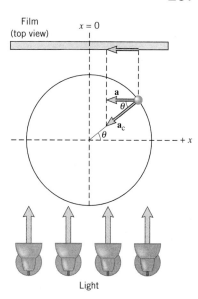

Figure 10.22 The acceleration **a** of the ball's shadow is the x component of the centripetal acceleration a_c of the ball on the reference circle.

EXAMPLE 8 • The Maximum Acceleration of a Loudspeaker Diaphragm

The diaphragm of a loudspeaker moves back and forth in simple harmonic motion to create sound, as in Figure 10.23. The frequency of the motion is $f = 1.0$ kHz, and the amplitude is $A = 2.0 \times 10^{-4}$ m. Find the maximum acceleration of the diaphragm.

Reasoning and Solution Using Equation 10.14 and the fact that $\omega = 2\pi f$, we find that

$$a_{max} = A\omega^2 = (2.0 \times 10^{-4} \text{ m})[2\pi(1.0 \times 10^3 \text{ Hz})]^2 = \boxed{7.9 \times 10^3 \text{ m/s}^2}$$

This acceleration is more than 800 times that due to gravity and is much larger than the acceleration experienced by astronauts during a rocket launch. The construction of the diaphragm must be sturdy enough to withstand such large accelerations.

The Physics of...
a loudspeaker diaphragm.

FREQUENCY OF VIBRATION

With the aid of Newton's second law ($\Sigma F = ma$), it is possible to determine the frequency at which an object of mass m vibrates on a spring. We assume that the mass of the spring itself is negligible and that the only force acting on the object in the horizontal direction is due to the spring, i.e., the Hooke's law restoring force. Thus, the net force is $\Sigma F = -kx$, and Newton's second law becomes $-kx = ma$, where a is the acceleration of the object. Using Equation 10.7 for the displacement x and Equation 10.13 for the acceleration a, we find that

$$-k(A \cos \omega t) = m(-A\omega^2 \cos \omega t)$$

which yields

$$\omega = \sqrt{\frac{k}{m}} \qquad (\omega \text{ in rad/s}) \qquad (10.15)$$

In this expression, the angular frequency ω must be in radians per second. Larger spring constants k and smaller masses m result in larger frequencies. Example 9 illustrates the effect of the mass on the frequency.

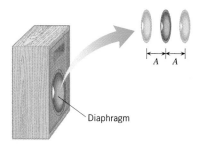

Figure 10.23 The diaphragm of a loudspeaker generates a 1.0-kHz sound by moving back and forth in simple harmonic motion.

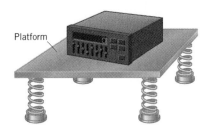

Platform

Figure 10.24 A piece of electronic equipment on a vibrating test platform.

The Physics of...
detecting and measuring small amounts of chemicals.

The Physics of...
a door-closing unit.

EXAMPLE 9 • A Vibrating Test Platform

Electronic equipment for high-performance aircraft like the Space Shuttle must be able to withstand large vibrations. Figure 10.24 shows a platform that can be used to test such equipment. The platform itself has a mass of 3.00 kg and is mounted on four springs, each of which has a spring constant of k = 1150 N/m. If electronic equipment is placed on the platform, find the frequency f at which the system vibrates when the electronic equipment has a mass of (a) 4.00 kg and (b) 7.00 kg.

Reasoning The vibrational frequency f is related to the angular frequency ω by $f = \omega/(2\pi)$ (Equation 10.10). The angular frequency of each spring is given by Equation 10.15 as $\omega = \sqrt{k/m}$, where m is the mass supported by each spring. Since there are four springs, each supports one-fourth of the total mass.

Solution

(a) The mass supported by each spring is $m = \frac{1}{4}(3.00 \text{ kg} + 4.00 \text{ kg}) = 1.75$ kg. The frequency of vibration is

$$f = \frac{1}{2\pi}\sqrt{\frac{k}{m}} = \frac{1}{2\pi}\sqrt{\frac{1150 \text{ N/m}}{1.75 \text{ kg}}} = \boxed{4.08 \text{ Hz}}$$

(b) The mass supported by each spring is now $m = \frac{1}{4}(3.00 \text{ kg} + 7.00 \text{ kg}) = 2.50$ kg, and the frequency is

$$f = \frac{1}{2\pi}\sqrt{\frac{k}{m}} = \frac{1}{2\pi}\sqrt{\frac{1150 \text{ N/m}}{2.50 \text{ kg}}} = \boxed{3.41 \text{ Hz}}$$

The frequency is lower here because of the greater inertia of the larger mass.

Example 9 indicates that the mass of the vibrating object influences the frequency of simple harmonic motion. Electronic sensors are being developed that take advantage of this effect in detecting and measuring small amounts of chemicals. These sensors utilize tiny quartz crystals that vibrate when an electric current passes through them. If the crystal is coated with a substance that absorbs a particular chemical, then its mass increases as the chemical is absorbed and, according to Equation 10.15, the frequency of the simple harmonic motion decreases. The change in frequency is detected electronically, and the sensor is calibrated to give the mass of the absorbed chemical as a function of the change in frequency.

10.5 ENERGY AND SIMPLE HARMONIC MOTION

ELASTIC POTENTIAL ENERGY

We saw in Chapter 6 that an object above the surface of the earth has gravitational potential energy. Therefore, when the object is allowed to fall, like the hammer of the pile driver in Figure 6.14, it can do work. A spring also has potential energy when the spring is stretched or compressed, which we refer to as ***elastic potential energy***. Because of elastic potential energy, a stretched or compressed spring can do work on an object that is attached to the spring. For instance, Figure 10.25 shows a door-closing unit that is often found on screen doors. When the door is opened, a spring inside the unit is compressed and has elastic potential energy. When the door is released, the compressed spring expands and does the work of closing the door.

To find an expression for the elastic potential energy, we will determine the work done by the spring force on an object. Figure 10.26 shows an object attached to one end of a stretched spring. When the object is released, the spring contracts and pulls the object from its initial position x_0 to its final position x_f. The work W done by a constant force is given by Equation 6.1 as $W = (F \cos \theta)s$, where F is the magnitude of the force, s is the magnitude of the displacement ($s = x_0 - x_f$), and θ is the angle between the force and the displacement. The magnitude of the spring force is not constant, however. Equation 10.6 gives the spring force as $F = -kx$, and as the spring contracts, the magnitude of this force changes from kx_0 to kx_f. In using Equation 6.1 to determine the work, we can account for the changing magnitude by using an average magnitude $\overline{F}$ in place of the constant magnitude F. Because the dependence of the spring force on x is linear, the magnitude of the average force is just one-half the sum of the initial and final values, or $\overline{F} = \frac{1}{2}(kx_0 + kx_f)$. The work W_{elastic} done by the average spring force is, then,

$$W_{\text{elastic}} = (\overline{F} \cos \theta)s = \tfrac{1}{2}(kx_0 + kx_f) \cos 0° (x_0 - x_f)$$

$$W_{\text{elastic}} = \underbrace{\tfrac{1}{2}kx_0^2}_{\substack{\text{Initial elastic} \\ \text{potential energy}}} - \underbrace{\tfrac{1}{2}kx_f^2}_{\substack{\text{Final elastic} \\ \text{potential energy}}} \quad (10.16)$$

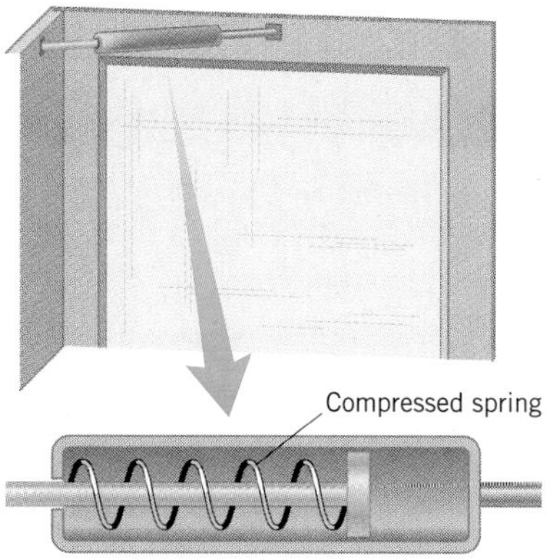

In the calculation above, θ is $0°$, since the spring force points in the same direction as the displacement. Equation 10.16 indicates that the work done by the spring force is equal to the difference between the initial and final values of the quantity $\frac{1}{2}kx^2$. The quantity $\frac{1}{2}kx^2$ is analogous to the quantity mgh, which we identified in Section 6.3 as the gravitational potential energy. Similarly, we identify the quantity $\frac{1}{2}kx^2$ as the elastic potential energy.

■ **DEFINITION OF ELASTIC POTENTIAL ENERGY**

The elastic potential energy PE_{elastic} is the energy that a spring has by virtue of being stretched or compressed. For an ideal spring that has a spring constant k and is stretched or compressed by an amount x relative to its unstrained length, the elastic potential energy is

$$PE_{\text{elastic}} = \tfrac{1}{2}kx^2 \quad (10.17)$$

SI Unit of Elastic Potential Energy: joule (J)

Equation 10.17 indicates that the elastic potential energy is a maximum for a fully stretched or compressed spring and zero for a spring that is neither stretched nor compressed ($x = 0$).

Figure 10.25 A door-closing unit. The elastic potential energy stored in the compressed spring is used to close the door.

Compressed spring

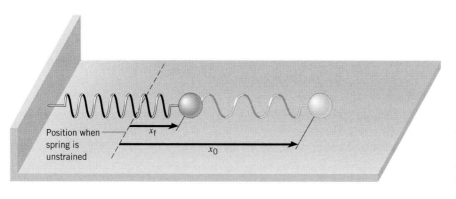

Position when spring is unstrained
x_f
x_0

Figure 10.26 When the object is released, its displacement changes from an initial value of x_0 to a final value of x_f.

Concepts at a Glance

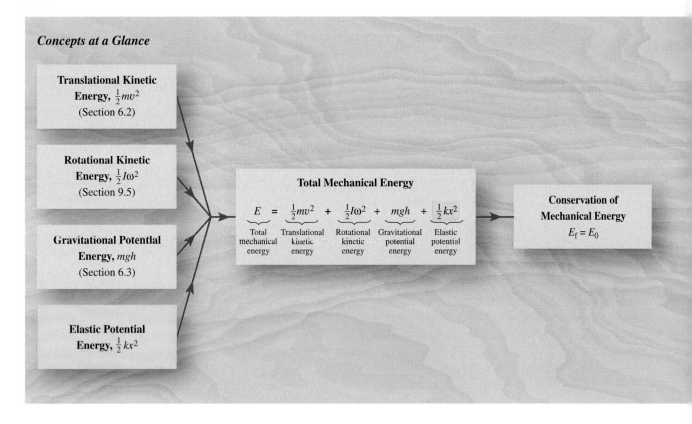

THE CONSERVATION OF MECHANICAL ENERGY

The idea of energy is a familiar one, for we have previously encountered two types: kinetic energy (including translational and rotational), and potential energy (gravitational). The total mechanical energy E is defined in Section 6.5 (see Figure 6.16) and Section 9.5 (see Figure 9.23) as the sum of these energies. We now expand this definition, as indicated by the concept chart in Figure 10.27, to include the elastic potential energy:

$$E \quad = \quad \underbrace{\tfrac{1}{2}mv^2}_{} \quad + \quad \underbrace{\tfrac{1}{2}I\omega^2}_{} \quad + \quad \underbrace{mgh}_{} \quad + \quad \underbrace{\tfrac{1}{2}kx^2}_{} \qquad (10.18)$$

| Total mechanical energy | Translational kinetic energy | Rotational kinetic energy | Gravitational potential energy | Elastic potential energy |

As Section 6.5 discusses, the total mechanical energy is conserved when external nonconservative forces (such as friction) do no net work. Then, the initial and final values of E are the same: $E_f = E_0$. The principle of conservation of total mechanical energy is the subject of the next example.

EXAMPLE 10 • An Object on a Horizontal Spring

Figure 10.28 shows the top view of an object ($m = 0.200$ kg) that is vibrating on a horizontal, frictionless table. The spring ($k = 545$ N/m) is stretched initially to $x_0 = 0.0450$ m and then released from rest (see part A of the drawing). Determine the final translational speed v_f of the object when the final displacement of the spring is (a) $x_f = 0.0225$ m and (b) $x_f = 0$.

Figure 10.27 *Concepts at a Glance*
The elastic potential energy is added to other energies to give the total mechanical energy. The total mechanical energy is conserved if the net work done by external nonconservative forces is zero. As this Bungee jumper oscillates up and down at the end of the elastic cord, the total mechanical energy remains constant, to the extent that the forces of friction and air resistance are negligible.

Reasoning The conservation of mechanical energy indicates that, in the absence of friction (a nonconservative force), the final and initial total mechanical energies are the same (see Figure 10.27):

$$E_f = E_0$$

$$\tfrac{1}{2}mv_f^2 + \tfrac{1}{2}I\omega_f^2 + mgh_f + \tfrac{1}{2}kx_f^2 = \tfrac{1}{2}mv_0^2 + \tfrac{1}{2}I\omega_0^2 + mgh_0 + \tfrac{1}{2}kx_0^2$$

Since the object is moving on a horizontal table, the final and initial heights are the same: $h_f = h_0$. The object is not rotating, so its angular speed is zero: $\omega_f = \omega_0 = 0$. And, as the problem states, the initial translational speed of the object is zero, $v_0 = 0$. With these substitutions, the equation above becomes

$$\tfrac{1}{2}mv_f^2 + \tfrac{1}{2}kx_f^2 = \tfrac{1}{2}kx_0^2$$

from which we can obtain v_f:

$$v_f = \sqrt{\frac{k}{m}\,(x_0^2 - x_f^2)}$$

Solution

(a) Since $x_0 = 0.0450$ m and $x_f = 0.0225$ m, the final translational speed is

$$v_f = \sqrt{\frac{545 \text{ N/m}}{0.200 \text{ kg}}\,[(0.0450 \text{ m})^2 - (0.0225 \text{ m})^2]} = \boxed{2.03 \text{ m/s}}$$

The total mechanical energy at this point is partly translational kinetic energy ($\tfrac{1}{2}mv_f^2 = 0.414$ J) and partly elastic potential energy ($\tfrac{1}{2}kx_f^2 = 0.138$ J). The total mechanical energy E is the sum of these two energies: $E = 0.414$ J $+ 0.138$ J $= 0.552$ J. Because the total mechanical energy remains constant during the motion, this value equals the initial total mechanical energy when the object is stationary and the energy is entirely elastic potential energy ($E_0 = \tfrac{1}{2}kx_0^2 = 0.552$ J).

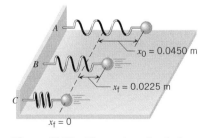

Figure 10.28 The total mechanical energy of this system is entirely elastic potential energy (*A*), partly elastic potential energy and partly kinetic energy (*B*), and entirely kinetic energy (*C*).

(b) When $x_0 = 0.0450$ m and $x_f = 0$, we have

$$v_f = \sqrt{\frac{k}{m}(x_0{}^2 - x_f{}^2)} = \sqrt{\frac{545 \text{ N/m}}{0.200 \text{ kg}}(0.0450 \text{ m})^2} = \boxed{2.35 \text{ m/s}}$$

Now the total mechanical energy is due entirely to the translational kinetic energy ($\frac{1}{2}mv_f{}^2 = 0.552$ J), since the elastic potential energy is zero. However, the total mechanical energy is the same as it is in part (a). In the absence of friction, the simple harmonic motion of a spring converts the different types of energy between one form and another, the total always remaining the same.

Conceptual Example 11 takes advantage of energy conservation to illustrate what happens to the maximum speed, amplitude, and angular frequency of a simple harmonic oscillator when its mass is changed suddenly at a certain point in the motion.

CONCEPTUAL EXAMPLE 11 • Changing the Mass of a Simple Harmonic Oscillator

Figure 10.29a shows a box of mass m attached to a spring that has a force constant k. The box rests on a horizontal, frictionless surface. The spring is initially stretched to $x = A$ and then released from rest. The box then executes simple harmonic motion that is characterized by a maximum speed v_{max}, an amplitude A, and an angular frequency ω. When the box is passing through the point where the spring is unstrained ($x = 0$), a second box of the same mass m and speed v_{max} is attached to it, as in part b of the drawing. Discuss what happens to (a) the maximum speed, (b) the amplitude, and (c) the angular frequency of the subsequent simple harmonic motion.

Reasoning and Solution

(a) The maximum speed of an object in simple harmonic motion occurs when the object is passing through the point where the spring is unstrained ($x = 0$), as in Figure 10.29b. Since the second box is attached at this point with the same speed, *the maximum speed of the two-box system remains the same as that of the one-box system.*

(b) At the same speed, the maximum kinetic energy of the two boxes is twice that of a single box, since the mass is twice as much. Subsequently, when the two boxes move to the left and compress the spring, their kinetic energy is converted into elastic potential energy. Since the two boxes have twice as much kinetic energy as one box alone, the two will have twice as much elastic potential energy when they come to a halt at the extreme left. Here, we are using the principle of conservation of me-

Figure 10.29 (*a*) A box of mass m, starting from rest at $x = A$, undergoes simple harmonic motion about $x = 0$. (*b*) When $x = 0$, a second box, with the same mass and speed, is attached.

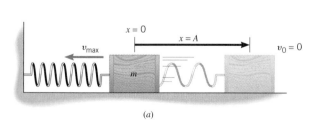

(a)

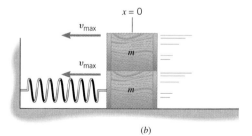

(b)

chanical energy, which applies since friction is absent. But the elastic potential energy is proportional to the amplitude squared (A^2) of the motion, so ***the amplitude of the two-box system is greater than that of the one-box system by a factor of*** $\sqrt{2}$.

(c) The angular frequency ω of a simple harmonic oscillator is given by Equation 10.15 as $\omega = \sqrt{k/m}$. Since the mass of the two-box system is twice that of the one-box system, ***the angular frequency of the two-box system is smaller than that of the one-box system by a factor of*** $\sqrt{2}$.

Related Homework Material: *Question 11, Problem 56*

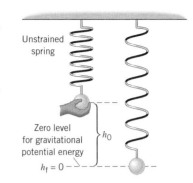

Figure 10.30 The ball is supported initially so that the spring is unstrained. After being released from rest, the ball falls through the distance h_0 before being momentarily stopped by the spring.

In the previous two examples, gravitational potential energy plays no role because the spring is horizontal. The next example illustrates that gravitational potential energy must be taken into account when a spring is oriented vertically.

EXAMPLE 12 • A Falling Ball on a Vertical Spring

A 0.20-kg ball is attached to a vertical spring, as in Figure 10.30. The spring constant of the spring is 28 N/m. The ball, supported initially so that the spring is neither stretched nor compressed, is released from rest. In the absence of air resistance, how far does the ball fall before being brought to a momentary stop by the spring?

Reasoning Since air resistance is absent, only the conservative forces of gravity and the spring act on the ball. Therefore, the principle of conservation of mechanical energy applies:

$$E_f = E_0$$
$$\tfrac{1}{2}mv_f^2 + \tfrac{1}{2}I\omega_f^2 + mgh_f + \tfrac{1}{2}kx_f^2 = \tfrac{1}{2}mv_0^2 + \tfrac{1}{2}I\omega_0^2 + mgh_0 + \tfrac{1}{2}kx_0^2$$

The problem states that the final and initial translational speeds of the ball are zero; $v_f = v_0 = 0$. The ball and spring do not rotate, so the final and initial angular speeds are also zero: $\omega_f = \omega_0 = 0$. As Figure 10.30 indicates, the initial height of the ball is h_0, while the final height is $h_f = 0$. In addition, the spring is unstrained ($x_0 = 0$) to begin with, and so it has no elastic potential energy initially. With these substitutions, the conservation of mechanical energy equation reduces to

$$\tfrac{1}{2}kx_f^2 = mgh_0$$

This result shows that the initial gravitational potential energy (mgh_0) is converted into elastic potential energy ($\tfrac{1}{2}kx_f^2$). When the ball falls to its lowest point, its displacement is $x_f = -h_0$, where the minus sign indicates that the displacement is downward. Substituting this result into the equation above and solving for h_0 yields $h_0 = 2mg/k$.

Solution The distance that the ball falls before coming to a momentary halt is

$$h_0 = \frac{2mg}{k} = \frac{2(0.20 \text{ kg})(9.8 \text{ m/s}^2)}{28 \text{ N/m}} = \boxed{0.14 \text{ m}}$$

• **PROBLEM SOLVING INSIGHT**
When evaluating the total mechanical energy E, always include a potential energy term for every conservative force acting on the system. In Example 12, there are two such terms, gravitational and elastic.

10.6 THE PENDULUM

As Figure 10.31 shows, a ***simple pendulum*** consists of a particle of mass m, attached to a frictionless pivot P by a cable of length L, whose mass is negligible. When the particle is pulled away from its equilibrium position by an angle θ and re-

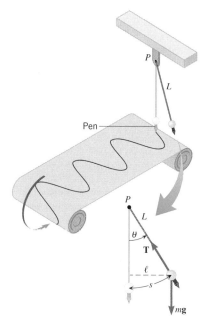

Figure 10.31 A simple pendulum swinging back and forth about the pivot P. If the angle θ is small, the swinging is approximately simple harmonic motion.

leased, the particle swings back and forth. By attaching a pen to the bottom of the swinging particle and moving a strip of paper beneath it at a steady rate, we can record the position of the particle as time passes. The graphical record reveals a pattern that is similar (but not identical) to the sinusoidal pattern for simple harmonic motion.

The force of gravity is responsible for the back-and-forth rotation about the axis at *P*. The rotation speeds up as the particle approaches the lowest point on the arc and slows down on the upward part of the swing. Eventually the angular speed is reduced to zero, and the particle swings back. As Section 9.4 discusses, a net torque is required to change the angular speed. The gravitational force *m***g** produces this torque. (The tension **T** in the cable creates no torque, because it points directly at the pivot *P* and, therefore, has a zero lever arm.) According to Equation 9.1, the magnitude of the torque τ is the product of the magnitude *mg* of the gravitational force and the lever arm ℓ, so that $\tau = -(mg)\ell$. The minus sign is included since the torque is a restoring torque: that is, it acts to reduce the angle θ (the angle θ is positive [counterclockwise], while the torque is negative [clockwise]). The lever arm ℓ is the perpendicular distance between the line of action of *m***g** and the pivot *P*. From Figure 10.31 it can be seen that ℓ is very nearly equal to the arc length *s* of the circular path when the angle θ is small (about 10° or less). Furthermore, if θ is expressed in radians, the arc length and the radius *L* of the circular path are related, according to $s = L\theta$. Under these conditions, it follows that $\ell \approx L\theta$, and the torque created by gravity is

$$\tau \approx \underbrace{-mgL}_{k'}\,\theta$$

In the equation above, the term *mgL* has a constant value k', independent of θ. *For small angles,* then, the torque that restores the pendulum to its vertical equilibrium position is proportional to the angular displacement θ. The expression $\tau = -k'\theta$ has the same form as the Hooke's law restoring force for a particle on a spring, $F = -kx$. Therefore, we expect the frequency of the back-and-forth movement of the pendulum to be given by an equation analogous to Equation 10.15 ($\omega = 2\pi f = \sqrt{k/m}$). In place of the spring constant *k*, the constant $k' = mgL$ will appear, and, as usual in rotational motion, in place of the mass *m*, the moment of inertia *I* will appear:

$$\omega = 2\pi f = \sqrt{\frac{mgL}{I}} \qquad \text{(small angles only)} \qquad (10.19)$$

The moment of inertia of a particle of mass *m*, rotating at a radius $r = L$ about an axis, is given by $I = mL^2$ (Equation 9.6). Substituting this expression for *I* into Equation 10.19 reveals for a simple pendulum that

Simple pendulum $\qquad \omega = 2\pi f = \sqrt{\dfrac{g}{L}} \qquad \text{(small angles only)} \qquad (10.20)$

The mass of the particle has been eliminated algebraically from this expression, so only the length *L* and the acceleration *g* due to gravity determine the frequency of a simple pendulum. If the angle of oscillation is large, the pendulum does not exhibit simple harmonic motion, and Equation 10.20 does not apply. Equation 10.20 provides the basis for using a pendulum to keep time, as Example 13 demonstrates.

EXAMPLE 13 • Keeping Time

Determine the length of a simple pendulum that will swing back and forth in simple harmonic motion with a period of 1.00 s.

Reasoning and Solution Equation 10.20 can be applied to find the length, if we remember that the relationship between frequency f and period T is $f = 1/T$:

$$2\pi f = \frac{2\pi}{T} = \sqrt{\frac{g}{L}}$$

$$L = \frac{T^2 g}{4\pi^2} = \frac{(1.00 \text{ s})^2(9.80 \text{ m/s}^2)}{4\pi^2} = \boxed{0.248 \text{ m}}$$

Figure 10.32 shows a clock that uses a pendulum to keep time.

It is not necessary that the object in Figure 10.31 be a simple particle. It may be an extended object, in which case the pendulum is called a ***physical pendulum.*** For small oscillations, Equation 10.19 still applies, but the moment of inertia I is no longer mL^2. The proper value for the rigid object must be used. (See Section 9.4 for a discussion of moment of inertia.) In addition, the length L for a physical pendulum is the distance between the axis at P and the center of gravity of the object.

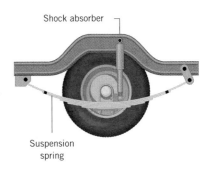

Figure 10.32 This pendulum clock keeps time as the pendulum swings back and forth.

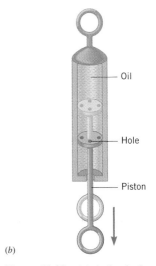

Figure 10.33 (*a*) A shock absorber mounted in the suspension system of an automobile and (*b*) a simplified, cutaway view of the shock absorber.

10.7 DAMPED HARMONIC MOTION

In simple harmonic motion, an object oscillates with a constant amplitude, because there is no mechanism for dissipating energy. In reality, however, friction or some other energy-dissipating mechanism is always present. In the presence of energy dissipation, the amplitude of oscillation decreases as time passes, and the motion is no longer simple harmonic motion. Instead, it is referred to as ***damped harmonic motion,*** the decrease in amplitude being called "damping."

One widely used application of damped harmonic motion is in the suspension system of an automobile. Figure 10.33*a* shows a shock absorber attached to a main suspension spring of a car. A shock absorber is designed to introduce damping forces, which reduce the vibrations associated with a bumpy ride. As part *b* of the drawing shows, a shock absorber consists of a piston in a reservoir of oil. When the piston moves in response to a bump in the road, holes in the piston head permit the piston to pass through the oil. Viscous forces that arise during this movement cause the damping.

Figure 10.34 illustrates the different degrees of damping that can exist. As applied to the example of a car's suspension system, these graphs show the vertical position of the chassis after it has been pulled upward by an amount A_0 at time $t_0 = 0$ and then released. Part *a* of the figure compares undamped or simple harmonic motion in curve 1 (red) to slightly damped motion in curve 2 (green). In damped harmonic motion, the chassis oscillates with decreasing amplitude until it eventually comes to rest. As the degree of damping is increased from curve 2 to curve 3 (gold), the car makes fewer oscillations before coming to a halt. Part *b* of the drawing shows that as the degree of damping is increased still further, there comes a point when the car does not oscillate at all after it is released but, rather, settles directly back to its equilibrium position, as in curve 4 (blue). The smallest degree of

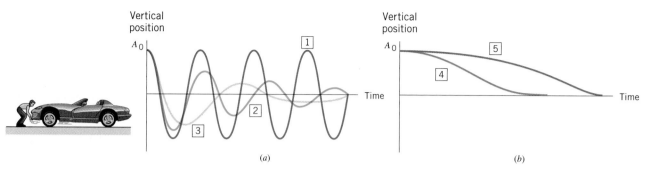

(a)   (b)

Figure 10.34 Damped harmonic motion. The degree of damping increases from curve 1 to curve 5. Curve 1 represents undamped or simple harmonic motion. Curves 2 and 3 show underdamped motion. Curve 4 represents critically damped harmonic motion. Curve 5 illustrates overdamped motion.

damping that completely eliminates the oscillations is termed "critical damping," and the motion is said to be *critically damped.*

Figure 10.34*b* also shows that the car takes the longest time to return to its equilibrium position in curve 5 (purple), where the degree of damping is increased above the value for critical damping. When the damping exceeds the critical value, the motion is said to be *overdamped.* In contrast, when the damping is less than the critical level, the motion is said to be *underdamped* (curves 2 and 3). Typical automobile shock absorbers are designed to produce underdamped motion somewhat like that in curve 3.

The Physics of...
a shock absorber.

10.8 DRIVEN HARMONIC MOTION AND RESONANCE

In damped harmonic motion, a mechanism such as friction dissipates the energy of an oscillating system, with the result that the amplitude of the motion decreases in time. This section discusses the opposite effect, namely, the increase in amplitude that results when energy is continually added to an oscillating system.

To set an object on an ideal spring into simple harmonic motion, some agent must apply a force that stretches or compresses the spring initially. Suppose that this force is applied at all times, not just for a brief initial moment. The force could be provided, for example, by a person who simply pushes and pulls the object back and forth. The resulting motion is known as ***driven harmonic motion,*** because the additional force drives or controls the behavior of the object to a large extent. The additional force is identified as the ***driving force.***

Figure 10.35 illustrates one particularly important example of driven harmonic motion. Here, the driving force has the same frequency as the spring system and always points in the direction of the object's velocity. The frequency of the spring system is $f = (1/2\pi)\sqrt{k/m}$ and is called a natural frequency, because it is the frequency at which the spring system naturally oscillates. Since the driving force and the velocity always have the same direction, positive work is done on the object at all times, and the total mechanical energy of the system increases. As a result, the amplitude of the vibration becomes larger and will increase without limit, if there is no damping force to dissipate the energy being added by the driving force. The situation depicted in Figure 10.35 is known as ***resonance.***

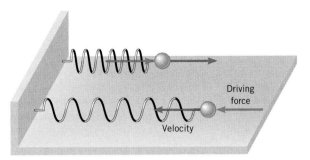

Figure 10.35 Resonance occurs when the frequency of the driving force matches a frequency at which the object naturally vibrates. The blue arrows represent the driving force, and the red arrows represent the velocity of the object.

■ **RESONANCE**

Resonance is the condition in which a time dependent force can transmit large amounts of energy to an oscillating object, leading to a large amplitude motion. In the absence of damping, resonance occurs when the frequency of the force matches a natural frequency at which the object will oscillate.

The role played by the frequency of a driving force is a critical one. The matching of this frequency with a natural frequency of vibration allows even a relatively weak force to produce a large amplitude vibration, because the effect of each push–pull cycle is cumulative.

Resonance can occur with any object that can oscillate, and springs need not be involved. The greatest tides in the world occur in the Bay of Fundy, which lies between the Canadian provinces of New Brunswick and Nova Scotia. Figure 10.36 shows the enormous difference between the water level at high and low tides, a difference that in some locations averages about 15 m. This phenomenon is partly due to resonance. The time, or period, that it takes for the tide to flow into and ebb out of a bay depends on the size of the bay, the topology of the bottom, and the configuration of the shoreline. The ebb and flow of the water in the Bay of Fundy has a period of 12.5 hours, which is very close to the lunar tidal period of 12.42 hours. Thus, each incoming ocean tide arrives just as the ebb of the previous tide in the bay is finished. The ocean tide then "drives" water into and out of the Bay of Fundy at a frequency that nearly matches the natural frequency of the bay. The result is the extraordinary high tide in the bay. (You can create a similar effect in a bathtub full of water by moving back and forth in synchronism with the waves you're causing.)

▶ **The Physics of...**
high tides at the Bay of Fundy.

Figure 10.36 The Bay of Fundy at (*a*) high tide and (*b*) low tide. In some places the water level changes by almost 15 m.

(*a*)

(*b*)

SUMMARY

The possible kinds of **elastic deformation** include stretch and compression, shear deformation, and volume deformation. The forces required to create them are given by Equations 10.1, 10.2, and 10.4 and are characterized with the aid of proportionality constants called, respectively, **Young's modulus,** the **shear modulus,** and the **bulk modulus.**

The **pressure** P is the magnitude F of the force acting perpendicular to a surface divided by the area A over which the force acts: $P = F/A$. The SI unit of pressure is N/m^2, a unit known as a pascal (Pa); $1\ Pa = 1\ N/m^2$.

Stress is the force per unit area applied to an object and causes **strain.** For stretch/compression and volume deformation, strain is the resulting fractional change in length or volume. For shear deformation, strain reflects the change in shape of an object. **Hooke's law** states that stress is directly proportional to strain, up to a limit called the **proportionality limit.**

The force F that must be applied to stretch or compress an **ideal spring** is $F = kx$, where k is the spring constant and x is the displacement of the spring from its unstrained length. A spring exerts a **restoring force** on an object attached to the spring. The restoring force produced by an ideal spring is $F = -kx$, where the minus sign indicates that the restoring force points opposite to the displacement of the spring.

Simple harmonic motion is the oscillatory motion that occurs when a restoring force of the form $F = -kx$ acts on an object. A graphical record of position versus time for an object in simple harmonic motion is sinusoidal. The **amplitude** A of the motion is the maximum distance that the object moves away from its equilibrium position. The **period** T is the time required to complete one cycle of the motion, while the **frequency** f is the number of cycles per second that occur. Frequency and period are related according to $f = 1/T$. The frequency f (in Hz) is related to the angular frequency ω (in rad/s) according to $\omega = 2\pi f$. For an object of mass m on a spring with spring constant k, the frequency

is determined by $2\pi f = \sqrt{k/m}$. The velocity and the acceleration in simple harmonic motion are continually changing with time; the maximum speed is $v_{max} = A\omega$, and the maximum acceleration is $a_{max} = A\omega^2$.

The **elastic potential energy** of an object attached to an ideal spring is $PE_{elastic} = \frac{1}{2}kx^2$. The total mechanical energy E of such a system is the sum of its translational and rotational kinetic energies, gravitational potential energy, and elastic potential energy; $E = \frac{1}{2}mv^2 + \frac{1}{2}I\omega^2 + mgh + \frac{1}{2}kx^2$. If external nonconservative forces like friction do no net work, the total mechanical energy of the system is conserved, $E_f = E_0$.

A **simple pendulum** is a particle of mass m attached to a frictionless pivot by a cable whose length is L and whose mass is negligible. The small-angle ($\leq 10°$) back-and-forth swinging of a simple pendulum is simple harmonic motion, while large-angle motion is not. The frequency of small-angle motion is given by $2\pi f = \sqrt{g/L}$. A **physical pendulum** consists of a rigid object, with moment of inertia I and mass m, suspended from a frictionless pivot. For small-angle displacements, the frequency of simple harmonic motion for a physical pendulum is determined by $2\pi f = \sqrt{mgL/I}$, where L is the distance between the axis of rotation and the center of gravity of the rigid object.

Damped harmonic motion is motion in which the amplitude of oscillation decreases as time passes. **Critical damping** is the minimum degree of damping that eliminates any oscillations in the motion as the object returns to its equilibrium position.

Driven harmonic motion occurs when an additional driving force is applied to an object along with the restoring force. **Resonance** is the condition under which a driving force can transmit large amounts of energy to an oscillating object, leading to large amplitude motion. In the absence of damping, resonance occurs when the frequency of the driving force matches a natural frequency at which the object oscillates.

CONCEPTUAL QUESTIONS

1. The drawing shows two cylinders. They are identical in all respects, except one is hollow. In a setup like that in Figure 10.3, identical forces are applied to the right end of each cylinder. Which cylinder, if either, stretches the most? Why?

2. Young's modulus for steel is thousands of times greater than

that for rubber. All other factors being equal, does this mean that steel stretches much more easily than rubber? Explain.

3. A trash compactor crushes empty aluminum cans, thereby reducing the total volume of the cans by 75%. Can the value given in Table 10.3 for the bulk modulus of aluminum be used to calculate the change in pressure generated in the trash compactor? Give a reason for your answer.

4. Both sides of the relation $F = S(\Delta X/L_0)A$ (Equation 10.2) can be divided by the area A to give F/A on the left side. Why can't this F/A term be called a pressure, such as the pressure that appears in $\Delta P = -B(\Delta V/V_0)$ (Equation 10.4)?

5. The block in the drawing rests on the ground. Which face, A, B, or C, experiences (a) the largest stress and (b) the smallest stress when the block is resting on it? Explain.

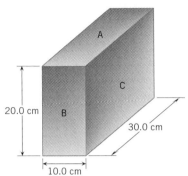

6. Two people pull on a horizontal spring that is attached to an immovable wall. Then, they detach it from the wall and pull on opposite ends of the horizontal spring. They pull just as hard in each case. In which situation, if either, does the spring stretch more? Account for your answer.

7. The drawing shows identical springs that are attached to a box in two different ways. Initially, the springs are unstrained. The box is then pulled to the right and released. In each case the initial displacement of the box is the same. At the moment of release, which box, if either, experiences the greater net force due to the springs? Provide a reason for your answer.

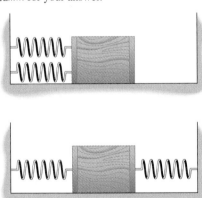

8. In Figures 10.19 and 10.22, the shadow moves in simple harmonic motion. Where on the shadow's path is (a) the velocity equal to zero and (b) the acceleration equal to zero?

9. A steel ball is dropped onto a concrete floor. Over and over again, it rebounds to its original height. Is the motion simple harmonic motion? Justify your answer.

10. Ignoring the damping introduced by the shock absorbers, explain why the number of passengers in a car affects the vibration frequency of the car's suspension system.

11. Review Conceptual Example 11 before answering this question. A block is attached to a horizontal spring and slides back and forth in simple harmonic motion on a frictionless horizontal sur-

face. A second identical block is suddenly attached to the first block. The attachment is accomplished by joining the blocks at one extreme end of the oscillation cycle. The velocities of the blocks are exactly matched at the instant of joining. Explain how (a) the amplitude, (b) the frequency, and (c) the maximum speed of the oscillation change.

12. A particle is oscillating in simple harmonic motion. The time required for the particle to travel through one complete cycle is equal to the period of the motion, no matter what the amplitude is. But how can this be, since larger amplitudes mean that the particle travels farther? Explain.

13. An electric saber saw consists of a blade that is driven back and forth by a pin mounted on the circumference of a rotating circular disk. As the disk rotates at a constant angular speed, the pin engages a slot and forces the blade back and forth, in the sequence (a), (b), (c), and (d) shown in the drawing. Is the motion of the blade simple harmonic motion? Explain.

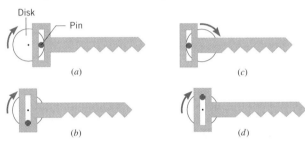

14. Is more elastic potential energy stored in a spring when the spring is compressed by one centimeter than when it is stretched by the same amount? Explain.

15. Suppose that a grandfather clock (a simple pendulum) is running slowly. That is, the time it takes to complete each cycle is longer than it should be. Should one shorten or lengthen the pendulum to make the clock keep the correct time? Why?

16. In principle, the motion of a simple pendulum and an object on an ideal spring can both be used to provide the basic time interval or period used in a clock. Which of the two kinds of clocks becomes more inaccurate when carried to the top of a high mountain? Justify your answer.

17. Suppose you were kidnapped and held prisoner by space invaders in a completely isolated room, with nothing but a watch and a pair of shoes (including shoelaces of known length). Explain how you might determine whether this room is on earth or on the moon.

18. Two people are sitting on playground swings. One person is pulled back 4° from the vertical and released, while the other is pulled back 8° from the vertical and released. If the two swings are started together, will they both come back to the starting points at the same time? Justify your answer.

19. A car travels over a road that contains a series of equally spaced bumps. Explain why a particularly jarring ride can result if the horizontal velocity of the car, the bump spacing, and the oscillation frequency of the car's suspension system are properly "matched."

PROBLEMS

Note: Unless otherwise indicated, the values for Young's modulus Y, the shear modulus S, and the bulk modulus B are given, respectively, in Table 10.1, Table 10.2, and Table 10.3.

ssm Solution is in the Student Solutions Manual. **www** Solution is available on the World Wide Web at http://www.wiley.com/college/cutnell ⚕ This icon represents a biomedical application.

Section 10.1 Elastic Deformation, Section 10.2 Stress, Strain, and Hooke's Law

1. ssm www A 3500-kg statue is placed on top of a cylindrical concrete ($Y = 2.3 \times 10^{10}$ N/m²) stand. The stand has a cross-sectional area of 7.3×10^{-2} m² and a height of 1.8 m. By how much does the statue compress the stand?

2. A 72-kg mountain climber hangs freely on a nylon rope (radius = 6.5 mm). If the rope stretches by 5.0×10^{-2} m, what is the unstretched length of the rope?

3. A rectangular solid block has the dimensions shown in the drawing for question 5 and a shear modulus of 7.00×10^9 N/m². A force of 4.2×10^4 N is applied as in Figure 10.5, with one face of the block fixed to an immovable horizontal surface. Determine the resulting shear deformation ΔX when the face fixed to the horizontal surface is (a) face A and (b) face C.

4. The drawing shows a 160-kg crate hanging from the end of a steel bar. The length of the bar is 0.10 m, and its cross-sectional area is 3.2×10^{-4} m². Neglect the weight of the bar itself and determine (a) the shear stress on the bar and (b) the vertical deflection ΔY of the right end of the bar.

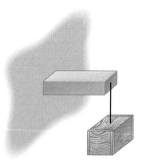

5. ssm A CD player is mounted on four cylindrical rubber blocks. Each cylinder has a height of 0.030 m and a cross-sectional area of 1.2×10^{-3} m², and the shear modulus for rubber is 2.6×10^6 N/m². If a horizontal force of magnitude 32 N is applied to the CD player, how far will the unit move sideways? Assume that each block is subjected to one-fourth of the force.

6. The pressure increases by 1.0×10^4 N/m² for every meter of depth beneath the surface of the ocean. At what depth does the volume of a Pyrex glass cube, 1.0×10^{-2} m on an edge at the ocean's surface, decrease by 1.0×10^{-10} m³?

7. An 1800-kg car, being lifted at a steady speed by a crane, hangs at the end of a cable whose radius is 6.0×10^{-3} m. The cable is 15 m in length and stretches by 8.0×10^{-3} m because of the weight of the car. Determine (a) the stress, (b) the strain, and (c) Young's modulus for the cable.

8. A copper cylinder and a brass cylinder are stacked end to end, as in the drawing. Each cylinder has a radius of 0.25 cm. A compressive force of $F = 6500$ N is applied to the right end of the brass cylinder. Find the amount by which the length of the stack decreases.

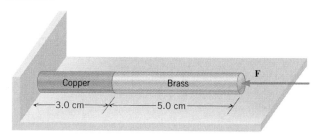

9. ssm Two metal beams are joined together by four rivets, as the drawing indicates. Each rivet has a radius of 5.0×10^{-3} m and is to be exposed to a shearing stress of no more than 5.0×10^8 Pa. What is the maximum tension **T** that can be applied to each beam, assuming that each rivet carries one-fourth of the total load?

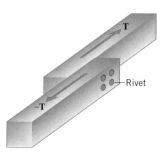

10. The shovel of a backhoe is controlled by hydraulic cylinders that are moved by oil under pressure. Determine the volume strain $\Delta V/V_0$ (including the algebraic sign) experienced by the oil, when the pressure increases from 1.8×10^5 Pa to 6.5×10^5 Pa while the shovel is digging a trench.

11. A piano wire ($Y = 2.0 \times 10^{10}$ N/m²) has a radius of 0.80 mm and a length of 0.76 m. One end of the wire is wrapped around a tuning peg whose radius is 1.8 mm. The other end is fixed in place. Initially, there is no tension in the wire. Find the tension when the tuning peg is turned through two revolutions.

12. A die is designed to punch holes of radii 1.00×10^{-2} m in a metal sheet that is 3.0×10^{-3} m thick, as the drawing illustrates. To punch through the sheet, the die must exert a shearing stress of 3.5×10^8 Pa. What force **F** must be applied to the die? *(Hint: Consider carefully which area is used in your calculation.)*

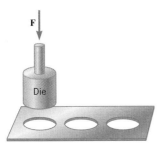

F

Die

*13. **ssm** A helicopter is lifting a 2100-kg jeep. The steel suspension cable is 48 m long and has a radius of 5.0×10^{-3} m. (a) Find the amount that the cable is stretched when the jeep is suspended motionless in the air. (b) What is the amount of cable stretch when the jeep is hoisted upward with an acceleration of 1.5 m/s²?

*14. A gymnast does a one-arm handstand. The humerus, which is the upper arm bone between the elbow and the shoulder joint, may be approximated as a 0.30-m-long cylinder with an outer radius of 1.00×10^{-2} m and a hollow inner core with a radius of 4.0×10^{-3} m. Excluding the arm, the mass of the gymnast is 63 kg. (a) What is the compressional strain of the humerus? (b) By how much is the humerus compressed?

*15. A block of copper is securely fastened to the floor. A force of 1800 N is applied to the top surface of the block, as the drawing shows. Find (a) the amount by which the height of the block is changed and (b) the shear deformation of the block.

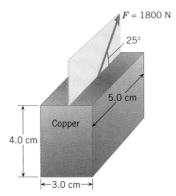

F = 1800 N

25°

5.0 cm

Copper

4.0 cm

←3.0 cm→

*16. The drawing shows one arm of a centrifuge. The arm is a solid cylinder of aluminum with a length of 0.10 m and a cross-sectional area of 1.3×10^{-4} m². At the end of the aluminum arm is a 0.58-kg specimen holder. By how much does the aluminum rod stretch when the centrifuge has an angular speed of 180 rad/s?

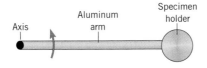

Specimen
holder

Aluminum
arm

Axis

*17. **ssm www** An 8.0-kg stone at the end of a steel wire is being whirled in a circle at a constant tangential speed of 12 m/s. The stone is moving on the surface of a frictionless horizontal

table. The wire is 4.0 m long and has a radius of 1.0×10^{-3} m. Find the strain in the wire.

*18. A 1.0×10^{-3}-kg spider is hanging vertically by a thread that has a Young's modulus of 4.5×10^{9} N/m² and a radius of 13×10^{-6} m. Suppose that a 95-kg person is hanging vertically on an aluminum wire. What is the radius of the wire that would exhibit the same strain as the spider's thread, when the thread is stressed by the full weight of the spider?

*19. A square plate is 1.0×10^{-2} m thick, measures 3.0×10^{-2} m on a side, and has a mass of 7.2×10^{-2} kg. The shear modulus of the material is 2.0×10^{10} N/m². One of the square faces rests on a flat horizontal surface, and the coefficient of static friction between the plate and the surface is 0.90. A force is applied to the top of the plate, as in Figure 10.5. Determine (a) the maximum possible amount of shear stress, (b) the maximum possible amount of shear strain, and (c) the maximum possible amount of shear deformation ΔX (see Figure 10.5) that can be created by the applied force just before the plate begins to move.

*20. Two solid cubes of the same size, one tungsten and one lead, are subjected to the same shearing force, as in the drawing that accompanies problem 78. Determine the ratio $\theta_{lead}/\theta_{tungsten}$ for the angles that characterize the resulting shear deformation, assuming that the angles are small.

21. **ssm www A solid brass sphere is subjected to a pressure of 1.0×10^{5} Pa due to the earth's atmosphere. On Venus the pressure due to the atmosphere is 9.0×10^{6} Pa. By what fraction $\Delta r/r_0$ (including the algebraic sign) does the radius of the sphere change when it is exposed to the Venusian atmosphere? Assume that the change in radius is very small relative to the initial radius.

**22. The drawing shows two crates that are connected by a steel wire that passes over a pulley. The unstretched length of the wire is 1.5 m, and its cross-sectional area is 1.3×10^{-5} m². The pulley is frictionless and massless. When the crates are accelerating, determine the change in length of the wire. Ignore the mass of the wire.

3.0 kg

5.0 kg

Section 10.3 The Ideal Spring and Simple Harmonic Motion

23. A spring has a spring constant of 248 N/m. Find the magnitude of the force needed (a) to stretch the spring by 3.00×10^{-2} m from its unstrained length and (b) to compress the spring by the same amount.

24. The graph shows the force *F* that an archer applies to the string of a long bow versus the string's displacement *x*. Drawing

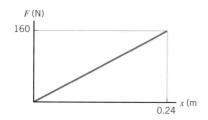

back this bow is analogous to stretching a spring. From the data in the graph determine the effective spring constant of the bow.

25. **ssm** A hand exerciser utilizes a coiled spring. A force of 89.0 N is required to compress the spring by 0.0191 m. Determine the force needed to compress the spring by 0.0508 m.

26. In an exercise apparatus, a spring stretches 0.24 m when a bodybuilder exerts a force of 410 N. When used vertically to support a 12-kg object, by how much does this spring compress?

27. In a room that is 2.44 m high, a spring (unstrained length = 0.30 m) hangs from the ceiling. From this spring a board hangs, so that its 1.98-m length is perpendicular to the floor, the lower end just extending to, but not touching, the floor. The board weighs 102 N. What is the spring constant of the spring?

28. A spring ($k = 830$ N/m) is hanging from the ceiling of an elevator, and a 5.0-kg object is attached to the lower end. By how much does the spring stretch (relative to its unstrained length) when the elevator is accelerating upward at $a = 0.60$ m/s^2?

29. **ssm** A car is hauling a 92-kg trailer, to which it is connected by a spring. The spring constant is 2300 N/m. The car accelerates with an acceleration of 0.30 m/s^2. By how much does the spring stretch?

*30. A 10.1-kg uniform board is wedged into a corner and held by a spring at a 50.0° angle, as the drawing shows. The spring has a spring constant of 176 N/m and is parallel to the floor. Find the amount by which the spring is stretched from its unstrained length.

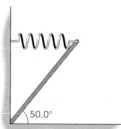

*31. To measure the static friction coefficient between a 2.3-kg block and a vertical wall, the setup shown in the drawing is used.

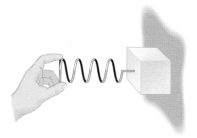

A spring (spring constant = 480 N/m) is attached to the block. Someone pushes on the end of the spring in a direction perpendicular to the wall until the block does not slip downward. If the spring in such a setup is compressed by 0.051 m, what is the coefficient of static friction?

*32. Review Conceptual Example 5 as an aid in solving this problem. An object is attached to the lower end of a 100-coil spring that is hanging from the ceiling. The spring stretches by 0.160 m. The spring is then cut into two identical springs of 50 coils each. As the drawing shows, each spring is attached between the ceiling and the object. By how much does each spring stretch?

33. **ssm A small ball is attached to one end of a spring that has an unstrained length of 0.200 m. The spring is held by the other end, and the ball is whirled around in a horizontal circle at a speed of 3.00 m/s. The spring remains nearly parallel to the ground during the motion and is observed to stretch by 0.010 m. By how much would the spring stretch if it were attached to the ceiling and the ball allowed to hang straight down, motionless?

**34. A 15.0-kg block rests on a horizontal table and is attached to one end of a massless, horizontal spring. By pulling horizontally on the other end of the spring, someone causes the block to accelerate uniformly and reach a speed of 5.00 m/s in 0.500 s. In the process, the spring is stretched by 0.200 m. The block is then pulled at a *constant speed* of 5.00 m/s, during which time the spring is stretched by only 0.0500 m. Find (a) the spring constant of the spring and (b) the coefficient of kinetic friction between the block and the table.

35. A 30.0-kg block is resting on a flat horizontal table. On top of this block is resting a 15.0-kg block, to which a horizontal spring is attached, as the drawing illustrates. The spring constant of the spring is 325 N/m. The coefficient of kinetic friction between the lower block and the table is 0.600, while the coefficient of static friction between the two blocks is 0.900. A horizontal force **F is applied to the lower block as shown. This force is increasing in such a way as to keep the blocks moving at a *constant*

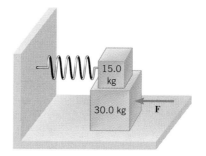

speed. At the point where the upper block begins to slip on the lower block determine (a) the amount by which the spring is compressed and (b) the magnitude of the force **F**.

Section 10.4 Simple Harmonic Motion and the Reference Circle

36. A small ball is attached to the outer edge of a stereo record that is revolving on a turntable at an angular speed of $\omega = 3.49$ rad/s. The record has a radius of 0.152 m. In a fashion similar to that in Figure 10.16, the shadow of the ball is projected onto a screen. For the simple harmonic motion of the shadow, what is the magnitude of (a) its maximum velocity and (b) its maximum acceleration?

37. **ssm** The shock absorbers in the suspension system of a car are in such bad shape that they have no effect on the behavior of the springs attached to the axles. Each of the identical springs attached to the front axle supports 320 kg. A person pushes down on the middle of the front end of the car and notices that it vibrates through five cycles in 3.0 s. Find the spring constant of either spring.

38. An 0.80-kg object is attached to one end of a spring, as in Figure 10.14, and the system is set into simple harmonic motion. The displacement x of the object as a function of time is shown in the drawing below. With the aid of this data, determine (a) the amplitude A of the motion, (b) the angular frequency ω, (c) the spring constant k, (d) the speed of the object at $t = 1.0$ s, and (e) the magnitude of the object's acceleration at $t = 1.0$ s.

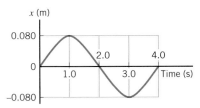

39. Refer to Conceptual Example 5 as an aid in solving this problem. A 100-coil spring has a spring constant of 420 N/m. It is cut into four shorter springs, each of which has 25 coils. One end of a 25-coil spring is attached to a wall. An object of mass 46 kg is attached to the other end of the spring, and the system is set into horizontal oscillation. What is the angular frequency of the motion?

40. A satellite circles the earth once every 5300 s in an orbit that passes over the north and south poles. The radius of the orbit is 6.5×10^6 m. The sun is shining and the satellite casts a shadow on the earth, much like the ball on the turntable in Figure 10.16 casts its shadow on the screen. What is the speed of the shadow as the satellite passes over the equator, where the suns' rays strike the earth perpendicularly?

41. **ssm www** Atoms in a solid are not stationary, but vibrate about their equilibrium positions. Typically, the frequency of vibration is about $f = 2.0 \times 10^{12}$ Hz, and the amplitude is about 1.1×10^{-11} m. For a typical atom, what is its (a) maximum speed and (b) maximum acceleration?

***42.** In Figure 10.17, the radius of the reference circle is 0.500 m.

Suppose the frequency of the simple harmonic motion of the shadow is 2.00 Hz. At time $t = 0.0500$ s calculate (a) the displacement x, (b) the magnitude of the velocity, and (c) the magnitude of the acceleration of the shadow.

***43.** Suppose that an object on a vertical spring oscillates up and down at a frequency of 5.00 Hz. By how much would this object, hanging at rest, stretch the spring?

***44.** When an object of mass m_1 is hung on a vertical spring and set into vertical simple harmonic motion, its frequency is 12.0 Hz. When another object of mass m_2 is hung on the spring along with m_1, the frequency of the motion is 4.00 Hz. Find the ratio m_2/m_1 of the masses.

***45.** **ssm** A spring stretches by 0.018 m when a 2.8-kg object is suspended from its end. How much mass should be attached to this spring so that its frequency of vibration is $f = 3.0$ Hz?

****46.** A copper rod (length = 2.0 m, radius = 3.0×10^{-3} m) hangs down from the ceiling. A 9.0-kg object is attached to the lower end of the rod. The rod acts as a "spring," and the object oscillates vertically with a small amplitude. Ignoring the rod's mass, find the frequency f of the simple harmonic motion.

Section 10.5 Energy and Simple Harmonic Motion

47. An archer pulls the bowstring back for a distance of 0.470 m before releasing the arrow. The bow and string act like a spring whose spring constant is 425 N/m. (a) What is the elastic potential energy of the drawn bow? (b) The arrow has a mass of 0.0300 kg. How fast is it traveling when it leaves the bow?

48. A horizontal spring ($k = 360$ N/m) is lying on a frictionless surface. One end of the spring is attached to a wall, and the other end is connected to an object of mass 2.8 kg. The spring is then compressed by 0.065 m and released from rest. What is the speed of the object at the instant when the spring is *stretched* by 0.048 m relative to its unstrained length?

49. **ssm** A 1.00×10^{-2}-kg block is resting on a horizontal frictionless surface and is attached to a horizontal spring whose spring constant is 124 N/m. The block is shoved parallel to the spring axis and is given an initial speed of 8.00 m/s, while the spring is initially unstrained. What is the amplitude of the resulting simple harmonic motion?

50. The spring constant for a spring in a dart gun is 1400 N/m. When the gun is cocked, the spring is compressed by 0.075 m. What is the speed of a 2.4×10^{-2}-kg dart when it leaves the gun horizontally?

51. In preparation for shooting a ball in a pinball machine, a spring ($k = 675$ N/m) is compressed by 0.0650 m relative to its unstrained length. The ball ($m = 0.0585$ kg) is at rest against the spring at point A. When the spring is released, the ball slides (without rolling) to point B, which is 0.300 m higher than point A. How fast is the ball moving at B?

52. A rifle fires a 2.10×10^{-2}-kg pellet straight upward, because the pellet rests on a compressed spring that is released when the trigger is pulled. The spring has a negligible mass and is compressed by 9.10×10^{-2} m from its unstrained length. The pellet rises to a height of 6.10 m above its position on the compressed spring. Ignoring air resistance, determine the spring constant.

53. **ssm** A 2.00-kg object is hanging from the end of a vertical spring. The spring constant is 50.0 N/m. The object is pulled 0.200 m downward and released from rest. Complete the table below by calculating the translational kinetic energy, the gravitational potential energy, the elastic potential energy, and the total mechanical energy E for each of the vertical positions indicated. The vertical positions h indicate distances above the point of release, where $h = 0$.

h (meters)	KE	PE (gravity)	PE (elastic)	E
0				
0.200				
0.400				

54. The drawing shows a block ($m = 1.7$ kg) and a spring ($k = 310$ N/m) on a frictionless incline. The spring is compressed by $x_0 = 0.31$ m relative to its unstrained position at $x = 0$ and then released. What is the speed of the block when the spring is still compressed by $x_f = 0.14$ m?

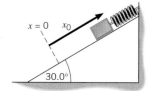
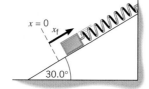

55. A heavy-duty stapling gun uses a 0.150-kg metal rod that rams against the staple to eject it. The rod is pushed by a stiff spring called a "ram spring" ($k = 34\ 000$ N/m). The mass of this spring may be ignored. Squeezing the handle of the gun first compresses the ram spring by 3.5×10^{-2} m from its unstrained length and then releases it. Assuming that the ram spring is oriented vertically and is still compressed by 1.0×10^{-2} m when the downward-moving ram hits the staple, find the speed of the ram at the instant of contact.

*56. Review Conceptual Example 11 before starting this problem. A block is attached to a horizontal spring and oscillates back and forth on a frictionless horizontal surface at a frequency of 3.00 Hz. The amplitude of the motion is 5.08×10^{-2} m. At the point where the block has its maximum speed, it suddenly splits into two identical parts, only one part remaining attached to the spring. (a) What is the amplitude and the frequency of the simple harmonic motion that exists after the block splits? (b) Repeat part (a), assuming that the block splits when it is at one of its extreme positions.

*57. **ssm** A 1.1-kg object is suspended from a vertical spring whose spring constant is 120 N/m. (a) Find the amount by which the spring is stretched from its unstrained length. (b) The object is pulled straight down by an additional distance of 0.20 m and released from rest. Find the speed with which the object passes through its original position on the way up.

*58. A 3.0-kg block is placed between two horizontal springs. Neither spring is strained when the block is located at the position labeled $x = 0$ in the drawing. The block is then displaced a dis-

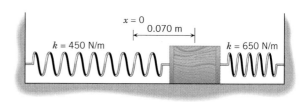

tance of 0.070 m from the position where $x = 0$ and released from rest. (a) What is the speed of the block when it passes back through the $x = 0$ position? (b) Determine the angular frequency ω of this system.

*59. A 14.6-kg block and a 29.2-kg block are resting on a horizontal frictionless surface. Between the two is squeezed a spring (spring constant = 1170 N/m). The spring is compressed by 0.152 m from its unstrained length. With what speed does each block move away when the mechanism keeping the spring squeezed is released?

**60. A 0.200-m uniform bar has a mass of 0.750 kg and is released from rest in the vertical position, as the drawing indicates. The spring is initially unstrained and has a spring constant of $k = 25.0$ N/m. Find the tangential speed with which end A strikes the horizontal surface.

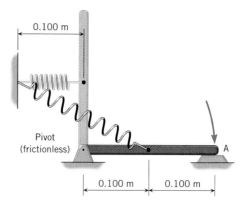

61. ssm A 70.0-kg circus performer is fired from a cannon that is elevated at an angle of 40.0° above the horizontal. The cannon uses strong elastic bands to propel the performer, much in the same way that a slingshot fires a stone. Setting up for this stunt involves stretching the bands by 3.00 m from their unstrained length. At the point where the performer flies free of the bands, his height above the floor is the same as that of the net into which he is shot. He takes 4.00 s to travel the horizontal distance of 50.0 m between this point and the net. Ignore friction and air resistance and determine the effective spring constant of the firing mechanism.

**62. A spring is mounted vertically on the floor. The mass of the spring is negligible. A certain object is placed on the spring to compress it. When the object is pushed further down by just a bit and then released, one up/down oscillation cycle occurs in 0.250 s. However, when the object is pushed down by 5.00×10^{-2} m to point P and then released, the object flies entirely off the spring. To what height above point P does the object rise in the absence of air resistance?

Section 10.6 The Pendulum

63. If the period of a simple pendulum is to be 2.0 s, what should be its length?

64. A spiral staircase winds up to the top of a tower in an old castle. To measure the height of the tower, a rope is attached to the top of the tower and hung down the center of the staircase. However, nothing is available with which to measure the length of the rope. Therefore, near the bottom of a rope a small object is attached so as to form a simple pendulum that just clears the floor. The period of the pendulum is measured to be 9.2 s. What is the height of the tower?

65. **ssm** Astronauts on a distant planet set up a simple pendulum of length 1.2 m. The pendulum executes simple harmonic motion and makes 100 complete vibrations in 280 s. What is the acceleration due to gravity?

66. Two playground swings start out together. After swing 1 has made 10 complete cycles, swing 2 has made only 8.5 cycles. What is the ratio L_1/L_2 of the lengths of the swings?

67. A wrecking ball is hanging at the end of a long cable on a crane. A student wants to estimate the length of the cable and, therefore, improvises by using a simple pendulum made from a 0.500-m length of string and a stone. The student observes that, in swinging back and forth over a small amplitude, the wrecking ball makes one complete oscillation cycle in the time it takes the stone to complete five cycles. What is the length of the cable?

*__68.__ A small object oscillates back and forth at the bottom of a frictionless hemispherical bowl, as the drawing illustrates. The radius of the bowl is R, and the angle θ is small enough that the object oscillates in simple harmonic motion. Derive an expression for the angular frequency ω of the motion. Express your answer in terms of R and g, the acceleration due to gravity.

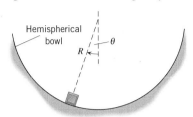

*__69.__ **ssm** Pendulum A is a physical pendulum made from a thin, rigid, and uniform rod whose length is d. One end of this rod is attached to the ceiling by a frictionless hinge, so the rod is free to swing back and forth. Pendulum B is a simple pendulum whose length is also d. Obtain the ratio T_A/T_B of their periods for small-angle oscillations.

**__70.__ A point on the surface of a solid sphere (radius = R) is attached directly to a pivot on the ceiling. The sphere swings back and forth as a physical pendulum with a small amplitude. What is the length of a simple pendulum that has the same period as this physical pendulum? Give your answer in terms of R.

ADDITIONAL PROBLEMS

71. The drawing shows how a piston in an automobile engine is attached to the crankshaft, which is rotating with an angular speed of $\omega = 126$ rad/s. If the shadow of point P could be projected onto a screen, the shadow would move in simple harmonic motion. Find (a) the amplitude, (b) the period, and (c) the maximum speed of the shadow's motion.

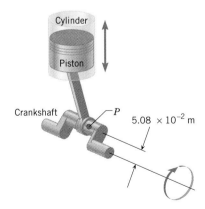

72. A pendulum clock can be approximated as a simple pendulum of length 1.00 m and keeps accurate time at a location where $g = 9.83$ m/s². In a location where $g = 9.78$ m/s², what must be the new length of the pendulum, such that the clock continues to keep accurate time (that is, its period remains the same)?

73. **ssm** The distance h_0 calculated in Example 12 is not the same as the distance d_0 (see Figure 10.15) at which the ball would hang motionless on the vertical spring. Determine d_0 to see whether it is greater or less than h_0, and explain why.

74. A 3.7-m length of aluminum cable is to be used in an attempt to pull a car out of a ditch. The maximum tension that the cable will have to sustain is 13 000 N, and, to be on the safe side, only 1.3×10^{-3} m of cable stretch will be tolerated. What is the radius of the thinnest cable that should be used?

75. A computer to be used in a satellite must be able to withstand accelerations of up to 25 times the acceleration due to gravity. In a test to see if it meets this specification, the computer is bolted to a frame that is vibrated back and forth in simple harmonic motion at a frequency of 9.5 Hz. What is the minimum amplitude of vibration that must be used in this test?

76. When the rubber band in a slingshot is stretched, it obeys Hooke's law. Suppose that the "spring constant" for the rubber band is $k = 44$ N/m. When the rubber band is pulled back with a force of 9.6 N, how far does it stretch?

77. **ssm** A piece of aluminum is surrounded by air at a pressure of 1.01×10^5 Pa. The aluminum is placed in a vacuum chamber where the pressure is reduced to zero. Determine the fractional change $\Delta V/V_0$ in the volume of the aluminum.

78. A copper cube, 0.30 m on a side, is subjected to two shearing forces, each of magnitude $F = 6.0 \times 10^6$ N (see the drawing). Find the angle θ (in degrees), which is one measure of how the shape of the block has been altered by shear deformation.

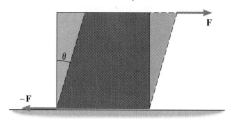

the drawing illustrates, so the ends of the wire make angles of 26° with respect to the horizontal. What is the radius of the wire?

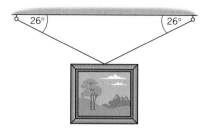

*79. A spring is compressed by 0.0800 m and is used to launch an object horizontally with a speed of 2.40 m/s. If the object were attached to the spring, at what angular frequency (in rad/s) would it oscillate?

*80. A piece of mohair from an Angora goat has a radius of 31×10^{-6} m. What is the least number of identical pieces of mohair that should be used to suspend a 75-kg person, so the strain $\Delta L/L_0$ experienced by each piece is less than 0.010? Assume that the tension is the same in all the pieces.

*81. **ssm** In 0.750 s, a 7.00-kg block is pulled through a distance of 4.00 m on a frictionless horizontal surface, starting from rest. The block has a constant acceleration and is pulled by means of a horizontal spring that is attached to the block. The spring constant of the spring is 415 N/m. By how much does the spring stretch?

*82. Suppose that F newtons of force are required to stretch a wire (circular cross section) by an amount ΔL. This wire is melted down, and a new wire (circular cross section) is formed that has half the length of the first. How much force is needed to stretch this second wire by the same amount ΔL? Express your answer in terms of F.

*83. The front spring of a car's suspension system has a spring constant of 1.50×10^6 N/m and supports a mass of 215 kg. The wheel has a radius of 0.400 m. The car is traveling on a bumpy road, and, due to resonance, the wheel starts to vibrate strongly when the car is traveling at a certain linear speed. What is this speed?

**84. A 1.00×10^{-2}-kg bullet is fired horizontally into a 2.50-kg wooden block attached to one end of a massless, horizontal spring ($k = 845$ N/m). The other end of the spring is fixed in place, and the spring is unstrained initially. The block rests on a horizontal, frictionless surface. The bullet strikes the block perpendicularly and quickly comes to a halt within it. As a result of this completely inelastic collision, the spring is compressed along its axis and causes the block/bullet to oscillate with an amplitude of 0.200 m. What is the speed of the bullet?

85. **ssm A steel wire is strung between two supports attached to a ceiling. Initially, there is no tension in the wire when it is horizontal. A 96-N picture is then hung from the center of the wire, as

**86. A tray is moved horizontally back and forth in simple harmonic motion at a frequency of $f = 2.00$ Hz. On this tray is an empty cup. Obtain the coefficient of static friction between the tray and the cup, given that the cup begins slipping when the amplitude of the motion is 5.00×10^{-2} m.

**87. A spring (spring constant = 80.0 N/m) is mounted on the floor and is oriented vertically. A 0.600-kg block is placed on top of this spring and pushed down to start it oscillating in simple harmonic motion. The block is not attached to the spring. (a) Obtain the frequency (in Hz) of the motion. (b) Determine the amplitude at which the block will lose contact with the spring.

**88. A cylindrically shaped piece of collagen (a substance found in the body in connective tissue) is being stretched by a force that increases from 0 to 3.0×10^{-2} N. The length and radius of the collagen are, respectively, 2.5 and 0.091 cm, and Young's modulus is 3.1×10^6 N/m². (a) If the stretching obeys Hooke's law, what is the spring constant k for collagen? (b) How much work is done by the variable force that stretches the collagen? (See Section 6.9 for a discussion of the work done by a variable force.)

89. **ssm The drawing shows a top view of a frictionless horizontal surface, where there are two springs with particles of mass m_1 and m_2 attached to them. Each spring has a spring constant of 120 N/m. The particles are pulled to the right and then released from the positions shown in the drawing. How much time passes before the particles are side-by-side for the first time at $x = 0$ if (a) $m_1 = m_2 = 3.0$ kg and (b) $m_1 = 3.0$ kg and $m_2 = 27$ kg?

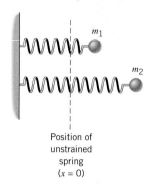

Position of unstrained spring ($x = 0$)

FLUIDS

This sea otter floats effortlessly, because the water exerts an
upward force, the "buoyant force," that balances its weight. This chapter
examines the forces and pressures that fluids, such as water and air, exert when
they are at rest and when they are in motion. Moving air, for instance,
gives rise to a force that lifts airplanes off the ground.

11.1 MASS DENSITY

Table 11.1
Mass Densities^a of Common Substances

Substance	Mass Density ρ (kg/m^3)
Solids	
Aluminum	2 700
Brass	8 470
Concrete	2 200
Copper	8 890
Diamond	3 520
Gold	19 300
Ice	917
Iron (steel)	7 860
Lead	11 300
Quartz	2 660
Silver	10 500
Wood (yellow pine)	550
Liquids	
Blood (whole, 37 °C)	1 060
Ethyl alcohol	806
Mercury	13 600
Oil (hydraulic)	800
Water (4 °C)	1.000×10^3
Gases	
Air	1.29
Carbon dioxide	1.98
Helium	0.179
Hydrogen	0.0899
Nitrogen	1.25
Oxygen	1.43

^{a} Unless otherwise noted, densities are given at 0 °C and 1 atm pressure.

Fluids are materials that can flow, and they include both gases and liquids. Air is the most familiar gaseous fluid, and we call it the wind when it flows from place to place. The most common liquid is water, which we use for generating hydroelectric power and white-water rafting. The *mass density* of a liquid or gas is an important factor that determines its behavior as a fluid. As indicated below, the mass density is the mass per unit volume and is denoted by the Greek letter rho (ρ).

■ **DEFINITION OF MASS DENSITY**

The mass density ρ is the mass m of a substance divided by its volume V:

$$\rho = \frac{m}{V} \qquad (11.1)$$

SI Unit of Mass Density: kg/m^3

Equal volumes of different substances generally have different masses, so the density depends on the nature of the material as Table 11.1 indicates. Gases have the smallest densities, because gas molecules are relatively far apart and a gas contains a large fraction of empty space. In contrast, the molecules are much more tightly packed in liquids and solids, and the tighter packing leads to larger densities. The density of a substance also depends on temperature and pressure. However, for the range of temperatures and pressures encountered in this text, the densities of liquids and solids do not differ much from the values in Table 11.1. On the other hand, the densities of gases are very sensitive to changes in temperature and pressure.

It is the mass of a substance, not its weight, that enters into the definition of density. In situations where weight is needed, it can be calculated from the mass density, the volume, and the acceleration of gravity, as Example 1 illustrates.

EXAMPLE 1 • Blood as a Fraction of Body Weight

The body of a man whose weight is about 690 N contains about 5.2×10^{-3} m^3 (5.5 qt) of blood. (a) Find the blood's weight and (b) express it as a percentage of the body weight.

Reasoning To find the weight W of the blood, we need the mass m, since $W = mg$, where g is the magnitude of the acceleration due to gravity. According to Table 11.1, the density of blood is 1060 kg/m^3, so the mass of the blood can be found by using the given volume of 5.2×10^{-3} m^3 in Equation 11.1.

Solution

(a) The mass and the weight of the blood are

$$m = \rho V = (1060 \text{ kg/m}^3)(5.2 \times 10^{-3} \text{ m}^3) = 5.5 \text{ kg} \qquad (11.1)$$

$$W = mg = (5.5 \text{ kg})(9.80 \text{ m/s}^2) = \boxed{54 \text{ N}} \qquad (4.5)$$

(b) The percentage of body weight contributed by the blood is

$$\text{Percentage} = \frac{54 \text{ N}}{690 \text{ N}} \times 100 = \boxed{7.8\%}$$

A convenient way to compare densities is to use the concept of *specific gravity.* The specific gravity of a substance is its density divided by the density of a standard reference material, usually chosen to be water at 4 °C.

$$\text{Specific gravity} = \frac{\text{Density of substance}}{\text{Density of water at 4 °C}} = \frac{\text{Density of substance}}{1.000 \times 10^3 \text{ kg/m}^3} \quad (11.2)$$

Being the ratio of two densities, specific gravity has no units. For example, Table 11.1 reveals that diamond has a specific gravity of 3.52, since the density of diamond is 3.52 times greater than the density of water at 4 °C.

The next two sections deal with the important concept of pressure. We will see that the density of a fluid is one factor determining the pressure that a fluid exerts.

11.2 PRESSURE

People who have fixed a flat tire know something about pressure. The final step in the job is to reinflate the tire to the proper pressure. The underinflated tire is soft, because it contains an insufficient number of air molecules to push outward against the rubber and give the tire that solid feel. When air is added from a pump, the number of molecules and the collective force they exert are increased. The air molecules within a tire are free to wander throughout its entire volume, and in the course of their wandering they collide with one another and with the inner walls of the tire. The collisions with the walls allow the air to exert a force against every part of the wall surface, as Figure 11.1 shows. The pressure P exerted by a fluid is defined in Section 10.1 as the magnitude F of the force acting perpendicular to a surface divided by the area A over which the force acts:

$$P = \frac{F}{A} \quad (11.3)$$

The SI unit for pressure is a newton/meter^2 (N/m^2), a combination that is referred to as a pascal (Pa). A pressure of 1 Pa is a very small amount. Many common situations involve pressures of approximately 10^5 Pa, an amount referred to as one *bar* of pressure. Alternatively, force can be measured in pounds and area in square inches, so another unit for pressure is pounds per square inch (lb/in.^{2}), often abbreviated as "psi."

Because of its pressure, the air in a tire applies a force to any surface with which it is in contact. Suppose, for instance, that a small cube is inserted inside the tire. As Figure 11.1 shows, the air pressure causes a force to act perpendicularly on each face of the cube. In a similar fashion, a liquid such as water also exerts pressure. A swimmer, for example, feels the water pushing perpendicularly inward everywhere on her body, as Figure 11.2 illustrates. In general, a static fluid cannot produce a force parallel to a surface, for if it did, the surface would apply a reaction force to the fluid, consistent with Newton's action–reaction law. In response, the fluid would flow and would not then be static.

While fluid pressure can generate a force, pressure itself is not a vector quantity, as is the force. In the definition of pressure, $P = F/A$, the symbol F refers only to the magnitude of the force, so that pressure has no directional characteristic. The force generated by the pressure of a static fluid is always perpendicular to the surface on which the fluid acts, as Example 2 illustrates.

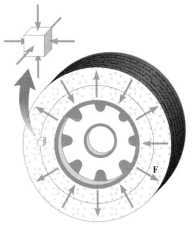

Figure 11.1 In colliding with the inner walls of the tire, the air molecules (blue dots) exert a force on every part of the wall surface. If a small cube were inserted inside the tire, the cube would experience forces acting perpendicular to each of its six faces.

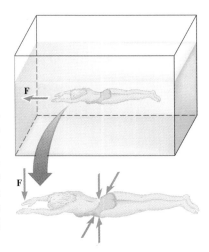

Figure 11.2 Water applies a force perpendicular to each surface within the water, including the walls and bottom of the swimming pool, and all parts of the swimmer's body.

EXAMPLE 2 • The Force on a Swimmer

Suppose the pressure acting on the back of a swimmer's hand is 1.2×10^5 Pa, a realistic value near the bottom of the diving end of a pool. The surface area of the back of the hand is 8.4×10^{-3} m². (a) Determine the magnitude of the force that acts on it. (b) Discuss the direction of the force.

Reasoning and Solution

(a) A pressure of 1.2×10^5 Pa is 1.2×10^5 N/m², and from the definition of pressure in Equation 11.3, we can see that the force is pressure times area:

$$F = PA = (1.2 \times 10^5 \text{ N/m}^2)(8.4 \times 10^{-3} \text{ m}^2) = \boxed{1.0 \times 10^3 \text{ N}}$$

This is a rather large force, about 230 lb.

(b) In Figure 11.2, the hand (palm downward) is oriented parallel to the bottom of the pool. Since the water pushes perpendicularly against the back of the hand, the force **F** is directed downward in the drawing. This downward-acting force is balanced by an upward-acting force on the palm, so that the hand is in equilibrium. If the hand were rotated by 90°, the direction of these forces would also be rotated by 90°, always being perpendicular to the hand.

A person need not be under water to experience the effects of pressure. Walking about on land, we are at the bottom of the earth's atmosphere, which is a fluid and pushes inward on our bodies just like the water in a swimming pool. As Figure 11.3 indicates, there is enough air above the surface of the earth to create the following pressure at sea level:

Atmospheric pressure at sea level 1.013×10^5 Pa = 1 atmosphere

This amount of pressure corresponds to 14.70 lb/in.² and is referred to as one *atmosphere (atm)*. One atmosphere of pressure is a significant amount. Look, for instance, in Figure 11.3 at the results of pumping out the air from within a gasoline can. With no internal air to push outward, the inward push of the external air is unbalanced and is strong enough to crumple the can.

Figure 11.4 shows one way that the force generated by fluid pressure is used. At limestone quarries workmen cut slots into the stone and then topple huge slices of it with the aid of air bags fitted into the slots. Pumps inflate the air bags, and the pressure of the air generates the force to push over the cut stone.

The Physics of...
cutting limestone.

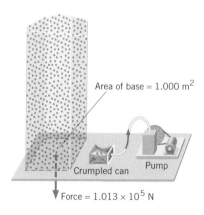

Area of base = 1.000 m²

Crumpled can   Pump

Force = 1.013×10^5 N

Figure 11.3 Atmospheric pressure at sea level is 1.013×10^5 Pa, which is sufficient to crumple a can if the inside air is pumped out.

Figure 11.4 Workmen at a limestone quarry topple a huge slice of stone with the aid of air bags fitted into slots and then inflated.

Figure 11.5 Lynx have large paws that act as natural snowshoes.

In contrast to the situation in Figure 11.4, it is sometimes useful to reduce the pressure. Lynx, for example, are especially well suited for hunting on snow because of their oversize paws (see Figure 11.5). The large paws function as snowshoes that distribute the weight over a large area. Thus, they reduce the weight per unit area or pressure that the cat applies to the surface, which helps to keep it from sinking into the snow.

The Physics of...
lynx paws.

11.3 PRESSURE AND DEPTH IN A STATIC FLUID

The pressure that an underwater swimmer experiences depends on how far beneath the surface he is. The deeper he goes, the more strongly the water pushes on his body. To help determine the relation between pressure and depth, we turn to Newton's second law. Our approach is outlined in Figure 11.6, which is an expanded version of the concept charts in Figures 4.26 and 10.12. In using the second law, we will focus on two types of external forces that act on a sample of the fluid. One is the gravitational force (the weight of the fluid). The other type is the collisional force that is responsible for fluid pressure, as mentioned in the previous section. Since the fluid is at rest, its acceleration is zero ($\mathbf{a} = 0$), and it is in equilibrium. By applying Newton's second law in the form $\Sigma \mathbf{F} = 0$, we will derive a relation between pressure and depth, as indicated in the upper-right portion of the concept chart in Figure 11.6. This relation is especially important, because it leads to two additional principles that are of great use in describing the properties of static fluids, Pascal's principle (Section 11.5) and Archimedes' principle (Section 11.6).

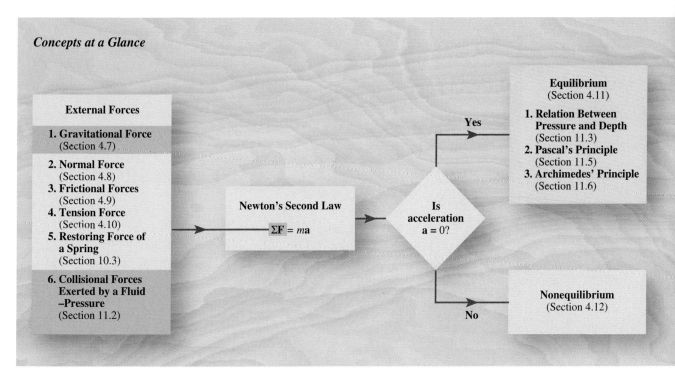

Concepts at a Glance

Figure 11.6 *Concepts at a Glance*
The gravitational force and the collisional forces exerted by a fluid are used with Newton's second law to determine the relation between pressure and depth in a static (**a** = 0) fluid. The deeper that this whale shark and snorkeler go, the greater will be the water pressure on them.

Figure 11.7 shows a container of fluid and focuses attention on one particular column of fluid. The free-body diagram in the figure shows all the vertical forces acting on the column. On the top face (area = A), the fluid pressure P_1 generates a downward force whose magnitude is $P_1 A$. Similarly, on the bottom face, the pressure P_2 generates an upward force of magnitude $P_2 A$. The pressure P_2 is greater than the pressure P_1, because the bottom face supports the weight of more fluid than the upper one does. In fact, the excess weight supported by the bottom face is exactly the weight of the fluid within the column. As the free-body diagram indi-

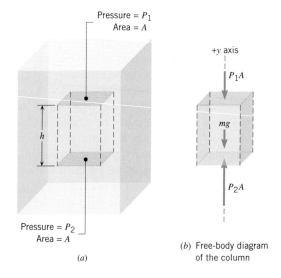

Figure 11.7 (*a*) A container of fluid in which one particular column of the fluid is outlined. The fluid is at rest. (*b*) The free-body diagram, showing the vertical forces acting on the column.

cates, this weight is mg, where m is the mass of the fluid and g is the magnitude of the acceleration due to gravity. Since the column of fluid is in equilibrium, we can set the sum of the vertical forces equal to zero and find that

$$\Sigma F_y = P_2 A - P_1 A - mg = 0 \quad \text{or} \quad P_2 A = P_1 A + mg$$

The mass m is related to the density ρ and the volume V of the column by $m = \rho V$. Since the volume is the cross-sectional area A times the vertical dimension h, we have $m = \rho A h$. With this substitution, the condition for equilibrium becomes $P_2 A = P_1 A + \rho A h g$. The area A can be eliminated algebraically from this expression, with the result that

$$P_2 = P_1 + \rho g h \tag{11.4}$$

Equation 11.4 indicates that if the pressure P_1 is known at a higher level, the larger pressure P_2 at a deeper level can be calculated by adding the increment $\rho g h$. In determining the pressure increment $\rho g h$, we assumed that the density ρ is the same at any vertical distance h. Another way of stating this assumption is to say that the fluid is incompressible. The assumption is reasonable for liquids, since the bottom layers of a liquid can support the upper layers with little compression. In a gas, however, the lower layers are compressed markedly by the weight of the upper layers, with the result that the density varies with vertical distance. For example, the density of our atmosphere is larger near the earth's surface than it is at higher altitudes. When applied to gases, the relation $P_2 = P_1 + \rho g h$ can be used only when h is small enough that any variation in ρ is negligible.

A significant feature of Equation 11.4 is that the pressure increment $\rho g h$ is affected by the vertical distance h, but not by any horizontal distance within the fluid. Conceptual Example 3 helps to clarify this feature.

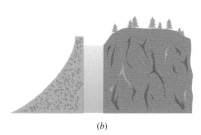

(a)

(b)

Figure 11.8 (*a*) The Hoover Dam in Nevada and Lake Mead behind it. (*b*) This drawing shows a hypothetical reservoir formed by removing most of the water from Lake Mead. Conceptual Example 3 compares the dam needed for this hypothetical reservoir with the Hoover Dam.

CONCEPTUAL EXAMPLE 3 • The Hoover Dam

Lake Mead is the largest wholly artificial reservoir in the United States and was formed after the completion of the Hoover Dam in 1936. As Figure 11.8*a* suggests, the water in the reservoir backs up behind the dam for a considerable distance (about 200 km or 120 miles). Suppose that all the water were removed, except for a relatively narrow vertical column in contact with the dam. Figure 11.8*b* shows a side view of this hypothetical situation, in which the water against the dam has the same depth as in Figure 11.8*a*. Would the Hoover Dam still be needed to contain the water in this hypothetical reservoir, or could a much less massive structure do the job?

Reasoning and Solution Since our hypothetical reservoir contains much less water than Lake Mead, it is tempting to say that a much less massive structure than the Hoover Dam would be needed. Not so, however. Hoover Dam would still be needed to contain our hypothetical reservoir. To see why, imagine a small square on the inner face of the dam, located beneath the water. The force on this square is the product of its area and the pressure of the water. But the pressure depends on the depth, that is, on the vertical distance *h* in Equation 11.4. The horizontal distance of the water behind the dam does not appear in this equation and, therefore, has no effect on the pressure. Consequently, at a given depth the small square experiences a force that is the same in both parts of Figure 11.8. Certainly the dam experiences greater forces generated by greater water pressures at greater depths. But no matter where on the inner face of the dam the square is, the force that the water applies to it depends only on the depth, not on the amount of water backed up behind the dam. ***Thus, the dam for our imaginary reservoir would sustain the same forces that the Hoover Dam sustains and would need to be equally large.***

The next example deals further with the relationship between pressure and depth given by Equation 11.4.

EXAMPLE 4 • The Swimming Hole

Figure 11.9 shows the cross section of a swimming hole. Points *A* and *B* are both located at a distance of *h* = 5.50 m below the surface of the water. Find the pressure at each of these two points.

Reasoning The pressure at point *B* is the same as that at point *A*, since both are located at the *same vertical distance* beneath the surface and only the vertical distance *h* affects the pressure increment ρgh in Equation 11.4. To understand this important feature more clearly, consider the path *AA′B′B* in Figure 11.9. The pressure de-

Figure 11.9 The pressures at points *A* and *B* are the same, since both points are located at the same vertical distance of 5.50 m beneath the surface of the water.

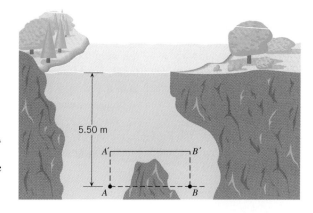

5.50 m

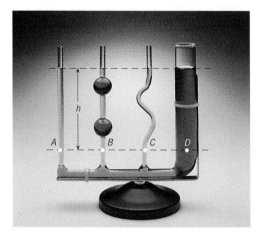

Figure 11.10 Since points A, B, C, and D are at the same distance h beneath the liquid surface, the pressure at each of them is the same.

creases on the way up along the vertical segment AA' and increases by the same amount on the way back down along segment $B'B$. Since no change in pressure occurs along the horizontal segment $A'B'$, the pressure is the same at A and B.

Solution The pressure acting on the surface of the water is the atmospheric pressure of 1.01×10^5 Pa. Using this value as P_1 in Equation 11.4, we can determine a value for the pressure P_2 at either point A or B, both of which are located 5.50 m under the water. Table 11.1 gives the density of water as 1.000×10^3 kg/m³.

$$P_2 = P_1 + \rho gh$$
$$P_2 = 1.01 \times 10^5 \text{ Pa} + (1.000 \times 10^3 \text{ kg/m}^3)(9.80 \text{ m/s}^2)(5.50 \text{ m}) = \boxed{1.55 \times 10^5 \text{ Pa}}$$

• **PROBLEM SOLVING INSIGHT**
The pressure at any point within a fluid depends on the vertical distance h of the point beneath the surface. However, for a given vertical distance, the pressure is the same, no matter where the point is located horizontally within the fluid.

Figure 11.10 shows an irregularly shaped container of liquid. Reasoning similar to that used in Example 4 leads to the conclusion that the pressure is the same at points A, B, C, and D, since each is at the same vertical distance h beneath the surface. In effect, the arteries in our bodies constitute an irregularly shaped "container" for the blood. The next example examines the blood pressure at different places in this "container."

EXAMPLE 5 • Blood Pressure

Blood in the arteries is flowing, but as a first approximation, the effects of this flow can be ignored and the blood can be treated as a static fluid. Estimate the amount by which the blood pressure P_2 in the anterior tibial artery at the foot exceeds the blood pressure P_1 in the aorta at the heart when the body is (a) reclining horizontally as in Figure 11.11a and (b) standing as in Figure 11.11b.

Reasoning and Solution

(a) When the body is horizontal, there is little or no vertical separation between the feet and the heart. Since $h = 0$,

$$P_2 - P_1 = \rho gh = \boxed{0} \qquad (11.4)$$

(b) When an adult is erect, the vertical separation between the feet and the heart is about 1.35 m, as Figure 11.11b indicates. Table 11.1 gives the density of blood as 1060 kg/m³, so that

$$P_2 - P_1 = \rho gh = (1060 \text{ kg/m}^3)(9.80 \text{ m/s}^2)(1.35 \text{ m}) = \boxed{1.40 \times 10^4 \text{ Pa}}$$

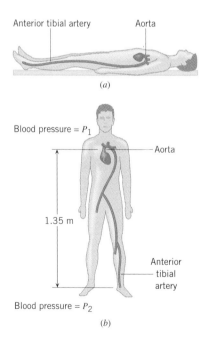

Anterior tibial artery   Aorta

(a)

Blood pressure = P_1

Aorta

1.35 m

Anterior tibial artery

Blood pressure = P_2

(b)

Figure 11.11 The blood pressure in the feet can exceed the blood pressure in the heart, depending on whether the body is (a) reclining horizontally or (b) standing erect.

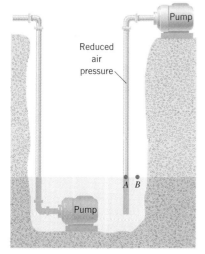

Figure 11.12 A water pump can be placed at the bottom of a well or at ground level. Conceptual Example 6 discusses how the depth of the well influences the choice of which method to use.

The Physics of...
pumping water from a well.

Sometimes fluid pressure places limits on how a job can be done. Conceptual Example 6 illustrates one way in which fluid pressure restricts the height to which water can be pumped.

CONCEPTUAL EXAMPLE 6 · Pumping Water

Figure 11.12 shows two methods for pumping water from a well. In one method, the pump is submerged in the water at the bottom of the well, while in the other, it is located at ground level. If the well is shallow, either technique can be used. However, if the well is very deep, only one of the methods works. Which is it?

Reasoning and Solution To answer this question, we need to examine the nature of the job done by each pump. The pump at the bottom of the well literally pushes water up the pipe. For a very deep well, the column of water becomes very tall, and the pressure at the bottom of the pipe becomes large, due to the pressure increment $\rho g h$ in Equation 11.4. However, as long as the pump can push with sufficient strength to overcome the large pressure, it can shove the next increment of water into the pipe, so the method can be used for very deep wells. In contrast, the pump at ground level does not push water at all. It removes air from the pipe. As the pump reduces the air pressure within the pipe (see point A), the greater air pressure outside (see point B) pushes water up into the pipe. But even the strongest pump can only remove *all* of the air from the pipe. Once the air is completely removed, an increase in pump strength does not increase the height to which the water is pushed by the external air pressure. ***Thus, a ground-level pump can only cause water to rise to a certain maximum height, so it cannot be used for very deep wells.*** (Incidentally, the pump at ground level works just the way you do when you drink through a straw. You draw the air out of the straw, and the external air pressure pushes the liquid up into it.)

Related Homework Material: Problems 24 and 26

11.4 PRESSURE GAUGES

One of the simplest pressure gauges is the mercury barometer used for measuring atmospheric pressure. This device is a tube sealed at one end, filled completely with mercury, and then inverted, so that the open end is under the surface of a pool of mercury (see Figure 11.13). Except for a negligible amount of mercury vapor, the space above the mercury in the tube is empty, and the pressure P_1 is nearly zero there. The pressure P_2 at point A at the bottom of the mercury column is the same as that at point B, namely, atmospheric pressure, for these two points are at the same level. With $P_1 = 0$ and $P_2 = P_{atm}$, it follows from Equation 11.4 that $P_{atm} = 0 + \rho g h$. Thus, the atmospheric pressure can be determined from the height h of the mercury in the tube, the density ρ of mercury, and the acceleration due to gravity. Usually weather forecasters report the pressure in terms of the height h, expressing it in millimeters or inches of mercury. For instance, using $P_{atm} = 1.013 \times 10^5$ Pa and $\rho = 13.6 \times 10^3$ kg/m^3 for the density of mercury, we find that $h = P_{atm}/(\rho g) = 760$ mm (29.9 inches).* Slight variations from this value occur, depending on weather conditions and altitude.

Figure 11.14 shows another kind of pressure gauge, the open-tube manometer. The phrase "open-tube" refers to the fact that one side of the U-tube is open to at-

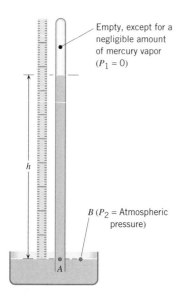

Figure 11.13 A mercury barometer.

---

* A pressure of one millimeter of mercury is sometimes referred to as one *torr*, to honor the inventor of the barometer, Evangelista Torricelli (1608–1647). Thus, one atmosphere of pressure is 760 torr.

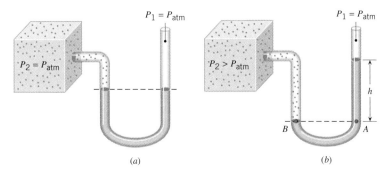

Figure 11.14 The U-shaped tube is called an open-tube manometer and can be used to measure the pressure P_2 in a container. In (a) P_2 equals the atmospheric pressure P_{atm}, while in (b) P_2 exceeds P_{atm}.

mospheric pressure. The tube contains a liquid, often mercury, and its other side is connected to the container whose pressure P_2 is to be measured. When the pressure in the container is equal to the atmospheric pressure, the liquid levels in both sides of the U-tube are the same, as Figure 11.14a indicates. When the pressure in the container is greater than atmospheric pressure, as in Figure 11.14b, the liquid in the tube is pushed downward on the left side and upward on the right side. The relation $P_2 = P_1 + \rho gh$ can be used to determine the container pressure. Atmospheric pressure exists at the top of the right column, so that $P_1 = P_{atm}$. The pressure P_2 is the same at points A and B, so we find that $P_2 = P_{atm} + \rho gh$, or

$$P_2 - P_{atm} = \rho gh$$

The height h is proportional to $P_2 - P_{atm}$, which is called the ***gauge pressure***. The gauge pressure is the amount by which the container pressure differs from atmospheric pressure. The actual value for P_2 is called the ***absolute pressure.***

The sphygmomanometer is a familiar device for measuring blood pressure. As Figure 11.15 illustrates, a squeeze bulb can be used to inflate the cuff with air, which cuts off the flow of blood through the artery below the cuff. When the release valve is opened, the cuff pressure drops. Blood begins to flow again when the pressure created by the heart at the peak of its beating cycle exceeds the cuff pressure. Using a stethoscope to listen for the initial flow, the operator can measure the corresponding cuff gauge pressure with an open-tube manometer. This cuff gauge pressure is called the *systolic* pressure. Eventually, there comes a point when even the pressure created by the heart at the low point of its beating cycle is sufficient to cause blood to flow. Identifying this point with the stethoscope, the operator can measure the corresponding cuff gauge pressure, which is referred to as the *diastolic* pressure. The systolic and diastolic pressures are reported in millimeters of mercury, and values of 120 and 80, respectively, are typical of a young, healthy heart.

11.5 PASCAL'S PRINCIPLE

As we have seen, the pressure in a fluid increases with depth, due to the weight of the fluid above the point of interest. A completely enclosed fluid may be subjected to an additional pressure by the application of an external force. For example, Figure 11.16a shows two interconnected cylindrical chambers. The chambers have different diameters and, together with the connecting tube, are completely filled with a liquid. The larger chamber is sealed at the top with a cap, while the smaller one on the left is fitted with a movable piston. We now follow the approach outlined earlier in Figure 11.6 and begin by asking what determines the pressure P_1 at a point immediately beneath the piston. According to the definition of pressure, it is the mag-

• PROBLEM SOLVING INSIGHT
When solving problems that deal with pressure, be sure to note the distinction between gauge pressure and absolute pressure.

The Physics of...
measuring blood pressure.

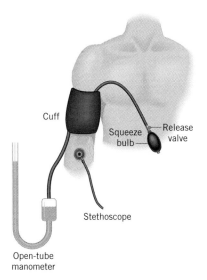

Figure 11.15 A sphygmomanometer is used to measure blood pressure.

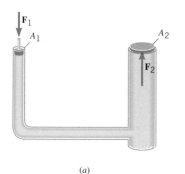

(a)

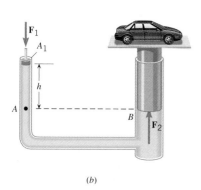

(b)

Figure 11.16 (a) An external force $\mathbf{F}_1$ is applied to the piston on the left. As a result, a force $\mathbf{F}_2$ is exerted on the cap on the chamber on the right. (b) The familiar hydraulic car lift.

The Physics of...
a hydraulic car lift.

• **PROBLEM SOLVING INSIGHT**
Note that the relation $F_1/A_1 = F_2/A_2$, which results from Pascal's principle, applies only when the points 1 and 2 lie at the same depth ($h = 0$) in the fluid.

nitude F_1 of the external force divided by the area A_1 of the piston: $P_1 = F_1/A_1$. If it is necessary to know the pressure P_2 *at any other place in the liquid,* we just add to the value of P_1 the increment $\rho g h$, which takes into account the depth h below the piston: $P_2 = P_1 + \rho g h$. The important feature here is this: The pressure P_1 adds to the pressure $\rho g h$ due to the depth of the liquid at any point, whether that point is in the smaller chamber, the connecting tube, or the larger chamber. Therefore, if the applied pressure P_1 is increased or decreased, the pressure at any other point within the confined liquid changes correspondingly. This behavior is described by *Pascal's principle.*

■ **PASCAL'S PRINCIPLE**

Any change in the pressure applied to a completely enclosed fluid is transmitted undiminished to all parts of the fluid and the enclosing walls.

The usefulness of the arrangement in Figure 11.16a becomes apparent when we calculate the force F_2 applied by the liquid to the cap on the right side. The area of the cap is A_2 and the pressure there is P_2. As long as the tops of the left and right chambers are at the same level, the pressure increment $\rho g h$ is zero, so that the relation $P_2 = P_1 + \rho g h$ becomes $P_2 = P_1$. Consequently, $F_2/A_2 = F_1/A_1$, and

$$F_2 = F_1 \left(\frac{A_2}{A_1} \right) \tag{11.5}$$

If area A_2 is larger than area A_1, a large force $\mathbf{F}_2$ can be applied to the cap on the right chamber, starting with a smaller force $\mathbf{F}_1$ on the left. Depending on the ratio of the areas A_2/A_1, the force $\mathbf{F}_2$ can be large indeed, as in the familiar hydraulic car lift shown in part b. In this device the force $\mathbf{F}_2$ is not applied to a cap that seals the larger chamber, but, rather, to a movable plunger that lifts a car. Example 7 deals with a hydraulic car lift.

EXAMPLE 7 • A Car Lift

In a hydraulic car lift, the input piston has a radius of $r_1 = 0.0120$ m and a negligible weight. The output plunger has a radius of $r_2 = 0.150$ m. The combined weight of the car and the plunger is $F_2 = 20\ 500$ N. The lift uses hydraulic oil that has a density of 8.00×10^2 kg/m³. What input force F_1 is needed to support the car and the output plunger when the bottom surfaces of the piston and plunger are at (a) the same level and (b) the levels shown in Figure 11.16b with $h = 1.10$ m?

Reasoning When the bottom surfaces of the piston and plunger are at the same levels, as in part (a), Equation 11.5 applies. However, this equation does not apply in part (b), where the bottom surface of the output plunger is $h = 1.10$ m below the input piston. Therefore, our solution in part (b) will take into account the pressure increment $\rho g h$. In either case, we will see that the input force is less than the combined weight of the plunger and car.

Solution

(a) Using $A = \pi r^2$ for the circular areas of the piston and plunger, we find from Equation 11.5 that

$$F_1 = F_2 \left(\frac{A_1}{A_2} \right) = F_2 \left(\frac{\pi r_1^2}{\pi r_2^2} \right) = (20\ 500\ \text{N}) \frac{(0.0120\ \text{m})^2}{(0.150\ \text{m})^2} = \boxed{131\ \text{N}}$$

(b) In Figure 11.16b, the bottom surface of the plunger at point B is at the same level as point A, which is at a depth h beneath the input piston. Therefore, we can apply $P_2 = P_1 + \rho gh$, with $P_2 = F_2/(\pi r_2^2)$ and $P_1 = F_1/(\pi r_1^2)$:

$$\frac{F_2}{(\pi r_2^2)} = \frac{F_1}{(\pi r_1^2)} + \rho gh$$

Solving for F_1 gives

$$F_1 = F_2\left(\frac{r_1^2}{r_2^2}\right) - \rho gh(\pi r_1^2)$$

$$= (20\ 500\ \text{N})\frac{(0.0120\ \text{m})^2}{(0.150\ \text{m})^2} - (8.00 \times 10^2\ \text{kg/m}^3)$$

$$\times\ (9.80\ \text{m/s}^2)(1.10\ \text{m})\ \pi\ (0.0120\ \text{m})^2 = \boxed{127\ \text{N}}$$

The answer here is less than in part (a), because the weight of the 1.10-m column of hydraulic oil provides some of the input force to support the car.

Figure 11.17 A backhoe uses a hydraulic fluid to generate a large output force, starting with a small input force.

In a device such as a hydraulic car lift, the same amount of work is done by both the input and output forces in the absence of friction. The larger output force $\mathbf{F}_2$ moves through a smaller distance, while the smaller input force $\mathbf{F}_1$ moves through a larger distance. The work, being the product of the magnitude of the force and the distance, is the same in either case since mechanical energy is conserved.

An enormous variety of clever devices use hydraulic fluids, just as the car lift does. In a backhoe, for instance, the fluids multiply a small input force into the large output force required for digging (see Figure 11.17).

11.6 ARCHIMEDES' PRINCIPLE

Anyone who has tried to push a beach ball under the water has felt how the water pushes back with a strong upward force. This upward force is called the **buoyant force,** and all fluids apply such a force to objects that are immersed in them. The buoyant force exists because fluid pressure is larger at greater depths.

In Figure 11.18a a cylinder of height h is being held under the surface of a liquid. The pressure P_1 on the top face generates the downward force P_1A, where A is the area of the face. Similarly, the pressure P_2 on the bottom face generates the upward force P_2A. Since the pressure is greater at greater depths, the upward force exceeds the downward force. Consequently, the liquid applies to the cylinder a net upward force, or buoyant force, whose magnitude F_B is

$$F_B = P_2A - P_1A = (P_2 - P_1)A$$

Substituting $P_2 - P_1 = \rho gh$ from Equation 11.4, we find that the buoyant force equals ρghA. In this result hA is the volume of liquid that the cylinder moves aside or displaces in being submerged, and ρ denotes the density of the liquid, not the density of the material from which the cylinder is made. Therefore, ρhA gives the mass m of the displaced fluid, so that the buoyant force equals mg, the weight of the displaced fluid. Part b of the drawing helps to clarify the meaning of the phrase "weight of the displaced fluid." This phrase refers to the weight of the fluid that would spill out, if the container were filled to the brim before the cylinder was inserted into the liquid. The buoyant force is not a new type of force. It is just the name given to the net upward force exerted by the fluid on the object.

The shape of the object in Figure 11.18 is not important. No matter what the shape, the buoyant force arises in a similar fashion, in accord with **Archimedes'**

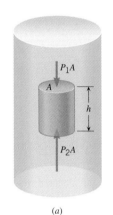

(a)

(b)

Figure 11.18 (a) The fluid applies an upward force P_2A to the bottom face of the submerged cylinder and a downward force P_1A to the top face. (b) The weight of the displaced fluid is the weight of the fluid that spills out as the cylinder is inserted into a completely filled container.

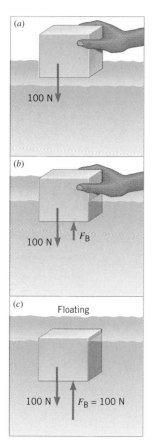

Figure 11.19 An object of weight 100 N is being inserted into a liquid. The deeper the object is, the more liquid it displaces, and the stronger the buoyant force is. In part *c*, the buoyant force matches the 100-N weight, so the object floats.

• **PROBLEM SOLVING INSIGHT**
When using Archimedes' principle to find the buoyant force F_B that acts on an object, be sure to use the density of the displaced fluid, not the density of the object.

principle. It was an impressive accomplishment that the Greek scientist Archimedes (ca. 287–212 B.C.) discovered the essence of this principle so long ago.

■ ARCHIMEDES' PRINCIPLE

Any fluid applies a buoyant force to an object that is partially or completely immersed in it; the magnitude of the buoyant force equals the weight of the fluid that the object displaces:

$$\underbrace{F_B}_{\substack{\text{Magnitude of} \\ \text{the buoyant} \\ \text{force}}} = \underbrace{W_{\text{fluid}}}_{\substack{\text{Weight of} \\ \text{the displaced} \\ \text{fluid}}} \qquad (11.6)$$

The effect that the buoyant force has depends on how strong it is compared with the other forces that are acting. For example, if the buoyant force is strong enough to balance the force of gravity, an object will float in a fluid. Figure 11.19 explores this possibility. In part *a*, a block that weighs 100 N is held above a liquid and displaces none of it, so the liquid exerts no buoyant force. In part *b*, the block displaces some liquid, and the liquid applies a buoyant force F_B to the block, according to Archimedes' principle. Nevertheless, if the block were released, it would fall further into the liquid, because the buoyant force is not sufficiently strong to balance the weight. Finally, in part *c*, enough of the block is submerged to provide a buoyant force that can balance the 100-N weight, so the block is in equilibrium and floats when released. If the buoyant force were not sufficiently large to balance the weight, even with the block completely submerged, the block would sink. Even if an object sinks, however, there is still a buoyant force acting on it; it's just that the buoyant force is not large enough to balance the weight. Example 8 provides additional insight into what determines whether an object will float or sink in a fluid.

EXAMPLE 8 • A Swimming Raft

A solid, square, pinewood raft measures 4.0 m on a side and is 0.30 m thick. (a) Determine whether the raft floats in water, and (b) if so, how much of the raft is beneath the surface (see the distance *h* in Figure 11.20).

Reasoning To determine whether the raft floats, we will compare the weight of the raft to the maximum possible buoyant force and see if there could be enough buoyant force to balance the weight. If so, then the value of the distance *h* can be obtained by utilizing the fact that the floating raft is in equilibrium, with the magnitude of the buoyant force equaling the raft's weight.

Solution

(a) The weight of the raft can be calculated from the density $\rho_{\text{pine}} = 550 \text{ kg/m}^3$ (Table 11.1), the volume of the wood, and the acceleration due to gravity. The volume of the wood is $V_{\text{pine}} = 4.0 \text{ m} \times 4.0 \text{ m} \times 0.30 \text{ m} = 4.8 \text{ m}^3$, so that

$$\begin{array}{c}\text{Weight}\\\text{of raft}\end{array} = (\rho_{\text{pine}} V_{\text{pine}})g = (550 \text{ kg/m}^3)(4.8 \text{ m}^3)(9.80 \text{ m/s}^2) = 26\,000 \text{ N}$$

The maximum possible buoyant force occurs when the entire raft is under the surface, displacing 4.8 m³ of water. The weight of this volume of water is the maximum buoyant force F_B^{max} and can be obtained using the density of water:

$$F_B^{\text{max}} = \rho_{\text{water}} V_{\text{pine}} g = (1.000 \times 10^3 \text{ kg/m}^3)(4.8 \text{ m}^3)(9.80 \text{ m/s}^2) = 47\,000 \text{ N}$$

Since the maximum possible buoyant force exceeds the 26 000-N weight of the raft, the raft will float only partially submerged at a distance h beneath the water.

(b) We now find the value of h. The buoyant force balances the raft's weight, so $F_{\text{B}} = 26\,000$ N. But according to Equation 11.6, the magnitude of the buoyant force is also the weight of the displaced fluid, so $F_{\text{B}} = W_{\text{fluid}} = 26\,000$ N. Using the density of water, we can also express the weight of the displaced water as $W_{\text{fluid}} = \rho_{\text{water}}V_{\text{water}}g$, where the volume is $V_{\text{water}} = 4.0\text{ m} \times 4.0\text{ m} \times h$. As a result,

$$W_{\text{fluid}} = \rho_{\text{water}}(4.0\text{ m} \times 4.0\text{ m} \times h)g = 26\,000\text{ N}$$

$$h = \frac{26\,000\text{ N}}{\rho_{\text{water}}(4.0\text{ m} \times 4.0\text{ m})g}$$

$$= \frac{26\,000\text{ N}}{(1.000 \times 10^3\text{ kg/m}^3)(4.0\text{ m} \times 4.0\text{ m})(9.80\text{ m/s}^2)} = \boxed{0.17\text{ m}}$$

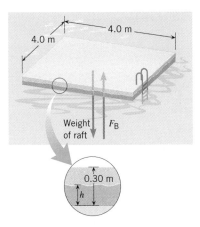

Figure 11.20 A raft floating with a distance h beneath the surface of the water.

In deciding whether the raft floats in part (a) of Example 8, we compared the raft's weight $[(\rho_{\text{pine}}V_{\text{pine}})g]$ to the maximum possible buoyant force $[(\rho_{\text{water}}V_{\text{pine}})g]$. The comparison depends only on the densities ρ_{pine} and ρ_{water}. The take-home message is this: Any object that is *solid throughout* will float in a liquid if the density of the object is less than or equal to the density of the liquid. For instance, at 0 °C ice has a density of 917 kg/m^3, while water has a density of 1000 kg/m^3. Therefore, ice floats in water.

Although a solid piece of a high-density material like steel will sink in water, such materials can, nonetheless, be used to make floating objects. A supertanker, for example, floats because it is *not* solid metal. It contains enormous amounts of empty space and, because of its shape, displaces enough water to balance its own large weight. Conceptual Example 9 focuses on the reason ships float.

CONCEPTUAL EXAMPLE 9 • How Much Water Is Needed to Float a Ship?

A ship floating in the ocean is a familiar sight. But is all that water really necessary? Can an ocean vessel float in the amount of water that a swimming pool contains, for instance?

Reasoning and Solution In principle, a ship can float in much less than the amount of water that a swimming pool contains; even a glass of water would do. To see why, look at Figure 11.21. Part *a* shows the ship floating in the ocean because it contains empty space within its hull and displaces enough water to balance its own weight. Part *b* shows the water that the ship displaces, which, according to Archimedes' principle, has a weight that equals the ship's weight. While this wedge-shaped portion of water represents the water *displaced* by the ship, it is not the amount that must be present to float the ship, as part *c* illustrates. This part of the drawing shows a canal, the cross section of which matches the shape in part *b*. **All that is needed, in principle, is a thin section of water that separates the hull of the floating ship from the sides of the canal.** This thin section of water could have a very small volume indeed and fit into a single glass!

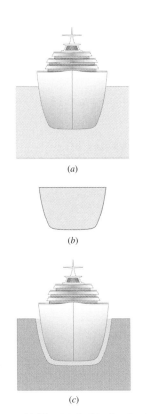

Figure 11.21 (*a*) A ship floating in the ocean. (*b*) This is the water that the ship displaces. (*c*) The ship floats here in a canal that has a cross section similar in shape to that in part *b*.

A useful application of Archimedes' principle can be found in car batteries. To alert the owner that recharging is necessary, some batteries include a state-of-charge indicator, such as the one illustrated in Figure 11.22. The battery includes a viewing port that looks down through a plastic rod, which extends into the battery acid. Attached to the end of this rod is a "cage," containing a green ball. The cage has holes

Figure 11.22 A state-of-charge indicator for a car battery.

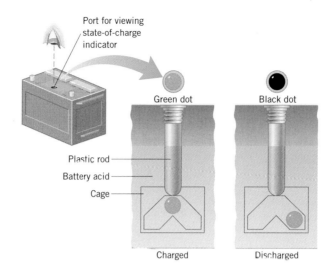

Port for viewing state-of-charge indicator

Green dot

Black dot

Plastic rod

Battery acid

Cage

Charged

Discharged

The Physics of...
a state-of-charge battery
indicator.

in it that allow the acid to enter. When the battery is charged, the density of the acid is large enough so that its buoyant force makes the ball rise to the top of the cage, to just beneath the plastic rod. The viewing port shows a green dot. As the battery discharges, the density of the acid decreases. Since the buoyant force is the weight of the acid displaced by the ball, the buoyant force also decreases. As a result, the ball sinks into one of the chambers oriented at an angle. With the ball no longer visible, the viewing port shows a dark or black dot, warning that the battery charge is low.

Archimedes' principle has allowed us to determine how an object can float in a liquid. This principle also applies to gases, as the next example illustrates.

EXAMPLE 10 • A Goodyear Airship

The Physics of...
a Goodyear airship.

Normally, a Goodyear airship, such as that in Figure 11.23, contains about 5.40×10^3 m^3 of helium whose density is 0.179 kg/m^3. Find the weight of the load W_L that the airship can carry in equilibrium at an altitude where the density of air is 1.20 kg/m^3.

Reasoning The airship and its load are in equilibrium. Thus, the buoyant force F_B applied to the airship by the surrounding air balances the weight W_{helium} of the helium and the weight W_L of the load, including the solid parts of the airship. The free-body diagram in Figure 11.23*b* shows these forces.

F_B

W_{helium}

W_L

Figure 11.23 (*a*) A helium-filled Goodyear airship. (*b*) The free-body diagram of the airship, including the load weight W_L.

(*a*)

(*b*) Free-body diagram of the airship

Solution Because the forces in the free-body diagram balance, we have

$$W_{\text{helium}} + W_{\text{L}} = F_{\text{B}} \quad \text{or} \quad W_{\text{L}} = F_{\text{B}} - W_{\text{helium}}$$

According to Archimedes' principle, the buoyant force is the weight of the displaced air: $F_{\text{B}} = W_{\text{air}} = \rho_{\text{air}}V_{\text{ship}}g$. The weight of the helium is $W_{\text{helium}} = \rho_{\text{helium}}V_{\text{ship}}g$, where we have assumed that the volume of the helium and the volume of the airship have nearly the same value of $V_{\text{ship}} = 5.40 \times 10^3 \text{ m}^3$. Thus, we find that

$$W_{\text{L}} = \rho_{\text{air}}V_{\text{ship}}g - \rho_{\text{helium}}V_{\text{ship}}g = (\rho_{\text{air}} - \rho_{\text{helium}})\,V_{\text{ship}}g$$

$$W_{\text{L}} = (1.20 \text{ kg/m}^3 - 0.179 \text{ kg/m}^3)(5.40 \times 10^3 \text{ m}^3)(9.80 \text{ m/s}^2) = \boxed{5.40 \times 10^4 \text{ N}}$$

11.7 FLUIDS IN MOTION

There are many examples of fluid motion or fluid flow. Water may flow smoothly and slowly in a quiet stream or violently over a waterfall. The air may form a gentle breeze or a raging tornado. To deal with such diversity, it helps to identify some of the basic types of fluid flow.

Fluid flow can be steady or unsteady. In *steady flow* the velocity of the fluid particles at any point is constant as time passes. For instance, in Figure 11.24 a fluid particle flows with a velocity of $\mathbf{v}_1 = +2$ m/s past point 1. In steady flow every particle passing through this point has this same velocity. At another location the velocity may be different, as in a river, which usually flows fastest near its center and slowest near its banks. Thus, at point 2, the fluid velocity is $\mathbf{v}_2 = +0.5$ m/s, and if the flow is steady, all particles passing through this point have a velocity of $+0.5$ m/s.

Unsteady flow exists whenever the velocity at a point in the fluid changes as time passes. *Turbulent flow* is an extreme kind of unsteady flow and occurs when there are sharp obstacles or bends in the path of a fast-moving fluid, as in the rapids in Figure 11.25. In turbulent flow, the velocity at any particular point changes erratically from moment to moment, both in magnitude and direction.

Fluid flow can be compressible or incompressible. Most liquids are nearly incompressible; that is, the density of a liquid remains almost constant as the pressure changes. To a good approximation, then, liquids flow in an incompressible manner. In contrast, gases are highly compressible. However, there are situations in which the density of a flowing gas remains constant enough that the flow can be considered incompressible.

Fluid flow can be viscous or nonviscous. A viscous fluid, such as honey, does not flow readily and is said to have a large viscosity. In contrast, water is less viscous and flows more readily; water has a smaller viscosity than honey. The flow of a viscous fluid is an energy-dissipating process. The viscosity hinders neighboring layers of fluid from sliding freely past one another. A fluid with zero viscosity flows in an unhindered manner with no dissipation of energy. Although no real fluid has zero viscosity at normal temperatures, many fluids have negligibly small viscosities. An incompressible, nonviscous fluid is called an *ideal fluid.*

Fluid flow can be rotational or irrotational. The flow is rotational when a part of the fluid has rotational as well as translational motion. To test for rotational flow, a small paddle wheel can be immersed in the fluid. If the wheel rotates, as in Figure 11.26a, the flow is rotational. If the paddle wheel does not rotate, as in part *b* of the figure, the flow is irrotational, and the fluid exhibits only translational motion.

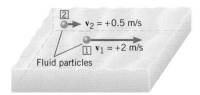

Figure 11.24 Two fluid particles in a stream. At different locations in the stream the particle velocities may be different, as indicated by $\mathbf{v}_1$ and $\mathbf{v}_2$.

Figure 11.25 The flow of water in a rapids is an example of turbulent flow.

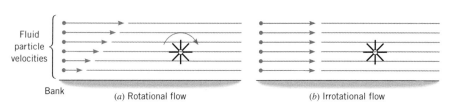

(a) Rotational flow   *(b) Irrotational flow*

Figure 11.26 (*a*) In a river, water near the bank usually has a smaller velocity than water near the center. This top view shows that a small paddle wheel placed near the bank would rotate, indicating rotational flow. (*b*) In an ideal fluid in which all portions of the fluid have the same velocity, a paddle wheel would not rotate, indicating irrotational flow.

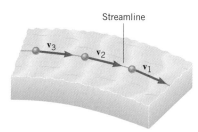

Figure 11.27 At any point along a streamline, the velocity vector of the fluid particle at that point is tangent to the streamline.

When the flow is steady, **streamlines** are often used to represent the trajectories of the fluid particles. A streamline is a line drawn in the fluid such that a tangent to the streamline at any point is parallel to the fluid velocity at that point. Figure 11.27 shows the velocity vectors at three points along a streamline. The fluid velocity can vary (in both magnitude and direction) from point to point along a streamline, but at any given point, the velocity is constant in time, as required by the condition of steady flow. In fact, steady flow is often called **streamline flow.**

Figure 11.28*a* illustrates a method for making streamlines visible by using small tubes to release a colored dye into the moving liquid. The dye does not immediately mix with the liquid and is carried along a streamline. In the case of a flowing gas, such as that in a wind tunnel, streamlines are often revealed by smoke streamers, as part *b* of the figure shows.

In steady flow, the pattern of streamlines is steady in time, and, as Figure 11.28*a* indicates, no two streamlines cross one another. If they did, every particle arriving at the crossing point could go one way or the other. This would mean that the velocity at the crossing point would change from moment to moment, a condition that does not exist in steady flow.

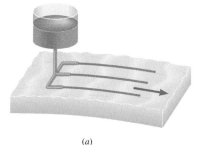

(a)

(b)

Figure 11.28 (*a*) In the steady flow of a liquid, a colored dye reveals the streamlines. (*b*) A smoke streamer reveals a streamline pattern for the air flowing around a car.

11.8 THE EQUATION OF CONTINUITY

Have you ever used your thumb to control the water flowing from the end of a hose, as in Figure 11.29? If so, you have seen that the water velocity increases when your thumb reduces the cross-sectional area of the hose opening. This kind of fluid behavior is described by the **equation of continuity,** which expresses the simple idea that if a fluid enters one end of a pipe at a rate of 5 kilograms per second, for example, then fluid must also leave at the same rate, assuming that there are no places between the entry and exit points to add or remove fluid. As we will see, it is because of this fact that water exits a hose more rapidly when you partially close off the hose opening. The mass of fluid per second (e.g., 5 kg/s) that flows through a tube is called the **mass flow rate.**

Figure 11.30 shows a small mass of fluid or fluid element moving along a tube. Upstream at position 2, where the tube has a cross-sectional area A_2, the fluid has a speed v_2 and a density ρ_2. Downstream at location 1, the corresponding quantities are A_1, v_1, and ρ_1. During a small time interval Δt, the fluid at point 2 moves a distance of $v_2\Delta t$, as the drawing shows. The volume of fluid that has flowed past this point is the cross-sectional area times this distance or $A_2 v_2 \Delta t$. The mass Δm of this fluid element is the product of the density and volume: $\Delta m = \rho_2 A_2 v_2 \Delta t$. Dividing

Δm by Δt gives the mass flow rate (the mass per second):

$$\text{Mass flow rate at position 2} = \frac{\Delta m}{\Delta t} = \rho_2 A_2 v_2 \qquad (11.7a)$$

Similar reasoning leads to the mass flow rate at position 1:

$$\text{Mass flow rate at position 1} = \rho_1 A_1 v_1 \qquad (11.7b)$$

Since no fluid can cross the sidewalls of the tube, the mass flow rates at positions 1 and 2 must be equal. But these positions were selected arbitrarily, so the mass flow rate has the same value everywhere in the tube, an important result known as the **equation of continuity.** The equation of continuity is an expression of the fact that mass is conserved (i.e., neither created nor destroyed) as it flows along the tube.

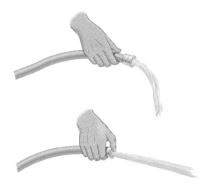

Figure 11.29 When the end of a hose is partially closed off, thus reducing its cross-sectional area, the fluid velocity increases.

■ **EQUATION OF CONTINUITY**

The mass flow rate ($\rho A v$) has the same value at every position along a tube that has a single entry and a single exit point for fluid flow. For two positions along such a tube

$$\rho_1 A_1 v_1 = \rho_2 A_2 v_2 \qquad (11.8)$$

where ρ = fluid density (kg/m^3)
     A = cross-sectional area of tube (m^2)
     v = fluid speed (m/s)

SI Unit of Mass Flow Rate: kg/s

The density of an incompressible fluid does not change during flow, so that $\rho_1 = \rho_2$, and the equation of continuity reduces to

Incompressible fluid $\qquad\qquad A_1 v_1 = A_2 v_2 \qquad (11.9)$

The quantity Av represents the volume of fluid per second that passes through the tube and is referred to as the **volume flow rate Q:**

$$Q = \text{volume flow rate} = Av \qquad (11.10)$$

Equation 11.9 shows that where the tube area is large, the fluid speed is small, and, conversely, where the tube area is small, the speed is large. Example 11 explores this behavior in more detail for the hose in Figure 11.29.

EXAMPLE 11 • A Garden Hose

A garden hose has an unobstructed opening with a cross-sectional area of $2.85 \times 10^{-4} \text{ m}^2$, from which water fills a bucket in 30.0 s. The volume of the bucket is $8.00 \times 10^{-3} \text{ m}^3$ (about two gallons). Find the speed of the water that leaves the

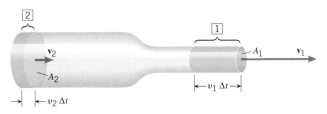

Figure 11.30 In general, a fluid flowing in a tube that has different cross-sectional areas A_1 and A_2 at positions 1 and 2 has different velocities $\mathbf{v}_1$ and $\mathbf{v}_2$ at these positions.

hose through (a) the unobstructed opening and (b) an obstructed opening that has only half as much area.

Reasoning If we can determine the volume flow rate Q, the speed of the water can be obtained from $Q = Av$ (Equation 11.10) as $v = Q/A$, since the area A is given. The volume flow rate can be found from the volume of the bucket and the time required to fill it.

Solution

(a) The volume flow rate is equal to the volume of the bucket divided by the time required to fill it:

$$Q = \frac{8.00 \times 10^{-3}\ \text{m}^3}{30.0\ \text{s}} = 2.67 \times 10^{-4}\ \text{m}^3/\text{s}$$

The speed of the water is

$$v = \frac{Q}{A} = \frac{2.67 \times 10^{-4}\ \text{m}^3/\text{s}}{2.85 \times 10^{-4}\ \text{m}^2} = \boxed{0.937\ \text{m/s}} \tag{11.10}$$

(b) Water can be considered incompressible, so the equation of continuity can be applied in the form $A_1 v_1 = A_2 v_2$. Since $A_2 = \frac{1}{2} A_1$, we find that

$$v_2 = \left(\frac{A_1}{A_2}\right) v_1 = \left(\frac{A_1}{\frac{1}{2}A_1}\right)(0.937\ \text{m/s}) = \boxed{1.87\ \text{m/s}} \tag{11.9}$$

The next example applies the equation of continuity to the flow of blood.

EXAMPLE 12 • A Clogged Artery

The Physics of...
a clogged artery.

In the condition known as atherosclerosis, a deposit or atheroma forms on the arterial wall and reduces the opening through which blood can flow. In the carotid artery in the neck, blood flows three times faster through a partially blocked region than it

Concepts at a Glance

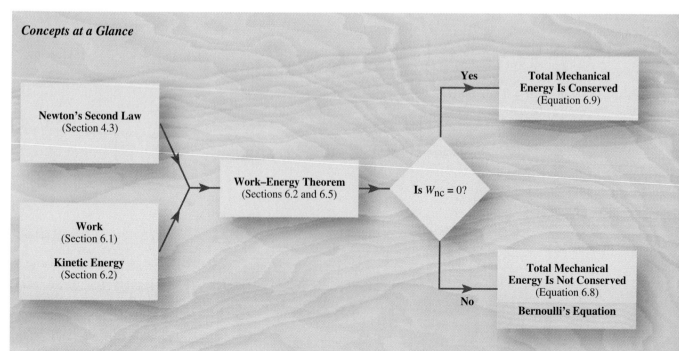

does through an unobstructed region. Determine the ratio of the effective radii of the artery at the two places.

Reasoning and Solution Blood, like most liquids, is incompressible, and the equation of continuity in the form of Equation 11.9 can be applied. Since the area of a circle is πr^2, it follows that

$$\underbrace{(\pi r_U{}^2)\, v_U}_{\substack{\text{Unobstructed} \\ \text{volume flow rate}}} = \underbrace{(\pi r_O{}^2)\, v_O}_{\substack{\text{Obstructed} \\ \text{volume flow rate}}}$$

The ratio of the radii is

$$\frac{r_U}{r_O} = \sqrt{\frac{v_O}{v_U}} = \sqrt{3} = \boxed{1.7}$$

The unobstructed artery has an effective radius that is 70% larger than the radius of the obstructed region.

• **PROBLEM SOLVING INSIGHT**
The equation of continuity in the form $A_1 v_1 = A_2 v_2$ applies only when the density of the fluid is constant. If the density is not constant, the equation of continuity is $\rho_1 A_1 v_1 = \rho_2 A_2 v_2$.

11.9 BERNOULLI'S EQUATION

For *steady, irrotational* flow, the speed, pressure, and elevation of an *incompressible, nonviscous* fluid are related by an equation discovered by Daniel Bernoulli (1700–1782). To derive ***Bernoulli's equation,*** we will use the work–energy theorem, which is introduced in Chapter 6. This theorem, as expressed by Equation 6.8, states that the net work W_{nc} done on an object by external nonconservative forces is equal to the change in the total mechanical energy of the object. As the concept chart in Figure 6.17 shows, if $W_{nc} = 0$, the total mechanical energy is conserved; otherwise, it is not. This chart is reproduced in Figure 11.31 and expanded to include Bernoulli's equation for fluid flow. As mentioned earlier, the pressure within a

Figure 11.31 *Concepts at a Glance* The work–energy theorem leads to Bernoulli's equation when the net work W_{nc} done by external nonconservative forces is not zero. Bernoulli's equation reveals that the pressure associated with moving air, such as this tornado in the movie *Twister,* is lower than that of stationary air, such as that in the house.

Figure 11.32 (*a*) In this horizontal pipe, the pressure in region 2 is greater than in region 1. The difference in pressure leads to the net force that accelerates the fluid to the right. (*b*) When the fluid changes elevation, the pressure at the bottom is greater than at the top, assuming the cross-sectional area of the pipe is constant.

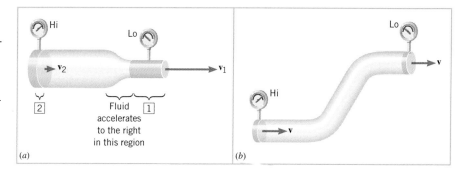

(*a*)   (*b*)

fluid is caused by collisional forces. Collisional forces are nonconservative. Therefore, when a fluid element is accelerated because of a difference in pressures, work is being done by nonconservative forces and W_{nc} is not zero. Consistent with the work–energy theorem, this work changes the total mechanical energy of the fluid element. We will now see how the work–energy theorem leads directly to Bernoulli's equation, as indicated in the lower right portion of Figure 11.31.

To begin with, let us make two observations about a moving fluid. First, whenever a fluid is flowing in a horizontal pipe and encounters a region of reduced cross-sectional area, the pressure of the fluid drops, as Figure 11.32*a* indicates. The reason for this follows from Newton's second law. When moving from the wider region 2 to the narrower region 1, the fluid speeds up or accelerates, consistent with the conservation of mass (as expressed by the equation of continuity). According to the second law, the accelerating fluid must be subjected to an unbalanced force. But there can be an unbalanced force only if the pressure in region 2 exceeds the pressure in region 1. We will see that the difference in pressures is given by Bernoulli's equation. Our second observation is that if the fluid moves to a higher elevation, the pressure at the lower level is greater than the pressure at the higher level, as in Figure 11.32*b*. The basis for this observation is our previous study of static fluids, and Bernoulli's equation will confirm it, provided that the cross-sectional area of the pipe does not change.

To derive Bernoulli's equation, consider Figure 11.33*a*. This drawing shows a fluid element of mass *m*, upstream in region 2 of a pipe. Both the cross-sectional area and the elevation are different at different places along the pipe. The speed, pressure, and elevation in this region are v_2, P_2, and y_2, respectively. Downstream

Figure 11.33 (*a*) A fluid element (dark blue) moving through a pipe whose cross-sectional area and elevation change. (*b*) The fluid element experiences a force −**F** on its top surface due to the fluid above it, and a force **F** + Δ**F** on its bottom surface due to the fluid below it.

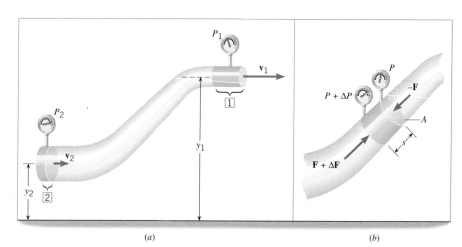

(*a*)   (*b*)

in region 1 these variables have the values v_1, P_1, and y_1. As Chapter 6 discusses, an object moving under the influence of gravity has a total mechanical energy E that is the sum of the kinetic energy KE and the gravitational potential energy PE: $E = \text{KE} + \text{PE} = \frac{1}{2}mv^2 + mgy$. When work W_{nc} is done on the fluid element by external nonconservative forces, the total mechanical energy changes. According to the work-energy theorem, the work equals the change in the total mechanical energy:

$$W_{nc} = E_1 - E_2 = \underbrace{(\tfrac{1}{2}mv_1{}^2 + mgy_1)}_{\substack{\text{Total mechanical}\\\text{energy in region 1}}} - \underbrace{(\tfrac{1}{2}mv_2{}^2 + mgy_2)}_{\substack{\text{Total mechanical}\\\text{energy in region 2}}} \qquad (6.8)$$

Figure 11.33b helps us understand how the work W_{nc} arises. On the top surface of the fluid element, the surrounding fluid exerts a pressure P. This pressure gives rise to a force of magnitude $F = PA$, where A is the cross-sectional area. On the bottom surface, the surrounding fluid exerts a slightly greater pressure, $P + \Delta P$, where ΔP is the pressure difference between the ends of the element. As a result, the force on the bottom surface is $F + \Delta F = (P + \Delta P)A$. The *net* force pushing the fluid element up the tube is $\Delta F = (\Delta P)A$. When the fluid element moves through its own length s, the work done is the product of the net force and the distance, according to Equation 6.1: Work $= (\Delta F)s = (\Delta P)As$. The quantity As is the volume V of the element, so the work is $(\Delta P)V$. The total work done on the fluid element in moving it from region 2 to region 1 is the sum of the small increments of work $(\Delta P)V$ done as the element moves along the tube. This sum amounts to $W_{nc} = (P_2 - P_1)V$, where $P_2 - P_1$ is the pressure difference between the two regions. With this expression for W_{nc}, the work-energy theorem becomes

$$W_{nc} = (P_2 - P_1)V = (\tfrac{1}{2}mv_1{}^2 + mgy_1) - (\tfrac{1}{2}mv_2{}^2 + mgy_2)$$

By dividing both sides of this result by the volume V, recognizing that m/V is the density ρ of the fluid, and rearranging terms, we obtain Bernoulli's equation.

■ BERNOULLI'S EQUATION

In the steady, irrotational flow of a nonviscous, incompressible fluid of density ρ, the pressure P, the fluid speed v, and the elevation y at any two points (1 and 2) are related by

$$P_1 + \tfrac{1}{2}\rho v_1{}^2 + \rho gy_1 = P_2 + \tfrac{1}{2}\rho v_2{}^2 + \rho gy_2 \qquad (11.11)$$

Since the points 1 and 2 were selected arbitrarily, the term $P + \frac{1}{2}\rho v^2 + \rho gy$ has a constant value at all positions in the flow. For this reason, Bernoulli's equation is sometimes expressed as $P + \frac{1}{2}\rho v^2 + \rho gy = \text{constant}$.

Equation 11.11 can be regarded as an extension of the earlier result that specifies how the pressure varies with depth in a static fluid ($P_2 = P_1 + \rho gh$), the terms $\frac{1}{2}\rho v_1{}^2$ and $\frac{1}{2}\rho v_2{}^2$ accounting for the effects of fluid speed. Bernoulli's equation reduces to the result for static fluids when the speed of the fluid is the same everywhere ($v_1 = v_2$), as it is when the cross-sectional area remains constant. Under such conditions, Bernoulli's equation is $P_1 + \rho gy_1 = P_2 + \rho gy_2$. After rearrangement, this result becomes

$$P_2 = P_1 + \rho g(y_1 - y_2) = P_1 + \rho gh$$

which is the result (Equation 11.4) for static fluids.

11.10 APPLICATIONS OF BERNOULLI'S EQUATION

When a moving fluid is contained in a horizontal pipe, all parts of it have the same elevation ($y_1 = y_2$), and Bernoulli's equation simplifies to

$$P_1 + \tfrac{1}{2}\rho v_1^2 = P_2 + \tfrac{1}{2}\rho v_2^2 \qquad (11.12)$$

Thus, the quantity $P + \tfrac{1}{2}\rho v^2$ remains constant throughout a horizontal pipe; if v increases, P decreases and vice versa. This is exactly the result that we deduced qualitatively from Newton's second law at the beginning of Section 11.9, and Conceptual Example 13 illustrates it.

CONCEPTUAL EXAMPLE 13 • Tarpaulins and Bernoulli's Equation

A tarpaulin is a piece of canvas that is used to cover a cargo, like that pulled by the truck in Figure 11.34. When the truck is stationary the tarpaulin lies flat, but it bulges outward when the truck is speeding down the highway. Account for this behavior.

Reasoning and Solution   The behavior is a direct consequence of the pressure changes that Bernoulli's equation describes for flowing fluids. When the truck is stationary, the air outside and inside the cargo area is stationary, so the air pressure is the same in both places. This pressure applies the same force to the outer and inner surfaces of the canvas, with the result that the tarpaulin lies flat. When the truck is moving, the outside air rushes over the top surface of the canvas. In accord with Bernoulli's equation (Equation 11.12), the moving air has a lower pressure than the stationary air within the cargo area. ***The greater inside pressure generates a greater force on the inner surface of the canvas, and the tarpaulin bulges outward.***

Related Homework Material: Problems 60, 61

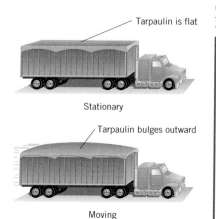

Tarpaulin is flat

Stationary

Tarpaulin bulges outward

Moving

Figure 11.34   The tarpaulin that covers the cargo is flat when the truck is stationary, but bulges outward when the truck is moving.

Example 14 applies Equation 11.12 to a dangerous physiological condition known as an aneurysm.

The Physics of... an aneurysm.

EXAMPLE 14 • An Enlarged Blood Vessel

An aneurysm is an abnormal enlargement of a blood vessel such as the aorta. Suppose that, because of an aneurysm, the cross-sectional area A_1 of the aorta increases to a value $A_2 = 1.7A_1$. The speed of the blood ($\rho = 1060$ kg/m^3) through a normal portion of the aorta is $v_1 = 0.40$ m/s. Assuming that the aorta is horizontal (the person is lying down), determine the amount by which the pressure P_2 in the enlarged region exceeds the pressure P_1 in the normal region.

Reasoning   According to the equation of continuity, the speed of the blood in the enlarged section of the artery is smaller than the speed in a healthy section. In turn, Bernoulli's equation for horizontal flow (Equation 11.12) indicates that the smaller speed leads to a greater pressure and may be used to find the pressure difference $P_2 - P_1$.

Solution   From the equation of continuity we find the speed v_2 of the blood in the aneurysm to be

$$v_2 = \left(\frac{A_1}{A_2}\right) v_1 = \left(\frac{A_1}{1.7A_1}\right)(0.40 \text{ m/s}) = 0.24 \text{ m/s} \qquad (11.9)$$

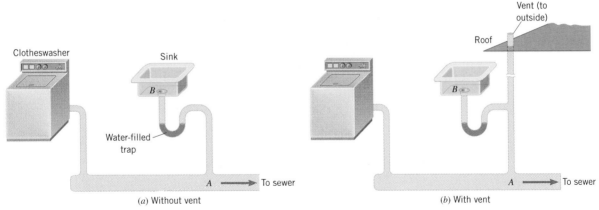

(a) Without vent

(b) With vent

From Bernoulli's equation for horizontal flow it follows that $P_1 + \frac{1}{2}\rho v_1^2 = P_2 + \frac{1}{2}\rho v_2^2$. Solving for $P_2 - P_1$, we find

$$P_2 - P_1 = \frac{1}{2}\rho(v_1^2 - v_2^2) = \frac{1}{2}(1060 \text{ kg/m}^3)[(0.40 \text{ m/s})^2 - (0.24 \text{ m/s})^2] = \boxed{54 \text{ Pa}}$$

This result is positive, indicating that P_2 is greater than P_1. The excess pressure puts added stress on the already weakened tissue of the arterial wall at the aneurysm.

Figure 11.35 In a household plumbing system, a vent is necessary to equalize the pressures at points A and B, thus preventing the trap from being emptied. An empty trap allows sewer gas to enter the house.

The impact of fluid flow on pressure is widespread. Figure 11.35, for instance, illustrates how household plumbing takes into account the implications of Bernoulli's equation. The U-shaped section of pipe beneath the sink is called a "trap," because it traps water, which serves as a barrier to prevent sewer gas from leaking into the house. Part a of the drawing shows poor plumbing. When water from the clotheswasher rushes through the sewer pipe, the high-speed flow causes the pressure at point A to drop. The pressure at point B in the sink, however, remains at the higher atmospheric pressure. As a result of this pressure difference, the water is pushed out of the trap and into the sewer line, leaving no protection against sewer gas. A correctly designed system is vented to the outside of the house, as in Figure 11.35b. The vent ensures that the pressure at A remains the same as that at B (atmospheric pressure), even when water from the clotheswasher is rushing through the pipe. Thus, the purpose of the vent is to prevent the trap from being emptied, not to provide an escape route for sewer gas.

The Physics of...
household plumbing.

One of the most spectacular examples of how fluid flow affects pressure is the dynamic lift on airplane wings. Figure 11.36a shows a wing (in cross section) moving to the right, with the air flowing past the wing to the left. Because of the shape

The Physics of...
airplane wings.

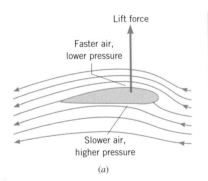

Lift force

Faster air,
lower pressure

Slower air,
higher pressure

(a)

(b)

Figure 11.36 (a) Streamlines of air flow around an airplane wing. The wing is moving to the right. (b) The end of this wing has roughly the shape indicated in part a.

Figure 11.37 A ski jumper curves his body to take advantage of the dynamic lift force due to the flow of air above and below him.

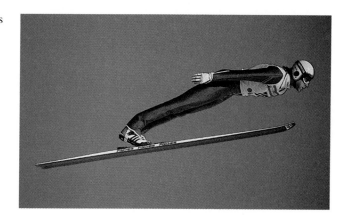

of the wing, the air travels faster over the curved upper surface than it does over the flatter lower surface. According to Bernoulli's equation, the pressure above the wing is lower (faster moving air), while the pressure below the wing is higher (slower moving air). Thus, the wing is lifted upward. Part *b* of the figure shows the wing of an airplane.

Ski jumpers use dynamic lift, just like airplane wings do, to help themselves stay in the air longer. Figure 11.37 shows a jumper who has curved his body to mimic the cross-sectional shape of an airplane wing.

The curve ball, one of the main weapons in the arsenal of a baseball pitcher, is another illustration of the effects of fluid flow. Figure 11.38*a* shows a baseball moving to the right with no spin. The view is from above, looking down toward the ground. In this situation, air flows with the same speed around both sides of the ball, and the pressure is the same on both sides. No net force exists to make the ball curve to either side. However, when the ball is given a spin, the air close to its surface is dragged around with it; the air on one half of the ball is speeded up (lower pressure), while that on the other half is slowed down (higher pressure). Part *b* of the picture illustrates the effects of a counterclockwise spin. The baseball experiences a net deflection force and curves on its way from the pitcher's mound to the plate, as part *c* shows.

As a final application of Bernoulli's equation, Figure 11.39*a* shows a large tank from which water is emerging through a small pipe near the bottom. Bernoulli's equation can be used to determine the speed (called the efflux speed) at which the water leaves the pipe, as the next example shows.

The Physics of...
ski jumping.

The Physics of...
a curve ball.

Figure 11.38 These views of a baseball are from above, looking down toward the ground, with the ball moving to the right. (*a*) Without spin, the ball does not curve to either side. (*b*) A spinning ball curves in the direction of the deflection force. (*c*) The spin in part *b* causes the ball to curve as shown here.

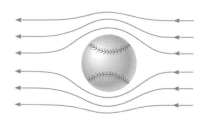

(*a*) Without spin

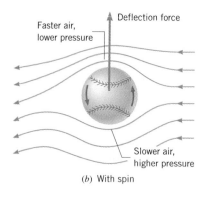

Deflection force

Faster air, lower pressure

Slower air, higher pressure

(*b*) With spin

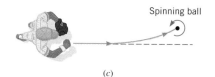

Spinning ball

(*c*)

EXAMPLE 15 • Efflux Speed

The tank in Figure 11.39a is open to the atmosphere at the top. Find an expression for the speed of the liquid leaving the pipe at the bottom.

Reasoning We assume that the liquid behaves as an ideal fluid. Therefore, we can apply Bernoulli's equation, and in preparation for doing so, we locate two points in the liquid in Figure 11.39a. Point 1 is just outside the efflux pipe, and point 2 is at the top surface of the liquid. The pressure at each of these points is equal to the atmospheric pressure, a fact that will be used to simplify Bernoulli's equation.

Solution Since the pressures at points 1 and 2 are the same, $P_1 = P_2$, Bernoulli's equation becomes $\frac{1}{2}\rho v_1^2 + \rho g y_1 = \frac{1}{2}\rho v_2^2 + \rho g y_2$. The density ρ can be eliminated algebraically from this result, which can then be solved for the square of the efflux speed v_1:

$$v_1^2 = v_2^2 + 2g(y_2 - y_1) = v_2^2 + 2gh$$

We have substituted $h = y_2 - y_1$ for the height of the liquid above the efflux tube. If the tank is very large, the liquid level changes only slowly, and the speed at point 2 can be set equal to zero, so that $\boxed{v_1 = \sqrt{2gh}}$.

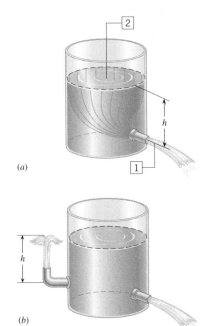

(a)

In Example 15 the liquid is assumed to be an ideal fluid, and the speed with which it leaves the pipe is the same as if the liquid had freely fallen through a height h (see Equation 2.9 with $x = h$ and $a = g$). This result is known as ***Torricelli's theorem.*** If the outlet pipe were pointed directly upward, as in part b of the drawing, the liquid would rise to a height h equal to the fluid level above the pipe. However, if the liquid is not an ideal fluid, its viscosity cannot be neglected. Then, the efflux speed would be less than that given by Bernoulli's equation, and the liquid would rise to a height less than h.

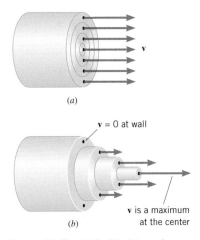

(b)

Figure 11.39 (*a*) Bernoulli's equation can be used to determine the speed of the liquid leaving the small pipe. (*b*) An ideal fluid (no viscosity) will rise to the fluid level in the tank after leaving a vertical outlet nozzle.

*11.11 VISCOUS FLOW

In an ideal fluid there is no viscosity to hinder the fluid layers as they slide past one another. Within a pipe of uniform cross section, every layer of an ideal fluid moves with the same velocity, even the layer next to the wall, as Figure 11.40a shows. When viscosity is present, the fluid layers have different velocities, as part b of the drawing illustrates. The fluid at the center of the pipe has the greatest velocity. In contrast, the fluid layer next to the wall surface does not move at all, because it is held tightly by intermolecular forces. So strong are these forces that if a solid surface moves, the adjacent fluid layer moves along with it and remains at rest *relative* to the moving surface.

To help introduce viscosity in a quantitative fashion, Figure 11.41a shows a viscous fluid between two parallel plates. The top plate is free to move while the bottom one is stationary. If the top plate is to move with a velocity **v** relative to the bottom plate, a force **F** is required. For a highly viscous fluid, like thick honey, a large force is needed; for a less viscous fluid, like water, a smaller force is necessary. As part b of the drawing suggests, we may imagine the fluid to be composed of many thin horizontal layers. When the top plate moves, the intermediate fluid layers slide over each other. The velocity of each layer is different, changing uniformly from **v** at the top plate to zero at the bottom plate. The resulting flow is called ***laminar***

Figure 11.40 (*a*) In ideal (nonviscous) fluid flow, all fluid particles across the pipe have the same velocity. (*b*) In viscous flow, the speed of the fluid is zero at the surface of the pipe and increases to a maximum along the center axis.

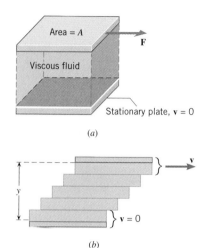

Figure 11.41 (*a*) A force **F** is applied to the top plate, which is in contact with a viscous fluid. (*b*) Because of the force **F**, the top plate and the adjacent layer of fluid move with a constant velocity **v**.

flow, since a thin layer is often referred to as a lamina. As each layer moves, it is subjected to viscous forces from its neighbors. The purpose of the force **F** is to compensate for the effect of these forces, so that any layer can move with a constant velocity.

The amount of force required in Figure 11.41*a* depends on several factors. Larger areas *A*, being in contact with more fluid, require larger forces, so that the force is proportional to the contact area ($F \propto A$). For a given area, greater speeds require larger forces, with the result that the force is proportional to the speed ($F \propto v$). The force is also inversely proportional to the perpendicular distance *y* between the top and bottom plates ($F \propto 1/y$). The larger the distance *y*, the smaller is the force required to achieve a given speed with a given contact area. These three proportionalities are related in the following manner: $F \propto Av/y$. Equation 11.13 expresses this relationship with the aid of a proportionality constant η (Greek letter *eta*), which is called the ***coefficient of viscosity*** or simply the ***viscosity.***

■ **FORCE NEEDED TO MOVE A LAYER OF VISCOUS FLUID WITH A CONSTANT VELOCITY**

The tangential force **F** required to move a fluid layer at a constant speed *v*, when the layer has an area *A* and is located a perpendicular distance *y* from an immobile surface, is given by

$$F = \frac{\eta A v}{y} \qquad (11.13)$$

where η is the coefficient of viscosity.

SI Unit of Viscosity: Pa · s

Common Unit of Viscosity: poise (P)

By solving this equation for the viscosity, $\eta = Fy/(vA)$, it can be seen that the SI unit for viscosity is N · m/[(m/s) · m²] = Pa · s. Another common unit for viscosity is the *poise* (P), which is used in the cgs system of units and is named after the French physician Jean Poiseuille (1797–1869, pronounced, approximately, as Pwah-zoy'). The following relation exists between the two units:

$$1 \text{ poise (P)} = 0.1 \text{ Pa} \cdot \text{s}$$

Values of viscosity depend on the nature of the fluid. Under ordinary conditions, the viscosities of gases are significantly *smaller* than those of liquids. Moreover, the viscosities of either liquids or gases depend markedly on temperature. Usually, the viscosities of liquids decrease as the temperature is increased. Anyone who has heated honey or oil, for example, knows that these fluids flow much more freely at an elevated temperature. In contrast, the viscosities of gases increase as the temperature is raised. In general, fluids having smaller values of η are more nearly ideal fluids, because they flow more readily with only relatively weak viscous forces impeding their movement; an ideal fluid has $\eta = 0$.

Viscous flow occurs in a wide variety of situations, such as oil moving through a pipeline or a liquid being forced through the needle of a hypodermic syringe. Figure 11.42 identifies the factors that determine the volume flow rate *Q* (in m³/s) of the fluid. First, a difference in pressure $P_2 - P_1$ must be maintained between any two locations along the pipe in order for the fluid to flow. In fact, *Q* is proportional to $P_2 - P_1$, a greater pressure difference leading to a larger flow rate. Second, a

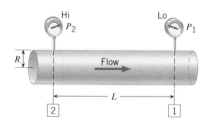

Figure 11.42 For viscous flow, the difference in pressure $P_2 - P_1$, the radius *R* and length *L* of the tube, and the viscosity η of the fluid influence the volume flow rate.

The Physics of...
pipeline pumping stations.

long pipe offers greater resistance to the flow than a short pipe does, and Q is inversely proportional to the length L. Because of this fact, long pipelines, such as the Alaskan pipeline, have pumping stations at various places along the line to compensate for a drop in pressure (see Figure 11.43). Third, high-viscosity fluids flow less readily than low viscosity fluids do, and Q is inversely proportional to the viscosity η. Finally, the volume flow rate is larger in a pipe of larger radius, other things being equal. The dependence on the radius R is a surprising one, Q being proportional to the fourth power of the radius, or R^4. If, for instance, the pipe radius is reduced to one-half of its original value, the volume flow rate is reduced to one-sixteenth of its original value, assuming the other variables remain constant. The mathematical relation for Q in terms of these parameters was discovered by Poiseuille and is known as **Poiseuille's law.**

■ **POISEUILLE'S LAW**

A fluid whose viscosity is η, flowing through a pipe of radius R and length L, has a volume flow rate Q given by

$$Q = \frac{\pi R^4 (P_2 - P_1)}{8\eta L} \qquad (11.14)$$

where P_1 and P_2 are the pressures at the ends of the pipe.

Example 16 illustrates the use of Poiseuille's law.

Figure 11.43 As oil flows along the Alaskan pipeline (top), the pressure drops because oil is a viscous fluid. Pumping stations, like this one (bottom), are located along the pipeline to compensate for the drop in pressure.

EXAMPLE 16 • Giving an Injection

A hypodermic syringe is filled with a solution whose viscosity is 1.5×10^{-3} Pa·s. As Figure 11.44 shows, the plunger area of the syringe is 8.0×10^{-5} m^2, and the length of the needle is 0.025 m. The internal radius of the needle is 4.0×10^{-4} m. The gauge pressure in a vein is 1900 Pa (14 mm of mercury). What force must be applied to the plunger, so that 1.0×10^{-6} m^3 of solution can be injected in 3.0 s?

Reasoning The necessary force is the pressure applied to the plunger times the area of the plunger. Since viscous flow is occurring, the pressure is different at different points along the syringe. However, the barrel of the syringe is so wide that little pressure difference is required to sustain the flow up to point 2, where the fluid encounters the narrow needle. Consequently, the pressure applied to the plunger is nearly equal to the pressure P_2 at point 2. To find this pressure, we apply Poiseuille's law to the needle. This law indicates that $P_2 - P_1 = 8\eta LQ/(\pi R^4)$. We note that the pressure P_1 is given as a gauge pressure, which, in this case, is the amount of pressure in excess of atmospheric pressure. This causes no difficulty, because we need to find the amount of force in excess of that applied to the plunger by the atmosphere. The volume flow rate Q can be obtained from the time needed to inject the known volume of solution.

Solution The volume flow rate is $Q = (1.0 \times 10^{-6}$ m$^3)/(3.0$ s$) = 3.3 \times 10^{-7}$ m^3/s. According to Poiseuille's law (Equation 11.14), the required pressure difference is

$$P_2 - P_1 = \frac{8\eta LQ}{\pi R^4} = \frac{8(1.5 \times 10^{-3} \text{ Pa·s})(0.025 \text{ m})(3.3 \times 10^{-7} \text{ m}^3/\text{s})}{\pi (4.0 \times 10^{-4} \text{ m})^4} = 1200 \text{ Pa}$$

Since $P_1 = 1900$ Pa, the pressure P_2 must be $P_2 = 1200$ Pa $+ 1900$ Pa $= 3100$ Pa. The force that must be applied to the plunger is this pressure times the plunger area:

$$F = (3100 \text{ Pa})(8.0 \times 10^{-5} \text{ m}^2) = \boxed{0.25 \text{ N}}$$

The Physics of...
a hypodermic syringe.

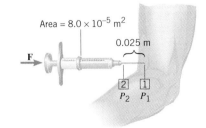

Area $= 8.0 \times 10^{-5}$ m^2

0.025 m

F

2   1
P_2 P_1

Figure 11.44 The difference in pressure $P_2 - P_1$ required to sustain the fluid flow through a hypodermic needle can be found with the aid of Poiseuille's law.

SUMMARY

Fluids are materials that can flow. The **mass density** ρ of any substance is its mass m divided by its volume V: $\rho = m/V$.

In the presence of gravity, the upper layers of a fluid push downward on the layers beneath, with the result that **fluid pressure is related to depth.** In an incompressible static fluid whose density is ρ, the relation is $P_2 = P_1 + \rho gh$, where P_1 is the pressure at one level and P_2 is the pressure at a level that is h meters deeper. The **gauge pressure** is the amount by which a pressure P differs from atmospheric pressure. The **absolute pressure** is the actual value for P.

According to **Pascal's principle,** any change in the pressure applied to a completely enclosed fluid is transmitted undiminished to all parts of the fluid and the enclosing walls.

The **buoyant force** is the net upward force that a fluid applies to any object that is immersed partially or completely in it. **Archimedes' principle** states that the magnitude of the buoyant force equals the weight of the fluid that the immersed object displaces.

The **mass flow rate** (in kg/s) of a fluid with a density ρ, flowing with a speed v in a pipe of cross-sectional area A, is the mass per second flowing past a point and is given by ρAv. The **equation of continuity** expresses the fact that mass is conserved; what flows into one end of a pipe flows out the other end, assuming there are no additional entry or exit points in between. In terms of the mass flow rate, the equation of continuity is $\rho_1 A_1 v_1 = \rho_2 A_2 v_2$, where the subscripts 1 and 2 denote two points along the pipe. If the fluid is incompressible, $\rho_1 = \rho_2$. The equation of continuity then becomes $A_1 v_1 = A_2 v_2$, where the product of the cross-sectional area and speed is called the **volume flow rate** Q; $Q = Av$.

Bernoulli's equation is a direct consequence of the work–energy theorem and describes the steady irrotational flow of an ideal fluid whose density is ρ. For any two points (1 and 2) in the fluid, this equation relates the pressure P, the speed v, and the elevation y: $P_1 + \frac{1}{2}\rho v_1^2 + \rho gy_1 = P_2 + \frac{1}{2}\rho v_2^2 + \rho gy_2$. When the flow is horizontal, Bernoulli's equation indicates that higher fluid speeds are associated with lower fluid pressures.

The **coefficient of viscosity** η is the proportionality constant that determines how much tangential force $\mathbf{F}$ is required to move a fluid layer at a constant speed v, when the layer has an area A and is located a perpendicular distance y from an immobile surface. The magnitude of the force is $F = \eta Av/y$. The SI unit of viscosity is Pa·s. A fluid whose viscosity is η, flowing through a pipe of radius R and length L, has a volume flow rate Q given by $Q = \pi R^4 (P_2 - P_1)/(8\eta L)$, where P_1 and P_2 are the pressures at the ends of the pipe. This equation is known as **Poiseuille's law.**

CONCEPTUAL QUESTIONS

1. A pile of empty aluminum cans has a volume of 1.0 m³. The density of aluminum is 2700 kg/m³. Explain why the mass of the pile of cans is *not* equal to $\rho_{Al} V = (2700 \text{ kg/m}^3)(1.0 \text{ m}^3) = 2700$ kg.

2. A method for resealing a partially full bottle of wine under a vacuum uses a specially designed rubber stopper to close the bottle. A simple pump is attached to the stopper, and to remove air from the bottle, the plunger of the pump is pulled up and then released. After about 15 pull-and-release cycles the wine is under a partial vacuum. On the 15th pull-and-release cycle, it is harder to pull up the plunger than on the first cycle. Why?

3. A person could not balance her entire weight on the pointed end of a single nail, because it would penetrate her skin. However, she can lie safely on a "bed of nails." A "bed of nails" is an array of many, many nails driven through a sheet of wood so that their pointed ends form a flat array. Why can the "bed of nails" trick be done safely?

4. ⚗ As you climb a mountain, your ears "pop" because of the changes in atmospheric pressure. In which direction does your eardrum move (a) as you climb up and (b) as you climb down? Give your reasoning.

5. A bottle of juice is sealed under partial vacuum, with a lid on which a red dot or "button" is painted. Around the button the following phrase is printed: "Button pops up when seal is broken." Explain why the button remains pushed in when the seal is intact.

6. A sealed container of water is full, except for a tube that is at-

tached to it, as the drawing shows. Water is poured into this tube, and before the tube is full the container bursts. Explain why.

7. A closed tank is completely filled with water. A valve is then opened at the bottom of the tank and water begins to flow out. When the water stops flowing, will the tank be completely empty, or will there still be a noticeable amount of water in it? Give a reason for your answer.

8. Could you use a straw to sip a drink on the moon? Justify your answer.

9. Why does the cork fly out with a loud "pop" when a bottle of champagne is opened? To answer this question on an exam a student says, "Because the gas pressure in the bottle is about 0.3×10^5 Pa." Explain whether the student is referring to absolute or gauge pressure.

10. A scuba diver is below the surface of the water when a storm approaches, dropping the air pressure above the water. Would a sufficiently sensitive pressure gauge attached to his wrist register this drop in air pressure? Give your reasoning.

11. A steel beam is suspended completely under water by a cable that is attached to one end of the beam, so it hangs vertically. Another identical beam is also suspended completely under water, but by a cable that is attached to the center of the beam so it hangs horizontally. Which beam, if either, experiences the greater buoyant force? Provide a reason for your answer. Neglect any change in water density with depth.

12. A glass beaker, filled to the brim with water, is resting on a scale. A block is placed in the water, causing some of it to spill over. The water that spills is wiped away, and the beaker is still filled to the brim. How do the initial and final readings on the scale compare if the block is made from (a) wood and (b) iron? In both cases, explain your answer, using data from Table 11.1 and Archimedes' principle.

13. On a distant planet the acceleration due to gravity is less than it is on earth. Would you float more easily in water on this planet than on earth? Account for your answer.

14. Put an ice cube in a glass, and fill the glass to the brim with water. When the ice cube melts and the temperature of the water returns to its initial value, would the water level drop, remain the same, or rise (causing water to spill out)? Explain what you would observe in terms of Archimedes' principle.

15. As a person dives toward the bottom of a swimming pool, the pressure increases noticeably. Does the buoyant force also increase? Justify your answer. Neglect any change in water density with depth.

16. Suppose that an orbiting space station of the future had a swimming pool in it. Would a buoyant force be exerted on a swimmer? Explain.

17. The drawing shows a side view of part of a system of gates that is being installed to protect the city of Venice, Italy, against flooding due to very high tides. When not in use the hollow, triangular-shaped gate is filled with seawater and rests on the ocean floor. The gate is anchored to the ocean floor by a hinge mechanism, about which it can rotate. When a dangerously high tide is

expected, water is pumped out of the gate, and the structure rotates into a position where it can protect against flooding. Explain why the gate rotates into position.

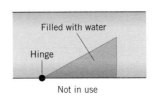 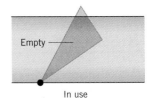

Filled with water

Hinge

Empty

Not in use

In use

18. In steady flow, the velocity **v** of a fluid particle at any point is constant in time. On the other hand, a fluid accelerates when it moves into a region of smaller cross-sectional area. (a) Explain what causes the acceleration. (b) Explain why the condition of steady flow does not rule out such an acceleration.

19. The cross-sectional area of a stream of water becomes smaller as the water falls from a faucet. Account for this phenomenon in terms of the equation of continuity. What would you expect to happen to the cross-sectional area when the water is shot upward, as it is in a fountain?

20. Have you ever had a large truck pass you from the opposite direction on a narrow two-lane road? You probably noticed that your car was pulled toward the truck as it passed. What can you conclude about the speed of the air between your car and the truck compared to that on the opposite side of the car? Provide a reason for your answer.

21. Can Bernoulli's equation describe the flow of water that is cascading down a rock-strewn spillway? Explain.

22. Hold two sheets of paper by adjacent corners, so that they hang downward. They should be parallel and slightly separated, so that you can see the floor through the gap between them. Blow air strongly down through the gap. Do the sheets move further apart, or do they come closer together? Discuss what you observe in terms of Bernoulli's equation.

23. Which way would you have to spin a baseball, so that it curves upward on its way to the plate? In describing the spin, state how you are viewing the ball. Justify your answer.

24. You are traveling on a train with your window open. As the train approaches its rather high operating speed, your ears "pop." You may have experienced a similar phenomenon in a plane climbing after takeoff or descending prior to landing. Your eardrums respond to a decrease or increase in the air pressure by "popping" outward or inward, respectively. Do your ears "pop" outward or inward on the train? Give your reasoning in terms of Bernoulli's equation.

25. A passenger is smoking in the backseat of a moving car. To remove the smoke, the driver opens a window just a bit. Explain why the smoke is drawn to and out of the driver's window.

26. The airport in Phoenix, Arizona, has occasionally been closed to large planes because of weather conditions that do not entail storms or low visibility. Instead, the conditions combine to create an unusually low air density. Using Bernoulli's equation, explain why such an air density would make it difficult for a large, heavy plane to take off, especially if the runway were not exceptionally long.

27. To change the oil in a car, you remove a plug beneath the engine and let the old oil run out. Your car has been sitting in the garage on a cold day. Before changing the oil, it is advisable to run the engine for a while. Why?

PROBLEMS

ssm Solution is in the Student Solutions Manual. **www** Solution is available on the World Wide Web at http://www.wiley.com/college/cutnell ⚕ This icon represents a biomedical application.

Section 11.1 Mass Density

1. **ssm** The ice on a lake is 0.010 m thick. The lake is circular, with a radius of 480 m. Find the mass of the ice.

2. Neutron stars consist only of neutrons and have unbelievably high densities. A typical mass and radius for a neutron star might be 2.7×10^{28} kg and 1.2×10^3 m. (a) Find the density of such a star. (b) If a dime ($V = 2.0 \times 10^{-7}$ m³) were made from this material, how much would it weigh (in pounds)?

3. A pirate in a movie is carrying a chest (0.30 m × 0.30 m × 0.20 m) that is supposed to be filled with gold. To see how ridiculous this is, determine the weight (in newtons) of the gold. To judge how large this weight is, remember that 1 N = 0.225 lb.

4. One end of a wire is attached to a ceiling, and a solid brass ball is tied to the lower end. The tension in the wire is 120 N. What is the radius of the brass ball?

5. **ssm** Accomplished silver workers in India can pound silver into incredibly thin sheets, as thin as 3.00×10^{-7} m (about one hundredth of the thickness of this sheet of paper). Find the area of such a sheet that can be formed from 1.00 kg of silver.

*6. A full can of soda has a mass of 0.416 kg. It contains 3.54×10^{-4} m³ of liquid. Assuming that the soda has the same density as water, find the volume of aluminum used to make the can.

*7. A gold prospector finds a solid rock that is composed solely of quartz and gold. The rock has a mass of 12.0 kg and a volume of 4.00×10^{-3} m³. What mass of gold is contained in the rock?

*8. Planners of an experiment are evaluating the design of a helium-filled (0 °C, 1 atm pressure) sphere of radius R. Ultrathin silver foil of thickness T is used to make the sphere, and the designers claim that the mass of helium in the sphere equals the mass of silver used. Assuming that T is much less than R, calculate the ratio T/R for such a sphere.

9. **ssm **www** An antifreeze solution is made by mixing ethylene glycol ($\rho = 1116$ kg/m³) with water. Suppose the specific gravity of such a solution is 1.0730. Assuming that the total volume of the solution is the sum of its parts, determine the volume percentage of ethylene glycol in the solution.

Section 11.2 Pressure

10. An airtight box has a removable lid of area 1.3×10^{-2} m² and negligible weight. The box is taken up a mountain where the

air pressure outside the box is 0.85×10^5 Pa. The inside of the box is completely evacuated. What is the magnitude of the force required to pull the lid off the box?

11. A rectangular table measures 0.80 m by 1.2 m. What is the magnitude and direction of the force that one atmosphere of pressure applies to the (a) top and (b) bottom surfaces of the table?

12. The circular top of a can of soda has a radius of 0.0320 m. The pull-tab has an area of 3.80×10^{-4} m². The absolute pressure of the carbon dioxide in the can is 1.40×10^5 Pa. Find the force that this gas generates (a) on the top of the can (including the pull-tab) and (b) on the pull-tab itself.

13. **ssm** High-heeled shoes can cause tremendous pressure to be applied to a floor. Suppose the radius of a heel is 6.00×10^{-3} m. At times during a normal walking motion, nearly the entire body weight acts perpendicular to the surface of such a heel. Find the pressure that is applied to the floor under the heel because of the weight of a 50.0-kg woman.

14. A storm approaches and the air pressure outside your apartment suddenly drops to 0.960×10^5 Pa from 1.013×10^5 Pa. Before the pressure inside your apartment has had time to change, what is the magnitude of the net force exerted on a window that measures 2.0 m × 3.1 m?

15. A brick weighs 17.8 N and is resting on the ground. Its dimensions are 0.203 m × 0.0890 m × 0.0570 m. A number of the bricks are then stacked on top of this one. What is the smallest number of whole bricks (including the one on the ground) that could be used, so that their weight creates a pressure of at least one atmosphere on the ground beneath the first brick? *(Hint: First decide which face of the brick is in contact with the ground.)*

*16. A cylinder is fitted with a piston, beneath which is a spring, as in the drawing. The cylinder is open at the top. Friction is ab-

sent. The spring constant of the spring is 3600 N/m. The piston has a negligible mass and a radius of 0.025 m. (a) When air beneath the piston is completely pumped out, how much does the atmospheric pressure cause the spring to compress? (b) How much work does the atmospheric pressure do in compressing the spring?

*17. **ssm** A cylinder (with circular ends) and a hemisphere are solid throughout and made from the same material. They are resting on the ground, the cylinder on one of its ends and the hemisphere on its flat side. The weight of each causes the same pressure to act on the ground. The cylinder is 0.500 m high. What is the radius of the hemisphere?

*18. A 58-kg skier is going down a 35° slope. The area of each ski in contact with the snow is 0.13 m². Determine the pressure that each ski exerts on the snow.

**19. A house has a roof (colored gray) with the dimensions shown in the drawing. Determine the magnitude and direction of the net force that the atmosphere applies to the roof, when the outside pressure rises suddenly by 10.0 mm of mercury before the pressure in the attic can adjust.

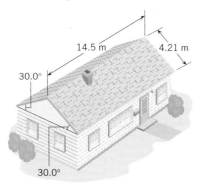

Section 11.3 Pressure and Depth in a Static Fluid, Section 11.4 Pressure Gauges

20. The deep end of a swimming pool has a depth of 3.00 m. The atmospheric pressure above the pool is 1.01×10^5 Pa. What is the pressure at the bottom of the pool?

21. **ssm** Some researchers believe that the dinosaur Barosaurus held its head erect on a long neck, much as a giraffe does. If so, fossil remains indicate that its heart would have been about 12 m below its brain. Assume that the blood has the density of water, and calculate the amount by which the blood pressure in the heart would have exceeded that in the brain. Size estimates for the single heart needed to withstand such a pressure range up to two tons. Alternatively, Barosaurus may have had a number of smaller hearts.

22. Suppose you are drinking a can of soda ($\rho = 1.0 \times 10^3$ kg/m³) using a straw. When you suck on the straw the gauge pressure inside your mouth and lungs is 1200 Pa. How high is the soda drawn up into the straw?

23. The Mariana trench is located in the Pacific Ocean at a depth of about 11 000 m below the surface of the water. The density of

seawater is 1025 kg/m³. (a) If an underwater vehicle were to explore such a depth, what force would the water exert on the vehicle's observation window (radius = 0.10 m)? (b) For comparison, determine the weight of a jetliner whose mass is 1.2×10^5 kg.

24. Review Conceptual Example 6 as an aid in understanding this problem. Consider the pump on the right-hand side of Figure 11.12, which acts to reduce the air pressure in the pipe. The air pressure outside the pipe is one atmosphere. Find the maximum depth from which this pump can extract water from the well.

25. **ssm** A water tower is a familiar sight in many towns. The purpose of such a tower is to provide storage capacity and to provide sufficient pressure in the pipes that deliver the water to customers. The drawing shows a spherical reservoir that contains 5.25×10^5 kg of water when full. The reservoir is vented to the atmosphere at the top. For a full reservoir, find the gauge pressure that the water has at the faucet in (a) house A and (b) house B. Ignore the diameter of the delivery pipes.

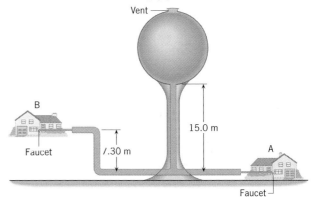

26. As background for this problem, review Conceptual Example 6. A submersible pump is put under the water at the bottom of a well and is used to push water up through a pipe. What minimum output gauge pressure must the pump generate to make the water reach the nozzle at ground level, 71 m above the pump?

27. The drawing shows an intravenous feeding. With the distance shown, nutrient solution ($\rho = 1030$ kg/m³) can just barely enter the blood in the vein. What is the gauge pressure of the venous blood? Express your answer in millimeters of mercury.

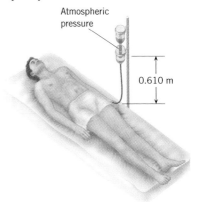

*28. A mercury barometer reads 747.0 mm on the roof of a building and 760.0 mm on the ground. Assuming a constant value of 1.29 kg/m³ for the density of air, determine the height of the building.

*29. **ssm www** Mercury is poured into a tall glass. Ethyl alcohol is then poured on top of the mercury until the height of the ethyl alcohol itself is 110 cm. The two fluids do not mix, and the air pressure at the top of the ethyl alcohol is one atmosphere. What is the absolute pressure at a point that is 7.10 cm below the ethyl alcohol–mercury interface?

*30. Two identical containers are open at the top and are connected at the bottom via a tube of negligible volume and a valve which is closed. Both containers are filled initially to the same height of 1.00 m, one with water, the other with mercury, as the drawing indicates. The valve is then opened. Water and mercury are immiscible. Determine the fluid level in the left container when equilibrium is reestablished.

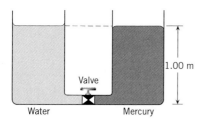

*31. The vertical face of a reservoir dam is 120 m wide and 11 m high. Find the total force that the water in a completely full reservoir exerts on this vertical face. *(Hint: The pressure varies linearly with depth, so you must use an average pressure.)*

**32. As the drawing illustrates, a pond has the shape of an inverted cone with the tip sliced off and has a depth of 5.00 m. The atmospheric pressure above the pond is 1.01×10^5 Pa. The circular top surface (radius $= R_2$) and circular bottom surface (radius $= R_1$) of the pond are both parallel to the ground. The magnitude of the force acting on the top surface is the same as the magnitude of the force acting on the bottom surface. Obtain (a) R_2 and (b) R_1.

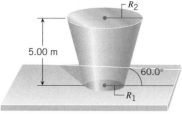

Section 11.5 Pascal's Principle

33. ssm The atmospheric pressure above a swimming pool changes from 755 to 765 mm of mercury. The bottom of the pool is a 12 m × 24 m rectangle. By how much does the force on the bottom of the pool increase?

34. In the process of changing a flat tire, a motorist uses a hydraulic jack. She begins by applying a force of $F_1 = 7.0$ N to the input piston. As a result, the output plunger applies a force of $F_2 = 420$ N to the car. The height difference between the input

piston and the output plunger can be neglected. What is the ratio A_2/A_1 of the plunger and piston areas?

35. In the hydraulic press used in a trash compactor, the radii of the input piston and the output plunger are 6.4×10^{-3} m and 5.1×10^{-2} m, respectively. If the height difference between the input piston and the output plunger can be neglected, what force is applied to the trash when the input force is 330 N?

36. The drawing shows a hydraulic chamber in which a spring (spring constant $= 1600$ N/m) is attached to the input piston, and a rock of mass 40.0 kg rests on the output plunger. The piston and plunger are at the same height, and each has a negligible mass. By how much is the spring compressed from its unstrained position?

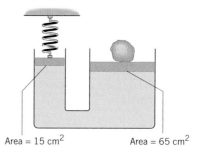

Area = 15 cm²   Area = 65 cm²

*37. **ssm** A dump truck uses a hydraulic cylinder, as the drawing illustrates. When activated by the operator, a pump injects hydraulic oil into the cylinder at an absolute pressure of 3.54×10^6 Pa and drives the output plunger, which has a radius of 0.150 m. Assuming the plunger remains perpendicular to the floor of the load bed, find the torque that the plunger creates about the axis identified in the drawing.

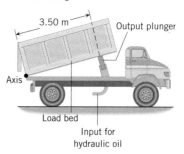

*38. The drawing shows a hydraulic system used with disc brakes. The force **F** is applied perpendicularly to the brake pedal.

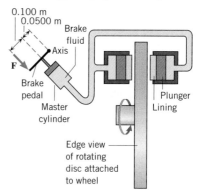

The brake pedal rotates about the axis shown in the drawing and causes a force to be applied perpendicularly to the input piston (radius = 9.50×10^{-3} m) in the master cylinder. The resulting pressure is transmitted by the brake fluid to the output plungers (radii = 1.90×10^{-2} m), which are covered with the brake linings. The linings are pressed against both sides of a disc attached to the rotating wheel. Suppose that the magnitude of **F** is 9.00 N. Assume that the input piston and the output plungers are at the same vertical level and find the force applied to each side of the revolving disc.

*39. Prepare for this problem by reviewing Example 7. The hydraulic oil in a car lift has a density of 8.00×10^2 kg/m^3. The weight of the input piston is negligible. The radii of the input piston and output plunger are 8.00×10^{-3} m and 0.140 m, respectively. What input force **F** is needed to support the 22 300-N combined weight of a car and the output plunger, when the bottom surfaces of the piston and plunger are at (a) the same level and (b) the levels shown in the drawing?

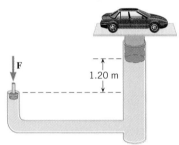

1.20 m

Section 11.6 Archimedes' Principle

40. A bodybuilder is holding a 29-kg steel barbell above her head. How much force would she have to exert if the barbell were lifted underwater?

41. **ssm** What is the radius of a hydrogen-filled balloon that would carry a load of 5750 N (in addition to the weight of the hydrogen) when the density of air is 1.29 kg/m^3?

42. A solid object floats on ethyl alcohol, with 68.2% of the object's volume submerged. Using Table 11.1, identify the substance from which the object is made.

43. What is the total mass of swimmers that the raft in Example 8 can carry and float with its top surface at water level?

44. An 80.0-kg person is wearing a life jacket and floating in water. The life jacket has a volume of 3.0×10^{-2} m^3 and is completely submerged under the water. The volume of the person's body that is underwater is 6.2×10^{-2} m^3. What is the density of the life jacket?

45. **ssm www** A person can change the volume of his body by taking air into his lungs. The amount of change can be determined by weighing the person under water. Suppose that under water a person weighs 20.0 N with partially full lungs and 40.0 N with empty lungs. Find the change in body volume.

*46. What is the smallest number of whole logs ($\rho = 725$ kg/m^3, radius = 0.0800 m, length = 3.00 m) that can be used to build a raft that will carry four people, each of whom has a mass of 80.0 kg?

*47. A spring has a spring constant of 578 N/m. When used to suspend an object in air, the spring stretches by 0.0640 m. When used to suspend the same object in water, the spring stretches by 0.0520 m. (a) What buoyant force acts on the object? (b) What is the volume of the part of the object that is underwater?

*48. To verify her suspicion that a rock specimen is hollow, a geologist weighs the specimen in air and in water. She finds that the specimen weighs twice as much in air as it does in water. The solid part of the specimen has a density of 5.0×10^3 kg/m^3. What fraction of the specimen's apparent volume is solid?

*49. **ssm** A 1967 Kennedy half-dollar has a mass of 1.150×10^{-2} kg. The coin is a mixture of silver and copper, and in water weighs 0.1011 N. Determine the mass of silver in the coin.

**50. One kilogram of glass ($\rho = 2.60 \times 10^3$ kg/m^3) is shaped into a hollow spherical shell that just barely floats in water. What are the inner and outer radii of the shell? Do not assume the shell is thin.

**51. A lighter-than-air balloon and its load of passengers and ballast are floating stationary above the earth. Ballast is weight (of negligible volume) that can be dropped overboard to make the balloon rise. The radius of this balloon is 6.25 m. Assuming a constant value of 1.29 kg/m^3 for the density of air, determine how much weight must be dropped overboard to make the balloon rise 105 m in 15.0 s.

**52. A spring is attached to the bottom of an empty swimming pool, with the axis of the spring oriented vertically. An 8.00-kg block of wood ($\rho = 840$ kg/m^3) is fixed to the top of the spring and compresses it. Then the pool is filled with water, completely covering the block. The spring is now observed to be stretched twice as much as it had been compressed. Determine the percentage of the block's total volume that is hollow. Ignore any air in the hollow space.

Section 11.8 The Equation of Continuity

53. **ssm** Water flows with a volume flow rate of 1.50 m^3/s in a pipe. Find the water speed where the pipe radius is 0.500 m.

54. Oil is flowing with a speed of 1.22 m/s through a pipeline with a radius of 0.305 m. How many gallons of oil (1 gal = 3.79×10^{-3} m^3) flow in one day?

55. Suppose that blood flows through the aorta with a speed of 0.35 m/s. The cross-sectional area of the aorta is 2.0×10^{-4} m^2. (a) Find the volume flow rate of the blood. (b) The aorta branches into tens of thousands of capillaries whose total cross-sectional area is about 0.28 m^2. What is the average blood speed through them?

56. Water flows straight down from an open faucet. The radius of the faucet is 0.75 cm and the speed of the water as it leaves is 0.85 m/s. At a distance of 0.10 m below the faucet, the speed of the water is 1.6 m/s. (a) What is the radius of the water stream at this point? (b) How long would it take to fill a bucket whose volume is 2.0×10^{-2} m^3?

*57. **ssm** A water line with an internal radius of 6.5×10^{-3} m is connected to a shower head that has 12 holes. The speed of the water in the line is 1.2 m/s. (a) What is the volume flow rate in the line? (b) At what speed does the water leave one of the holes (effective hole radius $= 4.6 \times 10^{-4}$ m) in the head?

*58. In an adjustable nozzle for a garden hose, a cylindrical plug is aligned along the axis of the hose and can be inserted into the hose opening. The purpose of the plug is to change the speed of the water leaving the hose. The speed of the water passing around the plug is to be three times greater than the speed of the water before it encounters the plug. Find the ratio of the plug radius to the inside hose radius.

Section 11.9 Bernoulli's Equation,
Section 11.10 Applications of Bernoulli's Equation

59. Prairie dogs are burrowing rodents. They do not suffocate in their burrows, because the effect of air speed on pressure creates sufficient air circulation. The animals maintain a difference in the shapes of two entrances to the burrow, and because of this difference, the air ($\rho = 1.29$ kg/m^3) blows past the openings at different speeds, as the drawing indicates. Assuming that the openings are at the same vertical level, find the difference in air pressure between the openings and indicate which way the air circulates.

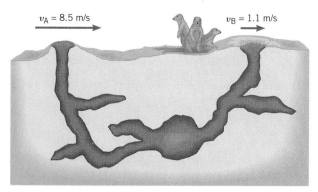

$v_A = 8.5$ m/s $v_B = 1.1$ m/s

60. Review Conceptual Example 13 before attempting this problem. The truck in that example is traveling at 27 m/s. The density of air is 1.29 kg/m^3. By how much does the pressure inside the cargo area beneath the tarpaulin exceed the outside pressure?

61. **ssm** Review Conceptual Example 13 as an aid in understanding this problem. Suppose that a 15-m/s wind is blowing across the roof of your house. The density of air is 1.29 kg/m^3. (a) Determine the reduction in pressure (below atmospheric pressure of stationary air) that accompanies this wind. (b) Explain why some roofs are "blown outward" during high winds.

62. Water is circulating through a closed pipe system in a two-floor apartment. On the first floor, the gauge pressure of the water is 9.7×10^4 Pa and its speed is 2.1 m/s. On the second floor, which is 4.0 m higher, the speed of the water is 3.7 m/s. The water speeds are different on the two floors, because the pipe diameters are different. (a) What is the gauge pressure of the water on the second floor? (b) When the water stops flowing, the gauge pressure of the water on the first floor becomes 1.1×10^5 Pa. What is the gauge pressure on the second floor?

63. A fountain sends a stream of water 5.00 m into the air. (a) Neglecting air resistance and any viscous effects, what must be

the speed of the water at the point where the water leaves the pipe feeding the fountain? At that point, the pressure is atmospheric pressure. (b) The effective cross-sectional area of the pipe is 5.00×10^{-4} m^2. How many gallons per minute are being used by the fountain? (1 gal $= 3.79 \times 10^{-3}$ m^3)

64. The water tower in the drawing is drained by a pipe that extends to the ground. The flow is nonviscous. (a) What is the absolute pressure at point 1 if the valve is *closed,* assuming that the top surface of the water at point 2 is at atmospheric pressure. (b) What is the absolute pressure at point 1 when the valve is opened and the water is flowing? Assume that the water speed at point 2 is negligible. (c) Assuming the effective cross-sectional area of the valve opening is 2.00×10^{-2} m^2, find the volume flow rate at point 1.

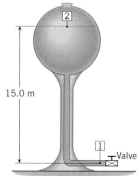

15.0 m

2

1 Valve

65. **ssm** An airplane wing is designed so that the speed of the air across the top of the wing is 251 m/s when the speed of the air below the wing is 225 m/s. The density of the air is 1.29 kg/m^3. What is the lifting force on a wing of area 24.0 m^2?

*66. A cylindrical water tunnel runs through a dam. The entrance to the tunnel has a radius of 1.30 m and is located at a depth of 22.0 m below the surface of the water. The other end of the tunnel has a radius of 0.840 m, is 46.0 m below the water's surface, and is open to the air. (a) What is the speed of the water when it exits the tunnel? (b) Determine the absolute pressure of the water just after it enters the tunnel.

*67. The construction of a flat rectangular roof (4.0 m × 5.5 m) allows it to withstand a maximum net outward force of 21 000 N. The density of the air is 1.29 kg/m^3. At what wind speed will this roof blow outward?

*68. Water is running out of a faucet, falling straight down, with an initial speed of 0.50 m/s. At what distance below the faucet is the radius of the stream reduced to one-half its value at the faucet?

*69. **ssm** A Venturi meter is a device for measuring the speed of a fluid within a pipe. The drawing shows a gas flowing at speed v_2 through a horizontal section of pipe whose cross-sectional area is

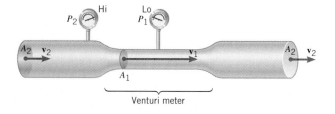

Hi   Lo
P_2 P_1

A_2 v_2 A_1 A_2 v_2

v_1

Venturi meter

$A_2 = 0.0700 \text{ m}^2$. The gas has a density of $\rho = 1.30 \text{ kg/m}^3$. The Venturi meter has a cross-sectional area of $A_1 = 0.0500 \text{ m}^2$ and has been substituted for a section of the larger pipe. The pressure difference between the two sections is $P_2 - P_1 = 120 \text{ Pa}$. Find (a) the speed v_2 of the gas in the larger original pipe and (b) the volume flow rate Q of the gas.

*70. In a very large closed tank, the absolute pressure of the air above the water is 6.01×10^5 Pa. The water leaves the bottom of the tank through a nozzle that is directed straight upward. The opening of the nozzle is 4.00 m below the surface of the water. (a) Find the speed at which the water leaves the nozzle. (b) Ignoring air resistance and viscous effects, determine the height to which the water rises.

*71. A pump and its horizontal intake pipe are located 12 m beneath the surface of a reservoir. The speed of the water in the intake pipe causes the pressure there to decrease, in accord with Bernoulli's principle. Assuming nonviscous flow, what is the maximum speed with which water can flow through the intake pipe?

**72. Two circular holes, one larger than the other, are cut in the side of a large water tank whose top is open to the atmosphere. The center of one of these holes is located twice as far beneath the surface of the water as the other. The volume flow rate of the water coming out of the holes is the same. (a) Decide which hole is located nearest the surface of the water. (b) Calculate the ratio of the radius of the larger hole to the radius of the smaller hole.

73. **ssm A uniform rectangular plate is hanging vertically downward from a hinge that passes along its left edge. By blowing air at 11.0 m/s *over the top of the plate only,* it is possible to keep the plate in a horizontal position, as illustrated in part *a* of the drawing. To what value should the air speed be reduced so that the plate is kept at a 30.0° angle with respect to the vertical, as in part *b* of the drawing? (Hint: Apply Bernoulli's equation in the form of Equation 11.12.)

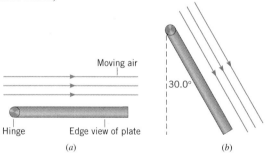

(a) Moving air / Hinge / Edge view of plate

(b) 30.0°

**74. The air speed of a plane can be measured with a Pitot-static tube, an example of which is illustrated in the drawing. The Pitot-static tube consists of two concentric tubes: The inner one is the static tube, and the outer one with holes in it is the Pitot tube. The difference in air pressure between the two is measured by the U-tube manometer in the drawing. The air speed inside the static tube is zero, because the closed tube presents an immovable obstacle to the flow of air. In contrast, the air rushing past the holes in the Pitot tube has a high speed. (a) By applying Bernoulli's equation, show that the speed of the air is $v = \sqrt{2(P_2 - P_1)/\rho}$, where $P_2 - P_1$ is the difference in pressure and ρ is the air density at the

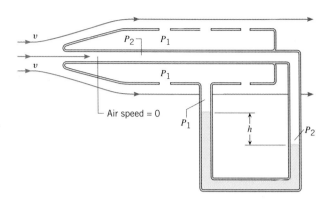

altitude of the plane. (b) By expressing $P_2 - P_1$ in terms of the height h and density ρ_0 of the fluid in the U-tube, show that the air speed can be expressed as $v = \sqrt{2gh\rho_0/\rho}$.

Section 11.11 Viscous Flow

75. A pressure difference of 1.1×10^3 Pa is needed to drive water ($\eta = 1.0 \times 10^{-3}$ Pa·s) through a pipe whose radius is 6.4×10^{-3} m. The volume flow rate of the water is 3.2×10^{-4} m^3/s. What is the length of the pipe?

76. ⌔ A blood vessel is 0.10 m in length and has a radius of 1.5×10^{-3} m. Blood ($\eta = 4 \times 10^{-3}$ Pa·s) flows at a rate of 1.0×10^{-7} m^3/s. Determine the difference in pressure that must be maintained between the two ends of the vessel.

77. **ssm www** A 1.3-m length of horizontal pipe has a radius of 6.4×10^{-3} m. Water flows with a volume flow rate of 9.0×10^{-3} m^3/s out of the right end of the pipe and into the air. What is the pressure in the flowing water at the left end of the pipe if the water behaves as (a) an ideal fluid and (b) a viscous fluid ($\eta = 1.00 \times 10^{-3}$ Pa·s)?

78. A cylindrical air duct in an air conditioning system has a length of 5.5 m and a radius of 7.2×10^{-2} m. A fan forces air ($\eta = 1.8 \times 10^{-5}$ Pa·s) through the duct, such that the air in a room (volume = 280 m^3) is replenished every ten minutes. Determine the difference in pressure between the ends of the air duct.

79. During a heavy rain, a 3.0-m × 4.6-m family room is flooded to a depth of 0.15 m. To remove the water ($\eta = 1.00 \times 10^{-3}$ Pa·s), a pump is used that does the job in two hours. The water flows out into the yard through a horizontal pipe of radius 6.4×10^{-3} m and length 6.7 m. What gauge pressure does the pump produce?

*80. Water ($\eta = 1.00 \times 10^{-3}$ Pa·s) is flowing through a horizontal pipe with a volume flow rate of 0.014 m^3/s. As the drawing

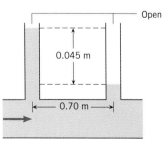

Open / 0.045 m / 0.70 m

shows, there are two vertical tubes that project from the pipe. From the data in the drawing, find the radius of the horizontal pipe.

*81. **ssm** When an object moves through a fluid, the fluid exerts a viscous force **F** on the object that tends to slow it down. For a small sphere of radius R, moving slowly with a speed v, the magnitude of the viscous force is given by Stokes' law, $F = 6\pi\eta Rv$, where η is the viscosity of the fluid. (a) What is the viscous force on a sphere of radius $R = 5.0 \times 10^{-4}$ m falling through water ($\eta = 1.00 \times 10^{-3}$ Pa·s) when the sphere has a speed of 3.0 m/s?

(b) The speed of the falling sphere increases until the viscous force balances the weight of the sphere. Thereafter, no net force acts on the sphere, and it falls with a constant speed called the terminal speed. If the sphere has a mass of 1.0×10^{-5} kg, what is its terminal speed?

*82. Two hoses are connected to the same outlet by means of a Y-connector. The other end of each hose is open to the atmosphere. One hose has a radius that is 1.50 times larger than the other, but each has the same length. Find the ratio of the average speed of the water in the larger hose to that in the smaller hose.

ADDITIONAL PROBLEMS

83. At what depth beneath the surface of a lake is the absolute pressure three times the atmospheric pressure of 1.01×10^5 Pa that acts on the lake's surface?

84. Water flows at a speed of 0.500 m/s through a horizontal hose with a radius of 0.0200 m. (a) At what speed does the water pass through a nozzle that has an effective radius of 3.00×10^{-3} m? (b) What must be the absolute pressure of the water entering the hose if the pressure at the nozzle is atmospheric pressure?

85. **ssm** A 0.10-m × 0.20-m × 0.30-m block is suspended from a wire and is completely under water. What buoyant force acts on the block?

86. A meat baster consists of a squeeze bulb attached to a plastic tube. When the bulb is squeezed and released, with the open end of the tube under the surface of the basting sauce, the sauce rises in the tube and can then be squirted over the meat. Suppose water rises 0.15 m in the tube when it is tested (see drawing). (a) Find the absolute pressure in the bulb, assuming that atmospheric pressure has the value of 1.013×10^5 Pa. (b) Would you expect this device to work better or worse on top of a mountain? Explain.

0.15 m

87. A two-lane highway is 150 km in length and has a width of 7.0 m. The solid concrete roadbed has a depth of 0.30 m. Find the weight of the concrete in the highway. (For comparison, the weight of a typical car is about 1.3×10^5 N.)

88. The *karat* is a dimensionless unit that is used to indicate the proportion of gold in a gold-containing alloy. An alloy that is one karat gold contains a weight of pure gold that is one part in twenty-four. What is the volume of gold in a 14.0-karat gold necklace whose weight is 1.27 N?

89. **ssm** A patient recovering from surgery is being given

fluid intravenously. The fluid has a density of 1030 kg/m³, and 9.5×10^{-4} m³ of it flows into the patient every six hours. Find the mass flow rate in kg/s.

90. Poiseuille's law remains valid as long as the fluid flow is laminar. For sufficiently high speed, however, the flow becomes turbulent, even if the fluid is moving through a smooth pipe with no restrictions. It is found experimentally that the flow is laminar as long as the *Reynolds number* Re is less than about 2000: Re = $2\bar{v}\rho R/\eta$. Here $\bar{v}$, ρ, and η are, respectively, the average speed, density, and viscosity of the fluid, and R is the radius of the pipe. Calculate the highest average speed that blood ($\rho = 1060$ kg/m³) could have and still remain in laminar flow when it flows through the aorta ($R = 8.0 \times 10^{-3}$ m).

91. A hydrometer is a device used to measure the density of a liquid. The hydrometer is a cylindrical tube that is weighted at one end, so that it floats with the heavier end downward. As the drawing illustrates, the hydrometer is contained inside a large "medicine dropper," into which the liquid is drawn using a squeeze bulb. The weighted hydrometer tube has a mass of 6.00×10^{-3} kg and a radius of 5.00×10^{-3} m. How far from the bottom of the tube should the mark be put that denotes (a) battery acid whose density is 1280 kg/m³ and (b) antifreeze solution whose density is 1073 kg/m³?

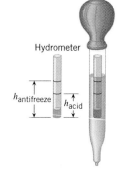

Hydrometer

$h_{antifreeze}$ h_{acid}

92. Water is flowing out of a horizontal pipe (radius = 1.9×10^{-2} m) onto the ground. The speed of the water in the pipe is 0.62 m/s. A nozzle (radius = 4.8×10^{-3} m) is attached to the end of the pipe. By how many pascals does the pressure of the water within the pipe increase?

93. **ssm** The main water line enters a house on the first floor.

The line has a gauge pressure of 1.90×10^5 Pa. (a) A faucet on the second floor, 6.50 m above the first floor, is turned off. What is the gauge pressure at this faucet? (b) How high could a faucet be before no water would flow from it, even if the faucet were open?

94. Only a small part of an iceberg protrudes above the water, while the bulk lies below the surface. The density of ice is 917 kg/m³ and that of seawater is 1025 kg/m³. Find the percentage of the iceberg's volume that lies below the surface.

*95. Three fire hoses are connected to a fire hydrant. Each hose has a radius of 0.020 m. Water enters the hydrant through an underground pipe of radius 0.080 m. In this pipe the water has a speed of 3.0 m/s. (a) How many kilograms of water are poured onto a fire in one hour? (b) Find the water speed in each hose.

*96. A hollow cubical box is 0.30 m on an edge. This box is floating in a lake with one-third of its height beneath the surface. The walls of the box have a negligible thickness. Water is poured into the box. What is the depth of the water in the box at the instant the box begins to sink?

*97. **ssm** A tube is sealed at both ends and contains a 0.0100-m long portion of liquid. The length of the tube is large compared to 0.0100 m. There is no air in the tube, and the vapor in the space above the liquid may be ignored. The tube is whirled around in a horizontal circle at a constant angular speed. The axis of the rotation passes through one end of the tube, and during the motion, the liquid collects at the other end. The pressure experienced by the liquid is the same as it would experience at the bottom of the tube, if the tube were completely filled with liquid and allowed to hang vertically. Find the angular speed (in rad/s) of the tube.

*98. A 1.00-m-tall container is filled to the brim, partway with mercury and the rest of the way with water. The container is open to the atmosphere. What must be the depth of the mercury, so the absolute pressure on the bottom of the container is twice the atmospheric pressure?

*99. An irregularly shaped chunk of concrete has a hollow spherical cavity inside. The mass of the chunk is 33 kg, and the volume enclosed by the outside surface of the chunk is 0.025 m³. What is the radius of the spherical cavity?

*100. A liquid is flowing through a horizontal pipe whose radius is 0.0200 m. The pipe bends straight upward through a height of 10.0 m and joins another horizontal pipe whose radius is 0.0400 m. What volume flow rate will keep the pressures in the two horizontal pipes the same?

101. ssm A solid cylinder (radius = 0.150 m, height = 0.120 m) has a mass of 7.00 kg. This cylinder is floating in water.

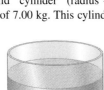

Then oil ($\rho = 725$ kg/m³) is poured on top of the water until the situation shown in the drawing results. How much of the height of the cylinder is in the oil?

102. The drawing shows a cylinder fitted with a piston that has a mass m_1 of 0.500 kg and a radius of 2.50×10^{-2} m. The top of the piston is open to the atmosphere. The pressure beneath the piston is maintained at a reduced (but constant) value by means of the pump. As shown, a rope of negligible mass is attached to the piston and passes over two massless pulleys. The other end of the rope is attached to a block that has a mass of $m_2 = 9.50$ kg. The block falls from rest down through a distance of 1.25 m in 3.30 s. Ignoring friction, find the absolute pressure beneath the piston.

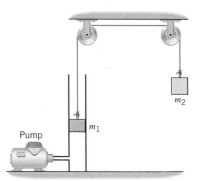

103. A siphon tube is useful for removing liquid from a tank. The siphon tube is first filled with liquid, and then one end is inserted into the tank. Liquid then drains out the other end, as the drawing illustrates. (a) Using reasoning similar to that employed in obtaining Torricelli's theorem, derive an expression for the speed v of the fluid emerging from the tube. This expression should give v in terms of the vertical height y and the acceleration due to gravity g. (Note that this speed does not depend on the depth d of the tube below the surface of the liquid.) (b) At what value of the vertical distance y will the siphon stop working? (c) Derive an expression for the absolute pressure at the highest point in the siphon (point A) in terms of the atmospheric pressure P_0, the fluid density ρ, g, and the heights h and y. (Note that the fluid speed at point A is the same as the speed of the fluid emerging from the tube, because the cross-sectional area of the tube is the same everywhere.)

TEMPERATURE AND HEAT

This volcanic eruption in Hawaii is spewing red-hot lava. The temperature is high because of the heat absorbed by the lava deep within the earth.

12.1 COMMON TEMPERATURE SCALES

To measure temperature we use a thermometer. Many thermometers make use of the fact that materials usually expand when their temperatures increase. Figure 12.1 shows the common mercury-in-glass thermometer, which utilizes the expansion of liquid mercury to indicate the temperature. The thermometer consists of a mercury-filled glass bulb connected to a capillary tube. When the mercury is heated, it expands into the capillary tube, the amount of expansion being proportional to the change in temperature. The outside of the glass is marked with an appropriate scale for reading the temperature.

A number of different temperature scales have been devised, two popular choices being the **Celsius** (formerly, centigrade) and **Fahrenheit scales.** Figure 12.1 illustrates these scales. Historically,* both scales were defined by assigning two temperature points on the scale and then dividing the distance between them into a number of equally spaced intervals. One point was chosen to be the temperature at which ice melts under one atmosphere of pressure (the "ice point"), and the other was the temperature at which water boils under one atmosphere of pressure (the "steam point"). On the Celsius scale, an ice point of 0 °C (0 degrees Celsius) and a steam point of 100 °C were selected. On the Fahrenheit scale, an ice point of 32 °F (32 degrees Fahrenheit) and a steam point of 212 °F were chosen. The Celsius scale is used worldwide, while the Fahrenheit scale is used mostly in the United States, often in home medical thermometers.

There is a subtle difference in the way the temperature of an object is reported, as compared to a *change* in its temperature. For example, the temperature of the human body is about 37 °C, where the symbol °C stands for "degrees Celsius." However, the *change* between two temperatures is specified in "Celsius degrees" (C°)—not in "degrees Celsius." Thus, if the body temperature rises to 39 °C, the change in temperature is 2 Celsius degrees or 2 C°, not 2 °C.

As Figure 12.1 indicates, the separation between the ice and steam points on the Celsius scale is divided into 100 Celsius degrees, while on the Fahrenheit scale the separation is divided into 180 Fahrenheit degrees. Therefore, the size of the Celsius degree is larger than that of the Fahrenheit degree by a factor of $\frac{180}{100}$, or $\frac{9}{5}$. Examples 1 and 2 illustrate how to convert between the Celsius and Fahrenheit scales using this factor.

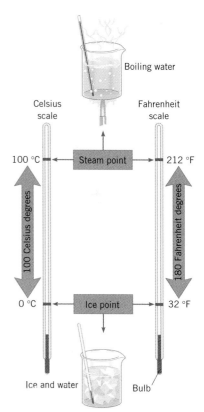

Figure 12.1 The Celsius and Fahrenheit temperature scales.

EXAMPLE 1 • Converting from a Fahrenheit to a Celsius Temperature

A healthy person has an oral temperature of 98.6 °F. What would this reading be on the Celsius scale?

Reasoning and Solution A temperature of 98.6 °F is 66.6 Fahrenheit degrees above the ice point of 32.0 °F. Since 1 C° = $\frac{9}{5}$ F°, the difference of 66.6 F° is equivalent to

$$(66.6 \text{ F}°)\left(\frac{1 \text{ C}°}{\frac{9}{5}\text{ F}°}\right) = 37.0 \text{ C}°$$

Thus, the person's temperature is 37.0 Celsius degrees above the ice point. Adding 37.0 Celsius degrees to the ice point of 0 °C on the Celsius scale gives a Celsius temperature of $\boxed{37.0 \text{ °C}}$.

---

* Today, the Celsius and Fahrenheit scales are defined in terms of the Kelvin temperature scale; Section 12.2 discusses the Kelvin scale.

EXAMPLE 2 • Converting From a Celsius to a Fahrenheit Temperature

A time and temperature sign on a bank indicates the outdoor temperature is −20.0 °C. Find the corresponding temperature on the Fahrenheit scale.

Reasoning and Solution The temperature of −20.0 °C is 20.0 Celsius degrees *below* the ice point of 0 °C. This number of Celsius degrees corresponds to

$$(20.0 \ \text{C}^\circ) \left(\frac{\frac{9}{5} \ \text{F}^\circ}{1 \ \text{C}^\circ} \right) = 36.0 \ \text{F}^\circ$$

The temperature, then, is 36.0 Fahrenheit degrees below the ice point. Subtracting 36.0 Fahrenheit degrees from the ice point of 32.0 °F on the Fahrenheit scale gives a Fahrenheit temperature of $\boxed{-4.0 \ ^\circ\text{F}}$.

The reasoning strategy used in Examples 1 and 2 for converting between the Celsius and Fahrenheit scales is summarized below.

REASONING STRATEGY

Converting Between Different Temperature Scales

1. Determine the magnitude of the difference between the stated temperature and the ice point on the initial scale.

2. Convert this number of degrees from one scale to the other scale by using the fact that $1 \ \text{C}^\circ = \frac{9}{5} \ \text{F}^\circ$.

3. Add or subtract the number of degrees on the new scale to or from the ice point on the new scale.

12.2 THE KELVIN TEMPERATURE SCALE

THE SCALE ITSELF

Although the Celsius and Fahrenheit scales are widely used, the ***Kelvin temperature scale*** has greater scientific significance. It was introduced by the Scottish physicist William Thompson (Lord Kelvin, 1824–1907), and in his honor each degree on the scale is called a kelvin (K). By international agreement, the symbol K is not written with a degree sign (°), nor is the word "degrees" used when quoting temperatures. For example, a temperature of 300 K (not 300 °K) is read as "three hundred kelvins," not "three hundred degrees kelvin." The kelvin is the SI base unit for temperature.

Figure 12.2 compares the Kelvin and Celsius scales. The size of one kelvin is identical to that of one Celsius degree, for there are one hundred divisions between the ice and steam points on both scales. As we will discuss shortly, experiments have shown that there exists a lowest possible temperature, below which no substance can be cooled. This lowest temperature is defined to be the zero point on the Kelvin scale, and is referred to as absolute zero. Moreover, the ice point (0 °C) occurs at 273.15 K on the Kelvin scale. Thus, the Kelvin temperature T and the Celsius temperature T_c are related by

$$T = T_c + 273.15 \tag{12.1}$$

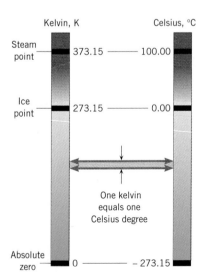

Figure 12.2 A comparison of the Kelvin and Celsius temperature scales.

THE CONSTANT-VOLUME GAS THERMOMETER

The number 273.15 in Equation 12.1 is an experimental result, obtained in studies that utilize a gas-based thermometer. When a gas is heated, its pressure increases, and when a gas is cooled, its pressure decreases, assuming the gas is confined to a fixed volume. For example, the air pressure in automobile tires can rise by as much as 20% after the car has been driven and the tires have become warm. The change in gas pressure with temperature is the basis for the ***constant-volume gas thermometer.***

A constant-volume gas thermometer consists of a gas-filled bulb to which a pressure gauge is attached, as in Figure 12.3. The gas is often hydrogen or helium at a low density, and the pressure gauge can be a U-tube manometer filled with mercury. The bulb is placed in thermal contact with the substance whose temperature is being measured. The volume of the gas is held constant by raising or lowering the *right* column of the U-tube manometer in order to keep the mercury level in the *left* column at the same reference level. The pressure of the gas is proportional to the height h of the mercury on the right. As the temperature changes, the pressure changes and can be used to indicate the temperature, once the constant-volume gas thermometer has been calibrated.

ABSOLUTE ZERO

Suppose the pressure of the gas in Figure 12.3 is measured at different temperatures. If the results are plotted on a pressure versus temperature graph, a straight line is obtained, as in Figure 12.4. If the straight line is extended or extrapolated to lower and lower temperatures, the line crosses the temperature axis at -273.15 °C. In reality, no gas can be cooled to this temperature, because all gases liquify before reaching it. However, helium and hydrogen liquify at such low temperatures that they are often used in the thermometer. This kind of graph can be obtained for different amounts and types of low-density gases. In all cases, it is found that the straight line extrapolates back to -273.15 °C on the temperature axis, which suggests that the value of -273.15 °C has fundamental significance. The significance of this number is that it is the ***absolute zero point*** for temperature measurement. The phrase "absolute zero" means that temperatures lower than -273.15 °C cannot be reached by continually cooling a gas or any other substance. If lower temperatures could be reached, then further extrapolation of the straight line in Figure 12.4 would suggest that negative absolute gas pressures could exist. Such a situation would be impossible, because a negative absolute gas pressure has no meaning. Thus, the Kelvin scale is chosen so that its zero temperature point is the lowest temperature attainable.

12.3 THERMOMETERS

All thermometers make use of the change in some physical property with temperature. A property that changes with temperature is called a ***thermometric property.*** For example, the thermometric property of the mercury thermometer is the length of the mercury column, while in the constant-volume gas thermometer it is the pressure of the gas. Several other thermometers and their thermometric properties will now be discussed.

The *thermocouple* is a thermometer used extensively in scientific laboratories.

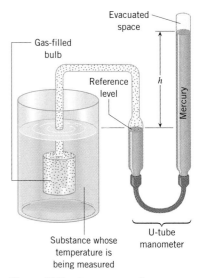

Figure 12.3 A constant-volume gas thermometer.

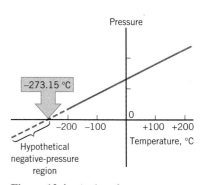

Figure 12.4 A plot of pressure versus temperature for a low-density gas at constant volume. The graph is a straight line and, when extrapolated (dashed line), crosses the temperature axis at -273.15 °C.

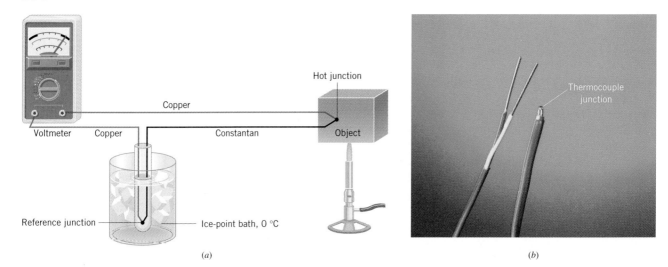

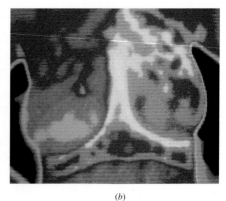

(a)   (b)

Figure 12.5 (*a*) A thermocouple is made from two different types of wires, copper and constantan in this case. (*b*) A thermocouple junction between two different wires.

The thermocouple consists of thin wires of different metals, welded together at the ends to form two junctions, as Figure 12.5 illustrates. Often the metals are copper and constantan (a copper–nickel alloy). One of the junctions, called the "hot" junction, is placed in thermal contact with the object whose temperature is being measured. The other junction, termed the "reference" junction, is kept at a known constant temperature (usually an ice–water mixture at 0 °C). The thermocouple generates a "voltage" that depends on the *difference in temperature* between the two junctions. This voltage is the thermometric property and is measured by a voltmeter, as the drawing indicates. With the aid of calibration tables, the temperature of the hot junction can be obtained from the voltage. Thermocouples are used to measure temperatures as high as 2300 °C or as low as −270 °C.

Most substances offer resistance to the flow of electricity. Because this electrical resistance changes with temperature, electrical resistance is another thermometric property. *Electrical resistance thermometers* are often made from platinum wire, because platinum has excellent mechanical and electrical properties in the temperature range from −270 °C to +700 °C. The electrical resistance of platinum wire is known as a function of temperature. Thus, the temperature of a substance can be determined by placing the resistance thermometer in thermal contact with the substance and measuring the resistance of the platinum wire.

Radiation emitted by an object can also be used to indicate the temperature of the object. At low to moderate temperatures, the predominant type of radiation

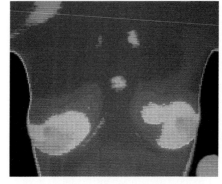

Figure 12.6 (*a*) Healthy breasts register a predominantly bluish color, indicating relatively cool temperatures. (*b*) The breast on the right in this thermograph has an invasive carcinoma (cancer) and registers colors from red to yellow/white, indicating markedly elevated temperatures.

(a)   (b) emitted is infrared radiation. As the temperature is raised, the intensity of the radiation increases substantially. In one interesting application, an infrared camera registers the intensity of the infrared radiation produced at different locations on the human body. The camera is connected to a color monitor that displays the different infrared intensities as different colors. This "thermal painting" of the body is called a *thermograph* or *thermogram*. Thermography is an important diagnostic tool in medicine. For example, breast cancer may be indicated in a thermograph by the elevated temperatures often associated with malignant tissue. Figure 12.6 shows typical thermographs used to diagnose breast cancer. Figure 12.7 shows a thermogram of the head and neck in which the sympathetic nerves on the patient's right side have been blocked by a local anesthetic. With the nerves blocked, the temperature rises by several Celsius degrees, as is evident from the red color in the thermogram.

Oceanographers and meteorologists use thermographs extensively to map the temperature distribution on the surface of the earth. For example, Figure 12.8 shows a satellite image of the eastern United States and Canada, and a portion of the Atlantic Ocean. The image vividly shows the temperature variations in the ocean, from a warm 80 °F in the south to a cool 40 °F in the north. The Gulf Stream is also evident in the picture, showing up as a deep red color. It moves up the southeastern coast of the United States and heads out into the Atlantic near the middle of the picture.

The Physics of... thermography.

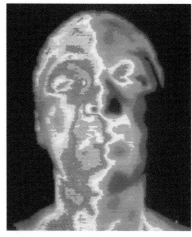

Figure 12.7 A thermogram of the face and neck after the sympathetic nerves on the patient's right side (the left side of the thermogram) have been blocked by a local anesthetic. The temperature in the red area is an average of 35.75 °C, and in the dark blue area it is 33.25 °C.

12.4 LINEAR THERMAL EXPANSION

NORMAL SOLIDS

Have you ever found the metal lid on a glass jar too tight to open? One solution is to run hot water over the lid, which loosens because the metal expands more than the glass does. To varying extents, most materials expand when heated and contract when cooled. The increase in any one dimension of a solid is called **linear expansion**, linear in the sense that the expansion occurs along a line. Figure 12.9 illustrates the linear expansion of a rod whose length is L_0 when the temperature is T_0. When the temperature increases to $T_0 + \Delta T$, the length becomes $L_0 + \Delta L$, where ΔT and ΔL are the magnitudes of the changes in temperature and length, respec-

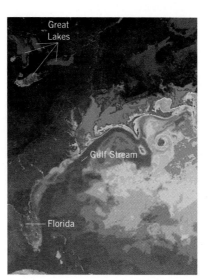

Figure 12.8 A thermograph of the eastern United States and Canada, and the Atlantic Ocean. The warmest water is red (about 80 °F), cooling to yellow, green, blue, and purple (about 40 °F) in the northern region.

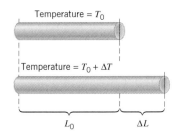

Figure 12.9 When the temperature of a rod is raised by an amount ΔT, the length of the rod increases by an amount ΔL.

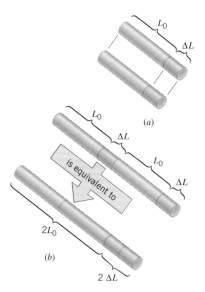

Figure 12.10 (*a*) Each of two identical rods expands by an amount ΔL when heated. (*b*) When the two rods are combined into a single rod of length $2L_0$, the "combined" rod expands by $2\,\Delta L$.

tively. Conversely, when the temperature decreases to $T_0 - \Delta T$, the length decreases to $L_0 - \Delta L$.

For modest temperature changes, experiments show that the change in length is directly proportional to the change in temperature ($\Delta L \propto \Delta T$). In addition, the change in length is proportional to the initial length of the rod, a fact that can be understood by considering Figure 12.10. Part *a* of the drawing shows two identical rods. Each rod has a length L_0 and expands by ΔL when the temperature increases by ΔT. Part *b* shows the two heated rods combined into a single rod, for which the total expansion is the sum of the expansions of each part, namely, $\Delta L + \Delta L = 2\,\Delta L$. Clearly, the amount of expansion doubles if the rod is twice as long to begin with. In other words, the change in length ΔL is directly proportional to the original length L_0 ($\Delta L \propto L_0$). Equation 12.2 expresses the fact that ΔL is proportional to both L_0 and ΔT ($\Delta L \propto L_0 \Delta T$) by using a proportionality constant α, which is called the ***coefficient of linear expansion.***

■ **LINEAR THERMAL EXPANSION OF A SOLID**

The length L_0 of an object changes by an amount ΔL when its temperature changes by an amount ΔT:

$$\Delta L = \alpha L_0 \Delta T \qquad (12.2)$$

where α is the coefficient of linear expansion.

Common Unit for the Coefficient of Linear Expansion: $\dfrac{1}{\text{C}^\circ} = (\text{C}^\circ)^{-1}$

Solving Equation 12.2 for α shows that $\alpha = \Delta L/(L_0 \Delta T)$. Since the length units of ΔL and L_0 cancel, the coefficient of linear expansion α has the unit of $(\text{C}^\circ)^{-1}$ when the temperature difference ΔT is expressed in Celsius degrees (C°). Different materials with the same initial length expand and contract by different amounts as the temperature changes, so the value of α depends on the nature of the material. Table 12.1 shows some typical values. Coefficients of linear expansion also vary somewhat depending on the range of temperatures involved, but the values in Table 12.1 are adequate approximations. Example 3 deals with a situation where a dramatic effect due to thermal expansion can be observed, even though the change in temperature is small.

EXAMPLE 3 • Buckling of a Sidewalk

A concrete sidewalk is constructed between two buildings on a day when the temperature is 25 °C. The sidewalk consists of two slabs, each three meters in length and of negligible thickness (Figure 12.11*a*). As the temperature rises to 38 °C, the slabs expand, but no space is provided for thermal expansion. The buildings do not move, so the slabs buckle upward. Determine the vertical distance y in part *b* of the drawing.

Reasoning and Solution The change in length of each slab can be determined by using the coefficient of linear expansion for concrete given in Table 12.1 and noting that the change in temperature is 13 C°:

$$\Delta L = \alpha L_0 \Delta T = [12 \times 10^{-6}\,(\text{C}^\circ)^{-1}](3.0\text{ m})(13\text{ C}^\circ) = 0.000\,47\text{ m} \quad (12.2)$$

The expanded length of each slab is 3.000 47 m. The vertical distance y can be obtained by applying the Pythagorean theorem to the right triangle in Figure 12.11*b*:

$$y = \sqrt{(3.000\,47\text{ m})^2 - (3.000\,00\text{ m})^2} = \boxed{0.053\text{ m}}$$

Table 12.1 Coefficients of Thermal Expansion for Solids and Liquids[a]

Substance	Coefficient of Thermal Expansion, $(C°)^{-1}$	
	Linear (α)	Volumetric (β)
Solids		
Aluminum	23×10^{-6}	69×10^{-6}
Brass	19×10^{-6}	57×10^{-6}
Concrete	12×10^{-6}	36×10^{-6}
Copper	17×10^{-6}	51×10^{-6}
Glass (common)	8.5×10^{-6}	26×10^{-6}
Glass (Pyrex)	3.3×10^{-6}	9.9×10^{-6}
Gold	14×10^{-6}	42×10^{-6}
Iron or steel	12×10^{-6}	36×10^{-6}
Lead	29×10^{-6}	87×10^{-6}
Nickel	13×10^{-6}	39×10^{-6}
Quartz (fused)	0.50×10^{-6}	1.5×10^{-6}
Silver	19×10^{-6}	57×10^{-6}
Liquids[b]		
Benzene	—	1240×10^{-6}
Carbon tetrachloride	—	1240×10^{-6}
Ethyl alcohol	—	1120×10^{-6}
Gasoline	—	950×10^{-6}
Mercury	—	182×10^{-6}
Methyl alcohol	—	1200×10^{-6}
Water	—	207×10^{-6}

[a] The values for α and β pertain to a temperature near 20 °C.
[b] Since liquids do not have fixed shapes, the coefficient of linear expansion is not defined for them.

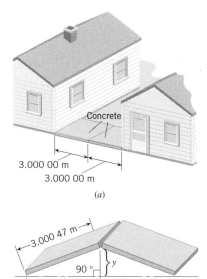

Figure 12.11 (*a*) Two concrete slabs completely fill the space between the buildings. (*b*) When the temperature increases, each slab expands, causing the sidewalk to buckle.

The buckling of a sidewalk is one consequence of not providing sufficient room for thermal expansion, and Figure 12.12*a* shows another. It is common for engineers to incorporate expansion joints or spaces at intervals along railroad tracks and bridge roadbeds to alleviate such problems. Part *b* of the figure shows such an expansion joint in a bridge.

(*a*)

(*b*)

Figure 12.12 (*a*) The rails buckled because inadequate allowance was made for thermal expansion. (*b*) An expansion joint in a bridge.

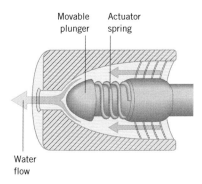

Figure 12.13   An antiscalding device.

While Example 3 shows how thermal expansion can cause problems, there are also times when it can be useful. For instance, each year thousands of children are taken to emergency rooms suffering from burns caused by scalding tap water. Such accidents can be reduced with the aid of the antiscalding device shown in Figure 12.13. This device screws onto the end of a faucet and quickly shuts off the flow of water when it becomes too hot. As the water temperature rises, the actuator spring expands and pushes the plunger forward, shutting off the flow. When the water cools, the spring contracts and the water flow resumes.

THERMAL STRESS

If the concrete slabs in Figure 12.11 had not buckled upward, they would have been subjected to immense forces from the buildings. The forces needed to keep a solid object from expanding must be strong enough to counteract any change in length that would occur due to a change in temperature. Although the change in temperature may be small, the forces—and hence the stresses—can be enormous, as Example 4 illustrates.

EXAMPLE 4 • The Stress on a Steel Beam

A steel beam is used in the roadbed of a bridge. The beam is mounted between two concrete supports when the temperature is 23 °C, with no room provided for thermal expansion (Figure 12.14). What compressional stress must the concrete supports apply to each end of the beam, if they are to keep the beam from expanding when the temperature rises to 42 °C?

Reasoning   Recall from Section 10.2 that the stress (force per unit cross-sectional area or F/A) required to change the length L_0 of an object by an amount ΔL is

$$\text{Stress} = Y \frac{\Delta L}{L_0} \qquad (10.1)$$

where Y is Young's modulus. If the steel beam were free to expand because of the change in temperature, the length would change by $\Delta L = \alpha L_0 \Delta T$. Because the concrete supports do not permit any expansion, they must supply a stress to compress the beam by an amount ΔL. Thus,

$$\text{Stress} = Y \frac{\Delta L}{L_0} = Y \frac{\alpha L_0 \Delta T}{L_0} = Y \alpha \Delta T$$

Solution   Young's modulus and the coefficient of linear expansion for steel are $Y = 2.0 \times 10^{11}$ N/m² (Table 10.1) and $\alpha = 12 \times 10^{-6}$ (C°)$^{-1}$ (Table 12.1), respectively. The change in temperature from 23 to 42 °C is $\Delta T = 19$ C°. The thermal stress is

Stress $= Y \alpha \Delta T = (2.0 \times 10^{11}$ N/m²$)[12 \times 10^{-6}$ (C°)$^{-1}](19$ C°$) =$

$$\boxed{4.6 \times 10^7 \text{ N/m}^2}$$

This stress is enormous. If the beam has a cross-sectional area of $A = 0.10$ m², the force applied to each end by a concrete support is $F = (\text{Stress})A = 4.6 \times 10^6$ N (over one million pounds).

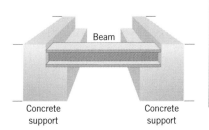

Figure 12.14   A steel beam is mounted between concrete supports with no room provided for thermal expansion of the beam.

THE BIMETALLIC STRIP

A *bimetallic strip* is made from two thin strips of metal that have *different* coefficients of linear expansion, as Figure 12.15a shows. Often brass [$\alpha = 19 \times$

$10^{-6}\,(\text{C}°)^{-1}]$ and steel $[\alpha = 12 \times 10^{-6}\,(\text{C}°)^{-1}]$ are selected. The two pieces are welded or riveted together. When the bimetallic strip is heated, the brass, having the larger value of α, expands more than the steel. Since the two metals are bonded together, the bimetallic strip bends into an arc as in part *b*, with the longer brass piece having a larger radius than the steel piece. When the strip is cooled, the bimetallic strip bends in the opposite direction, as in part *c*.

Bimetallic strips are frequently used as adjustable automatic switches in electrical appliances. Figure 12.16 shows an automatic coffee maker that turns off when the coffee is brewed to the selected strength. In part *a*, while the brewing cycle is on, electricity passes through the heating coil that heats the water. The electricity can flow, because the contact mounted on the bimetallic strip touches the contact mounted on the "strength" adjustment knob, thus providing a continuous path for the electricity. When the bimetallic strip gets hot enough to bend away, as in part *b* of the drawing, the contacts separate. The electricity stops, because it no longer has a continuous path along which to flow, and the brewing cycle is shut off. Turning the "strength" knob adjusts the brewing time by adjusting the distance through which the bimetallic strip must bend for the contact points to separate.

THE EXPANSION OF HOLES

An interesting example of linear expansion occurs when there is a hole in a piece of solid material. We know that the material itself expands when heated. But what about the hole? Does it expand, contract, or remain the same? Conceptual Example 5 provides some insight into the answer of this important question.

CONCEPTUAL EXAMPLE 5 • Do Holes Expand or Contract When the Temperature Increases?

Figure 12.17*a* shows eight square tiles that are arranged to form a square pattern with a hole in the center. If the tiles are heated, what happens to the size of the hole?

The Physics of...
an automatic coffee maker.

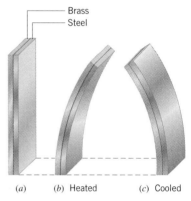

Brass
Steel

(*a*) (*b*) Heated (*c*) Cooled

Figure 12.15 (*a*) A bimetallic strip and how it behaves when (*b*) heated and (*c*) cooled.

Heating coil

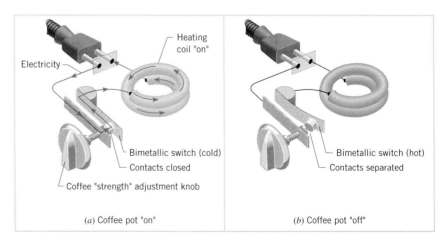

Electricity —

Heating coil "on"

Bimetallic switch (cold)
Contacts closed

Coffee "strength" adjustment knob

(*a*) Coffee pot "on"

Bimetallic switch (hot)
Contacts separated

(*b*) Coffee pot "off"

Figure 12.16 A bimetallic strip controls the brewing time on this automatic coffee maker. (*a*) When the bimetallic strip is cold, electricity flows through the heating coil. (*b*) When sufficiently hot, the bimetallic strip bends away from the adjustment knob contact, thus turning off the electricity.

Figure 12.17 (*a*) The tiles are arranged to form a square pattern with a hole in the center. (*b*) When the tiles are heated, the pattern, as well as the hole in the center, expands. (*c*) The expanded hole is the same size as a heated tile.

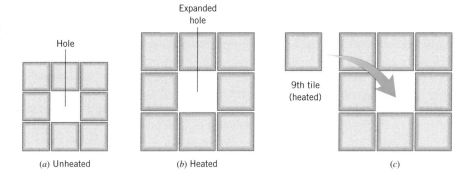

(*a*) Unheated   (*b*) Heated   (*c*)

Reasoning and Solution We can analyze this problem by disassembling the pattern into separate tiles, heating them, and then reassembling the pattern. Because each tile expands upon heating, it is evident from Figure 12.17*b* that the heated pattern expands and so does the hole in the center. In fact, if we had a ninth tile that was identical to and also heated like the others, it would fit exactly into the center hole, as Figure 12.17*c* indicates. Thus, not only does the hole in the pattern expand, but it expands exactly as much as one of the tiles. Since the ninth tile is made of the same material as the others, we see that *the hole expands just as if it were made of the material of the surrounding tiles.* The thermal expansion of the hole and the surrounding material is analogous to a photographic enlargement; in both situations everything is enlarged, including holes.

Related Homework Material: Problem 12

Instead of the separate tiles in Example 5, we could have used a square plate with a square hole in the center. The hole in the plate would have expanded just like the hole in the pattern of tiles. Furthermore, the same conclusion applies to a hole of any shape. Thus, it follows that **a hole in a piece of solid material expands when heated and contracts when cooled, just as if it were filled with the material that surrounds it.** If the hole is circular, the equation $\Delta L = \alpha L_0 \Delta T$ can be used to find the change in any linear dimension of the hole, such as its radius or diameter. Example 6 illustrates this type of linear expansion.

EXAMPLE 6 • A Heated Engagement Ring

A gold engagement ring has an inner diameter of 1.5×10^{-2} m and a temperature of 27 °C. The ring falls into a sink of hot water whose temperature is 49 °C. What is the change in the diameter of the hole in the ring?

Reasoning The hole expands as if it were filled with gold, so the change in the diameter is given by $\Delta L = \alpha L_0 \Delta T$, where $\alpha = 14 \times 10^{-6} (\text{C}°)^{-1}$ (Table 12.1) is the coefficient of linear expansion for gold and L_0 is the original diameter.

Solution The change in the ring's diameter is

$$\Delta L = \alpha L_0 \Delta T = [14 \times 10^{-6}\,(\text{C}°)^{-1}](1.5 \times 10^{-2}\,\text{m})(49\,°\text{C} - 27\,°\text{C}) =$$

$$\boxed{4.6 \times 10^{-6}\,\text{m}}$$

The previous two examples illustrate that holes expand like the surrounding material when heated. Therefore, holes in materials with larger coefficients of linear expansion expand more than those in materials with smaller coefficients of linear expansion. Conceptual Example 7 explores this aspect of thermal expansion.

CONCEPTUAL EXAMPLE 7 • Expanding Cylinders

Figure 12.18 shows a cross-sectional view of three cylinders, A, B, and C. Each is made from a different material; one is lead, one is brass, and one is steel. All three have the same temperature, and they barely fit inside each other. As the cylinders are heated to the same, but higher, temperature, cylinder C falls off, while cylinder A becomes tightly wedged to cylinder B. Which cylinder is made from which material?

Reasoning and Solution We need to consider how the outer and inner diameters of each cylinder change as the temperature is raised. With respect to the inner diameter, we will be guided by the fact that a hole expands as if it were filled with the surrounding material. According to Table 12.1, lead has the greatest coefficient of linear expansion, followed by brass, and then by steel. These data indicate that the outer and inner diameters of the lead cylinder change the most, while those of the steel cylinder change the least.

Since the steel cylinder expands the least, it cannot be the outer one, for if it were, the greater expansion of the middle cylinder would prevent the steel cylinder from falling off. The steel cylinder also cannot be the inner one, because then the greater expansion of the middle cylinder would allow the steel cylinder to fall out, contrary to what is observed. The only place left for the steel cylinder is in the middle, which leads to the two possibilities in Figure 12.18. In part *a*, lead is on the outside and will fall off as the temperature is raised, since lead expands more than steel. On the other hand, the inner brass cylinder expands more than the steel that surrounds it and becomes tightly wedged, as observed. Thus, one possibility is *A = brass, B = steel, and C = lead.*

In part *b* of the drawing, brass is on the outside. As the temperature is raised, brass expands more than steel, so the outer cylinder will again fall off. The inner lead cylinder has the greatest expansion and will be wedged against the middle steel cylinder. A second possible answer, then, is *A = lead, B = steel, and C = brass.*

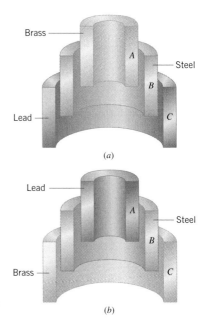

Figure 12.18 Conceptual Example 7 discusses the arrangements of the three cylinders shown in these cutaway views in parts *a* and *b*.

12.5 VOLUME THERMAL EXPANSION

NORMAL MATERIALS

The volume of a normal material increases as the temperature increases. Most solids and liquids behave in this fashion. By analogy with linear thermal expansion, the change in volume ΔV is proportional to the change in temperature ΔT and to the initial volume V_0, provided the change in temperature is not too large. These two proportionalities can be converted into Equation 12.3 below with the aid of a proportionality constant β, known as the *coefficient of volume expansion*. The algebraic form of this equation is similar to that for linear expansion, $\Delta L = \alpha L_0 \Delta T$.

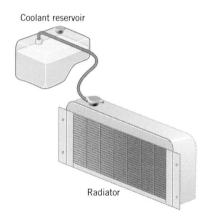

Coolant reservoir

Radiator

Figure 12.19 An automobile radiator and a coolant reservoir for catching the overflow from the radiator.

■ VOLUME THERMAL EXPANSION

The volume V_0 of an object changes by an amount ΔV when its temperature changes by an amount ΔT:

$$\Delta V = \beta V_0 \Delta T \qquad (12.3)$$

where β is the coefficient of volume expansion.

Common Unit for the Coefficient of Volume Expansion: $(C°)^{-1}$

The unit for β, like that for α, is $(C°)^{-1}$. Values for β depend on the nature of the material, and Table 12.1 lists some examples measured near 20 °C. The values of β for liquids are substantially larger than those for solids, because liquids typically expand more than solids, given the same initial volumes and temperature changes. Table 12.1 also shows that, for most solids, the coefficient of volume expansion is three times greater than the coefficient of linear expansion: $\beta = 3\alpha$. (See problem 39.)

If a cavity exists within a solid object, the volume of the cavity increases when the object expands, just as if the cavity were filled with the surrounding material. The expansion of the cavity is analogous to the expansion of a hole in a sheet of material. Accordingly, the change in volume of a cavity can be found using the relation $\Delta V = \beta V_0 \Delta T$, where β is the coefficient of volume expansion of the material that surrounds the cavity. Example 8 illustrates this point.

▶ **The Physics of...**
the overflow of an automobile radiator.

EXAMPLE 8 • An Automobile Radiator

A small plastic container, called the coolant reservoir, catches the radiator fluid that overflows when an automobile engine becomes hot (see Figure 12.19). The radiator is made of copper, and the coolant has a coefficient of volume expansion of $\beta = 410 \times 10^{-6} (C°)^{-1}$. If the radiator is filled to its 15-quart capacity when the engine is "cold" (6.0 °C), how much overflow from the radiator will spill into the reservoir when the coolant reaches its operating temperature of 92 °C?

Reasoning When the temperature increases, both the coolant and radiator expand. If they were to expand by the same amount, there would be no overflow. However, the liquid coolant expands more than the radiator, and the amount of overflow is the amount of coolant expansion *minus* the expansion of the radiator cavity.

Solution When the temperature increases by 86 C°, the coolant expands by an amount

$$\Delta V = \beta V_0 \Delta T = [410 \times 10^{-6} (C°)^{-1}](15 \text{ quarts})(86 \text{ C}°) = 0.53 \text{ quarts} \quad (12.3)$$

The volume of the radiator cavity expands as if it were filled with copper $[\beta = 51 \times 10^{-6} (C°)^{-1}$, see Table 12.1]. The expansion of the radiator cavity is

$$\Delta V = \beta V_0 \Delta T = [51 \times 10^{-6} (C°)^{-1}](15 \text{ quarts})(86 \text{ C}°) = 0.066 \text{ quarts}$$

The amount of coolant overflow is 0.53 quarts $-$ 0.066 quarts $= \boxed{0.46 \text{ quarts}}$.

● **PROBLEM SOLVING INSIGHT**
The way in which the level of a liquid in a container changes with temperature depends on both the change in volume of the liquid and the change in volume of the container.

THE ANOMALOUS BEHAVIOR OF WATER NEAR 4 °C

While most substances expand when heated, a few do not. For instance, if water at 0 °C is heated, its volume *decreases* until the temperature reaches 4 °C. Above 4 °C water behaves normally, and its volume increases as the temperature increases. Because a given mass of water has a minimum volume at 4 °C, the density (mass per unit volume) of water is greatest at 4 °C, as Figure 12.20 shows.

The fact that water has its greatest density at 4 °C, rather than at 0 °C, has important consequences for the way in which a lake freezes. When the air temperature drops, the surface layer of water is chilled. As the temperature of the surface layer drops toward 4 °C, this layer becomes more dense than the warmer water below. The denser water sinks and pushes up the deeper and warmer water, which in turn is chilled at the surface. This process continues until the temperature of the entire lake reaches 4 °C. Further cooling of the surface water below 4 °C makes it *less dense* than the deeper layers; consequently, the surface layer does not sink, but stays on top. Continued cooling of the top layer to 0 °C leads to the formation of ice that floats on the water, because ice has a smaller density than water at any temperature. Below the ice, however, the water temperature remains above 0 °C. The sheet of ice acts as an insulator that reduces the loss of heat from the lake, especially if the ice is covered with a blanket of snow, which is also an insulator. Furthermore, heat transferred from the ground beneath the lake helps to keep the water under the ice sheet from freezing. As a result, lakes usually do not freeze solid, even during prolonged cold spells, so fish and other aquatic life can survive during the winter.

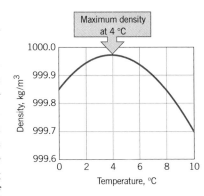

Figure 12.20 The density of water in the temperature range from 0 to 10 °C. Water has a maximum density of 999.973 kg/m^3 at 4 °C. (This value for the density is equivalent to the often-quoted density of 1.000 00 grams per milliliter.)

12.6 HEAT AND INTERNAL ENERGY

An object with a high temperature is said to be hot, and the word "hot" brings to mind the word "heat." *Heat* flows from a hotter object to a cooler object when the two are placed in contact. It is for this reason that a pot of hot coffee feels hot to the touch, while a glass of ice water feels cold. The temperature of hot coffee is higher than the normal body temperature of 37 °C, while the temperature of ice water is lower than 37 °C. When the person in Figure 12.21a touches the coffee cup, heat flows from the hotter cup into the cooler hand. When the person touches the glass in part *b* of the drawing, heat again flows from hot to cold, in this case from the warmer hand into the colder glass. The response of the nerves in the hand to the arrival or departure of heat prompts the brain to identify the pot as being hot and the glass as being cold.

But just what is heat? As the following definition indicates, heat is a form of energy, energy in transit from hot to cold.

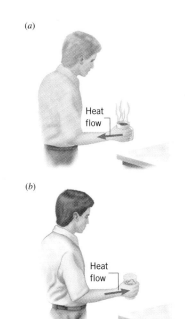

■ **DEFINITION OF HEAT**

Heat is energy that flows from a higher-temperature object to a lower-temperature object because of the difference in temperatures.

SI Unit of Heat: joule (J)

Being a kind of energy, heat is measured in the same units used for work, kinetic energy, and potential energy. Thus, the SI unit for heat is the joule.

The heat that flows from hot to cold in Figure 12.21 originates in the *internal energy* of the hot substance. The internal energy of a substance is the sum of the molecular kinetic energy (due to the random motion of the molecules), the molecular potential energy (due to forces that act between the atoms of a molecule and between molecules), and other kinds of molecular energy. When heat flows in circum-

Figure 12.21 Heat is energy in transit from hot to cold. (*a*) Heat flows from the hotter coffee cup to the colder hand. (*b*) Heat flows from the warmer hand to the colder glass of ice water.

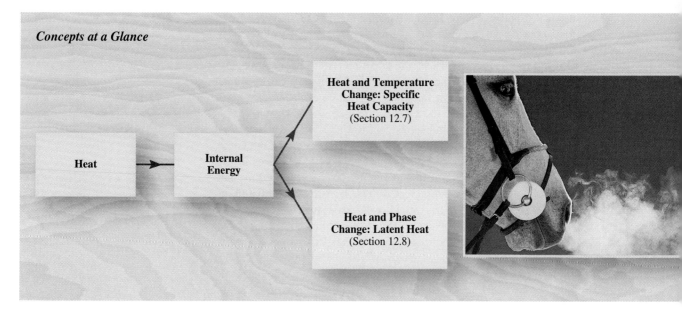

Concepts at a Glance

Figure 12.22 *Concepts at a Glance* When heat is added or removed from a substance, its internal energy can change. This change can cause a change in the temperature (Section 12.7) or a change in phase (Section 12.8). This horse exhales air and water vapor that have been heated by energy that originated in the horse's internal energy. Upon reaching the cold outside, the water vapor condenses to form a miniature cloud.

stances where the work done is negligible, the internal energy of the hot substance decreases and the internal energy of the cold substance increases. While heat may originate in the internal energy supply of a substance, *it is not correct to say that a substance contains heat.* The substance has internal energy, not heat. The word "heat" is used only when referring to the energy actually in transit from hot to cold.

In the next two sections, we will consider some of the effects of heat. For instance, when preparing spaghetti for dinner, the first thing that the cook does is to place a pot of water on the stove. Heat from the stove causes the internal energy of the water to increase, as the left-hand side of the concept chart in Figure 12.22 suggests. Associated with this increase in internal energy is a rise in temperature. After a while, however, the temperature reaches 100 °C and the water begins to boil. During boiling, any heat added to the water goes into producing steam, a process in which water changes from a liquid phase to a vapor phase. The next section investigates how the addition (or removal) of heat causes the temperature of a substance to change (see the upper-right portion of the chart in Figure 12.22). Then, Section 12.8 discusses the relationship between heat and phase changes (see the lower-right part of Figure 12.22), such as that which occurs when water boils.

12.7 HEAT AND TEMPERATURE CHANGE: SPECIFIC HEAT CAPACITY

SOLIDS AND LIQUIDS

Greater amounts of heat are needed to raise the temperature of solids or liquids to higher values. A greater amount of heat is also required to raise the temperature of a greater mass of material. Similar comments apply when the temperature is lowered, except that heat must be removed. For limited temperature ranges, experiment shows that the amount of heat Q is directly proportional to the change in temperature ΔT and to the mass m. These two proportionalities are expressed below in Equation 12.4, with the help of a proportionality constant c that is referred to as the ***specific heat capacity*** of the material.

■ **HEAT SUPPLIED OR REMOVED IN CHANGING THE TEMPERATURE OF A SUBSTANCE**

The heat Q that must be supplied or removed to change the temperature of a substance of mass m by an amount ΔT is

$$Q = cm\Delta T \qquad (12.4)$$

where c is the specific heat capacity of the substance.

Common Unit for Specific Heat Capacity: $J/(kg \cdot C°)$

Solving Equation 12.4 for the specific heat capacity shows that $c = Q/(m\,\Delta T)$, so the unit for specific heat capacity is $J/(kg \cdot C°)$. Table 12.2 reveals that the value of the specific heat capacity depends on the nature of the material. Example 9 applies Equation 12.4 to a hot jogger.

EXAMPLE 9 • A Hot Jogger

In a half hour, a 65-kg jogger can generate 8.0×10^5 J of heat. This heat is removed from the jogger's body by a variety of means, including the body's own temperature regulating mechanisms. If the heat were not removed, how much would the body temperature increase?

Reasoning and Solution Table 12.2 gives the average specific heat capacity of the human body as 3500 $J/(kg \cdot C°)$. With this value, Equation 12.4 shows that the temperature increase would be

$$\Delta T = \frac{Q}{cm} = \frac{8.0 \times 10^5 \text{ J}}{[3500 \text{ J}/(kg \cdot C°)](65 \text{ kg})} = \boxed{3.5 \text{ C}°}$$

Table 12.2 Specific Heat Capacities^a of Some Solids and Liquids

Substance	Specific Heat Capacity, c	
	$J/(kg \cdot C°)$	$kcal/(kg \cdot C°)^b$
Solids		
Aluminum	9.00×10^2	0.215
Copper	387	0.0924
Glass	840	0.20
Human body (37 °C, average)	3500	0.83
Ice (-15 °C)	2.00×10^3	0.478
Iron or steel	452	0.108
Lead	128	0.0305
Silver	235	0.0562
Liquids		
Benzene	1740	0.415
Ethyl alcohol	2450	0.586
Glycerin	2410	0.576
Mercury	139	0.0333
Water (15 °C)	4186	1.000

^a Except as noted, the values are for 25 °C and 1 atm of pressure.
^b The values given are the same in units of $cal/(g \cdot C°)$.

Figure 12.23 Wolves, such as this Gray wolf, often pant to get rid of excess heat.

An increase in body temperature of 3.5 °C could be life threatening. One way in which the jogger's body prevents it from occurring is to remove excess heat by perspiring. In contrast, wolves, such as the one in Figure 12.23, do not perspire but often pant to remove excess heat.

HEAT UNITS OTHER THAN THE JOULE

There are three heat units other than the joule in common use. One kilocalorie (1 kcal) was defined historically as the amount of heat needed to raise the temperature of one kilogram of water by one Celsius degree.* With $Q = 1.00$ kcal, $m = 1.00$ kg, and $\Delta T = 1.00$ C°, the equation $Q = cm\,\Delta T$ shows that such a definition is equivalent to a specific heat capacity for water of $c = 1.00$ kcal/(kg·C°). Similarly, one calorie (1 cal) was defined as the amount of heat needed to raise the temperature of one gram of water by one Celsius degree, which yields a value of $c = 1.00$ cal/(g·C°). (Nutritionists use the word "Calorie," with a capital C, to specify the energy content of foods; this use is unfortunate, since 1 Calorie = 1000 calories = 1 kcal.) The British thermal unit (Btu) is the other commonly used heat unit and was defined historically as the amount of heat needed to raise the temperature of one pound of water by one Fahrenheit degree.

It was not until the time of James Joule (1818–1889) that the relationship between energy in the form of work (in units of joules) and energy in the form of heat (in units of kilocalories) was firmly established. Joule's experiments revealed that the performance of mechanical work, like rubbing your hands together, can make the temperature of a substance rise, just as the absorption of heat can. His experiments and those of later workers have shown that

$$1 \text{ kcal} = 4186 \text{ joules} \quad \text{or} \quad 1 \text{ cal} = 4.186 \text{ joules}$$

Because of its historical significance, this conversion factor is known as the ***mechanical equivalent of heat.*** Example 10 illustrates the use of the joule and the kilocalorie as heat units.

EXAMPLE 10 • Taking a Hot Shower

Cold water at a temperature of 15 °C enters a heater, and the resulting hot water has a temperature of 61 °C. A person uses 120 kg of hot water in taking a shower. (a) Find the energy needed to heat the water. (b) Assuming that the power utility charges $0.10 per kilowatt·hour for electrical energy, determine the cost of heating the water.

Reasoning The amount Q of heat needed to raise the water temperature can be found from the relation $Q = cm\,\Delta T$, since the specific heat capacity, mass, and temperature change of the water are known. To determine the cost of this energy, we multiply the cost per unit of energy ($0.10 per kilowatt·hour) by the amount of energy used, expressed in energy units of kilowatt·hours.

Solution

(a) The amount of heat used in heating the water is

$$Q = cm\,\Delta T = [4186 \text{ J/(kg·C°)}](120 \text{ kg})(61 \text{ °C} - 15 \text{ °C}) = \boxed{2.3 \times 10^7 \text{ J}} \quad (12.4)$$

---

* From 14.5 to 15.5 °C.

(b) The kilowatt · hour (kWh) is the unit of energy that utility companies use in your electric bill. To calculate the cost, we need to determine the number of joules in one kilowatt · hour. Recall that 1 kilowatt is 1000 watts (1 kW = 1000 W), one watt is one joule per second (1 W = 1 J/s, see Equation 6.10), and one hour is equal to 3600 seconds (1 h = 3600 s). Thus,

$$1 \text{ kWh} = (1\text{kWh})\left(\frac{1000 \text{ W}}{1 \text{ kW}}\right)\left(\frac{1 \text{ J/s}}{1 \text{ W}}\right)\left(\frac{3600 \text{ s}}{1 \text{ h}}\right) = 3.60 \times 10^6 \text{ J}$$

The number of kilowatt · hours of energy used to heat the water is

$$(2.3 \times 10^7 \text{ J})\left(\frac{1 \text{ kWh}}{3.60 \times 10^6 \text{ J}}\right) = 6.4 \text{ kWh}$$

At a cost of $0.10 per kWh, the bill for the heat is $\boxed{\$0.64}$.

GASES

As we will see in Section 15.5, the value of the specific heat capacity depends on whether the pressure or volume is held constant while energy in the form of heat is added to or removed from a substance. The distinction between constant pressure and constant volume is usually not important for solids and liquids but is significant for gases. Different values are obtained when the specific heat capacity for a gas is measured under conditions of *constant pressure* as compared to conditions of *constant volume*. Table 12.3 illustrates the difference for several gases and indicates that the value c_P at constant pressure is greater than the value c_V at constant volume.

CALORIMETRY

In Section 6.8 we encountered the principle of conservation of energy, which states that energy can be neither created nor destroyed, but can only be converted from one form to another. There we dealt with kinetic and potential energies. In this chapter we have expanded our concept of energy to include heat, which is energy that flows from a higher-temperature object to a lower-temperature object because of the difference in temperature. No matter what its form, whether kinetic energy, potential energy, or heat, energy can be neither created nor destroyed. This fact governs the way objects at different temperatures come to an equilibrium temperature when they are placed in contact. If there is no heat loss to the external sur-

Table 12.3   Specific Heat Capacities^a of Gases

	Specific Heat Capacity	
Gas	Constant Pressure, c_P [J/(kg·C°)]	Constant Volume, c_V [J/(kg·C°)]
Ammonia	2190	1670
Carbon dioxide	833	638
Nitrogen	1040	739
Oxygen	912	651
Water vapor (100 °C)	2020	1520

^{a} Except as noted, the values are for 15 °C and 1 atm of pressure.

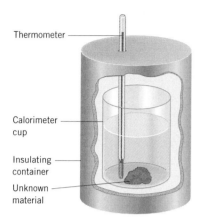

Thermometer

Calorimeter cup

Insulating container

Unknown material

Figure 12.24 A calorimeter can be used to measure the specific heat capacity of an unknown material.

roundings, the heat lost by the hotter objects equals the heat gained by the cooler ones, a process that is consistent with the conservation of energy. Just this kind of process occurs in a thermos. A perfect thermos would prevent any heat from leaking out or in. However, energy in the form of heat can flow *between* materials inside the thermos to the extent that they have different temperatures, for example, between ice cubes and warm tea. Such heat flow satisfies the conservation of energy, for the colder materials gain the energy that the hotter materials lose. The transfer of energy continues until a common temperature is reached at thermal equilibrium.

The kind of heat transfer that occurs in a thermos of iced tea also occurs in a calorimeter, which is the experimental apparatus used in a technique known as *calorimetry.* Figure 12.24 shows that, like a thermos, a calorimeter is essentially an insulated container. It can be used to determine the specific heat capacity of a substance, as the next example illustrates.

EXAMPLE 11 • Measuring the Specific Heat Capacity

The calorimeter cup in Figure 12.24 is made from 0.15 kg of aluminum and contains 0.20 kg of water. Initially, the water and the cup have a common temperature of 18.0 °C. An unknown material (m = 0.040 kg) is heated to a temperature of 97.0 °C and then added to the water. The temperature of the water, the cup, and the unknown material is 22.0 °C after thermal equilibrium is reestablished. Ignoring the small amount of heat gained by the thermometer, find the specific heat capacity of the unknown material.

Reasoning Since energy is conserved and there is negligible heat flow between the calorimeter and the outside surroundings, the heat gained by the cold water and the aluminum cup as they warm up is equal to the heat lost by the unknown material as it cools down. Each quantity of heat can be calculated using the relation $Q = cm\,\Delta T$, where we always write the change in temperature ΔT as the higher temperature minus the lower temperature. The equation "Heat gained = Heat lost" will then contain a single unknown quantity, the desired specific heat capacity.

Solution

$$\underbrace{(cm\,\Delta T)_{\text{aluminum}} + (cm\,\Delta T)_{\text{water}}}_{\substack{\text{Heat gained by}\\ \text{aluminum and water}}} = \underbrace{(cm\,\Delta T)_{\text{unknown}}}_{\substack{\text{Heat lost}\\ \text{by unknown material}}}$$

$$[9.00 \times 10^2 \text{ J/(kg} \cdot \text{C}°)](0.15 \text{ kg})(22.0 °\text{C} - 18.0 °\text{C})$$
$$+ [4186 \text{ J/(kg} \cdot \text{C}°)](0.20 \text{ kg})(22.0 °\text{C} - 18.0 °\text{C})$$
$$= c_{\text{unknown}}(0.040 \text{ kg})(97.0 °\text{C} - 22.0 °\text{C})$$

$$\boxed{c_{\text{unknown}} = 1300 \text{ J/(kg} \cdot \text{C}°)}$$

• **PROBLEM SOLVING INSIGHT**
In the equation "Heat gained = Heat lost," both sides must have the same algebraic sign. Therefore, when calculating heat contributions, always write any temperature changes as the higher minus the lower temperature.

12.8 HEAT AND PHASE CHANGE: LATENT HEAT

Surprisingly, there are situations in which the addition or removal of heat does not cause a temperature change. Consider a well-stirred glass of iced tea that has come to thermal equilibrium. Even though heat enters the glass from the warmer room, the temperature of the tea does not rise above 0 °C as long as ice cubes are present.

Apparently the heat is being used for some purpose other than raising the temperature. In fact, the heat is being used to melt the ice, and only when all of it is melted will the temperature of the liquid begin to rise.

An important point illustrated by the iced tea example is that there is more than one type or phase of matter. For instance, some of the water in the glass is in the solid phase (ice) and some in the liquid phase. The gas or vapor phase is the third familiar phase of matter. In the gas phase, water is referred to as water vapor or steam. All three phases of water are present in the scene depicted in Figure 12.25.

Matter can change from one phase to another, and heat plays a role in the change. Figure 12.26 summarizes the various possibilities for phase changes between solids, liquids, and gases. A solid can *melt* or *fuse* into a liquid if heat is added, while the liquid can *freeze* into a solid if heat is removed. Similarly, a liquid can *evaporate* into a gas if heat is supplied, while the gas can *condense* into a liquid if heat is taken away. Rapid evaporation, with the formation of vapor bubbles within the liquid, is called boiling. Finally, a solid can sometimes change directly into a gas if heat is provided. We say that the solid *sublimes* into a gas. Examples of sublimation are (1) solid carbon dioxide, CO_2 (dry ice), turning into gaseous CO_2 and (2) solid naphthalene (moth balls) turning into naphthalene fumes. Conversely, if heat is removed under the right conditions, the gas will condense directly into a solid.

Figure 12.27 displays a graph that indicates what typically happens when heat is added to a material that changes phase. The graph records temperature versus heat added and refers to water at the normal atmospheric pressure of 1.01×10^5 Pa. The water starts off as ice at the subfreezing temperature of $-30\ °C$. As heat is added, the temperature of the ice increases, in accord with the specific heat capacity of ice

Figure 12.25 The three phases of water: ice is floating in liquid water, while (invisible) water vapor is present in the air.

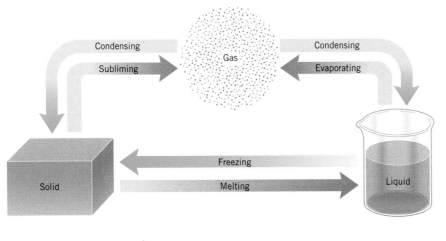

Figure 12.26 Three familiar phases of matter—solid, liquid, and gas—and the phase changes that can occur between any two of them.

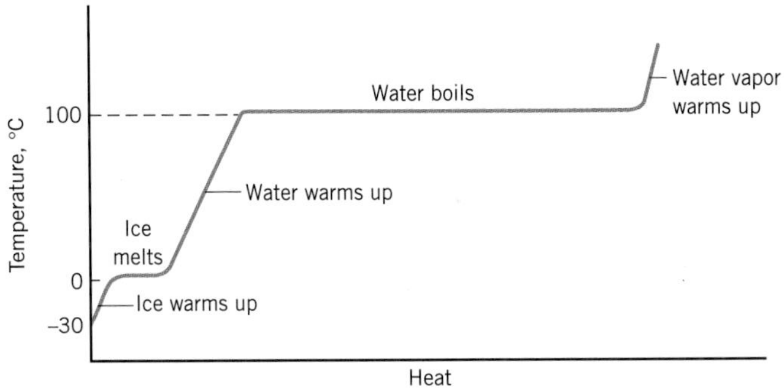

Figure 12.27 The graph shows the way the temperature of water changes as heat is added, starting with ice at $-30\ °C$. The pressure is atmospheric pressure.

[2000 J/(kg·C°)]. Not until the temperature reaches the normal melting/freezing point of 0 °C does the water begin to change phase. Then, when heat is added, the solid changes into the liquid, the temperature staying at 0 °C until *all the ice has melted.* Once all the material is in the liquid phase, additional heat causes the temperature to increase again, now in accord with the specific heat capacity of liquid water [4186 J/(kg·C°)]. When the temperature reaches the normal boiling/condensing point of 100 °C, the water begins to change from the liquid to the gas phase and continues to do so as long as heat is added. The temperature remains at 100 °C *until all liquid is gone.* When all of the material is in the gas phase, additional heat once again causes the temperature to rise, this time according to the specific heat capacity of water vapor at constant atmospheric pressure [2020 J/(kg·C°)]. Conceptual Example 12 applies the information in Figure 12.27 to a familiar situation.

CONCEPTUAL EXAMPLE 12 • Saving Energy

Suppose you are cooking spaghetti for dinner, and the instructions say to boil the noodles in water for ten minutes. To cook spaghetti in an open pot with the least amount of energy, should you turn up the burner to its fullest so the water vigorously boils, or should you turn down the burner so the water barely boils?

Reasoning and Solution The spaghetti needs to cook at a temperature of 100 °C for ten minutes. It doesn't matter whether the water is vigorously boiling or barely boiling, for its temperature is still 100 °C. Remember, as long as there is boiling water in the pot, and the pot is open to one atmosphere of pressure, no amount of additional heat will cause the water temperature to rise above 100 °C. Additional heat only vaporizes the water and produces steam, which is of no use in cooking the spaghetti. So, save your money and turn down the heat, for **the least amount of energy is expended when the water barely boils.**

When a substance changes from one phase to another, the amount of heat that must be added or removed depends on the type of material and the nature of the phase change. The heat per kilogram associated with a phase change is referred to as *latent heat:*

■ DEFINITION OF LATENT HEAT

The latent heat L is the heat per kilogram that must be added or removed when a substance changes from one phase to another at a constant temperature.

SI Unit of Latent Heat: J/kg

The *latent heat of fusion* L_f refers to the change between solid and liquid phases, the *latent heat of vaporization* L_v applies to the change between liquid and gas phases, and the *latent heat of sublimation* L_s refers to the change between solid and gas phases.

Table 12.4 gives some typical values of latent heats of fusion and vaporization. For instance, the latent heat of fusion for water is 3.35×10^5 J/kg. Thus, 3.35×10^5 J of heat must be supplied to melt one kilogram of ice at 0 °C into liquid water at 0 °C; conversely, this amount of heat must be removed from one kilogram of liquid water at 0°C to freeze the liquid into ice at 0 °C. In comparison, the latent heat of vaporization for water has the much larger value of 22.6×10^5 J/kg. When wa-

Table 12.4   Latent Heats^a of Fusion and Vaporization

Substance	Melting Point (°C)	Latent Heat of Fusion, L_f (J/kg)	Boiling Point (°C)	Latent Heat of Vaporization, L_v (J/kg)
Ammonia	−77.8	33.2×10^4	−33.4	13.7×10^5
Benzene	5.5	12.6×10^4	80.1	3.94×10^5
Copper	1083	20.7×10^4	2566	47.3×10^5
Ethyl alcohol	−114.4	10.8×10^4	78.3	8.55×10^5
Gold	1063	6.28×10^4	2808	17.2×10^5
Lead	327.3	2.32×10^4	1750	8.59×10^5
Mercury	−38.9	1.14×10^4	356.6	2.96×10^5
Nitrogen	−210.0	2.57×10^4	−195.8	2.00×10^5
Oxygen	−218.8	1.39×10^4	−183.0	2.13×10^5
Water	0.0	33.5×10^4	100.0	22.6×10^5

^{a} The values pertain to 1 atm pressure.

ter boils at 100 °C, 22.6×10^5 J of heat must be supplied for each kilogram of liquid turned into steam. And when steam condenses at 100 °C, this amount of heat is released from each kilogram of steam that changes back into liquid. Liquid water at 100 °C is hot enough by itself to cause a bad burn, and the additional effect of the large latent heat can cause severe tissue damage if condensation occurs on the skin. Examples 13 and 14 illustrate how to take into account the effect of latent heat when using the conservation of energy principle.

The Physics of... steam burns.

EXAMPLE 13 • Ice-cold Lemonade

Ice at 0 °C is placed in a Styrofoam cup containing 0.32 kg of lemonade at 27 °C. The specific heat capacity of lemonade is virtually the same as that of water. After the ice and lemonade reach an equilibrium temperature, some ice still remains. Ignore the specific heat capacity of the cup and any heat lost to the surroundings. Determine the mass of ice that has melted.

Reasoning   According to the principle of energy conservation, the heat gained by the melting ice equals the heat lost by the cooling lemonade. The heat gained by the melting ice is $Q = mL_f$, since the latent heat of fusion L_f is defined as the heat per kilogram that must be added when a solid changes to a liquid. The heat lost by the lemonade is given by $Q = cm \Delta T$, where ΔT is the higher temperature of 27 °C minus the lower equilibrium temperature. The equilibrium temperature is 0 °C, because there is some ice remaining, and ice is in equilibrium with liquid water when the temperature is 0 °C.

Solution

$$\underbrace{(mL_f)_{\text{ice}}}_{\substack{\text{Heat gained} \\ \text{by ice}}} = \underbrace{(cm \Delta T)_{\text{lemonade}}}_{\substack{\text{Heat lost} \\ \text{by lemonade}}}$$

The mass m_{ice} of ice that has melted is

$$m_{\text{ice}} = \frac{(cm \Delta T)_{\text{lemonade}}}{L_f} = \frac{[4186 \text{ J/(kg} \cdot \text{C}°)](0.32 \text{ kg})(27 \text{ °C} - 0 \text{ °C})}{33.5 \times 10^4 \text{ J/kg}} = \boxed{0.11 \text{ kg}}$$

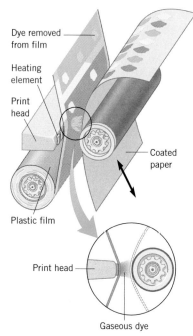

Figure 12.28 A dye sublimation printer. As the plastic film passes in front of the print head, the heat from a given heating element causes one of three pigments or dyes on the film to sublime from a solid to a gas. The gaseous dye is absorbed onto the coated paper as a dot of color. The size of the dots on the paper has been exaggerated for clarity.

The Physics of...
a dye-sublimation color
printer.

EXAMPLE 14 • Getting Ready for a Party

A 7.00-kg glass bowl [$c = 840$ J/(kg·C°)] contains 16.0 kg of punch at 25.0 °C. Two-and-a-half kilograms of ice [$c = 2.00 \times 10^3$ J/(kg·C°)] are added to the punch. The ice has an initial temperature of -20.0 °C, having been kept in a very cold freezer. The punch may be treated as if it were water [$c = 4186$ J/(kg·C°)], and it may be assumed that there is no heat flow between the punch bowl and the external environment. When thermal equilibrium is reached, all the ice has melted, and the final temperature of the mixture is above 0 °C. Determine this temperature.

Reasoning The final temperature can be found by using the conservation of energy, because the heat gained (a) by the ice in warming up to the melting point, (b) by the ice in changing phase from a solid to a liquid, and (c) by the liquid that results from the ice warming up to the final temperature is equal to the heat lost (d) by the punch and (e) by the bowl in cooling down. The heat gained or lost by each component in changing temperature can be determined from the relation $Q = cm\,\Delta T$. The heat gained when water changes phase from a solid to a liquid at 0 °C is $Q = mL_f$, where m is the mass of water and L_f is the latent heat of fusion.

Solution The heat gained or lost by each component is listed as follows:

(a) Heat gained when ice warms to 0.0 °C
$= [2.00 \times 10^3$ J/(kg·C°)](2.50 kg)[0.0 °C $- (-20.0$ °C)]

(b) Heat gained when ice melts at 0.0 °C
$= (2.50$ kg)(3.35 $\times 10^5$ J/kg)

(c) Heat gained when melted ice (liquid) warms to temperature T
$= [4186$ J/(kg·C°)](2.50 kg)($T - 0.0$ °C)

(d) Heat lost when punch cools to temperature T
$= [4186$ J/(kg·C°)](16.0 kg)(25.0 °C $- T$)

(e) Heat lost when bowl cools to temperature T
$= [840$ J/(kg·C°)](7.00 kg)(25.0 °C $- T$)

Setting the heat gained equal to the heat lost gives:

$$\underbrace{(a) + (b) + (c)}_{\text{Heat gained}} = \underbrace{(d) + (e)}_{\text{Heat lost}}$$

This equation can be solved to show that $\boxed{T = 11\ °C}$.

An interesting application of the phase change between a solid and a gas is found in one kind of color printer used with computers. A dye-sublimation printer uses a thin plastic film coated with separate panels of cyan (blue), yellow, and magenta pigment or dye. A full spectrum of colors is produced by using combinations of tiny spots of these dyes. As Figure 12.28 shows, the coated film passes in front of a print head that extends across the width of the paper and contains 2400 heating elements. When a heating element is turned on, the dye in front of it absorbs heat and goes from a solid to a gas—it sublimes—with no liquid phase in between. A coating on the paper absorbs the gaseous dye on contact, producing a small spot of color. The intensity of the spot is controlled by the heating element, since each element can produce 256 different temperatures; the hotter the element, the greater the amount of dye transferred to the paper. The paper makes three separate passes across the print head, once for each of the dyes. The final result is an image that is of near-photographic quality.

*12.9 EQUILIBRIUM BETWEEN PHASES OF MATTER

Under specific conditions of temperature and pressure, a substance can exist in equilibrium in more than one phase at the same time. Consider Figure 12.29, which shows a container kept at a constant temperature by a large reservoir of heated sand. Initially the container is evacuated, and part *a* shows it just after it has been partially filled with a liquid. A few fast-moving molecules escape the liquid and form a vapor phase, as part *b* suggests. These molecules pick up the required energy (the latent heat of vaporization) during collisions with neighboring molecules in the liquid. However, the reservoir of heated sand replenishes the energy carried away, thus maintaining the constant temperature. At first, the movement of molecules is predominantly from liquid to vapor, although some molecules in the vapor phase do reenter the liquid. As the molecules accumulate in the vapor, the number reentering the liquid eventually equals the number entering the vapor, and equilibrium becomes established, as in part *c*. From this point on, the concentration of molecules in the vapor phase does not change, and the vapor pressure remains constant. The pressure of the vapor that coexists in equilibrium with the liquid is called the *equilibrium vapor pressure* of the liquid.

The equilibrium vapor pressure does not depend on the volume of space above the liquid. If more space were provided, more liquid would vaporize, until equilibrium was reestablished at the same vapor pressure, assuming the same temperature is maintained. In fact, the equilibrium vapor pressure depends only on the temperature of the liquid; a higher temperature causes a higher pressure, as the graph in Figure 12.30 indicates for the specific case of water. Only when the temperature and vapor pressure correspond to a point on the curved line, which is called the *vapor pressure curve* or the *vaporization curve,* can liquid and vapor phases coexist in equilibrium.

To illustrate the use of a vaporization curve, let's see what happens when water boils in a pot that is *open to the air.* Assume that the air pressure acting on the water is 1.01×10^5 Pa (one atmosphere). When boiling occurs, bubbles of water vapor form throughout the liquid, rise to the surface, and break. For these bubbles to form and rise, the pressure of the vapor inside them must at least equal the air pressure acting on the surface of the water. According to Figure 12.30, a value of 1.01×10^5 Pa corresponds to a temperature of 100 °C. Consequently, water boils at 100 °C at

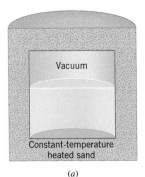

Vacuum

Constant-temperature heated sand

(a)

Constant-temperature heated sand

(b)

Constant-temperature heated sand

(c)

Figure 12.29 (*a*) Initially, only a liquid is in the evacuated container. (*b*) Some of the molecules begin entering the vapor phase. (*c*) Equilibrium is reached when the number of molecules entering the vapor phase equals the number returning to the liquid.

Figure 12.30 A plot of the equilibrium vapor pressure versus temperature is called the vapor pressure curve or the vaporization curve, the example shown being that for the liquid/vapor equilibrium of water.

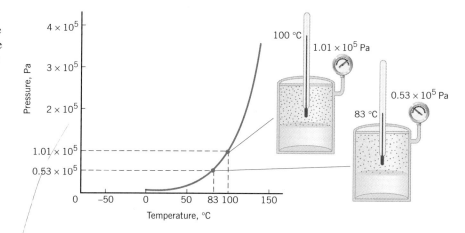

1 atmosphere of pressure. In general, a ***liquid boils at the temperature for which its vapor pressure equals the external pressure.*** Water will not boil, then, at sea level if the temperature is only 83 °C, because at this temperature the vapor pressure of water is only 0.53×10^5 Pa (see Figure 12.30), a value less than the external pressure of 1.01×10^5 Pa. However, water does boil at 83 °C on a mountain at an altitude of just under five kilometers, because the atmospheric pressure there is 0.53×10^5 Pa.

The fact that water can boil at a temperature less than 100 °C leads to an interesting phenomenon that Conceptual Example 15 discusses.

(a) Water boiling

CONCEPTUAL EXAMPLE 15 • How to Boil Water That Is Cooling Down

Figure 12.31a shows water boiling in an open flask. Shortly after the flask is removed from the burner, the boiling stops. A cork is then placed in the neck of the flask to seal it. To restart the boiling, should you pour hot (but not boiling) water or cold water over the neck of the flask, as in part b of the drawing?

Reasoning and Solution When the open flask is removed from the burner, the water begins to cool. Boiling stops, because water cannot boil when its temperature is less than 100 °C and the external pressure is one atmosphere (1.01×10^5 Pa). Certainly, pouring hot water over the corked flask can rewarm the water within it to some extent. But since the water being poured is not boiling, its temperature must be less than 100 °C. Therefore, it cannot reheat the water within the flask to 100 °C and restart the boiling. However, when cold water is poured over the corked flask, it causes some of the water vapor inside to condense. Consequently, the pressure above the liquid in the flask drops. When the pressure drops to a certain level, boiling restarts. This occurs when the pressure in the flask becomes equal to the vapor pressure of the water at its current temperature (which is now less than 100 °C). ***Thus, it is possible to restart the boiling by pouring cold water over the neck of the flask.***

(b) Water boiling again

Figure 12.31 (a) Water is boiling at a temperature of 100 °C and a pressure of one atmosphere. (b) The water boils at a temperature that is less than 100 °C, because the cool water reduces the pressure above the water in the flask.

The Physics of...
spray cans.

The operation of spray cans is based on the equilibrium between a liquid and its vapor. Figure 12.32a shows that a spray can contains a liquid propellant that is mixed with the product (such as hair spray). Inside the can, propellant vapor forms over the liquid. A propellant is chosen that has an equilibrium vapor pressure that is greater than atmospheric pressure at room temperature. Consequently, when the

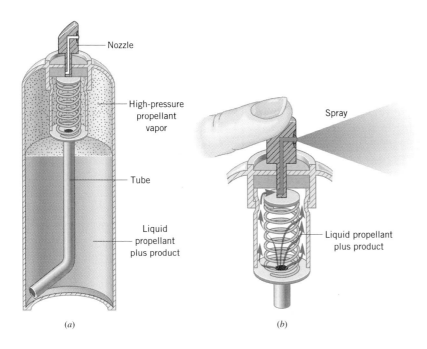

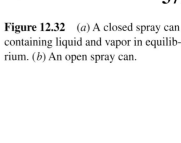

Figure 12.32 (*a*) A closed spray can containing liquid and vapor in equilibrium. (*b*) An open spray can.

nozzle of the can is pressed, as in part *b* of the drawing, the vapor pressure forces the liquid propellant and product up the tube in the can and out the nozzle as a spray. When the nozzle is released, the coiled spring reseals the can and the propellant vapor builds up once again to its equilibrium value.

As is the case for liquid/vapor equilibrium, a solid can be in equilibrium with its liquid phase only at specific conditions of temperature and pressure. For each temperature, there is a single pressure at which the two phases can coexist in equilibrium. A plot of the equilibrium pressure versus equilibrium temperature is referred to as the ***fusion curve,*** and Figure 12.33*a* shows a typical curve for a normal substance. A normal substance expands upon melting (e.g., carbon dioxide and sulfur). Since higher pressures make it more difficult for such materials to expand, a higher melting temperature is needed for a higher pressure, and the fusion curve slopes upward to the right. Part *b* of the picture illustrates the fusion curve for water, one of the few substances that contract when they melt. Higher pressures make it easier for such substances to melt. Consequently, a lower melting temperature is associated with a higher pressure, and the fusion curve slopes downward to the right.

It should be noted that just because two phases can coexist in equilibrium does not necessarily mean that they will. Other factors may prevent it. For example, water in an *open* bowl may never come into equilibrium with water vapor if air currents are present. Under such conditions the liquid, perhaps at a temperature of 25 °C, attempts to establish the corresponding equilibrium vapor pressure of 3.2×10^3 Pa. If air currents continually blow the water vapor away, however, equilibrium will never be established, and eventually the water will evaporate completely. Each kilogram of water that goes into the vapor phase takes along the latent heat of vaporization. Because of this heat loss, the remaining liquid would become cooler, except for the fact that the surroundings replenish the loss. In the case of the human body, water is exuded by the sweat glands and evaporates from a much larger area than the surface of a typical bowl of water. The removal of heat along with the water vapor is called evaporative cooling and is one mechanism that the body uses to maintain its constant temperature.

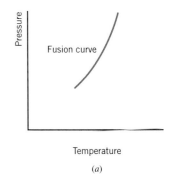

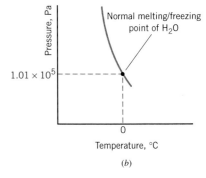

Figure 12.33 (*a*) The fusion curve for a normal substance that expands on melting. (*b*) The fusion curve for water, one of the few substances that contract on melting.

The Physics of...
evaporative cooling of the human body.

*12.10 HUMIDITY

Air is a mixture of gases, including nitrogen, oxygen, and water vapor. The total pressure of the mixture is the sum of the partial pressures of the component gases. The partial pressure of a gas is the pressure it would exert if it alone occupied the entire volume at the same temperature as the mixture. The partial pressure of water vapor in air depends on weather conditions. It can be as low as zero or as high as the equilibrium vapor pressure of water at the given temperature.

The Physics of...
relative humidity.

To provide an indication of how much water vapor is in the air, weather forecasters usually give the ***relative humidity.*** If the relative humidity is too low, the air contains such a small amount of water vapor that our skin and mucous membranes tend to dry out. If the relative humidity is too high, especially on a hot day, we become very uncomfortable and our skin feels "sticky." Under such conditions, the air holds so much water vapor that the water exuded by our sweat glands cannot evaporate efficiently. The relative humidity is defined as the ratio (expressed as a percentage) of the partial pressure of water vapor in the air to the equilibrium vapor pressure at a given temperature.

$$\text{Percent relative humidity} = \frac{\text{Partial pressure of water vapor}}{\text{Equilibrium vapor pressure of water at the existing temperature}} \times 100 \qquad (12.5)$$

The term in the denominator on the right of Equation 12.5 is given by the vaporization curve of water and is the pressure of the water vapor in equilibrium with the liquid. At a given temperature, the pressure of the water vapor in the air cannot exceed this value. If it did, the vapor would not be in equilibrium with the liquid and would condense as dew or rain to reestablish equilibrium.

When the partial pressure of the water vapor equals the equilibrium vapor pressure of water at a given temperature, the relative humidity is 100%. In such a situation, the vapor is said to be *saturated,* because it is present in the maximum amount, as it would be above a pool of liquid at equilibrium in a closed container. If the relative humidity is less than 100%, the water vapor is said to be *unsaturated.* Example 16 demonstrates how to find the relative humidity.

EXAMPLE 16 • Relative Humidities

One day, the partial pressure of water vapor in the air is 2.0×10^3 Pa. Using the vaporization curve for water in Figure 12.34, determine the relative humidity if the temperature is (a) 32 °C and (b) 21 °C.

Reasoning and Solution

(a) According to Figure 12.34, the equilibrium vapor pressure of water at 32 °C is 4.8×10^3 Pa. Equation 12.5 reveals that the relative humidity is

$$\text{Relative humidity at 32 °C} = \frac{2.0 \times 10^3 \text{ Pa}}{4.8 \times 10^3 \text{ Pa}} \times 100 = \boxed{42\%}$$

(b) A similar calculation shows that

$$\text{Relative humidity at 21 °C} = \frac{2.0 \times 10^3 \text{ Pa}}{2.5 \times 10^3 \text{ Pa}} \times 100 = \boxed{80\%}$$

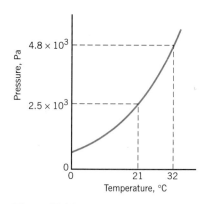

Figure 12.34 The vaporization curve of water.

When air containing a given amount of water vapor is cooled, a temperature is reached in which the partial pressure of the vapor equals the equilibrium vapor

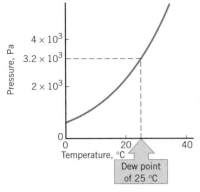

Figure 12.35 On the vaporization curve of water, the dew point is the temperature that corresponds to the actual partial pressure of water vapor in the air.

Figure 12.36 For fog to form, as it does in this forested region, the air temperature must drop below the dew point.

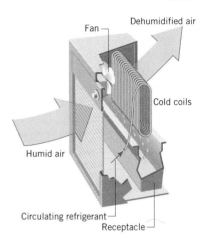

Figure 12.37 The cold coils of a dehumidifier cool the air blowing across them below the dew point, and water vapor condenses out of the air.

pressure. This temperature is known as the ***dew point.*** Figure 12.35 shows that if the partial pressure of water vapor is 3.2×10^3 Pa, the dew point is 25 °C. This partial pressure would correspond to a relative humidity of 100%, if the ambient temperature were equal to the dew-point temperature. Hence, the dew point is the temperature below which water vapor in the air condenses in the form of liquid drops (dew or fog). The closer the actual temperature is to the dew point, the closer the relative humidity is to 100%. Thus, for fog to form in the forested region shown in Figure 12.36, the air temperature must drop below the dew point. Similarly, water condenses on the outside of a cold glass when the temperature of the air next to the glass falls below the dew point. And the cold coils in a home dehumidifier (see Figure 12.37) function very much in the same way that the cold glass does. The coils are kept cold by a circulating refrigerant. When the air blown across them by the fan cools below the dew point, water vapor condenses in the form of droplets, which collect in a receptacle.

The Physics of...
fog formation.

The Physics of...
a home dehumidifier.

SUMMARY

On the **Celsius temperature scale,** there are 100 equal divisions between the ice point (0 °C) and the steam point (100 °C). On the **Fahrenheit temperature scale,** there are 180 equal divisions between the ice point (32 °F) and the steam point (212 °F). For most scientific work, the **Kelvin temperature scale** is the scale of choice. One kelvin is equal in size to one Celsius degree; however, the temperature T on the Kelvin scale differs from the temperature T_c on the Celsius scale by an additive constant of 273.15: $T = T_c + 273.15$. The lower limit of temperature is called **absolute zero** and is designated as 0 K on the Kelvin scale. The operation of any thermometer is based on the change in some physical property with temperature; this physical property is called a **thermometric property.**

Most substances expand when heated. For **linear expansion,** an object of length L_0 experiences a change in length

ΔL when the temperature changes by ΔT: $\Delta L = \alpha L_0 \Delta T$, where α is the **coefficient of linear expansion.** For **volume expansion,** the change in volume ΔV of an object of volume V_0 is given by $\Delta V = \beta V_0 \Delta T$, where β is the **coefficient of volume expansion.** When the temperature changes, a hole in a plate or a cavity in a piece of solid material expands or contracts as if the hole or cavity were filled with the surrounding material.

For an object held rigidly in place, a **thermal stress** can occur when the object attempts to expand or contract. The thermal stress can be extremely large, even when the temperature change is small.

The **internal energy** of a substance is the sum of the internal kinetic, potential, and other kinds of energy that the molecules of the substance have. **Heat** is energy that flows from a higher-temperature object to a lower-temperature object because of the difference in temperatures. The SI unit for heat is the joule.

The **specific heat capacity** c of a substance of mass m determines how much heat Q must be supplied to or removed from the substance to change its temperature by an amount ΔT: $Q = cm\,\Delta T$. When materials are placed in thermal contact within a perfectly insulated container, the **principle of energy conservation** requires that heat lost by warmer materials equals heat gained by cooler materials.

Heat must be supplied or removed to make a material change from one phase to another. The amount of heat per kilogram of material is called the latent heat of the phase change. The **latent heats of fusion, vaporization,** and **sublimation** refer, respectively, to the solid/liquid, liquid/vapor, and solid/vapor phase changes.

The **equilibrium vapor pressure** of a substance is the pressure of the vapor phase that is in equilibrium with the liquid phase. Vapor pressure depends only on temperature. For a liquid, a plot of the equilibrium vapor pressure versus temperature is called the **vapor pressure curve.** The **fusion curve** gives the combinations of temperature and pressure for equilibrium between solid and liquid phases.

The **relative humidity** is the ratio (expressed as a percentage) of the partial pressure of water vapor in the air to the equilibrium vapor pressure at the existing temperature. The **dew point** is the temperature below which the water vapor in the air condenses. On the vaporization curve of water, the dew point is the temperature that corresponds to the actual pressure of water vapor in the air.

CONCEPTUAL QUESTIONS

1. For the highest accuracy, would you choose an aluminum or a steel tape rule for year-round outdoor use? Why?

2. The first international standard of length was a metal bar kept at the International Bureau of Weights and Measures. One meter of length was defined to be the distance between two fine lines engraved near the ends of the bar. Why was it important that the bar be kept at a constant temperature?

3. A circular hole is cut through a flat aluminum plate. A spherical brass ball has a diameter that is slightly *smaller* than the diameter of the hole. The plate and the ball have the same temperature at all times. Should the plate and ball both be heated or both be cooled to *prevent* the ball from falling through the hole? Give your reasoning.

4. For added strength, many highways and buildings are constructed with reinforced concrete (concrete that is reinforced with embedded steel rods). Table 12.1 shows that the coefficient of linear expansion for concrete is the same as that for steel. Why is it important that these two coefficients be the same?

5. At a certain temperature, a rod is hung from an aluminum frame, as the drawing shows. A small gap exists between the rod and the floor. The frame and rod are heated uniformly. Explain whether the rod will ever touch the floor, assuming that the rod is made from (a) aluminum and (b) lead.

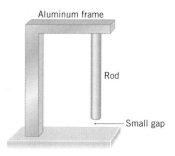

6. A simple pendulum is made using a long thin metal wire.

When the temperature drops, does the period of the pendulum increase, decrease, or remain the same? Account for your answer.

7. Two bimetallic strips have identical shapes. One is made from aluminum and steel, while the other is made from copper and silver. Which is likely to bend more for a given change in temperature? Justify your answer.

8. A hot steel ring fits snugly over a cold brass cylinder. The temperatures of the ring and cylinder are, respectively, above and below room temperature. Account for the fact that it is nearly impossible to pull the ring off the cylinder once the assembly has reached room temperature.

9. For glass baking dishes, Pyrex glass is used instead of common glass. A cold Pyrex dish taken from the refrigerator can be put directly into a hot oven without cracking from thermal stress. A dish made from common glass would crack. With the aid of Table 12.1, explain why Pyrex is better in this respect than common glass.

10. Suppose liquid mercury and glass both had the same coefficient of volume expansion. Explain why such a mercury-in-glass thermometer would not work.

11. Is the buoyant force provided by warm water (above 4 °C) greater than, less than, or equal to that provided by cold water (also above 4 °C)? Explain your reasoning.

12. When the bulb of a mercury-in-glass thermometer is inserted into boiling water, the mercury column first drops slightly before it begins to rise. Account for this phenomenon. *(Hint: Consider what happens initially to the glass.)*

13. Two different objects are supplied with equal amounts of heat. Give the reason(s) why their temperature changes would not necessarily be the same.

14. Two objects are made from the same material. They have different masses and temperatures. If the two are placed in contact, which object will experience the greater temperature change? Explain.

15. Two identical mugs contain hot coffee from the same pot.

One mug is full, while the other is only one-quarter full. Sitting on the kitchen table, which mug stays warmer longer? Explain.

16. Bicyclists often carry along a water bottle on hot days. When the bottle is wrapped in a wet sock, the water temperature stays near 75 °F for three or four hours, even when the air temperature is near 100 °F. Why does the water stay cool?

17. ⚕ To help lower the high temperature of a sick patient, an alcohol rub is sometimes used. Isopropyl alcohol is rubbed over the patient's back, arms, legs, etc., and allowed to evaporate. Why does the procedure work?

18. Your head feels colder under an air-conditioning vent when your hair is wet than when it is dry. Why?

19. ⚕ Suppose the latent heat of vaporization of H_2O were one-tenth its actual value. (a) Other things being equal, would it take the same time, a shorter time, or a longer time for a pot of water on a stove to boil away? (b) Would the evaporative cooling mechanism of the human body be as effective? Account for both answers.

20. Fruit blossoms are permanently damaged when the temperature drops below about −4 °C (a "hard freeze"). Orchard owners sometimes spray a film of water over the blossoms to protect them when a hard freeze is expected. From the point of view of phase changes, give a reason for the protection.

21. A camping stove is used to boil water on a mountain. Does it necessarily follow that the same stove can boil water at lower altitudes, such as at sea level? Provide a reason for your answer.

22. If a bowl of water is placed in a closed container and water vapor is pumped away rapidly enough, the remaining liquid will turn to ice. Explain why the ice appears.

23. ⚕ Medical instruments are sterilized under the hottest possible temperatures. Explain why they are sterilized in an autoclave, which is a device that is essentially a pressure cooker and heats the instruments in water under a pressure greater than one atmosphere.

24. A bottle of carbonated soda is left outside in subfreezing temperatures, although it remains in the liquid form. When the soda is brought inside and opened, it immediately freezes. Explain why this could happen.

25. A bowl of water is covered tightly and allowed to sit at a constant temperature of 23 °C for a long time. What is the relative humidity in the space between the surface of the water and the cover? Justify your answer.

26. Is it possible for dew to form on Tuesday night and not on Monday night, even though Monday night is the cooler night? Incorporate the idea of the dew point into your answer.

27. Two rooms in a house have the same temperature. One of the rooms contains an indoor swimming pool. On a cold day the windows of one of the rooms are "steamed up." Which room is it? Explain.

28. A jar is half filled with boiling water. The lid is then screwed on the jar. After the jar has cooled to room temperature, the lid is difficult to remove. Why?

PROBLEMS

Note: For problems in this set, use the values of α and β given in Table 12.1, and the values of c, L_f, and L_v given in Tables 12.2 and 12.4, unless stated otherwise.

ssm Solution is in the Student Solutions Manual. **www** Solution is available on the World Wide Web at http://www.wiley.com/college/cutnell ⚕ This icon represents a biomedical application.

Section 12.1 Common Temperature Scales, Section 12.2 The Kelvin Temperature Scale, Section 12.3 Thermometers

1. ssm What's your normal body temperature? It's probably not 98.6 °F, the oft-quoted average that was determined in the nineteenth century. A recent study has reported an average temperature of 98.2 °F. What is the *difference* between these averages, expressed in Celsius degrees?

2. A personal computer is designed to operate over the temperature range from 50.0 to 104 °F. To what do these temperatures correspond (a) on the Celsius scale and (b) on the Kelvin scale?

3. A comfortable temperature for most people is around 24 °C. What is this temperature (a) on the Fahrenheit scale and (b) on the Kelvin scale?

4. ⚕ Dermatologists often remove small precancerous skin lesions by freezing them quickly with liquid nitrogen, which has a temperature of 77 K. What is this temperature on the (a) Celsius and (b) Fahrenheit scales?

5. ssm A temperature of absolute zero occurs at −273.15 °C. What is this temperature on the Fahrenheit scale?

6. The greatest naturally occurring range in temperature of −90.0 °F to +98.0 °F was recorded in Verkhoyansk, Russia. What is the *difference* between the higher and lower temperature in (a) Celsius degrees and (b) kelvins?

*7. On the Rankine temperature scale, which is sometimes used in engineering applications, the ice point is at 491.67 °R and the steam point is at 671.67 °R. Determine a relationship (analogous to Equation 12.1) between the Rankine and Fahrenheit temperature scales.

*8. Space invaders land on earth. On the invaders' temperature scale, the ice point is at 25 °I (I = invader), and the steam point is at 156 °I. The invaders' thermometer shows the temperature on earth to be 58 °I. Using logic similar to that in Example 1 in the text, what would this temperature be on the Celsius scale?

9. **ssm A constant-volume gas thermometer (see Figures 12.3 and 12.4) has a pressure of 5.00×10^3 Pa when the gas temperature is 0.00 °C. What is the temperature (in °C) when the pressure is 2.00×10^3 Pa?

Section 12.4 Linear Thermal Expansion

10. A baking dish is taken from a refrigerator at 8 °C and put into an oven that is set for 155 °C. Each side of the cold dish has a length of 0.15 m. By how much does a side of the dish expand as it heats up, if the dish is made from (a) common glass and (b) Pyrex?

11. A steel aircraft carrier is 370 m long when moving through the icy North Atlantic at a temperature of 2.0 °C. By how much does the carrier lengthen when it is traveling in the warm Mediterranean Sea at a temperature of 21 °C?

12. Conceptual Example 5 provides background for this problem. A hole is drilled through a copper plate whose temperature is 11 °C. (a) When the temperature of the plate is increased, will the radius of the hole be larger or smaller than the radius at 11 °C? Why? (b) When the plate is heated to 110 °C, by what fraction $\Delta r/r_0$ will the radius of the hole change?

13. **ssm www** Find the approximate length of the Golden Gate bridge if it is known that the steel in the roadbed expands by 0.53 m when the temperature changes from +2 to +32 °C.

14. An aluminum baseball bat has a length of 0.86 m at a temperature of 25.0 °C. When the temperature of the bat is raised, the bat expands by 0.000 16 m. Determine the final temperature of the bat.

15. A steel beam is used in the construction of a skyscraper. By what fraction $\Delta L/L_0$ does the length of the beam increase when the temperature changes from that on a cold winter day (−15 °F) to that on a summer day (+105 °F)?

16. When a fused quartz rod and a lead rod experience the same temperature change, they change lengths by the same amount. If the initial length of the lead rod is 0.10 m, what is the initial length of the quartz rod?

17. **ssm** A rod made from a particular alloy is heated from 25.0 °C to the boiling point of water. Its length increases by 8.47×10^{-4} m. The rod is then cooled from 25.0 °C to the freezing point of water. By how much does the rod shrink?

18. When a copper rod is stretched, it will rupture when a tensile stress of 2.3×10^7 N/m² is applied at each end. A copper rod is fastened securely at both ends to an immovable support. The rod is then cooled. What is the change in temperature of the rod when it ruptures?

*19. A lead sphere has a diameter that is 0.050% larger than the inner diameter of a steel ring when each has a temperature of 70.0 °C. Thus, the ring will not slip over the sphere. At what common temperature will the ring just slip over the sphere?

*20. The brass bar and the aluminum bar in the drawing are each attached to an immovable wall. At 28 °C the air gap between the rods is 1.3×10^{-3} m. At what temperature will the gap be closed?

*21. **ssm www** A simple pendulum consists of a ball connected to one end of a thin brass wire. The period of the pendulum is 2.0000 s. The temperature rises by 140 C°, and the length of the wire increases. Determine the period of the heated pendulum.

*22. A cylindrical drinking glass is placed inside another cylindrical glass. The two just fit, because the outer radius (0.035 m) of the inside glass is equal to the inner radius of the outside glass. The glasses are made from common glass. The inside glass is then filled with cold water at 5 °C, while the outside glass is heated to 45 °C by running hot water over it. Find the separation distance between the two glasses.

*23. A steel ruler is accurate when the temperature is 25 °C. When the temperature drops to −15 °C, the ruler no longer reads correctly, but it can be made to read correctly if a stress is applied to each end of the ruler. (a) Should the stress be a compression or a tension? Why? (b) What is the magnitude of the necessary stress?

**24. A steel ruler is calibrated to read true at 20.0 °C. A draftsman uses the ruler at 40.0 °C to draw a line on a 40.0 °C copper plate. As indicated on the warm ruler, the length of the line is 0.50 m. To what temperature should the plate be cooled, such that the length of the line truly becomes 0.50 m?

25. **ssm A wire is made by attaching two segments together, end to end. One segment is made of aluminum and the other is steel. The effective coefficient of linear expansion of the two-segment wire is 19×10^{-6} (C°)$^{-1}$. What fraction of the length is aluminum?

**26. A steel bicycle wheel (without the rubber tire) is rotating freely with an angular speed of 18.00 rad/s. The temperature of the wheel changes from −100.0 to +300.0 °C. No net external torque acts on the wheel, and the mass of the spokes is negligible. (a) Does the angular speed increase or decrease as the wheel heats up? Why? (b) What is the angular speed at the higher temperature?

Section 12.5 Volume Thermal Expansion

27. A swimming pool contains 110 m³ of water. The sun heats the water from 17 to 27 °C. What is the change in the volume of the water?

28. A copper kettle contains water at 24 °C. When the water is heated to its boiling point, the volume of the kettle expands by 1.2×10^{-5} m³. Determine the volume of the kettle at 24 °C.

29. **ssm** When heated at constant pressure, a given volume of gas expands much more than an equal volume of liquid. The coefficient of volume expansion for air is about 3.7×10^{-3} (C°)$^{-1}$ at room temperature and one atmosphere of pressure. Find the ratio of the change in volume of air to the change in volume of water, assuming the same initial volumes and temperature changes.

30. A 1.5000-L (L = liter) flask is filled with a liquid at 97.0 °C. When the liquid is cooled to 15.0 °C, the flask is full only to the 1.3832-L mark. Neglect the contraction of the flask and use Table 12.1 to identify the liquid.

31. Many hot-water heating systems have a reservoir tank connected directly to the pipeline, so as to allow for expansion when the water becomes hot. The heating system of a house has 76 m of copper pipe whose inside radius is 9.5×10^{-3} m. When the water and pipe are heated from 24 to 78 °C, what must be the minimum volume of the reservoir tank to hold the overflow of water?

32. During an all-night cram session, a student heats up a one-half liter (0.50×10^{-3} m^3) glass (Pyrex) beaker of cold coffee. Initially, the temperature is 18 °C, and the beaker is filled to the brim. A short time later when the student returns, the temperature has risen to 92 °C. The coefficient of volume expansion of coffee is the same as that of water. How much coffee (in cubic meters) has spilled out of the beaker?

33. ssm Suppose you were selling apple cider for two dollars a gallon when the temperature is 4.0 °C. The coefficient of volume expansion of the cider is 280×10^{-6} (C°)$^{-1}$. If the expansion of the container is ignored, how much more money (in pennies) would you make per gallon by refilling the container on a day when the temperature is 26 °C?

34. An aluminum can is filled to the brim with 3.5×10^{-4} m^3 of soda at 5 °C. When the can and soda are heated to 78 °C, 3.6×10^{-6} m^3 of soda spills over. What is the coefficient of volume expansion of the soda?

35. A thin spherical shell of silver has an inner radius of 1.5×10^{-2} m when the temperature is 25 °C. The shell is heated to 135 °C. Find the change in the interior volume of the shell.

*__36.__ The density of mercury is 13 600 kg/m^3 at 0 °C. What would be its density at 166 °C?

*__37. ssm__ The bulk modulus of water is $B = 2.2 \times 10^{9}$ N/m^2. What change in pressure ΔP (in atmospheres) is required to keep water from expanding when it is heated from 15 to 25 °C?

*__38.__ A solid aluminum sphere has a radius of 0.50 m and a temperature of 75 °C. The sphere is then completely immersed in a pool of water whose temperature is 25 °C. The sphere cools, while the water temperature remains nearly at 25 °C, because the pool is very large. The sphere is weighed in the water immediately after being submerged (before it begins to cool) and then again after cooling to 25 °C. (a) Which weight is larger? Why? (b) Use Archimedes' principle to find the magnitude of the *difference* between the weights.

*__39.__ Each side of a cube has a length L_0. When the temperature of the cube is increased by ΔT, the enlarged volume of the cube becomes $(L_0 + \Delta L)^3$. Expand this expression for the enlarged volume and show that the change in volume ΔV is given approximately by $\Delta V = 3\alpha V_0 \Delta T$, where $V_0 = L_0^3$ is the initial volume and α is the coefficient of linear expansion. *(Hint: The terms involving $(\alpha \Delta T)^2$ and $(\alpha \Delta T)^3$ are much smaller than the term involving $\alpha \Delta T$ and, therefore, can be ignored; try substituting any reasonable numbers for α and ΔT to convince yourself of this fact.)* Comparing this expression with $\Delta V = \beta V_0 \Delta T$, we see that $\beta = 3\alpha$ for solids.

**__40.__ The column of mercury in a barometer (see Figure 11.13) has a height of 0.760 m when the pressure is one atmosphere and the temperature is 0.0 °C. Ignoring any change in the glass containing the mercury, what will be the height of the mercury column for the same one atmosphere of pressure when the tempera-

ture rises to 38.0 °C on a hot day? *(Hint: The pressure in the barometer is given by* Pressure $= \rho gh$, *and the density ρ of the mercury changes when the temperature changes.)*

**__41. ssm www__ Two identical thermometers made of Pyrex glass contain, respectively, identical volumes of mercury and methyl alcohol. If the expansion of the glass is taken into account, how many times greater is the distance between the degree marks on the methyl alcohol thermometer than that on the mercury thermometer?

Section 12.6 Heat and Internal Energy, Section 12.7 Heat and Temperature Change: Specific Heat Capacity

42. An ice chest at a beach party contains 12 cans of soda at 5.0 °C. Each can of soda has a mass of 0.35 kg and a specific heat capacity of 3800 J/(kg·C°). Someone adds a 6.5-kg watermelon at 27 °C to the chest. The specific heat capacity of watermelon is nearly the same as that of water. Ignore the specific heat capacity of the chest and determine the final temperature T of the soda and watermelon.

43. Blood can carry excess energy from the interior to the surface of the body, where the energy is dispersed in a number of ways. While a person is exercising, 0.6 kg of blood flows to the surface of the body and releases 2000 J of energy. The blood arriving at the surface has the temperature of the body interior, 37.0 °C. Assuming that blood has the same specific heat capacity as water, determine the temperature of the blood that leaves the surface and returns to the interior.

44. Two bars of identical mass are at 25 °C. One bar is made from glass and the other from another substance in Table 12.2. Identical amounts of heat are supplied to each. The glass bar reaches a temperature of 88 °C, while the other reaches 250.0 °C. What is the other substance?

45. ssm At a fabrication plant, a hot metal forging has a mass of 75 kg and a specific heat capacity of 430 J/(kg·C°). To harden it, the forging is quenched by immersion in 710 kg of oil that has a temperature of 32 °C and a specific heat capacity of 2700 J/(kg·C°). The final temperature of the oil and forging at thermal equilibrium is 47 °C. Assuming that heat flows only between the forging and the oil, determine the initial temperature of the forging.

46. Heat is supplied to an 0.850-kg copper pot at the rate of 203 joules per second. How much time does it take to raise the temperature of the pot from 21.0 °C to 95.0 °C if the pot (a) is empty and (b) contains 0.200 kg of water?

47. When resting, a person has a metabolic rate of about 4.2×10^{5} joules per hour. The person is submerged neck-deep into a tub containing 1.0×10^{3} kg of water at 27.00 °C. If the heat from the person goes only into the water, find the water temperature after half an hour.

48. When you take a bath, how many kilograms of hot water (49.0 °C) must you mix with cold water (13.0 °C) so that the temperature of the bath is 36.0 °C? The total mass of water (hot plus cold) is 191 kg. Ignore any heat flow between the water and its external surroundings.

49. **ssm** If the price of electrical energy is $0.10 per kilowatt · hour, what is the cost of using electrical energy to heat the water in a swimming pool (12.0 m × 9.00 m × 1.5 m) from 15 to 27 °C?

***50.** Heat is added to a silver bar of mass 0.039 kg and length 0.15 m. The bar expands by an amount of 4.3×10^{-4} m. How much heat was added?

***51.** The box of a well-known breakfast cereal states that one ounce of the cereal contains 110 Calories (1 food Calorie = 4186 J). If 2.0% of this energy could be converted by a weight lifter's body into work done in lifting a barbell, what is the heaviest barbell that could be lifted a distance of 2.1 m?

***52.** Water is moving with a speed of 5.00 m/s just before it passes over the top of a waterfall. At the bottom, 5.00 m below, the water flows away with a speed of 3.00 m/s. What is the largest amount by which the temperature of the water at the bottom could exceed the temperature of the water at the top?

***53.** **ssm** An electric hot water heater takes in cold water at 13.0 °C and delivers hot water. The hot water has a constant temperature of 45.0 °C, when the "hot" faucet is left open all the time and the volume flow rate is 5.0×10^{-6} m³/s. What is the minimum power rating of the hot water heater?

****54.** A steel rod ($\rho = 7860$ kg/m³) has a length of 2.0 m. It is bolted at both ends between immobile supports. Initially there is no tension in the rod, because the rod just fits between the supports. Find the tension that develops when the rod loses 3300 J of heat.

Section 12.8 Heat and Phase Change: Latent Heat

55. An ice cube tray holds 0.39 kg of water at 0 °C. How much heat must a freezer remove to make ice cubes at 0 °C?

56. Heat (5.64×10^5 J) is added to 1.70 kg of a substance in Table 12.4. The substance completely melts without any change in temperature. What is the substance?

57. **ssm** How much heat must be added to 0.45 kg of aluminum to change it from a solid at 130 °C to a liquid at 660 °C (its melting point)? The latent heat of fusion for aluminum is 4.0×10^5 J/kg.

58. A 10.0-kg block of ice has a temperature of −10.0 °C. The pressure is one atmosphere. The block absorbs 4.11×10^6 J of heat. What is the final temperature of the liquid water?

59. Suppose the amount of heat removed when 3.0 kg of water freezes at 0 °C were removed from ethyl alcohol at its freezing/melting point of −114 °C. How many kilograms of ethyl alcohol would freeze?

60. A woman finds the front windshield of her car covered with ice at −12.0 °C. The ice has a thickness of 4.50×10^{-4} m, and the windshield has an area of 1.25 m². The density of ice is 917 kg/m³. How much heat is required to melt the ice?

61. **ssm** Liquid nitrogen boils at a chilly −195.8 °C when the pressure is one atmosphere. A silver coin of mass 1.5×10^{-2} kg and temperature 25 °C is dropped into the liquid. What mass of nitrogen boils off as the coin cools to −195.8 °C?

62. ♨ The latent heat of vaporization of H_2O at body tempera-

ture (37.0 °C) is 2.42×10^6 J/kg. To cool the body of a 75-kg jogger [average specific heat capacity = 3500 J/(kg·C°)] by 1.5 C°, how many kilograms of water in the form of sweat have to be evaporated?

63. When it rains, water vapor in the air condenses into liquid water, and energy is released. (a) How much energy is released when 0.0254 m (one inch) of rain falls over an area of 2.59×10^6 m² (one square mile)? (b) If the average energy needed to heat one home for a year is 1.50×10^{11} J, how many homes can be heated with the energy determined in part (a)?

64. Water at 23 °C is sprayed on 0.180 kg of molten gold at 1063 °C (its melting point). The water boils away, leaving solid gold at 1063 °C. What minimum mass of water must be used?

***65.** **ssm** Ice at −10.0 °C and steam at 130 °C are brought together at atmospheric pressure in a perfectly insulated container. After thermal equilibrium is reached, the liquid phase at 50.0 °C is present. Ignoring the container and the equilibrium vapor pressure of the liquid at 50.0 °C, find the ratio of the mass of steam to the mass of ice.

***66.** To help keep his barn warm on cold days, a farmer stores 840 kg of solar-heated water in barrels. For how many hours would a 2.0-kW electric space heater have to operate to provide the same amount of heat as the water does, when it cools from 10.0 to 0.0 °C and completely freezes?

***67.** A 35-kg block of ice at 0 °C is sliding on a horizontal surface. The initial speed of the ice is 6.5 m/s and the final speed is 4.8 m/s. Assume that the part of the block that melts has a very small mass and that all the heat generated by kinetic friction goes into the block of ice, and determine the mass of ice that melts into water at 0 °C.

***68.** In the British engineering system, the latent heat of fusion of H_2O is 144 Btu/lb. Suppose an air conditioner can remove heat at a rate of 12 000 Btu/h. Determine the number of tons (1 ton = 2000 lb) of water at 0 °C that such an air conditioner could freeze into ice at 0 °C in 24 h. This is the number of "tons of air-conditioning" that the unit can provide, a phrase that is sometimes used to rate the cooling capacity of air conditioners.

***69.** **ssm** **www** An unknown material has a normal melting/freezing point of −25.0 °C, and the liquid phase has a specific heat capacity of 160 J/(kg·C°). One-tenth of a kilogram of the solid at −25.0 °C is put into a 0.150-kg aluminum calorimeter cup that contains 0.100 kg of glycerin. The temperature of the cup and the glycerin is initially 27.0 °C. All the unknown material melts, and the final temperature at equilibrium is 20.0 °C. The calorimeter loses no energy to the external environment. What is the latent heat of fusion of the unknown material?

***70.** Heat is extracted from a certain quantity of steam at 100.0 °C. As a result, the steam changes into ice at 0.0 °C. If this energy were used to accelerate the ice from rest, what would be the linear speed of the ice? For comparison, bullet speeds of about 700 m/s are common.

****71.** A locomotive wheel is 1.00 m in diameter. A 25.0-kg steel band has a temperature of 20.0 °C and a diameter that is 6.00×10^{-4} m less than that of the wheel. What is the smallest mass of water vapor at 100 °C that can be condensed on the steel band to heat it, so that it will fit onto the wheel? Do not ignore the water that results from the condensation.

Section 12.9 Equilibrium Between Phases of Matter

72. What pressure would be required for carbon dioxide to boil at a temperature of 30 °C? Use the vapor pressure curve that accompanies this problem.

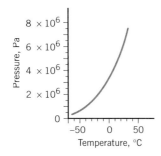

73. **ssm** Use the vapor pressure curve that accompanies problem 72 to determine the temperature at which liquid carbon dioxide exists in equilibrium with its vapor phase when the pressure is 3.5×10^6 Pa.

*74. A container is fitted with a movable piston of negligible mass and radius $r = 0.061$ m. Inside the container is liquid water in equilibrium with its vapor, as the drawing shows. The piston remains stationary with a 120-kg block on top of it. The air pressure acting on the top of the piston is one atmosphere. By using the vaporization curve for water in Figure 12.30, find the temperature of the water.

**75. A tall column of water is open to the atmosphere. At a depth of 10.3 m below the surface, the water is boiling. What is the temperature at this depth? Use the vaporization curve for water in Figure 12.30, as needed.

Section 12.10 Humidity

76. Using the vapor pressure curve for water that accompanies this problem, find the partial pressure of water vapor on a day when the weather forecast gives the relative humidity as 56.0% and the temperature as 30.0 °C.

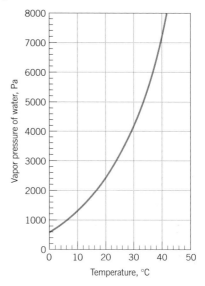

77. **ssm** What is the relative humidity on a day when the temperature is 30 °C and the dew point is 10 °C? Use the vapor pressure curve that accompanies problem 76.

78. The outdoor temperature is 15 °C, and the relative humidity is 45%. The partial pressure of water vapor inside a house is the same as outdoors, but the temperature in the house is 25 °C. Using the vapor pressure curve for water that accompanies problem 76, determine the relative humidity in the house.

*79. The temperature of the air in a room is 36 °C. A person turns on a dehumidifier and notices that when the cooling coils reach 30 °C, water begins to condense on them. What is the relative humidity in the room? Use the vapor pressure curve that accompanies problem 76.

*80. A woman has been outdoors where the temperature is 10 °C. She walks into a 25 °C house, and her glasses "steam up." Using the vapor pressure curve for water that accompanies problem 76, find the smallest possible value for the relative humidity of the room.

81. **ssm At a picnic, a glass contains 0.300 kg of tea at 30.0 °C, which is the air temperature. To make iced tea, someone adds 0.0670 kg of ice at 0.0 °C and stirs the mixture. When all the ice melts and the final temperature is reached, the glass begins to fog up, because water vapor condenses on the outer glass surface. Using the vapor pressure curve for water that accompanies problem 76, ignoring the specific heat capacity of the glass, and treating the tea as if it were water, estimate the relative humidity.

ADDITIONAL PROBLEMS

82. The Concorde is 62 m long when its temperature is 23 °C. In flight, the outer skin of this supersonic aircraft can reach 105 °C due to air friction. The coefficient of linear expansion of the skin is 2.0×10^{-5} (C °)$^{-1}$. Find the amount by which the Concorde expands.

83. Assume that the pressure is one atmosphere and determine the heat required to produce 2.00 kg of water vapor at 100.0 °C, starting with (a) 2.00 kg of water at 100.0 °C and (b) 2.00 kg of liquid water at 0.0 °C.

84. A commonly used method of fastening one part to another part is called "shrink fitting." A steel rod has a diameter of 2.0026 cm, and a flat plate contains a hole whose diameter is 2.0000 cm. The rod is cooled so that it just fits into the hole. When the rod warms up, the enormous thermal stress exerted by the plate holds the rod securely to the plate. By how many Celsius degrees should the rod be cooled?

85. ssm Find the mass of water that vaporizes when 2.10 kg of mercury at 205 °C is added to 0.110 kg of water at 80.0 °C.

86. Into a 0.200-kg copper cup (20.0 °C) is put 0.100 kg of aluminum at 50.0 °C and 0.250 kg of water at 85.0 °C. Assuming there is no heat flow between the cup and its outside environment, find the final equilibrium temperature. The final temperature is greater than 50.0 °C.

87. When the temperature of a coin is raised by 75 C°, the coin's diameter increases by 2.3×10^{-5} m. If the original diameter is 1.8×10^{-2} m, find the coefficient of linear expansion.

88. A precious-stone dealer wishes to find the specific heat capacity of a 0.030-kg gemstone. The specimen is heated to 95.0 °C and then placed in a 0.15-kg copper vessel that contains 0.080 kg of water at equilibrium at 25.0 °C. The loss of heat to the external environment is negligible. When equilibrium is established, the temperature is 28.5 °C. What is the specific heat capacity of the specimen?

89. ssm Suppose that the gas tank in your car is completely filled when the temperature is 17 °C. How many gallons will spill out of the twenty-gallon steel tank when the temperature rises to 35 °C?

90. The relative humidity is 35% when the temperature is 27 °C. Using the vapor pressure curve for water that accompanies problem 76, determine the dew point.

*91. A copper–constantan thermocouple generates a voltage of 4.75×10^{-3} volts when the temperature of the hot junction is 110.0 °C and the reference junction is kept at 0.0 °C. If the voltage is proportional to the difference in temperature between the junctions, what is the temperature of the hot junction when the voltage is 1.90×10^{-3} volts?

*92. An electrical resistance thermometer is made from platinum wire. The electrical resistance (measured in a unit called an *ohm*) changes from 0.40 ohms when the wire is at 0.00 °C to a value of 0.68 ohms when the wire is heated to 100.0 °C. Assume that the resistance varies linearly with temperature. What would be the resistance when the temperature is 25 °C?

*93. ssm It is claimed that if a lead bullet goes fast enough, it can melt completely when it comes to a halt suddenly, and all its kinetic energy is converted into heat via friction. Find the minimum speed of a lead bullet (initial temperature = 30.0 °C) for such an event to happen.

*94. A can is filled with a liquid to 97.0% of its capacity. The temperature of the can and the liquid is 0.0 °C. The material from which the can is made has a coefficient of volume expansion of

85×10^{-6} (C°)⁻¹. At a temperature of 100.0 °C, the can is observed to be filled to the brim. Determine the coefficient of volume expansion of the liquid.

*95. A 0.25-kg coffee mug is made from a material that has a specific heat capacity of 950 J/(kg·C°) and contains 0.30 kg of water. The cup and water are at 25 °C. To make a cup of coffee, a small electric heater is immersed in the water and brings it to a boil in two minutes. Assume that the cup and water always have the same temperature and determine the minimum power rating of this heater.

*96. The length of each edge of an aluminum cube is 0.050 m. The cube is heated to 145 °C from 25 °C. What is the increase in the surface area of the cube? (*Hint: The length of each edge of the cube expands when heated.*)

*97. ssm A rock of mass 0.20 kg falls from rest from a height of 15 m into a pail containing 0.35 kg of water. The rock and water have the same initial temperature. The specific heat capacity of the rock is 1840 J/(kg·C°). Ignore the heat absorbed by the pail itself, and determine the rise in the temperature of the rock and water.

**98. An aluminum wire of radius 3.0×10^{-4} m is stretched between the ends of a concrete block, as the drawing illustrates. When the system (wire and concrete) is at 35 °C, the tension in the wire is 50.0 N. What is the tension in the wire when the system is heated to 185 °C?

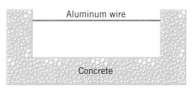

**99. An 85.0-N backpack is hung from the middle of an aluminum wire, as the drawing shows. The temperature of the wire drops by 20.0 C°. Find the tension in the wire at the lower temperature. Assume that the distance between the supports does not change, and ignore any thermal stress.

**100. A 1.5-kg steel sphere will not fit through a circular hole in a 0.85-kg aluminum plate, because the radius of the sphere is 0.10% larger than the radius of the hole. If both the sphere and the plate are always kept at the same temperature, how much heat must be put into the two so the ball just passes through the hole?

THE TRANSFER OF HEAT

Polar bears lose relatively small amounts of body heat, in part because of their fur, which consists of narrow, air-filled tubes. This chapter considers the processes by which heat is lost or gained.

Concepts at a Glance

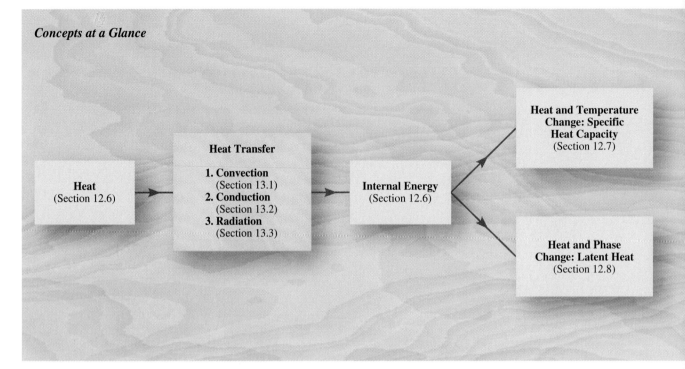

Figure 13.1 *Concepts at a Glance*
Heat is transferred from one place to another by three methods: convection, conduction, and radiation. As discussed in Chapter 12, heat can change the internal energy of a substance, which results in a change in the temperature or phase of the substance. The white-hot (over 1000 °C) cube shown in the photograph can be held between the fingers, because heat is transferred by conduction into the fingertips at a small rate.

13.1 CONVECTION

When heat is transferred to or from a substance, the internal energy of the substance can change as we saw in Chapter 12. This change in internal energy is accompanied by a change in temperature or a change in phase, as suggested by the concept chart in Figure 12.22. The transfer of heat from one place to another is important to our own well-being. For instance, within our home a furnace distributes heat on a cold day, and an air conditioner removes it on a hot day. Our bodies constantly transfer heat in one direction or another, to prevent the adverse effects of hypo- and hyperthermia. And virtually all our energy originates in the sun and is transferred to us over a distance of 150 million kilometers through the void of space. Today's sunlight provides the energy to drive photosynthesis in the plants that provide our food and, hence, metabolic energy. Ancient sunlight nurtured the organic matter that became the fossil fuels of oil, natural gas, and coal. This chapter examines the three processes by which heat is transferred: convection, conduction, and radiation. The concept chart in Figure 13.1, which is a modification of that in Figure 12.22, illustrates how these three methods of heat transfer are related to our previous study of heat and internal energy.

When part of a fluid is warmed, such as the air above a fire, the volume of the fluid expands, and the density decreases. According to Archimedes' principle (see Section 11.6), the surrounding cooler and denser fluid exerts a buoyant force on the warmer fluid and pushes it upward. As warmer fluid rises, the surrounding cooler fluid replaces it. This cooler fluid, in turn, is warmed and pushed upward. Thus, a continuous flow is established. The fluid flow carries along heat and is called a *convection current.* Whenever heat is transferred by the bulk movement of a gas or a liquid, the heat is said to be transferred by *convection.*

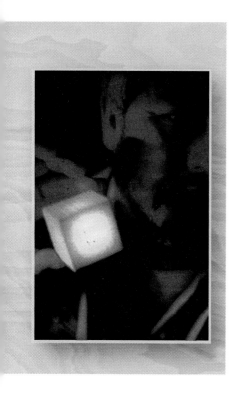

Figure 13.2 The smoke from a fire rises because of convection.

■ CONVECTION

Convection is the process in which heat is carried from place to place by the bulk movement of a fluid.

The smoke rising above the fire in Figure 13.2 is one visible result of convection. Figure 13.3 shows an example of convection currents in a pan of water being heated on a gas burner. The currents distribute the heat from the burning gas to all parts of the water. Conceptual Example 1 deals with some of the important roles that convection plays in the home.

Figure 13.3 Convection currents are set up when a pan of water is heated.

CONCEPTUAL EXAMPLE 1 • Hot Water Baseboard Heating and Refrigerators

Hot water baseboard heating units are frequently used in homes, where they are mounted on the wall next to the floor, as in Figure 13.4*a*. In contrast, the cooling coils in a refrigerator are mounted near the top of the refrigerator, as in part *b* of the drawing. The locations for these heating and cooling devices are different, yet each location is designed to maximize the production of convection currents. Explain how.

Reasoning and Solution An important goal for a heating system is to distribute heat throughout a room. The analogous goal for the cooling coils is to remove heat from all of the space within a refrigerator. In each case, the heating or cooling device is positioned so that convection makes the goal achievable. The air above the baseboard unit is heated, like the air above a fire. Buoyant forces from the surrounding cooler air push the warm air upward. Cooler air near the ceiling is displaced downward and then warmed by the baseboard heating unit, leading to the convection

The Physics of...
heating and cooling by convection.

Figure 13.4 (*a*) Air warmed by the baseboard heating unit is pushed to the top of the room by the cooler and denser air. (*b*) Air cooled by the cooling coils sinks to the bottom of the refrigerator. In both (*a*) and (*b*) a convection current is established.

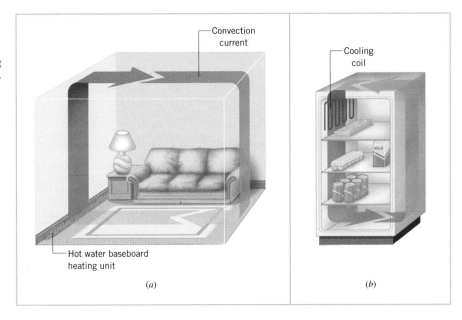

Convection current

Cooling coil

Hot water baseboard heating unit

(*a*)   (*b*)

current illustrated in Figure 13.4*a*. Had the heating unit been located near the ceiling, the warm air would have remained there, with very little convection to distribute the heat.

Within the refrigerator, as the air in contact with the top-mounted coils is cooled, its volume decreases and its density increases. The surrounding warmer and less dense air cannot provide sufficient buoyant force to support the colder air, which sinks downward. In the process, warmer air near the bottom is displaced upward and is then cooled by the coils, establishing the convection current shown in Figure 13.4*b*. Had the cooling coils been placed at the bottom of the refrigerator, stagnant, cool air would have collected there, with little convection to carry the heat from other parts of the refrigerator to the coils for removal.

Another example of convection occurs when the ground, heated by the sun's rays, warms the neighboring air. Surrounding cooler and denser air pushes the heated air upward. The resulting updraft or "thermal" can be quite strong, depending on the amount of heat that the ground can supply. As Figure 13.5 illustrates, these thermals can be used by glider pilots to gain considerable altitude. Birds such as eagles utilize thermals in a similar fashion.

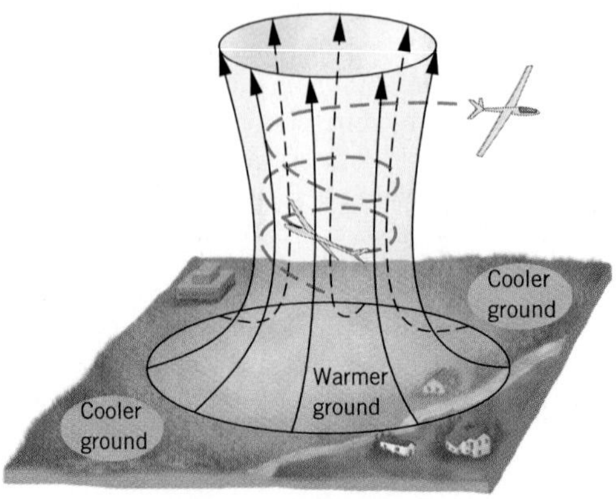

The Physics of...
"thermals."

Cooler ground

Cooler ground

Warmer ground

Figure 13.5 Updrafts, or thermals, are produced by the convective movement of air that has been warmed by the ground.

It is usual for air temperature to decrease with increasing altitude, and the resulting upward convection currents are important for dispersing pollutants from industrial sources and automobile exhaust systems. Sometimes, however, meteorological conditions cause a layer to form in the atmosphere where the temperature increases with increasing altitude. Such a layer is called an *inversion layer,* because its temperature profile is inverted compared to the usual situation. An inversion layer arrests the normal upward convection currents, causing a stagnant-air condition in which the concentration of pollutants increases substantially. This leads to a smog layer that can often be seen hovering over large cities (Figure 13.6).

We have been discussing **natural convection,** in which a temperature difference causes the density at one place in a fluid to be different than at another. Sometimes, natural convection is inadequate to transfer sufficient amounts of heat. In such cases *forced convection* is often used, and an external device such as a pump or a fan mixes the warmer and cooler portions of the fluid. Figure 13.7 shows an example of an automobile engine, where forced convection occurs in several ways. First, a pump circulates radiator fluid (water and antifreeze) through the engine to remove excess heat from the combustion process. Second, a radiator fan draws air through the radiator. Heat is transferred from the hotter radiator fluid to the cooler air, thereby cooling the fluid.

The Physics of...
cooling by forced convection.

13.2 CONDUCTION

Anyone who has fried a hamburger in an all-metal skillet knows that the metal handle becomes hot. Somehow, heat is transferred from the burner to the handle. Clearly, heat is not being transferred by the bulk movement of the metal or the surrounding air, so convection can be ruled out. Instead, heat is transferred directly through the metal by a process called *conduction.*

■ CONDUCTION

Conduction is the process whereby heat is transferred directly through a material, any bulk motion of the material playing no role in the transfer.

One mechanism for conduction occurs when the atoms or molecules in a hotter part of the material vibrate or move with greater energy than those in a cooler part. By means of collisions, the more energetic molecules pass on some of their energy to their less energetic neighbors. For example, imagine a gas that fills the space between two walls that face each other and are maintained at different temperatures. Molecules that strike the hotter wall absorb energy from it and rebound with a greater kinetic energy than when they arrived. As these more energetic molecules collide with their less energetic neighbors, they transfer some of their energy to them. Eventually, this energy is passed on until it reaches the molecules next to the cooler wall. These molecules, in turn, collide with the wall, giving up some of their energy to it in the process. Through such molecular collisions, heat is conducted from the hotter to the cooler wall.

A similar mechanism for the conduction of heat occurs in metals. Metals are different from most substances in having a pool of electrons that are more or less free to wander throughout the metal. These free electrons can transport energy and allow metals to transfer heat very well. The free electrons are also responsible for the excellent electrical conductivity that metals have.

Figure 13.6 In the absence of upward convection currents in the air, pollutants accumulate and form a smog layer in New York City.

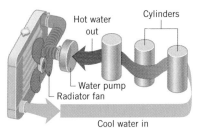

Figure 13.7 The forced convection generated by a pump circulates radiator fluid through an automobile engine to remove excessive heat.

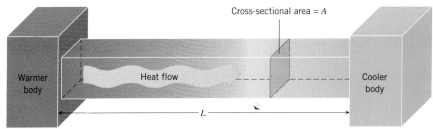

Figure 13.8 Heat is conducted through the bar when the ends of the bar are maintained at different temperatures. The heat flows from the warmer to the cooler end.

Those materials that conduct heat well are called ***thermal conductors,*** while those that conduct heat poorly are known as ***thermal insulators.*** Most metals, such as aluminum, copper, gold, and silver, are excellent thermal conductors, while wood, glass, and most plastics are common thermal insulators. Thermal insulators have many important applications. Virtually all new housing construction incorporates thermal insulation in attics and walls to reduce heating and cooling costs. And the wooden or plastic handles on many pots and pans reduce the flow of heat to the cook's hand.

To illustrate the factors that influence the conduction of heat, Figure 13.8 displays a rectangular bar. The ends of the bar are in thermal contact with two bodies, one of which is kept at a constant higher temperature, while the other is kept at a constant lower temperature. Although not shown for the sake of clarity, the sides of the bar are insulated, so the heat lost through them is negligible. The amount of heat Q conducted through the bar from the warmer end to the cooler end depends on a number of factors:

1. Q is proportional to the length of time t during which conduction takes place ($Q \propto t$). More heat flows in longer time periods.

2. Q is proportional to the temperature difference ΔT between the two ends of the bar ($Q \propto \Delta T$). A larger temperature difference causes more heat to flow. No heat flows when both ends have the same temperature, so that $\Delta T = 0$.

3. Q is proportional to the cross-sectional area A of the bar ($Q \propto A$). Figure 13.9 helps to explain this fact by showing two identical bars (insulated sides not shown) placed between the warmer and cooler bodies. Clearly, twice as much heat flows through two bars as through one, since the cross-sectional area has been doubled.

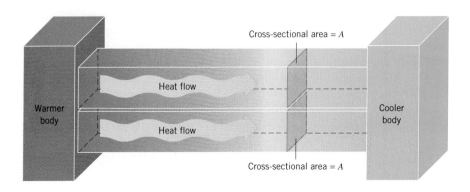

Figure 13.9 Twice as much heat flows through two identical bars as through one bar.

4. Q is inversely proportional to the length L of the bar ($Q \propto 1/L$). Greater lengths of material conduct less heat. To experience this effect, put two insulated mittens (the kind that cooks keep around the stove) on the *same hand*. Then, touch a hot pot and notice that it feels cooler than when you wear only one mitten, signifying that less heat passes through the greater thickness ("length") of material.

These proportionalities can be stated together as $Q \propto (A\,\Delta T)t/L$. Equation 13.1 expresses this result with the aid of a proportionality constant k, which is called the **thermal conductivity.**

■ **CONDUCTION OF HEAT THROUGH A MATERIAL**

The heat Q conducted during a time t through a bar of length L and cross-sectional area A is

$$Q = \frac{(kA\,\Delta T)t}{L} \qquad (13.1)$$

where ΔT is the temperature difference between the ends of the bar and k is the thermal conductivity of the material.

SI Unit of Thermal Conductivity: $J/(s \cdot m \cdot C°)$

Since $k = QL/(tA\,\Delta T)$, the SI unit for thermal conductivity is $J \cdot m/(s \cdot m^2 \cdot C°)$ or $J/(s \cdot m \cdot C°)$. The SI unit of power is the joule per second (J/s) or watt (W), so the thermal conductivity is also given in units of $W/(m \cdot C°)$.

Different materials have different thermal conductivities, and Table 13.1 gives some representative values. Because metals are such good thermal conductors, they have large thermal conductivities. In comparison, liquids and gases generally have small thermal conductivities. In fact, in most fluids the heat transferred by conduction is negligible compared to that transferred by convection when there are strong convection currents. Air, for instance, with its small thermal conductivity, is an excellent thermal insulator when confined to small spaces where no appreciable convection currents can be established. Goose down, Styrofoam, and wool derive their fine insulating properties in part from the small dead-air spaces within them, as Figure 13.10 illustrates. We also take advantage of dead-air spaces when we dress "in layers" during very cold weather and put on several layers of relatively thin clothing, rather than one thick layer. The air trapped between the layers acts as an excellent insulator.

Example 2 deals with the role that conduction through body fat plays in regulating body temperature.

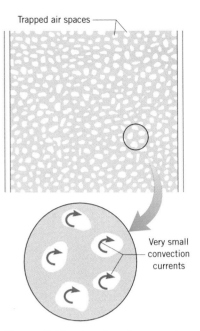

Figure 13.10 Styrofoam is an excellent thermal insulator because it contains many small, dead-air spaces. These small spaces inhibit heat transfer by convection currents, and air itself has a very low thermal conductivity.

▶ **The Physics of...**
dressing warm.

EXAMPLE 2 • Heat Transfer in the Human Body

When excessive heat is produced within the body, it must be transferred to the skin and dispersed if the temperature at the body interior is to be maintained at the normal value of 37.0 °C. One possible mechanism for transfer is conduction through body fat. Suppose that heat travels through 0.030 m of fat in reaching the skin, which has a total surface area of 1.7 m^2 and a temperature of 34.0 °C. Find the amount of heat that reaches the skin in half an hour (1800 s).

▶ **The Physics of...**
heat transfer in the human body.

Table 13.1 Thermal Conductivities[a] of Selected Materials

Substance	Thermal Conductivity, k [J/(s·m·C°)]
Metals	
Aluminum	240
Brass	110
Copper	390
Iron	79
Lead	35
Silver	420
Steel (stainless)	14
Gases	
Air	0.0256
Hydrogen (H$_2$)	0.180
Nitrogen (N$_2$)	0.0258
Oxygen (O$_2$)	0.0265
Other Materials	
Asbestos	0.090
Body fat	0.20
Concrete	1.1
Diamond	2450
Glass	0.80
Goose down	0.025
Ice (0 °C)	2.2
Styrofoam	0.010
Water	0.60
Wood (oak)	0.15
Wool	0.040

[a] Except as noted, the values pertain to temperatures near 20 °C.

Reasoning and Solution In Table 13.1 the thermal conductivity of body fat is given as $k = 0.20$ J/(s·m·C°). According to Equation 13.1,

$$Q = \frac{(kA\,\Delta T)t}{L}$$

$$Q = \frac{[0.20 \text{ J/(s·m·C°)}](1.7 \text{ m}^2)(37.0 \text{ °C} - 34.0 \text{ °C})(1800 \text{ s})}{0.030 \text{ m}} = \boxed{6.1 \times 10^4 \text{ J}}.$$

For comparison, a jogger can generate over ten times this amount of heat in a half hour. Thus, conduction through body fat is not a particularly effective way of removing excess heat. Heat transfer via blood flow to the skin is more effective and has the added advantage that the body can vary the blood flow as needed (see problem 9).

Virtually all homes contain insulation in the walls to reduce heat loss. Example 3 illustrates how to determine this loss with and without insulation.

EXAMPLE 3 • Layered Insulation

One wall of a house consists of 0.019-m-thick plywood backed by 0.076-m-thick insulation, as Figure 13.11 shows. The temperature at the inside surface is 25.0 °C, while the temperature at the outside surface is 4.0 °C, both being constant. The thermal conductivities of the insulation and plywood are, respectively, 0.030 and 0.080 J/(s·m·C°), and the area of the wall is 35 m². Find the heat conducted through the wall in one hour (a) with the insulation and (b) without the insulation.

Reasoning The temperature at the insulation–plywood interface (see drawing) must be determined before the heat conducted through the wall can be obtained. In calculating this temperature T, we use the fact that no heat is accumulating in the wall, for the inner and outer temperatures are constant. Therefore, the heat conducted through the insulation must equal the heat conducted through the plywood during the same time, that is, $Q_{\text{insulation}} = Q_{\text{plywood}}$. Each of the Q values can be expressed as $Q = (kA\,\Delta T)t/L$, according to Equation 13.1, leading to an expression that can be solved for the interface temperature T. Once a value for T is available, Equation 13.1 can be used to obtain the heat conducted through the wall.

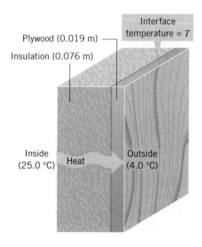

Plywood (0.019 m)
Insulation (0.076 m)
Interface temperature = T
Inside (25.0 °C) Heat Outside (4.0 °C)

Figure 13.11 Heat flows through the insulation and plywood from the warmer inside to the cooler outside. The temperature of the insulation–plywood interface is T.

Solution

(a) Using the fact that $Q_{\text{insulation}} = Q_{\text{plywood}}$ and Equation 13.1, we find that

$$\left[\frac{(kA\Delta T)t}{L}\right]_{\text{insulation}} = \left[\frac{(kA\Delta T)t}{L}\right]_{\text{plywood}}$$

$$\frac{[0.030 \text{ J/(s} \cdot \text{m} \cdot \text{C}°)]A(25.0 °\text{C} - T)t}{0.076 \text{ m}} = \frac{[0.080 \text{ J/(s} \cdot \text{m} \cdot \text{C}°)]A(T - 4.0 °\text{C})t}{0.019 \text{ m}}$$

Eliminating the area A and time t algebraically and solving this equation for T reveals that the temperature at the insulation–plywood interface is $T = 5.8 °\text{C}$.

The Physics of...
layered insulation.

The heat conducted through the wall is either $Q_{\text{insulation}}$ or Q_{plywood}, since the two quantities are equal. Choosing $Q_{\text{insulation}}$ and using $T = 5.8 °\text{C}$ in Equation 13.1, we find that

$$Q_{\text{insulation}} = \frac{[0.030 \text{ J/(s} \cdot \text{m} \cdot \text{C}°)](35 \text{ m}^2)(25.0 °\text{C} - 5.8 °\text{C})(3600 \text{ s})}{0.076 \text{ m}}$$

$$= \boxed{9.5 \times 10^5 \text{ J}}$$

• **PROBLEM SOLVING INSIGHT**
When heat is conducted through a multiple-layer material, and the high and low temperatures are constant, the heat conducted through each layer is the same.

(b) It is straightforward to use Equation 13.1 to calculate the amount of heat flowing through the plywood in one hour if the insulation were absent:

$$Q_{\text{plywood}} = \frac{[0.080 \text{ J/(s} \cdot \text{m} \cdot \text{C}°)](35 \text{ m}^2)(25.0 °\text{C} - 4.0 °\text{C})(3600 \text{ s})}{0.019 \text{ m}}$$

$$= \boxed{110 \times 10^5 \text{ J}}$$

Without insulation, the heat loss is increased by a factor of about 12.

Fruit growers sometimes protect their crops by spraying them with water when overnight temperatures are expected to drop below freezing. Some fruit, like the strawberries in Figure 13.12, can withstand temperatures down to freezing (0 °C) with minimal harm. However, as the temperature falls below freezing, the risk of tissue damage rises significantly. When water is sprayed on the strawberries, it can freeze and form a covering of ice. When the water freezes it releases heat (see Section 12.8), and some of this heat goes into warming the plant. In addition, both water and ice have relatively small thermal conductivities, as Table 13.1 indicates. Thus, they also protect the crop by acting as thermal insulators which reduce heat loss from the plants.

The Physics of...
protecting fruit plants from freezing.

Figure 13.12 Fruit growers sometimes spray water to protect their crops. After a subzero night, these berries are visible in their insulating jackets of ice.

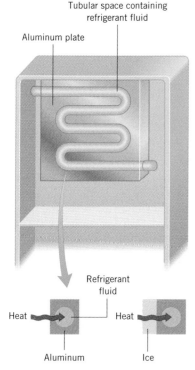

Tubular space containing refrigerant fluid

Aluminum plate

Refrigerant fluid

Heat → | Heat →

Aluminum | Ice

Figure 13.13 In a refrigerator, cooling is accomplished by a cold refrigerant fluid that circulates through a tubular space embedded within an aluminum plate. The arrangement works less well when coated with a layer of ice.

While a layer of ice may be beneficial to strawberry plants, it is not so desirable inside a refrigerator, as Conceptual Example 4 discusses.

CONCEPTUAL EXAMPLE 4 • An Iced-up Refrigerator

In a refrigerator heat is removed by a cold refrigerant fluid that circulates within a tubular space embedded within a metal plate, as Figure 13.13 illustrates. A good refrigerator cools food as quickly as possible. Decide whether the plate should be made from aluminum or stainless steel and whether the arrangement works better or worse when it becomes coated with a layer of ice.

Reasoning and Solution Figure 13.13 (see blowups) shows the metal cooling plate with and without a layer of ice. Without ice, heat passes by conduction through the metal to the refrigerant fluid within. For a given temperature difference across the thickness of metal, heat is transferred more quickly through the metal with the largest thermal conductivity. Table 13.1 indicates that the thermal conductivity of aluminum is more than 17 times larger than that of stainless steel. Therefore, *the plate should be made from aluminum.* When the plate becomes coated with ice, any heat that is removed by the refrigerant fluid must first be transferred by conduction through the ice before it even encounters the aluminum plate. But the conduction of heat through ice occurs much less readily than through aluminum, for Table 13.1 indicates that ice has a much smaller thermal conductivity. Moreover, Equation 13.1 indicates that the heat conducted per unit time (Q/t) is inversely proportional to the thickness L of the ice. Thus, as ice builds up, the heat removed per unit time by the cooling plate decreases. *When covered with ice, the cooling plate works less well.*

Related Homework Material: Problem 10

13.3 RADIATION

Energy from the sun is brought to earth by large amounts of visible light waves, as well as by substantial amounts of infrared and ultraviolet waves. These waves belong to a class of waves known as electromagnetic waves, a class that also includes the microwaves used for cooking and the radio waves used for AM and FM broadcasts. The sunbathers in Figure 13.14 feel hot, because their bodies absorb energy from the sun's electromagnetic waves. And anyone who has stood by a roaring fire or put a hand near an incandescent light bulb has experienced a similar effect. Thus, fires and light bulbs also emit electromagnetic waves, and when the energy of such waves is absorbed, it can have the same effect as heat.

The process of transferring energy via electromagnetic waves is called *radiation,* and, unlike convection or conduction, it does not require a material medium. Electromagnetic waves from the sun, for example, travel through the void of space during their journey to earth.

■ **RADIATION**

Radiation is the process in which energy is transferred by means of electromagnetic waves.

All bodies continuously radiate energy in the form of electromagnetic waves. Even an ice cube radiates energy, although so little of it is in the form of visible light that an ice cube cannot be seen in the dark. Likewise, the human body emits insufficient visible light to be seen in the dark. However, as we saw in Figures 12.6

Figure 13.14 Suntans are produced by ultraviolet rays.

and 12.7, the infrared waves radiating from the body can be detected in the dark by electronic cameras. Generally, an object does not emit much visible light until the temperature of the object exceeds about 1000 K. Then a characteristic red glow appears, like that of a heating coil on an electric stove. When its temperature reaches about 1700 K, an object begins to glow white-hot, like the tungsten filament in an incandescent light bulb.

In the transfer of energy by radiation, the absorption of electromagnetic waves is just as important as the emission. The surface of an object plays a significant role in determining how much radiant energy the object will absorb or emit. The two blocks in Figure 13.15, for example, are identical, except that one has a rough surface coated with lampblack (a fine black soot), while the other has a highly polished silver surface. A thermometer is inserted into each block, and the blocks are placed in direct sunlight. The temperature of the black block rises at a much faster rate than that of the silvery block, because lampblack absorbs about 97% of the incident radiant energy, while the silvery surface absorbs only about 10%. The remaining part of the incident energy is reflected in each case. We see the lampblack as black in color because it reflects so little of the light falling on it, while the silvery surface looks like a mirror because it reflects so much light. Since the color black is associated with nearly complete absorption of visible light, the term *perfect blackbody* or, simply, *blackbody* is used when referring to an object that absorbs *all* the electromagnetic waves falling on it.

All objects emit and absorb electromagnetic waves simultaneously. When a body has the same constant temperature as its surroundings, the amount of radiant energy being absorbed must balance the amount being emitted in a given interval of time. As Figure 13.16 illustrates, the block coated with lampblack absorbs and emits the same amount of radiant energy, and the silvery block does too. In either case, if absorption were greater than emission, the block would experience a net gain in energy. As a result, the temperature of the block would rise and not be constant. Similarly, if emission were greater than absorption, the temperature would fall. Since absorption and emission are balanced, *a material that is a good absorber, like lampblack, is also a good emitter, and a material that is a poor absorber, like polished silver, is also a poor emitter.* A perfect blackbody, being a perfect absorber, is also a perfect emitter.

The fact that a black surface is both a good absorber and a good emitter is the reason people are uncomfortable wearing dark clothes during the summer. Dark clothes absorb a large fraction of the sun's radiation and then reemit it in all directions. About one-half of the emitted radiation is directed inward toward the body

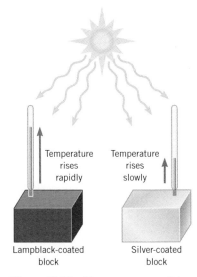

Figure 13.15 The temperature of the block coated with lampblack rises faster than the temperature of the block coated with silver, because the black surface absorbs radiant energy from the sun at a greater rate than does the silver surface.

The Physics of...
summer clothing.

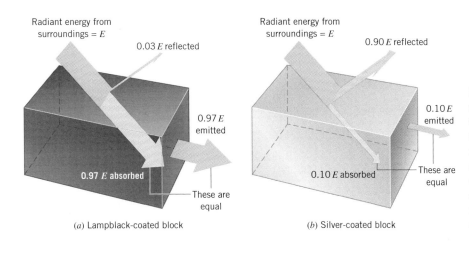

(a) Lampblack-coated block

(b) Silver-coated block

Figure 13.16 When a block and its surroundings have the same constant temperature, the block emits the same amount of radiant energy that it absorbs in a given time interval. This is true for a block coated with (a) lampblack and (b) silver. Although the block emits radiation in all directions, the emission is represented by a single arrow in these drawings.

and creates the sensation of warmth. Light-colored clothes, in contrast, are cooler to wear, since they absorb and reemit relatively little of the incident radiation.

The use of light colors for comfort also occurs in nature. Most lemurs, for instance, are nocturnal and have dark fur like the one shown in Figure 13.17a. Since they are active at night, the dark fur poses no disadvantage in absorbing excessive sunlight. Figure 13.17b shows a species of lemur called the white sifaka that lives in semiarid regions where there is little shade. The white color of the fur may help in thermoregulation, by reflecting sunlight during the hot part of the day. However, during the cool mornings, reflection of sunlight would be a hindrance in warming up. It is interesting to note that these lemurs have black skin and only sparse fur on their bellies, and that to warm up in the morning, they turn their dark bellies to the sun. The dark color enhances the absorption of sunlight.

The amount of radiant energy Q emitted by a perfect blackbody is proportional to the radiation time interval t ($Q \propto t$). The longer the time, the greater the amount of energy radiated. Experiment shows that Q is also proportional to the surface area A ($Q \propto A$). An object with a large surface area radiates more energy than one with a small surface area, other things being equal. Finally, experiment reveals that Q is proportional to the *fourth power of the Kelvin temperature* T ($Q \propto T^4$), so the emitted energy increases markedly with increasing temperature. If, for example, the Kelvin temperature of an object doubles, the object emits 2^4 or 16 times more energy. Combining these factors into a single proportionality, we see that $Q \propto T^4At$. This proportionality is converted into an equation by inserting a proportionality constant σ, known as the *Stefan–Boltzmann constant*. It has been found experimentally that $\sigma = 5.67 \times 10^{-8}$ J/(s·m²·K⁴):

$$Q = \sigma T^4 At$$

The relationship above holds only for a perfect emitter. Most objects are not perfect emitters, however. For instance, a dark-colored human body radiates only about 80% of the visible light energy that a perfect emitter would radiate, so Q (for dark skin) $= (0.80)\sigma T^4At$. The factor such as the 0.80 in this equation is called the **emissivity e** and is a dimensionless number between zero and one. The emissivity is the ratio of the energy an object actually radiates to the energy the object would radiate if it were a perfect emitter. For visible light, the value of e for the human body varies between about 0.65 and 0.80, the smaller values pertaining to lighter skin colors. For infrared radiation, e is nearly one for all skin colors. For a perfect blackbody emitter, $e = 1$. Including the factor e on the right side of the expression $Q = \sigma T^4At$ leads to the **Stefan–Boltzmann law of radiation.**

In Equation 13.2, the Stefan–Boltzmann constant σ is a universal constant in the sense that its value is the same for all bodies, regardless of the nature of their surfaces. The emissivity e, however, depends on the condition of the surface.

Example 5 shows how the Stefan–Boltzmann law can be used to determine the size of a star.

(a)

(b)

Figure 13.17 (a) Most lemurs, like this one, are nocturnal and have dark fur. (b) The species of lemur called the white sifaka, however, is active during the day and has white fur.

EXAMPLE 5 • A Supergiant Star

The supergiant star Betelgeuse has a surface temperature of about 2900 K and emits a radiant power (in joules per second or watts) of approximately 4×10^{30} W. The temperature is about one-half and the power about 10 000 greater than that of our sun. Assuming that Betelgeuse is a perfect emitter (emissivity $e = 1$) and spherical, find its radius.

Reasoning According to the Stefan–Boltzmann law, the power emitted is $Q/t = e\sigma T^4 A$. A star with a relatively small temperature T can have a relatively large radiant power Q/t only if the area A is large. As we will see, Betelgeuse indeed has a very large surface area, so its radius is enormous.

Solution Solving the Stefan–Boltzmann law for the area, we find

$$A = \frac{Q/t}{e\sigma T^4}$$

But the surface area of a sphere is $A = 4\pi r^2$, so $r = \sqrt{A/4\pi}$. Therefore, we have

$$r = \sqrt{\frac{Q/t}{4\pi e\sigma T^4}} = \sqrt{\frac{4 \times 10^{30} \text{ W}}{4\pi(1)[5.67 \times 10^{-8} \text{ J/(s} \cdot \text{m}^2 \cdot \text{K}^4)](2900 \text{ K})^4}}$$

$$= \boxed{3 \times 10^{11} \text{ m}}$$

For comparison, Mars orbits the sun at a distance of 2.28×10^{11} m. Betelgeuse is certainly a "supergiant."

> • **PROBLEM SOLVING INSIGHT**
> When solving an equation to arrive at a numerical answer, algebraically solve for the unknown variable in terms of the known variables. Then substitute in the numbers for the known variables, as this example shows.

The next example explains how to apply the Stefan–Boltzmann law when an object, such as a wood stove, simultaneously emits and absorbs radiant energy.

EXAMPLE 6 • A Wood-Burning Stove

> **The Physics of...**
> a wood-burning stove.

A wood-burning stove stands unused in a room where the temperature is 18 °C (291 K). A fire is started inside the stove. Eventually, the temperature of the stove surface reaches a constant 198 °C (471 K), and the room warms to a constant 29 °C (302 K). The stove has an emissivity of 0.900 and a surface area of 3.50 m². Determine the *net* radiant power generated by the stove when the stove (a) is unheated and has a temperature equal to room temperature and (b) has a temperature of 198 °C.

Reasoning The stove emits more radiant power when heated than when unheated. In both cases, however, the Stefan–Boltzmann law can be used to determine the amount of power emitted. Power is energy per unit time or Q/t. But in this problem we need to find the *net* power produced by the stove. The net power is the power the stove emits minus the power the stove absorbs. The power the stove absorbs comes from the walls, ceiling, and floor of the room, all of which emit radiation.

Solution

(a) Remembering that temperature must be expressed in kelvins when using the Stefan–Boltzmann law, we find that

> • **PROBLEM SOLVING INSIGHT**
> In the Stefan–Boltzmann law of radiation, the temperature T must be expressed in kelvins, not in degrees Celsius or degrees Fahrenheit.

$$\begin{array}{l} \text{Power emitted} \\ \text{by unheated} \\ \text{stove at 18 °C} \end{array} = \frac{Q}{t} = e\sigma T^4 A \qquad (13.2)$$

$$= (0.900)[5.67 \times 10^{-8} \text{ J/(s} \cdot \text{m}^2 \cdot \text{K}^4)](291 \text{ K})^4(3.50 \text{ m}^2) = 1280 \text{ W}$$

The fact that the unheated stove emits 1280 W of power and yet maintains a constant temperature means that the stove also absorbs 1280 W of radiant power from its surroundings. Thus, the *net* power generated by the unheated stove is zero:

$$
\begin{array}{c}
\text{Net power} \\
\text{generated by} \\
\text{stove at 18 °C}
\end{array}
=
\underbrace{1280\ \text{W}}_{\substack{\text{Power emitted} \\ \text{by stove at} \\ \text{18 °C}}}
-
\underbrace{1280\ \text{W}}_{\substack{\text{Power emitted by} \\ \text{room at 18 °C and} \\ \text{absorbed by stove}}}
= \boxed{0}
$$

(b) The hot stove (198 °C or 471 K) emits more radiant power than it absorbs from the cooler room. The radiant power the stove emits is

$$
\begin{array}{c}
\text{Power emitted} \\
\text{by stove at} \\
\text{198 °C}
\end{array}
= \frac{Q}{t} = e\sigma T^4 A
$$

$$
= (0.900)[5.67 \times 10^{-8}\ \text{J/(s·m}^2\text{·K}^4)](471\ \text{K})^4(3.50\ \text{m}^2) = 8790\ \text{W}
$$

The radiant power the stove absorbs from the room is identical to the power that the stove would emit at the constant room temperature of 29 °C (302 K). The reasoning here is exactly like that in part (a):

$$
\begin{array}{c}
\text{Power emitted by} \\
\text{room at 29 °C and} \\
\text{absorbed by stove}
\end{array}
= \frac{Q}{t} = e\sigma T^4 A
$$

$$
= (0.900)[5.67 \times 10^{-8}\ \text{J/(s·m}^2\text{·K}^4)](302\ \text{K})^4(3.50\ \text{m}^2) = 1490\ \text{W}
$$

The *net* radiant power the stove produces from the fuel it burns is

$$
\begin{array}{c}
\text{Net power} \\
\text{generated by} \\
\text{stove at 198 °C}
\end{array}
=
\underbrace{8790\ \text{W}}_{\substack{\text{Power emitted} \\ \text{by stove at} \\ \text{198 °C}}}
-
\underbrace{1490\ \text{W}}_{\substack{\text{Power emitted by} \\ \text{room at 29 °C and} \\ \text{absorbed by stove}}}
= \boxed{7300\ \text{W}}
$$

Example 6 illustrates that when an object has a higher temperature than its surroundings, the object emits a net radiant power $P_{\text{net}} = (Q/t)_{\text{net}}$. The net power is the power the object emits minus the power it absorbs. Applying the Stefan–Boltzmann law as in Example 6 leads to the following expression for P_{net} when the temperature of the object is T and the temperature of the environment is T_0:

$$
P_{\text{net}} = e\sigma A(T^4 - T_0^{\,4}) \tag{13.3}
$$

13.4 APPLICATIONS

To keep heating and air-conditioning bills to a minimum, it pays to use good thermal insulation in your home. Insulation inhibits convection between inner and outer walls and minimizes heat transfer by conduction. With respect to conduction, the logic behind home insulation comes directly from Equation 13.1. According to this equation, the heat per unit time Q/t flowing through a thickness of material is $Q/t = kA\,\Delta T/L$. Keeping the value for Q/t to a minimum means using materials that have small thermal conductivities k and large thicknesses L. Construction engineers, however, prefer to use Equation 13.1 in the slightly different form shown below:

$$
\frac{Q}{t} = \frac{A\,\Delta T}{L/k}
$$

Figure 13.18 The highly reflecting metal foil that covers the Hubble space telescope minimizes the temperature changes that would otherwise occur during each orbit around the earth.

The term L/k in the denominator is called the R value of the insulation. For a building material, it is convenient to talk about an R value, because it expresses in a single number the combined effects of thermal conductivity and thickness. Larger R values reduce the heat per unit time flowing through the material and, therefore, mean better insulation. It is also convenient to use R values to describe layered slabs formed by sandwiching together a number of materials with different thermal conductivities and different thicknesses. The R values for the individual layers can be added to give a single R value for the entire slab (see problem 42). It should be noted, however, that R values are expressed using units of feet, hours, F°, and BTU for thickness, time, temperature, and heat, respectively.

When it is in the earth's shadow, an orbiting satellite is shielded from the intense electromagnetic waves emitted by the sun. But when a satellite moves out of the earth's shadow, the satellite experiences the full effect of these waves. As a result, the temperature within a satellite would decrease and increase sharply during an orbital period and sensitive electronic circuitry would suffer, unless precautions are taken. To minimize temperature fluctuations, satellites are often covered with a highly reflecting and, hence, poorly absorbing metal foil, as Figure 13.18 shows. By reflecting much of the sunlight, the foil minimizes temperature rises. Being a poor absorber, the foil is also a poor emitter and reduces radiant energy losses. Reducing these losses keeps the temperature from falling excessively when the satellite is in the earth's shadow.

The design of solar collectors takes into account all three methods of heat transfer to capture radiant energy from the sun. As Figure 13.19 illustrates for one design, cool water is pumped into the collector, heated by solar energy, and then sent into the living quarters. The entire inside of the collector, including the water pipes, has a highly absorptive black coating, to capture as much radiant energy as possible. The copper from which the pipes are made has a large thermal conductivity, and, therefore, readily conducts the absorbed energy to the water. The glass cover minimizes the loss of heat due to air convection.

A thermos bottle, sometimes referred to as a Dewar flask, reduces the rate at which the hot liquids cool down or cold liquids warm up. A thermos usually consists of a double-walled glass vessel with silvered inner walls (see Figure 13.20) and accomplishes its job by minimizing heat transfer via convection, conduction, and radiation. The space between the walls is evacuated to minimize energy losses due to conduction and convection. The silvered surfaces reflect most of the radiant energy that would otherwise enter or leave the liquid in the thermos. Finally, little

The Physics of...
rating thermal insulation by
R values.

The Physics of...
regulating the temperature of an
orbiting satellite.

The Physics of...
a solar collector.

The Physics of...
a thermos bottle.

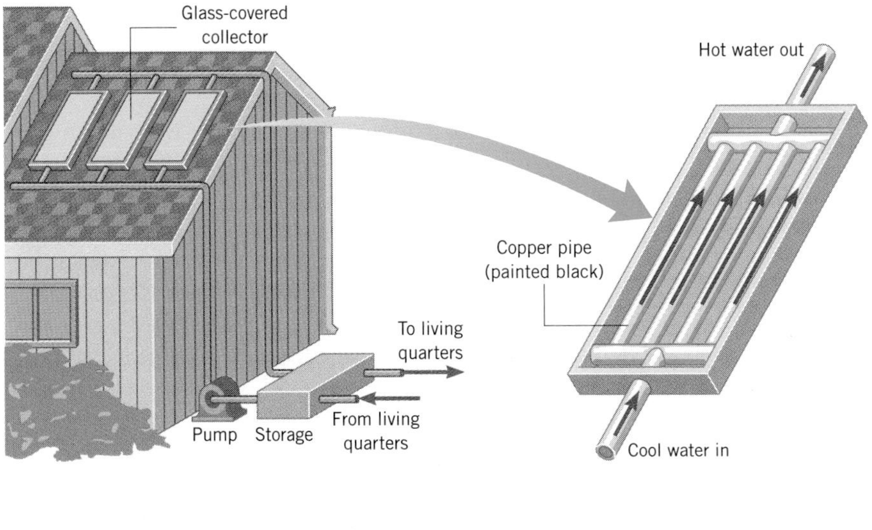

Figure 13.19 The design of a hot water solar collector takes into account heat transfer via convection, conduction, and radiation.

Figure 13.20 A thermos bottle or Dewar flask minimizes energy transfer due to convection, conduction, and radiation.

Figure 13.21 In a halogen cooktop, quartz–iodine lamps emit a large amount of electromagnetic energy that is absorbed directly by a pot or pan.

heat is lost through the glass or the rubberlike gaskets and stopper, since these materials have relatively small thermal conductivities.

The Physics of...
a halogen cooktop stove.

Halogen cooktops are a relatively new idea in cooking, and they use radiant energy to heat pots and pans. A halogen cooktop (see Figure 13.21) uses several quartz–iodine lamps, like the ones used for ultrabright automobile headlights. These lamps are electrically powered and are mounted below a ceramic top. They radiate a great deal of electromagnetic energy, which passes through the ceramic top and is absorbed directly by the bottom of the pot. Consequently, the pot heats up very quickly, rivaling the time of an open gas burner.

SUMMARY

Convection is the process in which heat is carried by the bulk movement of a fluid. During natural convection, the warmer, less dense part of a fluid is pushed upward by the buoyant force provided by the surrounding cooler and denser part. Forced convection occurs when an external device, such as a fan or a pump, causes the fluid to move.

Conduction is the process whereby heat is transferred directly through a material, any bulk motion of the material playing no role in the transfer. The heat Q conducted during a time t through a bar of length L and cross-sectional area A is expressed as $Q = (kA \Delta T)t/L$, where ΔT is the difference in temperature between the ends of the bar and k is the **thermal conductivity** of the material. Materials that have large values of k, such as most metals, are known as thermal conductors. Materials that have small values of k, such as Styrofoam and wood, are referred to as thermal insulators.

Radiation is the process in which energy is transferred by electromagnetic waves. All objects, regardless of their temperatures, simultaneously absorb and emit electromagnetic waves. A body that is at the same constant temperature as its surroundings absorbs and emits equal amounts of radiant energy per unit time. **Objects that are good absorbers of radiant energy are also good emitters, and objects that are poor absorbers are also poor emitters.** An object that absorbs all the radiation incident upon it is called a **perfect blackbody.** A perfect blackbody, being a perfect absorber, is also a perfect emitter.

The amount of radiant energy Q emitted during a time t by an object whose surface area is A and whose Kelvin temperature is T is given by the **Stefan–Boltzmann law,** $Q = e\sigma T^4 At$. In this equation, σ is the Stefan–Boltzmann constant $[\sigma = 5.67 \times 10^{-8} \text{ J/(s} \cdot \text{m}^2 \cdot \text{K}^4)]$ and e is the emissivity, a dimensionless number characterizing the surface of the object. The emissivity lies between 0 and 1, being zero for a nonemitting surface and one for a perfect blackbody. The net radiant power emitted by an object of temperature T located in an environment of temperature T_0 is $P_{net} = e\sigma A(T^4 - T_0^4)$.

CONCEPTUAL QUESTIONS

1. One often hears about heat transfer by convection in gases and liquids, but not in solids. Why?

2. A heavy drape, hung close to a cold window, reduces heat loss considerably by interfering primarily with one of the three processes of heat transfer. Explain which one.

3. The *windchill factor* is a term used by weather forecasters. Roughly speaking, it refers to the fact that you feel colder when the wind is blowing than when it is not, even though the air temperature is the same in either case. Which of the three processes for heat transfer plays the principal role in the windchill factor? Explain your reasoning.

4. Often the following warning sign is seen on bridges: "Caution—Bridge surface freezes before road surface." Account for the warning in terms of heat transfer processes. Note that, unlike the road, a bridge has both surfaces exposed to the air.

5. A piece of Styrofoam and a piece of wood are sandwiched together to form a layered slab. The two pieces have the same thickness and cross-sectional area. The exposed surfaces have constant temperatures. The temperature of the exposed Styrofoam surface is greater than the temperature of the exposed wood surface. Is the temperature of the Styrofoam–wood interface closer to the larger or smaller of the two temperatures? Give your reasoning, using Equation 13.1 and data from Table 13.1.

6. ♫ One way that heat is transferred from place to place inside the human body is by the flow of blood. Which one of the three heat transfer processes best describes this action of the blood? Justify your answer.

7. Some animals have hair, the strands of which are hollow, air-filled tubes. Other animals have hair that is composed of solid tubular strands. Which kind of hair would be more likely to give an animal an advantage for surviving in very cold climates? Why?

8. A poker used in a fireplace is held at one end, while the other end is in the fire. Why are pokers made of iron rather than copper? Ignore the fact that iron may be cheaper and stronger.

9. In Alaska, a lack of snow allowed the ground to freeze down to a depth of about one meter, causing buried water pipes to freeze and burst. Why did a lack of snow lead to this situation?

10. Concrete walls often contain steel reinforcement bars. Does the steel enhance or degrade the insulating value of the concrete? Explain.

11. Grandma says that it is quicker to bake a potato if you put a nail into it. In fact, she is right. Justify her baking technique in terms of one of the three processes of heat transfer.

12. Several days after a snowstorm, the roof on a house is uniformly covered with snow. On a neighboring house, however, the snow on the roof has completely melted. Which house is probably better insulated? Give your reasoning.

13. One car has a metal body, while another has a plastic body. On a cold winter day these cars are parked side by side. If you put a bare hand on each car, the metal body feels colder. Why?

14. A high-quality pot is designed so heat can enter readily and be distributed evenly, while the rate of energy loss from the pot is kept to a minimum. Many high-quality pots have copper bases and polished stainless steel sides. Based on conduction and radiation principles, explain why this design is better than all-copper or all-steel units.

15. Two objects have the same size and shape. Object A has an emissivity of 0.3, and object B has an emissivity of 0.6. Each radiates the same power. Is the Kelvin temperature of A twice that of B? Give your reasoning.

16. A concave mirror can be used to start a fire by directing sunlight onto a small spot on a piece of paper. Explain why the mirror does not get as hot as the paper.

17. Two strips of material, A and B, are identical, except they have emissivities of 0.4 and 0.7, respectively. The strips are heated to the same temperature and have a red glow. A brighter glow signifies that more energy per second is being radiated. Which strip has the brighter glow? Explain.

18. A pot of water is being heated on an electric stove. The diameter of the pot is smaller than the diameter of the heating element on which the pot rests. The exposed outer edges of the heating element are glowing cherry red. When you lift the pot, you see that the part of the heating element beneath it is not glowing cherry red, indicating that it is cooler than the outer edges. Why are the outer edges hotter?

19. To keep your hands as warm as possible during skiing, should you wear mittens or gloves? (Mittens, except for the thumb, do not have individual finger compartments.) Give a reason for your answer.

20. Two identical hot cups of cocoa are sitting on a table. One has a metal spoon in it and one does not. After five minutes, which cup is cooler? Explain in terms of heat transfer processes.

21. (a) Would a hot solid cube cool more rapidly if it were left intact or cut in half? Explain your answer in terms of one or more of the three heat transfer processes. (b) Using reasoning similar to that used in answering part (a), decide which cools faster, one pound of wide and flat lasagna noodles or one pound of spaghetti noodles. Assume that both kinds of noodles are made from the same pasta and start out with the same temperature.

22. One day during the winter the sun has been shining all day. Toward sunset a light snow begins to fall. It collects without melting on a cement playground, but it melts immediately upon contact with a black asphalt road adjacent to the playground. Account for the fact that the snow collects in one place but not in the other.

23. If you were stranded in the mountains in cold weather, it would help to minimize energy losses from your body by curling up into the tightest ball possible. Which of the factors in Equation 13.2 are you using to the best advantage by curling into a ball? Why?

PROBLEMS

Note: For problems in this set, use the values for thermal conductivities given in Table 13.1 unless stated otherwise.

ssm Solution is in the Student Solutions Manual. **www** Solution is available on the World Wide Web at http://www.wiley.com/college/cutnell ⚕ This icon represents a biomedical application.

Section 13.2 Conduction

1. ssm One end of an iron poker is placed in a fire where the temperature is 502 °C, and the other end is kept at a temperature of 26 °C. The poker is 1.2 m long and has a radius of 5.0×10^{-3} m. Ignoring the heat lost along the length of the poker, find the amount of heat conducted from one end of the poker to the other in 5.0 s.

2. A refrigerator has a surface area of 5.3 m². It is lined with 0.075-m-thick insulation whose thermal conductivity is 0.030 J/(s·m·C°). The interior temperature is kept at 5 °C, while the temperature at the outside surface is 25 °C. How much heat per second is being removed from the unit?

3. A closed box is filled with dry ice at a temperature of −78.5 °C, while the outside temperature is 21.0 °C. The box is cubical, measuring 0.350 m on a side, and the thickness of the walls is 3.00×10^{-2} m. In one day, 3.10×10^{6} J of heat is conducted through the six walls. Find the thermal conductivity of the box.

4. The temperature in an electric oven is 160 °C. The temperature at the outer surface in the kitchen is 50 °C. The oven (surface area = 1.6 m²) is insulated with material that has a thickness of 0.020 m and a thermal conductivity of 0.045 J/(s·m·C°). (a) How much energy is used to operate the oven for six hours? (b) At a price of $0.10 per kilowatt·hour for electrical energy, what is the cost of operating the oven?

5. ssm Due to a temperature difference ΔT, heat is conducted through an aluminum plate that is 0.035 m thick. The plate is then replaced by a stainless steel plate that has the same temperature difference and cross-sectional area. How thick should the steel plate be, so that the same amount of heat per second is conducted through it?

6. A hollow, cylindrical glass tube is filled with hydrogen (H_2) gas. The tube has a length of 0.25 m and an inner and outer radii of 0.020 and 0.023 m. One end of the tube is maintained at a temperature of 85 °C, while the other end is kept at 15 °C. There is no heat loss through the curved surface of the glass. Determine the heat that flows through the glass and hydrogen in a time of 55 s.

7. A skier wears a jacket filled with goose down that is 15 mm thick. Another skier wears a wool sweater that is 5.0 mm thick. Both have the same surface area. Assuming the temperature difference between the inner and outer surfaces of each garment is the same, calculate the ratio (wool/goose down) of the heat lost due to conduction during the same time interval.

8. Heat is conducted by two bars, one made from silver and the other from brass. The bars have the same cross-sectional area and the same length. The temperature difference between the ends of each bar is the same. Ignore any heat lost through the sides of the bars. What percentage of the total power, or energy per second, is conducted by the silver bar?

9. ssm www ⚕ In the conduction equation $Q = (kA\,\Delta T)t/L$, the combination of factors kA/L is called the *conductance*. The human body has the ability to vary the conductance of the tissue beneath the skin by means of vasoconstriction and vasodilation, in which the flow of blood to the veins and capillaries underlying the skin is decreased and increased, respectively. The conductance can be adjusted over a range such that the tissue beneath the skin is equivalent to a thickness of 0.080 mm of Styrofoam or 3.5 mm of air. By what factor can the body adjust the conductance?

*10. Review Conceptual Example 4 before attempting this problem. To illustrate the effect of ice on the aluminum cooling plate, consider the drawing below and the data contained therein. Ignore any limitations due to significant figures. (a) Calculate the heat per second per square meter that is conducted through the ice–aluminum sandwich. (b) Calculate the heat per second per square meter that would be conducted through the aluminum if the ice were not present. Notice how much larger the answer is in (b) as compared to (a).

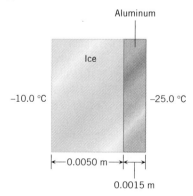

*11. One end of a brass bar is maintained at 295 °C, while the other end is kept at a constant but lower temperature. The cross-sectional area of the bar is 4.0×10^{-4} m². Because of insulation, there is negligible heat loss through the sides of the bar. Heat flows through the bar, however, at the rate of 2.7 J/s. What is the temperature of the bar at a point 0.20 m from the hot end?

*12. Two identical bars are attached end to end, as part *a* of the drawing illustrates. In a time of 44 minutes, a certain amount of heat is conducted from the left end to the right end. Suppose, instead, that the two bars were placed on top of one another, as in part *b* of the drawing, where the difference in temperature between the ends of the bars is the same as in part *a*. How long (in minutes) would it take for the same amount of heat to be conducted from left to right along this combination?

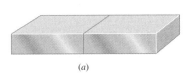

(a)   (b)

*13. **ssm** Two rods, one of aluminum and the other of copper, are joined end to end. The cross-sectional area of each is 4.0×10^{-4} m^2, and the length of each is 0.040 m. The free end of the aluminum rod is kept at 302 °C, while the free end of the copper rod is kept at 25 °C. The loss of heat through the sides of the rods may be ignored. (a) What is the temperature at the aluminum–copper interface? (b) How much heat is conducted through the unit in 2.0 s? (c) What is the temperature in the aluminum rod at a distance of 0.015 m from the hot end?

*14. Three building materials, plasterboard [$k = 0.30$ J/(s·m·C°)], brick [$k = 0.60$ J/(s·m·C°)], and wood [$k = 0.10$ J/(s·m·C°)], are sandwiched together as the drawing illustrates. The temperatures at the inside and outside surfaces are 27 °C and 0 °C, respectively. Each material has the same thickness and cross-sectional area. Find the temperature (a) at the plasterboard–brick interface and (b) at the brick–wood interface.

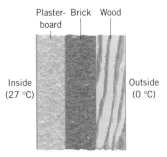

Plaster- Brick Wood
board

Inside
(27 °C)

Outside
(0 °C)

*15. In an aluminum pot, 0.20 kg of water at 100 °C boils away in five minutes. The bottom of the pot is 2.5×10^{-3} m thick and has a surface area of 0.020 m^2. To prevent the water from boiling too rapidly, a stainless steel plate has been placed between the pot and the heating element. The plate is 1.2×10^{-3} m thick, and its area matches that of the pot. Assuming that heat is conducted into the water only through the bottom of the pot, find the temperature at (a) the aluminum–steel interface and (b) the steel surface in contact with the heating element.

**16. A 0.30-m-thick sheet of ice covers a lake. The air temperature at the ice surface is −15 °C. In five minutes, the ice thickens by a small amount. Assume that no heat flows from the ground below into the water and that the added ice is very thin compared to 0.30 m. Find the number of millimeters by which the ice thickens.

17. **ssm www Two cylindrical rods have the same mass. One is made of silver (density = 10 500 kg/m^3), and one is made of iron (density = 7860 kg/m^3). Both rods conduct the same amount of heat per second when the same temperature difference is maintained across their ends. What is the ratio (silver-to-iron) of (a) the lengths and (b) the radii of these rods?

Section 13.3 Radiation

18. The filament of a light bulb has a temperature of 3.0×10^3 °C and radiates sixty watts of power. The emissivity of the filament is 0.36. Find the surface area of the filament.

19. An object emits 30 W of radiant power. If it were a perfect blackbody, other things being equal, it would emit 90 W of radiant power. What is the emissivity of the object?

20. Assume that the earth has an average surface temperature of 22 °C and is a perfect emitter of radiation ($e = 1$). Find the energy per second per square meter that the earth radiates into space.

21. **ssm** A car parked in the sun absorbs energy at a rate of 560 watts per square meter of surface area. The car reaches a temperature at which it radiates energy at this same rate. Treating the car as a perfect radiator ($e = 1$), find the temperature.

22. The energy per second, or power, radiated by the sun is 3.9×10^{26} W. (a) Imagine that the surface of the sun extends all the way out to the earth, a radial distance of about 1.5×10^{11} m. If this imaginary sun were a perfect blackbody radiator, what would be its surface temperature? (b) Repeat part (a), assuming that the sun extends out to the planet Pluto, a radial distance of 5.9×10^{12} m.

23. By what factor should the Kelvin temperature of an object be increased to double the radiant energy per second emitted by the object?

24. One day, the temperature of a radiator has to be 68 °C to keep the surrounding walls of a room at 24 °C. The next day is warmer outside, so the temperature of the radiator needs to be only 49 °C to keep the walls at 24 °C. Assuming the room is heated only by radiation, determine the ratio of the *net* power radiated by the unit on the colder day to that radiated on the warmer day.

25. **ssm** Review Example 6 before attempting this problem. Suppose the stove in that example had a surface area of only 2.00 m^2. What would its temperature (in kelvins) have to be so that it still generated a net power of 7300 W?

26. ⚡ Suppose the skin temperature of a naked person is 34 °C when the person is standing inside a room whose temperature is 25 °C. The skin area of the individual is 1.5 m^2. (a) Assuming the emissivity is 0.80, find the net loss of radiant power from the body. (b) Determine the number of food Calories of energy (1 food Calorie = 4186 J) that is lost in one hour due to the net loss rate obtained in part (a). Metabolic conversion of food into energy replaces this loss.

*27. Our sun has a radius of 6.96×10^8 m. Sirius B is a white dwarf star that emits only 0.04 times the radiant power of the sun, even though it has a surface temperature that is four times that of the sun. (a) What is the radius of Sirius B? (b) The mass of Sirius B is 2×10^{30} kg, nearly the same as the sun's. What is the mass density of this white dwarf star? For comparison, the density of water is 1000 kg/m^3.

*28. Liquid helium is stored at its boiling-point temperature of 4.2 K in a spherical container ($r = 0.30$ m). The container is a perfect blackbody radiator. The container is surrounded by a spherical shield whose temperature is 77 K. A vacuum exists in the space between the container and shield. The latent heat of vaporization for helium is 2.1×10^4 J/kg. What mass of liquid helium boils away through the venting valve in one hour?

29. **ssm A solid cylinder is radiating power. It has a length that is ten times its radius. It is cut into a number of smaller cylinders, each of which has the same length. Each small cylinder has the same temperature as the original cylinder. The total radiant power emitted by the pieces is twice that emitted by the original cylinder. How many smaller cylinders are there?

**30. A small sphere (emissivity = 0.90, radius = r_1) is located at the center of a spherical asbestos shell (thickness = 1.0 cm, outer radius = r_2). The thickness of the shell is small compared to the inner and outer radii of the shell. The temperature of the small sphere is 800.0 °C, while the temperature of the inner surface of the shell is 600.0 °C, both temperatures remaining constant. Assuming that $r_2/r_1 = 10.0$ and ignoring any air inside the shell, find the temperature of the outer surface of the shell.

ADDITIONAL PROBLEMS

31. In an electrically heated home, the temperature of the ground in contact with a concrete basement wall is 12.8 °C. The temperature at the inside surface of the wall is 20.0 °C. The wall is 0.10 m thick and has an area of 9.0 m². Assume one kilowatt · hour of electrical energy costs $0.10. How many hours are required for one dollar's worth of energy to be conducted through the wall?

32. Two light bulbs are identical, except that one has a filament that operates at 2700 K, while the other operates at 2100 K. Find the ratio of the power radiated by the hotter bulb to that radiated by the cooler bulb.

33. **ssm** A person's body is covered with 1.6 m² of wool clothing. The thickness of the wool is 2.0×10^{-3} m. The temperature at the outside surface of the wool is 11 °C, and the skin temperature is 36 °C. How much heat per second does the person lose due to conduction?

34. A solar collector is placed in direct sunlight where it absorbs energy at the rate of 880 J/s for each square meter of its surface. The emissivity of the solar collector is $e = 0.75$. What equilibrium temperature does the collector reach? Assume that the only energy loss is due to the emission of radiation.

35. A rod, made from an unknown metal, is 1.2 m long and has a cross-sectional area of 3.0×10^{-4} m². Except for its ends, the rod is insulated, so that no heat escapes through the sides of the rod. The rod conducts heat at a rate equal to or greater than 3.7 J/s, when a temperature difference of 75 C° is maintained between its ends. From which one or more of the metals listed in Table 13.1 could the rod be made?

36. The concrete wall of a building is 0.10 m thick. The temperature inside the building is 20.0 °C, while the temperature outside is 0.0 °C. Heat is conducted through the wall. When the building is unheated, the inside temperature falls to 0.0 °C, and heat conduction ceases. However, the wall does emit radiant energy when its temperature is 0.0 °C. The radiant energy emitted per second per square meter is the same as the heat lost per second per square meter due to conduction. What is the emissivity of the wall?

37. **ssm www** How many days does it take for a perfect blackbody cube (0.0100 m on a side, 30.0 °C) to radiate the same amount of energy that a one-hundred-watt light bulb uses in one hour?

*38. At a picnic, a Styrofoam cup contains lemonade and ice at 0 °C. The thickness of the cup is 2.0×10^{-3} m, and the area is

0.016 m². The temperature at the outside surface of the cup is 35 °C. The latent heat of fusion for ice is 3.35×10^5 J/kg. What mass of ice melts in one hour?

*39. Two pots are identical except that the flat bottom of one is aluminum, whereas that of the other is copper. Water in these pots is boiling away at 100 °C at the same rate. The temperature of the heating element on which the aluminum bottom is sitting is 155 °C. Assume that heat enters the water only through the bottoms of the pots and find the temperature of the heating element on which the copper bottom rests.

*40. A solid sphere has a temperature of 773 K. The sphere is melted down and recast into a cube that has the same emissivity and emits the same radiant power as the sphere. What is the cube's temperature?

41. **ssm www The drawing shows a solid cylindrical rod made from a center cylinder of lead and an outer concentric jacket of copper. Except for its ends, the rod is insulated (not shown), so that the loss of heat from the curved surface is negligible. When a temperature difference is maintained between its ends, this rod conducts one-half the amount of heat that it would conduct if it were solid copper. Determine the ratio of the radii r_1/r_2.

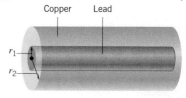

**42. The drawing shows a layered slab formed by sandwiching together two pieces of different insulating materials that have thermal conductivities of k_1 and k_2. The thicknesses L_1 and L_2 are different, and the temperature T_H is greater than the temperature T_C. Show that the heat conducted through the slab in a time t is given by $Q = A(T_H - T_C)t/[(L_1/k_1) + (L_2/k_2)]$, so that the effective R value of the slab is $(L_1/k_1) + (L_2/k_2)$.

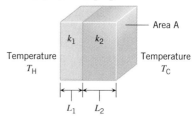

THE IDEAL GAS LAW AND KINETIC THEORY

Balloon fish can inflate their bodies with air, so that the spines in the
skin become erect. Their internal air pressure is determined, in part, by the depth
of the water in which the fish live.

14.1 MOLECULAR MASS, THE MOLE, AND AVOGADRO'S NUMBER

ATOMIC AND MOLECULAR MASSES

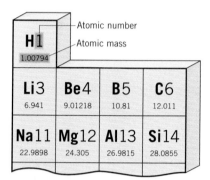

Figure 14.1 A portion of the periodic table showing the atomic number and atomic mass of each element. In the periodic table it is customary to omit the symbol "u" denoting atomic mass units.

The relative masses of the atoms of different elements can be expressed in terms of their **atomic masses,*** which indicate how massive one atom is compared to another. The atomic masses of all the elements are listed in the periodic table, part of which is shown in Figure 14.1. The complete periodic table is given on the inside of the back cover. In general, the masses listed are average values and take into account the various types or isotopes† of an element that exist naturally. The periodic table indicates, for example, that the lithium atom (Li) has an average atomic mass of 6.941, while the corresponding average value for the magnesium atom (Mg) is 24.305; thus, atomic magnesium is more massive than atomic lithium by a factor of 24.305/6.941 = 3.502.

To set up the atomic mass scale, a value (along with a unit) must be chosen for one of the elements. The unit on this scale is called an **atomic mass unit** (u). The reference element has been chosen to be the most abundant isotope of carbon, called carbon-12, and its atomic mass is defined to be exactly twelve atomic mass units, or 12 u. Thus, one atomic mass unit is exactly one-twelfth the mass of a carbon-12 atom. In Figure 14.1 the atomic mass of carbon (C) is given as 12.011 u, rather than exactly 12 u, because a small amount (about 1%) of the naturally occurring material is an isotope called carbon-13. The value 12.011 u is an average that reflects the small contribution of carbon-13.

The **molecular mass** of a molecule is the sum of the atomic masses of its atoms. For instance, hydrogen and oxygen have atomic masses of 1.007 94 and 15.9994 u, respectively, so that the molecular mass of a water molecule (H_2O) is $M =$ 2(1.007 94 u) + 15.9994 u = 18.0153 u.

THE MOLE AND AVOGADRO'S NUMBER

Macroscopic amounts of materials contain large numbers of atoms or molecules. Even in a small volume of gas, 1 cm^3, for example, the number is enormous. It is convenient to express such large numbers in terms of a single unit, the **gram-mole,** or simply the **mole** (symbol: *mol*). **One gram-mole of a substance contains as many particles (atoms or molecules) as there are atoms in 12 grams of the isotope carbon-12.** Experiment shows that 12 grams of carbon-12 contain 6.022×10^{23} atoms. This number is called **Avogadro's number** N_A, after the Italian scientist Amedeo Avogadro (1776–1856). Although defined in terms of carbon atoms, the concept of a mole can be applied to any collection of objects by noting that one mole is Avogadro's number of objects. Thus, one mole of atomic sulfur contains 6.022×10^{23} sulfur atoms, one mole of water contains 6.022×10^{23} H_2O molecules, and one mole of golf balls contains 6.022×10^{23} golf balls. Just as the meter is the SI base unit for length, the mole is the SI base unit for expressing "the amount of a substance."

One mole of any substance contains Avogadro's number of particles. However, a mole of one substance does not, in general, have the same mass as a mole of another substance. For example, one mole of carbon-12 atoms is defined to have a

---

* In chemistry the expression "atomic weight" is frequently used in place of "atomic mass."
† Isotopes are discussed in Section 31.1.

mass of 12 grams. However, one mole of aluminum atoms has a mass of 26.9815 grams for the following reason. An aluminum atom is more massive than a carbon-12 atom by the ratio of their atomic masses, (26.9815 u)/(12 u). Since Avogadro's number (1 mole) of carbon-12 atoms has a mass of 12 grams, then Avogadro's number (1 mole) of aluminum atoms must have a mass of

$$\left(\frac{26.9815 \text{ u}}{12 \text{ u}}\right)(12 \text{ grams}) = 26.9815 \text{ grams}$$

(a)

Thus, one mole of aluminum has a mass that is equal to its atomic mass, expressed in grams. This reasoning can be extended to any other atom or molecule, with the result that *one mole of a substance has a mass in grams that is equal to the atomic or molecular mass of the substance.*

Example 1 illustrates how to use the concepts of the mole, atomic mass, and Avogadro's number to determine the number of atoms and molecules present in two famous gemstones.

EXAMPLE 1 • The Hope Diamond and the Rosser Reeves Ruby

Figure 14.2*a* shows the Hope diamond (44.5 carats), which is almost pure carbon. Figure 14.2*b* shows the Rosser Reeves ruby (138 carats), which is primarily aluminum oxide (Al_2O_3). One carat is equivalent to a mass of 0.200 g. Determine (a) the number of carbon atoms in the diamond and (b) the number of Al_2O_3 molecules in the ruby.

Reasoning The number of atoms (or molecules) in a known mass of material is the number of moles of material times the number of atoms (or molecules) per mole (Avogadro's number). We can find the number of moles by dividing the known mass by the atomic mass of the substance.

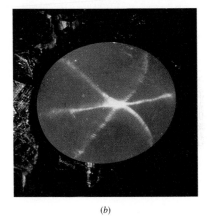

(b)

Figure 14.2 (*a*) The oval-shaped Hope diamond surrounded by 16 smaller diamonds. (*b*) The Rosser Reeves ruby. Both gems are on display at the Smithsonian Institution in Washington, D.C.

Solution

(a) The mass of the Hope diamond is (44.5 carats)[(0.200 g)/(1 carat)] = 8.90 g. Since the average atomic mass of naturally occurring carbon is 12.011 u, one mole of it has a mass of 12.011 g. The number of moles of carbon in the diamond is

$$(8.90 \text{ g})\left(\frac{1 \text{ mol}}{12.011 \text{ g}}\right) = 0.741 \text{ mol}$$

Because one mole contains Avogadro's number N_A of carbon atoms, the number of atoms in the diamond is $(0.741)N_A$, or

$$(0.741 \text{ mol})\left(\frac{6.022 \times 10^{23} \text{ atoms}}{1 \text{ mol}}\right) = \boxed{4.46 \times 10^{23} \text{ atoms}}$$

The Physics of... gemstones.

(b) The mass of the Rosser Reeves ruby is (138 carats)[(0.200 g)/(1 carat)] = 27.6 g. The molecular mass of aluminum oxide (Al_2O_3) is the sum of the atomic masses of its atoms, which are 26.9815 u for aluminum and 15.9994 u for oxygen:

$$\text{Molecular mass} = \underbrace{2(26.9815 \text{ u})}_{\substack{\text{Mass of 2} \\ \text{aluminum} \\ \text{atoms}}} + \underbrace{3(15.9994 \text{ u})}_{\substack{\text{Mass of 3} \\ \text{oxygen} \\ \text{atoms}}} = 101.9612 \text{ u}$$

Thus, one mole of Al_2O_3 has a mass of 101.9612 g. Calculations like those in part (a) reveal that the ruby contains 0.271 mol or $\boxed{1.63 \times 10^{23} \text{ molecules of } Al_2O_3}$.

14.2 THE IDEAL GAS LAW

An ideal gas is an idealized model for real gases that have sufficiently low densities. The condition of low density means that the molecules of the gas are so far apart that they do not interact (except during collisions that are effectively elastic). The ideal gas law expresses the relationship between the absolute pressure, the Kelvin temperature, the volume, and the number of moles of the gas.

In discussing the constant volume gas thermometer, Section 12.2 has already explained the relationship between the absolute pressure and Kelvin temperature of a low-density gas. This thermometer utilizes a small amount of gas (e.g., hydrogen or helium) placed inside a bulb of constant volume. Since the density is kept low, the gas behaves as an ideal gas. Experiment reveals that a plot of gas pressure versus temperature is a straight line, as in Figure 12.4. This plot is redrawn in Figure 14.3, with the change that the temperature axis is now labeled in kelvins rather than in degrees Celsius. The graph indicates that the absolute pressure P is directly proportional to the Kelvin temperature T ($P \propto T$), for a fixed volume and a fixed number of molecules.

The relation between absolute pressure and the number of molecules of an ideal gas is simple. Experience indicates that it is possible to increase the pressure of a gas by adding more molecules; this is exactly what happens when a tire is pumped up. When the volume and temperature of a low-density gas are kept constant, doubling the number of molecules doubles the pressure. Thus, the absolute pressure of an ideal gas is proportional to the number of molecules or, equivalently, to the number of moles n of the gas ($P \propto n$).

To see how the absolute pressure of a gas depends on the volume of the gas, look at the partially filled balloon in Figure 14.4a. This balloon is "soft," because the pressure of the air is too low to expand the balloon to its fullest. However, if all the air in the balloon is squeezed into a small "bubble," as in part b of the figure, the "bubble" has a very tight feel. This tightness indicates that the pressure in the smaller volume is high enough to stretch the rubber substantially. Thus, it is possible to increase the pressure of a gas by reducing its volume, and if the number of molecules and the temperature are kept constant, the absolute pressure of an ideal gas is inversely proportional to its volume V ($P \propto 1/V$).

The three relations just discussed for the absolute pressure of an ideal gas can be expressed as a single proportionality, $P \propto nT/V$. This proportionality can be written as an equation by inserting a proportionality constant R, called the ***universal gas***

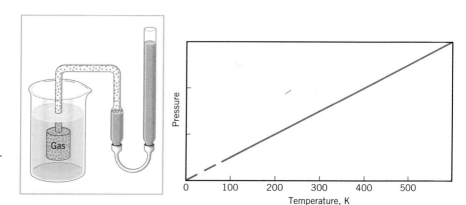

Figure 14.3 The pressure inside a constant-volume gas thermometer is directly proportional to the Kelvin temperature, which is characteristic of an ideal gas.

constant. The value of R has been determined experimentally to be 8.31 J/(mol·K) for any real gas whose density is sufficiently low to ensure ideal gas behavior. The resulting equation is known as the ***ideal gas law.***

■ **IDEAL GAS LAW**

The absolute pressure P of an ideal gas is directly proportional to the Kelvin temperature T and the number of moles n of the gas and is inversely proportional to the volume V of the gas: $P = R(nT/V)$. In other words,

$$PV = nRT \qquad (14.1)$$

where R is the universal gas constant and has the value of 8.31 J/(mol·K) in SI units.

Sometimes, it is convenient to express the ideal gas law in terms of the total number of particles N, instead of the number of moles n. To obtain such an expression, we multiply and divide the right side of the ideal gas law by Avogadro's number $N_A = 6.022 \times 10^{23}$ particles/mol* and recognize that the product nN_A is equal to the total number N of particles:

$$PV = nRT = nN_A \left(\frac{R}{N_A} \right) T = N \left(\frac{R}{N_A} \right) T$$

The constant term R/N_A is referred to as ***Boltzmann's constant,*** in honor of the Austrian physicist Ludwig Boltzmann (1844–1906), and is represented by the symbol k:

$$k = \frac{R}{N_A} = \frac{8.31 \text{ J/(mol·K)}}{6.022 \times 10^{23} \text{ mol}^{-1}} = 1.38 \times 10^{-23} \text{ J/K}$$

With this substitution, the ideal gas law becomes

$$PV = NkT \qquad (14.2)$$

Example 2 presents an application of the ideal gas law.

(a)

(b)

Figure 14.4 (a) The air pressure in the partially filled balloon can be increased by decreasing the volume of the balloon, as illustrated in (b).

The Physics of... ✎
oxygen in the lungs.

EXAMPLE 2 • Oxygen in the Lungs

In the lungs, the respiratory membrane separates tiny sacs of air (absolute pressure = 1.00×10^5 Pa) from the blood in the capillaries. These sacs are called alveoli, and it is from them that oxygen enters the blood. The average radius of the alveoli is 0.125 mm, and the air inside contains 14% oxygen. Assuming that the air behaves as an ideal gas at body temperature (310 K), find the number of oxygen molecules in one of the sacs.

Reasoning and Solution The volume of a sac is $V = \frac{4}{3}\pi r^3$. The form of the ideal gas law given in Equation 14.2 is convenient to use here, because it explicitly contains the total number N of molecules:

$$N = \frac{PV}{kT} = \frac{(1.00 \times 10^5 \text{ Pa})[\frac{4}{3}\pi(0.125 \times 10^{-3} \text{ m})^3]}{(1.38 \times 10^{-23} \text{ J/K})(310 \text{ K})} = 1.9 \times 10^{14}$$

The number of oxygen molecules is 14% of this value or $0.14N = \boxed{2.7 \times 10^{13}}$.

• **PROBLEM SOLVING INSIGHT**
In the ideal gas law, the temperature T must be expressed on the Kelvin scale. The Celsius and Fahrenheit scales cannot be used.

---

* "Particles" is not an SI unit and is often omitted. Then, particles/mol = 1/mol = mol^{-1}.

With the aid of the ideal gas law, it can be shown (see problem 12) that one mole of an ideal gas occupies a volume of 22.4 liters at a temperature of 273 K (0 °C) and a pressure of one atmosphere (1.013×10^5 Pa). These conditions of temperature and pressure are known as ***standard temperature and pressure (STP).*** Conceptual Example 3 discusses another interesting application of the ideal gas law.

The Physics of...
rising beer bubbles.

CONCEPTUAL EXAMPLE 3 • Beer Bubbles on the Rise

The next time you get a chance, watch the bubbles rise in a glass of beer. If you look carefully, you'll see them grow in size as they move upward, often doubling in volume by the time they reach the surface. You might also notice that their speed increases as they rise. In Figure 14.5, for example, the bubbles near the bottom are smaller and more closely spaced, while those higher up are larger and farther apart. Why does a bubble grow and move faster as it ascends?

Reasoning and Solution Beer bubbles contain mostly carbon dioxide (CO_2), a gas that is in the beer because of the fermentation process. The volume V of gas in a bubble is related to its temperature T, pressure P, and the number n of moles of CO_2 by the ideal gas law: $V = nRT/P$. Thus, one or more of these variables must be responsible for the growth of a bubble. Temperature can be eliminated immediately, since it is constant throughout the beer. What about the pressure? As a bubble rises, its depth decreases, and so does the fluid pressure. Since the volume is inversely proportional to pressure, part of the bubble growth is due to the decreasing pressure of the surrounding beer. Recall, however, that some bubbles double in volume on the way up. To account for the doubling, there would have to be two atmospheres of pressure at the bottom of the glass, compared to the one atmosphere at the top. The pressure increment due to depth is $\rho g h$ according to Equation 11.4, so the extra pressure of one atmosphere at the bottom would mean 1.01×10^5 Pa $= \rho g h$. Solving for h with ρ equal to the density of water reveals that $h = 10.3$ m. Since most beer glasses are only about 0.2 m tall, we can rule out a change in pressure as the major cause of the change in volume. Therefore, we are left with only one variable, the number of moles of CO_2 gas in the bubble. In fact, the number of moles does increase as the bubble rises. Each bubble acts as a nucleation site for other CO_2 molecules, so as the bubble moves upward, it accumulates carbon dioxide from the surrounding beer and grows larger.

Now consider the speed of the bubble. As a bubble grows bigger, it displaces more of the surrounding beer. According to Archimedes' principle (see Section 11.6), the upward-acting buoyant force becomes larger as a result. This increasing buoyant force more than offsets the increasing weight of the bubble and any downward-acting resistive force from the surrounding beer. Thus, there is a net upward force on the bubble. According to Newton's second law, the net force produces an acceleration that results in the bubble moving faster and faster as it approaches the top of the beer.

Related Homework Material: Problem 22

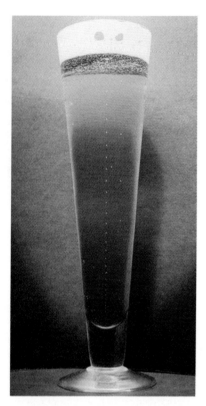

Figure 14.5 The bubbles in a glass of beer grow larger and travel with a greater speed as they move upward.

Historically the work of several investigators led to the formulation of the ideal gas law. The Irish scientist Robert Boyle (1627–1691) discovered that at a constant temperature, the absolute pressure of a fixed mass (fixed number of moles) of a low-density gas is inversely proportional to its volume ($P \propto 1/V$). This fact is often called Boyle's law and can be derived from the ideal gas law by noting that $P = nRT/V = $ constant$/V$ when n and T are constants. Alternatively, if an ideal gas changes from an initial pressure and volume (P_i, V_i) to a final pressure and volume (P_f, V_f), it is possible to write $P_i V_i = nRT$ and $P_f V_f = nRT$. Since the right sides of

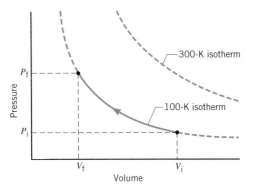

Figure 14.6 A pressure-versus-volume plot for an ideal gas at a constant temperature. Each isotherm is a plot of the equation $P = nRT/V = $ constant$/V$, one with $T = 100$ K and the other with $T = 300$ K.

these equations are equal, we may equate the left sides to give the following concise way of expressing **Boyle's law:**

Constant T, constant n $$P_i V_i = P_f V_f \qquad (14.3)$$

Figure 14.6 illustrates how pressure and volume change according to Boyle's law for a fixed number of moles of an ideal gas at a constant temperature of 100 K. The gas begins with an initial pressure and volume of P_i and V_i and is compressed. The pressure increases as the volume decreases, according to $P = nRT/V$, until the final pressure and volume of P_f and V_f are reached. The curve that passes through the initial and final points is called an *isotherm,* meaning "same temperature." If the temperature had been 300 K, rather than 100 K, the compression would have occurred along the 300-K isotherm. Different isotherms do not intersect. Example 4 deals with an application of Boyle's law to scuba diving.

(a)

EXAMPLE 4 • Scuba* Diving

In scuba diving, a greater water pressure acts on a diver at greater depths. The air pressure inside the body cavities (e.g., lungs, sinuses) must be maintained at the same pressure as that of the surrounding water; otherwise they would collapse. A special valve automatically adjusts the pressure of the air breathed from a scuba tank to ensure that the air pressure equals the water pressure at all times. The scuba gear in Figure 14.7a consists of a 0.0150-m^3 tank filled with compressed air at an absolute pressure of 2.02×10^7 Pa. Assuming that air is consumed at a rate of 0.0300 m^3 per minute and that the temperature is the same at all depths, determine how long the diver can stay under seawater at a depth of (a) 10.0 m and (b) 30.0 m.

Reasoning The time (in minutes) that a scuba diver can remain under water is equal to the volume of air that is available divided by the volume per minute consumed by the diver. The available volume is the volume of air at the pressure P_2 breathed by the diver. This pressure is determined by the depth h beneath the surface, according to $P_2 = P_1 + \rho g h$ (Equation 11.4), where $P_1 = 1.01 \times 10^5$ Pa is the atmospheric pressure at the surface (see Figure 14.7b). Since we know the pressure and volume of air in the scuba tank, and since the temperature is constant, we can use Boyle's law to find the volume of air available at the pressure P_2.

Solution

(a) Using $\rho = 1025$ kg/m^3 for the density of seawater, we find that the absolute pressure P_2 at the depth of $h = 10.0$ m is

$$P_2 = P_1 + \rho g h = 1.01 \times 10^5 \text{ Pa} + (1025 \text{ kg/m}^3)(9.80 \text{ m/s}^2)(10.0 \text{ m})$$
$$= 2.01 \times 10^5 \text{ Pa}$$

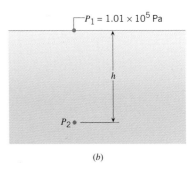

(b)

Figure 14.7 (a) The air pressure inside the body cavities of scuba divers must be maintained at the same pressure as that of the surrounding water. (b) The pressure P_2 at a depth h below the surface of the ocean is given by $P_2 = P_1 + \rho g h$, where ρ is the density of seawater.

The Physics of...
scuba diving.

---
* The word is an acronym for **S**elf-**C**ontained **U**nderwater **B**reathing **A**pparatus.

The pressure and volume of the air in the tank are $P_i = 2.02 \times 10^7$ Pa and $V_i = 0.0150$ m^3, respectively. According to Boyle's law, the volume of air V_f available at a pressure of $P_f = 2.01 \times 10^5$ Pa is

$$V_f = \frac{P_i V_i}{P_f} = \frac{(2.02 \times 10^7 \text{ Pa})(0.0150 \text{ m}^3)}{2.01 \times 10^5 \text{ Pa}} = 1.51 \text{ m}^3 \qquad (14.3)$$

Of this volume, only 1.51 m^3 − 0.0150 m^3 = 1.50 m^3 is available for breathing, because 0.0150 m^3 of air always remains in the tank. At a consumption rate of 0.0300 m^3/min, the compressed air will last for

$$t = \frac{1.50 \text{ m}^3}{0.0300 \text{ m}^3/\text{min}} = \boxed{50.0 \text{ min}}$$

(b) The calculation here is like that in part (a). Equation 11.4 indicates that at a depth of 30.0 m, the absolute water pressure is 4.02×10^5 Pa. Because this pressure is twice that at the 10.0-m depth, Boyle's law reveals that the volume of air provided by the tank is now only $V_f = 0.754$ m^3. The air available for use is 0.754 m^3 − 0.0150 m^3 = 0.739 m^3. At a consumption rate of 0.0300 m^3/min, the air will last for $\boxed{t = 24.6 \text{ min}}$, so the deeper dive must have a shorter duration.

Another investigator whose work contributed to the formulation of the ideal gas law was the Frenchman, Jacques Charles (1746–1823). He discovered that at a constant pressure, the volume of a fixed mass (fixed number of moles) of a low-density gas is directly proportional to the Kelvin temperature ($V \propto T$). This relationship is known as Charles' law and can be obtained from the ideal gas law by noting that $V = nRT/P = (\text{constant})T$, if n and P are constant. Equivalently, when an ideal gas changes from an initial volume and temperature (V_i, T_i) to a final volume and temperature (V_f, T_f), it is possible to write $V_i/T_i = nR/P$ and $V_f/T_f = nR/P$. Thus, one way of stating **Charles' law** is

Constant P, constant n $\qquad \dfrac{V_i}{T_i} = \dfrac{V_f}{T_f}$ $\qquad\qquad$ (14.4)

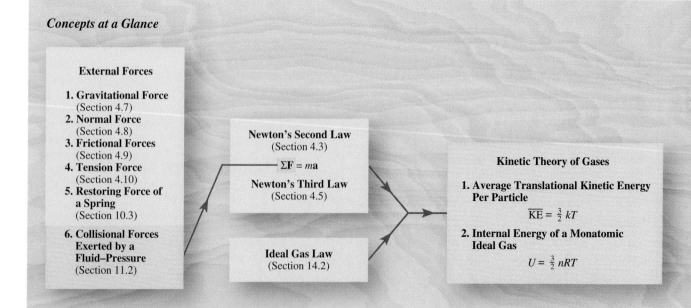

Concepts at a Glance

External Forces

1. **Gravitational Force** (Section 4.7)
2. **Normal Force** (Section 4.8)
3. **Frictional Forces** (Section 4.9)
4. **Tension Force** (Section 4.10)
5. **Restoring Force of a Spring** (Section 10.3)
6. **Collisional Forces Exerted by a Fluid–Pressure** (Section 11.2)

Newton's Second Law (Section 4.3)
$\Sigma \mathbf{F} = m\mathbf{a}$
Newton's Third Law (Section 4.5)

Ideal Gas Law (Section 14.2)

Kinetic Theory of Gases

1. **Average Translational Kinetic Energy Per Particle**
$\overline{\text{KE}} = \frac{3}{2} kT$

2. **Internal Energy of a Monatomic Ideal Gas**
$U = \frac{3}{2} nRT$

14.3 KINETIC THEORY OF GASES

As useful as it is, the ideal gas law provides no insight as to how pressure and temperature are related to properties of the molecules themselves, such as their speeds. To show how such microscopic properties are related to the pressure and temperature of an ideal gas, this section examines the dynamics of molecular motion. As the concept chart in Figure 14.8 illustrates, by bringing Newton's second and third laws of motion together with the ideal gas law, we will be able to show that the average translational kinetic energy $\overline{\text{KE}}$ of a particle in an ideal gas is $\overline{\text{KE}} = \frac{3}{2}kT$, where k is Boltzmann's constant and T is the Kelvin temperature. In the process, we will also be able to show that the internal energy U of a monatomic ideal gas is $U = \frac{3}{2}nRT$, where n is the number of moles and R is the universal gas constant.

THE DISTRIBUTION OF MOLECULAR SPEEDS

A macroscopic container filled with a gas at standard temperature and pressure contains a large number of particles (atoms or molecules). These particles are in constant, random motion, colliding with each other and with the walls of the container. In the course of one second, a particle undergoes many collisions, and each one changes the particle's speed and direction of motion. As a result, the atoms or molecules, in general, have different speeds. It is possible, however, to speak about an average particle speed. At any given instant, some particles have speeds less than, some near, and some greater than the average. For conditions of low gas density, the distribution of speeds within a large collection of molecules at a constant temperature was calculated by the Scottish physicist James Clerk Maxwell (1831–1879). Figure 14.9 displays the Maxwell speed distribution curves for O_2 gas at two different temperatures. When the temperature is 300 K, the maximum in the curve indicates that the most probable speed is about 400 m/s. At a temperature

Figure 14.8 *Concepts at a Glance* To formulate the kinetic theory of gases, Newton's second and third laws are brought together with the ideal gas law. One of the predictions of the theory is that the internal energy of an ideal gas is proportional to the Kelvin temperature of the gas. Since air behaves approximately as an ideal gas, the internal energy of the air inside a hot-air balloon increases as the temperature rises.

Figure 14.9 The Maxwell distribution curves for molecular speeds in oxygen gas at temperatures of 300 and 1200 K.

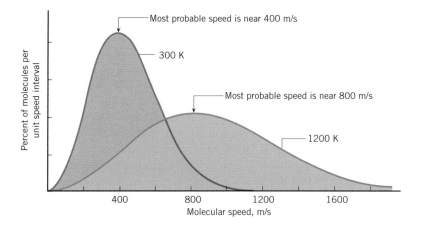

of 1200 K, the distribution curve shifts to the right, and the most probable speed increases to about 800 m/s.

KINETIC THEORY

If a ball is thrown against a wall, it exerts a force on the wall. As Figure 14.10 suggests, gas particles do the same thing, except that their masses are smaller and their speeds are greater. The number of particles is so great and they strike the wall so often that the effect of their individual impacts appears as a continuous force. Dividing the magnitude of this force by the area of the wall gives the pressure exerted by the gas.

To calculate the force, consider an ideal gas composed of N identical particles in a cubical container whose sides have length L. Except for elastic* collisions, these particles do not interact. Figure 14.11 focuses attention on one particle of mass m as it strikes the right wall perpendicularly and rebounds elastically. While approaching the wall, the particle has a velocity $+v$ and linear momentum $+mv$ (see Section 7.1 for a review of linear momentum). The particle rebounds with a velocity $-v$ and momentum $-mv$, travels to the left wall, rebounds again, and heads back toward the right. The time t between collisions with the right wall is the round-trip distance $2L$ divided by the speed of the particle, that is, $t = 2L/v$. According to Newton's second law of motion, in the form of the impulse–momentum theorem, the average force exerted on the particle by the wall is given by the change in the particle's momentum per unit time:

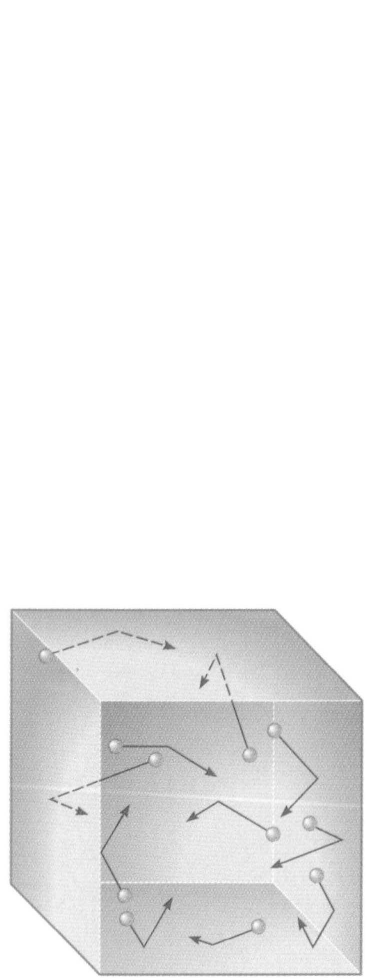

$$\text{Average force} = \frac{\text{Final momentum } - \text{ Initial momentum}}{\text{Time between successive collisions}} \qquad (7.4)$$

$$= \frac{(-mv) - (+mv)}{2L/v} = \frac{-mv^2}{L}$$

Figure 14.10 The pressure that a gas exerts is caused by the impact of its molecules on the walls of the container.

According to Newton's third law of action–reaction, the force applied to the wall by the particle is equal in magnitude to this value, but oppositely directed (i.e., $+mv^2/L$). The magnitude F of the *total* force exerted on the right wall is equal to the number of particles that collide with the wall during the time t multiplied by the

---

* The term "elastic" is used here to mean that *on the average,* in a large number of particles, there is no gain or loss of translational kinetic energy because of collisions.

average force exerted by each particle. Since the N particles move randomly in three dimensions, one-third of them on the average strike the right wall during the time t. Therefore, the total force is

$$F = \left(\frac{N}{3}\right)\left(\frac{m\overline{v^2}}{L}\right)$$

In this result v^2 has been replaced by $\overline{v^2}$, *the average* value of the squared speed. The collection of particles possesses a Maxwell distribution of speeds, so an average value for v^2 must be used, rather than a value for any individual particle. The square root of the quantity $\overline{v^2}$ is called the **root-mean-square speed,** or, for short, the *rms speed;* $v_{rms} = \sqrt{\overline{v^2}}$. With this substitution, the total force becomes

$$F = \left(\frac{N}{3}\right)\left(\frac{mv_{rms}^2}{L}\right)$$

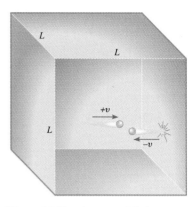

Figure 14.11 A gas particle is shown colliding elastically with the right wall of the container and rebounding from it.

Pressure is force per unit area, so the pressure P acting on a wall of area L^2 is

$$P = \frac{F}{L^2} = \left(\frac{N}{3}\right)\left(\frac{mv_{rms}^2}{L^3}\right)$$

Since the volume of the box is $V = L^3$, the equation above can be written as

$$PV = \tfrac{2}{3}N(\tfrac{1}{2}mv_{rms}^2) \tag{14.5}$$

Equation 14.5 relates the macroscopic properties of the gas, its pressure and volume, to the microscopic properties of the constituent particles, their mass and speed. Since the term $\tfrac{1}{2}mv_{rms}^2$ is the average translational kinetic energy $\overline{KE}$ of an individual particle, it follows that

$$PV = \tfrac{2}{3}N(\overline{KE})$$

This result is similar to the ideal gas law, $PV = NkT$. Both equations have identical terms on the left, so the terms on the right must be equal: $\tfrac{2}{3}N(\overline{KE}) = NkT$. Therefore,

$$\overline{KE} = \tfrac{1}{2}mv_{rms}^2 = \tfrac{3}{2}kT \tag{14.6}$$

Equation 14.6 is significant, for it allows us to interpret temperature in terms of the motion of gas particles. This equation indicates that the Kelvin temperature is directly proportional to the average translational kinetic energy per particle in an ideal gas, no matter what the pressure and volume are. On the average, the particles have greater kinetic energies when the gas is hotter than when it is cooler. Conceptual Example 5 discusses a common misconception about the relation between kinetic energy and temperature.

CONCEPTUAL EXAMPLE 5 • Does a Single Particle Have a Temperature?

Each particle in a gas has a kinetic energy. Furthermore, the equation $\tfrac{1}{2}mv_{rms}^2 = \tfrac{3}{2}kT$ establishes the relationship between the average kinetic energy per particle and temperature of an ideal gas. Is it valid, then, to conclude that a single particle has a temperature?

Reasoning and Solution We know that a gas contains an enormous number of particles that are traveling with a distribution of speeds, such as those indicated by the graphs in Figure 14.9. Therefore, the particles do not all have the same kinetic energy, but possess a distribution of kinetic energies ranging from very nearly zero to extremely large values. If each particle had a temperature that was associated with its kinetic energy, there would be a whole range of different temperatures within the gas. This is not so, for a gas at thermal equilibrium has only one temperature (see Section 15.2), a temperature that would be registered by a thermometer placed in the gas. Thus, temperature is a property that characterizes the gas as a whole, a fact that is inherent in the relation $\frac{1}{2}mv_{rms}^2 = \frac{3}{2}kT$. The term v_{rms} is a kind of *average particle speed.* Therefore, $\frac{1}{2}mv_{rms}^2$ is the *average kinetic energy* per particle and is characteristic of the gas as a whole. Since the Kelvin temperature is proportional to $\frac{1}{2}mv_{rms}^2$, it is also a characteristic of the gas as a whole and cannot be ascribed to each gas particle individually. Thus, *a single gas particle does not have a temperature.*

If two ideal gases have the same temperature, the relation $\frac{1}{2}mv_{rms}^2 = \frac{3}{2}kT$ indicates that the average kinetic energy of each kind of gas particle is the same. In general, however, the rms speeds of the different particles are not the same, for the masses may be different. The next example illustrates these facts and shows how rapidly gas particles move at normal temperatures.

EXAMPLE 6 • The Speed of Molecules in Air

Air is primarily a mixture of nitrogen N_2 (molecular mass = 28.0 u) and oxygen O_2 (molecular mass = 32.0 u). Assume that each behaves as an ideal gas and determine the rms speed of the nitrogen and oxygen molecules when the temperature of the air is 293 K.

Reasoning The rms speed can be obtained from Equation 14.6 as $v_{rms} = \sqrt{2(\overline{KE})/m}$. In this expression, the average kinetic energy is $\overline{KE} = \frac{3}{2}kT$ and is the same for both types of molecules at the same temperature. However, the masses of the nitrogen and oxygen molecules are different. Since Avogadro's number of particles (1 mol) has a mass in grams equal to the molecular mass, we can determine the mass of each type of particle.

Solution The average kinetic energy per particle for both nitrogen and oxygen is

$$\overline{KE} = \tfrac{3}{2}kT = \tfrac{3}{2}(1.38 \times 10^{-23}\text{ J/K})(293\text{ K}) = \boxed{6.07 \times 10^{-21}\text{ J}} \qquad (14.6)$$

Using Avogadro's number and the molecular mass of nitrogen, we find a particle mass of

$$m = \frac{28.0\text{ g/mol}}{6.022 \times 10^{23}\text{ mol}^{-1}} = 4.65 \times 10^{-23}\text{ g} = 4.65 \times 10^{-26}\text{ kg}$$

Similarly, we find the particle mass for oxygen to be 5.31×10^{-26} kg. The calculations of the rms speeds are shown below:

Nitrogen $\qquad v_{rms} = \sqrt{\dfrac{2(\overline{KE})}{m}} = \sqrt{\dfrac{2(6.07 \times 10^{-21}\text{ J})}{4.65 \times 10^{-26}\text{ kg}}} = \boxed{511\text{ m/s}}$

Oxygen $\qquad v_{rms} = \sqrt{\dfrac{2(\overline{KE})}{m}} = \sqrt{\dfrac{2(6.07 \times 10^{-21}\text{ J})}{5.31 \times 10^{-26}\text{ kg}}} = \boxed{478\text{ m/s}}$

For comparison, the speed of sound at a temperature of 293 K is 343 m/s (767 mi/h).

• **PROBLEM SOLVING INSIGHT**
The average translational kinetic energy is the same for all ideal-gas molecules at the same temperature, regardless of the molecular mass. The rms translational speed of the molecules is not the same, however, but depends on the mass.

As we have seen in Example 6, the rms speeds of the nitrogen and oxygen molecules in air are greater than the speed of sound in air. Extremely small molecules have even larger speeds, however, and Conceptual Example 7 considers one consequence of this fact.

CONCEPTUAL EXAMPLE 7 • Why Is There No Hydrogen in Our Atmosphere?

Hydrogen is the most plentiful element in the universe and must have been abundant in the atmosphere during the early stages of the earth's evolution. Yet today, our atmosphere contains mostly nitrogen and oxygen. Why doesn't it contain hydrogen?

Reasoning and Solution The hydrogen present in the early stages of the earth's atmosphere must have escaped into outer space. But how can hydrogen escape if it, like any other object, is pulled toward the earth by the gravitational force? The answer lies in the concept of *escape speed*. If an object is moving away from the earth with a speed that is equal to or greater than the escape speed, it will leave the earth and never return. The escape speed is about 11 200 m/s, so if a hydrogen molecule has a speed equal to or greater than this, it will escape. We have seen in Example 6 that the rms speed is inversely proportional to the square root of the mass, and that nitrogen (N_2) has an rms speed of 511 m/s at a temperature of 293 K. Hydrogen (H_2) has a mass that is about 14 times smaller than nitrogen, so its rms speed is about $\sqrt{14}$ times greater, or about 1900 m/s. Although the rms speed of hydrogen is only 17% of the escape speed, some molecules move at speeds many times the rms speed, according to a Maxwell speed distribution, and therefore escape into space. In the five, or so, billion years since the formation of the earth, there has been time for essentially all of the hydrogen in the primordial atmosphere to escape.

Related Homework Material: Problem 32

The equation $\overline{KE} = \frac{3}{2}kT$ has also been applied to particles much larger than atoms or molecules. The English botanist Robert Brown (1773–1858) observed through a microscope that pollen grains suspended in water move on very irregular, zigzag paths. This Brownian motion can also be observed with other particle suspensions, such as fine smoke particles in air. In 1905, Albert Einstein (1879–1955) showed that Brownian motion could be explained as a response of the large suspended particles to impacts from the moving molecules of the fluid medium (e.g., water or air). As a result of the impacts, the suspended particles have the same average translational kinetic energy as the fluid molecules, namely, $\overline{KE} = \frac{3}{2}kT$. But unlike the molecules, the particles are large enough to be seen through a microscope and, because of their relatively large mass, have a comparatively small average speed.

THE INTERNAL ENERGY OF A MONATOMIC IDEAL GAS

Chapter 15 deals with the science of thermodynamics, in which the concept of internal energy plays an important role. Using the results just developed for the average translational kinetic energy, we conclude this section by expressing the internal energy of a monatomic ideal gas in a form that is suitable for use later on.

The internal energy of a substance is the sum of the various kinds of energy that the atoms or molecules of the substance possess. A monatomic ideal gas is composed of single atoms. These atoms are assumed to be so small that the mass is concentrated at a point, with the result that the moment of inertia I about the center of mass is negligible. Thus, the rotational kinetic energy $\frac{1}{2}I\omega^2$ is also negligible. Vibrational kinetic and potential energies are absent, because the atoms are not connected by chemical bonds and, except for elastic collisions, do not interact. As a result, the internal energy U is the total translational kinetic energy of the N atoms that constitute the gas: $U = N(\frac{1}{2}mv_{rms}^2)$. Since $\frac{1}{2}mv_{rms}^2 = \frac{3}{2}kT$ according to Equation 14.6, the internal energy can be written in terms of the Kelvin temperature as

$$U = N(\tfrac{3}{2}kT)$$

Usually, U is expressed in terms of the number of moles n, rather than the number of atoms N. Using the fact that Boltzmann's constant is $k = R/N_A$, where R is the universal gas constant and N_A is Avogadro's number, and realizing that $N/N_A = n$, we find that

Monatomic ideal gas $\qquad\qquad U = \tfrac{3}{2}nRT \qquad\qquad$ (14.7)

*14.4 DIFFUSION

You can smell the fragrance of a perfume at some distance from an open bottle, because perfume molecules evaporate from the liquid, where they are relatively concentrated, and spread out into the air, where they are less concentrated. During their journey, they collide with other molecules, so their paths resemble the zigzag paths characteristic of Brownian motion. The process in which molecules move from a region of higher concentration to one of lower concentration is called **diffusion.** Diffusion also occurs in liquids (Figure 14.12 shows ink diffusing through water, for instance) and solids. However, compared to the rate of diffusion in gases, the rate is generally smaller in liquids and even smaller in solids. The host medium, such as the air or water in the examples above, is referred to as the **solvent,** while the diffusing substance, like the perfume molecules or the ink in Figure 14.12, is known as the **solute.** Relatively speaking, diffusion is a slow process, even in a gas. Conceptual Example 8 illustrates why.

CONCEPTUAL EXAMPLE 8 • Why Diffusion Is Relatively Slow

We have seen in Example 6 that a gas molecule has a translational rms speed of hundreds of meters per second at room temperature. At such a speed, a molecule could travel across an ordinary room in just a fraction of a second. Yet, it often takes several seconds, and sometimes minutes, for the fragrance of a perfume to reach the other side of a room. Why does it take so long?

Reasoning and Solution When a perfume molecule diffuses through air, it makes millions of collisions each second with air molecules. As Figure 14.13 illustrates, the speed and direction of motion change abruptly as a result of each collision. Between collisions, the perfume molecule moves in a straight line at a constant speed. Although a perfume molecule does move very fast between collisions, it wanders only slowly away from the bottle because of the zigzag path resulting from the collisions. It would take a long time for a molecule to diffuse in this manner across a room. Usually, however, convection currents are present and carry the fragrance across the room in a matter of seconds or minutes.

Related Homework Material: Problems 44, 46, and 47

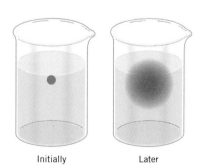

Initially   Later

Figure 14.12 A drop of ink placed in water eventually becomes completely dispersed because of diffusion.

The diffusion process can be described in terms of the arrangement in Figure 14.14a. A hollow channel of length L and cross-sectional area A is filled with a fluid. The left end of the channel is connected to a container in which the solute concentration C_2 is relatively high, while the right end is connected to a container in which the solute concentration C_1 is lower. These concentrations are defined as the total mass of the solute molecules divided by the volume of the solution (e.g., 0.1 kg/m^3). Because of the difference in concentration between the ends of the channel, $\Delta C = C_2 - C_1$, there is a net diffusion of the solute from the left end to the right end.

Figure 14.14a is similar to Figure 13.8 for the conduction of heat along a bar, which, for convenience, is reproduced in Figure 14.14b. When the ends of the bar are maintained at different temperatures, T_2 and T_1, the heat Q conducted along the bar in a time t is

$$Q = \frac{(kA\,\Delta T)t}{L} \tag{13.1}$$

where $\Delta T = T_2 - T_1$, and k is the thermal conductivity. Whereas conduction is the flow of heat from a region of higher temperature to a region of lower temperature, diffusion is the mass flow of solute from a region of higher concentration to a region of lower concentration. By analogy with Equation 13.1, it is possible to write an equation for diffusion: (1) replace Q by the mass m of solute that is diffusing through the channel, (2) replace $\Delta T = T_2 - T_1$ by the difference in concentrations $\Delta C = C_2 - C_1$, and (3) replace k by a constant known as the diffusion constant D. The resulting equation, first formulated by the German physiologist Adolf Fick (1829–1901), is referred to as *Fick's law of diffusion.*

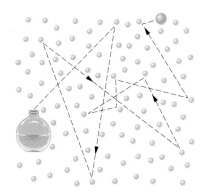

Figure 14.13 A perfume molecule collides with millions of air molecules during its journey, so the path has a zigzag shape. Although the air molecules are shown as stationary, they are also moving.

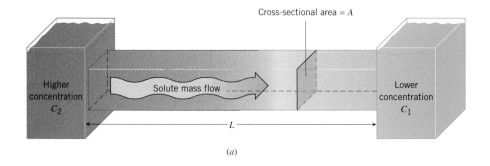

(a)

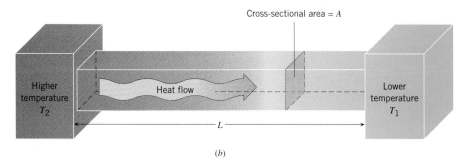

(b)

Figure 14.14 (a) Solute diffuses through the channel from the region of higher concentration to the region of lower concentration. (b) Heat is conducted along a bar whose ends are maintained at different temperatures.

■ **FICK'S LAW OF DIFFUSION**

The mass m of solute that diffuses in a time t through a solvent contained in a channel of length L and cross-sectional area A is

$$m = \frac{(DA\,\Delta C)t}{L} \tag{14.8}$$

where ΔC is the concentration difference between the ends of the channel and D is the diffusion constant.

SI Unit for the Diffusion Constant: m^2/s

It can be verified from Equation 14.8 that the diffusion constant has units of m^2/s, the exact value depending on the nature of the solute and the solvent. For example, the diffusion constant for ink in water is different than that for ink in benzene. Example 9 illustrates an important application of Fick's law.

EXAMPLE 9 • Water Given Off by Plant Leaves

The Physics of...
water loss from plant leaves.

Large amounts of water can be given off by plants. It has been estimated, for instance, that a single sunflower plant can lose up to a pint of water a day during the growing season. Figure 14.15 shows a cross-sectional view of a leaf. Inside the leaf, water passes from the liquid phase to the vapor phase at the walls of the mesophyll cells. The water vapor then diffuses through the intercellular air spaces and eventually exits the leaf through small openings, called stomatal pores. The diffusion constant for water vapor in air is $D = 2.4 \times 10^{-5}$ m^2/s. A stomatal pore has a cross-sectional area of about $A = 8.0 \times 10^{-11}$ m^2 and a length of about $L = 2.5 \times 10^{-5}$ m. The concentration of water vapor on the interior side of a pore is roughly $C_2 = 0.022$ kg/m^3, while that on the outside is approximately $C_1 = 0.011$ kg/m^3. Determine the mass of water vapor that passes through a stomatal pore in one hour.

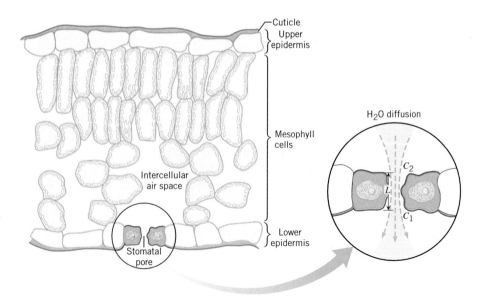

Figure 14.15 A cross-sectional view of a leaf. Water vapor diffuses out of the leaf through a stomatal pore.

Reasoning and Solution Fick's law of diffusion shows that

$$m = \frac{(DA \, \Delta C)t}{L} \qquad (14.8)$$

$$m = \frac{(2.4 \times 10^{-5} \text{ m}^2/\text{s})(8.0 \times 10^{-11} \text{ m}^2)(0.022 \text{ kg/m}^3 - 0.011 \text{ kg/m}^3)(3600 \text{ s})}{2.5 \times 10^{-5} \text{ m}}$$

$$= \boxed{3.0 \times 10^{-9} \text{ kg}}$$

This amount of water may not seem significant. However, a single leaf may have a million or so stomatal pores, so the water lost by an entire plant can be substantial.

SUMMARY

Each element in the periodic table is assigned an **atomic mass.** One **atomic mass unit** (u) is exactly one-twelfth the mass of an atom of carbon-12. The **molecular mass** of a molecule is the sum of the atomic masses of its atoms. One **mole** of a substance contains **Avogadro's number** N_A of particles, where $N_A = 6.022 \times 10^{23}$ particles per mole. The mass in grams of one mole of a substance is equal to the atomic or molecular mass of its particles.

The **ideal gas law** relates the absolute pressure P, the volume V, the number of moles n, and the Kelvin temperature T of an ideal gas according to $PV = nRT$, where $R = 8.31$ J/(mol·K) is the universal gas constant. An alternative form of the ideal gas law is $PV = NkT$, where N is the number of particles and $k = R/N_A$ is Boltzmann's constant. A real gas behaves as an ideal gas when the density of the real gas is low enough that its particles do not interact, except via elastic collisions.

The distribution of particle speeds in an ideal gas at constant temperature is the **Maxwell speed distribution.** According to the **kinetic theory of gases,** an ideal gas consists of a large number of particles (atoms or molecules) that are in constant random motion. The particles are far apart compared to their dimensions, so they do not interact except when elastic collisions occur. The pressure on the walls of a container is produced by the impacts of the particles with the walls. According to the kinetic theory of gases, the Kelvin temperature T of an ideal gas is a measure of the average translational kinetic energy $\overline{KE}$ per particle through the relation $\overline{KE} = \frac{3}{2}kT$. The **internal energy** U of n moles of a monatomic ideal gas is $U = \frac{3}{2}nRT$.

Diffusion is the process whereby solute molecules move through a solvent from a region of higher concentration to a region of lower concentration. **Fick's law of diffusion** states that the mass m of solute that diffuses in a time t through a solvent contained in a channel of length L and cross-sectional area A is given by $m = (DA \, \Delta C)t/L$, where ΔC is the concentration difference between the ends of the channel and D is the diffusion constant.

CONCEPTUAL QUESTIONS

1. (a) Which, if either, contains a greater number of molecules, a mole of hydrogen (H_2) or a mole of oxygen (O_2)? (b) Which one has more mass? Give reasons for your answers.

2. Suppose two different substances, A and B, have the same mass densities. (a) In general, does one mole of substance A have the same mass as one mole of substance B? (b) Does 1 m^3 of substance A have the same mass as 1 m^3 of substance B? Justify each answer.

3. A tightly sealed house has a large ceiling fan that blows air out of the house and into the attic. This fan is turned on, and the owners forget to open any windows or doors. What happens to the air pressure in the house after the fan has been on for a while, and does it become easier or harder for the fan to do its job? Explain.

4. Above the liquid in a can of hair spray is a gas at a relatively high pressure. The label on the can includes the warning "Do not store at high temperatures." Use the ideal gas law and explain why the warning is given.

5. Assuming that air behaves like an ideal gas, explain what happens to the pressure in a tightly sealed house when the electric furnace turns on for a while.

6. 🔊 When you climb a mountain, your eardrums "pop" outward as the air pressure decreases. When you come down, they "pop" inward as the pressure increases. At the sea coast, there is a cave that can be entered only by swimming through a completely submerged passage and emerging into a pocket of air within the cave. The cave is not vented to the external atmosphere. As the tide comes in, the water level in the cave rises, and your eardrums "pop." Is this "popping" analogous to what happens as you climb up or down a mountain? Give your reasoning.

7. Airplanes have emergency oxygen bags that fall from the ceiling when the air pressure in the cabin suddenly drops. At high altitudes, where the air pressure outside the plane is relatively low, the oxygen bags expand to their fullest. However, at low altitudes, the bags expand only partially. Why don't the bags expand to their fullest at low altitudes?

8. Atmospheric pressure decreases with increasing altitude. With this fact in mind, explain why helium-filled weather balloons are underinflated when they are launched from earth.

9. A slippery cork is being pressed into a very full (but not 100% full) bottle of wine. When released, the cork slowly slides back out. However, if some wine is removed from the bottle before the cork is inserted, the cork does not slide out. Account for these observations in terms of the ideal gas law.

10. A commonly used packing material consists of "bubbles" of air trapped between bonded layers of plastic, as the photograph shows. Using the ideal gas law, explain why this packing material offers less protection on cold days than on warm days.

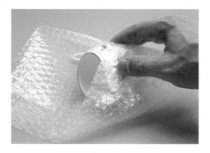

11. The kinetic theory of gases assumes that a gas molecule rebounds with the same speed after colliding with the wall of a container. Suppose the speed after the collision is less than that before the collision. Would the pressure of the gas be greater than, equal to, or less than that predicted by kinetic theory? Account for your answer.

12. If the translational speed of each molecule in an ideal gas were tripled, would the Kelvin temperature also triple? If not, by what factor would the Kelvin temperature increase? Account for your answer.

13. If the temperature of an ideal gas is doubled from 50 to 100 °C, does the average translational kinetic energy of each particle double? If not, why not?

14. It is possible for both the pressure and volume of a monatomic ideal gas to change *without* causing the internal energy of the gas to change. Explain how this could occur.

15. Suppose that the atoms in a container of helium (He) have the same translational rms speed as the molecules in a container of argon (Ar). Treating each gas as an ideal gas, explain which, if either, has the greater temperature.

16. The ionosphere is the uppermost part of the earth's atmosphere. The temperature of the ionized gas in the ionosphere is about 1000 K. However, the density of the gas is extremely low, being on the order of 10^{11} molecules/m^3. Although the temperature is very high, an astronaut in the ionosphere would not burn up. Why?

17. ⚕ In the lungs, oxygen in very small sacs called alveoli diffuses into the blood. The diffusion occurs directly through the walls of the sacs. The walls are very thin, so the oxygen diffuses over a distance L that is quite small. Because there are so many alveoli, the effective area A across which diffusion occurs is very large. Use this information, together with Fick's law of diffusion, to explain why the mass of oxygen per second that diffuses into the blood is large.

Problems

Note: The pressures referred to in these problems are absolute pressures, unless indicated otherwise.

ssm Solution is in the Student Solutions Manual. **www** Solution is available on the World Wide Web at http://www.wiley.com/college/cutnell ⚕ This icon represents a biomedical application.

Section 14.1 Molecular Mass, the Mole, and Avogadro's Number

1. ssm Hemoglobin has a molecular mass of 64 500 u. Find the mass (in kg) of one molecule of hemoglobin.

2. The artificial sweetener NutraSweet is a chemical called aspartame ($C_{14}H_{18}N_2O_5$). What is (a) its molecular mass (in atomic mass units) and (b) the mass (in kg) of an aspartame molecule?

3. The chlorophyll-*a* molecule, $C_{55}H_{72}MgN_4O_5$, is important in photosynthesis. (a) Determine its molecular mass (in atomic mass units). (b) What is the mass (in grams) of 3.00 moles of chlorophyll-*a* molecules?

4. A glass of water has a volume of 6.0×10^{-4} m^3. How many moles of water molecules are there in the glass?

5. ssm A mass of 135 g of an element is known to contain 30.1×10^{23} atoms. What is the element?

6. ⚕ Manufacturers of headache remedies routinely claim that their own brands are more potent pain relievers than the competing brands. The best way of making the comparison is to compare the number of molecules in the standard dosage. Tylenol uses 325 mg of acetaminophen ($C_8H_9NO_2$) as the standard dose, while Advil uses 2.00×10^2 mg of ibuprofen ($C_{13}H_{18}O_2$). Find the number of molecules of pain reliever in the standard doses of (a) Tylenol and (b) Advil.

*7. What mass of chlorine gas (Cl_2) must be added to four grams of hydrogen gas (H_2), if all the molecules are used to produce hydrochloric acid (HCl)?

*8. Ethyl alcohol (C_2H_5OH) has a density of 806 kg/m^3 and a volume of 2.00×10^{-3} m^3. (a) Determine the mass (in kg) of a molecule of ethyl alcohol, and (b) find the number of molecules in the liquid.

9. **ssm www At the normal boiling point of a material, the liquid phase has a density of 958 kg/m^3, and the vapor phase has a density of 0.598 kg/m^3. Determine the ratio of the distance between neighboring molecules in the gas phase to that in the liquid phase. *(Hint: Assume that the volume of each phase is filled with many cubes, with one molecule at the center of each cube.)*

Section 14.2 The Ideal Gas Law

10. An ultrahigh vacuum pump can reduce the absolute pressure to 1.2×10^{-7} Pa. In a volume of 2.0 m^3 and at a temperature of 3.0×10^2 K, how many molecules of gas are present at this pressure?

11. At the start of a trip, a driver adjusts the absolute pressure in her tires to be 2.81×10^5 Pa when the outdoor temperature is 284 K. At the end of the trip she measures the pressure to be 3.01×10^5 Pa. Ignoring the expansion of the tires, find the air temperature inside the tires at the end of the trip.

12. Find the volume occupied by one mole of an ideal gas at STP conditions. Express your answer in liters (1 liter = 1000 cm^3 = 10^{-3} m^3).

13. **ssm** In a diesel engine, the piston compresses air at 305 K to a volume that is one-sixteenth of the original volume and a pressure that is 48.5 times the original pressure. What is the temperature of the air after the compression?

14. A 0.030-m^3 container is initially evacuated, and then 4.0 g of water is placed in it. After a time, all the water evaporates, and the temperature is 388 K. Find the pressure.

15. ☟ Oxygen for hospital patients is kept in special tanks, where the oxygen has a pressure of 65.0 atmospheres and a temperature of 288 K. The tanks are stored in a separate room, and the oxygen is pumped to the patient's room, where it is administered at a pressure of 1.00 atmosphere and a temperature of 297 K. What volume does 1.00 m^3 of oxygen in the tanks occupy at the conditions in the patient's room?

16. In a portable oxygen system, the oxygen (O_2) is contained in a cylinder whose volume is 0.0028 m^3. A full cylinder has an absolute pressure of 1.5×10^7 Pa when the temperature is 296 K. Find the mass (in kg) of oxygen in the cylinder.

17. **ssm** A young male adult takes in about 5.0×10^{-4} m^3 of fresh air during a normal breath. Fresh air contains approximately 21% oxygen. Assuming that the pressure in the lungs is

1.0×10^5 Pa and air is an ideal gas at a temperature of 310 K, find the number of oxygen molecules in a normal breath.

18. A frictionless gas-filled cylinder is fitted with a movable piston, as the drawing shows. The block resting on the top of the piston determines the constant pressure that the gas has. The height h is 0.120 m when the temperature is 273 K and increases as the temperature increases. What is the value of h when the temperature reaches 318 K?

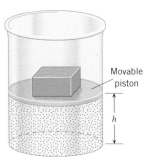

Movable piston

h

19. On the sunlit surface of Venus, the atmospheric pressure is 9.0×10^6 Pa, and the temperature is 740 K. On the earth's surface the atmospheric pressure is 1.0×10^5 Pa, while the surface temperature can reach 320 K. These data imply that Venus has a "thicker" atmosphere at its surface than does the earth, which means that the number of molecules per unit volume (N/V) is greater on the surface of Venus than on the earth. Find the ratio $(N/V)_{\text{Venus}}/(N/V)_{\text{Earth}}$.

20. Two ideal gases have the same mass density and the same absolute pressure. One of the gases is helium (He), and its temperature is 175 K. The other gas is neon (Ne). What is the temperature of the neon?

*21. **ssm** The relative humidity is 55% on a day when the temperature is 30.0 °C. Using the graph that accompanies problem 76 in Chapter 12, determine the number of moles of water vapor per cubic meter of air.

*22. Review Conceptual Example 3 before starting this problem. A bubble, located 0.200 m beneath the surface of the beer, rises to the top. The air pressure at the top is 1.01×10^5 Pa. Assume that the density of beer is the same as that of fresh water. If the temperature and number of moles of CO_2 remain constant as the bubble rises, find the ratio of its volume at the top to that at the bottom.

*23. One assumption of the ideal gas law is that the atoms or molecules themselves occupy a negligible volume. Verify that this assumption is reasonable by considering gaseous argon (Ar). Argon has an atomic radius of 0.70×10^{-10} m. For STP conditions, calculate the percentage of the total volume occupied by the atoms.

*24. Assume that the pressure in a room remains constant at 1.01×10^5 Pa and the air is composed only of nitrogen (N_2). The volume of the room is 60.0 m^3. When the temperature increases from 289 to 302 K, what mass of air (in kg) escapes from the room?

*25. **ssm www** A primitive diving bell consists of a cylindrical tank with one end open and one end closed. The tank is lowered into a freshwater lake, open end downward. Water rises into the tank, compressing the trapped air, whose temperature remains constant during the descent. The tank is brought to a halt when the distance between the surface of the water in the tank and the surface of the lake is 40.0 m. Atmospheric pressure at the surface of the lake is 1.01×10^5 Pa. Find the fraction of the tank's volume that is filled with air.

*26. The drawing shows two thermally insulated tanks. They are connected by a valve that is initially closed. Each tank contains neon gas at the pressure, temperature, and volume indicated in the drawing. When the valve is opened, the contents of the two tanks mix, and the pressure becomes constant throughout. (a) What is the final temperature? Ignore any change in temperature of the tanks themselves. *(Hint: The heat gained by the gas in one tank is equal to that lost by the other.)* (b) What is the final pressure?

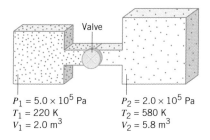

$P_1 = 5.0 \times 10^5$ Pa
$T_1 = 220$ K
$V_1 = 2.0$ m^3

$P_2 = 2.0 \times 10^5$ Pa
$T_2 = 580$ K
$V_2 = 5.8$ m^3

**27. A spherical balloon is made from a material whose mass is 3.00 kg. The thickness of the material is negligible compared to the 1.50-m radius of the balloon. The balloon is filled with helium (He) at a temperature of 305 K and just floats in air, neither rising nor falling. The density of the surrounding air is 1.19 kg/m^3. Find the absolute pressure of the helium gas.

**28. A gas fills the right portion of a horizontal cylinder whose radius is 5.00 cm. The initial pressure of the gas is 1.01×10^5 Pa. A frictionless movable piston separates the gas from the left portion of the cylinder that is evacuated and contains an ideal spring, as the drawing shows. The piston is initially held in place by a pin. The spring is initially unstrained, and the length of the gas-filled portion is 20.0 cm. When the pin is removed and the gas is allowed to expand, the length of the gas-filled chamber doubles. The initial and final temperatures are equal. Determine the spring constant of the spring.

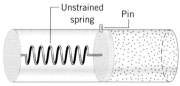

Unstrained
spring   Pin

29. **ssm A cylindrical glass beaker of height 1.520 m rests on a table. The bottom half of the beaker is filled with a gas, and the top half is filled with liquid mercury that is exposed to the atmosphere. The gas and mercury do not mix, because they are separated by a frictionless, movable piston of negligible mass and thickness. The initial temperature is 273 K. The temperature is increased until a value is reached when one-half of the mercury has spilled out. Ignore the thermal expansion of the glass and mercury, and find this temperature.

Section 14.3 Kinetic Theory of Gases

30. The surface of the sun has a temperature of about 6.0×10^3 K. This hot gas contains hydrogen atoms ($m = 1.67 \times 10^{-27}$ kg). Find the rms speed of these atoms.

31. Two moles of nitrogen (N_2) gas are placed in a container whose volume is 8.5×10^{-3} m^3. The pressure of the gas is 4.5×10^5 Pa. What is the average translational kinetic energy of a nitrogen molecule?

32. See Conceptual Example 7 as an aid in solving this problem. At what temperature is the translational rms speed of hydrogen molecules (H_2) equal to the escape speed?

33. **ssm** The average value of the squared speed $\overline{v^2}$ does not equal the square of the average speed $(\overline{v})^2$. To verify this fact, consider three particles with the following speeds: $v_1 = 3.0$ m/s, $v_2 = 7.0$ m/s, and $v_3 = 9.0$ m/s. Calculate (a) $\overline{v^2} = \frac{1}{3}(v_1^2 + v_2^2 + v_3^2)$ and (b) $(\overline{v})^2 = [\frac{1}{3}(v_1 + v_2 + v_3)]^2$.

34. The temperature in the ionosphere (the uppermost part of the earth's atmosphere) is about 1.0×10^3 K. What is the ratio of the translational rms speed of an oxygen molecule at this temperature to that of an oxygen molecule near the earth where the temperature is 290 K?

35. Initially, the translational rms speed of a molecule of an ideal gas is 463 m/s. The pressure and volume of this gas are kept constant, while the number of molecules is doubled. What is the final translational rms speed of the molecules?

36. At what temperature would the translational rms speed of hydrogen molecules (H_2) be equal to that of oxygen molecules (O_2) at 3.0×10^2 K?

37. **ssm** Suppose that a tank contains 680 m^3 of neon at an absolute pressure of 1.01×10^5 Pa. The temperature is changed from 293.2 to 294.3 K. What is the increase in the internal energy of the neon?

*38. Helium (He), a monatomic gas, fills a 0.010-m^3 container. The pressure of the gas is 6.2×10^5 Pa. How long would a 0.25-hp engine have to run (1 hp = 746 W) to produce an amount of energy equal to the internal energy of this gas?

*39. The temperature near the surface of the earth is 295 K. An argon atom (atomic mass = 39.948 u) has a kinetic energy equal to the average translational kinetic energy and is moving straight up. If the atom does not collide with any other atoms or molecules, how high up would it go before coming to rest? Assume that the acceleration due to gravity is constant throughout the ascent.

*40. The pressure of sulfur dioxide (SO_2) is 2.12×10^4 Pa. There are 421 moles in a volume of 50.0 m^3. Find the translational rms speed of the sulfur dioxide molecules.

41. **ssm **www** In a TV, electrons with a speed of 8.4×10^7 m/s strike the screen from behind, causing it to glow. The electrons come to a halt after striking the screen. Each electron has a mass of 9.11×10^{-31} kg, and there are 6.2×10^{16} electrons per second hitting the screen over an area of 1.2×10^{-7} m². What is the pressure that the electrons exert on the screen?

Section 14.4 Diffusion

42. A tube has a length of 0.015 m and a cross-sectional area of 7.0×10^{-4} m². The tube is filled with a solution of sucrose in water. The diffusion constant of sucrose in water is 5.0×10^{-10} m²/s. A difference in concentration of 3.0×10^{-3} kg/m³ is maintained between the ends of the tube. How much time is required for 8.0×10^{-13} kg of sucrose to be transported through the tube?

43. It is found that the amino acid glycine diffuses through water at a mass rate of $m/t = 6.00 \times 10^{-14}$ kg/s. The diffusion constant is 10.6×10^{-10} m²/s. A tube of water has a radius of 1.40 cm. What difference in concentration per unit length of the tube $\Delta C/L$ must be maintained to give this flow?

44. Review Conceptual Example 8 before starting this problem. Ammonia, which has a strong smell, is diffusing through air contained in a tube of length L and cross-sectional area of 4.0×10^{-4} m². The diffusion constant for ammonia in air is 4.2×10^{-5} m²/s. In a certain time, 8.4×10^{-8} kg of ammonia diffuses through the air when the difference in ammonia concentration between the ends of the tube is 3.5×10^{-2} kg/m³. Find the speed at which ammonia diffuses through the air, that is, the length of the tube divided by the time to travel this length.

45. ssm www Carbon tetrachloride (CCl_4) is diffusing through benzene (C_6H_6), as the drawing illustrates. The concentration of CCl_4 at the left end of the tube is maintained at 1.00×10^{-2} kg/m³, and the diffusion constant is 20.0×10^{-10} m²/s. The CCl_4 enters the tube at a mass rate of 5.00×10^{-13} kg/s. Using this data and that shown in the drawing, find (a) the mass of CCl_4 per second that passes point A and (b) the concentration of CCl_4 at point A.

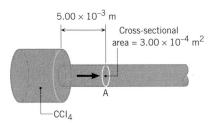

5.00 × 10⁻³ m

Cross-sectional
area = 3.00 × 10⁻⁴ m²

A

CCl_4

46. Review Conceptual Example 8 before working this problem. For water vapor in air at 293 K, the diffusion constant is $D = 2.4 \times 10^{-5}$ m²/s. As outlined in problem 47(a), the time required for the first solute molecules to traverse a channel of length L is $t = L^2/(2D)$, according to Fick's law. Find the time t for $L = 0.010$ m. (b) For comparison, how long would a water molecule take to travel $L = 0.010$ m at the translational rms speed of water molecules (assumed to be an ideal gas) at a temperature of 293 K? (c) Explain why the answer to part (a) is so much longer than the answer to part (b).

***47.** Review Conceptual Example 8 as background for this problem. It is possible to convert Fick's law into a form that is useful when the concentration is zero at one end of the diffusion channel ($C_1 = 0$ in Figure 14.14a). (a) Noting that AL is the volume V of the channel and that m/V is the average concentration of solute in the channel, show that Fick's law becomes $t = L^2/(2D)$. This form of Fick's law can be used to estimate the time required for the first solute molecules to traverse the channel. (b) A bottle of perfume is opened in a room where convection currents are absent. Assuming that the diffusion constant for perfume in air is 1.0×10^{-5} m²/s, estimate the minimum time required for the perfume to be smelled 2.5 cm away.

****48.** The drawing shows a container that is partially filled with 2.0 grams of water. The temperature is maintained at a constant 20 °C. The space above the liquid contains air that is completely saturated with water vapor. A tube of length 0.15 m and cross-sectional area 3.0×10^{-4} m² connects the water vapor at one end to air that remains completely dry at the other end. The diffusion constant for water vapor in air is 2.4×10^{-5} m²/s. How long does it take for the water in the container to evaporate completely? (*Hint: Refer to problem 76 in Chapter 12 to find the pressure of the water vapor.*)

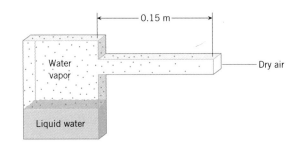

0.15 m

Water
vapor

Dry air

Liquid water

ADDITIONAL PROBLEMS

49. ssm A bicycle tire whose volume is 4.1×10^{-4} m³ has a temperature of 296 K and an absolute pressure of 4.8×10^5 Pa. A cyclist brings the pressure up to 6.2×10^5 Pa without changing the temperature or volume. How many moles of air must be pumped into the tire?

50. A Goodyear blimp typically contains 5400 m³ of helium (He) at an absolute pressure of 1.1×10^5 Pa. The temperature of the helium is 280 K. What is the mass (in kg) of the helium in the blimp?

51. A can of soda has a mass of 354 g. Assume that the soda is mainly water, and find the number of water molecules in the can.

52. What is the density (in kg/m³) of nitrogen gas (molecular mass = 28 g/mol) at a pressure of 2.0 atmospheres and a temperature of 310 K?

53. ssm The diffusion constant of ethanol in water is 12.4×10^{-10} m²/s. A cylinder has a cross-sectional area of 4.00 cm² and a length of 2.00 cm. A difference in ethanol concentration of 1.50 kg/m³ is maintained between the ends of the cylinder. In one hour, what mass of ethanol diffuses through the cylinder?

54. If the translational rms speed of the water vapor molecules (H_2O) in air is 648 m/s, what is the translational rms speed of the carbon dioxide molecules (CO_2) in the same air? Both gases are at the same temperature.

55. Very fine smoke particles are suspended in air. The translational rms speed of a smoke particle is 4.9×10^{-3} m/s, and the temperature is 295 K. Find the mass of a particle.

56. A closed cylindrical tank has a height of 0.80 m. The tank is initially filled with air at a pressure of 2.0 atm. Water is pumped into the tank until the air pressure reaches 6.0 atm. The temperature is constant. Determine the height of the compressed air.

***57. ssm** A drop of water has a radius of 9.00×10^{-4} m. How many water molecules are in the drop?

***58.** Compressed air can be pumped underground into huge caverns as a form of energy storage. The volume of a cavern is 5.6×10^5 m³ and the pressure of the air in it is 7.7×10^6 Pa. Assume that air is a diatomic ideal gas whose internal energy U is given by $U = \frac{5}{2}nRT$. If one home uses 30.0 kW·h of energy per day, how many homes could this internal energy serve for one day?

***59.** An apartment has a living room whose dimensions are 3.3 m × 4.5 m × 8.1 m. Assume that the air in the room is composed of 79% nitrogen (N_2) and 21% oxygen (O_2). At STP conditions, what is the mass (in grams) of the air?

****60.** In 10.0 s, 200 bullets strike and embed themselves in a wall. The bullets strike the wall perpendicularly. Each bullet has a mass of 5.0×10^{-3} kg and a speed of 1200 m/s. (a) What is the average change in momentum per second for the bullets? (b) Determine the average force exerted on the wall. (c) Assuming the bullets are spread out over an area of 3.0×10^{-4} m², obtain the average pressure they exert on this region of the wall.

****61. ssm** Estimate the spacing between the centers of neighboring atoms in a piece of solid aluminum, based on a knowledge of the density (2700 kg/m³) and atomic mass (26.9815 u) of aluminum. *(Hint: Assume that the volume of the solid is filled with many small cubes, with one atom at the center of each.)*

****62.** The mass of a hot-air balloon and its occupants is 320 kg (excluding the hot air inside the balloon). The air outside the balloon has a pressure of 1.01×10^5 Pa and a density of 1.29 kg/m³. To lift off, the air inside the balloon is heated. The volume of the heated balloon is 650 m³. The pressure of the heated air remains the same as that of the outside air. To what temperature (in kelvins) must the air be heated so that the balloon just lifts off? The molecular mass of air is 29 g/mol.

THERMODYNAMICS

This geyser is shooting out hot water that has been heated and pushed to the surface from wells deep within the earth. The laws that govern heat and work form the basis of thermodynamics, the subject of this chapter.

15.1 THERMODYNAMIC SYSTEMS AND THEIR SURROUNDINGS

Figure 15.1 The hot air in a balloon is one example of a thermodynamic system.

We have studied heat and work as separate topics. Often, however, they occur simultaneously. Our society relies heavily, for instance, on engines, such as the internal combustion engine used in automobiles. Here, fuel is burned at a relatively high temperature, and some of its internal energy is used for doing the work of driving the pistons up and down. Excess heat is generated in the combustion process and is removed by the cooling system to prevent overheating. *Thermodynamics* is the branch of physics that is built upon the fundamental laws that heat and work obey.

In thermodynamics the collection of objects upon which attention is being focused is called the *system,* while everything else in the environment is called the *surroundings.* For example, the system in an automobile engine could be the burning gasoline, while the surroundings would then include the pistons, the exhaust system, the radiator, and the outside air. The system and its surroundings are separated by walls of some kind. Walls that permit heat to flow through them, such as those of the engine block, are called *diathermal walls.* Perfectly insulating walls that do not permit heat to flow between the system and its surroundings are called *adiabatic walls.*

To understand what the laws of thermodynamics have to say about the relationship between heat and work, it is necessary to describe the physical condition or *state of a system.* We might be interested, for instance, in the hot air within one of the balloons in Figure 15.1. The hot air itself would be the system, and the skin of the balloon provides the walls that separate this system from the surrounding cooler air. The state of the system would be specified by giving values for the pressure, volume, and temperature of the hot air, assuming that its mass is known.

As this chapter discusses, there are four laws of thermodynamics. We begin with the one known as the zeroth law and then consider the remaining three.

15.2 THE ZEROTH LAW OF THERMODYNAMICS

The zeroth law of thermodynamics deals with the concept of *thermal equilibrium.* Two systems are said to be in thermal equilibrium if there is no net flow of heat between them when they are brought into thermal contact. For instance, you are *not* in thermal equilibrium with the water in Lake Michigan in January. Just dive into it, and you will find out how quickly your body loses heat to the frigid water. To help explain the central idea of the zeroth law of thermodynamics, Figure 15.2*a* shows

Figure 15.2 (*a*) Systems A and B are surrounded by adiabatic walls and register the same temperature on the thermometer. (*b*) When A is put into thermal contact with B through diathermal walls, no net flow of heat occurs between the systems.

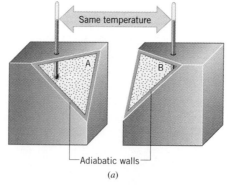
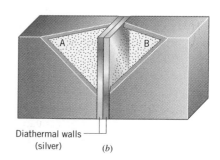

Same temperature

Adiabatic walls
(*a*)

Diathermal walls
(silver)
(*b*)

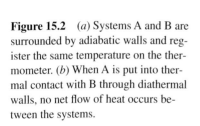

two systems labeled A and B. Each is within a container whose adiabatic walls are made from insulation that prevents the flow of heat, and each has the same temperature, as indicated by the thermometers. In part *b*, one wall of each container is replaced by a thin silver sheet, and the two sheets are touched together. Silver has a large thermal conductivity, so heat flows through it readily and the silver sheets behave as diathermal walls. Even though the diathermal walls would permit it, no net flow of heat occurs in part *b*, indicating that the two systems are in thermal equilibrium. There is no net flow of heat, because the two systems have the same temperature. We see, then, that *temperature is the indicator of thermal equilibrium in the sense that there is no net flow of heat between two systems in thermal contact that have the same temperature.*

In Figure 15.2 the thermometer plays an important role. System A is in equilibrium with the thermometer, and so is system B. In each case, the thermometer registers the same temperature, thereby indicating that the two systems are equally hot. Consequently, systems A and B are found to be in thermal equilibrium with each other. In effect, the thermometer is a third system. The fact that system A and system B are each in thermal equilibrium with this third system means that they are in thermal equilibrium with each other. This finding is an example of the *zeroth law of thermodynamics.*

■ **THE ZEROTH LAW OF THERMODYNAMICS**
Two systems individually in thermal equilibrium with a third system are in thermal equilibrium with each other.

The zeroth law establishes temperature as the indicator of thermal equilibrium and implies that all parts of a system must be in thermal equilibrium if the system is to have a definable single temperature. In other words, there can be no flow of heat within a system in thermal equilibrium.

15.3 THE FIRST LAW OF THERMODYNAMICS

In Chapter 6 we saw that forces can do work, which, in turn, can change the kinetic and potential energy of an object. For example, the atoms and molecules of a substance exert forces on one another. As a result, they have kinetic and potential energy. These and other kinds of molecular energy constitute the internal energy of a substance. In Chapter 12 we learned that heat is the transfer of energy from a higher-temperature object to a lower-temperature object because of the difference in temperatures. When a substance participates in a process involving energy in the form of work and heat, the internal energy of the substance can change. The relationship between work, heat, and changes in the internal energy is known as the first law of thermodynamics and is illustrated in the concept chart in Figure 15.3. We will now see that the first law of thermodynamics is an expression of the conservation of energy.

Suppose that a system gains heat Q and that this is the only effect occurring. Consistent with the law of conservation of energy, the internal energy of the system increases from an initial value of U_i to a final value of U_f, the change being $\Delta U = U_f - U_i = Q$. In writing this equation, we use the convention that *heat Q is positive when the system gains heat and negative when the system loses heat.* The

Concepts at a Glance

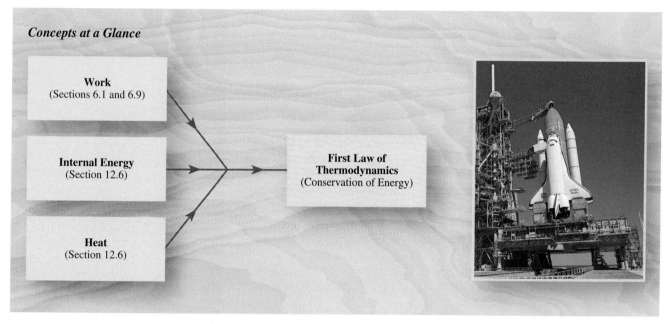

Figure 15.3 *Concepts at a Glance* The concepts of work, heat, and the internal energy of a system are related by the first law of thermodynamics, which is an expression of the law of conservation of energy. The photograph shows the Space Shuttle being readied for a launch. The onboard fuel contains internal energy. During liftoff, part of this internal energy is used to heat the exhaust gas. The high-speed hot exhaust gas provides the force that propels the spacecraft into orbit, doing work on it in the process.

internal energy of a system can also change because of work. If a system does work W on its surroundings and there is no heat flow, energy conservation indicates that the internal energy of the system decreases from U_i to U_f, the change now being $\Delta U = U_f - U_i = -W$. The minus sign is included because we follow the convention that **work is positive when it is done by the system and negative when it is done on the system.** If a system gains energy in the form of heat Q and simultaneously loses energy by doing work W, the change in internal energy due to both factors is given below by Equation 15.1. Thus, the **first law of thermodynamics** is just the conservation of energy principle applied to heat, work, and the change in the internal energy.

■ THE FIRST LAW OF THERMODYNAMICS

The internal energy of a system changes from an initial value U_i to a final value of U_f due to heat Q and work W:

$$\Delta U = U_f - U_i = Q - W \qquad (15.1)$$

Q is positive when the system gains heat and negative when it loses heat. W is positive when work is done by the system and negative when work is done on the system.

Example 1 illustrates the use of Equation 15.1 and the sign conventions for Q and W.

EXAMPLE 1 • Positive and Negative Work

Figure 15.4 illustrates a system and its surroundings. In part *a*, the system gains 1500 J of heat from its surroundings, and 2200 J of work is done *by* the system on the surroundings. In part *b*, the system also gains 1500 J of heat, but 2200 J of work is done *on* the system by the surroundings. In each case, determine the change in the internal energy of the system.

Reasoning In Figure 15.4*a* the system loses more energy in doing work than it gains in the form of heat, so the internal energy of the system decreases. Thus, we expect the change in the internal energy, $\Delta U = U_f - U_i$, to be negative. In part *b* of the drawing, the system gains energy both in the form of heat and in the form of work. The internal energy of the system increases, and we expect ΔU to be positive.

Solution

(a) The heat is positive, $Q = +1500$ J, since it is gained by the system. The work is positive, $W = +2200$ J, since it is done *by* the system. According to the first law of thermodynamics

$$\Delta U = Q - W = (+1500 \text{ J}) - (+2200 \text{ J}) = \boxed{-700 \text{ J}} \qquad (15.1)$$

The minus sign for ΔU indicates that the internal energy has decreased, as expected.
(b) The heat is positive, $Q = +1500$ J, since it is gained by the system. But the work is negative, $W = -2200$ J, since it is done *on* the system. Thus, the change in the internal energy of the system is

$$\Delta U = Q - W = (+1500 \text{ J}) - (-2200 \text{ J}) = \boxed{+3700 \text{ J}} \qquad (15.1)$$

The plus sign for ΔU indicates that the internal energy has increased, as expected.

> • **PROBLEM SOLVING INSIGHT**
> **When using the first law of thermodynamics, as expressed by Equation 15.1, be careful to adhere to the proper sign conventions for the heat Q and the work W.**

In the first law of thermodynamics, the internal energy U, heat Q, and work W are energy quantities, and each is expressed in energy units such as joules. However, there is a fundamental difference between U, on the one hand, and Q and W on the other. The next example sets the stage for explaining this difference.

EXAMPLE 2 • An Ideal Gas

The temperature of three moles of a monatomic ideal gas is reduced from $T_i = 540$ K to $T_f = 350$ K by two different methods. In the first method 5500 J of heat flows into the gas, while in the second method, 1500 J of heat flows into it. In each case find (a) the change in the internal energy of the gas and (b) the work done by the gas.

Reasoning Since the internal energy of a monatomic ideal gas is $U = \frac{3}{2}nRT$ (Equation 14.7) and since the number of moles n is fixed, only a change in temperature T can alter the internal energy. Since the change in T is the same in both methods, the change in U is also the same. From the given temperature values, the change in internal energy ΔU can be determined. Then, the first law of thermodynamics can be used with ΔU and the given heat values to calculate the work.

Solution

(a) Using Equation 14.7 for the internal energy of a monatomic ideal gas, we find for each method of adding heat that

$$\Delta U = \frac{3}{2}nR(T_f - T_i) = \frac{3}{2}(3.0 \text{ mol})[8.31 \text{ J/(mol·K)}](350 \text{ K} - 540 \text{ K}) = \boxed{-7100 \text{ J}}$$

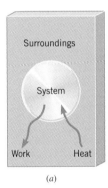

(a)

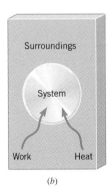

(b)

Figure 15.4 (*a*) The system gains energy in the form of heat, but loses energy because work is done by the system. (*b*) The system gains energy in the form of heat and also gains energy because work is done on the system.

(b) Since ΔU is now known and the heat is given in each method, Equation 15.1 can be used to determine the work:

1st method $W = Q - \Delta U = 5500 \text{ J} - (-7100 \text{ J}) = \boxed{12\ 600 \text{ J}}$

2nd method $W = Q - \Delta U = 1500 \text{ J} - (-7100 \text{ J}) = \boxed{8600 \text{ J}}$

In each method the gas does work, but it does more work in the first method.

To understand the difference between U, on the one hand, and Q and W on the other, consider the value for ΔU in Example 2. In both methods ΔU is the same. The value of ΔU is determined once the initial and final temperatures are specified, because the internal energy of an ideal gas depends only on the Kelvin temperature. Temperature is one of the variables (along with pressure and volume) that define the state of a system. ***The internal energy depends only on the state of a system, not on the method by which the system arrives at a given state.*** In recognition of this characteristic, internal energy is referred to as a ***function of state.**** In contrast, heat and work are not functions of state, because they have different values for each different method used to make the system change from one state to another, as in Example 2.

15.4 THERMAL PROCESSES

A system can interact with its surroundings in many ways, and the heat and work that come into play always obey the first law of thermodynamics. This section introduces four common thermal processes. In each case, the process is assumed to be ***quasi-static,*** which means that it occurs slowly enough that a uniform pressure and temperature exist throughout all regions of the system at all times.

   An isobaric process is one that occurs at constant pressure. For instance, Figure 15.5 shows a substance (solid, liquid, or gas) contained in a chamber fitted with a movable frictionless piston. The pressure P experienced by the substance is always the same and is determined by the external atmosphere and by the weight of the piston and the block resting on it. Heating the substance makes it expand and do work W in lifting the piston and block through the displacement **s**. The work can be calculated from $W = Fs$ (Equation 6.1), where F is the magnitude of the force and s is the magnitude of the displacement. The force is generated by the pressure P acting on the bottom surface of the piston, according to $F = PA$ (Equation 10.3), where A is the cross-sectional area of the piston. With this substitution for F, the work becomes $W = (PA)s$. But the product $A \cdot s$ is the change in volume of the material, $\Delta V = V_f - V_i$, where V_f and V_i are the final and initial volumes, respectively. Thus, the expression for the work is

Isobaric process $$W = P\Delta V = P(V_f - V_i) \qquad (15.2)$$

Consistent with our sign convention for work, this result predicts a positive value for the work done *by a system* when the system expands isobarically (V_f exceeds V_i). Equation 15.2 also applies to an isobaric compression (V_f less than V_i). Then, the work is negative, again consistent with the sign convention, since work must be

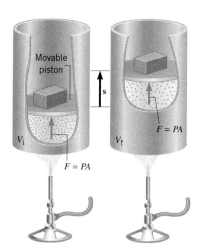

Figure 15.5 The substance in the chamber is expanding isobarically, because the pressure is held constant by the external atmosphere and the weight of the movable piston and the block.

---

* The fact that an ideal gas is used in Example 2 does not restrict our conclusion here. Had a real (non-ideal) gas or other material been used, the only difference would have been that the expression for the internal energy would have been more complicated. It might have involved the volume V, as well as the temperature T, for instance.

done *on the system* to compress it. Example 3 emphasizes that $W = P\Delta V$ applies to any system, solid, liquid, or gas, as long as the pressure remains constant while the volume changes.

EXAMPLE 3 • Isobaric Expansion of Water

One gram of water is placed in the cylinder in Figure 15.5, and the pressure is maintained at 2.0×10^5 Pa. The temperature of the water is raised by 31 C°. In one case, the water is in the liquid phase and expands by the small amount of 1.0×10^{-8} m³. In another case, the water is in the gas phase and expands by the much greater amount of 7.1×10^{-5} m³. For the water in each case, find (a) the work done and (b) the change in the internal energy.

Reasoning In both cases the material expands, so work is done by the system (water) on the surroundings and is, therefore, positive. The work is done at a constant pressure, so it can be found from $W = P\Delta V$. Once the work is known, the first law of thermodynamics, $\Delta U = Q - W$, can be used to find the change in internal energy, provided a value for the heat Q can be found. The heat needed to raise the temperature of each material can be obtained from $Q = cm\Delta T$ (Equation 12.4), where the specific heat capacity of liquid water is $c = 4186$ J/(kg·C°) and the specific heat capacity of water vapor at constant pressure is $c_P = 2020$ J/(kg·C°).

Solution

(a) In both cases, the process is isobaric, so the work done is given by Equation 15.2:

$$W_{\text{liquid}} = P\Delta V = (2.0 \times 10^5 \text{ Pa})(1.0 \times 10^{-8} \text{ m}^3) = \boxed{0.0020 \text{ J}}$$
$$W_{\text{gas}} = P\Delta V = (2.0 \times 10^5 \text{ Pa})(7.1 \times 10^{-5} \text{ m}^3) = \boxed{14 \text{ J}}$$

Note that under comparable conditions, liquids (and solids) change volume much less than gases do, which leads to a much smaller work of expansion or compression for liquids than for gases.

(b) Using $\Delta U = Q - W$ and $Q = cm\Delta T$, we find that $\Delta U = cm\Delta T - W$. Therefore,

$$\Delta U_{\text{liquid}} = [4186 \text{ J/(kg·C°)}](0.0010 \text{ kg})(31 \text{ C°}) - 0.0020 \text{ J}$$
$$= 130 \text{ J} - 0.0020 \text{ J} = \boxed{130 \text{ J}}$$
$$\Delta U_{\text{gas}} = [2020 \text{ J/(kg·C°)}](0.0010 \text{ kg})(31 \text{ C°}) - 14 \text{ J}$$
$$= 63 \text{ J} - 14 \text{ J} = \boxed{49 \text{ J}}$$

Virtually all the 130 J of heat added to the liquid serves to change the internal energy, since the volume change and the corresponding work of expansion are so small. In contrast, a significant fraction of the 63 J of heat added to the vapor causes work of expansion to be done, so that only 49 J is left for the internal energy change.

It is often convenient to display thermal processes graphically. For instance, Figure 15.6 shows a plot of pressure versus volume for an isobaric expansion. Since the pressure is constant, the graph is a horizontal straight line, beginning at the initial volume V_{i} and ending at the final volume V_{f}. In terms of such a plot, the work $W = P(V_{\text{f}} - V_{\text{i}})$ is the area under the graph, which is the shaded rectangle of height P and width $V_{\text{f}} - V_{\text{i}}$.

Another common thermal process is an ***isochoric process, one that occurs at constant volume.*** Figure 15.7a illustrates an isochoric process in which a substance (solid, liquid, or gas) is heated. The substance would expand if it could, but the

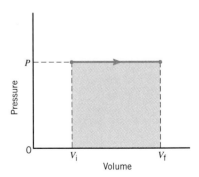

Figure 15.6 For an isobaric process, a pressure-versus-volume plot is a horizontal straight line, and the work done $[W = P(V_{\text{f}} - V_{\text{i}})]$ is the colored rectangular area under the graph.

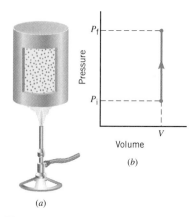

(a)

(b)

Figure 15.7 (*a*) The substance in the chamber is being heated isochorically, because the rigid chamber keeps the volume constant. (*b*) The pressure–volume plot for an isochoric process is a vertical straight line. The area under the graph is zero, indicating that no work is done.

rigid container keeps the volume constant, so the pressure–volume plot shown in Figure 15.7*b* is a vertical straight line. The expansion of the container itself is negligible. Because the volume is constant, the pressure inside rises, and the substance exerts more and more force on the walls. While enormous forces can be generated in the closed container, no work is done, since the walls do not move. Consistent with zero work being done and the interpretation of work as the area under the pressure–volume graph, the area under the vertical straight line in Figure 15.7*b* is zero. Since no work is done, the first law of thermodynamics indicates that the heat in an isochoric process serves only to change the internal energy of the system: $\Delta U = Q - W = Q$.

A third important thermal process is an ***isothermal process, one that takes place at constant temperature.*** The next section illustrates the important features of an isothermal process when the system is an ideal gas.

Last, there is the ***adiabatic process, one that occurs without the transfer of heat.*** Since there is no heat transfer, Q equals zero, and the first law indicates that $\Delta U = Q - W = -W$. Thus, when work is done by a system adiabatically, the internal energy of the system decreases by exactly the amount of the work done. When work is done on a system adiabatically, the internal energy increases correspondingly. The next section discusses an adiabatic process for an ideal gas.

A process may be complex enough that no part of it is recognizable as one of the four discussed above. For instance, Figure 15.8 shows a process for a gas in which the pressure, volume, and temperature are changed along the straight line from X to Y. With the aid of integral calculus, it can be shown that ***the area under a pressure–volume graph is the work for any kind of process,*** so the area representing the work has been colored in the drawing. The volume increases, and work is done by the gas. This work is positive by convention, as is the area. In contrast, if a process reduces the volume, work is done on the gas, and this work is negative by convention. Correspondingly, the area under the pressure–volume graph would be assigned a negative value. In Example 4, we determine the work for the case shown in Figure 15.8.

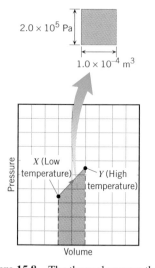

Figure 15.8 The thermal process that occurs from X to Y is not isobaric, isochoric, isothermal, or adiabatic. Nonetheless, the work done by the gas is given by the colored area.

EXAMPLE 4 • Work and the Area Under a Pressure–Volume Graph

Determine the work for the process in which the pressure, volume, and temperature of a gas are changed along the straight line from X to Y in Figure 15.8.

Reasoning The work is given by the area (in color) under the straight line between X and Y. Since the volume increases, work is done by the gas on the surroundings, so the work is positive. The area can be found by counting squares in Figure 15.8 and multiplying by the area per square.

Solution We estimate that there are 8.8 colored squares in the drawing. The area of one square is $(2.0 \times 10^5 \text{ Pa})(1.0 \times 10^{-4} \text{ m}^3) = 2.0 \times 10^1 \text{ J}$, so the work is

$$W = +(8.8 \text{ squares})(2.0 \times 10^1 \text{ J/square}) = \boxed{+180 \text{ J}}$$

The determination of work as the area under the graph of pressure versus volume is very similar to the procedure followed in Section 6.9, where we also determined the work done by a variable force as the area under a graph (see Figure 6.25). However, the graph used in Section 6.9 was a graph of the force component in the direction of the displacement plotted against the magnitude of the displacement.

15.5 THERMAL PROCESSES THAT UTILIZE AN IDEAL GAS

ISOTHERMAL EXPANSION OR COMPRESSION

When a system performs work isothermally, the temperature remains constant. In Figure 15.9a, for instance, a metal cylinder contains n moles of an ideal gas, and the large mass of hot water maintains the cylinder and gas at a constant Kelvin temperature T. The piston is held in place initially so the volume of the gas is V_i. As the external force applied to the piston is reduced quasi-statically, the gas expands to the final volume V_f. Figure 15.9b gives a plot of pressure ($P = nRT/V$) versus volume for the process. The solid red line in the graph is called an isotherm (meaning "constant temperature"), because it represents the relation between pressure and volume when the temperature is held constant. The work W done by the gas is *not* given by $W = P\,\Delta V = P(V_f - V_i)$, because the pressure is not constant. Nevertheless, the work is equal to the area under the graph. The techniques of integral calculus lead to the following result* for W:

Isothermal expansion or compression of an ideal gas

$$W = nRT \ln\left(\frac{V_f}{V_i}\right) \qquad (15.3)$$

Where does the energy for this work originate? Since the internal energy of an ideal gas is proportional to the Kelvin temperature ($U = \frac{3}{2}nRT$ for a monatomic ideal gas), the internal energy remains constant throughout an isothermal process, and the change in internal energy is zero. The first law of thermodynamics becomes $\Delta U = 0 = Q - W$. In other words, $Q = W$, and the energy for the work originates in the hot water. Heat flows into the gas from the water, as Figure 15.9a illustrates. If the gas is compressed isothermally, Equation 15.3 still applies, and heat flows out of the gas into the water. The following example deals with the isothermal expansion of an ideal gas.

EXAMPLE 5 • Isothermal Expansion of an Ideal Gas

Two moles of the monatomic gas argon expand isothermally at 298 K, from an initial volume of $V_i = 0.025 \text{ m}^3$ to a final volume of $V_f = 0.050 \text{ m}^3$. Assuming that argon is

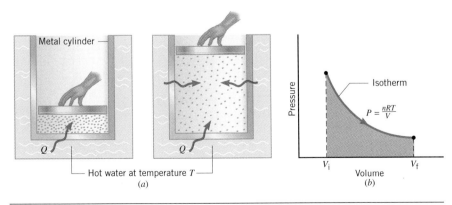

Metal cylinder

Q

Hot water at temperature T

(a)

Pressure

Isotherm

$P = \frac{nRT}{V}$

V_i V_f

Volume

(b)

Figure 15.9 (a) The ideal gas in the cylinder is expanding isothermally at temperature T. The force holding the piston in place is reduced slowly, so the expansion occurs quasi-statically. (b) The work done by the gas is given by the colored area.

* In this result, "ln" denotes the natural logarithm to the base $e = 2.71828$. The natural logarithm is related to the common logarithm to the base ten by $\ln(V_f/V_i) = 2.303 \log(V_f/V_i)$.

Figure 15.10 (*a*) The ideal gas in the cylinder is expanding adiabatically. The force holding the piston in place is reduced slowly, so the expansion occurs quasi-statically. (*b*) A plot of pressure versus volume yields the adiabatic curve shown in red, which intersects the isotherms (blue) at the initial temperature T_i and the final temperature T_f. The work done by the gas is given by the colored area.

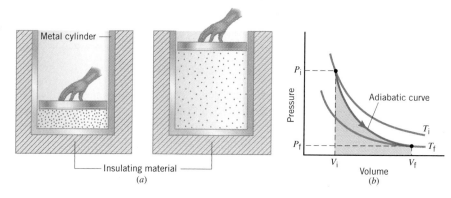

an ideal gas, find (a) the work done by the gas, (b) the change in the internal energy of the gas, and (c) the heat supplied to the gas.

Reasoning and Solution

(a) The work done by the gas can be found from Equation 15.3:

$$W = nRT \ln\left(\frac{V_f}{V_i}\right) = (2.0 \text{ mol})[8.31 \text{ J/(mol} \cdot \text{K)}](298 \text{ K}) \ln\left(\frac{0.050 \text{ m}^3}{0.025 \text{ m}^3}\right)$$

$$= \boxed{+3400 \text{ J}}$$

(b) The internal energy of a monatomic ideal gas is $U = \frac{3}{2}nRT$ and does not change when the temperature is constant. Therefore, $\boxed{\Delta U = 0}$.

(c) The heat Q added to the gas can be determined from the first law of thermodynamics:

$$Q = \Delta U + W = 0 + 3400 \text{ J} = \boxed{+3400 \text{ J}} \tag{15.1}$$

ADIABATIC EXPANSION OR COMPRESSION

When a system performs work adiabatically, no heat flows into or out of the system. Figure 15.10*a* shows an arrangement in which n moles of an ideal gas do work under adiabatic conditions, expanding quasi-statically from an initial volume V_i to a final volume V_f. The arrangement is similar to that in Figure 15.9 for isothermal expansion. However, a different amount of work is done here, because the cylinder is now surrounded by insulating material that prevents the flow of heat, so $Q = 0$. According to the first law of thermodynamics, the change in internal energy is $\Delta U = Q - W = -W$. Since the internal energy of an ideal monatomic gas is $U = \frac{3}{2}nRT$, it follows that $\Delta U = U_f - U_i = \frac{3}{2}nR(T_f - T_i)$, where T_i and T_f are the initial and final Kelvin temperatures. With this substitution, the relation $\Delta U = -W$ becomes

Adiabatic expansion or compression of a monatomic ideal gas

$$W = \tfrac{3}{2}nR(T_i - T_f) \tag{15.4}$$

When an ideal gas expands adiabatically, it does positive work, so W is positive in Equation 15.4. Therefore, the term $T_i - T_f$ is also positive, and the final temperature of the gas must be less than the initial temperature. The internal energy of the gas is reduced to provide the necessary energy to do the work, and because the internal energy is proportional to the Kelvin temperature, the temperature decreases. Figure 15.10b shows a plot of pressure versus volume for the adiabatic process. The adiabatic curve (red) intersects the isotherms (blue) at the higher initial temperature $[T_i = P_iV_i/(nR)]$ and the lower final temperature $[T_f = P_fV_f/(nR)]$. The colored area under the adiabatic curve represents the work done.

The reverse of an adiabatic expansion is an adiabatic compression (W is negative), and Equation 15.4 indicates that the final temperature exceeds the initial temperature. The energy provided by the agent doing the work increases the internal energy of the gas. As a result, the gas becomes hotter.

The equation that gives the adiabatic curve (red) between the initial pressure and volume (P_i, V_i) and the final pressure and volume (P_f, V_f) in Figure 15.10b can be derived using integral calculus. The result is

Adiabatic expansion or compression of an ideal gas

$$P_iV_i^\gamma = P_fV_f^\gamma \tag{15.5}$$

where the exponent γ is the ratio of the specific heat capacities at constant pressure and constant volume, $\gamma = c_P/c_V$. Equation 15.5 applies in conjunction with the ideal gas law, for *each point* on the adiabatic curve satisfies the relation $PV = nRT$.

15.6 SPECIFIC HEAT CAPACITIES AND THE FIRST LAW OF THERMODYNAMICS

In this section the first law of thermodynamics is used to gain an understanding of the factors that determine the specific heat capacity of a material. Remember, when the temperature of a substance changes as a result of heat flow, the change in temperature ΔT and the amount of heat Q are related according to $Q = cm \, \Delta T$ (Equation 12.4). In this expression c denotes the specific heat capacity in units of J/(kg·C°), and m is the mass in kilograms. It is more convenient now to express the amount of material as the number of moles n, rather than the number of kilograms. Therefore, we replace the expression $Q = cm \, \Delta T$ with the following analogous expression:

$$Q = Cn \, \Delta T \tag{15.6}$$

where the capital letter C (as opposed to the lowercase c) refers to the **molar specific heat capacity** in units of J/(mol·K). In addition, the unit for measuring the temperature change ΔT is the Kelvin (K) rather than the Celsius degree (C°). For gases it is necessary to distinguish between the molar specific heat capacities C_P and C_V, which apply, respectively, to conditions of constant pressure and constant volume. With the help of the first law of thermodynamics and an ideal gas as an example, it is possible to see why C_P and C_V differ.

To determine the molar specific heat capacities, we must first calculate the heat Q needed to raise the temperature of an ideal gas from T_i to T_f. According to the first law, $Q = \Delta U + W$. We also know that the internal energy of a monatomic ideal gas is $U = \frac{3}{2}nRT$ (Equation 14.7). As a result, $\Delta U = U_f - U_i = \frac{3}{2}nR(T_f - T_i)$. When the heating process occurs at constant pressure, the work done is given by Equation 15.2: $W = P \, \Delta V = P(V_f - V_i)$. For an ideal gas $PV = nRT$, so the work becomes $W = nR(T_f - T_i)$. On the other hand, when the volume is constant, $\Delta V = 0$, and the work done is zero. The calculation of the heat is summarized below:

$$Q = \Delta U + W$$
$$Q_{\text{constant pressure}} = \tfrac{3}{2}nR(T_f - T_i) + nR(T_f - T_i)$$
$$Q_{\text{constant volume}} = \tfrac{3}{2}nR(T_f - T_i) + 0$$

The molar specific heat capacities can now be determined, since Equation 15.6 indicates that $C = Q/[n(T_f - T_i)]$:

Constant pressure
for a monatomic $C_P = \tfrac{3}{2}R + R = \tfrac{5}{2}R$ (15.7)
ideal gas

Constant volume
for a monatomic $C_V = \tfrac{3}{2}R$ (15.8)
ideal gas

The ratio γ of the specific heats is

Monatomic
ideal gas $\gamma = \dfrac{C_P}{C_V} = \dfrac{\tfrac{5}{2}R}{\tfrac{3}{2}R} = \dfrac{5}{3}$ (15.9)

For real monatomic gases near room temperature, experimental values of C_P and C_V give ratios very close to the theoretical value of $\tfrac{5}{3}$.

The difference between C_P and C_V arises because work is done when the gas expands in response to the addition of heat under conditions of constant pressure, whereas no work is done under conditions of constant volume. For a monatomic ideal gas, C_P exceeds C_V by an amount equal to R, the ideal gas constant:

$$C_P - C_V = R \qquad (15.10)$$

In fact, it can be shown that Equation 15.10 applies to any kind of ideal gas—monatomic, diatomic, etc.

Concepts at a Glance

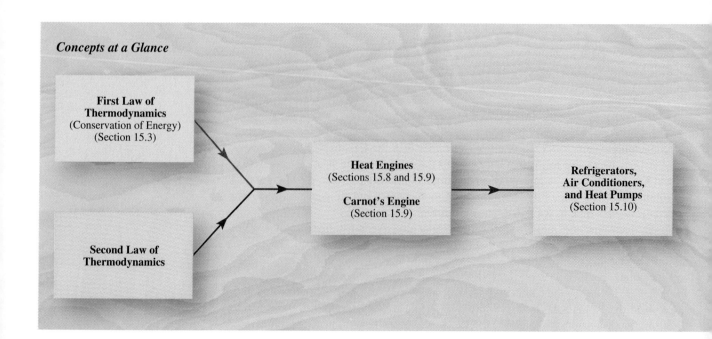

15.7 THE SECOND LAW OF THERMODYNAMICS

Ice cream melts when left out on a warm day. A cold can of soda warms up on a hot day at a picnic. This is always the case, because ice cream and soda never become colder when left in a hot environment. Heat always flows spontaneously from hot to cold, and never from cold to hot. The spontaneous flow of heat is the focus of one of the most profound laws in all of science, the *second law of thermodynamics.*

■ **THE SECOND LAW OF THERMODYNAMICS: THE HEAT FLOW STATEMENT**

Heat flows spontaneously from a substance at a higher temperature to a substance at a lower temperature and does not flow spontaneously in the reverse direction.

It is important to realize that the second law of thermodynamics deals with a different aspect of nature than does the first law of thermodynamics. The second law is a statement about the natural tendency of heat to flow from hot to cold, whereas the first law deals with energy conservation and focuses on both heat and work. A number of important devices depend on heat and work in their operation, and to understand such devices both laws of thermodynamics are needed. For instance, an automobile engine is a type of heat engine, because it uses heat to produce work. In discussing heat engines in Sections 15.8 and 15.9, we will bring together the first and second laws to analyze engine efficiency, as the concept chart in Figure 15.11 indicates. Then, in Section 15.10 we will see that refrigerators, air conditioners, and heat pumps are also devices that utilize heat and work and that they are closely related to heat engines. The way in which these three appliances operate also depends on both the first and second laws of thermodynamics.

Figure 15.11 *Concepts at a Glance* The first and second laws of thermodynamics are used to evaluate the performance of heat engines, as well as refrigerators, air conditioners, and heat pumps. As the photograph illustrates, the Space Shuttle uses heat engines to lift itself into orbit.

15.8 HEAT ENGINES

A *heat engine* is any device that uses heat to perform work. To illustrate features common to all heat engines, Figure 15.12 shows a simplified steam engine. The boiler receives heat from a high-temperature source, such as an oil or gas burner. In the boiler, liquid water is converted into steam at a high temperature and pressure. The steam passes through the intake valve into the cylinder and expands against the piston, forcing it to move. The piston may be coupled to a wheel in order to turn it, for example. The work done on the moving piston comes from the internal energy of the steam, so the temperature of the steam in the cylinder drops. As the piston returns to its original position, the cooler steam is forced out the exhaust valve and into the condenser, where the steam is changed back into a liquid. The heat of condensation is released to a heat sink, which may be the atmosphere. The water is then pumped back to the boiler, and the cycle is repeated. The combination of steam and water is known as the *working substance* of the engine, for it is the agent that does the work. Heat engines, such as the steam engine, operate in repeating cycles. In other words, the conclusion of one cycle is the start of another, so the working substance is in the same state at the end of one cycle and the beginning of the next.

The essential features of a steam engine are listed below, because they are shared by all types of heat engines:

1. Heat is supplied to the engine at a relatively high temperature.
2. Part of the input heat is used to perform work.
3. The remainder of the input heat is rejected at a temperature lower than the input temperature.

Figure 15.13 emphasizes these essential features in a schematic fashion. In this drawing the symbol Q_H refers to the magnitude of the input heat. The subscript H stands for "hot," and the place from which the input heat comes is called the "hot

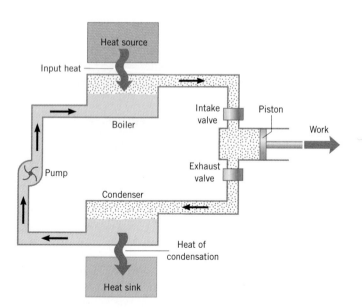

Figure 15.12   A steam engine.

reservoir." The symbol Q_C denotes the magnitude of the rejected heat. The subscript C means "cold," and the place in the environment where the rejected heat goes is known as the "cold reservoir." The symbol W denotes the magnitude of the work done by the engine. The symbols Q_H, Q_C, and W refer to magnitudes only, without reference to algebraic signs, so negative values are not used for these quantities in the equations in which they appear.

To be highly efficient, a heat engine must produce a relatively large amount of work from as little input heat as possible. Thus, the *efficiency e* of a heat engine is defined as the ratio of the work W done by the engine to the input heat Q_H:

$$e = \frac{\text{Work done}}{\text{Input heat}} = \frac{W}{Q_H} \qquad (15.11)$$

If the input heat were converted entirely into work, the engine would have an efficiency of 1.00, since $W = Q_H$; such an engine would be 100% efficient. *Efficiencies are often quoted as percentages obtained by multiplying the ratio W/Q_H by a factor of 100.* Thus, an efficiency of 68% would mean that a value of 0.68 is used for the value of the efficiency in Equation 15.11.

An engine, like any device, must obey the principle of conservation of energy. Some of the engine's input heat Q_H is converted into work W, and the remainder Q_C is rejected to the cold reservoir. If there are no other losses in the engine, the principle of energy conservation requires that

$$Q_H = W + Q_C \qquad (15.12)$$

Solving this equation for W and substituting the result into Equation 15.11 leads to the following alternative expression for the efficiency e of a heat engine:

$$e = \frac{Q_H - Q_C}{Q_H} = 1 - \frac{Q_C}{Q_H} \qquad (15.13)$$

Example 6 illustrates how the concepts of efficiency and energy conservation are applied to a heat engine.

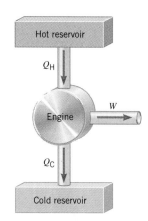

Figure 15.13 This schematic representation of a heat engine shows the input heat (magnitude = Q_H) that originates from the hot reservoir, the work (magnitude = W) that the engine does, and the heat (magnitude = Q_C) that the engine rejects to the cold reservoir.

EXAMPLE 6 • An Automobile Engine

An automobile engine has an efficiency of 22.0% and produces 2510 J of work. How much heat is rejected by the engine?

Reasoning Energy conservation indicates that the amount of heat rejected to the cold reservoir is the part of the input heat that is not converted into work. The amount Q_C rejected is $Q_C = Q_H - W$, according to Equation 15.12. To use this equation, however, we need a value for the input heat Q_H, so we begin our solution by finding this value.

Solution From Equation 15.11 for the efficiency e, we find that

$$Q_H = \frac{W}{e} = \frac{2510 \text{ J}}{0.220} = 11\,400 \text{ J}$$

The rejected heat is

$$Q_C = Q_H - W = 11\,400 \text{ J} - 2510 \text{ J} = \boxed{8900 \text{ J}} \qquad (15.12)$$

• **PROBLEM SOLVING INSIGHT**
When efficiency is stated as a percentage (e.g., 22.0%), it must be converted to a decimal fraction (e.g., 0.220) before being used in an equation.

In Example 6, less than one-quarter of the input heat is converted into work, because the efficiency of the automobile engine is only 22.0%. If the engine were 100% efficient, all the input heat would be converted into work. Unfortunately, nature does not permit 100% efficient heat engines to exist, as the next section discusses.

15.9 CARNOT'S PRINCIPLE AND THE CARNOT ENGINE

CARNOT'S PRINCIPLE

What is it that allows a heat engine to operate with maximum efficiency? The French engineer Sadi Carnot (1796–1832) proposed that a heat engine has maximum efficiency when the processes within the engine are reversible. The essence of a *reversible process* is illustrated in Figure 15.14, where a gas supports a frictionless piston. In part *a* of the drawing, the gas has a pressure, volume, and temperature of P_1, V_1, and T_1. When 100 J of heat are added to the gas, as in part *b*, the gas expands, pushes the piston upward, and thereby does work. When the gas stops expanding, its pressure, volume, and temperature are P_2, V_2, and T_2. In part *c*, the piston falls back to its original position and does work in compressing the gas as heat is removed. If the process is reversible, 100 J of heat flows out of the gas and back into the environment as the gas returns to its original state (P_1, V_1, and T_1). Thus, *a reversible process is one in which both the system (the gas) and its environment (the piston and the rest of the universe) can be returned to exactly the states they were in before the process occurred.*

In a reversible process, *both* the system and its environment can be returned to their initial states. Therefore, a process that involves an energy-dissipating mechanism, such as friction, cannot be reversible, because the energy wasted due to friction would alter the system or the environment or both. There are also reasons other than friction why a process may not be reversible. For instance, the spontaneous flow of heat from a hot substance to a cold substance is irreversible, even though friction is not present. For heat to flow in the reverse direction, work must be done, as we will see in Section 15.10. The agent doing such work must be located in the environment of the hot and cold substances, and, therefore, the environment must change while the heat is moved back from cold to hot. Since the system and the environment cannot *both* be returned to their initial states, the process of spontaneous heat flow is irreversible. In fact, all spontaneous processes are irreversible, such as the explosion of an unstable chemical or the bursting of a bubble. When the word

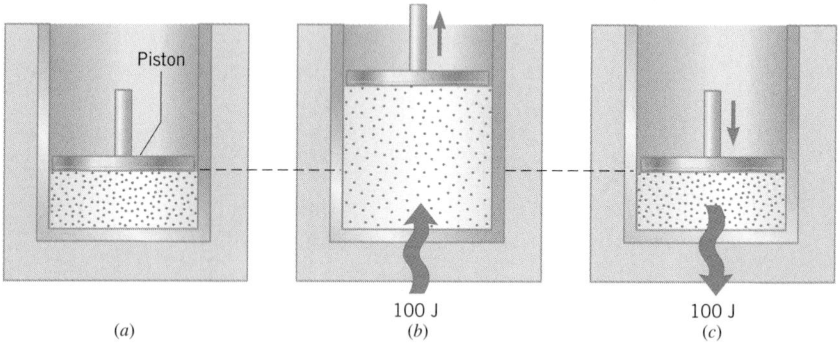

Figure 15.14 (*a*) The frictionless piston is supported by the gas. (*b*) When 100 J of heat is added, the gas expands and does work in lifting the piston. (*c*) If the process is reversible, the gas and its environment can both be returned to their initial states.

"reversible" is used in connection with engines, it does not just mean a gear that allows the engine to operate a device in reverse. All cars have a reverse gear, for instance, but no automobile engine is thermodynamically reversible, since friction exists no matter which way the car moves.

Today, the idea that the efficiency of a heat engine is a maximum when the engine operates reversibly is referred to as *Carnot's principle.*

■ CARNOT'S PRINCIPLE: AN ALTERNATIVE STATEMENT OF THE SECOND LAW OF THERMODYNAMICS

No irreversible engine operating between two reservoirs at constant temperatures can have a greater efficiency than a reversible engine operating between the same temperatures. Furthermore, all reversible engines operating between the same temperatures have the same efficiency.

Carnot's principle is quite remarkable, for no mention is made of the working substance of the engine. It does not matter if the working substance is a gas, a liquid, or a solid. As long as the process is reversible, the efficiency of the engine is a maximum. Carnot's principle does *not* state, or even imply, that a reversible engine has an efficiency of 100%.

It can be shown that if Carnot's principle were not valid, it would be possible for heat to flow spontaneously from a cold substance to a hot substance, in violation of the second law of thermodynamics. In effect, then, Carnot's principle is another way of expressing the second law.

THE CARNOT ENGINE

No real engine operates reversibly. Nonetheless, the idea of a reversible engine provides a useful standard for judging the performance of real engines. Figure 15.15 shows a reversible engine, called a *Carnot engine,* that is particularly useful as an idealized model. An important feature of the Carnot engine is that all input heat Q_H originates from a hot reservoir *at a single temperature* T_H and all rejected heat Q_C goes into a cold reservoir *at a single temperature* T_C.

Carnot's principle implies that the efficiency of a reversible engine is independent of the working substance of the engine, and therefore can depend only on the temperatures of the hot and cold reservoirs. Since Efficiency = $1 - Q_C/Q_H$ according to Equation 15.13, the ratio Q_C/Q_H can depend only on the reservoir temperatures. This observation led Lord Kelvin to propose a *thermodynamic temperature scale.* He proposed that the thermodynamic temperatures of the cold and hot reservoirs be defined such that the ratio of these temperatures is equal to Q_C/Q_H. Thus, the thermodynamic temperature scale is related to the heats absorbed and rejected by a Carnot engine, and is independent of the working substance. If a reference point on the thermodynamic temperature scale is properly chosen, it can be shown that the scale is identical to the Kelvin temperature scale introduced in Section 12.2 and used in the ideal gas law. As a result, the ratio of the rejected heat Q_C to the input heat Q_H is

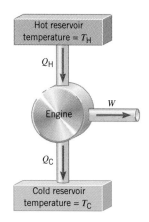

Figure 15.15 A Carnot engine is a reversible engine in which all input heat Q_H originates from a hot reservoir at a single temperature T_H, and all rejected heat Q_C goes into a cold reservoir at a single temperature T_C. The work done by the engine is W.

$$\frac{Q_C}{Q_H} = \frac{T_C}{T_H} \qquad (15.14)$$

where the temperatures T_C and T_H *must be expressed in kelvins.*

The efficiency of a Carnot engine can be written in a particularly useful way by substituting Equation 15.14 into the relation Efficiency $= 1 - Q_C/Q_H$:

$$\begin{matrix} \text{Efficiency of a} \\ \text{Carnot engine} \end{matrix} = 1 - \frac{T_C}{T_H} \qquad (15.15)$$

This relation gives the maximum possible efficiency for a heat engine operating between two Kelvin temperatures T_C and T_H, and the next example illustrates its application.

EXAMPLE 7 • A Tropical Ocean as a Heat Engine

> Water near the surface of a tropical ocean has a temperature of 298.2 K (25.0 °C), while water 700 m beneath the surface has a temperature of 280.2 K (7.0 °C). It has been proposed that the warm water be used as the hot reservoir and the cool water as the cold reservoir of a heat engine. (a) Find the maximum possible efficiency for such an engine. (b) Determine the minimum input heat Q_H that would be needed if a number of these engines were to produce an amount of work equal to the 9.3×10^{19} J of energy that the United States consumed in 1994.
>
> *Reasoning*   The maximum possible efficiency is the efficiency that a Carnot engine would have (Equation 15.15) operating between temperatures of $T_H = 298.2$ K and $T_C = 280.2$ K. Once the maximum efficiency is found, the minimum input heat Q_H needed for the engines to produce $W = 9.3 \times 10^{19}$ J of work can be obtained with the aid of Equation 15.11 ($Q_H = W/$Efficiency).
>
> *Solution*
>
> (a) Using $T_H = 298.2$ K and $T_C = 280.2$ K in Equation 15.15, we find that
>
> $$\begin{matrix} \text{Efficiency of a} \\ \text{Carnot Engine} \end{matrix} = 1 - \frac{T_C}{T_H} = 1 - \frac{280.2 \text{ K}}{298.2 \text{ K}} = \boxed{0.060 \ (6.0\%)}$$
>
> (b) According to Equation 15.11, the input heat Q_H needed for the engines to produce 9.3×10^{19} J of work is
>
> $$Q_H = \frac{W}{\text{Efficiency}} = \frac{9.3 \times 10^{19} \text{ J}}{0.060} = \boxed{1.6 \times 10^{21} \text{ J}}$$
>
> Real engines, being less efficient than Carnot engines, would require an input energy that is greater than 1.6×10^{21} J to produce the same amount of work.

In Example 7 the maximum possible efficiency is only 6.0%. The small efficiency arises because the Kelvin temperatures of the hot and cold reservoirs are so close. A greater efficiency is possible only when there is a greater difference between the reservoir temperatures. However, there are limits on how large the efficiency of a heat engine can be, as Conceptual Example 8 discusses.

CONCEPTUAL EXAMPLE 8 • Natural Limits on the Efficiency of a Heat Engine

> Consider a hypothetical engine that receives 1000 J of heat as input from a hot reservoir and delivers 1000 J of work, rejecting no heat to a cold reservoir whose temperature is above 0 K. Decide whether this engine violates the first or the second law of thermodynamics, or both.

The Physics of...
extracting work from a warm ocean.

• PROBLEM SOLVING INSIGHT
When determining the efficiency of a Carnot engine, be sure the temperatures T_C and T_H of the cold and hot reservoirs are expressed in kelvins; degrees Celsius or degrees Fahrenheit will not do.

Reasoning and Solution As we know, the first law is an expression of energy conservation. From the point of view of energy conservation, there is nothing wrong with an engine that converts 1000 J of heat into 1000 J of work. Energy has been neither created nor destroyed; it has only been transformed from one form (heat) to another (work). This engine does, however, violate the second law of thermodynamics. Since all of the input heat is converted into work, the efficiency of the engine is 1, or 100%. But Equation 15.15, which is based on the second law, indicates that the maximum possible efficiency is $1 - T_C/T_H$, where T_C and T_H are the temperatures of the cold and hot reservoirs, respectively. Since we know that T_C is above 0 K, it is clear that the ratio T_C/T_H is greater than zero, so the maximum possible efficiency must be less than 1, or less than 100%. It is important to understand here that the first and second laws of thermodynamics address different aspects of nature. *It is the second law, not the first law, that limits the efficiencies of heat engines to values less than 100%.*

Example 8 has emphasized that *even a perfect heat engine has an efficiency that is less than 1.0 or 100%.* In this regard, we note that the maximum possible efficiency, as given by Equation 15.15, approaches 1.0 when T_C approaches absolute zero (0 K). However, experiments have shown that it is not possible to cool a substance to absolute zero (see Section 15.12), so nature does not permit the existence of a 100% efficient heat engine. As a result, there will always be heat rejected to a cold reservoir whenever a heat engine is used to do work, even if friction and other irreversible processes are eliminated completely.

15.10 REFRIGERATORS, AIR CONDITIONERS, AND HEAT PUMPS

The natural tendency of heat is to flow from hot to cold, as indicated by the second law of thermodynamics. However, if work is used, heat can be *made* to flow from cold to hot, against its natural tendency. Refrigerators, air conditioners, and heat pumps are, in fact, devices that do just that. As Figure 15.16 illustrates, these devices use work W to extract an amount of heat Q_C from the cold reservoir and deposit an amount of heat Q_H into the hot reservoir. Generally speaking, such a process is called a *refrigeration process.* A comparison of this drawing with Figure 15.15 shows that the directions of the arrows symbolizing heat and work in a refrigeration process are opposite to those in an engine process. Nonetheless, energy is conserved during a refrigeration process, just as it is in an engine process, so $Q_H = W + Q_C$. Moreover, if the process occurs reversibly, we have ideal devices that are called Carnot refrigerators, Carnot air conditioners, and Carnot heat pumps. For these ideal devices, the relation $Q_C/Q_H = T_C/T_H$ applies, just as it does for the Carnot engine.

In a *refrigerator,* the interior of the unit is the cold reservoir, while the warmer exterior is the hot reservoir. As Figure 15.17 illustrates, the refrigerator takes heat from the food inside and deposits it into the kitchen, along with the energy needed to do the work of making the heat flow from cold to hot. For this reason, the outside surfaces (usually the sides and back) of most refrigerators are warm to the touch while the units operate. Thus, a refrigerator warms the kitchen.

An *air conditioner* is like a refrigerator, except that the room itself is the cold

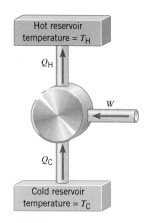

Figure 15.16 In a refrigeration process, work W is used to remove heat Q_C from the cold reservoir and deposit heat Q_H into the hot reservoir.

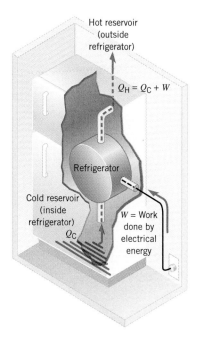

Figure 15.17 A refrigerator.

The Physics of...
a refrigerator.

Figure 15.18 A window air conditioner removes heat from a room, which is the cold reservoir, and deposits heat outdoors, which is the hot reservoir.

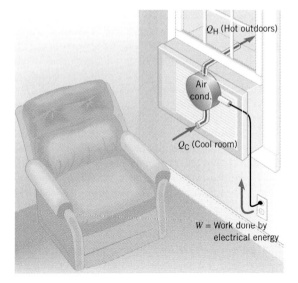

The Physics of...
an air conditioner.

reservoir and the outdoors is the hot reservoir. Figure 15.18 shows a window unit, which cools a room by removing heat and depositing it outside, along with the work used to make the heat flow from cold to hot. Conceptual Example 9 considers a common misconception about refrigerators and air conditioners.

CONCEPTUAL EXAMPLE 9 • You Can't Beat the Second Law of Thermodynamics

Is it possible to cool your kitchen by leaving the refrigerator door open or cool your bedroom by putting a window air conditioner on the floor by the bed?

Reasoning and Solution Whatever heat Q_C is removed from the air directly in front of the open refrigerator is deposited directly back into the kitchen at the rear of the unit. Moreover, according to the second law, work W is needed to move that heat from cold to hot, and the energy from this work is also deposited into the kitchen as additional heat. Thus, the open refrigerator puts into the kitchen an amount of heat $Q_H = Q_C + W$, which is more than it removes. ***Rather than cooling the kitchen, the open refrigerator warms it up.*** Putting a window air conditioner on the floor to cool your bedroom is similarly a no-win game. The heat pumped out the back of the air conditioner and into the bedroom is greater than the heat pulled into the front of the unit. Consequently, ***the air conditioner actually warms the bedroom.***

Related Homework Material: Problem 72

The performance of a refrigerator or air conditioner is rated according to its coefficient of performance. Such appliances exhibit a high performance when they can remove a relatively large amount of heat Q_C from the cold reservoir with as little work W as possible. Therefore, the coefficient of performance is defined as the ratio of Q_C to W, and the greater this ratio is, the better the performance is:

Refrigerator
or
air conditioner

$$\text{Coefficient of performance} = \frac{Q_C}{W} \qquad (15.16)$$

Commercially available refrigerators and air conditioners have coefficients of performance in the range 2–6, depending on the temperatures involved. The coefficients of performance for these real devices are less than those for ideal, or Carnot, refrigerators and air conditioners.

In a sense, refrigerators and air conditioners operate like pumps. They pump heat "uphill" from a lower temperature to a higher temperature, just as a water pump forces water uphill from a lower elevation to a higher elevation. It would be appropriate to call them heat pumps. However, the name "heat pump" is reserved for the device illustrated in Figure 15.19, which is a home heating appliance. The **heat pump** uses work W to make heat Q_C from the wintry outdoors (the cold reservoir) flow up the temperature "hill" into a warm house (the hot reservoir). According to the conservation of energy, the heat pump deposits inside the house an amount of heat $Q_H = Q_C + W$. The air conditioner and the heat pump do closely related jobs. The air conditioner refrigerates the inside of the house and heats up the outdoors, while the heat pump refrigerates the outdoors and heats up the inside. These jobs are so closely related that most heat pump systems serve in a dual capacity, being equipped with a switch that converts them from heaters in the winter into air conditioners in the summer.

Heat pumps are popular for home heating in today's energy-conscious world, and it is easy to understand why. Suppose 1000 J of energy is available to use for home heating. Figure 15.20*a* shows that a conventional electric heating system uses this 1000 J to heat a coil of wire, just as in a toaster. A fan blows air across the hot coil, and forced convection carries the 1000 J of heat into the room. Part *b* of the drawing shows that a heat pump does not use the 1000 J directly as heat. Instead, it uses the 1000 J to do the work of pumping heat Q_C from the cooler outdoors into the warmer room and delivers to the inside an amount of energy $Q_H = Q_C + 1000$ J. Obviously, the heat pump puts more than 1000 J of heat into the room, whereas the conventional electric heating system provides only 1000 J. The next example shows how the basic relations $Q_H = W + Q_C$ and $Q_C/Q_H = T_C/T_H$ are used with heat pumps.

The Physics of...
a heat pump.

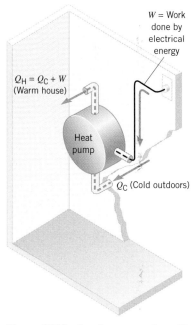

W = Work done by electrical energy

$Q_H = Q_C + W$
(Warm house)

Heat pump

Q_C (Cold outdoors)

Figure 15.19 In a heat pump the cold reservoir is the wintry outdoors, and the hot reservoir is the inside of the house.

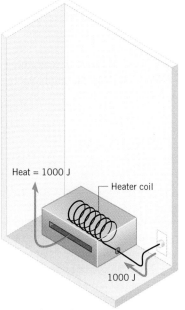

Heat = 1000 J

Heater coil

1000 J

(*a*) Conventional electric heating

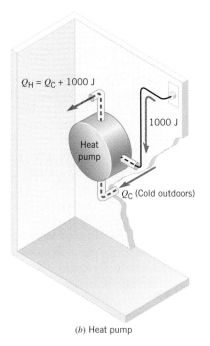

$Q_H = Q_C + 1000$ J

1000 J

Heat pump

Q_C (Cold outdoors)

(*b*) Heat pump

Figure 15.20 (*a*) This conventional electric heating system is delivering 1000 J of heat to the living room. (*b*) In a heat pump, 1000 J of energy is used as work to obtain heat Q_C from the outdoors, the amount of energy delivered to the room being $Q_H = Q_C + 1000$ J.

EXAMPLE 10 • A Heat Pump

An ideal or Carnot heat pump is used to heat a house to a temperature of $T_H = 294$ K (21 °C). How much work must be done by the pump to deliver $Q_H = 3350$ J of heat into the house when the outdoor temperature T_C is (a) 273 K (0 °C) and (b) 252 K (-21 °C)?

Reasoning The conservation of energy ($Q_H = W + Q_C$) applies to the heat pump. Thus, the work can be determined from $W = Q_H - Q_C$, provided we can obtain a value for Q_C, the heat taken by the pump from the outside. To determine Q_C, we use the fact that the pump is a Carnot heat pump and operates reversibly. Therefore, the relation $Q_C/Q_H = T_C/T_H$ (Equation 15.14) applies. Solving it for Q_C, we obtain $Q_C = Q_H(T_C/T_H)$. Using this result we find that

$$W = Q_H - Q_C = Q_H - Q_H\left(\frac{T_C}{T_H}\right) = Q_H\left(1 - \frac{T_C}{T_H}\right)$$

Solution

(a) At an indoor temperature of $T_H = 294$ K and an outdoor temperature of $T_C = 273$ K, the work needed is

$$W = Q_H\left(1 - \frac{T_C}{T_H}\right) = (3350 \text{ J})\left(1 - \frac{273 \text{ K}}{294 \text{ K}}\right) = \boxed{240 \text{ J}}$$

(b) This solution is identical to that in part (a), except that it is now cooler outside, so $T_C = 252$ K. The necessary work is $W = \boxed{479 \text{ J}}$, which is more than in part (a). More work must be done because the heat is pumped up a greater temperature "hill" when the outside is colder than when it is warmer.

It is also possible to specify a coefficient of performance for heat pumps. However, unlike refrigerators and air conditioners, the job of a heat pump is to heat, not to cool. As a result, the coefficient of performance of a heat pump is the ratio of the heat Q_H delivered into the house to the work W required to deliver it:

Heat pump $\qquad$ Coefficient of performance $= \dfrac{Q_H}{W}$ $\qquad\qquad$ (15.17)

The coefficient of performance depends on the indoor and outdoor temperatures. Commercially available units have coefficients of about 3–4 under favorable conditions.

15.11 ENTROPY AND THE SECOND LAW OF THERMODYNAMICS

ENTROPY

A Carnot engine has the maximum possible efficiency for its operating conditions, because the processes occurring within it are reversible. Irreversible processes, such as friction, cause real engines to operate at less than maximum efficiency, for they reduce our ability to use heat to perform work. As an extreme example, imagine that a hot object is placed in thermal contact with a cold object, so heat flows spontaneously, and hence irreversibly, from hot to cold. Eventually both objects reach the same temperature, and $T_C = T_H$. A Carnot engine using these two objects as heat reservoirs is unable to do work, because the efficiency of the engine is zero [Efficiency $= 1 - (T_C/T_H) = 0$]. In general, irreversible processes cause us to lose some, but not necessarily all, of the ability to perform work. This partial loss can be expressed in terms of a concept called *entropy.*

To introduce the idea of entropy we recall the relation $Q_C/Q_H = T_C/T_H$ that applies to a Carnot engine. This equation can be rearranged as $Q_C/T_C = Q_H/T_H$, which focuses attention on the heat Q divided by the Kelvin temperature T. The quantity Q/T is called the change in the entropy ΔS:

$$\Delta S = \left(\frac{Q}{T}\right)_R \qquad (15.18)$$

In this expression the temperature T must be in kelvins, and the subscript R refers to the word "reversible." It can be shown that Equation 15.18 applies to any process in which heat Q enters or leaves a system reversibly at a constant temperature. Such is the case for the heat that flows into and out of the reservoirs of a Carnot engine. Equation 15.18 indicates that the SI unit for entropy is a joule per kelvin (J/K).

Entropy, like internal energy, is a function of the state or condition of the system. Only the state of a system determines the entropy S that a system has. Therefore, the change in entropy ΔS is equal to the entropy of the final state of the system minus the entropy of the initial state.

We can now describe what happens to the entropy of a Carnot engine. As the engine operates, the entropy of the hot reservoir decreases, since heat Q_H departs at a Kelvin temperature T_H. The change in the entropy of the hot reservoir is $\Delta S_H = -Q_H/T_H$, where the minus sign is needed to indicate a decrease in entropy, since the symbol Q_H denotes only the magnitude of the heat. In contrast, the entropy of the cold reservoir increases by an amount $\Delta S_C = +Q_C/T_C$, for the rejected heat enters the cold reservoir at a Kelvin temperature T_C. The total change in entropy is

$$\Delta S_C + \Delta S_H = \frac{Q_C}{T_C} - \frac{Q_H}{T_H} = 0$$

because $Q_C/T_C = Q_H/T_H$ according to Equation 15.14.

The fact that the total change in entropy is zero for a Carnot engine is a specific illustration of a general result. It can be proved that when *any* reversible process occurs, the change in the entropy of the universe is zero; $\Delta S_{universe} = 0$ for a reversible process. The word "universe" means that $\Delta S_{universe}$ takes into account the entropy changes of all parts of the system and all parts of the environment. **Reversible processes, then, do not alter the total entropy of the universe.** To be sure, the entropy of one part of the universe may change because of a reversible process, but if so, the entropy of another part must change in the opposite way by the same amount.

To understand what happens to the entropy of the universe when an *irreversible* process occurs is more difficult, for the expression $\Delta S = (Q/T)_R$ does not apply directly to such a process. However, if a system changes irreversibly from an initial state to a final state, this expression can be used to calculate ΔS indirectly, as Figure 15.21 indicates. We imagine a hypothetical reversible process that causes the system to change between *the same initial and final states* and then find ΔS for this reversible process. The value obtained for ΔS also applies to the irreversible process that actually occurs, since only the nature of the initial and final states, and not the path between them, determines ΔS. Example 11 illustrates this indirect method and shows that spontaneous (irreversible) processes cause the entropy of the universe to increase.

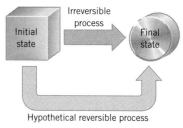

$$\Delta S \text{ for irreversible} = \Delta S \text{ for hypothetical}$$
$$\text{process} \qquad \text{reversible process}$$

Figure 15.21 Although the relation $\Delta S = (Q/T)_R$ applies to reversible processes, it can be used as part of an indirect procedure to find the entropy change for an irreversible process. This drawing illustrates the procedure discussed in the text.

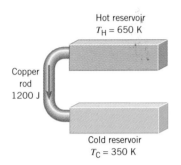

Figure 15.22 Heat flows sponta-neously from a hot reservoir to a cold reservoir.

EXAMPLE 11 • The Entropy of the Universe Increases

Figure 15.22 shows 1200 J of heat flowing spontaneously through a copper rod from a hot reservoir at 650 K to a cold reservoir at 350 K. Determine the amount by which this irreversible process changes the entropy of the universe, assuming that no other changes occur.

Reasoning The hot-to-cold heat flow is irreversible, so the relation $\Delta S = (Q/T)_R$ is applied to a hypothetical process whereby the 1200 J of heat is taken reversibly from the hot reservoir and added reversibly to the cold reservoir.

Solution The total entropy change of the universe is the algebraic sum of the entropy changes for each reservoir:

$$\Delta S_{\text{universe}} = -\underbrace{\frac{1200 \text{ J}}{650 \text{ K}}}_{\substack{\text{Entropy lost} \\ \text{by the hot} \\ \text{reservoir}}} + \underbrace{\frac{1200 \text{ J}}{350 \text{ K}}}_{\substack{\text{Entropy gained} \\ \text{by the cold} \\ \text{reservoir}}} = \boxed{+1.6 \text{ J/K}}$$

The irreversible process causes the entropy of the universe to increase by 1.6 J/K.

Example 11 is a specific illustration of a general result: ***any irreversible process increases the entropy of the universe.*** In other words, $\Delta S_{\text{universe}} > 0$ for an irreversible process. Reversible processes do not alter the entropy of the universe, whereas irreversible processes cause the entropy to increase. Therefore, the entropy of the universe continually increases, like time itself, and entropy is sometimes called "time's arrow." It can be shown that the behavior of the entropy of the universe constitutes a completely general statement of the second law of thermodynamics, which applies not only to heat flow but also to all kinds of other processes.

■ **THE SECOND LAW OF THERMODYNAMICS STATED IN TERMS OF ENTROPY**

The total entropy of the universe does not change when a reversible process occurs ($\Delta S_{\text{universe}} = 0$) and increases when an irreversible process occurs ($\Delta S_{\text{universe}} > 0$).

ENERGY THAT IS UNAVAILABLE FOR DOING WORK

When an irreversible process occurs and the entropy of the universe increases, the energy available for doing work decreases, as the next example illustrates.

EXAMPLE 12 • Energy Unavailable for Doing Work

Suppose that 1200 J of heat is used as input for an engine under two different conditions. In Figure 15.23a the heat is supplied by a hot reservoir whose temperature is 650 K. In part *b* of the drawing, the heat flows irreversibly through a copper rod into a second reservoir whose temperature is 350 K and then enters the engine. In either case, a 150-K reservoir is used as the cold reservoir. For each case, determine the maximum amount of work that can be obtained from the 1200 J of heat.

Reasoning The work obtained from the engine is the product of its efficiency and the input heat: $W = (\text{Efficiency})Q_H = (\text{Efficiency}) \times (1200 \text{ J})$. The maximum amount of work is obtained when the efficiency is a maximum, that is, when the engine is a Carnot engine. The efficiency of a Carnot engine is given by Equation 15.15 as Efficiency = $1 - T_C/T_H$. Therefore, the efficiency may be determined from the Kelvin temperatures of the hot and cold reservoirs.

Solution

Before irreversible heat flow

$$\text{Efficiency} = 1 - \frac{T_C}{T_H} = 1 - \frac{150 \text{ K}}{650 \text{ K}} = 0.77$$

$$W = (\text{Efficiency})(1200 \text{ J}) = (0.77)(1200 \text{ J}) = \boxed{920 \text{ J}}$$

After irreversible heat flow

$$\text{Efficiency} = 1 - \frac{T_C}{T_H} = 1 - \frac{150 \text{ K}}{350 \text{ K}} = 0.57$$

$$W = (\text{Efficiency})(1200 \text{ J}) = (0.57)(1200 \text{ J}) = \boxed{680 \text{ J}}$$

When the 1200 J of input heat is taken from the 350-K reservoir instead of the 650-K reservoir, the efficiency of the Carnot engine is smaller. As a result, less work (680 J versus 920 J) can be extracted from the input heat.

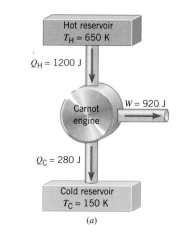

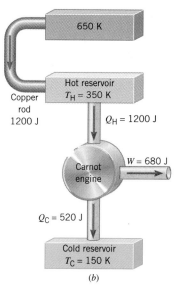

Figure 15.23 Heat in the amount of 1200 J is used as input for an engine under two different conditions in parts *a* and *b*.

Example 12 shows that 240 J less work (920 J − 680 J) can be performed when the input heat is obtained from the reservoir with the smaller temperature. In other words, the irreversible process of heat flow through the copper rod causes energy to become unavailable for doing work in the amount of $W_{\text{unavailable}} = 240$ J. Example 11 shows that this irreversible process also causes the entropy of the universe to increase by an amount $\Delta S_{\text{universe}} = +1.6$ J/K. These values for $W_{\text{unavailable}}$ and $\Delta S_{\text{universe}}$ are in fact related, because if you multiply $\Delta S_{\text{universe}}$ by 150 K, which is the lowest Kelvin temperature in Example 12, you obtain $W_{\text{unavailable}} = (150 \text{ K}) \times (1.6 \text{ J/K}) = 240$ J. This is one illustration of the following general result:

$$W_{\text{unavailable}} = T_0 \Delta S_{\text{universe}} \tag{15.19}$$

where T_0 is the Kelvin temperature of the coldest heat reservoir. Since irreversible processes cause the entropy of the universe to increase, they cause energy to be degraded, for part of the energy becomes unavailable for the performance of work. In contrast, there is no penalty when reversible processes occur, because for them $\Delta S_{\text{universe}} = 0$, and there is no loss of work. However, the entropy of the universe is always increasing, so the amount of energy that is unavailable for work is also increasing.

ORDER AND DISORDER

Entropy can also be interpreted in terms of order and disorder. As an example, consider a block of ice (Figure 15.24) with each of its H_2O molecules fixed rigidly in place in a highly structured and ordered arrangement. In comparison, the puddle of

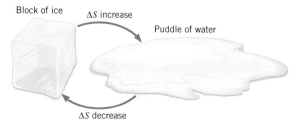

Figure 15.24 A block of ice is an example of an ordered system relative to a puddle of water.

Figure 15.25 Demolition experts are causing this building to go from an ordered state (lower entropy) to a disordered state (higher entropy).

water into which the ice melts is disordered and unorganized, for the molecules in a liquid are free to move from place to place. Heat is required to melt the ice and produce the disorder. Moreover, heat flow into a system increases the entropy of the system, according to $\Delta S = (Q/T)_R$. We associate an increase in entropy, then, with an increase in disorder. Conversely, we associate a decrease in entropy with a decrease in disorder or a greater degree of order. Example 13 illustrates an order-to-disorder change and the increase of entropy that accompanies it.

EXAMPLE 13 • Order to Disorder

Find the change in entropy that results when a 2.3-kg block of ice melts slowly (reversibly) at 273 K (0 °C).

Reasoning and Solution Since the phase change occurs reversibly at a constant temperature, the change in entropy can be found by using $\Delta S = (Q/T)_R$, where Q is the heat absorbed by the melting ice. This heat can be determined by using the relation $Q = mL_f$ (see Section 12.8), where m is the mass and $L_f = 3.35 \times 10^5$ J/kg is the latent heat of fusion of water. The change in entropy is

$$\Delta S = \left(\frac{Q}{T}\right)_R = \frac{mL_f}{T} = \frac{(2.3 \text{ kg})(3.35 \times 10^5 \text{ J/kg})}{273 \text{ K}} = \boxed{+2.8 \times 10^3 \text{ J/K}}$$

a result that is positive, since the ice absorbs heat as it melts.

Figure 15.25 shows another order-to-disorder change that can be described in terms of entropy.

15.12 THE THIRD LAW OF THERMODYNAMICS

To the zeroth, first, and second laws of thermodynamics we add the third (and last) law. The ***third law of thermodynamics*** indicates that it is impossible to reach a temperature of absolute zero.

> ■ **THE THIRD LAW OF THERMODYNAMICS**
> It is not possible to lower the temperature of any system to absolute zero in a finite number of steps.

This law, like the second law, can be expressed in a number of ways, but a discussion of them is beyond the scope of this text. The third law is needed to explain a number of experimental observations that cannot be explained by the other laws of thermodynamics.

SUMMARY

Thermodynamics is the branch of physics built upon the laws obeyed by energy in the form of work and heat. A thermodynamic **system** is the collection of objects on which attention is being focused, and the **surroundings** are everything else. The **state of a system** is the physical condition of the system, as described by values for physical parameters, often pressure, volume, and temperature.

Two systems are in **thermal equilibrium** if there is no net flow of heat between them when they are brought into thermal contact. The **zeroth law of thermodynamics** states that two systems individually in thermal equilibrium with a third system are in thermal equilibrium with each other. **Temperature** is the indicator of thermal equilibrium in the sense that there is no net flow of heat between two systems in thermal contact that have the same temperature.

The **first law of thermodynamics** states that due to heat Q and work W, the internal energy of a system changes from its initial value of U_i to a final value of U_f, according to $\Delta U = U_f - U_i = Q - W$. The first law is the conservation of energy principle applied to heat, work, and change in the internal energy. The internal energy is called a **function of state,** because it depends only on the state of the system and not on the method by which the system came to be in a given state. Heat and work are not functions of state, because they depend on how the system is changed from one state to another.

Thermal processes are **quasi-static** when they occur slowly enough that a uniform pressure and temperature exist throughout the system. An **isobaric process** is one that occurs at constant pressure. The work W done when a system changes at a constant pressure P from an initial volume V_i to a final volume V_f is $W = P(V_f - V_i)$. An **isochoric process** is one that takes place at constant volume, and no work is done in such a process. An **isothermal process** is one that occurs at constant temperature. An **adiabatic process** is one that takes place without the transfer of heat. The work done in a quasi-static thermal process is given by the area under the pressure-versus-volume graph for the process.

When n moles of an ideal gas change quasi-statically from an initial volume V_i to a final volume V_f at a constant Kelvin temperature T, the work done is $W = nRT \ln(V_f/V_i)$. When n moles of a monatomic ideal gas change quasi-statically and adiabatically from an initial temperature T_i to a final temperature T_f, the work done is $W = \frac{3}{2}nR(T_i - T_f)$. Along with the ideal gas law, an ideal gas also obeys the relation $P_iV_i^{\gamma} = P_fV_f^{\gamma}$ in an adiabatic process, where $\gamma = C_P/C_V$ is the ratio of the specific heat capacities at constant pressure and constant volume.

The **molar specific heat capacity** C of a substance determines how much heat Q is added or removed when the temperature of n moles of the substance changes by an amount ΔT: $Q = Cn\,\Delta T$. For a monatomic ideal gas, the molar specific heat capacities at constant pressure and constant volume are, respectively, $C_P = \frac{5}{2}R$ and $C_V = \frac{3}{2}R$, where R is the ideal gas constant. For any type of ideal gas $C_P - C_V = R$.

A **reversible process** is one in which both the system and its environment can be returned to exactly the initial states they were in before the process occurred. All spontaneous processes (such as the conduction of heat) and any process involving friction are irreversible.

The **second law of thermodynamics** can be stated in a number of equivalent forms. In terms of heat flow, the second law declares that heat flows spontaneously from a substance at a higher temperature to a substance at a lower temperature and does not flow spontaneously in the reverse direction. In the form known as **Carnot's principle,** the second law states that no irreversible engine operating between two reservoirs at constant temperatures can have a greater efficiency than a reversible engine operating between the same temperatures. Furthermore, all reversible engines operating between the same temperatures have the same efficiency. In terms of **entropy,** the second law states that the total entropy of the universe does not change when a reversible process occurs and increases when an irreversible process occurs.

A **heat engine** operates in cycles and produces work W from input heat Q_H that is extracted from a heat reservoir at a relatively high temperature. The engine rejects heat Q_C into a reservoir at a relatively low temperature. The **efficiency** of a heat engine is defined as Efficiency $= W/Q_H$. In addition, the principle of the conservation of energy requires that $Q_H = W + Q_C$.

A **Carnot engine** is a reversible engine in which all input heat Q_H originates from a hot reservoir at a single Kelvin temperature T_H and all rejected heat Q_C goes into a cold reservoir at a single Kelvin temperature T_C. For a Carnot engine, $Q_C/Q_H = T_C/T_H$. The **efficiency of a Carnot engine** is the maximum efficiency that an engine operating between two fixed temperatures can have: Efficiency of a Carnot engine $= 1 - T_C/T_H$.

Refrigerators, air conditioners, and **heat pumps** are devices that utilize work W to make heat Q_C flow from a lower Kelvin temperature T_C to a higher Kelvin temperature T_H. In the process (the refrigeration process) they deposit an amount of heat Q_H at the higher temperature. The principle of the conservation of energy requires that $Q_H = W + Q_C$. If the refrigeration process is perfect, in the sense that it occurs reversibly, the devices are called Carnot devices and the relation $Q_C/Q_H = T_C/T_H$ holds. The **coefficient of performance of a refrigerator or an air conditioner** is Q_C/W, while the **coefficient of performance of a heat pump** is Q_H/W.

The **change in entropy** ΔS for a process in which heat Q enters or leaves a system reversibly at a constant Kelvin temperature T is $\Delta S = (Q/T)_R$, where R stands for "reversible." Irreversible processes cause energy to be degraded in the sense that part of the energy becomes unavailable for the performance of work. The **energy that is unavailable for doing work** because of an irreversible process is $W_{\text{unavailable}} = T_0 \Delta S_{\text{universe}}$, where $\Delta S_{\text{universe}}$ is the total entropy change of the universe and T_0 is the Kelvin temperature of the coldest reservoir into which heat is rejected. Increased entropy is associated with a greater degree of disorder and decreased entropy with a lesser degree of disorder (more order).

The **third law of thermodynamics** states that it is not possible to lower the temperature of any system to absolute zero in a finite number of steps.

CONCEPTUAL QUESTIONS

1. Ignore friction and assume that air behaves as an ideal gas. The plunger of a bicycle tire pump is pushed down rapidly with the end of the pump sealed so that no air escapes and there is little time for heat to flow through the cylinder wall. Explain why the cylinder of the pump becomes warm to the touch.

2. One hundred joules of heat is added to a gas, and the gas expands at constant pressure. Is it possible that the internal energy increases by 200 J? Account for your answer with the aid of the first law of thermodynamics.

3. A gas is compressed isothermally, and its internal energy increases. Is the gas an ideal gas? Justify your answer.

4. Listed below are five values of heat and work that result when a system interacts with its surroundings. In each case, state whether the internal energy of the system increases, decreases, or remains the same, and justify your choices using the first law of thermodynamics.

   (a) $W = -500$ J and $Q = 0$

   (b) $W = 0$ and $Q = -200$ J

   (c) $W = +100$ J and $Q = +100$ J

   (d) $W = -100$ J and $Q = -100$ J

   (e) $W = +300$ J and $Q = +500$ J

5. (a) Is it possible for the temperature of a substance to rise without heat flowing into it? (b) Does the temperature of a substance necessarily have to change because heat flows into or out of it? In each case, give your reasoning and use the example of an ideal gas.

6. The drawing shows a pressure–volume graph in which a gas expands at constant pressure from A to B, and then goes from B to C at constant volume. Complete the table below by deciding whether each of the remaining four entries should be positive $(+)$, negative $(-)$, or zero (0). Give a reason for each answer that you choose.

	ΔU	Q	W
$A \rightarrow B$	$+$		
$B \rightarrow C$		$+$	

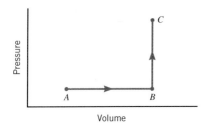

7. The drawing shows an arrangement for an adiabatic free expansion or "throttling" process. The process is adiabatic because the entire arrangement is contained within perfectly insulating walls. The gas in chamber A rushes suddenly into chamber B through a hole in the partition. Chamber B is evacuated, so the gas expands there under zero external pressure and the work $W = P \Delta V$ is zero. Assume that the gas is an ideal gas and explain how the final temperature of the gas after expansion compares to its initial temperature.

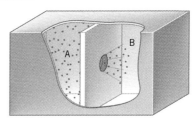

8. Suppose a material contracts when it is heated. Follow the same line of reasoning used in the text to reach Equations 15.7 and 15.8 and deduce which specific heat capacity for the material is larger, C_P or C_V.

9. When a solid melts at constant pressure, the volume of the resulting liquid does not differ much from the volume of the solid. Using what you know about the first law of thermodynamics and about the latent heat of fusion, describe how the internal energy of the liquid compares to the internal energy of the solid.

10. Suppose that you want to heat a gas so that its temperature will be as high as possible. Would you heat it under conditions of constant pressure or constant volume? Why?

11. Consider a hypothetical device that takes 10 000 J of heat from a hot reservoir and 5000 J of heat from a cold reservoir and produces 15 000 J of work. (a) Does this device violate the first law of thermodynamics? Explain. (b) Does the device violate the second law of thermodynamics? Explain.

12. If you saw an advertisement for an automobile that claimed the same gas mileage with and without the air conditioner operating, would you be suspicious? Explain, using what the second law of thermodynamics has to say about work and the direction of heat flow in an air conditioner.

13. The second law of thermodynamics, in the form of Carnot's principle, indicates that the most efficient heat engine operating between two temperatures is a reversible one. Does this mean that a reversible engine operating between the temperatures of 600 and 400 K must be more efficient than an *irreversible* engine operating between 700 and 300 K? Provide a reason for your answer.

14. Three reversible engines, A, B, and C, use the same cold reservoir for their exhaust heats. However, they use different hot reservoirs that have the following temperatures: (A) 1000 K, (B) 1100 K, and (C) 900 K. Rank these engines in order of increasing efficiency (smallest efficiency first). Account for your answer.

15. Suppose you wish to improve the efficiency of a Carnot engine. Compare the improvement to be realized via each of the following alternatives: (a) lower the Kelvin temperature of the cold reservoir by a factor of four, (b) raise the Kelvin temperature of the hot reservoir by a factor of four, (c) cut the Kelvin temperature of the cold reservoir in half and double the Kelvin temperature of the hot reservoir. Give your reasoning.

16. A refrigerator is kept in a garage that is not heated in the cold winter or air-conditioned in the hot summer. Does it cost more for this refrigerator to make a kilogram of ice cubes in the winter or in the summer? Give your reasoning.

17. Is it possible for a Carnot heat pump to have a coefficient of performance that is less than one? Justify your answer.

18. Air conditioners and refrigerators both remove heat from a cold reservoir and deposit it in a hot reservoir. Why, then, does an air conditioner cool the inside of a house while a refrigerator warms the house?

19. It has been said that heat pumps can't possibly deliver more energy into your house than they consume in operating. Can they or can't they? Explain.

20. A refrigerator is advertised as being easier to "live with" during the summer, because it puts into your kitchen only the heat that it removes from the food. Does this advertising claim violate the second law of thermodynamics? Account for your answer.

21. On a summer day a window air conditioner cycles on and off, according to how the temperature within the room changes. Are you more likely to be able to fry an egg on the outside part of the unit when the unit is on or when it is off? Explain.

22. An event happens somewhere in the universe and, as a result, the entropy of an object changes by −5 J/K. Which one (or more) of the following is a possible value for the entropy change for the rest of the universe: −5 J/K, 0 J/K, +5 J/K, +10 J/K? Account for your choice(s) in terms of the second law of thermodynamics.

23. When water freezes from a less-ordered liquid to a more-ordered solid, its entropy decreases. Why doesn't this decrease in entropy violate the second law of thermodynamics?

24. In each of the following cases, which has the greater entropy: (a) a handful of popcorn kernels or the popcorn that results from them, (b) a salad before or after it has been tossed, (c) a messy apartment with clothes strewn all over or a neat apartment? Why?

25. A glass of water contains a teaspoon of dissolved sugar. After a while, the water evaporates, leaving behind sugar crystals. The entropy of the sugar crystals is less than the entropy of the dissolved sugar, because the sugar crystals are in a more ordered state. Why doesn't this process violate the second law of thermodynamics?

26. A builder uses lumber to construct a building, which is unfortunately destroyed in a fire. Thus, the lumber existed at one time or another in three different states: (1) as unused building material, (2) as a building, and (3) as a burned-out shell of a building. Rank these three states in order of decreasing entropy (largest first). Provide a reason for the ranking.

PROBLEMS

ssm Solution is in the Student Solutions Manual. **www** Solution is available on the World Wide Web at http://www.wiley.com/college/cutnell ⚲ This application represents a biomedical application.

Section 15.3 The First Law of Thermodynamics

1. ssm When one gallon of gasoline is burned in a car engine, 1.19×10^8 J of internal energy is released. Suppose that 1.00×10^8 J of this energy flows directly into the surroundings (engine block and exhaust system) in the form of heat. If 6.0×10^5 J of work is required to make the car go one mile, how many miles can the car travel on one gallon of gas?

2. Suppose that a runner loses 4.7×10^5 J of heat and does 2.8×10^5 J of work during a workout. Determine each of the following quantities (including the algebraic sign): (a) Q, (b) W, and (c) ΔU.

3. The work done to compress one mole of a monatomic ideal gas is 6200 J. The temperature of the gas changes from 350 to 550 K. (a) How much heat flows between the gas and its surroundings? (b) Determine whether the heat flows into or out of the gas.

4. The drawing shows two objects, A and B. The arrows in the drawing symbolize the flow of heat into or out of an object, as well as the work done on or by an object. Find the change in the

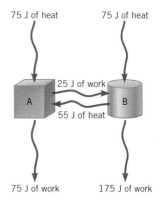

75 J of heat   75 J of heat

25 J of work

A   B

55 J of heat

75 J of work   175 J of work

internal energy of the system when it is (a) object A only, (b) object B only, and (c) objects A and B together. In each case, specify whether the change is a decrease or an increase.

5. ssm The internal energy of a system changes because the system gains 165 J of heat and performs 312 J of work. In returning to its initial state, the system loses 114 J of heat. During this return process, (a) how much work is involved, and (b) is work done by the system or is work done on the system?

*6. In exercising, a weight lifter loses 0.150 kg of water through evaporation, the heat required to evaporate the water coming from the weight lifter's body. The work done in lifting weights is 1.40×10^5 J. (a) Assuming that the latent heat of vaporization of perspiration is 2.42×10^6 J/kg, find the change in the internal energy of the weight lifter. (b) Determine the minimum number of nutritional calories of food (1 nutritional calorie = 4186 J) that must be consumed to replace the loss of internal energy.

Section 15.4 Thermal Processes

7. A gas is compressed under isobaric conditions, and its volume changes from 7.0×10^{-3} to 2.0×10^{-3} m^3. The pressure of the gas is 1.5×10^5 Pa. (a) Determine the work done (including the algebraic sign). (b) How much work is done (including the algebraic sign) if the gas then expands from 2.0×10^{-3} to 8.0×10^{-3} m^3 under isobaric conditions?

8. The internal energy of a system increases by 210 J during an adiabatic process. Determine (a) whether work is done on or by the system, and (b) find the magnitude of the work.

9. ssm A system gains 1500 J of heat, while the internal energy of the system increases by 4500 J and the volume decreases by 0.010 m^3. Assume the pressure is constant and find its value.

10. (a) Using the data presented in the accompanying pressure-versus-volume graph, estimate the magnitude of the work done when the system changes from A to B to C along the path shown. (b) Determine whether the work is done by the system or on the system and, hence, whether the work is positive or negative.

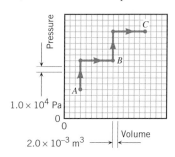

11. The internal energy of a system increases by 1350 J when the system gains 1150 J of heat at a constant pressure of 1.01×10^5 Pa. (a) What is the change in the volume of the gas? (b) Does the volume increase or decrease?

12. A gas is contained in a chamber such as that in Figure 15.5.

Suppose the region outside the chamber is evacuated and the total mass of the block and the movable piston is 135 kg. When 2050 J of heat flows into the gas, the internal energy of the gas increases by 1730 J. What is the distance s through which the piston rises?

13. ssm When a .22-caliber rifle is fired, the expanding gas from the burning gunpowder creates a pressure behind the bullet. This pressure causes the force that pushes the bullet through the barrel. The barrel has a length of 0.61 m and an opening whose radius is 2.8×10^{-3} m. A bullet (mass = 2.6×10^{-3} kg) has a speed of 370 m/s after passing through this barrel. Ignore friction and determine the average pressure of the expanding gas.

14. The pressure and volume of a gas are changed along the path ABCA. Determine the work done (including the algebraic sign) in each segment of the path: (a) A to B, (b) B to C, and (c) C to A.

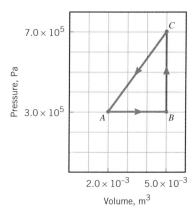

*15. A piece of aluminum has a volume of 1.4×10^{-3} m^3. The coefficient of volume expansion for aluminum is $\beta = 69 \times 10^{-6}$ $(C°)^{-1}$. The temperature of this object is raised from 20 to 320 °C. How much work is done by the expanding aluminum if the air pressure is 1.01×10^5 Pa?

*16. When a monatomic ideal gas expands at a constant pressure of 2.6×10^5 Pa, the volume of the gas increases by 6.2×10^{-3} m^3. (a) Determine the heat that flows into or out of the gas. (b) Specify the direction of the flow.

*17. ssm www A monatomic ideal gas expands isobarically. Using the first law of thermodynamics, prove that the heat Q is positive, so that it is impossible for heat to flow out of the gas.

*18. The latent heat of sublimation for zinc (atomic mass = 65.4 u) at 6.00×10^2 K is 1.99×10^6 J/kg. Assume that the zinc vapor can be treated as a monatomic ideal gas and that the volume of one kilogram of solid is negligible compared to that of the vapor. What percentage of the latent heat serves to change the internal energy during sublimation?

**19. Water is heated in an open pan where the air pressure is one atmosphere. The water remains a liquid, which expands by a small amount as it is heated. Determine the ratio of the work done by the water to the heat absorbed by the water.

Section 15.5 Thermal Processes That Utilize an Ideal Gas

20. The temperature of three moles of an ideal gas is 373 K. How much work does the gas do in expanding isothermally to four times its initial volume?

21. ssm Three moles of an ideal gas are compressed from 5.5×10^{-2} to 2.5×10^{-2} m^3. During the compression, 6.1×10^3 J of work is done on the gas, and heat is removed in order to keep the temperature of the gas constant at all times. Find (a) ΔU, (b) Q, and (c) the temperature of the gas.

22. Five moles of oxygen expands isothermally from 0.100 to 0.400 m^3. To maintain the constant temperature, 2.50×10^4 J of heat is added to the system. Assuming oxygen to be an ideal gas, determine the temperature.

23. Six grams of helium (molecular mass = 4.0 u) expands isothermally at 370 K and does 9600 J of work. Assuming that helium is an ideal gas, determine the ratio of the final volume of the gas to the initial volume.

24. A monatomic ideal gas ($\gamma = \frac{5}{3}$) is compressed adiabatically, and its volume is reduced by a factor of two. Determine the factor by which its pressure increases.

25. ssm A monatomic ideal gas has an initial temperature of 405 K. This gas expands and does the same amount of work whether the expansion is adiabatic or isothermal. When the expansion is adiabatic, the final temperature of the gas is 245 K. What is the ratio of the final to the initial volume when the expansion is isothermal?

*26. A bubble from the tank of a scuba diver in a lake contains 3.5×10^{-4} mol of gas. The bubble expands as it rises to the surface from a freshwater depth of 10.3 m. Assuming that the gas is an ideal gas and the temperature remains constant at 291 K, find the amount of heat that flows into the bubble.

*27. A diesel engine does not use spark plugs to ignite the fuel and air in the cylinders. Instead, the temperature required to ignite the fuel occurs because the pistons compress the air in the cylinders. Suppose air at an initial temperature of 27 °C is compressed adiabatically to a temperature of 681 °C. Assume the air to be an ideal gas for which $\gamma = \frac{7}{5}$. Find the compression ratio, which is the ratio of the initial volume to the final volume.

*28. The drawing refers to one mole of a monatomic ideal gas and shows a process that has four steps, two isobaric (A to B, C to

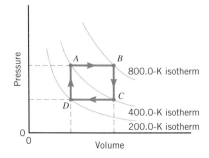

D) and two isochoric (B to C, D to A). Complete the following table by calculating ΔU, W, and Q (including the algebraic signs) for each of the four steps.

	ΔU	W	Q
A to B			
B to C			
C to D			
D to A			

*29. Using the relationship for an adiabatic expansion or compression of an ideal gas ($P_i V_i^\gamma = P_f V_f^\gamma$) together with the ideal gas law, derive an expression similar to the one above, but involving only volume, temperature, and γ.

*30. The pressure and volume of an ideal monatomic gas change from A to B to C, as the drawing shows. The curved line between A and C is an isotherm. (a) Determine the total heat for the process and (b) state whether the flow of heat is into or out of the gas.

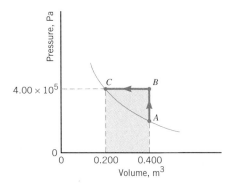

**31. ssm The work done by one mole of a monatomic ideal gas ($\gamma = \frac{5}{3}$) in expanding adiabatically is 825 J. The initial temperature and volume of the gas are 393 K and 0.100 m^3. Obtain (a) the final temperature and (b) final volume of the gas.

**32. One mole of a monatomic ideal gas has an initial pressure, volume, and temperature of P_0, V_0, and 438 K, respectively. It undergoes an isothermal expansion that triples the volume of the gas. Then, the gas undergoes an isobaric compression back to its original volume. Finally, the gas undergoes an isochoric increase in pressure, so that the final pressure, volume, and temperature are P_0, V_0, and 438 K, respectively. Find the total heat for this three-step process, and state whether it is absorbed by or given off by the gas.

15.6 Specific Heat Capacities and the First Law of Thermodynamics

33. ssm How much heat is required to change the temperature of 1.5 mol of a monatomic ideal gas by 77 K, if the pressure is held constant?

34. Suppose 750 J of heat is removed from five moles of a monatomic ideal gas. What change in temperature occurs when the energy is removed under conditions of (a) constant volume and (b) constant pressure?

35. Argon is a monatomic gas whose molecular mass is 39.9 u. The temperature of eight grams of argon is raised by 75 K under conditions of constant pressure. Assuming that argon is an ideal gas, how much heat is required?

36. Three moles of a monatomic ideal gas are heated at a constant volume of 1.50 m³. The amount of heat added is 5.24×10^3 J. (a) What is the change in the temperature of the gas? (b) Find the change in its internal energy. (c) Determine the change in pressure.

37. **ssm** The temperature of 2.5 mol of a monatomic ideal gas is 350 K. The internal energy of this gas is doubled by the addition of heat. How much heat is needed when it is added at (a) constant volume and (b) constant pressure?

38. Heat Q is added to a monatomic ideal gas at constant pressure. As a result, the gas does work W. Find the ratio Q/W.

*39. A ten-watt heater is used to heat a monatomic ideal gas at a constant pressure of 2.50×10^5 Pa. During the process, the 1.00×10^{-3}-m³ volume of the gas increases by 20.0%. How long was the heater on?

*40. Suppose that 31.4 J of heat is added to an ideal gas. The gas expands at a constant pressure of 1.40×10^4 Pa while changing its volume from 3.00×10^{-4} to 8.00×10^{-4} m³. The gas is not monatomic, so the relation $C_P = \frac{5}{2}R$ does not apply. (a) Determine the change in the internal energy of the gas. (b) Calculate its molar specific heat capacity C_P.

*41. **ssm** A monatomic ideal gas expands at constant pressure. (a) What percentage of the heat being supplied to the gas is used to increase the internal energy of the gas? (b) What percentage is used for doing the work of expansion?

**42. One mole of neon, a monatomic gas, starts out at conditions of standard temperature and pressure. The gas is heated at constant volume until its pressure is tripled, then further heated at constant pressure until its volume is doubled. Assume that neon behaves as an ideal gas. For the entire process, find the heat added to the gas.

15.8 Heat Engines

43. The input heat for an engine is 2.41×10^4 J, and the rejected heat is 5.86×10^3 J. Find the work done by the engine.

44. An automobile engine has an efficiency of 14% before a tune-up. After the tune-up, the efficiency increases to 20.0%. For a given amount of input energy, determine the ratio of the work done by the engine after the tune-up to that before the tune-up.

45. **ssm** In doing 16 600 J of work, an engine rejects 9700 J of heat. What is the efficiency of the engine?

46. An engine has an efficiency of 64% and produces 5500 J of work. Determine (a) the input heat and (b) the rejected heat.

*47. A hiker of mass 58 kg climbs up a mountain through a vertical distance of 950 m. To make the climb, her body generates an extra 4.5×10^6 J of energy. (a) How much work does she do in changing her gravitational potential energy? (b) What is her efficiency (expressed as a fraction between 0 and 1) in doing the work in part (a)?

*48. Engine A discards 72% of its input heat into a cold reservoir. Engine B has twice the efficiency as engine A. What percentage of its input heat does engine B discard?

49. **ssm **www** Engine A receives three times more input heat, produces five times more work, and rejects two times more heat than engine B. Find the efficiency of (a) engine A and (b) engine B.

Section 15.9 Carnot's Principle and the Carnot Engine

50. An engine has a hot reservoir temperature of 950 K and a cold reservoir temperature of 620 K. The engine operates at three-fifths maximum efficiency. What is the efficiency of the engine?

51. Five thousand joules of heat is put into a Carnot engine whose hot and cold reservoirs have temperatures of 500 and 200 K, respectively. How much heat is converted into work?

52. A fossil-fuel generating plant operates with an efficiency of 40%, and the temperature of its cold reservoir is 300 K. What is the maximum temperature of the steam produced by the fuel, which is the temperature of the hot reservoir?

53. **ssm** A Carnot engine has an efficiency of 0.700, and the temperature of its cold reservoir is 378 K. (a) Determine the temperature of its hot reservoir. (b) If 5230 J of heat is rejected to the cold reservoir, what amount of heat is put into the engine?

54. The ratio of the input heat to the discarded heat of a Carnot engine is 1.5. (a) What is the efficiency of the engine? (b) What is the ratio of the Kelvin temperature of the hot reservoir to the Kelvin temperature of the cold reservoir?

55. A Carnot heat engine has an efficiency of 0.18 when its hot reservoir has a temperature of 112 °C. What must be the reservoir's temperature (in kelvins) to produce an efficiency of 0.34? Assume that the temperature of the cold reservoir is the same in both cases.

56. A heat engine uses a hot reservoir consisting of a large amount of boiling water and a cold reservoir consisting of a large tub of ice and water. When 6800 J of heat is put into the engine and the engine produces work, how many kilograms of ice in the tub is melted due to the heat delivered to the cold reservoir?

*57. **ssm** A power plant taps steam superheated by geothermal energy to 505 K (the temperature of the hot reservoir) and uses the steam to do work in turning the turbine of an electric generator. The steam is then converted back into water in a condenser at 323 K (the temperature of the cold reservoir), after which the water is pumped back down into the earth where it is heated again. The output power (work per unit time) of the plant is 84 000 kilowatts. Determine (a) the maximum efficiency at which this plant can operate and (b) the minimum amount of rejected heat that must be removed from the condenser every twenty-four hours.

*58. From a hot reservoir at a temperature of T_1, Carnot engine A takes an input heat of 5550 J, delivers 1750 J of work, and rejects heat to a cold reservoir that has a temperature of 503 K. This cold reservoir at 503 K also serves as the hot reservoir for engine B, which uses the rejected heat of the first engine as input heat. Engine B also delivers 1750 J of work, while rejecting heat to an even colder reservoir that has a temperature of T_2. Find the temperatures (a) T_1 and (b) T_2.

*59. Suppose the gasoline in a car engine burns at 631 °C, while the exhaust temperature (the temperature of the cold reservoir) is 139 °C and the outdoor temperature is 27 °C. Assume that the engine can be treated as a Carnot engine (a gross oversimplification). In an attempt to increase mileage performance, an inventor builds a second engine that functions between the exhaust and outdoor temperatures and uses the exhaust heat to produce additional work. Assume that the inventor's engine can also be treated as a Carnot engine. Determine the ratio of the total work produced by both engines to that produced by the first engine alone.

**60. The hot and cold reservoirs of a Carnot engine have temperatures of 845 and 395 K, respectively. The engine does the work of lifting a 15.0-kg block straight up from rest, so that at a height of 5.00 m the block has a speed of 8.50 m/s. How much heat must be put into the engine?

**61. ssm A nuclear-fueled electric power plant utilizes a so-called "boiling water reactor." In this type of reactor, nuclear energy causes water under pressure to boil at 285 °C (the temperature of the hot reservoir). After the steam does the work of turning the turbine of an electric generator, the steam is converted back into water in a condenser at 40 °C (the temperature of the cold reservoir). To keep the condenser at 40 °C, the rejected heat must be carried away by some means, for example, by water from a river. The plant operates at three-fourths of its Carnot efficiency, and the electrical output power of the plant is 1.2×10^9 watts. A river with a water flow rate of 1.0×10^5 kg/s is available to remove the rejected heat from the plant. Find the number of Celsius degrees by which the temperature of the river rises.

Section 15.10 Refrigerators, Air Conditioners, and Heat Pumps

62. What is the coefficient of performance of an air conditioner that uses 3500 J of electrical energy to remove 11 200 J of heat from a room?

63. A refrigerator does 2500 J of work in order to remove 8500 J of heat from the cold reservoir. (a) Determine the coefficient of performance of the refrigerator. (b) How much heat is deposited into the kitchen (the hot reservoir)?

64. The water in a deep underground well is used as the cold reservoir of a Carnot heat pump that maintains the temperature of a house at 301 K. To deposit 14 200 J of heat in the house, the heat pump requires 800 J of work. Determine the temperature of the well water.

65. ssm www The temperatures indoors and outdoors are 299 and 312 K, respectively. A Carnot air conditioner deposits 6.12×10^5 J of heat outdoors. How much heat is removed from the house?

66. The coefficient of performance of a refrigerator is 4.6. How much electrical energy is used in removing 4100 J of heat from the food inside?

67. A heat pump removes 2090 J of heat from the outdoors and delivers 3140 J of heat to the inside of a house. (a) How much work does the heat pump need? (b) What is the coefficient of performance of the heat pump?

68. A Carnot engine has an efficiency of 0.70. If this engine were run backward as a heat pump, what would be the coefficient of performance?

*69. ssm www A Carnot refrigerator transfers heat from its inside (6.0 °C) to the room air outside (20.0 °C). (a) Find the coefficient of performance of the refrigerator. (b) Determine the magnitude of the minimum work needed to cool 5.00 kg of water from 20.0 to 6.0 °C when it is placed in the refrigerator.

*70. An engine is run in reverse as a heat pump. An identical engine (with the same values of Q_H, Q_C, and W as the first engine) is run in reverse as a refrigerator. The coefficient of performance of the heat pump is three times greater than the coefficient of performance of the refrigerator. Obtain (a) the coefficient of performance of the refrigerator, (b) the coefficient of performance of the heat pump, and (c) the efficiency of the engine.

*71. How long would a 3.00-kW space heater have to run to put into a kitchen the same amount of heat as a refrigerator (coefficient of performance = 3.00) does when it freezes 1.50 kg of water at 20.0 °C into ice at 0.0 °C?

*72. Review Conceptual Example 9 before attempting this problem. A window air conditioner has an average coefficient of performance of 2.0. This unit has been placed on the floor by the bed, in a futile attempt to cool the bedroom. During this attempt 7.6×10^4 J of heat is pulled in the front of the unit. The room is sealed and contains 3800 mol of air. Assuming that the molar specific heat capacity of the air is $C_V = \frac{5}{2}R$, determine the rise in temperature caused by operating the air conditioner in this manner.

**73. ssm A Carnot engine uses hot and cold reservoirs that have temperatures of 1684 and 842 K, respectively. The input heat for this engine is Q_H. The work delivered by the engine is used to operate a Carnot heat pump. The pump removes heat from the 842-K reservoir and puts it into a hot reservoir at a temperature T'. The amount of heat removed from the 842-K reservoir is also Q_H. Find the temperature T'.

Section 15.11 Entropy and the Second Law of Thermodynamics

74. The inside of a house is at 20.0 °C and the outside is at -15 °C. The house loses 1.30×10^5 J of heat. Find $\Delta S_{universe}$, the change in entropy of the universe.

75. Four kilograms of carbon dioxide sublimes from solid "dry ice" to a gas at a pressure of one atmosphere and a temperature of

194.7 K. The latent heat of sublimation is 5.77×10^5 J/kg. Find the change in entropy of the carbon dioxide.

76. Heat Q flows spontaneously from a reservoir at 394 K into a reservoir that has a lower temperature T. Because of the spontaneous flow, thirty percent of Q is rendered unavailable for work when a Carnot engine operates between the reservoir at temperature T and a reservoir at 248 K. Find the temperature T.

77. ssm Find the change in entropy of the H_2O molecules when (a) three kilograms of ice melts into water at 273 K and (b) three kilograms of water changes into steam at 373 K. (c) On the basis of the answers to parts (a) and (b), discuss which change creates more disorder in the collection of H_2O molecules.

*78. (a) Find the equilibrium temperature that results when one kilogram of liquid water at 373 K is added to two kilograms of liquid water at 283 K in a perfectly insulated container. (b) When heat is added to or removed from a solid or liquid of mass m and specific heat capacity c, the change in entropy can be shown to be

$\Delta S = mc \ln(T_f/T_i)$, where T_i and T_f are the initial and final Kelvin temperatures. Use this equation to calculate the entropy change for each amount of water. Then combine the two entropy changes algebraically to obtain the total entropy change of the universe. Note that the process is irreversible, so the total entropy change of the universe is greater than zero. (c) Assuming that the coldest reservoir at hand has a temperature of 273 K, determine the amount of energy that becomes unavailable for doing work because of the irreversible process.

*79. (a) Five kilograms of water at 80.0 °C is mixed in a perfect thermos with 2.00 kg of ice at 0.0 °C, and the mixture is allowed to reach equilibrium. Using the expression $\Delta S = mc \ln(T_f/T_i)$ [see problem 78] and the change in entropy for melting, find the change in entropy that occurs. (b) Should the entropy of the universe increase or decrease as a result of the mixing process? Give your reasoning and state whether your answer in part (a) is consistent with your answer here.

ADDITIONAL PROBLEMS

80. Five moles of a monatomic gas expand adiabatically, and its temperature decreases from 370 to 290 K. Determine (a) the work done (including the algebraic sign) by the gas, and (b) the change in its internal energy.

81. ssm One-half mole of a monatomic ideal gas absorbs 1200 J of heat while 2500 J of work is done by the gas. (a) What is the temperature change of the gas? (b) Is the change an increase or a decrease?

82. The amount of heat that an engine rejects to the cold reservoir is three times the amount of work that it does. What is the efficiency of the engine?

83. The specific heat capacity of a material is 1100 J/(kg·C°). The temperature of 2.0 kg of this solid material is raised by 6.0 C°. Ignoring the work that corresponds to the small change in the volume of the material, determine the change in the internal energy of the material.

84. A Carnot refrigerator maintains the food inside it at 276 K, while the temperature of the kitchen is 298 K. The refrigerator removes 3.00×10^4 J of heat from the food. How much heat is delivered to the kitchen?

85. ssm A process occurs in which the entropy of a system increases by 125 J/K. During the process, the energy that becomes unavailable for doing work is zero. (a) Is this process reversible or irreversible? Give your reasoning. (b) Determine the change in the entropy of the surroundings.

86. The temperature of 2.5 mol of helium (a monatomic gas) is lowered by 35 K under conditions of constant volume. Assuming that helium behaves as an ideal gas, how much heat is removed from the gas?

87. A Carnot heat pump operates between an outdoor tempera-

ture of 265 K and an indoor temperature of 298 K. Find its coefficient of performance.

88. A Carnot engine operates between temperatures of 650 and 350 K. To improve the efficiency of the engine, it is decided either to raise the temperature of the hot reservoir by 40 K or to lower the temperature of the cold reservoir by 40 K. Which change gives the greatest improvement? Justify your answer by calculating the efficiency in each case.

89. ssm The volume of a gas is changed along the curved line between A and B in the drawing. Do not assume that the curved line is an isotherm or that the gas is ideal. (a) Find the magnitude of the work for the process, and (b) determine whether the work is positive or negative.

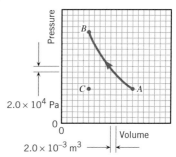

*90. Refer to the drawing in problem 89, where the curve between A and B is now an isotherm. An ideal gas begins at A and is changed along the horizontal line from A to C and then along the vertical line from C to B. (a) Find the heat for the process ACB and (b) determine whether it flows into or out of the gas.

*91. Two kilograms of liquid water at 0 °C is put into the freezer compartment of a Carnot refrigerator. The temperature of the compartment is −15 °C, and the temperature of the kitchen is 27 °C. If the cost of electrical energy is ten cents per kilowatt · hour, how much does it cost to make two kilograms of ice at 0 °C?

*92. Refer to the drawing that accompanies problem 10. When a system changes from A to B along the path shown on the pressure-versus-volume graph, it gains 2700 J of heat. What is the change in the internal energy of the system?

*93. **ssm** A monatomic ideal gas ($\gamma = \frac{5}{3}$) is contained within a perfectly insulated cylinder that is fitted with a movable piston. The initial pressure of the gas is 1.50×10^5 Pa. The piston is pushed so as to compress the gas, with the result that the Kelvin temperature doubles. What is the final pressure of the gas?

*94. Suppose a monatomic ideal gas is contained within a vertical cylinder that is fitted with a movable piston. The piston is frictionless and has a negligible mass. The area of the piston is 3.14×10^{-2} m^2, and the pressure outside the cylinder is 1.01×10^5 Pa. Heat (2093 J) is removed from the gas. Through what distance does the piston drop?

*95. 🔊 Even at rest, the human body generates heat. The heat arises because of the body's metabolism, that is, the chemical reactions that are always occurring in the body to generate energy. In rooms designed for use by large groups, adequate ventilation or air conditioning must be provided to remove this heat. Consider a classroom containing 200 students. Assume that the metabolic rate of generating heat is 130 W for each student and that the heat accumulates during a fifty-minute lecture. In addition, assume that the air has a molar specific heat of $C_V = \frac{5}{2}R$ and that the room (volume = 1200 m^3, initial pressure = 1.01×10^5 Pa, and initial temperature = 21 °C) is sealed shut. If all the heat generated by the students were absorbed by the air, by how much would the air temperature rise during a lecture?

*96. The hot reservoir for a Carnot engine has a temperature of 890 K, while the cold reservoir has a temperature of 670 K. The heat input for this engine is 4800 J. The 670-K reservoir also serves as the hot reservoir for a second Carnot engine. This second engine uses the rejected heat of the first engine as input and extracts additional work from it. The rejected heat from the second engine goes into a reservoir that has a temperature of 420 K. Find the total work delivered by the two engines.

97. **ssm An engine has an efficiency e_1. The engine takes input heat Q_H from a hot reservoir and delivers work W_1. The heat rejected by this engine is used as input heat for a second engine, which has an efficiency e_2 and delivers work W_2. The overall efficiency of this two-engine device is the total work delivered ($W_1 + W_2$) divided by the input heat Q_H. Find an expression for the overall efficiency e in terms of e_1 and e_2.

**98. The drawing shows an adiabatically isolated cylinder that is divided initially into two identical parts by an adiabatic partition. Both sides contain one mole of a monatomic ideal gas ($\gamma = \frac{5}{3}$), with the initial temperature being 525 K on the left and 275 K on the right. The partition is then allowed to move slowly (i.e., quasi-statically) to the right, until the pressures on each side of the partition are the same. Find the final temperatures on the (a) left and (b) right.

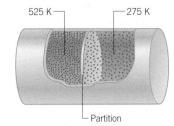

525 K   275 K

Partition

WAVES AND SOUND

**Dolphins communicate with each other by transmitting audible "squeaks"
and "clicks" through the water. They also gain information about their environment
by monitoring the echoes of the sounds that reflect from objects.**

16.1 THE NATURE OF WAVES

Water waves have two features common to all waves:

1. A wave is a traveling disturbance.
2. A wave carries energy from place to place.

In Figure 16.1 the wave created by the motorboat travels across the lake and disturbs the fisherman. It should be noted that *there is no bulk flow of water* outward from the motorboat. The wave is not a bulk movement of water such as a river, but, rather, a disturbance traveling on the surface of the lake. Part of the wave's energy in Figure 16.1 is transferred to the fisherman and his boat.

We will consider two basic types of waves, transverse and longitudinal. Figure 16.2 illustrates how a transverse wave can be generated using a Slinky, a remarkable toy that is a long, loosely coiled spring. If one end of the Slinky is jerked up and down, as in part *a*, an upward pulse is sent traveling toward the right. If the end is then jerked down and up, as in part *b*, a downward pulse is generated and also moves to the right. If the end is continually moved up and down in simple harmonic motion, an entire wave is produced. As part *c* illustrates, the wave consists of a series of alternating upward and downward sections that propagate to the right, disturbing the vertical position of the Slinky in the process. To focus attention on the disturbance, a colored dot is attached to the Slinky in part *c* of the drawing. As the wave advances, the dot is displaced up and down in simple harmonic motion. The motion of the dot occurs perpendicular, or transverse, to the direction in which the wave travels. This example shows that ***a transverse wave is one in which the disturbance is perpendicular to the direction of travel of the wave.*** Radio waves, light waves, and microwaves are transverse waves. Transverse waves also travel on the strings of instruments such as guitars and banjos.

A longitudinal wave can also be generated with a Slinky, and Figure 16.3 demonstrates how. When one end of the Slinky is pushed forward along its length (i.e., longitudinally) and then returned to its starting point, as in part *a*, a region where the coils are squeezed together or compressed is sent traveling to the right. If the end is pulled backward and then returned to its starting point, as in part *b*, a region where the coils are pulled apart or stretched is formed and also moves to the right. If the end is continually moved back and forth in simple harmonic motion, an

Figure 16.1 The wave created by the motorboat travels across the lake and disturbs the fisherman.

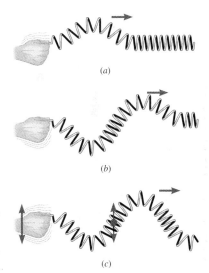

Figure 16.2 (*a*) An upward pulse moves to the right, followed by (*b*) a downward pulse. (*c*) When the end of the Slinky is moved up and down continuously, a transverse wave is produced.

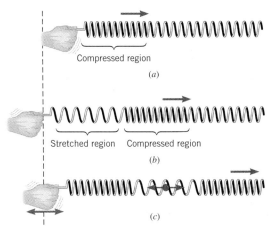

Compressed region

(*a*)

Stretched region   Compressed region

(*b*)

(*c*)

Figure 16.3 (*a*) A compressed region moves to the right, followed by (*b*) a stretched region. (*c*) When the end of the Slinky is moved back and forth continuously, a longitudinal wave is produced.

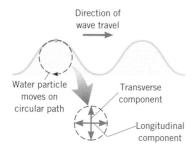

Figure 16.4 A water wave is neither transverse nor longitudinal, since water particles at the surface move clockwise on nearly circular paths as the wave moves from left to right.

entire wave is created. As part *c* shows, the wave consists of a series of alternating compressed and stretched regions that travel to the right and disturb the separation between adjacent coils. A colored dot is once again attached to the Slinky to emphasize the vibratory nature of the disturbance. In response to the wave, the dot moves back and forth in simple harmonic motion along the line of travel of the wave. Thus, *a longitudinal wave is one in which the disturbance is parallel to the line of travel of the wave.* A sound wave is a longitudinal wave.

Some waves are neither transverse nor longitudinal. For instance, in a water wave the motion of the water particles is not strictly perpendicular or strictly parallel to the line along which the wave travels. Instead, the motion includes both transverse and longitudinal components, since the water particles at the surface move on nearly circular paths, as Figure 16.4 indicates.

16.2 PERIODIC WAVES

The transverse and longitudinal waves that we have been discussing are called *periodic waves,* because they consist of patterns that are produced over and over again by the source. In Figures 16.2 and 16.3 the repetitive patterns occur as a result of the simple harmonic motion of the left end of the Slinky. Every segment of the Slinky vibrates in simple harmonic motion. Therefore, some of the terminology (cycle, amplitude, period, and frequency) used to describe periodic waves is the same as that given in Chapter 10 for simple harmonic motion, as the concept chart in Figure 16.5 emphasizes. Figure 16.6 uses a graphical representation of a transverse wave on a Slinky to introduce this terminology. One *cycle* of the wave is shaded in color in both parts of the drawing. A wave is a series of many cycles. In part *a* the vertical position of the Slinky is plotted on the vertical axis, while the corresponding distance along the length of the Slinky is plotted on the horizontal axis. Such a graph is equivalent to a photograph of the wave taken at one instant in time and shows the disturbance that exists at each point along the Slinky's length. As marked on this graph, the *amplitude A* is the maximum excursion of a particle

Figure 16.5 *Concepts at a Glance* For periodic waves, the terms cycle, amplitude, period, and frequency have the same meaning as they do in simple harmonic motion. Sound, such as that produced by this elegant pipe organ in Tokyo, Japan, is a periodic wave and is described using this terminology.

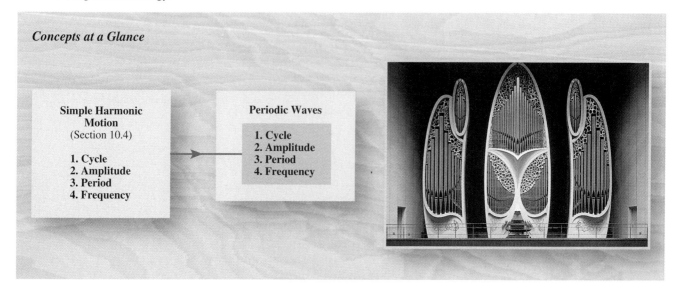

Concepts at a Glance

Simple Harmonic Motion
(Section 10.4)

1. Cycle
2. Amplitude
3. Period
4. Frequency

Periodic Waves

1. Cycle
2. Amplitude
3. Period
4. Frequency

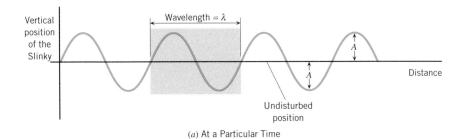

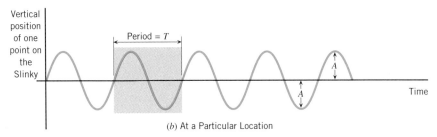

Figure 16.6 In parts *a* and *b*, one cycle of the wave is shaded in color, and the amplitude of the wave is denoted as *A*.

of the medium from the particle's undisturbed position. The amplitude is the distance between a crest, or highest point on the wave pattern, and the undisturbed position; it is also the distance between a trough, or lowest point on the wave pattern, and the undisturbed position. The *wavelength* λ is the horizontal length of one cycle of the wave, as shown in Figure 16.6*a*. The wavelength is also the horizontal distance between two successive crests, two successive troughs, or any two successive equivalent points on the wave.

Part *b* of Figure 16.6 shows a graph in which time, rather than distance, is plotted on the horizontal axis. This graph is obtained by observing a single point on the Slinky. As the wave passes, the point under observation oscillates up and down in simple harmonic motion. As indicated on the graph, the *period T* is the time required for one complete up/down cycle, just as it is for an object on a spring. Equivalently, the period is the time required for the wave to travel a distance of one wavelength. The period *T* is related to the *frequency f*, just as it is for any example of simple harmonic motion:

$$f = \frac{1}{T} \tag{10.9}$$

The period is commonly measured in seconds, and frequency is measured in cycles per second or hertz (Hz). If, for instance, one cycle of a wave takes one-tenth of a second to pass an observer, then ten cycles pass per second, as Equation 10.9 indicates [$f = 1/(0.1\text{ s}) = 10\text{ cycles/s} = 10$ Hz].

A simple relation exists between the period, the wavelength, and the speed of a wave, a relation that Figure 16.7 helps to introduce. Imagine waiting at a railroad crossing, while a freight train moves by at a constant speed *v*. The train consists of a long line of identical boxcars, each of which has a length λ and requires a time *T* to pass, so the speed is $v = λ/T$. This same equation applies for a wave and relates the speed of the wave to the wavelength λ and the period *T*. Since the frequency of

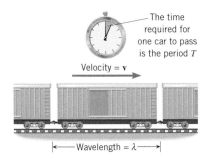

Figure 16.7 A train moving at a constant speed serves as an analogy for a traveling wave.

a wave is $f = 1/T$, the expression for the speed is

$$v = \frac{\lambda}{T} = f\lambda \qquad (16.1)$$

The terminology just discussed and the fundamental relations $f = 1/T$ and $v = f\lambda$ apply to longitudinal as well as to transverse waves. Example 1 illustrates how the wavelength of a wave is determined by the wave speed and the frequency established by the source.

EXAMPLE 1 • **The Wavelengths of Radio Waves**

AM and FM radio waves are transverse waves that consist of electric and magnetic disturbances. These waves travel at a speed of 3.00×10^8 m/s. A station broadcasts an AM radio wave whose frequency is 1230×10^3 Hz (1230 kHz on the dial) and an FM radio wave whose frequency is 91.9×10^6 Hz (91.9 MHz on the dial). Find the distance between adjacent crests in each wave.

Reasoning The distance between adjacent crests is the wavelength λ. Since the speed of each wave is $v = 3.00 \times 10^8$ m/s and the frequencies are known, the relation $v = f\lambda$ can be used to determine the wavelengths.

Solution

AM
$$\lambda = \frac{v}{f} = \frac{3.00 \times 10^8 \text{ m/s}}{1230 \times 10^3 \text{ Hz}} = \boxed{244 \text{ m}}$$

FM
$$\lambda = \frac{v}{f} = \frac{3.00 \times 10^8 \text{ m/s}}{91.9 \times 10^6 \text{ Hz}} = \boxed{3.26 \text{ m}}$$

Notice that the wavelength of an AM radio wave is longer than two and one-half football fields!

> • **PROBLEM SOLVING INSIGHT**
> The equation $v = f\lambda$ applies to any kind of periodic wave.

16.3 THE SPEED OF A WAVE ON A STRING

The properties of the material* or medium through which a wave travels determine the speed of the wave. For example, Figure 16.8 shows a transverse wave on a string and draws attention to four string particles that have been drawn as colored dots. As the wave moves to the right, each particle is displaced, one after the other, from its undisturbed position. In the drawing, particles 1 and 2 have already been displaced upward, while particles 3 and 4 are not yet affected by the wave. Particle 3 will be next to move, because the section of string immediately to its left (i.e., particle 2) will pull it upward.

Figure 16.8 leads us to conclude that the speed with which the wave moves to the right depends on how quickly one particle of the string is accelerated upward in response to the net pulling force exerted by its adjacent neighbors. In accord with Newton's second law, a stronger net force results in a greater acceleration, and,

---

* Electromagnetic waves can move through a vacuum, as well as through materials such as glass and water.

Figure 16.8 As a transverse wave moves to the right with speed v, each string particle is displaced, one after the other, from its undisturbed position.

thus, a faster-moving wave. The ability of one particle to pull on its neighbors depends on how tightly the string is stretched, that is, on the tension (see Section 4.10 for a review of tension). The greater the tension, the greater the pulling force the particles exert on each other, and the faster the wave travels, other things being equal. Along with the tension, a second factor influences the wave speed. According to Newton's second law, the inertia or mass of particle 3 in Figure 16.8 also affects how quickly it responds to the upward pull of particle 2. For a given net pulling force, a smaller mass has a greater acceleration than a larger mass. Therefore, other things being equal, a wave travels faster on a string whose particles have a small mass, or, as it turns out, on a string that has a small mass per unit length. The mass per unit length is called the *linear density* of the string. The effects of the tension F and the mass per unit length m/L are evident in the following expression for the speed v of a small-amplitude wave on a string:

$$v = \sqrt{\frac{F}{m/L}} \qquad (16.2)$$

The motion of transverse waves along a string is important in the operation of musical instruments, such as the guitar, the violin, and the piano. In these instruments, the strings are either plucked, bowed, or struck to produce transverse waves. Example 2 discusses the speed of the waves on the strings of a guitar.

EXAMPLE 2 • Waves Traveling on Guitar Strings

Transverse waves travel on the strings of an electric guitar after the strings are plucked. (See Figure 16.9.) The length of each string between its two fixed ends is 0.628 m, and the mass is 0.208 g for the highest pitched E string and 3.32 g for the lowest pitched E string. Each string is under a tension of 226 N. Find the speeds of the waves on the two strings.

Reasoning and Solution The tension F, the mass m, and the length L are known for each string, and the speeds of the waves are given by Equation 16.2:

High-pitched E $\quad v = \sqrt{\frac{F}{m/L}} = \sqrt{\frac{226\ \text{N}}{(0.208 \times 10^{-3}\ \text{kg})/(0.628\ \text{m})}} = \boxed{826\ \text{m/s}}$

Low-pitched E $\quad v = \sqrt{\frac{F}{m/L}} = \sqrt{\frac{226\ \text{N}}{(3.32 \times 10^{-3}\ \text{kg})/(0.628\ \text{m})}} = \boxed{207\ \text{m/s}}$

Notice how fast the waves move; the speeds correspond to 1850 mi/h and 463 mi/h.

Transverse vibration of the string

Figure 16.9 Transverse waves are generated on a guitar string by plucking it.

The Physics of... waves on guitar strings.

Conceptual Example 3 offers additional insight into the nature of a wave as a traveling disturbance.

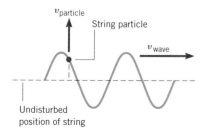

Figure 16.10 A transverse wave on a string is moving to the right with a speed v_{wave}. A string particle moves up and down in simple harmonic motion about the undisturbed position of the string. The speed of the particle $v_{particle}$ changes from moment to moment as the wave passes.

CONCEPTUAL EXAMPLE 3 • Wave Speed Versus Particle Speed

Is the speed of a transverse wave on a string the same as the speed at which a particle on the string moves (see Figure 16.10)?

Reasoning and Solution The speed v_{wave} of a transverse wave on a string specifies how fast the disturbance travels along the string. According to Equation 16.2, this speed is determined by the tension F and the linear density m/L of the string. Thus, the speed of the wave depends solely on the *properties of the string,* and, if these properties do not change, the speed remains constant.

The particle speed $v_{particle}$ specifies how fast the particle is moving as it oscillates up and down, and it is different than the wave speed. If the source of the wave (e.g., the hand in Figure 16.2c) vibrates in simple harmonic motion, each string particle vibrates in a like manner, with the same amplitude and frequency as the source. Moreover, the particle speed, unlike the wave speed, is not constant. As for any object in simple harmonic motion, the particle speed is greatest when the particle is passing through the undisturbed position of the string and zero when the particle is at its maximum displacement. The speed depends on the amplitude and frequency of the motion, as expressed by Equation 10.11. Thus, the speed of a string particle is determined by the *properties of the source* creating the wave and not on the properties of the string itself. In contrast, the speed of the wave is determined by the properties of the string. We see, then, that *the two speeds, v_{wave} and $v_{particle}$, are not the same.*

Related Homework Material: *Question 4, Problems 20 and 104*

*16.4 THE MATHEMATICAL DESCRIPTION OF A WAVE

When a wave travels through a medium, it displaces the particles of the medium from their undisturbed positions. Suppose a particle is located at a distance x from a coordinate origin. We would like to know the displacement y of this particle from its undisturbed position at any time t as the wave passes. For periodic waves that result from simple harmonic motion of the source, the expression for the displacement involves a sine or cosine, a fact that is not surprising. After all, in Chapter 10 simple harmonic motion is described using sinusoidal equations, and the graphs for a wave in Figure 16.6 look like a plot of displacement versus time for an object oscillating on a spring (see Figure 10.14).

Our tack will be to present the expression for the displacement and then show graphically that it gives a correct description. Equation 16.3 represents the displacement of a particle caused by a wave traveling in the $+x$ direction (to the right) that has an amplitude A, frequency f, and wavelength λ. Equation 16.4 applies to a wave moving in the $-x$ direction (to the left).

Wave motion toward $+x$
$$y = A \sin\left(2\pi f t - \frac{2\pi x}{\lambda}\right) \tag{16.3}$$

Wave motion toward $-x$
$$y = A \sin\left(2\pi f t + \frac{2\pi x}{\lambda}\right) \tag{16.4}$$

These equations apply to transverse or longitudinal waves and assume that $y = 0$ when $x = 0$ and $t = 0$.

Consider a transverse wave moving in the $+x$ direction along a string. The term

$(2\pi ft - 2\pi x/\lambda)$ in Equation 16.3 is called the *phase angle* of the wave. A string particle located at the origin ($x = 0$) exhibits simple harmonic motion with a phase angle of $2\pi ft$; that is, its displacement as a function of time is $y = A \sin (2\pi ft)$. A particle located at a distance x also exhibits simple harmonic motion, but its phase angle is

$$2\pi ft - \frac{2\pi x}{\lambda} = 2\pi f\left(t - \frac{x}{f\lambda}\right) = 2\pi f\left(t - \frac{x}{v}\right)$$

The quantity x/v is the time needed for the wave to travel the distance x. In other words, the simple harmonic motion that occurs at x is delayed by the time interval x/v compared to the motion at the origin.

Figure 16.11 shows the displacement y plotted as a function of position x along the string at a series of time intervals separated by one-fourth of the period $T(t = 0, \frac{1}{4}T, \frac{2}{4}T, \frac{3}{4}T, T)$. These graphs are constructed by substituting the corresponding value for t into Equation 16.3, remembering that $f = 1/T$, and then calculating y at a series of values for x. The graphs are like photographs taken at various times as the wave moves to the right. For reference, the colored square on each graph marks the place on the wave that is located at $x = 0$ when $t = 0$. As time passes, the colored square moves to the right, along with the wave. It should be noted that the phase angle ($2\pi ft - 2\pi x/\lambda$) is measured in *radians,* not degrees. Therefore, **when using a calculator to evaluate the function sin ($2\pi ft - 2\pi x/\lambda$), the calculator must be set to its radian mode.** In a similar manner, it can be shown that Equation 16.4 represents a wave moving in the $-x$ direction.

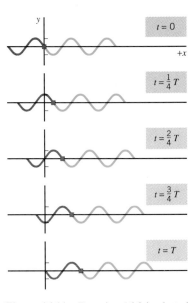

Figure 16.11 Equation 16.3 is plotted here at a series of times separated by one-fourth of the period T. The colored square in each of the graphs marks the place on the wave that is located at $x = 0$ when $t = 0$. As time passes, the wave moves to the right.

16.5 THE NATURE OF SOUND

LONGITUDINAL SOUND WAVES

Sound is a longitudinal wave that is created by a vibrating object, such as a guitar string, the human vocal cords, or the diaphragm of a loudspeaker. Moreover, sound can be created or transmitted only in a medium, such as a gas, liquid, or solid. As we will see, the particles of the medium must be present for the disturbance of the wave to move from place to place. Sound cannot exist in a vacuum.

To see how sound waves are produced and why they are longitudinal, consider the vibrating diaphragm of a loudspeaker. When the diaphragm moves outward, it compresses the air directly in front of it, as in Figure 16.12a. This compression causes the air pressure to rise slightly. The region of increased pressure is called a ***condensation*** and travels away from the speaker at the speed of sound. The conden-

The Physics of...
how a loudspeaker diaphragm produces sound.

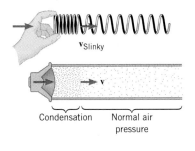

Condensation   Normal air pressure

v_{Slinky}

(a)

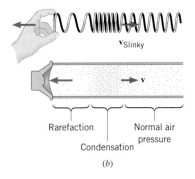

v_{Slinky}

Rarefaction   Normal air pressure

Condensation

(b)

Figure 16.12 (a) When the speaker diaphragm moves outward, it creates a condensation. (b) When the diaphragm moves inward, it creates a rarefaction. The condensation and rarefaction on the Slinky are included for comparison. In reality, the velocity of the wave on the Slinky v_{Slinky} is much smaller than the velocity of sound in air v. For simplicity, the two waves are shown here to have the same velocity.

Figure 16.13 Both the wave on the Slinky and the sound wave are longitudinal waves. The colored dots attached to the Slinky and to an air molecule vibrate back and forth parallel to the line of travel of the wave.

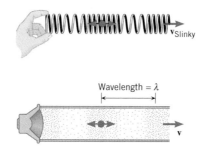

Wavelength = λ

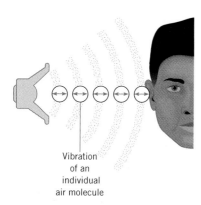

Vibration
of an
individual
air molecule

Figure 16.14 Although the condensations and rarefactions travel from the speaker to the listener, the individual air molecules do not move with the wave. A given molecule vibrates back and forth about a fixed location.

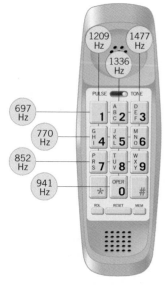

Figure 16.15 A push-button telephone and a schematic showing the two pure tones produced when each button is pressed.

sation is analogous to the compressed region of coils in a longitudinal wave on a Slinky, which is included in Figure 16.12*a* for comparison. After producing a condensation, the diaphragm reverses its motion and moves inward, as in part *b* of the drawing. The inward motion produces a region known as a ***rarefaction,*** where the air pressure is slightly less than normal. The rarefaction is similar to the stretched region of coils in a longitudinal Slinky wave. Following immediately behind the condensation, the rarefaction also travels away from the speaker at the speed of sound. Figure 16.13 further emphasizes the similarity between a sound wave and a longitudinal Slinky wave. As the wave passes, the colored dots attached both to the Slinky and to an air molecule execute simple harmonic motion about their undisturbed positions. The colored arrows on either side of the dots indicate that the simple harmonic motion occurs parallel to the line of travel. The drawing also shows that the wavelength λ is the distance between the centers of two successive condensations; λ is also the distance between the centers of two successive rarefactions.

Figure 16.14 illustrates a sound wave spreading out in space after being produced by a loudspeaker. When the condensations and rarefactions arrive at the ear, they force the eardrum to vibrate at the same frequency as the speaker diaphragm. The vibratory motion of the eardrum is interpreted by the brain as sound. It should be emphasized that sound is not a mass movement of air, like the wind. As the condensations and rarefactions of the sound wave travel outward from the vibrating diaphragm in Figure 16.14, for example, the individual air molecules are not carried along with the wave. Rather, each molecule executes simple harmonic motion about a fixed location. In doing so, one molecule collides with its neighbor and passes the condensations and rarefactions forward. The neighbor, in turn, repeats the process.

THE FREQUENCY OF A SOUND WAVE

Each cycle of a sound wave includes one condensation and one rarefaction, and the ***frequency*** is the number of cycles per second that passes by a given location. For example, if the diaphragm of a speaker vibrates back and forth in simple harmonic motion at a frequency of 1000 Hz, then 1000 condensations, each followed by a rarefaction, are generated every second, thus forming a sound wave whose frequency is also 1000 Hz. A sound with a single frequency is called a ***pure tone.*** Experiments have shown that a healthy young person hears all sound frequencies from approximately 20 to 20 000 Hz (20 kHz). The ability to hear the high frequencies decreases with age, however, and a normal middle-aged adult hears frequencies only up to 12–14 kHz.

Pure tones are used in push-button telephones, such as that shown in Figure 16.15. These phones simultaneously produce two pure tones when each button is pressed, a different pair of tones for each different button. The tones are transmitted electronically to the central telephone office, where they activate switching circuits that complete the call. For example, the drawing indicates that pressing the "5" button produces pure tones of 770 and 1336 Hz simultaneously, while the "9" button generates tones of 852 and 1477 Hz.

Sound can be generated whose frequency lies below 20 Hz or above 20 kHz, although humans normally do not hear it. Sound waves with frequencies below 20 Hz are said to be *infrasonic,* while those with frequencies above 20 kHz are referred to as *ultrasonic.* Rhinoceroses use infrasonic frequencies as low as 5 Hz to call one another (Figure 16.16), while bats use ultrasonic frequencies up to 100 kHz for locating their food sources and navigating (Figure 16.17).

Frequency is an objective property of a sound wave, because frequency can be measured with an electronic frequency counter. A listener's perception of frequency, however, is subjective. The brain interprets the frequency detected by the ear primarily in terms of the subjective quality called *pitch.* A pure tone with a large (high) frequency is interpreted as a high-pitched sound, while a pure tone with a small (low) frequency is interpreted as a low-pitched sound. A piccolo produces high-pitched sounds, and a tuba produces low-pitched sounds.

THE PRESSURE AMPLITUDE OF A SOUND WAVE

Figure 16.18 illustrates a pure-tone sound wave traveling in a tube. Attached to the tube is a series of gauges that indicate the pressure variations along the wave. The graph shows that the air pressure varies sinusoidally along the length of the tube. Although this graph has the appearance of a transverse wave, remember that the sound itself is a longitudinal wave. The graph also shows the *pressure amplitude* of the wave, which is the magnitude of the maximum change in pressure, measured relative to the undisturbed or atmospheric pressure. The pressure fluctuations in a sound wave are normally very small. For instance, in a typical conversation between two people the pressure amplitude is about 3×10^{-2} Pa, certainly a small amount compared with the atmospheric pressure of $1.01 \times 10^{+5}$ Pa. The ear is remarkable in being able to detect such small changes.

Loudness is an attribute of sound that depends primarily on the amplitude of the wave: the larger the amplitude, the louder the sound. The pressure amplitude is an objective property of a sound wave, since it can be measured. Loudness, on the other hand, is subjective. Each individual determines what is loud, depending on the acuteness of his or her hearing.

The Physics of...
push-button telephones.

Figure 16.16   Rhinoceroses call to one another using infrasonic sound waves.

Figure 16.17   Bats use ultrasonic sound waves for navigating and locating food sources.

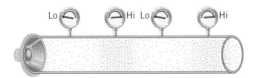

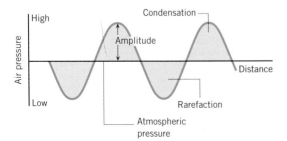

Figure 16.18   A sound wave is a series of alternating condensations and rarefactions. The graph shows that the condensations are regions of higher-than-normal air pressure, and the rarefactions are regions of lower-than-normal air pressure.

16.6 THE SPEED OF SOUND

GASES

Table 16.1 Speed of Sound in Gases, Liquids, and Solids

Substance	Speed (m/s)
Gases	
Air (0 °C)	331
Air (20 °C)	343
Carbon dioxide (0 °C)	259
Oxygen (0 °C)	316
Helium (0 °C)	965
Liquids	
Chloroform (20 °C)	1004
Ethanol (20 °C)	1162
Mercury (20 °C)	1450
Fresh water (20 °C)	1482
Solids	
Copper	5010
Glass (Pyrex)	5640
Lead	1960
Steel	5960

Sound travels through gases, liquids, and solids at considerably different speeds, as Table 16.1 reveals. Near room temperature, the speed of sound in air is 343 m/s (767 mi/h) and is markedly greater in liquids and solids. For example, sound travels more than four times faster in water and more than seventeen times faster in steel than it does in air. In general, sound travels slowest in gases, faster in liquids, and fastest in solids.

Like the speed of a wave on a guitar string, the speed of sound depends on the properties of the medium. In a gas, it is only when molecules collide that the condensations and rarefactions of a sound wave can move from place to place. It is reasonable, then, to expect the speed of sound in a gas to have the same order of magnitude as the average molecular speed between collisions. For an ideal gas this average speed is the translational rms-speed given by Equation 14.6: $v_{rms} = \sqrt{3kT/m}$, where T is the Kelvin temperature, m is the mass of a molecule, and k is Boltzmann's constant. Although the expression for v_{rms} overestimates the speed of sound, it does give the correct dependence on Kelvin temperature and particle mass. Careful analysis shows that the speed of sound in an ideal gas is given by

$$\textit{Ideal gas} \qquad v = \sqrt{\frac{\gamma kT}{m}} \qquad (16.5)$$

where $\gamma = c_P/c_V$ is the ratio of the specific heat capacity at constant pressure c_P to the specific heat capacity at constant volume c_V.

The factor γ is introduced in Section 15.5, where the adiabatic compression and expansion of an ideal gas is discussed. It appears in Equation 16.5 because the condensations and rarefactions of a sound wave are formed by adiabatic compressions and expansions of the gas. The regions that are compressed (the condensations) become slightly warmed, and the regions that are expanded (the rarefactions) become slightly cooled. However, no appreciable heat flows from a condensation to an adjacent rarefaction, because the distance between the two (half a wavelength) is relatively large for most audible sound waves and a gas is a poor thermal conductor. Thus, the compression and expansion process is adiabatic. Example 4 illustrates the use of Equation 16.5.

The Physics of...
an ultrasonic ruler.

EXAMPLE 4 • An Ultrasonic Ruler

Figure 16.19 shows an ultrasonic ruler that is used to measure the distance between itself and a target, such as a wall. To initiate the measurement, the ruler generates a pulse of ultrasonic sound that travels to the wall and, like an echo, reflects from it. The reflected pulse returns to the ruler, which measures the time it takes for the round-trip. Using a preset value for the speed of sound, the unit determines the distance to the wall and displays it on a digital readout. Suppose the round-trip travel time is 20.0×10^{-3} s on a day when the air temperature is 296 K (23 °C). Assuming that air is an ideal gas for which $\gamma = 1.40$ and that the average molecular mass of an air molecule is 28.9 u, find the distance x to the wall.

Reasoning The distance between the ruler and the wall is $x = vt$, where v is the speed of sound and t is the time for the sound pulse to reach the wall. The time t is one-half the round-trip time, so $t = 10.0 \times 10^{-3}$ s. The speed of sound in air can be

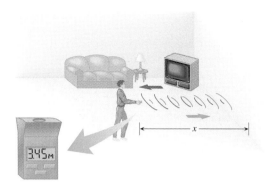

Figure 16.19 An ultrasonic ruler uses sound with a frequency greater than 20 kHz to measure the distance x to the wall. The blue arcs and blue arrow denote the outgoing sound wave, while the red arcs and red arrow denote the wave reflected from the wall.

obtained directly from Equation 16.5, provided that the mass of an air molecule can be found. The mass of an air molecule is the average molecular mass of air (expressed in kilograms) divided by Avogadro's number N_A (see Section 14.1).

Solution The mass of an air molecule is the mass of one mole of air, expressed in kilograms, divided by Avogadro's number, N_A:

$$m = \frac{28.9 \times 10^{-3}\,\text{kg/mol}}{N_A} = \frac{28.9 \times 10^{-3}\,\text{kg/mol}}{6.022 \times 10^{23}\,\text{mol}^{-1}} = 4.80 \times 10^{-26}\,\text{kg}$$

For the speed of sound, we find

$$v = \sqrt{\frac{\gamma k T}{m}} = \sqrt{\frac{(1.40)(1.38 \times 10^{-23}\,\text{J/K})(296\,\text{K})}{4.80 \times 10^{-26}\,\text{kg}}} = 345\,\text{m/s} \quad (16.5)$$

The distance to the wall is

$$x = vt = (345\,\text{m/s})(10.0 \times 10^{-3}\,\text{s}) = \boxed{3.45\,\text{m}}$$

• **PROBLEM SOLVING INSIGHT**
When using the equation $v = \sqrt{\gamma k T/m}$ to calculate the speed of sound in an ideal gas, be sure to express the temperature T in kelvins and not in degrees Celsius or Fahrenheit.

Conceptual Example 5 illustrates how the speed of sound in air can be used to estimate the distance to a thunderstorm, using a handy rule of thumb.

CONCEPTUAL EXAMPLE 5 • Thunder, Lightning, and a Rule of Thumb

There is a rule of thumb for estimating how far away a thunderstorm is. After you see a flash of lightning, count off the seconds until the thunder is heard. Divide the number of seconds by five. The result gives the approximate distance (in miles) to the thunderstorm. Why does this rule work?

Reasoning and Solution Figure 16.20 shows a lightning bolt from a thunderstorm and a person who is standing one mile (1.6×10^3 m) away. When the lightning oc-

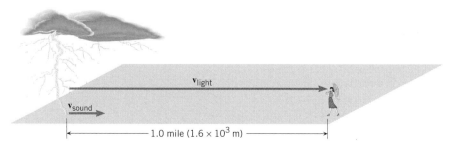

Figure 16.20 A lightning bolt from a thunderstorm generates a flash of light and sound (thunder). The speed of light is much greater than the speed of sound. Therefore, the light reaches the person first, followed about 5 seconds later by the sound.

curs, light and sound (thunder) are produced very nearly at the same instant. Light travels so rapidly ($v_{\text{light}} = 3.0 \times 10^8$ m/s) that it reaches the observer almost instantaneously. Its travel time is only $(1.6 \times 10^3 \text{ m})/(3.0 \times 10^8 \text{ m/s}) = 5 \times 10^{-6}$ s. In comparison, sound travels very slowly ($v_{\text{sound}} = 343$ m/s). The time for the thunder to reach the person is $(1.6 \times 10^3 \text{ m})/(343 \text{ m/s}) = 5$ s. Thus, the time interval between seeing the flash and hearing the thunder is about 5 seconds for every mile of travel. ***This rule of thumb works because the speed of light is so much greater than the speed of sound*** that the time needed for the light to reach the observer is negligible compared to the time needed for the sound.

Related Homework Material: *Problem 42*

LIQUIDS

In a liquid, the speed of sound depends on the density ρ and the *adiabatic* bulk modulus B_{ad} of the liquid:

Liquid
$$v = \sqrt{\frac{B_{\text{ad}}}{\rho}} \qquad (16.6)$$

The bulk modulus is introduced in Section 10.1 in a discussion of the volume deformation of liquids and solids. There it is tacitly assumed that the temperature remains constant while the volume of the material changes; i.e., the compression or expansion is isothermal. However, the condensations and rarefactions in a sound wave occur under *adiabatic* rather than isothermal conditions. Thus, the adiabatic bulk modulus B_{ad} must be used when calculating the speed of sound in liquids. Values of B_{ad} will be provided as needed in this text. The next example emphasizes that sound travels much faster in a liquid than in a gas.

EXAMPLE 6 • An Ultrasonic Ruler, Revisited

Example 4 discusses an ultrasonic ruler that measures the distance between two points using a pulse of ultrasonic sound. The ruler is calibrated to measure distances in air. Will it correctly measure distances under the ocean, where the adiabatic bulk modulus and density of seawater are $B_{\text{ad}} = 2.31 \times 10^9$ Pa and $\rho = 1025$ kg/m^3? If not, determine the factor by which it is in error.

Reasoning Since the ultrasonic sound travels much faster in seawater than it does in air, the reflected pulse returns in a much shorter time. This quicker return time fools the ultrasonic ruler into believing the object is much closer than it actually is. Therefore, the ruler will not measure underwater distances correctly. The error depends on the speed of sound in seawater.

Solution The speed of sound in seawater can be determined from Equation 16.6:

$$v = \sqrt{\frac{B_{\text{ad}}}{\rho}} = \sqrt{\frac{2.31 \times 10^9 \text{ Pa}}{1025 \text{ kg/m}^3}} = 1500 \text{ m/s}$$

In comparison, the speed of sound in air is about 343 m/s. Thus, the ultrasonic ruler calculates erroneously that the object is $(1500 \text{ m/s})/(343 \text{ m/s}) = \boxed{4.4}$ times closer underwater than it actually is.

SOLID BARS

When sound travels through a long slender solid bar, the speed of the sound depends on the properties of the medium according to

Long slender solid bar
$$v = \sqrt{\frac{Y}{\rho}}$$
(16.7)

where Y is Young's modulus (defined in Section 10.1) and ρ is the density.

16.7 SOUND INTENSITY

Sound waves carry energy that can be used to do work, like forcing the eardrum to vibrate. Or, in an extreme case such as a sonic boom, the energy can be sufficient to cause damage to windows and buildings. The amount of energy transported per second by a sound wave is called the ***power*** of the wave and is measured in SI units of joules per second (J/s) or watts (W).

When a sound wave leaves a source, such as the loudspeaker in Figure 16.21, the power spreads out and passes through surfaces that have increasingly larger areas. The concept chart in Figure 16.22 indicates how we will take this spreading-out effect into account. We will bring together the ideas of power and the area through which the power passes, in the process formulating the concept of sound intensity. This concept of intensity will recur in Chapter 24 when we discuss electromagnetic waves. The ***sound intensity I*** is defined as the sound power P that passes perpendicularly through a surface divided by the area A of that surface:

$$I = \frac{P}{A}$$
(16.8)

The unit of sound intensity is power per unit area, or W/m². The next example illustrates how the sound intensity changes as the distance from a loudspeaker changes.

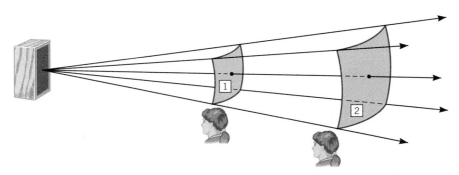

Figure 16.21 The power carried by a sound wave spreads out after leaving a source, such as a loudspeaker. Thus, the power passes perpendicularly through surface 1 and then through surface 2, which has the larger area.

Concepts at a Glance

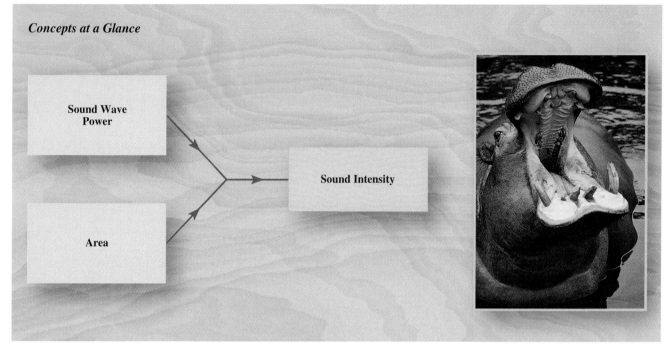

Figure 16.22 *Concepts at a Glance* The concept of sound intensity takes into account both the sound power and the area through which the power passes. A bellowing bull hippopotamus produces a sound intensity nearly as large as that at the loudest rock concert.

• PROBLEM SOLVING INSIGHT
Sound intensity *I* and sound power *P* are different concepts. They are related, however, since intensity expresses the idea of power per unit area.

EXAMPLE 7 • Sound Intensities

In Figure 16.21, 12×10^{-5} W of sound power passes perpendicularly through the surfaces labeled 1 and 2. These surfaces have areas of $A_1 = 4.0$ m^2 and $A_2 = 12$ m^2. Determine the sound intensity at each surface and discuss why listener 2 hears a quieter sound than listener 1.

Reasoning and Solution   The sound intensity at each surface can be computed from Equation 16.8:

Surface 1 $I_1 = \dfrac{P}{A_1} = \dfrac{12 \times 10^{-5} \text{ W}}{4.0 \text{ m}^2} = \boxed{3.0 \times 10^{-5} \text{ W/m}^2}$

Surface 2 $I_2 = \dfrac{P}{A_2} = \dfrac{12 \times 10^{-5} \text{ W}}{12 \text{ m}^2} = \boxed{1.0 \times 10^{-5} \text{ W/m}^2}$

The sound intensity is less at the more-distant surface, where the same power passes through a threefold greater area. The ear of a listener, with its fixed area, intercepts less power where the intensity, or power per unit area, is smaller. Thus, listener 2 intercepts less of the sound power than listener 1. With less power striking the ear, the sound is quieter.

For a 1000-Hz tone, the smallest sound intensity that the human ear can detect is about 1×10^{-12} W/m^2; this intensity is called the ***threshold of hearing.*** On the other extreme, continuous exposure to intensities greater than 1 W/m^2 can be painful and result in permanent hearing damage. The human ear is remarkable for the wide range of intensities to which it is sensitive.

If a source emits sound *uniformly in all directions,* the sound intensity depends on distance in a simple way. Figure 16.23 shows such a source at the center of an imaginary sphere (for clarity only a hemisphere is shown). The radius of the sphere is r. Since all the radiated sound power P passes through the spherical surface of area $A = 4\pi r^2$, the intensity at a distance r is

Spherically uniform radiation $\qquad I = \dfrac{P}{4\pi r^2}$ $\qquad\qquad\qquad$ (16.9)

From this we see that the intensity of a source that radiates sound uniformly in all directions varies as $1/r^2$. For example, if the distance increases by a factor of two, the sound intensity decreases by a factor of $2^2 = 4$. Example 8 illustrates the effect of the $1/r^2$ dependence of intensity on distance.

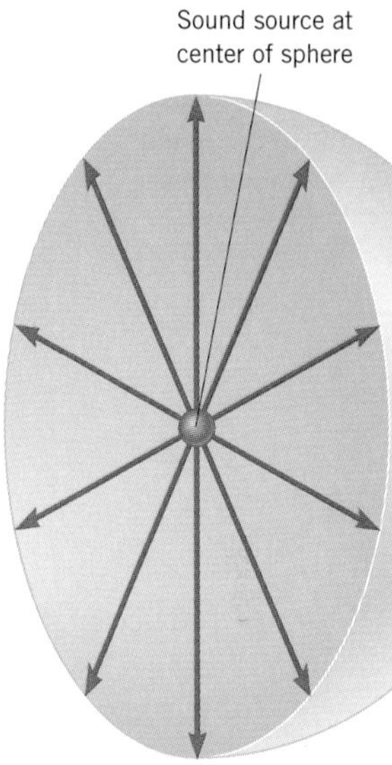

EXAMPLE 8 • Fireworks

During a fireworks display, a rocket explodes high in the air, as Figure 16.24 illustrates. Assume that the sound spreads out uniformly in all directions and that reflections from the ground can be ignored. When the sound reaches listener 2, who is $r_2 = 640$ m away from the explosion, the sound has an intensity of $I_2 = 0.10$ W/m^2. What is the sound intensity detected by listener 1, who is $r_1 = 160$ m away from the explosion?

Reasoning Listener 1 is four times closer to the explosion than listener 2. Therefore, the sound intensity detected by listener 1 is $4^2 = 16$ times greater than that detected by listener 2.

Solution The ratio of the sound intensities can be found using Equation 16.9:

$$\frac{I_1}{I_2} = \frac{\dfrac{P}{4\pi r_1^2}}{\dfrac{P}{4\pi r_2^2}} = \frac{r_2^2}{r_1^2} = \frac{(640\ \text{m})^2}{(160\ \text{m})^2} = 16$$

As a result, $I_1 = (16)I_2 = (16)(0.10\ \text{W/m}^2) = \boxed{1.6\ \text{W/m}^2}$.

Equation 16.9 is valid only when no walls, ceilings, floors, etc. are present to reflect the sound and cause it to pass through the same surface more than once. Conceptual Example 9 demonstrates why this is so.

Figure 16.23 The sound source at the center of the sphere emits sound uniformly in all directions. In this drawing, only a hemisphere is shown, for clarity.

• **PROBLEM SOLVING INSIGHT**
Equation 16.9 can be used only when the sound spreads out uniformly in all directions and there are no reflections of the sound waves.

Sound source at center of sphere

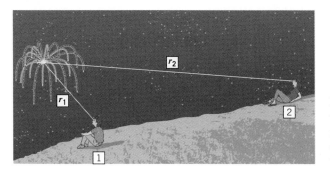

Figure 16.24 If an explosion in a fireworks display radiates sound uniformly in all directions, the intensity at any distance r is $I = P/(4\pi r^2)$, where P is the sound power of the explosion.

Figure 16.25 When someone sings in the shower, the sound power passing through the imaginary spherical surface is the sum of the direct sound power and the reflected sound power.

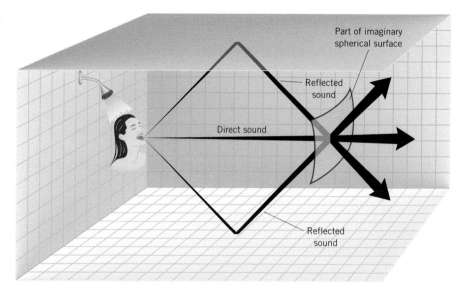

CONCEPTUAL EXAMPLE 9 • Reflected Sound and Sound Intensity

Suppose the person singing in the shower in Figure 16.25 produces a sound power P. Sound reflects from the surrounding shower stall. At a distance r in front of the person, does Equation 16.9 $[I = P/(4\pi r^2)]$ underestimate, overestimate, or give the correct sound intensity?

Reasoning and Solution In arriving at Equation 16.9, it was assumed that the sound spreads out uniformly from the source and passes only once through the imaginary surface that surrounds it (see Figure 16.23). Only part of this imaginary surface is shown in Figure 16.25. The drawing illustrates three paths by which the sound passes through the surface. The "direct" sound travels directly along a path from its source to the surface. It is the intensity of the direct sound that is given by $I = P/(4\pi r^2)$. The remaining paths are two of the many that characterize the sound reflected from the shower stall. The total sound power that passes through the surface is the sum of the direct and reflected powers. Thus, the total sound intensity at a distance r from the source is greater than that of the direct sound alone, and ***the relation $I = P/(4\pi r^2)$ underestimates the sound intensity from the singing, because it does not take into account the reflected sound.*** People like to sing in the shower, because their voices sound so much louder due to the enhanced intensity caused by the reflected sound.

Related Homework Material: *Problems 56 and 105*

16.8 DECIBELS

COMPARING SOUND INTENSITIES

The ***decibel*** (dB) is a measurement unit used when comparing two sound intensities. The simplest method of comparison would be to compute the ratio of the intensities. For instance, we could compare $I = 8 \times 10^{-12}$ W/m^2 to $I_0 = 1 \times 10^{-12}$ W/m^2 by computing $I/I_0 = 8$ and stating that I is eight times greater than I_0.

However, because of the way in which the human hearing mechanism responds to intensity, it is more appropriate to use a logarithmic scale for the comparison. For this purpose, the *intensity level β* (expressed in decibels) is defined as follows:

$$\beta \text{ (in decibels)} = 10 \log\left(\frac{I}{I_0}\right) \tag{16.10}$$

where "log" denotes the logarithm to the base ten. I_0 is the intensity of the reference level to which I is being compared and is often the threshold of hearing, $I_0 = 1.00 \times 10^{-12}$ W/m². With the aid of a calculator, the intensity level can be evaluated for the values of I and I_0 given above:

$$\beta = 10 \log\left(\frac{8 \times 10^{-12} \text{ W/m}^2}{1 \times 10^{-12} \text{ W/m}^2}\right) = 10 \log 8 = 10(0.9) = 9 \text{ dB}$$

This result indicates that I is 9 decibels greater than I_0.

Although β is called the "intensity level," it is *not* an intensity and does *not* have intensity units of W/m². In fact, the decibel, like the radian, is unitless, since it is the product of the number 10 and a logarithm, both of which are pure numbers without units.

Notice that if both I and I_0 are at the threshold of hearing, then $I = I_0$, and the intensity level is 0 dB according to Equation 16.10:

$$\beta = 10 \log\left(\frac{I_0}{I_0}\right) = 10 \log 1 = 0$$

since log 1 = 0. Thus, *an intensity level of zero decibels does not mean that the sound intensity I is zero; it means that I = I₀.*

Table 16.2 lists the intensities I and the associated intensity levels β for some common sounds, using the threshold of hearing as the reference level. Intensity levels can be measured with a sound level meter, such as the one in Figure 16.26. The intensity level β is displayed on its scale, assuming that the threshold of hearing is 0 dB.

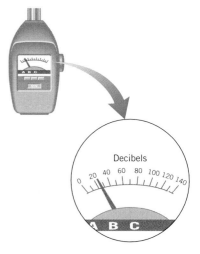

Figure 16.26 A sound level meter and a close-up view of its decibel scale.

INTENSITY LEVEL CHANGES AND LOUDNESS CHANGES

When a sound wave reaches a listener's ear, the sound is interpreted by the brain as loud or soft, depending on the intensity of the wave. Greater intensities give rise to

Table 16.2 Typical Sound Intensities and Intensity Levels Relative to the Threshold of Hearing

	Intensity I (W/m²)	Intensity Level β (dB)
Threshold of hearing	1.0×10^{-12}	0
Rustling leaves	1.0×10^{-11}	10
Whisper	1.0×10^{-10}	20
Normal conversation (1 meter)	3.2×10^{-6}	65
Inside car in city traffic	1.0×10^{-4}	80
Car without muffler	1.0×10^{-2}	100
Live rock concert	1.0	120
Threshold of pain	10	130

louder sounds. However, the relation between intensity and loudness is not a simple proportionality, for doubling the intensity does *not* double the loudness, as we will now see.

Suppose you are sitting in front of a stereo system that is producing an intensity level of 90 dB. If the volume control on the amplifier is turned up slightly to produce a 91 dB level, you would just barely notice the change in loudness. ***Hearing tests have revealed that a one-decibel (1-dB) change in the intensity level is approximately the smallest change in loudness that an average listener can detect.*** Since 1 dB is the smallest perceivable increment in loudness, a change of 3 dB, say, from 90 to 93 dB, is still a rather small change in loudness. Example 10 determines the factor by which the sound intensity must be changed to achieve this small change.

EXAMPLE 10 • Comparing Sound Intensities

Audio system 1 produces an intensity level of $\beta_1 = 90.0$ dB, while system 2 produces an intensity level of $\beta_2 = 93.0$ dB. The corresponding intensities (in W/m^2) are I_1 and I_2. Determine the ratio I_2/I_1.

Reasoning Intensity levels are related to intensities by logarithms (see Equation 16.10), and it is a property of logarithms (see Appendix D) that $\log A - \log B = \log (A/B)$. Subtracting the two intensity levels and using this property, we find that

$$\beta_2 - \beta_1 = 10 \log \left(\frac{I_2}{I_0} \right) - 10 \log \left(\frac{I_1}{I_0} \right) = 10 \log \left(\frac{I_2/I_0}{I_1/I_0} \right) = 10 \log \left(\frac{I_2}{I_1} \right)$$

Solution Using the result just obtained, we find

$$93.0 \text{ dB} - 90.0 \text{ dB} = 10 \log \left(\frac{I_2}{I_1} \right)$$

$$0.30 = \log \left(\frac{I_2}{I_1} \right) \quad \text{or} \quad \frac{I_2}{I_1} = 10^{0.30} = \boxed{2.0}$$

Since doubling the intensity changes the loudness by only a small amount (3 dB) and does not double it, there is no simple proportionality between intensity and loudness.

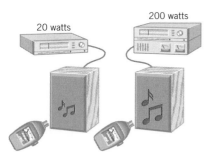

20 watts

200 watts

Figure 16.27 In spite of its tenfold greater power, the 200-watt audio system has only about double the loudness of the 20-watt system, when both are set for maximum volume.

To double the loudness of a sound, the intensity must be increased by more than a factor of two. ***Experiment shows that if the intensity level increases by 10 dB, the new sound seems approximately twice as loud as the original sound.*** For instance, a 70-dB intensity level sounds about twice as loud as a 60-dB level, and an 80-dB intensity level sounds about twice as loud as a 70-dB level. The factor by which the sound intensity must be increased to double the loudness can be determined by the method used in Example 10:

$$\beta_2 - \beta_1 = 10.0 \text{ dB} = 10 \left[\log \left(\frac{I_2}{I_0} \right) - \log \left(\frac{I_1}{I_0} \right) \right]$$

Solving this equation reveals that $I_2/I_1 = 10.0$. Thus, increasing the sound intensity by a factor of ten will double the perceived loudness. Consequently, with both audio systems in Figure 16.27 set at maximum volume, the 200-watt system will sound only twice as loud as the much cheaper 20-watt system.

16.9 APPLICATIONS OF SOUND

SONAR

Sonar (**so**und **na**vigation **r**anging) is a technique for determining water depth and locating underwater objects, such as reefs, submarines, and schools of fish. The core of a sonar unit consists of an ultrasonic transmitter and receiver mounted on the bottom of a ship, as Figure 16.28 illustrates. The transmitter emits a short pulse of ultrasonic sound, and at a later time the reflected pulse returns and is detected by the receiver. The water depth is determined from the electronically measured round-trip time of the pulse and a knowledge of the speed of sound in water; the depth registers automatically on an appropriate meter. Such a depth measurement is similar to the distance measurement discussed for the ultrasonic ruler in Example 4.

ULTRASOUND IN MEDICINE

When ultrasonic waves are used in medicine for diagnostic purposes, high-frequency sound pulses are produced by a transmitter and directed into the body. As in sonar, reflections occur. They occur each time a pulse encounters a boundary between two tissues that have different densities or a boundary between a tissue and the adjacent fluid. By scanning ultrasonic waves across the body and detecting the echoes generated from various locations within the body, it is possible to obtain a "picture" or sonogram of the inner anatomy. Ultrasonic waves are employed extensively in obstetrics to examine the developing fetus (Figure 16.29). The fetus, surrounded by the amniotic sac, can be distinguished from other anatomical features so that fetal size, position, and possible abnormalities can be detected.

Ultrasound is also used in other medically related areas. For instance, malignancies in the liver, kidney, brain, and pancreas can be detected with ultrasound. In cases where internal hemorrhaging occurs, it is possible to identify the bleeding area and even obtain a gross estimate of blood loss using ultrasonic techniques. Yet another application involves monitoring the real-time movement of pulsating structures, such as heart valves ("echocardiography") and large blood vessels.

The Physics of... sonar.

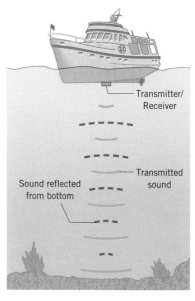

Figure 16.28 Sonar uses ultrasonic sound to measure water depth.

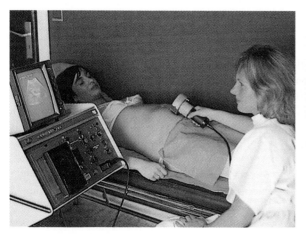 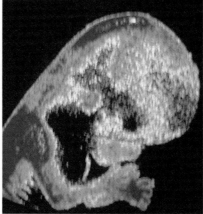

Figure 16.29 An ultrasonic scanner can be used to produce images of the fetus as it develops in the uterus.

The Physics of...
ultrasonic imaging.

The Physics of...
the cavitron ultrasonic surgical
aspirator.

When ultrasound is used to locate internal anatomical features or foreign objects in the body, the wavelength of the sound wave must be about the same size, or smaller, than the object to be located. Therefore, high frequencies in the range from 1 to 15 MHz (1 MHz = 1 megahertz = 1×10^6 Hz) are the norm. For instance, the wavelength of 5-MHz ultrasound is $\lambda = v/f = 0.3$ mm, if a value of 1540 m/s is used for the speed of sound through tissue. A sound wave with a frequency higher than 5 MHz and a correspondingly shorter wavelength is required for locating objects smaller than 0.3 mm.

Ultrasound also has applications other than locating objects in the body. Neurosurgeons use a device called a **c**avitron **u**ltrasonic **s**urgical **a**spirator (CUSA) to remove brain tumors once thought to be inoperable. Ultrasonic sound waves cause the slender tip of the CUSA probe (see Figure 16.30) to vibrate at approximately 23 kHz. The probe shatters any section of the tumor that it touches, and the fragments are flushed out of the brain with a saline solution. Because the tip of the probe is small, the surgeon can selectively remove small bits of malignant tissue without damaging the surrounding healthy tissue.

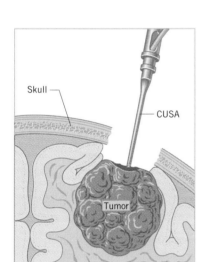

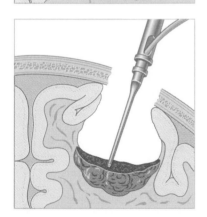

Figure 16.30 Neurosurgeons use a cavitron ultrasonic surgical aspirator (CUSA) to "cut out" brain tumors without adversely affecting the surrounding healthy tissue.

16.10 THE DOPPLER EFFECT

Have you ever heard an approaching fire truck and noticed the distinct change in the sound of the siren as the truck passes? The effect is similar to what you get when you put the two syllables "eee" and "yow" together to produce "eee-yow." While the truck approaches, the pitch of the siren is relatively high ("eee"), but as the truck passes and moves away, the pitch suddenly drops ("yow"). Something similar, but less familiar, occurs when an observer moves toward or away from a stationary source of sound. Such phenomena were first identified in 1842 by the Austrian physicist Christian Doppler (1803–1853) and are referred to as Doppler effects. To explain them, we will take the approach outlined in the concept chart in Figure 16.31. We will incorporate the effect of the velocities of the source and observer of the sound into the definitions of wavelength and frequency. In so doing, we will learn that the **_Doppler effect_** is the change in pitch or frequency of the sound detected by an observer because the sound source and the observer have different velocities with respect to the medium of sound propagation.

MOVING SOURCE

To see how the Doppler effect arises, consider the sound emitted by a siren on the stationary fire truck in Figure 16.32a. Like the truck, the air is assumed to be stationary with respect to the earth. Each solid blue arc in the drawing represents a condensation of the sound wave. Since the sound pattern is symmetrical, listeners standing in front of or behind the truck detect the same number of condensations per second and, consequently, hear the same frequency. Once the truck begins to move, the situation changes, as part b of the picture illustrates. Ahead of the truck, the condensations are now closer together, resulting in a decrease in the wavelength of the sound. This "bunching-up" occurs because the moving truck "gains ground" on a previously emitted condensation before emitting the next one. Since the condensations are closer together, the observer standing in front of the truck senses more of them arriving per second than she does when the truck is stationary. The increased rate of arrival corresponds to a greater sound frequency, which the observer hears as a higher pitch. Behind the moving truck, the condensations are far-

Concepts at a Glance

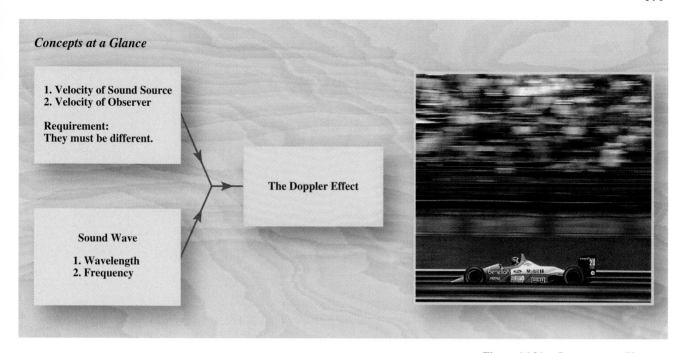

1. Velocity of Sound Source
2. Velocity of Observer

Requirement:
They must be different.

The Doppler Effect

Sound Wave

1. Wavelength
2. Frequency

Figure 16.31 *Concepts at a Glance* The Doppler effect arises when the source and the observer of a sound wave have different velocities with respect to the medium through which the sound travels. At an Indy-car racing event, the Doppler effect creates the characteristic sound that you hear when the cars pass by. It is similar to that heard when a train blowing its horn passes you at a high speed.

ther apart than they are when the truck is stationary. This increase in the wavelength occurs because the truck pulls away from condensations emitted toward the rear. Consequently, fewer condensations per second arrive at the ear of an observer behind the truck, corresponding to a smaller sound frequency or lower pitch.

If the stationary siren in Figure 16.32*a* emits a condensation at the time $t = 0$, it will emit the next one at time T, where T is the period of the wave. The distance between these two condensations is the wavelength λ of the sound produced by the stationary source, as Figure 16.33*a* indicates. When the truck is moving with a speed v_s (the subscript "s" stands for the "source" of sound) toward a stationary observer, the siren also emits condensations at $t = 0$ and at time T. However, prior to emitting the second condensation, the truck moves closer to the observer by a distance $v_s T$, as Figure 16.33*b* shows. As a result, the distance between successive condensations is no longer the wavelength λ created by the stationary siren, but, rather, a wavelength λ' that is shortened by the amount $v_s T$:

$$\lambda' = \lambda - v_s T$$

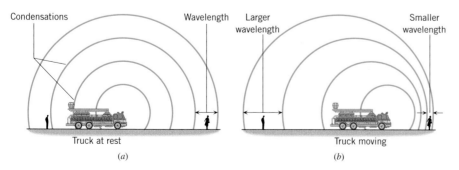

Condensations   Wavelength Larger wavelength   Smaller wavelength

Truck at rest

(a)

Truck moving

(b)

Figure 16.32 *(a)* When the truck is stationary, the wavelength of the sound is the same in front of and behind the truck. *(b)* When the truck is moving, the wavelength in front of the truck becomes smaller, while the wavelength behind the truck becomes larger.

Figure 16.33 (*a*) When the fire truck is stationary, the distance between successive condensations is one wavelength λ. (*b*) When the truck moves with a speed v_s, the wavelength of the sound in front of the truck is shortened to λ'.

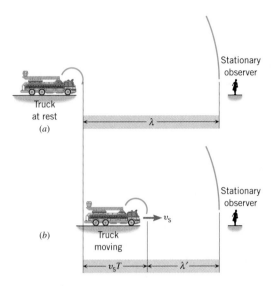

The frequency f', as perceived by the stationary observer, is the speed of sound v divided by the shortened wavelength λ', according to Equation 16.1:

$$f' = \frac{v}{\lambda'} = \frac{v}{\lambda - v_\text{s}T}$$

But for the stationary siren, $\lambda = v/f$ and $T = 1/f$, where f is the frequency at which the source emits the sound (not the frequency f' perceived by the observer). With the aid of these substitutions for λ and T, the expression for f' can be arranged to give the following result:

Source moving toward stationary observer

$$f' = f\left(\frac{1}{1 - \dfrac{v_\text{s}}{v}}\right) \qquad (16.11)$$

Since the term $1 - v_\text{s}/v$ is in the denominator in Equation 16.11 and is less than one, the frequency f' heard by the observer is *greater* than the frequency f emitted by the source. The difference between these two frequencies, $f' - f$, is called the **Doppler shift,** and its magnitude depends on the ratio of the speed of the source v_s to the speed of sound v.

When the siren moves away from, rather than toward, the observer, the wavelength λ' becomes *greater* than λ according to

$$\lambda' = \lambda + v_\text{s}T$$

Notice the presence of the "+" sign in this equation, in contrast to the "−" sign that appeared earlier. The same reasoning that led to Equation 16.11 can be used to obtain an expression for the observed frequency f':

Source moving away from stationary observer

$$f' = f\left(\frac{1}{1 + \dfrac{v_\text{s}}{v}}\right) \qquad (16.12)$$

The denominator $1 + v_\text{s}/v$ in Equation 16.12 is greater than one, so the frequency f' heard by the observer is *less* than the frequency f emitted by the source. The next example illustrates how large the Doppler shift is in a familiar situation.

EXAMPLE 11 • The Sound of a Passing Train

A high-speed train is traveling at a speed of 44.7 m/s (100 mi/h) when the engineer sounds the 415-Hz warning horn. The speed of sound is 343 m/s. What are the frequency and wavelength of the sound, as perceived by a person standing at a crossing, when the train is (a) approaching and (b) leaving the crossing?

Reasoning When the train approaches, the person at the crossing hears a sound whose frequency is greater than 415 Hz, because of the Doppler effect. As the train moves away, the person hears a frequency that is less than 415 Hz. We may use Equations 16.11 and 16.12, respectively, to determine these frequencies. In either case, the observed wavelength can be obtained according to Equation 16.1 as the speed of sound divided by the observed frequency.

Solution

(a) When the train approaches, the observed frequency is

$$f' = f\left(\frac{1}{1 - \dfrac{v_s}{v}}\right) = (415 \text{ Hz})\left(\frac{1}{1 - \dfrac{44.7 \text{ m/s}}{343 \text{ m/s}}}\right) = \boxed{477 \text{ Hz}} \quad (16.11)$$

The observed wavelength is

$$\lambda' = \frac{v}{f'} = \frac{343 \text{ m/s}}{477 \text{ Hz}} = \boxed{0.719 \text{ m}} \quad (16.1)$$

(b) When the train leaves the crossing, the observed frequency is

$$f' = f\left(\frac{1}{1 + \dfrac{v_s}{v}}\right) = (415 \text{ Hz})\left(\frac{1}{1 + \dfrac{44.7 \text{ m/s}}{343 \text{ m/s}}}\right) = \boxed{367 \text{ Hz}} \quad (16.12)$$

In this case, the observed wavelength is

$$\lambda' = \frac{v}{f'} = \frac{343 \text{ m/s}}{367 \text{ Hz}} = \boxed{0.935 \text{ m}}$$

MOVING OBSERVER

Figure 16.34 shows how the Doppler effect arises when the sound source is stationary and the observer moves, again assuming the air is stationary. The observer moves with a speed v_o ("o" stands for observer) toward the stationary source and covers a distance $v_o t$ in a time t. During this time, the moving observer encounters all the condensations that he would if he were stationary, *plus an additional*

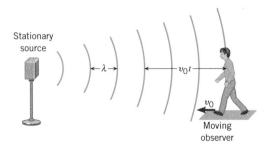

Stationary source

Moving observer

Figure 16.34 An observer moving with a speed v_o toward the stationary source intercepts more wave condensations per unit of time than does a stationary observer.

number. The additional number of condensations encountered is the distance $v_{\mathrm{o}}t$ divided by the distance λ between successive condensations, or $v_{\mathrm{o}}t/\lambda$. Thus, the additional number of condensations encountered per second is v_{o}/λ. Since a stationary observer would hear a frequency f emitted by the source, the moving observer hears a higher frequency f' given by

$$f' = f + \frac{v_{\mathrm{o}}}{\lambda} = f\left(1 + \frac{v_{\mathrm{o}}}{f\lambda}\right)$$

Using the fact that $v = f\lambda$, we find that

Observer moving toward stationary source
$$f' = f\left(1 + \frac{v_{\mathrm{o}}}{v}\right) \tag{16.13}$$

An observer moving *away from* a stationary source moves in the same direction as the sound wave and, as a result, intercepts *fewer* condensations per second than a stationary observer does. In this case, the moving observer hears a smaller frequency f' that is given by

Observer moving away from stationary source
$$f' = f\left(1 - \frac{v_{\mathrm{o}}}{v}\right) \tag{16.14}$$

The physical mechanism producing the Doppler effect in the case of the moving observer is different from that in the case of the moving source. When the source moves and the observer is stationary, the wavelength of the sound changes, giving rise to the frequency f' heard by the observer. On the other hand, when the observer moves and the source is stationary, the *wavelength of the sound does not change.* Instead, a moving observer intercepts a different number of wave condensations per second than does a stationary observer and, therefore, detects a different frequency f'.

GENERAL CASE

It is possible for both the sound source and the observer to move with respect to the medium of sound propagation. If the medium is stationary, Equations 16.11–16.14 may be combined to give the observed frequency f' as

Source and observer both moving
$$f' = f\left(\frac{1 \pm \dfrac{v_{\mathrm{o}}}{v}}{1 \mp \dfrac{v_{\mathrm{s}}}{v}}\right) \tag{16.15}$$

In the numerator, the plus sign applies when the observer moves toward the source, and the minus sign applies when the observer moves away from the source. In the denominator, the minus sign is used when the source moves toward the observer, and the plus sign is used when the source moves away from the observer. The symbols v_{o}, v_{s}, and v denote numbers without an algebraic sign, because the direction of travel has been taken into account by the plus and minus signs that appear directly in this equation.

The Physics of...
the Doppler flow meter.

DOPPLER FLOW METER

The Doppler flow meter is an interesting medical application of the Doppler effect. This device measures the speed of blood flow, using transmitting and receiving elements that are placed directly on the skin, as in Figure 16.35. The transmitter emits a continuous sound whose frequency is typically about 5 MHz. When the sound is reflected from the red blood cells, its frequency is changed in a kind of Doppler effect, because the cells are moving. The receiving element detects the reflected sound, and an electronic counter measures its frequency, which is Doppler-shifted relative to the transmitter frequency. From the change in frequency the speed of the blood flow can be determined. Typically, the change in frequency is around 600 Hz for flow speeds of about 0.1 m/s. The Doppler flow meter can be used to locate regions where blood vessels have narrowed, since greater flow speeds occur in the narrowed regions, according to the equation of continuity (see Section 11.8). In addition, the Doppler flow meter can be used to detect the motion of a fetal heart as early as 8–10 weeks after conception.

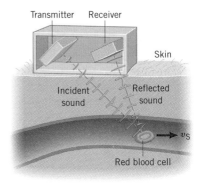

Figure 16.35 A Doppler flow meter measures the speed of red blood cells.

NEXRAD

NEXRAD stands for **Nex**t Generation Weather **Rad**ar and is a new nationwide system used by the National Weather Service to provide dramatically improved early warning of severe storms, such as the tornado in Figure 16.36*a*. The system is based on radar waves, which are a type of electromagnetic wave (see Chapter 24) and, like sound waves, can exhibit the Doppler effect. The Doppler effect is at the heart of NEXRAD. As Figure 16.36*b* illustrates, a tornado is a swirling mass of air and water droplets. Radar pulses are sent out by a NEXRAD unit, whose protective covering is shaped like a soccer ball. The waves reflect from the water droplets and return to the unit, where the radar frequency is measured and compared to the outgoing frequency. For instance, droplets at point *A* in the drawing are moving toward the unit, and the radar waves reflected from them have their frequency Doppler-shifted to higher values. Droplets at point *B*, however, are moving away from the unit. The frequency of the waves reflected from these droplets is Doppler-shifted to lower values. Computer processing of the Doppler frequency shifts leads to color-enhanced views on display screens (see Figure 16.37). These views reveal the direction and magnitude of the wind velocity and can identify, from distances up to 140 mi, the swirling air masses that are likely to spawn tornadoes. The equations

The Physics of...
Next Generation Weather Radar.

Figure 16.36 (*a*) A tornado is one of nature's most dangerous storms. (*b*) The National Weather Service uses the NEXRAD system, which is based on Doppler-shifted radar, to identify the storms that are likely to spawn tornadoes.

(*a*)

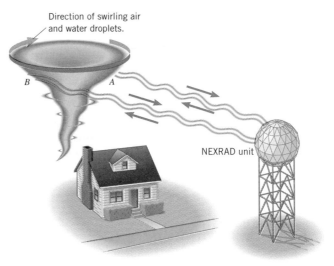

(*b*)

Figure 16.37 In this color-enhanced NEXRAD view of a tornado shades of green denote winds moving toward the NEXRAD station, and shades of red denote winds moving away from the station, which is located below and to the right of the figure. The white dot marks the center of the storm, and the arrow indicates how the winds are circulating.

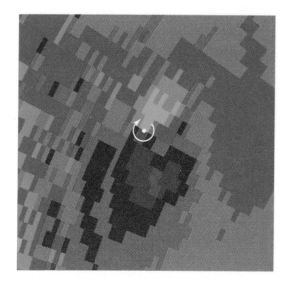

that specify the Doppler frequency shifts are different than those given for sound waves by Equation 16.15. The reason for the difference is that radar waves propagate from one place to another by a different mechanism than that of sound waves (see Section 24.5).

*16.11 THE SENSITIVITY OF THE HUMAN EAR

The Physics of... hearing.

Although the ear is capable of detecting sound intensities as small as 1×10^{-12} W/m^2, it is *not* equally sensitive to all frequencies, as Figure 16.38 shows. This figure displays a series of graphs that are known as the *Fletcher–Munson curves,* after H. Fletcher and M. Munson, who first determined them in 1933. In these graphs the audible sound frequencies are plotted on the horizontal axis, while the sound intensity levels (in decibels) are plotted on the vertical axis. Each curve is a *constant loudness* curve, in the sense that it shows the sound intensity level needed at each frequency to make the sound appear to have the same loudness. For example, the lowest (red) curve represents the threshold of hearing. It shows the intensity levels at which sounds of different frequencies just become audible. The graph indicates

Figure 16.38 Each curve represents the intensity levels at which sounds of various frequencies have the same loudness. The curves are labeled by their intensity levels at 1000 Hz. These curves are known as the Fletcher–Munson curves.

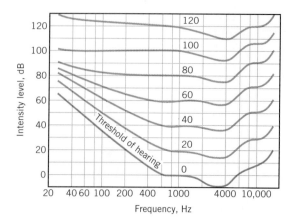

that the intensity level of a 100-Hz sound must be about 37 dB greater than the intensity level of a 1000-Hz sound to be at the threshold of hearing. Therefore, the ear is *less sensitive* to a 100-Hz sound than it is to a 1000-Hz sound. In general, Figure 16.38 reveals that the ear is most sensitive in the range from about 1 to 5 kHz, and becomes progressively less sensitive at higher and lower frequencies.

Each curve in Figure 16.38 represents a different loudness, and each is labeled according to its intensity level at 1000 Hz. For instance, the curve labeled "60" represents all sounds that have the same loudness as that of a 1000-Hz sound whose intensity level is 60 dB. These constant-loudness curves become "flatter" as the loudness increases, the relative flatness indicating that the ear is nearly equally sensitive to all frequencies when the sound is loud. Thus, when you play your stereo at very loud levels, you hear the low frequencies, the middle frequencies, and the high frequencies about equally well. However, when you turn down the volume control on the receiver so the sound becomes quiet, the high and low frequencies seem to "disappear," for the ear is relatively insensitive to these frequencies under such conditions.

SUMMARY

A **wave** is a traveling disturbance and carries energy from place to place. In a **transverse wave,** the disturbance is perpendicular to the direction of travel of the wave. In a **longitudinal wave,** the disturbance is parallel to the line along which the wave travels.

In a **periodic wave,** the pattern of the disturbance is produced over and over again by the source of the wave. The **amplitude** of the wave is the maximum excursion of a particle of the medium from the particle's undisturbed position. The **wavelength** λ is the distance along the length of the wave between two successive equivalent points, such as two crests or two troughs. The **period** T is the time required for the wave to travel a distance of one wavelength. The **frequency** f (in hertz) is the number of wave cycles per second that pass an observer and is the reciprocal of the period (in seconds): $f = 1/T$. The **speed** v of a wave is related to the wavelength and the frequency according to $v = f\lambda$.

The **speed of a wave** depends on the properties of the medium in which the wave travels. For a transverse wave on a string that has a tension F and a mass per unit length m/L, the wave speed is $v = \sqrt{F/(m/L)}$.

Sound is a longitudinal wave that consists of alternating regions of greater-than-normal pressure (condensations) and less-than-normal pressure (rarefactions). Each cycle of a sound wave includes one condensation and one rarefaction. A sound wave with a large frequency is interpreted by the brain as a high-pitched sound, while one with a small frequency is interpreted as a low-pitched sound. The **pressure amplitude** of a sound wave is the magnitude of the maximum change in pressure, measured relative to the undisturbed pressure. The larger the pressure amplitude, the louder the sound.

The **speed of sound** v depends on the properties of the medium. In an ideal gas, the speed of sound is $v = \sqrt{\gamma kT/m}$, where $\gamma = c_P/c_V$ is the ratio of the specific heat capacities at constant pressure and constant volume, k is Boltzmann's constant, T is the Kelvin temperature, and m is the mass of a molecule of the gas. In a liquid, the speed of sound is $v = \sqrt{B_{ad}/\rho}$, where B_{ad} is the adiabatic bulk modulus and ρ is the mass density. For a solid in the shape of a long slender bar, the expression for the speed of sound is $v = \sqrt{Y/\rho}$, where Y is Young's modulus.

The **intensity** I of a sound wave is the power P that passes perpendicularly through a surface divided by the area A of the surface: $I = P/A$. The SI unit of intensity is watts per square meter (W/m^2). The smallest sound intensity that humans can detect is known as the **threshold of hearing** and is about 1×10^{-12} W/m^2 for a 1-kHz sound. When a source emits sound uniformly in all directions and no reflections are present, the intensity of the sound is inversely proportional to the square of the distance from the source.

The **intensity level** β (in decibels) is used to compare a sound intensity I to the sound intensity I_0 of a reference level: $\beta = 10 \log (I/I_0)$.

The **Doppler effect** is the change in frequency detected

by an observer because the sound source and the observer have different velocities with respect to the medium of sound propagation. If the observer and source move with speeds v_o and v_s, respectively, and if the medium is stationary, the frequency f' detected by the observer is

$$f' = f\left(\frac{1 \pm v_o/v}{1 \mp v_s/v}\right)$$

where f is the frequency of the sound emitted by the source, and v is the speed of sound. In the numerator, the plus sign applies when the observer moves toward the source, and the minus sign applies when the observer moves away from the source. In the denominator, the minus sign is used when the source moves toward the observer, and the plus sign is used when the source moves away from the observer.

CONCEPTUAL QUESTIONS

1. Considering the nature of a water wave (see Figure 16.4), describe the motion of a fishing float on the surface of a lake when a wave passes beneath the float. Is it really correct to say that the float bobs straight "up and down"? Explain.

2. "Domino Toppling" is one entry in the *Guiness Book of World Records*. The event consists of lining up an incredible number of dominoes and then letting them topple, one after another. Is the disturbance that propagates along the line of dominoes transverse, longitudinal, or partly both? Explain.

3. Suppose that a longitudinal wave moves along a Slinky at a speed of 5 m/s. Does one coil of the Slinky move through a distance of 5 m in one second? Justify your answer.

4. Examine Conceptual Example 3 before addressing this question. A wave moves on a string with a constant velocity. Does this mean that the particles of the string always have zero acceleration? Justify your answer.

5. A wire is strung tightly between two immovable posts. Discuss how an increase in temperature affects the speed of a transverse wave on this wire. Give your reasoning, ignoring any change in the mass per unit length of the wire.

6. A rope of mass m is hanging down from the ceiling. Nothing is attached to the loose end of the rope. A transverse wave is traveling on the rope. As the wave travels up the rope, does the speed of the wave increase, decrease, or remain the same? Give a reason for your choice.

7. One end of each of two identical strings is attached to a wall. Each string is being pulled tight by someone at the other end. A transverse pulse is sent traveling along one of the strings. A bit later an identical pulse is sent traveling along the other string. What, if anything, can be done to make the second pulse catch up with and pass the first pulse? Account for your answer.

8. In Section 4.10 the concept of a "massless" rope is discussed. Would it take any time for a transverse wave to travel the length of a massless rope? Justify your answer.

9. In a sound wave, are there any particles that are *always* at rest as the wave passes by? Justify your answer.

10. Do you expect an echo to return to you more quickly or less quickly on a hot day as compared to a cold day, other things being equal? Account for your answer.

11. A loudspeaker produces a sound wave. Does the wavelength of the sound increase, decrease, or remain the same, when the wave travels from air into water? Justify your answer. *(Hint: The frequency does not change as the sound enters the water.)*

12. A person is making a bowl of JELL-O. It starts out as a liquid and then sets to a gel. What would you expect to happen to the speed of sound in this material as the JELL-O sets? Does it increase, decrease, or remain the same? Give your reasoning.

13. Some animals rely on an acute sense of hearing for survival, and the visible part of the ear on such animals is often relatively large. Explain how this anatomical feature helps to increase the sensitivity of the animal's hearing for low-intensity sounds.

14. A source is emitting sound uniformly in all directions. There are no reflections anywhere. A *flat* surface faces the source. Is the sound intensity the same at all points on the surface? Give your reasoning.

15. If the sound intensity level is measured relative to the threshold of hearing, what does it mean for the intensity level to be negative, e.g., −20 dB?

16. If two people talk simultaneously and each creates an intensity level of 65 dB at a certain point, does the total intensity level at this point equal 130 dB? Account for your answer.

17. Two cars, one behind the other, are traveling in the same direction at the same speed. Does either driver hear the other's horn at a frequency that is different from that heard when both cars are at rest? Justify your answer.

18. A source of sound produces the same frequency underwater as it does in air. This source has the same velocity in air as it does underwater. The observer of the sound is stationary, both in air and underwater. Is the Doppler effect greater in air or underwater when the source (a) approaches and (b) moves away from the observer? Explain.

19. A music fan at a swimming pool is listening to a radio on a diving platform. The radio is playing a constant frequency tone when this fellow, clutching his radio, jumps. Describe the Doppler effect heard by (a) a person left behind on the platform and (b) a person down below floating on a rubber raft. In each case, specify (1) whether the observed frequency is greater or smaller than the frequency produced by the radio, (2) whether the observed frequency is constant, and (3) how the observed frequency changes during the fall, if it does change. Give your reasoning.

20. When a car is at rest, its horn emits a frequency of 600 Hz. A person standing in the middle of the street hears the horn with a frequency of 580 Hz. Should the person jump out of the way? Account for your answer.

21. The text discusses how the Doppler effect arises when (1) the observer is stationary and the source moves and (2) the observer moves and the source is stationary. A car is speeding toward a large wall and sounds the horn. Is the Doppler effect present in the echo that the driver hears? If it is present, from which of the above situations does it arise, (1) or (2) or both? Explain.

PROBLEMS

Section 16.1 The Nature of Waves, Section 16.2 Periodic Waves

1. ssm A person standing in the ocean notices that after a wave crest passes by, ten more crests pass in a time of 120 s. What is the frequency of the wave?

2. To navigate, a porpoise emits a sound wave that has a wavelength of 2.5 cm. The speed at which sound travels in seawater is 1470 m/s. Find the period of the wave.

3. Consider the freight train in Figure 16.7. Suppose 15 boxcars pass by in a time of 12.0 s and each has a length of 14.0 m. (a) What is the frequency at which each boxcar passes? (b) What is the speed of the train?

4. A light wave travels through air at a speed of 3.0×10^8 m/s. Red light has a wavelength of about 6.6×10^{-7} m. What is the frequency of red light?

5. ssm In Figure 16.2 the hand moves the end of the Slinky up and down through two complete cycles in one second. The wave moves along the Slinky at a speed of 0.50 m/s. Find the distance between two adjacent crests on the wave.

6. A longitudinal wave with a frequency of 3.0 Hz takes 1.7 s to travel the length of a 2.5-m Slinky (see Figure 16.3). Determine the wavelength of the wave.

7. A wave has a frequency of 45 Hz and a speed of 22 m/s. Determine, if possible, (a) its period, (b) its wavelength, and (c) its amplitude. If it is not possible to determine any of these quantities, then so state.

8. A person lying on an air mattress in the ocean rises and falls through one complete cycle every five seconds. The crests of the wave causing the motion are 20.0 m apart. Determine (a) the frequency and (b) the speed of the wave.

9. ssm Suppose the amplitude and frequency of the transverse wave in Figure 16.2c are, respectively, 1.3 cm and 5.0 Hz. Find the *total vertical distance* (in cm) through which the colored dot moves in 3.0 s.

*10. The speed of a transverse wave on a string is 450 m/s, while the wavelength is 0.18 m. The amplitude of the wave is 2.0 mm. How much time is required for a particle of the string to move through a total distance of 1.0 km?

*11. A 3.49 rad/s ($33\frac{1}{3}$ rpm) record has a 5.00-kHz tone cut in the

groove. If the groove is located 0.100 m from the center of the record (see drawing), what is the "wavelength" in the groove?

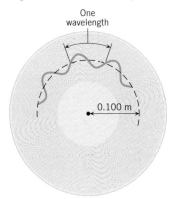

One wavelength

0.100 m

**12. A water-skier is moving at a speed of 12.0 m/s. When she skis in the same direction as a traveling wave, she springs upward every 0.600 s because of the wave crests. When she skis in the direction opposite to that in which the wave moves, she springs upward every 0.500 s in response to the crests. The speed of the skier is greater than the speed of the wave. Determine (a) the speed and (b) the wavelength of the wave.

Section 16.3 The Speed of a Wave on a String

13. ssm The linear density of the A string on a violin is 7.8×10^{-4} kg/m. A wave on the string has a frequency of 440 Hz and a wavelength of 65 cm. What is the tension in the string?

14. A transverse wave is traveling with a speed of 300 m/s on a horizontal string. If the tension in the string is increased by a factor of four, what is the speed of the wave?

15. A 0.75-m string is stretched so the tension is 2.3 N. A transverse wave with a frequency of 150 Hz and a wavelength of 0.40 m travels on the string. What is the mass of the string?

16. A wire is stretched between two posts. Another wire is stretched between two posts that are twice as far apart. The wires have the same mass. If the same tension exists in both wires, find the ratio of the speed of a transverse wave on the shorter wire to that on the longer wire.

17. ssm www The middle C string on a piano is under a tension of 944 N. The period and wavelength of a wave on this string are 3.82×10^{-3} s and 1.26 m, respectively. Find the linear density of the string.

18. Two wires are parallel, and one is directly above the other. Each has a length of 50.0 m and a mass per unit length of 0.020 kg/m. However, the tension in wire A is 6.00×10^2 N, while the tension in wire B is 3.00×10^2 N. Transverse wave pulses are generated simultaneously, one at the left end of wire A and one at the right end of wire B. The pulses travel toward each other. How much time does it take until the pulses pass each other?

*19. To measure the acceleration due to gravity on a distant planet, an astronaut hangs a 0.085-kg ball from the end of a wire. The wire has a length of 1.5 m and a linear density of 3.1×10^{-4} kg/m. Using electronic equipment, the astronaut measures the time for a transverse pulse to travel the length of the wire and obtains a value of 0.083 s. The mass of the wire is negligible compared to the mass of the ball. Determine the acceleration due to gravity.

*20. Review Conceptual Example 3 before starting this problem. A horizontal wire is under a tension of 315 N and has a mass per unit length of 6.50×10^{-3} kg/m. A transverse wave with an amplitude of 2.50 mm and a frequency of 585 Hz is traveling on this wire. As the wave passes, a particle of the wire moves up and down in simple harmonic motion. Obtain (a) the speed of the wave and (b) the maximum speed with which the particle moves up and down.

*21. ssm www The drawing shows a frictionless incline and pulley. The two blocks are connected by a wire (mass per unit length = 0.0250 kg/m) and remain stationary. A transverse wave on the wire has a speed of 75.0 m/s. Neglecting the weight of the wire relative to the tension in the wire, find the masses m_1 and m_2 of the blocks.

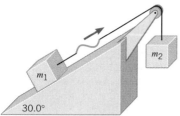

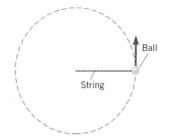

**22. The drawing shows a 15.0-kg ball being whirled in a circular path on the end of a string. The motion occurs on a frictionless,

horizontal table. The angular speed of the ball is $\omega = 12.0$ rad/s. The string has a mass of 0.0230 kg. How much time does it take for a wave on the string to travel from the center of the circle to the ball?

**23. A copper wire, whose cross-sectional area is 1.1×10^{-6} m², has a linear density of 7.0×10^{-3} kg/m and is strung between two walls. At the ambient temperature, a transverse wave travels with a speed of 46 m/s on this wire. The coefficient of linear expansion for copper is 17×10^{-6} (C°)$^{-1}$, and Young's modulus for copper is 1.1×10^{11} N/m². What will be the speed of the wave when the temperature is lowered by 14 C°? Ignore any change in the linear density caused by the change in temperature.

Section 16.4 The Mathematical Description of a Wave

(Note: The phase angles $(2\pi ft - 2\pi x/\lambda)$ and $(2\pi ft + 2\pi x/\lambda)$ are measured in radians, not degrees.)

24. A wave has a displacement (in meters) of $y = (0.45) \sin (8.0\pi t + \pi x)$, where t and x are expressed in seconds and meters, respectively. (a) Find the amplitude, the frequency, the wavelength, and the speed of the wave. (b) Is this wave traveling in the $+x$ or $-x$ direction?

25. ssm A wave has the following properties: amplitude = 0.37 m, period = 0.77 s, wave speed = 12 m/s. The wave is traveling in the $-x$ direction. What is the mathematical expression (similar to Equation 16.3 or 16.4) for the wave?

26. The drawing shows two graphs that represent a transverse wave on a string. The wave is moving in the $+x$ direction. Using the information contained in these graphs, write the mathematical expression (similar to Equation 16.3 or 16.4) for the wave.

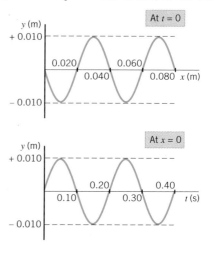

27. The displacement (in meters) of a wave is $y = (0.26) \sin (\pi t - 3.7\pi x)$, where t is in seconds and x is in meters. (a) Is the wave traveling in the $+x$ or $-x$ direction? (b) What is the displacement y when $t = 38$ s and $x = 13$ m?

*28. Refer to the graphs that accompany problem 26. From the data in these graphs, determine the speed of the wave.

*29. **ssm** A transverse wave is traveling on a string. The displacement y of a particle from its equilibrium position is given by $y = (0.021 \text{ m}) \sin (25t - 2.0x)$. Note that the phase angle $(25t - 2.0x)$ is in radians, t is in seconds, and x is in meters. The linear density of the string is 1.6×10^{-2} kg/m. What is the tension in the string?

**30. A transverse wave on a string has an amplitude of 0.20 m and a frequency of 175 Hz. Consider the particle of the string at $x = 0$. It begins with a displacement of $y = 0$ when $t = 0$, according to Equation 16.3 or 16.4. How much time passes between the first two instants when this particle has a displacement of $y = 0.10$ m?

Section 16.5 The Nature of Sound,
Section 16.6 The Speed of Sound

31. A rhinoceros is calling to her mate using infrasonic sound with a frequency of 5.0 Hz. Her mate is 480 m away. The speed of sound is 343 m/s. How many cycles of the sound wave are between the two animals?

32. The wavelength of a sound wave in air is 2.74 m at 20 °C. What is the wavelength of this sound wave in fresh water at 20 °C? (*Hint: The frequency of the sound is the same in both media.*)

33. ssm The speed of a sound in a container of hydrogen at 201 K is 1220 m/s. What would be the speed of sound if the temperature were raised to 405 K? Assume that hydrogen behaves like an ideal gas.

34. Have you ever listened for an approaching train by kneeling next to a railroad track and putting your "ear to the rail?" Young's modulus for steel is $Y = 2.0 \times 10^{11}$ N/m², and the density of steel is $\rho = 7860$ kg/m³. On a day when the temperature is 20 °C, how many times greater is the speed of sound in the rail than in the air?

35. A bat emits sound whose frequency is 91 kHz. The air temperature is 35 °C, so the speed of sound is not 343 m/s. Find the wavelength of the sound.

36. A sound wave is incident on a pool of fresh water. The sound enters the water perpendicularly and travels a distance of 0.45 m before striking a 0.15-m-thick copper block lying on the bottom. The sound passes through the block, reflects from the bottom surface of the block, and returns to the top of the water along the same path. How much time elapses between when the sound enters and leaves the water?

37. ssm At 20 °C the densities of fresh water and ethanol are, respectively, 998 and 789 kg/m³. Find the ratio of the adiabatic bulk modulus of fresh water to the adiabatic bulk modulus of ethanol at 20 °C.

38. An explosion occurs at the end of a pier. The sound reaches the other end of the pier by traveling through three media: air, fresh water, and a slender handrail of solid steel. The speeds of sound in air, water, and the handrail are 343, 1482, and 5040 m/s, respectively. The sound travels a distance of 125 m in each medium. (a) Through which medium does the sound arrive first, second, and third? (b) After the first sound arrives, how much later do the second and third sounds arrive?

39. At what temperature is the speed of sound in helium (atomic mass = 4.003 u) the same as the speed of sound in oxygen at 0 °C? Consider helium to be a monatomic, ideal gas ($\gamma = 1.67$).

40. As the drawing illustrates, a siren can be made by blowing a jet of air through 20 equally spaced holes in a rotating disk. The time it takes for successive holes to move past the air jet is the period of the sound. The siren is to produce a 2200-Hz tone. What must be the angular speed ω (in rad/s) of the disk?

Air jet

*41. **ssm** A long slender bar is made from an unknown material. The length of the bar is 0.83 m, its cross-sectional area is 1.3×10^{-4} m², and its mass is 2.1 kg. A sound wave travels from one end of the bar to the other end in 1.9×10^{-4} s. From which one of the materials listed in Table 10.1 is the bar most likely to be made?

*42. Review the rule of thumb in Conceptual Example 5. Suppose the speed of light were 961 m/s, instead of 3.0×10^8 m/s, but the speed of sound remained at 343 m/s. The new rule of thumb is stated as follows:

> After you see a flash of lightning, count off the seconds until the thunder is heard. Divide the number of seconds by the number X. The result gives the approximate distance (in miles) to the thunderstorm.

What is the number X?

*43. A sonar unit on a boat works like the ultrasonic ruler discussed in Examples 4 and 6 and is capable of detecting the return of an ultrasonic pulse during times up to 1.500 s after the pulse is transmitted. (a) Assuming that the freshwater temperature is 20 °C, what is the maximum water depth that can be measured? (b) If the sonar unit measures time with an error of ±0.004 s, what is the error (in meters) in the depth measurement?

*44. At a height of ten meters above the surface of a lake, a sound pulse is generated. The echo from the bottom of the lake returns to the point of origin 0.140 s later. The air and water temperatures are 20 °C. How deep is the lake?

*45. **ssm** A hunter is standing on flat ground between two vertical cliffs that are directly opposite one another. He is closer to one cliff than the other. He fires a gun and, after a while, hears three echoes. The second echo arrives 1.6 s after the first, and the third echo arrives 1.1 s after the second. Assuming that the speed of sound is 343 m/s and that there are no reflections of sound from the ground, find the distance between the cliffs.

* 46. A monatomic ideal gas ($\gamma = 1.67$) is contained within a box whose volume and pressure are 2.5 m³ and 3.5×10^5 Pa, respectively. The total mass of the gas is 2.3 kg. Find the speed of sound in the gas.

* 47. Both krypton (Kr) and neon (Ne) can be approximated as monatomic ideal gases. The atomic mass of krypton is 83.8 u, while that of neon is 20.2 u. A loudspeaker produces a sound whose wavelength in krypton is 1.25 m. If the loudspeaker were used to produce sound of the same frequency in neon at the same temperature, what would be the wavelength?

** 48. A jet is flying horizontally, as the drawing shows. When the plane is directly overhead at *B*, a person on the ground hears the sound coming from *A* in the drawing. The average temperature of the air is 20 °C. If the speed of the plane at *A* is 164 m/s, what is its speed at *B*, assuming that it has a constant acceleration?

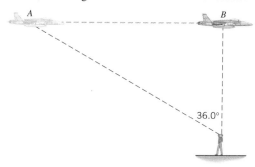

** 49. **ssm www** As a prank, someone drops a water-filled balloon out of a window. The balloon is released from rest at a height of 10.0 m above the ears of a man who is the target. Because of a guilty conscience, however, the prankster shouts a warning after the balloon is released. The warning will do no good, however, if shouted after the balloon reaches a certain point, even if the man could react infinitely quickly. Assuming that the air temperature is 20 °C and ignoring the effect of air resistance on the balloon, determine how far above the man's ears this point is.

** 50. Civil engineers use a transit theodolite when surveying. A modern version of this device determines distance by measuring the time required for an ultrasonic pulse to reach a target, reflect from it, and return. Effectively, such a theodolite is calibrated properly when it is programmed with the speed of sound appropriate for the ambient air temperature. (a) Suppose the round-trip time for the pulse is 0.580 s on a day when the air temperature is 293 K, the temperature for which the instrument is calibrated. How far is the target from the theodolite? (b) Assume that air behaves as an ideal gas. If the air temperature were 298 K, rather than the calibration temperature of 293 K, what percentage error would there be in the distance measured by the theodolite?

Section 16.7 Sound Intensity

51. A typical adult ear has a surface area of 2.1×10^{-3} m². The sound intensity during a normal conversation is about 3.2×10^{-6} W/m² at the listener's ear. Assume the sound strikes the sur-

face of the ear perpendicularly. How much power is intercepted by the ear?

52. The average sound intensity inside a busy restaurant is 3.2×10^{-5} W/m². How much energy goes into each ear (area = 2.1×10^{-3} m²) during a one-hour meal?

53. **ssm www** At a distance of 3.8 m from a siren, the sound intensity is 3.6×10^{-2} W/m². Assuming that the siren radiates sound uniformly in all directions, find the total power radiated.

54. Suppose that sound is emitted uniformly in all directions by a public address system. The intensity at a location 22 m away from the sound source is 3.0×10^{-4} W/m². What is the intensity at a spot that is 78 m away?

55. A source of sound radiates power uniformly in all directions. Two spheres are centered on this source. Sphere A has a radius of 0.80 m, while sphere B has a radius of 1.1 m. Consider the sound power that passes through a 0.17-m² patch of surface on sphere A. What is the area of the surface on sphere B through which this same power passes?

56. Suppose in Conceptual Example 9 (see Figure 16.25) that the person is producing 0.50 watt of sound power. Some of the sound is reflected from the floor and ceiling. The intensity of this reflected sound at a distance of 3.0 m from the source is 2.1×10^{-3} W/m². What is the total sound intensity due to both the direct and reflected sounds, at this point?

57. **ssm** A loudspeaker has a circular opening with a radius of 0.0950 m. The electrical power needed to operate the speaker is 25.0 W. The average sound intensity at the opening is 17.5 W/m². What percentage of the electrical power is converted by the speaker into sound power?

* 58. When a helicopter is hovering 1450 m directly overhead, an observer on the ground measures a sound intensity *I*. Assume that sound is radiated uniformly from the helicopter and that ground reflections are negligible. How far must the helicopter fly in a straight line parallel to the ground before the observer measures a sound intensity of $\frac{1}{4}I$?

* 59. A dish of lasagna is being heated in a microwave oven. The effective area of the lasagna that is exposed to the microwaves is 1.6×10^{-2} m². The mass of the lasagna is 0.25 kg, and its specific heat capacity is 3400 J/(kg·C°). The temperature rises by 80.0 C° in 7.0 minutes. What is the intensity of the microwaves in the oven?

* 60. Two sources each emit sound power uniformly in all directions. There are no reflections. Both sources are located on the *x* axis, one at the origin and the other at $x = +123$ m. The source at the origin emits four times more power than the other source. Where on the *x* axis is the intensity of each source equal? Note that there are two answers.

** 61. **ssm** A rocket, starting from rest, travels straight up with an acceleration of 58.0 m/s². When the rocket is at a height of 562 m, it produces sound that eventually reaches a ground-based monitoring station directly below. The sound is emitted uniformly in all directions. The monitoring station measures a sound intensity *I*.

Later on, the station measures an intensity $\frac{1}{3}I$. Assuming the speed of sound is 343 m/s, find the time that has elapsed between the two measurements.

Section 16.8 Decibels

62. The bellow of a territorial bull hippopotamus has been measured at 115 dB above the threshold of hearing. What is the sound intensity?

63. An amplified guitar has a sound intensity level that is 14 dB greater than the same unamplified sound. What is the ratio of the amplified intensity to the unamplified intensity?

64. One of the important specifications of a cassette tape deck is its signal-to-noise rating. This specification indicates how much of the sound intensity created when playing a tape is due to the musical tones (the signal) and how much is due to the hissing sound produced by the moving tape (the noise). Suppose the sound intensity due to the musical tones is 3.0×10^{-5} W/m^2, while that due to the tape hiss is 4.8×10^{-11} W/m^2. What is the signal-to-noise rating, which is the number of *decibels* by which the signal exceeds the noise?

65. ssm When a person wears a hearing aid, the sound intensity level increases by 30.0 dB. By what factor does the sound intensity increase?

66. The sound intensity level of a jet engine is 138 dB, while at a rock concert it is 115 dB. Find the ratio of the sound intensity of the jet engine to the sound intensity at the rock concert.

67. The equation $\beta = 10 \log (I/I_0)$, which defines the decibel, is sometimes written in terms of power P (in watts) rather than intensity I (in watts/meter^2). The form $\beta = 10 \log (P/P_0)$ can be used to compare two power levels in terms of decibels. Suppose that stereo amplifier A is rated at $P = 250$ watts per channel, while amplifier B has a rating of $P_0 = 45$ watts per channel. (a) Expressed in decibels, how much more powerful is A compared to B? (b) Will A sound more than twice as loud as B? Justify your answer.

68. For information, read problem 67 before working this problem. Stereo manufacturers express the power output of a stereo amplifier using the decibel, abbreviated as dBW, where the "W" indicates that a reference power level of $P_0 = 1.00$ W has been used. If an amplifier has a power rated at 17.5 dBW, how many watts of power can this amplifier deliver?

69. ssm A listener doubles his distance from a source that emits sound uniformly in all directions. By how many decibels does the sound intensity level change?

*__70.__ Two identical rifles are shot at the same time, and the sound intensity level is 80.0 dB. What would be the sound intensity level if only one rifle were shot? (*Hint: The answer is not 40.0 dB.*)

*__71.__ A portable radio is sitting at the edge of a balcony 5.1 m above the ground. The unit is emitting sound uniformly in all directions. By accident, it falls from rest off the balcony and continues to play on the way down. A gardener is working in a flower bed directly below the falling unit. From the instant the unit begins to fall, how much time is required for the sound intensity level heard by the gardener to increase by 10.0 dB?

*__72.__ At an outdoor party, Arnold is talking at a distance of 1.5 m from the beer keg. A number of other people at a distance of 3.5 m from the keg are also talking. Each individual, including Arnold, is producing the same sound power, which is assumed to spread out uniformly in all directions. The sound intensity at the beer keg due to all the others is at least as large as that produced by Arnold. What is the minimum number of other people talking?

73. ssm Suppose that when a certain sound intensity level (in dB) triples, the sound intensity (in W/m^2) also triples. Determine this sound intensity level.

74. A source emits sound uniformly in all directions. A radial line is drawn from this source. On this line, determine the positions of two points, 1.00 m apart, such that the intensity level at one point is 2.00 dB greater than that at the other.

Section 16.10 The Doppler Effect

75. At a football game, a stationary spectator is watching the halftime show. A trumpet player in the band is playing a 784-Hz tone while marching directly toward the spectator at a speed of 0.83 m/s. On a day when the speed of sound is 343 m/s, what frequency does the spectator hear?

76. A hawk is flying directly away from a bird watcher at a speed of 11.0 m/s. The hawk produces a shrill cry whose frequency is 865 Hz. The speed of sound is 343 m/s. What is the frequency that the bird watcher hears?

77. ssm From a vantage point very close to the track at a stock car race, you hear the sound emitted by a moving car. You detect a frequency that is 0.86 times smaller than that emitted by the car when it is stationary. The speed of sound is 343 m/s. What is the speed of the car?

78. Suppose you are stopped for a traffic light, and an ambulance approaches you from behind with a speed of 18 m/s. The siren on the ambulance produces sound with a frequency of 955 Hz. The speed of sound in air is 343 m/s. What is the wavelength of the sound reaching your ears?

79. A train is blowing its whistle while traveling at a speed of 33.0 m/s. The speed of sound is 343 m/s. Observer A is directly in front of the train, while observer B is directly behind it. Find the ratio of the whistle frequency heard by A to that heard by B.

*__80.__ A bicyclist and a car have the same speed and are moving toward each other. The car emits a sound of frequency 508 Hz, and the bicyclist hears the frequency as 542 Hz. The temperature of the air is 20 °C. How fast is each going?

*__81.__ **ssm** An aircraft carrier has a speed of 13.0 m/s relative to the water. A jet is catapulted from the deck and has a speed of 67.0 m/s relative to the water. The engines produce a 1550-Hz whine, and the speed of sound is 343 m/s. What is the frequency of the sound heard by the crew on the ship?

*__82.__ A microphone is moving in air toward a stationary source of sound (speed of sound = 343 m/s). The detected frequency is

83.0 Hz greater than the emitted frequency. When the microphone moves at the same speed toward the same stationary source in a liquid, the detected frequency is only 23.0 Hz greater than the emitted frequency. What is the speed of sound in the liquid?

*83. Two trucks travel at the same speed. They are far apart on adjacent lanes and approach each other essentially head-on. One driver hears the horn of the other truck at a frequency that is 1.20 times the frequency he hears when the trucks are stationary. The speed of sound is 343 m/s. At what speed is each truck moving?

*84. A bungee jumper jumps from rest and screams with a frequency of 589 Hz. The air temperature is 20 °C. What is the frequency heard by the people on the ground below when she has fallen a distance of 11.0 m? Assume that the bungee cord has not yet taken effect, so she is in free-fall.

*85. **ssm** A motorcycle starts from rest and accelerates along a straight line at 2.81 m/s². The speed of sound is 343 m/s. A siren at the starting point remains stationary. How far has the motorcycle gone when the driver hears the frequency of the siren at 90.0% of the value it has when the motorcycle is stationary?

**86. A microphone is attached to a spring that is suspended from the ceiling, as the drawing indicates. Directly below on the floor is a stationary 440-Hz source of sound. The microphone vibrates up and down in simple harmonic motion with a period of 2.0 s. The difference between the maximum and minimum sound frequencies detected by the microphone is 2.1 Hz. Ignoring any reflections of sound in the room and using 343 m/s for the speed of sound, determine the amplitude of the simple harmonic motion.

Sound source

ADDITIONAL PROBLEMS

87. The volume control on a stereo amplifier is adjusted so the sound intensity level increases from 23 to 61 dB. What is the ratio of the final sound intensity to the original sound intensity?

88. The magnetic tape of a cassette deck moves with a speed of 0.048 m/s ($1\frac{7}{8}$ inches per second). The recording head records a 15 000-Hz tone on the tape. What is the wavelength λ of the magnetized regions?

89. **ssm** Argon (molecular mass = 39.9 u) is a monatomic gas. Assuming that it behaves like an ideal gas at 298 K (γ = 1.67), find (a) the rms-speed of argon atoms and (b) the speed of sound in argon.

90. A vibrator moves one end of a rope up and down to generate a wave. The tension in the rope is 58 N. The frequency is then doubled. To what value must the tension be adjusted, so the new wave has the same wavelength as the old one?

91. The security alarm on a parked car goes off and produces a frequency of 960 Hz. The speed of sound is 343 m/s. As you drive toward this parked car, pass it, and drive away, you observe the frequency to change by 95 Hz. At what speed are you driving?

92. A person fishing from a pier observes that four wave crests pass by in 7.0 s and estimates the distance between two successive crests as 4.0 m. The timing starts with the first crest and ends with the fourth. What is the speed of the wave?

93. **ssm** Humans can detect a difference in sound intensity levels as small as 1.0 dB. What is the ratio of the sound intensities?

94. A rocket in a fireworks display explodes high in the air. The sound spreads out uniformly in all directions. The intensity of the sound is 2.0×10^{-6} W/m² at a distance of 120 m from the explosion. Find the distance from the source at which the intensity is 0.80×10^{-6} W/m².

95. The distance between a loudspeaker and the left ear of a listener is 2.70 m. (a) Calculate the time required for sound to travel this distance if the air temperature is 20 °C. (b) Assuming that the sound frequency is 523 Hz, how many wavelengths of sound are contained in this distance?

96. A wave is moving in the +x direction. Assuming that the wave has the following properties, write the equation of the wave (similar to Equation 16.3 or 16.4): speed = 7.1 m/s, amplitude = 0.15 m, wavelength = 0.28 m.

97. **ssm** A speeder looks in his rearview mirror. He notices that a police car has pulled behind him and is matching his speed of 38 m/s. The siren on the police car has a frequency of 860 Hz when the police car and the listener are stationary. The speed of sound is 343 m/s. What frequency does the speeder hear when the siren is turned on in the moving police car?

98. The right-most key on a piano produces a sound wave that has a frequency of 4185.6 Hz. Assuming that the speed of sound in air is 343 m/s, find the corresponding wavelength.

99. A rocket engine emits 2.0×10^5 J of sound energy every second. The sound is emitted uniformly in all directions. What is the sound intensity level, measured relative to the threshold of hearing, at a distance of 85 m away from the engine?

*100. Sound is coming through an open window whose dimensions are 1.1 m × 0.75 m. The sound intensity level is 95 dB above the threshold of hearing. How much sound *energy* comes through the window in one hour?

*101. **ssm** Two blocks are connected by a wire that has a mass per unit length of 8.50×10^{-4} kg/m. One block has a mass of 19.0 kg, while the other has a mass of 42.0 kg. These blocks are being pulled across a horizontal frictionless floor by a horizontal force **P** that is applied to the less massive block. A transverse wave travels on the wire between the blocks with a speed of 352 m/s (relative to the wire). The mass of the wire is negligible compared to the mass of the blocks. Find the magnitude of **P**.

*102. In Figure 16.3c the colored "dot" exhibits simple harmonic motion as the longitudinal wave passes. The wave has an amplitude of 5.4×10^{-3} m and a frequency of 4.0 Hz. Find the maximum acceleration of the dot.

*103. A steel cable, of cross-sectional area 2.83×10^{-3} m², is kept under a tension of 1.00×10^{4} N. The density of steel is 7860 kg/m³ (this is *not* the linear density). At what speed does a transverse wave move along the cable?

*104. Review Conceptual Example 3 before starting this problem. The amplitude and frequency of a transverse wave on a string are, respectively, 7.8×10^{-3} m and 6.0 Hz. Remembering that a particle of the string exhibits simple harmonic motion as the wave passes, determine the maximum speed of a particle.

*105. **ssm** Review Conceptual Example 9 as background for this problem. A loudspeaker is generating sound in a room. At a certain point, the sound waves coming directly from the speaker (without reflecting from the walls) create an intensity level of 75.0 dB. The waves reflected from the walls create, by themselves, an intensity level of 72.0 dB at the same point. These levels are relative to the threshold of hearing. Relative to the threshold of hearing, what is the total intensity level? *(Hint: The answer is not 147.0 dB.)*

*106. The sound intensity level of a person speaking normally is about 65 dB above the threshold of hearing. What is the minimum number of people speaking simultaneously, each with this intensity level, that is necessary to produce a sound intensity level at least 78 dB above the threshold of hearing?

**107. In a mixture of argon (atomic mass = 39.9 u) and neon (atomic mass = 20.2 u), the speed of sound is 363 m/s at 3.00×10^{2} K. Assume that both monatomic gases behave as ideal gases. Find the percentage of the atoms that are argon and the percentage that are neon.

**108. The sonar unit on a boat is designed to measure the depth of fresh water ($\rho = 1.00 \times 10^{3}$ kg/m³, $B_{ad} = 2.28 \times 10^{9}$ Pa). When the boat moves into salt water ($\rho = 1025$ kg/m³, $B_{ad} = 2.20 \times 10^{9}$ Pa), the sonar unit is no longer calibrated properly. In salt water, the sonar unit indicates the water depth to be 10.0 m. What is the actual depth of the water?

THE PRINCIPLE OF LINEAR SUPERPOSITION AND INTERFERENCE PHENOMENA

Wind and stringed instruments, such as those shown here, use standing waves to produce their sound. Standing waves are one manifestation of the principle of linear superposition, the subject of this chapter.

17.1 THE PRINCIPLE OF LINEAR SUPERPOSITION

As we have seen in the last chapter, sound is a pressure wave. Often, two or more sound waves are present at the same place at the same time, such as when everyone is talking at a party or when music plays from the speakers of a stereo system. To illustrate what happens when several waves pass simultaneously through the same region, let's first consider Figures 17.1 and 17.2, which show two transverse pulses of equal heights moving toward each other along a Slinky. In Figure 17.1 both pulses are "up," while in Figure 17.2 one is "up" and the other is "down." Part *b* of each drawing shows the two pulses beginning to overlap. The pulses merge, and the Slinky assumes a shape that is *the sum of the shapes of the individual pulses.* Thus, when the two "up" pulses overlap completely, as in Figure 17.1*c*, the Slinky has a pulse height that is twice the height of an individual pulse. Likewise, when the "up" pulse and the "down" pulse overlap exactly, as in Figure 17.2*c*, they momentarily cancel, and the Slinky becomes straight. In either case, the two pulses move apart after overlapping, and the Slinky once again conforms to the shapes of the individual pulses.

The adding together of individual pulses to form a resultant pulse is an example of a more general concept called the ***principle of linear superposition.***

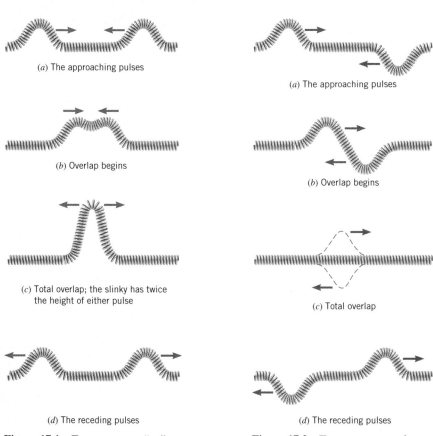

(a) The approaching pulses

(b) Overlap begins

(c) Total overlap; the slinky has twice the height of either pulse

(d) The receding pulses

Figure 17.1 Two transverse "up" pulses passing through each other.

(a) The approaching pulses

(b) Overlap begins

(c) Total overlap

(d) The receding pulses

Figure 17.2 Two transverse pulses, one "up" and one "down," passing through each other.

> ### ■ THE PRINCIPLE OF LINEAR SUPERPOSITION
>
> When two or more waves are present simultaneously at the same place, the resultant wave is the sum of the individual waves.

This principle can be applied to all types of waves, including sound waves, water waves, and electromagnetic waves such as light. It embodies one of the most important concepts in all of physics, and the remainder of this chapter deals with examples related to it.

17.2 CONSTRUCTIVE AND DESTRUCTIVE INTERFERENCE OF SOUND WAVES

Suppose that the sounds from two speakers overlap in the middle of a listening area, as in Figure 17.3, and that each speaker produces a sound wave of the same amplitude and frequency. For convenience, the wavelength of the sound is chosen to be $\lambda = 1$ m. In addition, assume the diaphragms of the speakers vibrate in phase; that is, they move outward together and inward together. If the distance of each speaker from the overlap point is the same (3 m in the drawing), the condensations (C) of one wave always meet the condensations of the other when the waves come together; similarly, rarefactions (R) always meet rarefactions. Figure 17.4 shows the pressure patterns of the individual sounds, as well as the combined pressure pattern at the overlap point. According to the principle of linear superposition, the combined pattern is the sum of the individual patterns. As a result, the pressure fluctuations at the overlap point have twice the amplitude that the individual waves have, and a listener at this spot hears a louder sound than that coming from either speaker alone. When two waves always meet condensation-to-condensation and rarefaction-to-rarefaction (or crest-to-crest and trough-to-trough), they are said to be *exactly in phase* and to exhibit *constructive interference.*

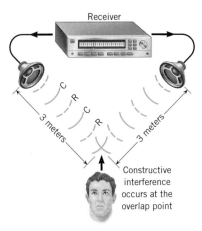

Figure 17.3 As a result of constructive interference between the two sound waves, a loud sound is heard at an overlap point located equally distant from two in-phase speakers (C, condensation; R, rarefaction).

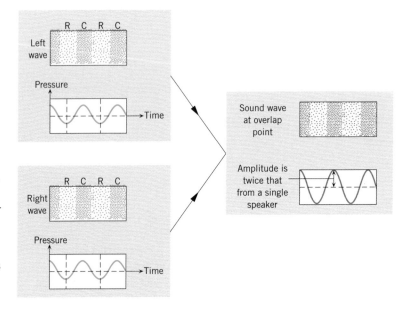

Figure 17.4 When sound waves from the left and right speakers in Figure 17.3 arrive at the overlap point, condensations (C) always meet condensations and rarefactions (R) always meet rarefactions, leading to constructive interference and a loud sound. The drawings in red are obtained by adding the drawings in blue.

Now consider what happens if one of the speakers is moved. The result is surprising. In Figure 17.5, the left speaker is moved away* from the overlap point by a distance equal to one-half of the wavelength, or 0.5 m. Therefore, at the overlap point, a condensation arriving from the left meets a rarefaction arriving from the right. Likewise, a rarefaction arriving from the left meets a condensation arriving from the right. According to the principle of linear superposition, the net effect is a mutual cancellation of the two waves, as Figure 17.6 emphasizes. The condensations from one wave offset the rarefactions from the other, leaving only a *constant air pressure.* A constant air pressure, devoid of condensations and rarefactions, means that a listener detects no sound. When two waves always meet condensation-to-rarefaction (or crest-to-trough), they are said to be ***exactly out of phase*** and to exhibit ***destructive interference.***

When two waves meet, they interfere constructively if they always meet exactly in phase and destructively if they always meet exactly out of phase. In either case, the wave patterns do not shift relative to one another as time passes. Sources that produce waves in this fashion are called ***coherent sources.***

Destructive interference is the basis of a useful technique for reducing the loudness of undesirable sounds. For instance, Figure 17.7 shows a pair of noise-canceling headphones. Small microphones are mounted inside the headphones and detect noise such as the engine noise that an airplane pilot would hear. The headphones also contain circuitry to process the electronic signals from the microphones and reproduce the noise in a form that is exactly out of phase compared to the original. This out-of-phase version is played back through the headphone speakers and, because of destructive interference, combines with the original noise to produce a quieter background.

It should be apparent that if the left speaker in Figure 17.5 were moved away from the overlap point by *another* one-half wavelength ($3\frac{1}{2}$ m + $\frac{1}{2}$ m = 4 m), the

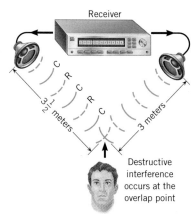

Figure 17.5 The speakers in this drawing vibrate in phase. However, the left speaker is one-half of a wavelength ($\frac{1}{2}$ m) farther from the overlap point than the right speaker. Because of destructive interference, no sound is heard at the overlap point (C, condensation; R, rarefaction).

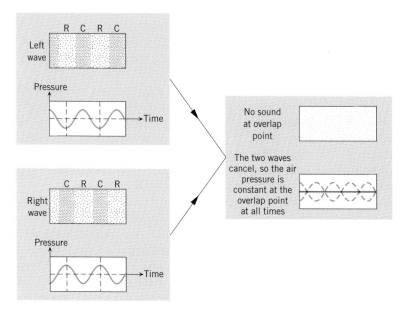

Figure 17.6 The sound waves from the left and right speakers in Figure 17.5 arrive at the overlap point in such a way that the condensations (C) from one speaker always meet the rarefactions (R) from the other, leading to destructive interference and an absence of sound. The drawings in red are obtained by adding the drawings in blue.

---

* When the left speaker is moved back, its sound intensity and, hence, its pressure amplitude decrease at the overlap point. In this chapter assume that the power delivered to the left speaker by the receiver is increased slightly to keep the amplitudes equal at the overlap point.

Figure 17.7 Noise-canceling head-phones utilize destructive interference.

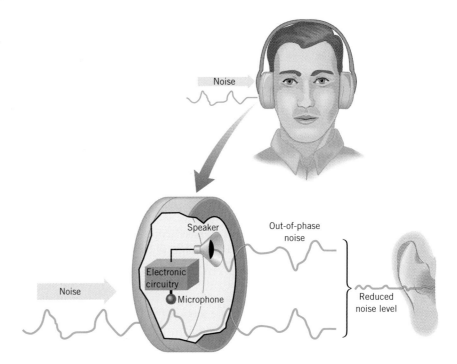

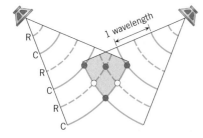

The Physics of...
noise-canceling headphones.

two waves would again be in phase, and constructive interference would occur. The listener would hear a loud sound, because the left wave travels one whole wavelength ($\lambda = 1$ m) farther than the right wave and, at the overlap point, condensation meets condensation and rarefaction meets rarefaction. In general, the important issue is the *difference* in the distances traveled by each wave in reaching the overlap point.

For two wave sources vibrating in phase, a difference in path lengths that is an integer number (1, 2, 3, . . .) of wavelengths leads to constructive interference; a difference in path lengths that is a half-integer number ($\frac{1}{2}$, $1\frac{1}{2}$, $2\frac{1}{2}$, . . .) of wavelengths leads to destructive interference.

Interference effects can also be detected if the two speakers are fixed in position and the listener moves about the room. Consider Figure 17.8, where the sound waves spread outward from each speaker, as indicated by the concentric circular arcs. Each solid arc represents the center of a condensation, while each dashed arc represents the center of a rarefaction. Where the two waves overlap, there are places of constructive interference and places of destructive interference. Constructive interference occurs at any spot where two condensations or two rarefactions intersect, and the drawing shows four such places as solid dots. A listener stationed at any one of these locations hears a loud sound. On the other hand, destructive interference occurs at any place where a condensation and a rarefaction intersect, such as the two open dots in the picture. A listener situated at a point of destructive interference hears no sound. At locations where neither constructive nor destructive interference occurs, the two waves partially reinforce or partially cancel, depending on the position relative to the speakers. Thus, it is possible for a listener to walk about the overlap region and hear marked variations in loudness.

The individual sound waves from the speakers in Figure 17.8 carry energy, and the resultant wave in the overlap region has the same energy as the sum of the energies of the individual waves. This fact is consistent with the principle of conserva-

Figure 17.8 The overlapping of two sound waves produces interference effects in the shaded region. The solid lines denote the centers of condensations (C), while the dashed lines denote the centers of rarefactions (R). Constructive interference occurs at each solid dot (●), and destructive interference occurs at each open dot (○).

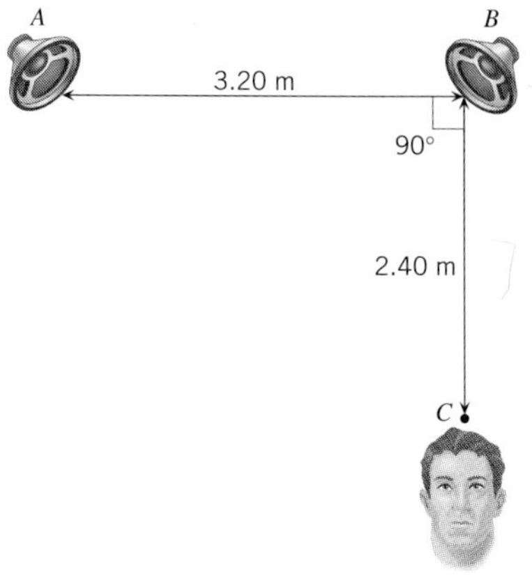

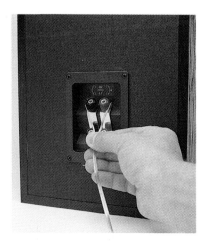

Figure 17.9 Example 1 assumes that the speakers are vibrating in phase and discusses whether this setup leads to constructive or destructive interference at point C for 214-Hz sound waves.

tion of energy, which we first encountered in Section 6.8. This principle states that energy can neither be created nor destroyed, but can only be converted from one form to another. One of the interesting consequences of interference is that the energy is redistributed, so there are places within the overlap region where the sound is loud and other places where there is no sound at all. Interference, so to speak, "robs Peter to pay Paul," but energy is always conserved in the process. Example 1 illustrates how to decide what a listener hears.

EXAMPLE 1 • What Does a Listener Hear?

In Figure 17.9 two in-phase loudspeakers, A and B, are separated by 3.20 m. A listener is stationed at point C, which is 2.40 m in front of speaker B. The triangle ABC is a right triangle. Both speakers are playing identical 214-Hz tones, and the speed of sound is 343 m/s. Does the listener hear a loud sound or no sound?

Reasoning The listener will hear either a loud sound or no sound, depending upon whether the interference occurring at point C is constructive or destructive. To determine which it is, we need to find the difference in the distances traveled by the two sound waves that reach point C and see whether the difference is an integer or half-integer number of wavelengths. In either event, the wavelength can be found from Equation 16.1 ($\lambda = v/f$).

Solution Since the triangle ABC is a right triangle, the distance AC is given by the Pythagorean theorem as $\sqrt{(3.20 \text{ m})^2 + (2.40 \text{ m})^2} = 4.00$ m. The distance BC is given as 2.40 m. Thus, we find that the difference in the travel distances for the waves is 4.00 m − 2.40 m = 1.60 m. The wavelength of the sound is

$$\lambda = \frac{v}{f} = \frac{343 \text{ m/s}}{214 \text{ Hz}} = 1.60 \text{ m} \qquad (16.1)$$

Since the difference in the distances is one wavelength, constructive interference occurs at point C, and ***the listener hears a loud sound.***

Up to this point, we have been assuming that the speaker diaphragms vibrate synchronously or in phase; that is, they move outward together and inward together. This may not be the case, however, and Conceptual Example 2 considers what happens then.

CONCEPTUAL EXAMPLE 2 • Out-of-Phase Speakers

To make a speaker operate, two wires must be connected between the speaker and the receiver (amplifier), as in Figure 17.10. To ensure that the diaphragms of two speakers vibrate in phase, it is necessary to make these connections in exactly the same way. If the wires for one speaker are not connected just as they are for the other speaker, the two diaphragms will vibrate out of phase. Whenever one diaphragm moves outward, the other will move inward, and vice versa. Suppose that in Figures 17.3 and 17.5 the connections are made so that the speaker diaphragms vibrate out of phase, everything else remaining the same. In each case, what kind of interference would result at the overlap point?

Reasoning and Solution Since the diaphragms are vibrating out of phase, one of them is now moving exactly opposite to the way it was moving originally. Let us as-

Figure 17.10 A loudspeaker is connected to a receiver (amplifier) by two wires.

sume that it is the one on the left in the drawings. The effect of this change is that every condensation originating from the left speaker becomes a rarefaction and every rarefaction becomes a condensation. As a result, in Figure 17.3 a rarefaction from the left now meets a condensation from the right at the overlap point, and ***destructive interference results.*** In Figure 17.5, a condensation from the left now meets a condensation from the right at the overlap point, ***leading to constructive interference.*** These results are opposite to those shown in the figures.

Instructions for connecting stereo systems specifically warn users to avoid out-of-phase vibration of the speaker diaphragms. One way to check your system is to play some music that is rich in low-frequency bass tones. Use the monaural mode on your receiver so the same sound is coming from each speaker. While the music plays, slide the speakers toward each other. If the diaphragms are moving in phase, the bass sound will either remain the same or slightly improve as the speakers come together. If the diaphragms are moving out of phase, the bass sound will decrease noticeably because of destructive interference when the speakers are right next to each other. In this event, simply interchange the wires to the terminals on one (but not both) of the speakers.

Related Homework Material: *Problem 8*

The Physics of...
wiring stereo speakers.

The phenomena of constructive and destructive interference are exhibited by all types of waves, not just sound waves. We will encounter interference effects again in Chapter 27, in connection with light waves.

17.3 DIFFRACTION

Section 16.5 discusses the fact that sound is a pressure wave created by a vibrating object, such as a loudspeaker. The previous two sections of this chapter have examined what happens when two sound waves are present simultaneously at the same place; according to the principle of linear superposition, a resultant wave is formed from the sum of the individual waves. This principle reveals that overlapping sound waves exhibit interference effects, whereby the sound energy is redistributed within the overlap region. As the concept chart in Figure 17.11 shows, we will now use the principle of linear superposition to explore another interference effect, that of diffraction.

When a wave encounters an obstacle or the edges of an opening, it bends around them. For instance, a sound wave produced by a stereo system bends around the edges of an open doorway, as Figure 17.12*a* illustrates. If such bending did not occur, sound could be heard outside the room only at locations directly in front of the doorway, as part *b* of the drawing suggests. (It is assumed that no sound is transmitted directly through the walls.) The bending of a wave around an obstacle or the edges of an opening is called ***diffraction.*** All kinds of waves exhibit diffraction.

To demonstrate how the bending of waves arises, Figure 17.13 shows an expanded view of Figure 17.12. When the sound wave reaches the doorway, the air in the doorway is set into longitudinal vibration. In effect, each air molecule in the doorway then becomes a source of a sound wave in its own right, and, for purposes of illustration, the drawing shows two of the molecules. Each molecule produces a sound wave that expands outward in three dimensions, much like a water wave does in two dimensions when a stone is dropped into a pond. The sound waves generated by all the molecules in the doorway must be added together to obtain the total

Concepts at a Glance

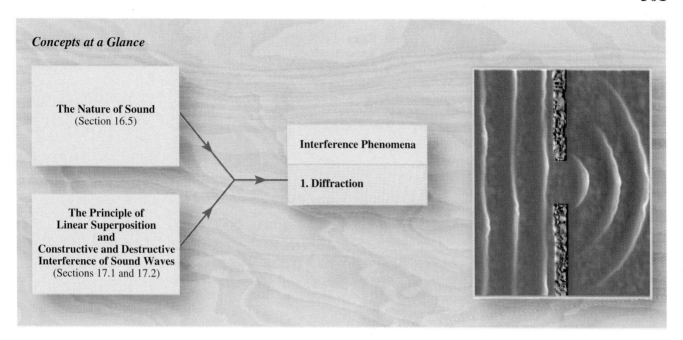

The Nature of Sound
(Section 16.5)

Interference Phenomena

1. Diffraction

The Principle of
Linear Superposition
and
Constructive and Destructive
Interference of Sound Waves
(Sections 17.1 and 17.2)

sound intensity at any location outside the room, in accord with the principle of linear superposition. However, even considering only the waves from the two molecules in the picture, it is clear that the expanding wave patterns reach locations off to either side of the doorway. The net effect is a "bending" or diffraction of the sound around the edges of the opening. Further insight into the origin of diffraction can be obtained with the aid of Huygens' principle (see Section 27.5).

When the sound waves generated by every molecule in the doorway are added together, it is found that there are places where the intensity is a maximum and places where it is zero, in a fashion similar to that discussed in the previous section. Analysis shows that at a great distance from the doorway the intensity is a maximum directly opposite the center of the opening. As the distance to either side of the center increases, the intensity decreases and reaches zero, then rises again to a maximum, falls again to zero, rises back to a maximum, and so on. Only the maximum at the center is a strong one. The other maxima are weak and become progres-

Figure 17.11 *Concepts at a Glance*
By combining our knowledge of sound waves with the principle of linear superposition, we can understand the interference phenomenon of diffraction. The drawing shows a top view of a water wave traveling to the right and passing through an opening in a breakwater. The bending of the wave in the region to the right of the breakwater is an example of diffraction.

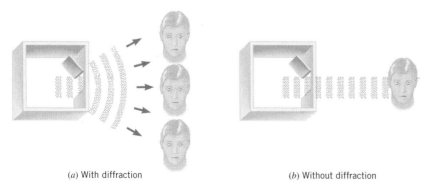

(a) With diffraction   (b) Without diffraction

Figure 17.12 (a) The bending of a sound wave around the edges of the doorway is an example of diffraction. The source of the sound within the room is not shown. (b) If diffraction did not occur, the sound wave would not bend as it passes through the doorway.

Figure 17.13 Each vibrating air molecule in the doorway generates a sound wave that expands outward and bends, or diffracts, around the edges of the doorway. Because of interference effects among the sound waves produced by all the molecules in the doorway, the sound intensity is mostly confined to the region defined by the angle θ on either side of the doorway.

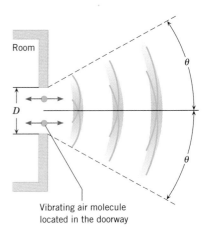

Diffraction horn

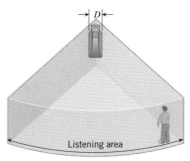

Listening area

(a) Correct mounting

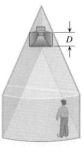

(b) Incorrect mounting

Figure 17.14 To achieve a wide dispersion of sound into the listening area, a diffraction horn loudspeaker must be mounted as in part a and not as in part b.

sively weaker at greater distances from the center. In Figure 17.13 the angle θ defines the location of the first minimum intensity point on either side of the center. Equation 17.1 gives θ in terms of the wavelength λ and the width D of the doorway and assumes that the doorway can be treated like a slit whose height is very large compared to its width:

Single slit—first minimum $$\sin \theta = \frac{\lambda}{D} \tag{17.1}$$

Waves also bend around the edges of openings other than single slits. Particularly important is the diffraction of sound by a circular opening. In this case, the angle θ is related to the wavelength λ and the diameter D of the opening by

Circular opening
—first minimum $$\sin \theta = 1.22 \frac{\lambda}{D} \tag{17.2}$$

An important point to remember about Equations 17.1 and 17.2 is that the extent of the diffraction depends on the ratio of the wavelength to the size of the opening. If the ratio λ/D is small, then θ is small. Little diffraction occurs, and the waves are beamed in the forward direction as they leave an opening, much like the light from a flashlight. Such sound waves are said to have "narrow dispersion." Since high-frequency sound has a relatively small wavelength, it tends to have a narrow dispersion. On the other hand, for larger values of the ratio λ/D, the angle θ is larger. The waves spread out over a larger region and are said to have a "wide dispersion." Low-frequency sound, with its relatively large wavelength, typically has a wide dispersion.

Loudspeaker designers, for instance, utilize a large value of λ/D in the type of speaker known as a diffraction horn. They choose the width D of the horn (see Figure 17.14) to be much smaller than the wavelengths of the sounds that are to be reproduced. Figure 17.14a shows a diffraction horn mounted so that diffraction spreads the sound into the widest possible listening area. If diffraction is to be an effective spreading mechanism, the width D must be parallel to the floor, as in part a of the drawing, and not perpendicular to the floor as in part b. The orientation in part b is incorrect, because it assumes that the wide flaring of the horn spreads out the sound, which is not the case in a diffraction horn.

Diffraction horns are only one type of loudspeaker. Other types allow the sound to emerge through circular openings, and Example 3 illustrates one way in which wide dispersion is achieved in such speakers.

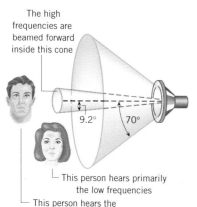

The high frequencies are beamed forward inside this cone

9.2°   70°

This person hears primarily the low frequencies

This person hears the high and low frequencies

Figure 17.15 Because the dispersion of high frequencies is less than that of low frequencies, you should be directly in front of the speaker to hear both the high and low frequencies equally well.

Tweeter

Figure 17.16 Small-diameter speakers, called tweeters, are used to produce high-frequency sound. The small diameter helps to promote a wider dispersion of the sound.

EXAMPLE 3 • Designing a Loudspeaker for Wide Dispersion

(a) A 1500-Hz sound and a 8500-Hz sound each come from a loudspeaker whose diameter is 0.30 m. Assuming that the speed of sound in air is 343 m/s, find the diffraction angle θ for each sound. (b) A second speaker is to be designed for the 8500-Hz sound. The goal is to produce this sound with the same wide dispersion that the 0.30-m speaker gives to the 1500-Hz sound. What should be the diameter of the second speaker?

Reasoning The diffraction angle θ for each sound wave is given by $\sin \theta = 1.22(\lambda/D)$. In part (a), we will use this equation to find the angles. In part (b), knowing the angle, we will use the equation to find the diameter. In either case, it is first necessary to calculate the wavelengths of the sounds from $\lambda = v/f$ (Equation 16.1).

Solution

(a) The wavelengths of the two sounds are

$$\lambda_{1500} = \frac{343 \text{ m/s}}{1500 \text{ Hz}} = 0.23 \text{ m} \quad \text{and} \quad \lambda_{8500} = \frac{343 \text{ m/s}}{8500 \text{ Hz}} = 0.040 \text{ m}$$

The diffraction angles can now be determined:

1500-Hz sound $\sin \theta = 1.22 \dfrac{\lambda_{1500}}{D} = 1.22 \left(\dfrac{0.23 \text{ m}}{0.30 \text{ m}} \right) = 0.94$ (17.2)

$\theta = \sin^{-1} 0.94 = \boxed{70°}$

8500-Hz sound $\sin \theta = 1.22 \dfrac{\lambda_{8500}}{D} = 1.22 \left(\dfrac{0.040 \text{ m}}{0.30 \text{ m}} \right) = 0.16$ (17.2)

$\theta = \sin^{-1} 0.16 = \boxed{9.2°}$

Figure 17.15 illustrates these results.

(b) To find the speaker diameter that will give a $\theta = 70°$ dispersion to the 8500-Hz sound, we again use the relation $\sin \theta = 1.22(\lambda/D)$:

$$D = \frac{1.22 \, \lambda_{8500}}{\sin \theta} = \frac{1.22(0.040 \text{ m})}{\sin (70°)} = \boxed{0.052 \text{ m}}$$

This result shows that high-frequency sound can have a wide dispersion, *provided the diameter of the speaker is small enough*. Accordingly, loudspeaker designers use a small diameter speaker called a tweeter to generate high-frequency sound, as Figure 17.16 indicates.

• **PROBLEM SOLVING INSIGHT**
When a wave passes through an opening, the extent of diffraction is greater when the ratio λ/D is greater, where λ is the wavelength of the wave and D is the width or diameter of the opening.

The Physics of...
tweeter loudspeakers.

As we have seen, diffraction is an interference effect, one in which some of the wave's energy is directed into regions that would otherwise not be accessible. Energy, of course, is conserved during this process, for energy is only redistributed during diffraction; no energy is created or destroyed.

17.4 BEATS

Constructive and destructive interference can occur when two sounds with the *same frequency* overlap. Now we consider what happens when the two overlapping waves have *slightly different frequencies*. As the concept chart in Figure 17.17 illustrates, this gives rise to the phenomenon of beats, which can also be understood with the aid of the principle of linear superposition.

Consider Figure 17.18, which shows sound waves coming from two tuning forks placed side by side. A tuning fork has the property of producing a single-frequency sound wave when struck with a sharp blow. The two tuning forks in the drawing are identical, and each is designed to produce a 440-Hz tone. However, a small piece of putty has been attached to one fork, whose frequency is lowered to 438 Hz because of the added mass. When the forks are sounded simultaneously, the loudness of the resulting sound rises and falls periodically—faint, then loud, then faint, then loud, and so on. The periodic variations in loudness are called *beats* and result from the interference between two sound waves with slightly different frequencies.

For clarity, Figure 17.18 shows the condensations and rarefactions of the sound waves separately. In reality, however, the waves spread out and overlap. In accord with the principle of linear superposition, the ear detects the combined total of the two. Notice that there are places where the waves interfere constructively and places where they interfere destructively. When a region of constructive interference reaches the ear, a loud sound is heard. When a region of destructive interference arrives, the sound intensity drops to zero (assuming each of the waves has the same amplitude). The number of times per second that the loudness rises and falls

Figure 17.17 *Concepts at a Glance*
This figure, which is an extension of the concept chart in Figure 17.11, shows that the principle of linear superposition can be used to explain the phenomenon of beats. Piano tuners and other musicians routinely use beats in the process of tuning their instruments.

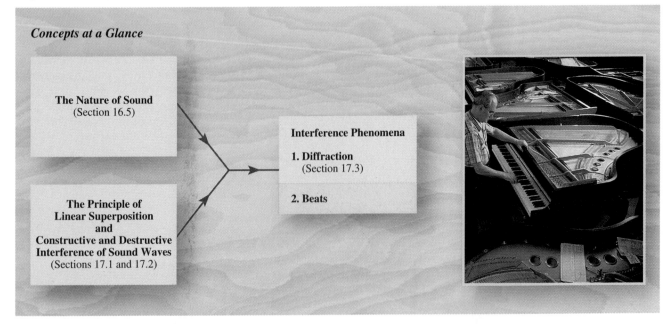

Concepts at a Glance

The Nature of Sound
(Section 16.5)

The Principle of
Linear Superposition
and
Constructive and Destructive
Interference of Sound Waves
(Sections 17.1 and 17.2)

Interference Phenomena

1. Diffraction
(Section 17.3)

2. Beats

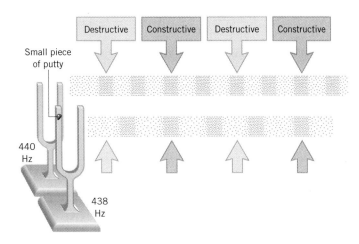

Figure 17.18 Two tuning forks have slightly different frequencies of 440 and 438 Hz. The phenomenon of beats occurs when the forks are sounded simultaneously. The sound waves are not drawn to scale.

is the ***beat frequency*** and is the ***difference*** between the two sound frequencies. Thus, in the situation illustrated in Figure 17.18, an observer hears the sound loudness rise and fall at the rate of 2 times per second (440 Hz − 438 Hz).

Figure 17.19 helps to explain why the beat frequency is the difference between the two frequencies. The drawing displays graphical representations of the pressure patterns of a 10-Hz wave and a 12-Hz wave, along with the pressure pattern that results when the two overlap. These frequencies have been chosen for convenience, even though they lie below the audio range and are inaudible. Audible sound waves behave in exactly the same way. The top two drawings in blue show the pressure variations in a one-second interval of each wave. The third drawing in red shows the result of adding together the blue patterns according to the principle of linear superposition. Notice that the amplitude in the red drawing is not constant, as it is in the individual waves. Instead, the amplitude changes from a minimum to a maximum, back to a minimum, etc. When such pressure variations reach the ear and occur in the audible frequency range, they produce a loud sound when the amplitude is a maximum, and a faint sound when the amplitude is a minimum. Two loud-faint cycles, or beats, occur in the one-second interval shown in the drawing, corresponding to a beat frequency of 2 Hz. This is consistent with our earlier statement that the

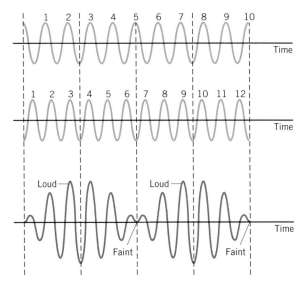

Figure 17.19 A 10-Hz sound wave and a 12-Hz sound wave, when added together, produce a wave with a beat frequency of 2 Hz. The drawings show the pressure patterns (in blue) of the individual waves and the pressure pattern (in red) that results when the two overlap. The time interval shown is one second.

beat frequency is the difference between the frequencies of the individual waves, that is, 12 Hz − 10 Hz = 2 Hz.

Musicians often tune their instruments by listening to a beat frequency. For instance, a guitar player sounds an out-of-tune string along with a tone from a source known to have the correct frequency. The guitarist adjusts the tension in the string until the beats vanish, ensuring that the string is vibrating at the correct frequency.

17.5 TRANSVERSE STANDING WAVES

A standing wave is another interference effect that can occur when two waves overlap. Standing waves can arise with transverse waves, such as those on a guitar string, and with longitudinal sound waves, such as those in a flute. As the concept chart in Figure 17.20 indicates, standing waves are another phenomenon that can be explained by using the principle of linear superposition.

Figure 17.21 shows some of the essential features of transverse standing waves. In this figure the left end of each string is vibrated back and forth, while the right end is attached to a wall. Regions of the string move so fast that they appear only as a "blur" in the photographs. Each of the patterns shown is called a *transverse standing wave pattern.* Notice that the patterns include special places called nodes and antinodes. The *nodes* are places that do not vibrate at all, and the *antinodes* are places where maximum vibration occurs. To the right of each photograph is a series of superimposed drawings that help us to visualize the motion of the string as it vibrates in a standing wave pattern. These drawings freeze the shape of the string at various times and emphasize the maximum vibration that occurs at an antinode with the aid of a red "dot" attached to the string.

Each standing wave pattern is produced at a unique frequency of vibration. These frequencies form a series, the smallest frequency f_1 corresponding to the one-loop pattern and the larger frequencies being integer multiples of f_1, as Figure 17.21 indicates. Thus, if f_1 is 10 Hz, the frequency needed to establish the 2-loop pattern

Figure 17.20 *Concepts at a Glance* This chart, which is a continuation of those in Figures 17.11 and 17.17, indicates that transverse and longitudinal standing waves are related to the principle of linear superposition. Jamey Turner entertains a crowd with a glass harp. The harp is a collection of glasses that are partially filled with water. The sound arises when a longitudinal standing wave is established in the air above the water in each glass.

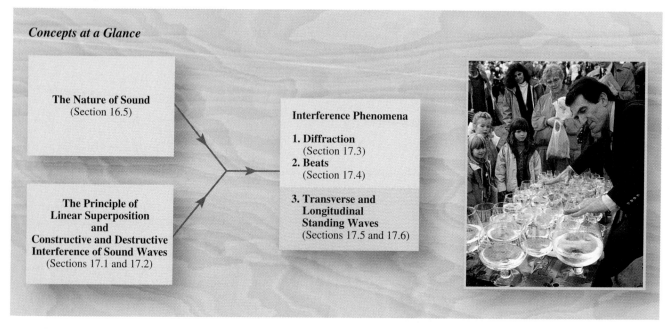

Concepts at a Glance

The Nature of Sound
(Section 16.5)

The Principle of
Linear Superposition
and
Constructive and Destructive
Interference of Sound Waves
(Sections 17.1 and 17.2)

Interference Phenomena

1. **Diffraction**
   (Section 17.3)
2. **Beats**
   (Section 17.4)

3. **Transverse and Longitudinal Standing Waves**
   (Sections 17.5 and 17.6)

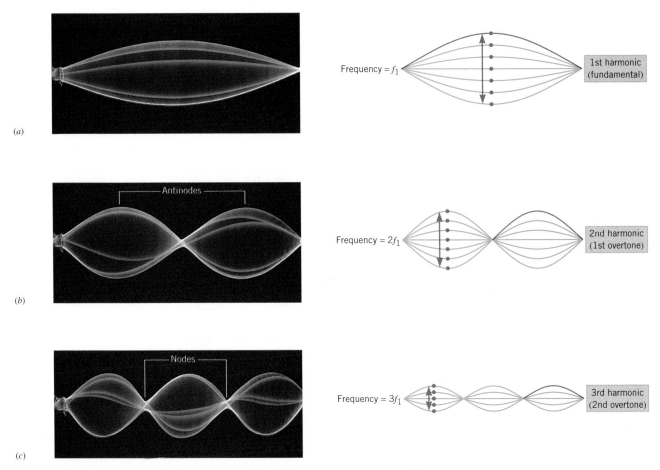

(a)

(b)

(c)

Figure 17.21 Vibrating a string at certain unique frequencies sets up transverse standing wave patterns, such as the three shown in the photographs on the left. Each illustration on the right shows the various shapes that the string assumes at various times as it vibrates. The red dots attached to the strings focus attention on the maximum vibration that occurs at an antinode. In each of the drawings on the right, one-half of a wave cycle is outlined in red.

is $2f_1$ or 20 Hz, while that needed to create the 3-loop pattern is $3f_1$ or 30 Hz, and so on. The frequencies in this series (f_1, $2f_1$, $3f_1$, etc.) are called **harmonics.** The lowest frequency f_1 is called the first harmonic, and the higher frequencies are labeled as the second harmonic ($2f_1$), the third harmonic ($3f_1$), and so forth. The harmonic number (1st, 2nd, 3rd, etc.) corresponds to the number of loops in the standing wave pattern. The frequencies in this series are also referred to as the fundamental frequency, the first overtone, the second overtone, and so on. Thus, frequencies above the fundamental are **overtones** (see Figure 17.21).

Standing waves arise because identical waves travel on the string in *opposite directions* and combine in accord with the principle of linear superposition. A standing wave is said to be "standing," since it does not travel in one direction or the other, as do the individual waves that produce it. Figure 17.22 shows why there are waves traveling in both directions on the string. At the top of the picture, one-half of a wave cycle (the remainder of the wave is omitted for clarity) is moving toward the wall on the right. When the half-cycle reaches the wall, it causes the string to pull upward on the wall. Consistent with Newton's action–reaction law, the wall

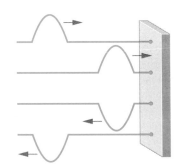

Figure 17.22 In reflecting from the wall, a forward-traveling half-cycle becomes a backward-traveling half-cycle that is inverted.

pulls downward on the string, and a downward-pointing half-cycle is sent back toward the left. Thus, the wave reflects from the wall. Upon arriving back at the point of origin, the wave reflects again, this time from the hand vibrating the string. The hand is essentially fixed (the vibration amplitude of the hand is assumed to be small) and behaves as the wall does in causing reflections. Repeated reflections at both ends of the string create a multitude of wave cycles traveling in both directions.

As each new cycle is formed by the vibrating hand, previous cycles that have reflected from the wall arrive and reflect again from the hand. Unless the timing is right, however, the new cycles and the reflected cycles tend to offset one another, and the formation of a standing wave is inhibited. Think about pushing someone on a swing and timing your pushes so that the effect of one push reinforces that of another. Such reinforcement in the case of the wave cycles leads to a large amplitude standing wave. Suppose the string has a length L and its left end is being vibrated at a frequency f_1. The time required to create a new wave cycle is the period T of the wave, where $T = 1/f_1$. On the other hand, the time needed for a cycle to travel from the hand to the wall and back, a distance of $2L$, is $2L/v$, where v is the wave speed. Reinforcement between new and reflected cycles occurs if these two times are equal; that is, if $1/f_1 = 2L/v$. Thus, a standing wave is established when the string is vibrated with a frequency of $f_1 = v/(2L)$.

Repeated reinforcement between newly created and reflected cycles causes a large amplitude standing wave to develop on the string, *even though the hand itself vibrates with only a small amplitude.* Thus, the motion of the string is a resonance effect, analogous to that discussed in Section 10.8 for an object attached to a spring. The frequency f_1 at which resonance occurs is sometimes called a ***natural frequency*** of the string, similar to the frequency at which an object oscillates on a spring.

There is a difference between the resonance of the string and the resonance of a spring system, however. An object on a spring has only a single natural frequency, whereas the string has a *series* of natural frequencies. The series arises because a reflected wave cycle need not return to its point of origin in time to reinforce *every* newly created cycle. Reinforcement can occur, for instance, on *every other* new cycle, as it does if the string is vibrated at twice the frequency f_1, or $f_2 = 2f_1$. Likewise, if the vibration frequency is $f_3 = 3f_1$, reinforcement occurs on *every third* new cycle. Similar arguments apply for any frequency $f_n = nf_1$, where n is an integer. As a result, the series of natural frequencies that lead to standing waves on a string fixed at both ends is given by

String fixed at both ends
$$f_n = n\left(\frac{v}{2L}\right) \qquad n = 1, 2, 3, 4, \ldots \qquad (17.3)$$

• PROBLEM SOLVING INSIGHT
The distance between two successive nodes (or between two successive antinodes) of a standing wave is equal to one-half of a wavelength.

It is also possible to obtain Equation 17.3 in another way. In Figure 17.21, one-half of a wave cycle is outlined in red for each of the harmonics, to show that each "loop" in a standing wave pattern corresponds to one-half a wavelength. Since the two fixed ends of the string are nodes, the length L of the string must contain an integer number n of half-wavelengths: $L = n(\frac{1}{2}\lambda_n)$ or $\lambda_n = 2L/n$. Using this result for the wavelength in the relation $f_n\lambda_n = v$ shows that $f_n(2L/n) = v$, which can be rearranged to give Equation 17.3.

Standing waves on a string play an important role in the way many musical instruments produce sound. For instance, a guitar string is stretched between two supports and, when plucked, vibrates according to the series of natural frequencies given by Equation 17.3. The next two examples deal with how this series of frequencies governs the playing and the design of a guitar.

EXAMPLE 4 • Playing a Guitar

The heaviest string on an electric guitar has a linear density of $m/L = 5.28 \times 10^{-3}$ kg/m and is stretched with a tension of $F = 226$ N. This string produces the musical note E when vibrating along its entire length in a standing wave at the fundamental frequency of 164.8 Hz. (a) Find the length L of the string between its two fixed ends (see Figure 17.23*a*). (b) A guitar player wants the string to vibrate at a fundamental frequency of 2×164.8 Hz = 329.6 Hz, as it must if the musical note E is to be sounded one octave higher in pitch. To accomplish this, he presses the string against the proper fret and then plucks the string (see part *b* of the drawing). Find the distance L between the fret and the bridge of the guitar.

Reasoning The fundamental frequency f_1 is given by Equation 17.3 with $n = 1$: $f_1 = v/(2L)$. Since f_1 is known in both parts (a) and (b), the length L in each case can be calculated directly from this expression, once the speed v is known. The speed, in turn, is related to the tension F and the linear density m/L according to Equation 16.2.

Solution

(a) The speed is

$$v = \sqrt{\frac{F}{m/L}} = \sqrt{\frac{226 \text{ N}}{5.28 \times 10^{-3} \text{ kg/m}}} = 207 \text{ m/s} \qquad (16.2)$$

According to $f_1 = v/(2L)$, the length of the string is

$$L = \frac{v}{2f_1} = \frac{207 \text{ m/s}}{2(164.8 \text{ Hz})} = \boxed{0.628 \text{ m}}$$

(b) The distance L that locates the fret can be determined exactly as in part (a) by using the wave speed $v = 207$ m/s and noting that the frequency is now $f_1 = 329.6$ Hz: $\boxed{L = 0.314 \text{ m}}$. This length is exactly half that determined in part (a), because the frequencies have a ratio of 2:1.

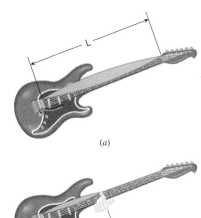

(a)

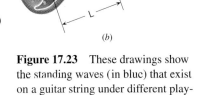

(b)

Figure 17.23 These drawings show the standing waves (in blue) that exist on a guitar string under different playing conditions.

CONCEPTUAL EXAMPLE 5 • The Frets on a Guitar

Figure 17.24 shows the frets on the neck of a guitar. They allow the player to produce a complete sequence of musical notes using a single string. Starting with the fret at the top of the neck, each successive fret indicates where the player should press to get the next note in the sequence. Musicians call the sequence the chromatic scale, and every thirteenth note in it corresponds to one octave or a doubling of the sound frequency. The spacing between the frets is greatest at the top of the neck and decreases with each additional fret further on down. The spacing eventually becomes smaller than the width of a finger, limiting the number of frets that can be used. Why does the spacing between the frets decrease going down the neck?

Reasoning and Solution Our reasoning is based on Equation 17.3, with $n = 1$ [$f_1 = v/(2L)$]. The value of n is 1, because a string vibrates mainly at its fundamental frequency when plucked, as mentioned in Example 4. This equation indicates that the fundamental frequency f_1 is inversely proportional to the length L between a

The Physics of...
the frets on a guitar.

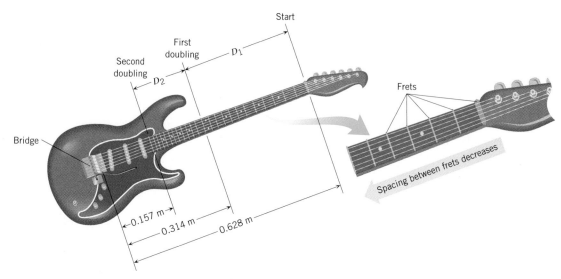

Figure 17.24 The spacing between the frets on the neck of a guitar decreases going down the neck toward the bridge.

given fret and the bridge of the guitar. Thus, in Example 4 we found that the E-string had a length of 0.628 m corresponding to a frequency of $f_1 = 164.8$ Hz. We also found that the length between the bridge and the fret that must be pressed to double this frequency to 329.6 Hz is one-half of 0.628 m, or 0.314 m. To understand why the spacing between frets decreases going down the neck, consider the fret that must be pressed to double the frequency again, from 329.6 Hz to 659.2 Hz. The length between the bridge and this fret would be one-half of 0.314 m, or 0.157 m. The three lengths of 0.628 m, 0.314 m, and 0.157 m are shown in Figure 17.24. The distances D_1 and D_2 are also shown and give the spacings between the starting fret at the top of the neck, the fret for the first doubling, and the fret for the second doubling. Clearly, D_1 is greater than D_2, and the frets near the top of the neck have more space between them than those further on down. All the frets on a guitar have locations that are determined by the musical notes associated with them.

Related Homework Material: *Problem 36*

17.6 LONGITUDINAL STANDING WAVES

Standing wave patterns can also be formed from longitudinal waves. For example, when sound reflects from a wall, the forward- and backward-going waves can produce a standing wave. Figure 17.25 illustrates the vibrational motion in a longitudinal standing wave on a Slinky. As in a transverse standing wave, there are nodes and antinodes. At the nodes the coils of the Slinky do not vibrate at all, that is, they have no displacement. At the antinodes the coils vibrate with maximum amplitude, as indicated by the red dots in the picture. At an antinode the coils have a maximum

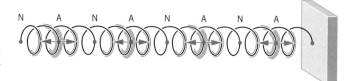

Figure 17.25 A longitudinal standing wave on a Slinky showing the displacement nodes (N) and antinodes (A).

displacement. The vibration occurs along the line of travel of the individual waves, as is to be expected for longitudinal waves. In a standing wave of sound, the molecules behave as the red dots do.

Musical instruments in the wind family depend on longitudinal standing waves in producing sound. Since wind instruments (trumpet, flute, clarinet, pipe organ, etc.) are modified tubes or columns of air, it is useful to examine the standing waves that can be set up in such tubes. Figure 17.26 shows two cylindrical columns of air that are open at both ends. Sound waves, originating from a tuning fork, travel up and down within each tube, since they reflect from the ends of the tubes, even though the ends are open. If the frequency f of the tuning fork matches one of the natural frequencies of the air column, the downward- and upward-traveling waves combine to form a standing wave, and the sound of the tuning fork becomes markedly louder. To emphasize the longitudinal nature of the standing wave patterns, the left side of the drawing replaces the air in the tubes with Slinkies, on which the nodes and antinodes are indicated with red dots. As an additional aid in visualizing the standing waves, the right side of the drawing shows blurred blue patterns within each tube. These patterns symbolize the amplitude of the vibrating air molecules at various locations. Wherever the pattern is widest, the amplitude of vibration is greatest (a displacement antinode), and wherever the pattern is narrowest there is no vibration (a displacement node).

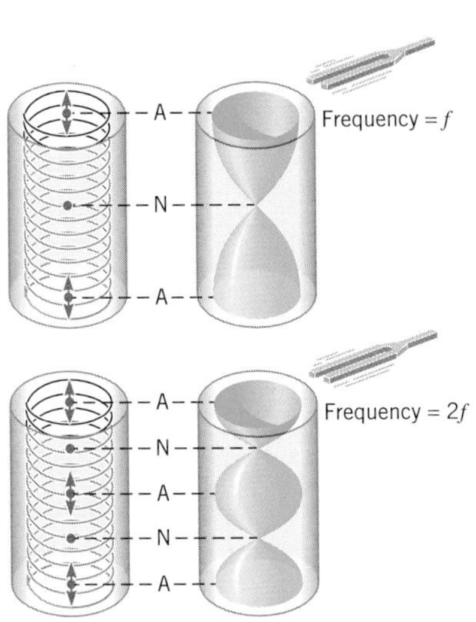

To determine the natural frequencies of the air columns in Figure 17.26, notice that there is a displacement antinode at each end of the tube, because the air molecules there are free to move.* As in a transverse standing wave, the distance between two successive antinodes is one-half of a wavelength, so the length L of the tube must be an integer number n of half-wavelengths: $L = n(\frac{1}{2}\lambda_n)$ or $\lambda_n = 2L/n$. Using this wavelength in the relation $f_n = v/\lambda_n$ shows that the natural frequencies f_n of the tube are

Tube open at both ends $$f_n = n\left(\frac{v}{2L}\right) \quad n = 1, 2, 3, 4, \ldots \qquad (17.4)$$

At these frequencies, large amplitude standing waves develop within the tube due to resonance. Example 6 illustrates how Equation 17.4 is involved when a flute is played.

Figure 17.26 A pictorial representation of longitudinal standing waves on a Slinky (left side) and in a tube of air (right side) that is open at both ends (A, antinode; N, node).

EXAMPLE 6 • Playing a Flute

When all the holes are closed on one type of flute, the lowest note it can sound is a middle C, whose fundamental frequency is 261.6 Hz. (a) The air temperature is 293 K, and the speed of sound is 343 m/s. Assuming the flute is a cylindrical tube open at both ends, determine the distance L in Figure 17.27, that is, the distance from the mouthpiece to the end of the tube. (This distance is only approximate, since the antinode does not occur exactly at the mouthpiece.) (b) A flautist can alter the length of the flute by adjusting the extent to which the head joint is inserted into the main stem of the instrument. If the air temperature rises to 305 K, to what length must a flute be adjusted to play a middle C?

Reasoning The fundamental frequency f_1 is given by Equation 17.4 with $n = 1$: $f_1 = v/(2L)$. This expression can be used to calculate the length as $L = v/(2f_1)$.

Figure 17.27 The length L of a flute between the mouthpiece and the end of the instrument determines the fundamental frequency of the lowest playable note.

---

* In reality, the antinode does not occur exactly at the open end. However, if the tube's diameter is small compared to its length, little error is made in assuming that the antinode is located right at the end.

The Physics of...
a flute.

When the speed of sound v changes, as it does when the temperature changes, the length of the flute must be changed. The effect of temperature on the speed of sound in air is given by $v = \sqrt{\gamma k T/m}$ (Equation 16.5), assuming air behaves as an ideal gas. Thus, the speed is proportional to the square root of the Kelvin temperature ($v \propto \sqrt{T}$), a fact that we can use to find the speed at the higher temperature.

Solution

(a) At a temperature of 293 K, when the speed of sound is v = 343 m/s, the length of the flute is

$$L = \frac{v}{2f_1} = \frac{343 \text{ m/s}}{2(261.6 \text{ Hz})} = \boxed{0.656 \text{ m}}$$

(b) Since $v \propto \sqrt{T}$, it follows that

$$\frac{v_{305 \text{ K}}}{v_{293 \text{ K}}} = \frac{\sqrt{305 \text{ K}}}{\sqrt{293 \text{ K}}} = 1.02$$

As a result, $v_{305 \text{ K}} = 1.02(v_{293 \text{ K}}) = 1.02(343 \text{ m/s}) = 3.50 \times 10^2$ m/s. The adjusted flute length is

$$L = \frac{v}{2f_1} = \frac{3.50 \times 10^2 \text{ m/s}}{2(261.6 \text{ Hz})} = \boxed{0.669 \text{ m}}$$

Thus, to play in tune at the higher temperature, a flautist must lengthen the flute by 0.013 m.

Standing waves can also exist in a tube with only one end open, as the patterns in Figure 17.28 indicate. Note the difference between these patterns and those in Figure 17.26. Here the standing waves have a displacement antinode at the open end and a displacement node at the closed end, where the air molecules are not free to move. Since the distance between a node and an adjacent antinode is one-fourth of a wavelength, the length L of the tube must be an odd number of quarter-wavelengths: $L = 1(\frac{1}{4}\lambda)$ and $L = 3(\frac{1}{4}\lambda)$ for the two standing wave patterns in Figure 17.28. In general, then, $L = n(\frac{1}{4}\lambda_n)$, where n is any odd integer (n = 1, 3, 5, . . .). From this result it follows that $\lambda_n = 4L/n$, and the natural frequencies f_n can be obtained from the relation $f_n = v/\lambda_n$:

Tube open at only one end $$f_n = n\left(\frac{v}{4L}\right) \qquad n = 1, 3, 5, \ldots \qquad (17.5)$$

A tube open at only one end can develop standing waves only at the odd harmonic frequencies f_1, f_3, f_5, etc. In contrast, a tube open at both ends can develop standing waves at all harmonic frequencies f_1, f_2, f_3, etc. Moreover, the fundamental frequency f_1 of a tube open at only one end (Equation 17.5) is one-half that of a tube open at both ends (Equation 17.4). In other words, a tube open only at one end needs to be only one-half as long as a tube open at both ends in order to produce the *same* fundamental frequency.

Energy is also conserved when a standing wave is produced, either on a string or in a tube of air. The energy of the standing wave is the sum of the energies of the individual waves that comprise the standing wave. Once again, interference redistributes the energy of the individual waves to create locations of greatest energy (displacement antinodes) and locations of no energy (displacement nodes).

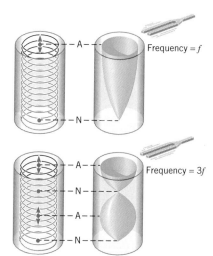

Frequency = f

Frequency = $3f$

Figure 17.28 A pictorial representation of the longitudinal standing waves on a Slinky (left side) and in a tube of air (right side) that is open only at one end (A, antinode; N, node).

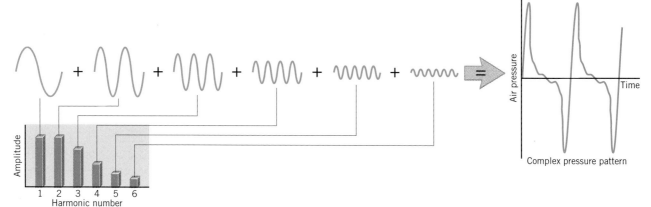

Figure 17.29 The graph on the right shows the pattern of pressure fluctuations such as a singer might produce. The pattern is the sum of the first six harmonics. The relative amplitudes of the harmonics correspond to the heights of the vertical bars in the bar graph.

*17.7 COMPLEX SOUND WAVES

Musical instruments produce sound in a way that depends on standing waves. Examples 4 and 5 illustrate the role of transverse standing waves on the string of an electric guitar, while Example 6 stresses the role of longitudinal standing waves in the air column within a flute. In each example sound is produced at the fundamental frequency of the instrument.

In general, however, a musical instrument does not produce just the fundamental frequency when it plays a note, but generates a number of harmonics as well. Different instruments, such as a violin and a trumpet, generate harmonics to different extents, and the harmonics give the instruments their characteristic sound qualities or timbres. Suppose, for instance, that a violinist and a trumpet player both sound concert A, a note whose fundamental frequency is 440 Hz. Even though both instruments are playing the same note, most people can distinguish the sound of the violin from that of the trumpet. The instruments sound different because the relative amplitudes of the harmonics (880 Hz, 1320 Hz, etc.) that the instruments create are different.

The sound wave corresponding to a note produced by a musical instrument or a singer is called a ***complex sound wave,*** because it consists of a mixture of the fundamental and harmonic frequencies. The pattern of pressure fluctuations in a complex wave can be obtained by using the principle of linear superposition, as Figure 17.29 indicates. This drawing shows a bar graph in which the heights of the bars give the relative amplitudes of the harmonics contained in a note such as a singer might produce. When the individual pressure patterns for each of the six harmonics are added together, they yield the complex pressure pattern shown on the right side of the picture.*

In practice, a bar graph such as that in Figure 17.29 is determined with the aid of an electronic instrument known as a spectrum analyzer. When the note is produced, the complex sound wave is detected by a microphone that converts the wave into an electrical signal. The electrical signal, in turn, is fed into the spectrum analyzer, as Figure 17.30 illustrates. The spectrum analyzer then determines the amplitude and

The Physics of...
a spectrum analyzer.

---

* In carrying out the addition, we assume that each individual pattern begins at zero at the origin when the time equals zero.

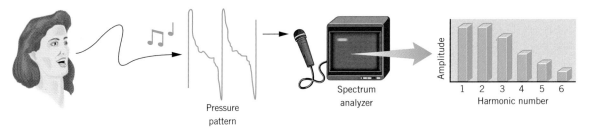

Figure 17.30 A microphone detects a complex sound wave, and a spectrum analyzer determines the amplitude and frequency of each harmonic present in the wave.

frequency of each harmonic present in the complex wave and displays the results on its screen. Conceptual Example 7 deals with the complex wave that corresponds to the sound of a normal voice and stresses the role that longitudinal standing waves play in determining voice quality.

CONCEPTUAL EXAMPLE 7 • Deep Sea Diving and Voice Quality

Divers working in underwater chambers at great depths must deal with the problem of nitrogen narcosis (the "bends"), in which nitrogen dissolves into the blood at toxic levels. One way to avoid the problem is for divers to breathe a mixture containing only helium and oxygen. Helium has an interesting effect on the voice, giving it a high-pitched quality, like that of Donald Duck's voice. For simplicity, assume that the voice is generated by the vocal cords vibrating above a gas-filled tube open at only one end. Explain why helium raises the pitch or frequency of the fundamental and higher harmonics that characterize the voice.

Reasoning and Solution For a tube of length L open at only one end, Equation 17.5 gives the natural frequencies as $f_n = nv/(4L)$, where n is an odd integer and v is the speed of sound. We see that each natural frequency is proportional to the speed of sound. It is because of this proportionality to the speed of sound that the Donald Duck effect occurs. Table 16.1 indicates that the speed of sound in helium is nearly three times greater than the speed in air. Consequently, ***the effect of the helium is to increase the speed of the sound,*** thereby raising substantially the frequency of the fundamental and higher harmonics that characterize the voice.

Related Homework Material: *Problem 44*

The Physics of...
deep sea diving and voice quality.

SUMMARY

The **principle of linear superposition** states that when two or more waves are present simultaneously at the same place, the resultant wave is the sum of the individual waves. **Constructive interference** occurs at a point when two waves meet there crest-to-crest and trough-to-trough, thus reinforcing each other. **Destructive interference** occurs when the waves meet crest-to-trough and cancel each other. When the waves meet crest-to-crest and trough-to-trough, they are **exactly in phase.** When they meet crest-to-trough, they are **exactly out of phase.**

Diffraction is the bending of a wave around an obstacle or the edges of an opening. The angle through which the wave bends depends on the ratio of the wavelength λ of the

wave to the width D of the opening; the greater the ratio λ/D, the greater the angle.

Beats are the periodic variations in amplitude that arise from the linear superposition of two waves that have slightly different frequencies. When the waves are sound waves, the variations in amplitude cause the loudness to vary at the **beat frequency,** which is the difference between the frequencies of the waves.

A transverse or longitudinal **standing wave** is the pattern of disturbance that results when oppositely traveling waves of the same frequency and amplitude pass through each other. A standing wave has places of minimum and maximum vibration called, respectively, **nodes** and **anti-**

nodes. Under resonant conditions, standing waves can be established only at certain frequencies f_n, known as the **natural frequencies.** For a string that is fixed at both ends and has a length L, the natural frequencies are $f_n = n(v/2L)$, where v is the speed of the wave on the string and n is any positive integer, $n = 1, 2, 3, \ldots$. For a gas in a cylindrical tube open at both ends, the natural frequencies are given by the same expression, where v is the speed of sound in the gas and $n = 1, 2, 3, \ldots$. However, if the cylindrical tube is open only at one end, the natural frequencies are $f_n = n(v/4L)$, where $n = 1, 3, 5, \ldots$.

A **complex sound wave** consists of a mixture of a fundamental frequency and overtone frequencies.

CONCEPTUAL QUESTIONS

1. Does the principle of linear superposition imply that two sound waves, passing through the same place at the same time, always create a louder sound than either wave alone? Explain.

2. Suppose you are sitting at the overlap point between the two speakers in Figure 17.5. Because of destructive interference, you hear no sound, even though both speakers are emitting identical sound waves. One of the two speakers is suddenly shut off. Describe what you would hear.

3. Review Figure 17.3. As you walk along a line that is perpendicular to the line between the speakers and passes through the overlap point, you do not observe the loudness to change from loud to faint to loud. However, as you walk along a line through the overlap point and parallel to the line between the speakers, you do observe the loudness to alternate between faint and loud. Explain why your observations are different in the two cases.

4. At an open-air rock concert the music is played through speakers that are on the stage. The speakers point straight forward and not toward the sides. You are wandering about and notice that the sounds of the female vocalists can be heard in front of the stage but not off to the sides. The rhythmic bass, however, can be heard both in front of the stage and off to the sides. How does the phenomenon of diffraction explain your observations?

5. Refer to Example 1 in Section 16.2. Which type of radio wave, AM or FM, diffracts more readily around a given obstacle? Give your reasoning.

6. A tuning fork has a frequency of 440 Hz. The string of a violin and this tuning fork, when sounded together, produce a beat frequency of 1 Hz. From these two pieces of information alone, is it possible to determine the exact frequency of the violin string? Explain.

7. When the regions of constructive and destructive interference in Figure 17.18 move past a listener's ear, a beat frequency of 2 Hz is heard. Suppose that the tuning forks in the drawing were sounded underwater and that the listener is also underwater. The forks vibrate at 438 and 440 Hz, just as they do in air. However, sound travels four times faster in water than in air. Explain why the listener hears a beat frequency of 2 Hz, just as he does in air, even though the regions of constructive and destructive interference are moving past his ear four times faster.

8. The tension in a guitar string is doubled. Does the frequency of oscillation also double? If not, by what factor does the frequency change? Specify whether the change is an increase or a decrease.

9. A string is attached to a wall and vibrates back and forth, as in Figure 17.21. The vibration frequency and length of the string are fixed. The tension in the string is changed, and it is observed that at certain values of the tension a standing wave pattern develops. Account for the fact that no standing waves are observed once the tension is increased beyond a certain value.

10. A string is vibrating back and forth as in Figure 17.21a. The tension in the string is decreased by a factor of four, with the frequency and the length of the string remaining the same. Draw the new standing wave pattern that develops on the string. Give your reasoning.

11. By blowing across the top of an empty soda bottle, a relatively loud sound can be created, because a standing wave develops in the bottle. Explain why the pitch of the sound becomes higher when the bottle is partially filled with water.

12. In Figure 17.26 the tubes are filled with air. Suppose, instead, that the tube at the bottom of the drawing is filled with a gas in which the speed of sound is twice what it is in air. The frequency of the tuning fork remains unchanged. Draw the standing wave pattern that would exist in the tube. Justify your drawing.

13. Standing waves can ruin the acoustics of a concert hall if there is excessive reflection of the sound waves that the performers generate. For example, suppose a performer generates a 2093-Hz tone. If a large amplitude standing wave is present, it is possible for a listener to move a distance of only 4.1 cm and hear the loudness of the tone change from loud to faint. Account for this observation in terms of standing waves, pointing out why the distance is 4.1 cm.

14. The natural frequencies of a wind instrument change when it is brought inside from the cold outdoors. Why do the frequencies change, and is the change an increase or a decrease? Justify your answers.

15. The tones produced by a typical orchestra are complex sound waves, and most have fundamental frequencies less than 5000 Hz. However, a high-quality stereo system must be able to reproduce frequencies up to 20 000 Hz accurately. Explain why.

PROBLEMS

ssm Solution is in the Student Solutions Manual. **www** Solution is available on the World Wide Web at
http://www.wiley.com/college/cutnell ⚕ This icon represents a biomedical application.

**Section 17.1 The Principle of Linear Superposition,
Section 17.2 Constructive and Destructive Interference
of Sound Waves**

1. **ssm** The drawing shows a string on which two rectangular
pulses are traveling at a constant speed of 1 cm/s at time $t = 0$.
Using the principle of linear superposition, draw the shape of the
string at $t = 1$ s, 2 s, 3 s, and 4 s.

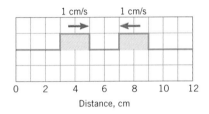

2. Repeat problem 1, assuming that the pulse on the right is
pointing downward, rather than upward.

3. Repeat problem 1, assuming that the pulses have the shape
(half up and half down) shown in the drawing.

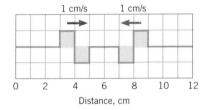

4. In Figure 17.9, the two speakers are vibrating in phase and are
separated by 3.20 m. Both are reproducing identical 214-Hz tones.
The speed of sound is 343 m/s. Suppose point C is 6.00 m directly in front of speaker B, instead of the 2.40 m shown in the
drawing. Does constructive or destructive interference occur at
point C? Why?

5. **ssm** Two loudspeakers are vibrating in phase. They are set
up as in Figure 17.9, and point C is located as shown there. The
speed of sound is 343 m/s. The speakers play the same tone. What
is the smallest frequency that will produce destructive interference
at point C?

6. The sound produced by the loudspeaker in the drawing has a
frequency of 12 000 Hz and arrives at the microphone via two different paths. The sound travels through the left tube *LXM,* which
has a fixed length. Simultaneously, the sound travels through the
right tube *LYM,* the length of which can be changed by moving
the sliding section. At *M,* the sound waves coming from the two
paths interfere. As the length of the path *LYM* is changed, the
sound loudness detected by the microphone changes. When the
sliding section is pulled out by 0.020 m, the loudness changes

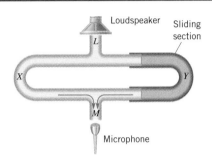

from a maximum to a minimum. Find the speed at which sound
travels through the gas in the tube.

✱7. Two loudspeakers on a concert stage are vibrating in phase. A
listener is 51.4 m from the left speaker and 30.0 m from the right
one. The listener can respond to all frequencies from 20 to
20 000 Hz, and the speed of sound is 343 m/s. What are the two
lowest frequencies that can be heard loudly due to constructive interference?

✱8. Review Conceptual Example 2 in preparation for this problem. Assume that the two loudspeakers in Figure 17.9 are vibrating *out of phase* instead of in phase. The speed of sound is
343 m/s. What is the smallest frequency that will produce destructive interference at point C?

✱9. **ssm** **www** The drawing shows a loudspeaker A and point
C, where a listener is positioned. A second loudspeaker B is located somewhere to the right of A. Both speakers vibrate in phase
and are playing a 68.6-Hz tone. The speed of sound is 343 m/s.
What is the closest to speaker A that speaker B can be located, so
that the listener hears no sound?

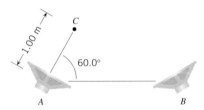

✱✱10. Speakers A and B are vibrating in phase. They are directly
facing each other, are 7.80 m apart, and are each playing a 73.0-
Hz tone. The speed of sound is 343 m/s. On the line *between* the
speakers there are three points where constructive interference occurs. What are the distances of these three points from speaker A?

Section 17.3 Diffraction

11. The width D of a diffraction horn loudspeaker is 0.050 m.
The speed of sound is 343 m/s. At what frequency is the diffraction angle θ equal to 45°?

12. The width D of a diffraction horn and the diameter of a circular speaker are equal. The sound produced by the two speakers

has the same diffraction angle θ. What is the ratio of the wavelength of the sound produced by the diffraction horn to that produced by the circular speaker?

13. **ssm** A speaker has a diameter of 0.30 m. (a) Assuming that the speed of sound is 343 m/s, find the diffraction angle θ for a 2.0-kHz tone. (b) What speaker diameter D should be used to generate a 6.0-kHz tone whose diffraction angle is as wide as that for the 2.0-kHz tone in part (a)?

14. A person is sitting at an angle θ off to the side of a diffraction horn loudspeaker that has a width D of 0.060 m. The speed of sound is 343 m/s. This individual does not hear a sound wave that has a frequency of 8100 Hz. When she is sitting at an angle $\theta/2$, there is a different frequency that she does not hear. What is it?

*15. A row of seats is parallel to a stage at a distance of 8.7 m from it. At the center and front of the stage is a diffraction horn loudspeaker that has a width D of 0.075 m. The speaker is playing a tone that has a frequency of 1.0×10^4 Hz. The speed of sound is 343 m/s. What is the separation between two seats, located on opposite sides of the center of the row, at which the tone cannot be heard?

*16. A 3.00-kHz tone is being produced by a speaker with a diameter of 0.175 m. The air temperature changes from 0 to 29 °C. Assuming air to be an ideal gas, find the *change* in the diffraction angle θ.

Section 17.4 Beats

17. **ssm** Two ultrasonic sound waves combine and form a beat frequency that is in the range of human hearing. The frequency of one of the ultrasonic waves is 70 kHz. What is (a) the smallest possible and (b) the largest possible value for the frequency of the other ultrasonic wave?

18. A tuning fork vibrates at a frequency of 524 Hz. An out-of-tune piano string vibrates at 529 Hz. How much time separates successive beats?

19. Two guitars are slightly out of tune. When they play the same note simultaneously, the sounds they produce have wavelengths of 0.776 and 0.769 m. On a day when the speed of sound is 343 m/s, what beat frequency is heard?

20. When a guitar string is sounded along with a 440-Hz tuning fork, a beat frequency of 5 Hz is heard. When the same string is sounded along with a 436-Hz tuning fork, the beat frequency is 9 Hz. What is the frequency of the string?

21. **ssm** Two pure tones are sounded together. The drawing

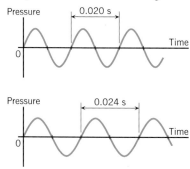

shows the pressure variations of the two sound waves, measured with respect to atmospheric pressure. What is the beat frequency?

*22. A sound wave is traveling in seawater, where the adiabatic bulk modulus and density are 2.31×10^9 Pa and 1025 kg/m^3, respectively. The wavelength of the sound is 3.35 m. A tuning fork is struck underwater and vibrates at 440.0 Hz. What would be the beat frequency heard by an underwater swimmer?

*23. One car is approaching and another is moving away from a bystander. Each car is moving with a speed of 6.00 m/s, and each blows a horn that has a frequency of 392 Hz. The speed of sound is 343 m/s. What is the beat frequency heard by the bystander?

**24. Two loudspeakers are mounted on a merry-go-round whose radius is 9.01 m. When stationary, the speakers both play a tone whose frequency is 100.0 Hz. As the drawing illustrates, they are situated at opposite ends of a diameter. The speed of sound is 343.00 m/s, and the merry-go-round revolves once every 20.0 s. What is the beat frequency that is detected by the listener when the merry-go-round is near the position shown?

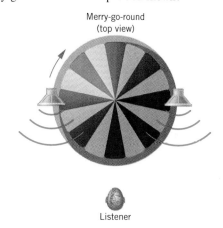

Merry-go-round
(top view)

Listener

Section 17.5 Transverse Standing Waves

25. **ssm** The A string on a string bass is tuned to vibrate at a fundamental frequency of 55.0 Hz. If the tension in the string were increased by a factor of four, what would be the new fundamental frequency?

26. If the string in Figure 17.21 is vibrating at a frequency of 4.0 Hz and the distance between two successive nodes is 0.30 m, what is the speed of the waves on the string?

27. The G string on a guitar has a fundamental frequency of 196 Hz and a length of 0.62 m. This string is pressed against the proper fret to produce the note C, whose fundamental frequency is 262 Hz. What is the distance L between the fret and the end of the string at the bridge of the guitar (see Figure 17.23b)?

28. The lowest note on a piano has a fundamental frequency of 27.5 Hz and is produced by a wire that has a length of 1.18 m. The speed of sound in air is 343 m/s. Determine the ratio of the wavelength of the sound wave to the wavelength of the waves that travel on the wire.

29. ssm On a cello, the string with the largest linear density $(1.56 \times 10^{-2}$ kg/m$)$ is the C string. This string produces a fundamental frequency of 65.4 Hz and has a length of 0.800 m between the two fixed ends. Find the tension in the string.

30. A string of length 2.50 m is fixed at both ends. When the string vibrates at a frequency of 85.0 Hz, a standing wave with five loops is formed. (a) What is the wavelength of the waves that travel on the string? (b) What is the speed of the waves? (c) What would be the fundamental frequency of this string?

31. Electric guitars often come equipped with a "whammy bar," which is a lever that allows the performer to adjust the tension in the strings. In this way, the performer can "dive bomb." To dive bomb means to produce a tone that begins with a high frequency and ends with a low frequency and to go through all the frequencies in between. Suppose that a performer dive bombs the frequency in half. By what factor has the whammy bar reduced the tension in the strings?

32. Sometimes, when a wind blows across a long wire, a low-frequency "moaning" sound is produced. This sound arises because a standing wave is set up on the wire, like a standing wave on a guitar string. Assume that a wire (linear density = 0.0140 kg/m) sustains a tension of 323 N, because the wire is stretched between two poles that are 19.0 m apart. The lowest frequency that a human ear can detect is about 20.0 Hz. What is the lowest harmonic number n that could be responsible for the "moaning" sound?

*33. ssm The E-string on an electric bass guitar has a length of 0.628 m and, when producing the note E, vibrates at a fundamental frequency of 41.2 Hz. Players sometimes add to their instruments a device called a "D-tuner." This device allows the E-string to be used to produce the note D, which has a fundamental frequency of 36.7 Hz. The D-tuner works by extending the length of the string, keeping all other factors the same. By how much does a D-tuner extend the length of the E-string?

*34. Two strings have different lengths and linear densities, as the drawing shows. They are joined together and stretched so that the tension in each string is 190.0 N. The free ends of the joined string are fixed in place. Find the lowest frequency that permits standing waves in both strings with a node at the junction. The standing wave pattern in each string may have a different number of loops.

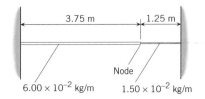

3.75 m   1.25 m

Node

6.00×10^{-2} kg/m 1.50×10^{-2} kg/m

*35. A copper block is suspended in air from a wire. As the drawing shows, a container of mercury is then raised up around the block, until the fundamental frequency of the wire is reduced by a factor of two. Determine the ratio h/h_0 that gives the fraction of

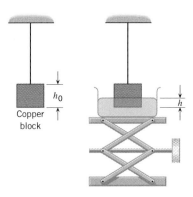

h_0

Copper block

h

the block immersed in the mercury. *(Hint: See Table 11.1 for density values and take advantage of Archimedes' principle.)*

**36. Review Conceptual Example 5 before attempting this problem. As the drawing shows, the length of a guitar string is 0.628 m. Note that the frets are numbered for convenience. A performer can play a musical scale on a single string, because the spacing *between the frets* is designed according to the following rule: When the string is pushed against any fret j, the fundamental frequency of the shortened string is larger by a factor of the twelfth root of two $(\sqrt[12]{2})$ than it is when the string is pushed against the fret $j - 1$. Assuming that the tension in the string is the same for any note, find the spacing (a) between fret 1 and fret 0 and (b) between fret 7 and fret 6.

0.628 m

7 6 5 4 3 2 1 0

**37. ssm www The note that is three octaves above middle C is supposed to have a fundamental frequency of 2093 Hz. On a certain piano the steel wire that produces this note has a cross-sectional area of 7.85×10^{-7} m². The wire is stretched between two pegs. When the piano is tuned properly to produce the correct frequency at 25.0 °C, the wire is under a tension of 818.0 N. Suppose the temperature drops to 20.0 °C. In addition, as an approximation, assume that the wire is kept from contracting as the temperature drops. Consequently, the tension in the wire changes. What beat frequency is produced when this piano and another instrument (properly tuned) sound the note simultaneously?

Section 17.6 Longitudinal Standing Waves, Section 17.7 Complex Sound Waves

38. A cylindrical tube sustains standing waves at the following frequencies: 500, 700, and 900 Hz. There are no standing waves at frequencies of 600 and 800 Hz. (a) What is the fundamental frequency? (b) Is the tube open at both ends or at only one end?

39. A piccolo and a flute can be approximated as cylindrical tubes with both ends open. The lowest fundamental frequency produced by one kind of piccolo is 587.3 Hz, while that produced by a flute is 261.6 Hz. What is the ratio of the length of the piccolo to that of the flute?

40. The range of human hearing is roughly from twenty hertz to twenty kilohertz. Based on these limits and a value of 343 m/s for the speed of sound, what are the lengths of the longest and shortest pipes (open at both ends and producing sound at their fundamental frequencies) that you expect to find in a pipe organ?

41. **ssm** A tube of air is open at only one end and has a length of 1.5 m. This tube sustains a standing wave at its third harmonic. What is the distance between one node and the adjacent antinode?

42. One method for measuring the speed of sound uses standing waves. A cylindrical tube is open at both ends, and one end is placed against a loudspeaker. A movable plunger is inserted into the other end. The distance between the loudspeaker and the plunger is L. When the loudspeaker generates a 485-Hz tone, the smallest value of L for which a standing wave is formed is 0.264 m. What is the speed of sound in the gas in the tube? *(Hint: The plunger closes one end of the tube.)*

43. The fundamental frequencies of two air columns are the same. Column A is open at both ends, while column B is open at only one end. The length of column A is 0.60 m. What is the length of column B?

44. Refer to Conceptual Example 7 as background for this problem. Suppose that someone's lungs are first filled with pure helium at 36 °C and then air at 20 °C. Assume that both helium and air are ideal gases. The nth harmonic frequency of the person's voice is f_n^{helium} and f_n^{air}. Find the ratio $f_n^{\text{helium}}/f_n^{\text{air}}$.

45. **ssm** Both neon (Ne) and helium (He) are monatomic gases and can be assumed to be ideal gases. The fundamental frequency of a tube of neon is 268 Hz. What is the fundamental frequency of the tube if the tube is filled with helium, all other factors remaining the same?

*46. A tube, open at both ends, contains an unknown ideal gas for which $\gamma = 1.40$. At 293 K, the shortest tube in which a standing wave can be set up with a 294-Hz tuning fork has a length of 0.248 m. Find the mass of a gas molecule.

*47. A cylindrical pipe is *closed at both ends*. Derive an expression for the frequencies of the allowed standing waves, similar in form to Equations 17.4 and 17.5, in terms of the speed of sound v, the length of the pipe L, and the harmonic number n. State which integer values of n are allowed.

*48. A person hums into the top of a well and finds that standing waves are established at frequencies of 42, 70.0, and 98 Hz. The frequency of 42 Hz is not necessarily the fundamental frequency. The speed of sound is 343 m/s. How deep is the well?

*49. **ssm www** A vertical tube is closed at one end and open to air at the other end. The air pressure is 1.01×10^5 Pa. The tube has a length of 0.75 m. Mercury (mass density = 13 600 kg/m³) is poured into it to shorten the effective length for standing waves. What is the absolute pressure at the bottom of the mercury column, when the fundamental frequency of the shortened, air-filled tube is equal to the third harmonic of the original tube?

**50. A tube, open at only one end, is cut into two shorter (nonequal) lengths. The piece open at both ends has a fundamental frequency of 425 Hz, while the piece open only at one end has a fundamental frequency of 675 Hz. What is the fundamental frequency of the original tube?

ADDITIONAL PROBLEMS

51. A stretched rubber band has a length of 0.10 m and a fundamental frequency of 440 Hz. What is the speed at which waves travel on the rubber band?

52. The fundamental frequency of a vibrating system is 400 Hz. For each of the following systems, give the three lowest frequencies (excluding the fundamental) at which standing waves can occur: (a) a string fixed at both ends, (b) a cylindrical pipe with both ends open, and (c) a cylindrical pipe with only one end open.

53. **ssm** Review Example 1 in the text. Speaker A is moved further to the left, while ABC remains a right triangle. What is the separation between the speakers when constructive interference occurs again at point C?

54. When a tuning fork is sounded together with a 492-Hz tone, a beat frequency of 2 Hz is heard. Then a small piece of putty is stuck to the tuning fork, and the tuning fork is again sounded along with the 492-Hz tone. The beat frequency decreases. What is the frequency of the tuning fork?

55. Two pulses are traveling toward each other, each having a speed of 1 cm/s. At $t = 0$, their positions are shown in the drawing. When $t = 1$ s, what is the height of the resultant pulse at (a) $x = 3$ cm and at (b) $x = 4$ cm?

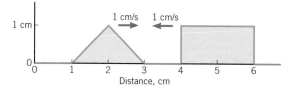

56. A tube with a cap on one end, but open at the other end, produces a standing wave whose fundamental frequency is 130.8 Hz. The speed of sound is 343 m/s. (a) If the cap is removed, what is the new fundamental frequency? (b) How long is the tube?

57. **ssm** Ideally, the strings on a violin are stretched with the same tension. Each has the same length between its two fixed ends. The musical notes and corresponding fundamental frequencies of two of these strings are G (196.0 Hz) and E (659.3 Hz). The linear density of the E string is 3.47×10^{-4} kg/m. What is the linear density of the G string?

*58. The graph shows a transverse standing wave on a string. (a) What is the wavelength of each wave that combines to form the standing wave? (b) If the velocity of one of the component waves is +12.0 cm/s, what is the velocity of the other? (c) What is the frequency of each component wave? (d) Suppose that a "dot" is attached to the string at $x = 2.0$ cm. Determine the maximum speed (in cm/s) of this dot as it vibrates up and down.

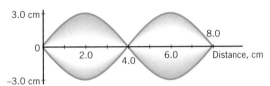

*59. Two loudspeakers face each other, vibrate in phase, and produce identical 440-Hz tones. A listener walks from one speaker toward the other at a constant speed and hears the loudness change (loud–soft–loud) at a frequency of 3.0 Hz. The speed of sound is 343 m/s. What is the walking speed?

**60. Two tuning forks X and Y have different frequencies and produce an 8-Hz beat frequency when sounded together. When X is sounded along with a 392-Hz tone, a 3-Hz beat frequency is detected. When Y is sounded along with the 392-Hz tone, a 5-Hz beat frequency is heard. What are the frequencies f_X and f_Y when (a) f_X is greater than f_Y and (b) f_X is less than f_Y?

**61. ssm www The drawing shows an arrangement in which a block (mass = 15.0 kg) is held in position on a frictionless incline by a cord (length = 0.600 m). The mass per unit length of the cord is 1.20×10^{-2} kg/m, so the mass of the cord is negligible compared to the mass of the block. The cord is being vibrated at a frequency of 165 Hz (vibration source not shown in the drawing). What are the values of the angle θ between 15.0° and 90.0° at which a standing wave exists on the cord?

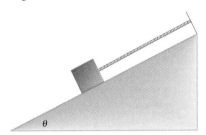

APPENDIX A POWERS OF TEN AND SCIENTIFIC NOTATION

In science, very large and very small decimal numbers are conveniently expressed in terms of powers of ten, some of which are listed below:

$$10^3 = 10 \times 10 \times 10 = 1000 \qquad 10^{-3} = \frac{1}{10 \times 10 \times 10}$$
$$= 0.001$$

$$10^2 = 10 \times 10 = 100 \qquad 10^{-2} = \frac{1}{10 \times 10} = 0.01$$

$$10^1 = 10 \qquad 10^{-1} = \frac{1}{10} = 0.1$$
$$10^0 = 1$$

Using powers of ten, we can write the radius of the earth in the following way, for example:

$$\text{Earth radius} = 6\,380\,000 \text{ m} = 6.38 \times 10^6 \text{ m}$$

The factor of ten raised to the sixth power is ten multiplied by itself six times, or one million, so the earth's radius is 6.38 million meters. Alternatively, the factor of ten raised to the sixth power indicates that the decimal point in the term 6.38 is to be moved six places *to the right* to obtain the radius as a number without powers of ten.

For numbers less than one, negative powers of ten are used. For instance, the Bohr radius of the hydrogen atom is

$$\text{Bohr radius} = 0.000\,000\,000\,0529 \text{ m} = 5.29 \times 10^{-11} \text{ m}$$

The factor of ten raised to the minus eleventh power indicates that the decimal point in the term 5.29 is to be moved eleven places *to the left* to obtain the radius as a number without powers of ten. Numbers expressed with the aid of powers of ten are said to be in *scientific notation.*

Calculations that involve the multiplication and division of powers of ten are carried out as in the following examples:

$$(2.0 \times 10^6)(3.5 \times 10^3) = (2.0 \times 3.5) \times 10^{6+3} = 7.0 \times 10^9$$

$$\frac{9.0 \times 10^7}{2.0 \times 10^4} = \left(\frac{9.0}{2.0}\right) \times 10^7 \times 10^{-4}$$
$$= \left(\frac{9.0}{2.0}\right) \times 10^{7-4} = 4.5 \times 10^3$$

The general rules for such calculations are

$$\frac{1}{10^n} = 10^{-n} \tag{A-1}$$

$$10^n \times 10^m = 10^{n+m} \qquad \text{(Exponents added)} \tag{A-2}$$

$$\frac{10^n}{10^m} = 10^{n-m} \qquad \text{(Exponents subtracted)} \tag{A-3}$$

where n and m are any positive or negative number.

Scientific notation is convenient because of the ease with which it can be used in calculations. Moreover, scientific notation provides a convenient way to express the significant figures in a number, as Appendix B discusses.

APPENDIX B SIGNIFICANT FIGURES

The number of *significant figures* in a number is the number of digits whose values are known with certainty. For instance, a person's height is measured to be 1.78 m, with the measurement error being in the third decimal place. All three digits are known with certainty, so that the number contains three significant figures. If a zero is given as the last digit to the right of the decimal point, the zero is presumed to be significant. Thus, the number 1.780 m contains four significant figures. As another example, consider a distance of 1500 m. This number contains only two significant figures, the one and the five. The zeros immediately to the left of the unexpressed decimal point are not counted as significant figures. However, zeros located between significant figures are significant, so a distance of 1502 m contains four significant figures.

Scientific notation is particularly convenient from the point of view of significant figures. Suppose it is known that a certain distance is fifteen hundred meters, to four significant figures. Writing the number as 1500 m presents a problem, because it implies that only two significant figures are known. In contrast, the scientific notation of 1.500×10^3 m has the advantage of indicating that the distance is known to four significant figures.

When two or more numbers are used in a calculation, the number of significant figures in the answer is limited by the number of significant figures in the original data. For instance, a rectangular garden with sides of 9.8 m and 17.1 m has an area of (9.8 m)(17.1 m). A calculator gives 167.58 m² for this product. However, one of the original lengths is known only to two significant figures, so the final answer is limited to only two significant figures and should be rounded off to 170 m². In general, *when numbers are multiplied or divided, the number of significant figures in the final answer equals the smallest number of significant figures in any of the original factors.*

The number of significant figures in the answer to an addition or a subtraction is also limited by the original data. Consider the total distance along a biker's trail that consists of three segments with the distances shown below:

$$
\begin{array}{rl}
2.5 & \text{km} \\
11 & \text{km} \\
\underline{5.26} & \text{km} \\
\text{Total} \quad 18.76 & \text{km}
\end{array}
$$

The distance of 11 km contains no significant figures to the right of the decimal point. Therefore, neither does the sum of the three distances, and the total distance should not be reported as 18.76 km. Instead, the answer is rounded off to 19 km. In general, *when numbers are added or subtracted, the last significant figure in the answer occurs in the last column (counting from left to right) containing a number that results from a combination of digits that are all significant.* In the answer of 18.76 km, the eight is the sum of $2 + 1 + 5$, each digit being significant. However, the seven is the sum of $5 + 0 + 2$, and the zero is not significant, since it comes from the 11-km distance, which contains no significant figures to the right of the decimal point.

APPENDIX C ALGEBRA

C1 PROPORTIONS AND EQUATIONS

Physics deals with physical variables and the relations between them. Typically, variables are represented by the letters of the English and Greek alphabets. Sometimes, the relation between variables is expressed as a proportion or inverse proportion. Other times, however, it is more convenient or necessary to express the relation by means of an equation, which is governed by the rules of algebra.

If two variables are *directly proportional* and one of them doubles, then the other variable also doubles. Similarly, if one variable is reduced to one-half its original value, then the other is also reduced to one-half its original value. In general, if x is directly proportional to y, then increasing or decreasing one variable by a given factor causes the other variable to change in the same way by the same factor. This kind of relation is expressed as $x \propto y$, where the symbol $\propto$ means "is proportional to."

Since the proportional variables x and y always increase and decrease by the same factor, the ratio of x to y must have a constant value, or $x/y = k$, where k is a constant, independent of the values for x and y. Consequently, a proportionality such as $x \propto y$ can also be expressed in the form of an equation: $x = ky$. The constant k is referred to as a *proportionality constant.*

If two variables are *inversely proportional* and one of them increases by a given factor, then the other decreases by the same factor. An inverse proportion is written as $x \propto 1/y$. This kind of proportionality is equivalent to the following equation: $xy = k$, where k is a proportionality constant, independent of x and y.

C2 SOLVING EQUATIONS

Some of the variables in an equation typically have known values, and some do not. It is often necessary to solve the equation so that a variable whose value is unknown is expressed in terms of the known quantities. *In the process of solving an equation, it is permissible to manipulate the equation in any way, as long as a change made on one side of the equals sign is also made on the other side.* For example, consider the equation $v = v_0 + at$. Suppose values for v, v_0, and a are available, and the value of t is required. To solve the equation for t, we begin by subtracting v_0 from *both* sides:

$$
\begin{array}{rcl}
v & = & v_0 + at \\
\underline{-v_0} & = & \underline{-v_0} \\
v - v_0 & = & at
\end{array}
$$

Next, we divide both sides of $v - v_0 = at$ by the quantity a:

$$
\frac{v - v_0}{a} = \frac{at}{a} = (1)t
$$

On the right side, the a in the numerator divided by the a in the denominator equals one, so that

$$t = \frac{v - v_0}{a}$$

It is always possible to check the correctness of the algebraic manipulations performed in solving an equation by substituting the answer back into the original equation. In the previous example, we substitute the answer for t into $v = v_0 + at$:

$$v = v_0 + a\left(\frac{v - v_0}{a}\right) = v_0 + (v - v_0) = v$$

The result $v = v$ implies that our algebraic manipulations were done correctly.

Algebraic manipulations other than addition, subtraction, multiplication, and division may play a role in solving an equation. The same basic rule applies, however: Whatever is done to the left side of an equation must also be done to the right side. As another example, suppose it is necessary to express v_0 in terms of v, a, and x, where $v^2 = v_0^2 + 2ax$. By subtracting $2ax$ from both sides, we isolate v_0^2 on the right:

$$
\begin{array}{rcl}
v^2 & = & v_0^2 + 2ax \\
-2ax & = & -2ax \\
\hline
v^2 - 2ax & = & v_0^2
\end{array}
$$

To solve for v_0, we take the positive and negative square root of *both* sides of $v^2 - 2ax = v_0^2$:

$$v_0 = \pm\sqrt{v^2 - 2ax}$$

C3 SIMULTANEOUS EQUATIONS

When more than one variable in a single equation is unknown, additional equations are needed if solutions are to be found for all of the unknown quantities. Thus, the equation $3x + 2y = 7$ cannot be solved by itself to give unique values for both x and y. However, if x and y simultaneously obey the equation $x - 3y = 6$, then both unknowns can be found.

There are a number of methods by which such simultaneous equations can be solved. One method is to solve one equation for x in terms of y and substitute the result into the other equation to obtain an expression containing only the single unknown variable y. The equation $x - 3y = 6$, for instance, can be solved for x by adding $3y$ to each side, with the result that $x = 6 + 3y$. The substitution of this expression for x into the equation $3x + 2y = 7$ is shown below:

$$3x + 2y = 7$$
$$3(6 + 3y) + 2y = 7$$
$$18 + 9y + 2y = 7$$

We find, then, that $18 + 11y = 7$, a result that can be solved for y:

$$
\begin{array}{rcr}
18 + 11y = & & 7 \\
-18 & & -18 \\
\hline
11y = & & -11
\end{array}
$$

Dividing both sides of this result by 11 shows that $y = -1$. The value of $y = -1$ can be substituted in either of the original equations to obtain a value for x:

$$
\begin{array}{rcl}
x - 3y & = & 6 \\
x - 3(-1) & = & 6 \\
x + 3 & = & 6 \\
-3 & & -3 \\
\hline
x & = & 3
\end{array}
$$

C4 THE QUADRATIC FORMULA

Equations occur in physics that include the square of a variable. Such equations are said to be *quadratic* in that variable and often can be put into the following form:

$$ax^2 + bx + c = 0 \qquad \text{(C-1)}$$

where a, b, and c are constants independent of x. This equation can be solved to give the **quadratic formula,** which is

$$x = \frac{-b \pm \sqrt{b^2 - 4ac}}{2a} \qquad \text{(C-2)}$$

The $\pm$ in the quadratic formula indicates that there are two solutions. For instance, if $2x^2 - 5x + 3 = 0$, then $a = 2$, $b = -5$, and $c = 3$. The quadratic formula gives the two solutions as follows:

Solution 1:
Plus sign

$$x = \frac{-b + \sqrt{b^2 - 4ac}}{2a}$$

$$= \frac{-(-5) + \sqrt{(-5)^2 - 4(2)(3)}}{2(2)}$$

$$= \frac{+5 + \sqrt{1}}{4} = \frac{3}{2}$$

Solution 2:
Minus sign

$$x = \frac{-b - \sqrt{b^2 - 4ac}}{2a}$$

$$= \frac{-(-5) - \sqrt{(-5)^2 - 4(2)(3)}}{2(2)}$$

$$= \frac{+5 - \sqrt{1}}{4} = 1$$

APPENDIX **D** EXPONENTS AND LOGARITHMS

Appendix A discusses powers of ten, such as 10^3, which means ten multiplied by itself three times, or $10 \times 10 \times 10$. The three is referred to as an **exponent.** The use of exponents extends beyond powers of ten. In general, the term y^n means the factor y is multiplied by itself n times. For example, y^2, or y squared, is familiar and means $y \times y$. Similarly, y^5 means $y \times y \times y \times y \times y$.

The rules that govern algebraic manipulations of exponents are the same as those given in Appendix A (see Equations A-1, A-2, and A-3) for powers of ten:

$$\frac{1}{y^n} = y^{-n} \qquad \text{(D-1)}$$

$$y^n y^m = y^{n+m} \qquad \text{(Exponents added)} \qquad \text{(D-2)}$$

$$\frac{y^n}{y^m} = y^{n-m} \qquad \text{(Exponents subtracted)} \qquad \text{(D-3)}$$

To the three rules above we add two more that are useful. One of these is

$$y^n z^n = (yz)^n \qquad \text{(D-4)}$$

The following example helps to clarify the reasoning behind this rule:

$$3^2 5^2 = (3 \times 3)(5 \times 5) = (3 \times 5)(3 \times 5) = (3 \times 5)^2$$

The other additional rule is

$$(y^n)^m = y^{nm} \qquad \text{(Exponents multiplied)} \qquad \text{(D-5)}$$

To see why this rule applies, consider the following example:

$$(5^2)^3 = (5^2)(5^2)(5^2) = 5^{2+2+2} = 5^{2\times3}$$

Roots, such as a square root or a cube root, can be represented with fractional exponents. For instance,

$$\sqrt{y} = y^{1/2} \quad \text{and} \quad \sqrt[3]{y} = y^{1/3}$$

In general, the nth root of y is given by

$$\sqrt[n]{y} = y^{1/n} \qquad \text{(D-6)}$$

The rationale for Equation D-6 can be explained using the fact that $(y^n)^m = y^{nm}$. For instance, the fifth root of y is the number that, when multiplied by itself five times, gives back y. As shown below, the term $y^{1/5}$ satisfies this definition:

$$(y^{1/5})(y^{1/5})(y^{1/5})(y^{1/5})(y^{1/5}) = (y^{1/5})^5 = y^{(1/5)\times5} = y$$

Logarithms are closely related to exponents. To see the connection between the two, note that it is possible to express any number y as another number B raised to the exponent x. In other words,

$$y = B^x \qquad \text{(D-7)}$$

The exponent x is called the **logarithm** of the number y. The number B is called the **base number.** One of two choices for the base number is usually used. If $B = 10$, the logarithm is known as the *common logarithm,* for which the notation "log" applies:

Common logarithm $\qquad y = 10^x \quad \text{or} \quad x = \log y \qquad \text{(D-8)}$

If $B = e = 2.718 \ldots$, the logarithm is referred to as the *natural logarithm,* and the notation "ln" is used:

Natural logarithm $\qquad y = e^z \quad \text{or} \quad z = \ln y \qquad \text{(D-9)}$

The two kinds of logarithms are related by

$$\ln y = 2.3026 \log y \qquad \text{(D-10)}$$

Both kinds of logarithms are often given on calculators.

The logarithm of the product or quotient of two numbers A and C can be obtained from the logarithms of the individual numbers according to the rules below. These rules are illustrated here for natural logarithms, but they are the same for any kind of logarithm.

$$\ln(AC) = \ln A + \ln C \qquad \text{(D-11)}$$

$$\ln\left(\frac{A}{C}\right) = \ln A - \ln C \qquad \text{(D-12)}$$

Thus, the logarithm of the product of two numbers is the sum of the individual logarithms, and the logarithm of the quotient of two numbers is the difference between the individual logarithms. Another useful rule concerns the logarithm of a number A raised to an exponent n:

$$\ln A^n = n \ln A \qquad \text{(D-13)}$$

Rules D-11, D-12, and D-13 can be derived from the definition of the logarithm and the rules governing exponents.

APPENDIX $\mathbf{E}$ GEOMETRY AND TRIGONOMETRY

E1 GEOMETRY

Angles

Two angles are equal if

1. They are vertical angles (see Figure E1).

2. Their sides are parallel (see Figure E2).

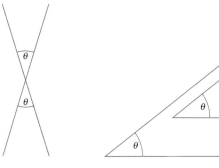

| Figure E1 | Figure E2 |

3. Their sides are mutually perpendicular (see Figure E3).

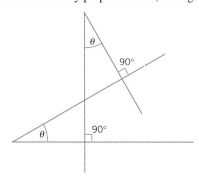

Figure E3

Triangles

1. The *sum of the angles* of any triangle is 180° (see Figure E4).

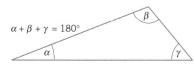

Figure E4

2. A *right triangle* has one angle that is 90°.

3. An *isosceles triangle* has two sides that are equal.

4. An *equilateral triangle* has three sides that are equal. Each angle of an equilateral triangle is 60°.

5. Two triangles are *similar* if two of their angles are equal (see Figure E5). The corresponding sides of similar triangles are proportional to each other:

$$\frac{a_1}{a_2} = \frac{b_1}{b_2} = \frac{c_1}{c_2}$$

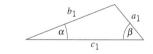
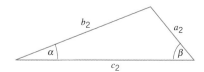

Figure E5

6. Two similar triangles are *congruent* if they can be placed on top of one another to make an exact fit.

Circumferences, Areas, and Volumes of Some Common Shapes

1. Triangle of base b and altitude h (see Figure E6):

$$\text{Area} = \tfrac{1}{2}bh$$

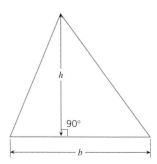

Figure E6

2. Circle of radius r:

$$\text{Circumference} = 2\pi r$$
$$\text{Area} = \pi r^2$$

3. Sphere of radius r:

$$\text{Surface area} = 4\pi r^2$$
$$\text{Volume} = \tfrac{4}{3}\pi r^3$$

4. Right circular cylinder of radius r and height h (see Figure E7):

$$\text{Surface area} = 2\pi r^2 + 2\pi rh$$
$$\text{Volume} = \pi r^2 h$$

Figure E7

E2 TRIGONOMETRY

Basic Trigonometric Functions

1. For a right triangle, the sine, cosine, and tangent of an angle θ are defined as follows (see Figure E8):

$$\sin\theta = \frac{\text{Side opposite }\theta}{\text{Hypotenuse}} = \frac{h_o}{h}$$

$$\cos\theta = \frac{\text{Side adjacent to }\theta}{\text{Hypotenuse}} = \frac{h_a}{h}$$

$$\tan\theta = \frac{\text{Side opposite }\theta}{\text{Side adjacent to }\theta} = \frac{h_o}{h_a}$$

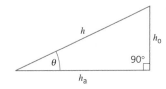

Figure E8

2. The secant ($\sec\theta$), cosecant ($\csc\theta$), and cotangent ($\cot\theta$) of an angle θ are defined as follows:

$$\sec\theta = \frac{1}{\cos\theta} \qquad \csc\theta = \frac{1}{\sin\theta} \qquad \cot\theta = \frac{1}{\tan\theta}$$

Triangles and Trigonometry

1. The **Pythagorean theorem** states that the square of the hypotenuse of a right triangle is equal to the sum of the squares of the other two sides (see Figure E8):

$$h^2 = h_o{}^2 + h_a{}^2$$

2. The **law of cosines** and the **law of sines** apply to any triangle, not just a right triangle, and they relate the angles and the lengths of the sides (see Figure E9):

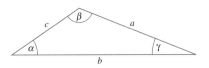

Figure E9

Law of cosines $c^2 = a^2 + b^2 - 2ab\cos\gamma$

Law of sines $\dfrac{a}{\sin\alpha} = \dfrac{b}{\sin\beta} = \dfrac{c}{\sin\gamma}$

Other Trigonometric Identities

1. $\sin(-\theta) = -\sin\theta$
2. $\cos(-\theta) = \cos\theta$
3. $\tan(-\theta) = -\tan\theta$
4. $(\sin\theta)/(\cos\theta) = \tan\theta$
5. $\sin^2\theta + \cos^2\theta = 1$
6. $\sin(\alpha \pm \beta) = \sin\alpha\cos\beta \pm \cos\alpha\sin\beta$

If $\alpha = 90°$, $\sin(90° \pm \beta) = \cos\beta$
If $\alpha = \beta$, $\sin 2\beta = 2\sin\beta\cos\beta$

7. $\cos(\alpha \pm \beta) = \cos\alpha\cos\beta \mp \sin\alpha\sin\beta$

If $\alpha = 90°$, $\cos(90° \pm \beta) = \mp\sin\beta$
If $\alpha = \beta$, $\cos 2\beta = \cos^2\beta - \sin^2\beta = 1 - 2\sin^2\beta$

APPENDIX F SELECTED ISOTOPES[a]

Atomic No. Z	Element	Symbol	Atomic Mass No. A	Atomic Mass u	% Abundance, or Decay Mode If Radioactive	Half-life (If Radioactive)
0	(Neutron)	n	1	1.008 665	β^-	10.37 min
1	Hydrogen	H	1	1.007 825	99.985	
	Deuterium	D	2	2.014 102	0.015	
	Tritium	T	3	3.016 050	β^-	12.33 yr

[a] Data for atomic masses are taken from *Handbook of Chemistry and Physics*, 66th ed., CRC Press, Boca Raton, FL. The masses are those for the neutral atom, including the Z electrons. Data for percent abundance, decay mode, and half-life are taken from E. Browne and R. Firestone, *Table of Radioactive Isotopes*, V. Shirley, Ed., Wiley, New York, 1986. α = alpha particle emission, β^- = negative beta emission, β^+ = positron emission, γ = γ-ray emission, EC = electron capture.

APPENDIX F Selected Isotopes *(continued)*

Atomic No. Z	Element	Symbol	Atomic Mass No. A	Atomic Mass u	% Abundance, or Decay Mode If Radioactive	Half-life (If Radioactive)
2	Helium	He	3	3.016 030	0.000 138	
			4	4.002 603	≈ 100	
3	Lithium	Li	6	6.015 121	7.5	
			7	7.016 003	92.5	
4	Beryllium	Be	7	7.016 928	EC, γ	53.29 days
			9	9.012 182	100	
5	Boron	B	10	10.012 937	19.9	
			11	11.009 305	80.1	
6	Carbon	C	11	11.011 432	β^+, EC	20.39 min
			12	12.000 000	98.90	
			13	13.003 355	1.10	
			14	14.003 241	β^-	5730 yr
7	Nitrogen	N	13	13.005 738	β^+, EC	9.965 min
			14	14.003 074	99.634	
			15	15.000 108	0.366	
8	Oxygen	O	15	15.003 065	β^+, EC	122.2 s
			16	15.994 915	99.762	
			18	17.999 160	0.200	
9	Fluorine	F	18	18.000 937	EC, β^+	1.8295 h
			19	18.998 403	100	
10	Neon	Ne	20	19.992 435	90.51	
			22	21.991 383	9.22	
11	Sodium	Na	22	21.994 434	β^+, EC, γ	2.602 yr
			23	22.989 767	100	
			24	23.990 961	β^-, γ	14.659 h
12	Magnesium	Mg	24	23.985 042	78.99	
13	Aluminum	Al	27	26.981 539	100	
14	Silicon	Si	28	27.976 927	92.23	
			31	30.975 362	β^-, γ	2.622 h
15	Phosphorus	P	31	30.973 762	100	
			32	31.973 907	β^-	14.282 days
16	Sulfur	S	32	31.972 070	95.02	
			35	34.969 031	β^-	87.51 days
17	Chlorine	Cl	35	34.968 852	75.77	
			37	36.965 903	24.23	
18	Argon	Ar	40	39.962 384	99.600	
19	Potassium	K	39	38.963 707	93.2581	
			40	39.963 999	β^-, EC, γ	1.277×10^9 yr
20	Calcium	Ca	40	39.962 591	96.941	
21	Scandium	Sc	45	44.955 910	100	
22	Titanium	Ti	48	47.947 947	73.8	
23	Vanadium	V	51	50.943 962	99.750	
24	Chromium	Cr	52	51.940 509	83.789	
25	Manganese	Mn	55	54.938 047	100	

APPENDIX F Selected Isotopes *(continued)*

Atomic No. Z	Element	Symbol	Atomic Mass No. A	Atomic Mass u	% Abundance, or Decay Mode If Radioactive	Half-life (If Radioactive)
26	Iron	Fe	56	55.934 939	91.72	
27	Cobalt	Co	59	58.933 198	100	
			60	59.933 819	β^-, γ	5.271 yr
28	Nickel	Ni	58	57.935 346	68.27	
			60	59.930 788	26.10	
29	Copper	Cu	63	62.939 598	69.17	
			65	64.927 793	30.83	
30	Zinc	Zn	64	63.929 145	48.6	
			66	65.926 034	27.9	
31	Gallium	Ga	69	68.925 580	60.1	
32	Germanium	Ge	72	71.922 079	27.4	
			74	73.921 177	36.5	
33	Arsenic	As	75	74.921 594	100	
34	Selenium	Se	80	79.916 520	49.7	
35	Bromine	Br	79	78.918 336	50.69	
36	Krypton	Kr	84	83.911 507	57.0	
			89	88.917 640	β^-, γ	3.16 min
			92	91.926 270	β^-, γ	1.840 s
37	Rubidium	Rb	85	84.911 794	72.165	
38	Strontium	Sr	86	85.909 267	9.86	
			88	87.905 619	82.58	
			90	89.907 738	β^-	28.5 yr
			94	93.915 367	β^-, γ	1.235 s
39	Yttrium	Y	89	88.905 849	100	
40	Zirconium	Zr	90	89.904 703	51.45	
41	Niobium	Nb	93	92.906 377	100	
42	Molybdenum	Mo	98	97.905 406	24.13	
43	Technecium	Tc	98	97.907 215	β^-, γ	4.2×10^6 yr
44	Ruthenium	Ru	102	101.904 348	31.6	
45	Rhodium	Rh	103	102.905 500	100	
46	Palladium	Pd	106	105.903 478	27.33	
47	Silver	Ag	107	106.905 092	51.839	
			109	108.904 757	48.161	
48	Cadmium	Cd	114	113.903 357	28.73	
49	Indium	In	115	114.903 880	95.7; β^-	4.41×10^{14} yr
50	Tin	Sn	120	119.902 200	32.59	
51	Antimony	Sb	121	120.903 821	57.3	
52	Tellurium	Te	130	129.906 229	38.8; β^-	2.5×10^{21} yr
53	Iodine	I	127	126.904 473	100	
			131	130.906 114	β^-, γ	8.040 days
54	Xenon	Xe	132	131.904 144	26.9	
			136	135.907 214	8.9	
			140	139.921 620	β^-, γ	13.6 s

APPENDIX F Selected Isotopes *(continued)*

Atomic No. Z	Element	Symbol	Atomic Mass No. A	Atomic Mass u	% Abundance, or Decay Mode If Radioactive	Half-life (If Radioactive)
55	Cesium	Cs	133	132.905 429	100	
			134	133.906 696	β^-, EC, γ	2.062 yr
56	Barium	Ba	137	136.905 812	11.23	
			138	137.905 232	71.70	
			141	140.914 363	β^-, γ	18.27 min
57	Lanthanum	La	139	138.906 346	99.91	
58	Cerium	Ce	140	139.905 433	88.48	
59	Praseodymium	Pr	141	140.907 647	100	
60	Neodymium	Nd	142	141.907 719	27.13	
61	Promethium	Pm	145	144.912 743	EC, α, γ	17.7 yr
62	Samarium	Sm	152	151.919 729	26.7	
63	Europium	Eu	153	152.921 225	52.2	
64	Gadolinium	Gd	158	157.924 099	24.84	
65	Terbium	Tb	159	158.925 342	100	
66	Dysprosium	Dy	164	163.929 171	28.2	
67	Holmium	Ho	165	164.930 319	100	
68	Erbium	Er	166	165.930 290	33.6	
69	Thulium	Tm	169	168.934 212	100	
70	Ytterbium	Yb	174	173.938 859	31.8	
71	Lutetium	Lu	175	174.940 770	97.41	
72	Hafnium	Hf	180	179.946 545	35.100	
73	Tantalum	Ta	181	180.947 992	99.988	
74	Tungsten (wolfram)	W	184	183.950 928	30.67	
75	Rhenium	Re	187	186.955 744	62.60; β^-	4.6×10^{10} yr
76	Osmium	Os	191	190.960 920	β^-, γ	15.4 days
			192	191.961 467	41.0	
77	Iridium	Ir	191	190.960 584	37.3	
			193	192.962 917	62.7	
78	Platinum	Pt	195	194.964 766	33.8	
79	Gold	Au	197	196.966 543	100	
			198	197.968 217	β^-, γ	2.6935 days
80	Mercury	Hg	202	201.970 617	29.80	
81	Thallium	Tl	205	204.974 401	70.476	
			208	207.981 988	β^-, γ	3.053 min
82	Lead	Pb	206	205.974 440	24.1	
			207	206.975 872	22.1	
			208	207.976 627	52.4	
			210	209.984 163	α, β^-, γ	22.3 yr
			211	210.988 735	β^-, γ	36.1 min
			212	211.991 871	β^-, γ	10.64 h
			214	213.999 798	β^-, γ	26.8 min

APPENDIX F Selected Isotopes *(continued)*

Atomic No. Z	Element	Symbol	Atomic Mass No. A	Atomic Mass u	% Abundance, or Decay Mode If Radioactive	Half-life (If Radioactive)
83	Bismuth	Bi	209	208.980 374	100	
			211	210.987 255	α, β^-, γ	2.14 min
			212	211.991 255	β^-, α, γ	1.0092 h
84	Polonium	Po	210	209.982 848	α, γ	138.376 days
			212	211.988 842	α, γ	45.1 s
			214	213.995 176	α, γ	163.69 μs
			216	216.001 889	α, γ	150 ms
85	Astatine	At	218	218.008 684	α, β^-	1.6 s
86	Radon	Rn	220	220.011 368	α, γ	55.6 s
			222	222.017 570	α, γ	3.825 days
87	Francium	Fr	223	223.019 733	α, β^-, γ	21.8 min
88	Radium	Ra	224	224.020 186	α, γ	3.66 days
			226	226.025 402	α, γ	1.6×10^3 yr
			228	228.031 064	β^-, γ	5.75 yr
89	Actinium	Ac	227	227.027 750	α, β^-, γ	21.77 yr
			228	228.031 015	β^-, γ	6.13 h
90	Thorium	Th	228	228.028 715	α, γ	1.913 yr
			231	231.036 298	β^-, γ	1.0633 days
			232	232.038 054	100; α, γ	1.405×10^{10} yr
			234	234.043 593	β^-, γ	24.10 days
91	Protactinium	Pa	231	231.035 880	α, γ	3.276×10^4 yr
			234	234.043 303	β^-, γ	6.70 h
			237	237.051 140	β^-, γ	8.7 min
92	Uranium	U	232	232.037 130	α, γ	68.9 yr
			233	233.039 628	α, γ	1.592×10^5 yr
			235	235.043 924	0.7200; α, γ	7.037×10^8 yr
			236	236.045 562	α, γ	2.342×10^7 yr
			238	238.050 784	99.2745; α, γ	4.468×10^9 yr
			239	239.054 289	β^-, γ	23.47 min
93	Neptunium	Np	239	239.052 933	β^-, γ	2.355 days
94	Plutonium	Pu	239	239.052 157	α, γ	2.411×10^4 yr
			242	242.058 737	α, γ	3.763×10^5 yr
95	Americium	Am	243	243.061 375	α, γ	7.380×10^3 yr
96	Curium	Cm	245	245.065 483	α, γ	8.5×10^3 yr
97	Berkelium	Bk	247	247.070 300	α, γ	1.38×10^3 yr
98	Californium	Cf	249	249.074 844	α, γ	350.6 yr
99	Einsteinium	Es	254	254.088 019	α, γ, β^-	275.7 days
100	Fermium	Fm	253	253.085 173	EC, α, γ	3.00 days
101	Mendelevium	Md	255	255.091 081	EC, α	27 min
102	Nobelium	No	255	255.093 260	EC, α	3.1 min
103	Lawrencium	Lr	257	257.099 480	α, EC	646 ms
104	Rutherfordium	Rf	261	261.108 690	α	1.08 min
105	Hahnium	Ha	262	262.113 760	α	34 s

ANSWERS TO ODD-NUMBERED PROBLEMS

CHAPTER 1

1. (a) 5×10^{-3} g   (b) 5 mg
    (c) 5×10^3 μg
3. (a) 5700 s   (b) 86 400 s
5. 1.37 lb
7. (a) correct   (b) not correct
    (c) not correct   (d) correct
    (e) correct
9. 3.13×10^8 m^3
11. 713 m
13. 80.1 km, 25.9° south of west
15. 3.73 m
17. 99° opposite the 190-cm side
    3.0×10^1 degrees opposite the 95-cm side
    51° opposite the 150-cm side
19. 35.3°
21. (a) 5.31 km, south
    (b) 5.31 km, north
23. 200 N due east or 600 N due west
25. (a) 5.70×10^2 newtons
    (b) 33.6° south of west
27. (a) 67.9 units, 45° south of west
    (b) 67.9 units, 45° north of west
29. (a) 5600 newtons
    (b) along the dashed line
31. (a) 6.00 units
    (b) 36.9° north of west
    (c) 6.00 units
    (d) 36.9° south of west
33. (a) C   (b) B
35. (a) 147 km   (b) 47.9 km
37. 222 m, 55.8° below the $-x$ axis
39. (a) 58°   (b) 240 newtons
41. (a) 322 newtons
    (b) 209 newtons
    (c) 279 newtons
43. 268 km, 38.5° north of east
45. 7.1 m, 9.9° north of east
47. 30.2 m, 10.2°
49. (a) 10.4 units
    (b) 12.0 units
51. (a) 371 units
    (b) 354 units
53. 6.88 km, 26.9°
55. (a) 2.96 km, 33.2° north of east
    (b) 2.96 km, 33.2° north of west
57. (a) 15.8 m/s   (b) 6.37 m/s

59. 3.00 m, 42.8° above the $-x$ axis
61. 0.90 km, 56° north of west
63. (a) 6.43 m, 70.0° above the $-x$ axis
    (b) 7.66 m, 20.0° below the $-x$ axis
65. (a) 78 newtons   (b) 34°
67. 288 units due west and 156 units due north

CHAPTER 2

1. (a) 12.4 km
    (b) 8.8 km, due east
3. 2.5 m/s
5. 0.80 s
7. 60 s
9. 7.2×10^3 m
11. 34 km/h, due north
13. (a) 1.5 m/s^2 (b) 1.5 m/s^2
    (c) yes, 76 m
15. (a) 4.6 m/s   (b) 4.6 m/s
17. 3.44 m/s, due west
19. 26.0 m/s
21. 13 m/s
23. (a) 1.6 m/s^2 (b) 2.0×10^1 m
25. 3.1 m/s^2, opposite to the direction of the motion
27. 4.6 m/s^2, opposite to the direction of the motion
29. 0.74 m/s
31. 17.7 m
33. 1.37 s
35. 0.81 m/s^2, in the same direction as the velocity
37. 14 s
39. 18 m
41. 44.1 m/s
43. 3.1 s
45. 1.1 s
47. (a) -7.9 m/s   (b) 3.2 m
49. 1.7 s
51. 12 m
53. 0.40 s
55. 0.932 m/s
57. 10.6 m
59. 2.5 m
61. The answer is in graphical form.
63. $a_A = 1.9$ m/s^2, $a_B = 0$ m/s^2, $a_C = 3.3$ m/s^2
65. -8.3 km/h^2

67. (a) 6.6 s   (b) 5.3 m/s
69. 91.5 m/s
71. 2 m
73. (a) 4.0 s   (b) 4.0 s
75. 4.5 m/s^2
77. (a) 2.67×10^4 m
    (b) 6.74 m/s, due north
79. $+35$ m/s
81. 2.0×10^1 m
83. 0.81 km
85. (a) 13 m/s   (b) 0.93 m/s^2

CHAPTER 3

1. 8.8×10^2 m
3. 27.4 m
5. 242 m/s
7. (a) 0.33 m/s^2 (b) 0.16 m/s^2
    (c) -0.29 m/s^2
9. 27.0°
11. (a) 2.99×10^4 m/s
    (b) 2.69×10^4 m/s
13. 14.6 s
15. 4.42 s
17. 14.1 m/s
19. (a) 85 m   (b) 610 m
21. (a) 1.78 s   (b) 20.8 m/s
23. 39 m/s
25. (a) 239 m/s, 57.1° with respect to the horizontal
    (b) 239 m/s, 57.1° with respect to the horizontal
27. 24 buses
29. (a) 1.1 s   (b) 1.3 s
31. 5.2 m
33. 14.9 m
35. 17.8 m/s
37. 42°
39. 23 m/s, 31° above the horizontal
41. 11 m/s
43. 5.8 m/s
45 $D = 850$ m, $H = 31$ m
47. The answer is a proof.
49. (a) 41 m/s, due east
    (b) 41 m/s, due west
51. 24 s
53. (a) 2.0×10^3 s
    (b) 1.8×10^3 m
55. 6.3 m/s, 18° north of east

57. 5.2 m/s, 52° west of south
59. 7.8 m/s, 174.8° east of north
61. 3.05 m/s, 14.8° north of west
63. 4.90 m
65. 30.0 m
67. (a) 6490 m/s (b) 7070 m/s
69. 1.7 s
71. 15.7°
73. 63.9 m/s, 85° west of south
75. (a) 5.5 s (b) 42 m/s
77. 0.141° and 89.860°

CHAPTER 4

1. 37 N
3. 93 N
5. 130 N
7. 2.61×10^5 N
9. 9.0
11. 1.20 m/s^2, left
13. 30.9 m/s^2, 27.2° above the $+x$ axis
15. 0.78 m, 21° south of east
17. 18.4 N, 68° north of east
19. 8.70×10^{-12} N
21. (a) $m = 115$ kg, $W = 1.13 \times 10^3$ N
     (b) $m = 115$ kg, $W = 0$ N
23. (a) 5.67×10^{-5} N to the right
     (b) 3.49×10^{-5} N to the right
     (c) 9.16×10^{-5} N to the left
25. (a) 3.75 m/s^2 (b) 2.4×10^2 N
27. 1.76×10^{24} kg
29. 0.223 m/s^2
31. 0.0050
33. 4.7 kg
35. $0.414 L$ or $-2.414 L$
37. 7.3×10^2 N
39. 267 N
41. (a) 447 N (b) 241 N
43. (a) 550 N (b) 7.2 m/s
45. (a) 0.980 m/s^2, opposite to the
     skater's motion
     (b) 29.5 m
47. 4.0×10^2 N
49. 58.8 N
51. (a) 1400 N (b) 2400 N
53. 9.70 N
55. (a) 57 600 N (b) 20 600 N
57. 0.444
59. 251 N
61. 1.9×10^2 N
63. 40.0 N, 59.8° above the horizontal
65. 406 N
67. 220 N, 64° north of east
69. 1730 N, due west
71. (a) 45 N (b) 37 N
73. 6.6 m/s
75. (a) 1610 N (b) 2640 N

77. (a) 1.3 N (b) 6.5 N
79. (a) 4.5 m/s^2 (b) 1200 N
81. 0.14 m/s^2
83. 7.1 m/s^2
85. 0.265 m
87. 49.1 m
89. 1.2 s
91. (a) 1.00×10^2 N (b) 41.6 N
93. (a) 10.5 m/s^2
     (b) 1.07 times greater
95. 929 N
97. 1.00×10^2 N, 53.1° south of east
99. (a) 914 N (b) 822 N
101. 4290 N
103. (a) 0.0640 m/s^2
     (b) 1.58×10^4 km
105. (a) 3.56 m/s^2 (b) 281 N
107. 0.200
109. 8.7 s
111. 3.9 m/s^2
113. 16.3 N
115. 38 m
117. 0.665

CHAPTER 5

1. 160 s
3. 320 m
5. 332 m
7. 24 m/s^2
9. 2.2
11. 426 N
13. (a) 1.2×10^4 N
     (b) 1.7×10^4 N
15. 3.3 m
17. 3500 N
19. (a) 5.1×10^3 N
     (b) 15 m/s
21. 2.0×10^1 m/s
23. (a) 19 m/s (b) 23 m/s
25. 2.12×10^6 N
27. 4.20×10^4 m/s
29. 1.33×10^4 m/s
31. 27.5 days
33. 2.45×10^4 N
35. 0.125
37. 4.72×10^3 m
39. (a) 5.7 m/s (b) 4.8 m/s
41. 2.9×10^4 N
43. 9.65 m/s
45. 0.68 m/s
47. 594 N
49. 0.71
51. 4.4×10^{-6} N
53. (a) 1.70×10^3 N
     (b) 1.66×10^3 N
55. 105 m

57. 23 N at 19.0 m/s and 77 N at
     38.0 m/s
59. 28°

CHAPTER 6

1. 2.2×10^3 J
3. (a) 24 300 J (b) $-10 400$ J
5. 42.8°
7. (a) 5.4×10^4 J
     (b) -2.3×10^4 J
9. 1.5
11. 203 N
13. (a) 3.1×10^3 J
     (b) 2.2×10^2 J
15. (a) 913 J (b) 913 J
17. 6.4×10^5 J
19. 9×10^3 m/s
21. 18%
23. (a) 25 m (b) 5.6 s
25. 7.1×10^{10} J
27. 2.39×10^5 J
29. (a) -3.0×10^4 J
     (b) no
31. 6.38×10^2 N
33. 2.3×10^4 J
35. (a) 22.2 m/s (b) 22.2 m/s
     (c) 22.2 m/s
37. At $h = 20.0$ m, KE $= 0$ J, PE $=$
     392 J, and $E = 392$ J. At $h = 10.0$ m,
     KE $= 196$ J, PE $= 196$ J, and $E =$
     392 J. At $h = 0$ m, KE $= 392$ J, PE $=$
     0 J, and $E = 392$ J.
39. 4.8 m/s
41. 1.7 m/s
43. 3.1 m/s
45. 6.33 m
47. 0.327 m
49. -4.51×10^4 J
51. (a) -1086 J
     (b) 2.01 m below the starting point
53. (a) 4.0 N (b) 4.3 N
55. (a) 3.1 J (b) 24 N
57. 4.17 m/s
59. 3.0×10^1 W
61. 3.6×10^6 J
63. (a) 6.4×10^4 W
     (b) 8.5×10^4 W
65. 6.7×10^2 N
67. 2.0°
69. (a) bow 1 (b) 25 J
71. (a) 93 J (b) 0 J
     (c) 2.3 m/s
73. 4.13 m
75. 23 m/s
77. -2.6×10^6 J
79. (a) 990 W (b) 240 W

81. -1.21×10^6 J
83. 2340 N
85. 13.5 m

CHAPTER 7

1. 11 N·s, parallel to the average force
3. 165 N, upward
5. -8.7 kg·m/s
7. $+69$ N
9. 3.7 N·s
11. 9.00×10^2 N
13. (a) 5.56 m/s
   (b) -2.83 m/s (1.50-kg ball),
      $+2.73$ m/s (4.60-kg ball)
   (c) 0.409 m (1.50-kg ball),
      0.380 m (4.60-kg ball)
15. $+7.1 \times 10^5$ m/s
17. (a) -1.5 m/s   (b) $+1.1$ m/s
19. (a) $+6.71$ m/s   (b) 0.559
21. 1.5×10^{-10} m/s
23. (a) $+9.0$ m/s
   (b) -3.7×10^4 N·s
   (c) 5.2 m
25. 0.097 m
27. 84 kg
29. 3.00 m
31. $+182$ m/s
33. (a) -0.400 m/s (5.00-kg ball),
      $+1.60$ m/s (7.50-kg ball)
   (b) $+0.800$ m/s
35. 0.19 s
37. (a) 73.0°   (b) 4.28 m/s
39. 8 bounces
41. 4.67×10^6 m
43. 6.46×10^{-11} m
45. $+9.3$ m/s
47. 4.24 kg·m/s, 45.0° south of east
49. (a) Bonzo, since he has the smaller
      recoil velocity
   (b) 1.7
51. $+4.9 \times 10^3$ kg·m/s
53. $+4500$ m/s, in the same direction as
      the rocket before the explosion
55. 0.330 m/s, opposite the horizontal
      velocity component of the stone
57. $m_1 = 1.00$ kg, $m_2 = 1.00$ kg
59. (a) 86.5 m/s   (b) 50.0 m/s
61. $+547$ m/s

CHAPTER 8

1. 63.7 grad
3. 4π rad
5. 13 rad/s
7. (a) 9.4×10^{-4} s
   (b) 0.13 m
9. 157.3 rad/s

11. 0.28 rad
13. 336 m/s
15. 25 rev
17. (a) 4.00×10^1 rad
   (b) 15.0 rad/s
19. 1.1×10^5 rad
21. (a) 1.2×10^4 rad
   (b) 1.1×10^2 s
23. (a) 4.60×10^3 rad
   (b) 2.00×10^2 rad/s^2
25. 12.5 s
27. 3.20×10^4 rad
29. 7.37 s
31. (a) 0.332 rad/s   (b) 128 m
33. 157 m/s
35. 0.34 m
37. (a) 1.25 m/s   (b) 7.98 rev/s
39. 18.6°
41. 380 m/s^2
43. 16
45. (a) 2.5 m/s^2 (b) 3.1 m/s^2
47. 1.88 m
49. 1.00 rad
51. 1450 rad
53. 8.71 rad/s^2
55. 208 rad
57. 0.300 m/s
59. 13.5 m/s
61. (a) 9.00 m/s^2
   (b) Radially inward, toward the center
      of the track
63. 35 rev/s
65. 492 rad/s
67. 3.50×10^4 m/s
69. 28.0 rad/s
71. $1/\sqrt{3}$
73. 2 revolutions

CHAPTER 9

1. 0.25 m
3. (a) 84 N·m, counterclockwise about
      corner A
   (b) 150 N·m, clockwise about corner
      B
5. 4.2 N·m
7. 824 N
9. 0.667 m
11. (a) 487 N   (b) 209 N
13. 0.591
15. (a) 194 N   (b) 32.3 N
17. (a) 27 N, to the left
   (b) 27 N, to the right
   (c) 27 N, to the right
   (d) 143 N, downward and to the left
      (79° below the horizontal)
19. 1.0×10^3 N, to the left

21. (a) 7.40×10^2 N
   (b) 0.851 m
23. 228 N
25. 37.6°
27. 17.5 N
29. 1.25 kg·m^2
31. 1.1 N·m
33. (a) 0.131 kg·m^2
   (b) 3.6×10^{-4} kg·m^2
   (c) 0.149 kg·m^2
35. (a) 8.80 rad/s^2 (b) 99.7 N
37. (a) 2.67 kg·m^2 (b) 1.16 m
39. 0.34 N
41. 0.78 N
43. 22.0 kg
45. 6.1×10^5 rev/min
47. 432 J
49. (a) $v_{T1} = 12.0$ m/s, $v_{T2} = 9.00$ m/s,
      $v_{T3} = 18.0$ m/s
   (b) 1.08×10^3 J
   (c) 60.0 kg·m^2
   (d) 1.08×10^3 J
51. 2/5
53. 3/4
55. 1.08×10^7 m
57. 4.4 kg·m^2
59. 0.34 rad/s
61. 8% increase
63. 0.17 m
65. 1.3
67. (a) 0.14 rad/s
   (b) An external torque must be applied
      in a direction opposite to the
      angular deceleration.
69. 0.060 kg·m^2
71. 755 m
73. (a) 1.21×10^3 N
   (b) 1.01×10^3 N, downward
75. 1.26 m/s
77. 51.4 N

CHAPTER 10

1. 3.7×10^{-5} m
3. (a) 4.0×10^{-5} m
   (b) 1.0×10^{-5} m
5. 7.7×10^{-5} m
7. (a) 1.6×10^8 N/m^2
   (b) 5.3×10^{-4}
   (c) 3.0×10^{11} N/m^2
9. 1.6×10^5 N
11. 1.2×10^3 N
13. (a) 6.3×10^{-2} m
   (b) 7.3×10^{-2} m
15. (a) 1.8×10^{-7} m
   (b) 1.0×10^{-6} m
17. 4.6×10^{-4}

19. (a) 710 N/m^2
   (b) 3.5×10^{-8}
   (c) 3.5×10^{-10} m
21. -4.4×10^{-5}
23. (a) 7.44 N   **(b)** 7.44 N
25. 237 N
27. 640 N/m
29. 0.012 m
31. 0.92
33. 2.29×10^{-3} m
35. (a) 0.407 m   **(b)** 397 N
37. 3.5×10^4 N/m
39. 6.0 rad/s
41. (a) 140 m/s
   (b) 1.7×10^{15} m/s^2
43. 9.93×10^{-3} m
45. 4.3 kg
47. (a) 46.9 J   **(b)** 55.9 m/s
49. 7.18×10^{-2} m
51. 6.55 m/s
53.

h (meters)	KE	PE (gravity)	PE (elastic)	E
0	0	0	8.76 J	8.76 J
0.200	1.00 J	3.92 J	3.84 J	8.76 J
0.400	0	7.84 J	0.92 J	8.76 J

55. 16 m/s
57. (a) 9.0×10^{-2} m
   (b) 2.1 m/s
59. 0.556 m/s (29.2-kg block), 1.11 m/s (14.6-kg block)
61. 2.36×10^3 N/m
63. 0.99 m
65. 6.0 m/s^2
67. 12.5 m
69. 0.816
71. (a) 5.08×10^{-2} m
   (b) 5.00×10^{-2} s
   (c) 6.40 m/s
73. 0.070 m
75. 0.069 m
77. 1.4×10^{-6}
79. 3.00×10^1 rad/s
81. 0.240 m
83. 33.4 m/s
85. 4.0×10^{-5} m
87. (a) 1.84 Hz   **(b)** 7.35×10^{-2} m
89. (a) 0.25 s   **(b)** 0.75 s

CHAPTER 11

1. 6.6×10^6 kg
3. 3400 N
5. 317 m^2
7. 1.57 kg
9. 63%

11. (a) 9.7×10^4 N, downward
   (b) 9.7×10^4 N, upward
13. 4.33×10^6 Pa
15. 29 bricks
17. 0.750 m
19. 1.41×10^5 N, straight down
21. 1.2×10^5 Pa
23. (a) 3.5×10^6 N
   (b) 1.2×10^6 N
25. (a) 2.45×10^5 Pa
   (b) 1.73×10^5 Pa
27. 46.2 mm Hg
29. 1.19×10^5 Pa
31. 7.1×10^7 N
33. 3.8×10^5 N
35. 2.1×10^4 N
37. 8.50×10^5 N·m
39. (a) 72.8 N   **(b)** 74.7 N
41. 4.89 m
43. 2.2×10^3 kg
45. 2.04×10^{-3} m^3
47. (a) 6.9 N   **(b)** 7.0×10^{-4} m^3
49. 6.3×10^{-3} kg
51. 1120 N
53. 1.91 m/s
55. (a) 7.0×10^{-5} m^3/s
   (b) 2.5×10^{-4} m/s
57. (a) 1.6×10^{-4} m^3/s
   (b) 2.0×10^1 m/s
59. 46 Pa, air enters at B and exits at A
61. (a) 150 Pa
   (b) The pressure inside the roof is greater than that outside the roof, so there is a net outward force.
63. (a) 9.90 m/s   **(b)** 78.4 gal/min
65. 1.92×10^5 N
67. 38 m/s
69. (a) 14 m/s   **(b)** 0.98 m^3/s
71. 21 m/s
73. 7.78 m/s
75. 2.3 m
77. (a) 1.01×10^5 Pa
   (b) 1.19×10^5 Pa
79. 2.9×10^3 Pa
81. (a) 2.8×10^{-5} N
   (b) 1.0×10^1 m/s
83. 20.6 m
85. 59 N
87. 6.8×10^9 N
89. 4.5×10^{-5} kg/s
91. (a) 0.0597 m   **(b)** 0.0712 m
93. (a) 1.26×10^5 Pa
   (b) 19.4 m
95. (a) 2.2×10^5 kg
   (b) 16 m/s
97. 31.3 rad/s

99. 0.13 m
101. 7.6×10^{-2} m
103. (a) $v = \sqrt{2gy}$
   (b) $y = 0$
   (c) $P_A = P_0 - \rho g(y + h)$

CHAPTER 12

1. 0.2 C°
3. (a) 75 °F   **(b)** 297 K
5. -459.67 °F
7. $T_R = T_F + 459.67$
9. -164 °C
11. 0.084 m
13. 1500 m
15. 8.0×10^{-4}
17. -2.82×10^{-4} m
19. 41 °C
21. 2.0027 s
23. (a) Since the ruler shrinks as the temperature decreases, a tension is needed to stretch the ruler.
   (b) 9.6×10^7 N/m^2
25. 0.6
27. 0.23 m^3
29. 18
31. 1.8×10^{-4} m^3
33. one penny
35. 8.9×10^{-8} m^3
37. 45 atm
39. The answer is a proof.
41. 6.9
43. 36.2 °C
45. 940 °C
47. 27.05 °C
49. \$230
51. 4.4×10^3 N
53. 6.7×10^2 W
55. 1.3×10^5 J
57. 3.9×10^5 J
59. 9.3 kg
61. 3.9×10^{-3} kg
63. (a) 1.49×10^{14} J
   (b) 993 homes
65. 0.223
67. 1.0×10^{-3} kg
69. 1.9×10^4 J/kg
71. 0.237 kg
73. 0 °C
75. 123 °C
77. 33%
79. 76%
81. 28%
83. (a) 4.52×10^6 J
   (b) 5.36×10^6 J
85. 9.49×10^{-3} kg
87. 1.7×10^{-5} (C°)$^{-1}$

89. 0.33 gal
91. 44.0 °C
93. 3.50×10^2 m/s
95. 930 W
97. 0.016 C°
99. 996 N

CHAPTER 13

1. 12 J
3. 1.47×10^{-2} J/(s·m·C°)
5. 2.0×10^{-3} m
7. 4.8
9. 17
11. 283 °C
13. (a) 130 °C   (b) 830 J
   (c) 237 °C
15. (a) 100.78 °C   (b) 107.2 °C
17. (a) 2.0   (b) 0.61
19. 0.3
21. 320 K
23. 1.19
25. 532 K
27. (a) 9×10^6 m
   (b) 7×10^8 kg/m^3
29. 12
31. 14 h
33. 8.0×10^2 J/s
35. aluminum, copper, or silver
37. 14.5 da
39. 134 °C
41. 0.74

CHAPTER 14

1. 1.07×10^{-22} kg
3. (a) 893.51 u   (b) 2680 g
5. aluminum
7. 141 g
9. 11.7
11. 304 K
13. 925 K
15. 67.0 m^3
17. 2.5×10^{21} molecules
19. 39
21. 0.93 mol/m^3
23. 3.9×10^{-3}%
25. 0.205
27. 6.19×10^5 Pa
29. 308 K
31. 4.8×10^{-21} J
33. (a) 46.3 m^2/s^2
   (b) 40.1 m^2/s^2
35. 327 m/s
37. 3.9×10^5 J
39. 9400 m
41. 4.0×10^1 Pa
43. 0.0919 kg/m^4

45. (a) 5.00×10^{-13} kg/s
   (b) 5.8×10^{-3} kg/m^3
47. (a) The answer is a proof.
   (b) 31 s
49. 2.3×10^{-2} mol
51. 1.18×10^{25} molecules
53. 1.34×10^{-7} kg
55. 5.1×10^{-16} kg
57. 1.02×10^{20} molecules
59. 1.5×10^5 g
61. 2.6×10^{-10} m

CHAPTER 15

1. 32 miles
3. (a) -3700 J
   (b) Heat flows out of the gas.
5. (a) -261 J
   (b) Work is done on the system.
7. (a) -7.5×10^2 J
   (b) $+9.0 \times 10^2$ J
9. 3.0×10^5 Pa
11. (a) -2.0×10^{-3} m^3
   (b) The volume decreases.
13. 1.2×10^7 Pa
15. 2.9 J
17. The answer is a proof.
19. 4.99×10^{-6}
21. (a) 0   (b) -6.1×10^3 J
   (c) 310 K
23. 8.0
25. 1.81
27. 18.0
29. $T_i V_i^{\gamma-1} = T_f V_f^{\gamma-1}$
31. (a) 327 K   (b) 0.132 m^3
33. 2400 J
35. 310 J
37. (a) 1.1×10^4 J
   (b) 1.8×10^4 J
39. 12.5 s
41. (a) 60.0%   (b) 40.0%
43. 1.82×10^4 J
45. 0.631
47. (a) 5.4×10^5 J
   (b) 0.12
49. (a) 5/9   (b) 1/3
51. 3000 J
53. (a) 1260 K
   (b) 1.74×10^4 J
55. 478 K
57. (a) 0.360
   (b) 1.3×10^{13} J
59. 1.23
61. 5.7 C°
63. (a) 3.4   (b) 1.10×10^4 J
65. 5.86×10^5 J
67. (a) 1050 J   (b) 2.99

69. (a) 2.0×10^1
   (b) 1.5×10^4 J
71. 279 s
73. 1.26×10^3 K
75. $+1.19 \times 10^4$ J/K
77. (a) $+3.68 \times 10^3$ J/K
   (b) $+1.82 \times 10^4$ J/K
   (c) The vaporization process creates
     more disorder.
79. (a) $+541$ J/K
   (b) The entropy of the universe
     increases.
81. (a) -2.1×10^2 K
   (b) Decrease
83. 13 000 J
85. (a) Reversible   (b) -125 J/K
87. 9.03
89. (a) 3100 J   (b) Negative
91. 0.30 cents
93. 8.49×10^5 Pa
95. 75 K
97. $e = e_1 + e_2 - e_1 e_2$

CHAPTER 16

1. 0.083 Hz
3. (a) 1.25 Hz   (b) 17.5 m/s
5. 0.25 m
7. (a) 0.022 s
   (b) 0.49 m
   (c) The amplitude cannot be
     determined.
9. 78 cm
11. 6.98×10^{-5} m
13. 64 N
15. 4.8×10^{-4} kg
17. 8.68×10^{-3} kg/m
19. 1.2 m/s^2
21. $m_1 = 28.7$ kg, $m_2 = 14.3$ kg
23. 79 m/s
25. $y = 0.37 \sin(2.6\pi t + 0.22\pi x)$, where
   x and y are in meters and t is in
   seconds.
27. (a) $+x$ direction
   (b) -0.080 m
29. 2.5 N
31. 7.0 cycles
33. 1730 m/s
35. 3.9×10^{-3} m
37. 2.06
39. 28.8 K
41. tungsten
43. (a) 1112 m   (b) ± 3 m
45. 650 m
47. 2.55 m
49. 0.404 m
51. 6.7×10^{-9} W

53. 6.5 W
55. 0.32 m^2
57. 1.98%
59. 1.0 × 10^4 W/m^2
61. 2.6 s
63. 25
65. 1000
67. (a) 7.4 dB
　　(b) No, because an increase of 10 dB is needed to double the loudness.
69. − 6.0 dB
71. 0.84 s
73. 2.39 dB
75. 786 Hz
77. 56 m/s
79. 1.21
81. 1350 Hz
83. 31 m/s
85. 209 m
87. 6300
89. (a) 431 m/s　　(b) 322 m/s
91. 17 m/s
93. 1.3
95. (a) 7.87 × 10^{-3} s
　　(b) 4.12 wavelengths
97. 860 Hz
99. 120 dB
101. 153 N
103. 21.2 m/s
105. 76.8 dB
107. 57% argon, 43% neon

CHAPTER 17

1. The answer is a series of drawings.
3. The answer is a series of drawings.
5. 107 Hz
7. 32.1 Hz and 48.1 Hz
9. 3.89 m
11. 9700 Hz
13. (a) 44°　　(b) 0.10 m
15. 8.9 m
17. (a) 50 kHz　　(b) 90 kHz
19. 4 Hz
21. 8 Hz
23. 14 Hz
25. 110.0 Hz
27. 0.46 m
29. 171 N
31. 4
33. 0.077 m
35. 0.49
37. 12 Hz
39. 0.445
41. 0.5 m
43. 0.30 m

45. 602 Hz
47. $f_n = n\left(\dfrac{v}{2L}\right)$, where $n = 1, 2, 3, \ldots$
49. 1.68 × 10^5 Pa
51. 88 m/s
53. 5.06 m
55. (a) 2 cm　　(b) 1 cm
57. 3.93 × 10^{-3} kg/m
59. 1.2 m/s
61. 20.8° and 53.1°

CHAPTER 18

1. 1.5 × 10^{13} electrons
3. 3.1 × 10^{13} electrons
5. (a) + 1.5 q　　(b) + 4 q
　　(c) + 4 q
7. 26 N
9. (a) 0.83 N
　　(b) Attractive, because the spheres carry charges of opposite sign.
11. 0.14 N
13. 0.38 N, 49° below the − x axis.
15. 9
17. (a) 4.56 × 10^{-8} C
　　(b) 3.25 × 10^{-6} kg
19. 2.6 × 10^{12} electrons
21. 92.0 N/m
23. 3.5 × 10^{-5} C
25. (a) 15.4°　　(b) 0.813 N
27. 1.8 N, due east
29. 0.45 N, due east
31. − 2.5 × 10^{-5} C
33. 1.3 m
35. 7.1 × 10^{-2} m^2
37. 6.5 × 10^3 N/C, downward
39. 2.8 × 10^5 N/C, − x direction
41. $q_1 = 0.716\, q$, $q_2 = 0.0895\, q$
43. 0.364
45. 61°
47. 3.25 × 10^{-8} C
49. (a) 350 N · m^2/C
　　(b) 460 N · m^2/C
51. 1.8 × 10^3 N · m^2/C
53. (a) The flux through the face in the x, z plane at $y = 0$ is − 6.0 × 10^1 N · m^2/C. The flux through the face parallel to the x, z plane at $y = 0.20$ m is + 6.0 × 10^1 N · m^2/C. The flux through each of the remaining four faces is zero.
　　(b) Flux = 0
55. The answer is a proof.
57. 120 N
59. 18 N/C

61. (a) Positive, so that the electrostatic force points upward
　　(b) 2.53 × 10^7 protons
63. (a) 1.4 N, − x direction
　　(b) 1.4 N, + x direction
65. (a) both positive or both negative
　　(b) 1.7 × 10^{-16} C
67. 5.53 × 10^{-2} m
69. − 3.3 × 10^{-6} C
71. (a) 8.2 × 10^{-8} C
　　(b) 8.2 × 10^{-3} N

CHAPTER 19

1. 1.1 × 10^{-20} J
3. 4.7 × 10^7
5. 9.4 × 10^7 m/s
7. 7.0 × 10^1 hp
9. 339 V
11. 18.7 m/s
13. + 3.6 × 10^{-9} C
15. No work is done.
17. − 4.7 × 10^{-2} J
19. − $q/\sqrt{2}$
21. 0.0342 m
23. 6.1 × 10^{-14} m
25. − 0.746 J
27. (a) − 2q/3　　(b) − 2q
29. 18 000 V
31. 2.1 × 10^3 V
33. + 9.0 × 10^3 V
35. (a) 159 V　　(b) 125 V
　　(c) 135 V
37. 0.213 J
39. 12 V
41. 5.3
43. 7.0 × 10^{13}
45. (a) 1.3 × 10^{-12} C
　　(b) 8.1 × 10^6
47. (a) 41 J
　　(b) 8200 W
49. 52 V
51. 1.0 × 10^{-4} C
53. The answer is a proof.
55. + 45 V, Point B is at the higher potential.
57. 1.1 × 10^3 V
59. 8.0 × 10^{-5} C
61. 4.2 × 10^3 V/m, directed from A to B
63. 1.7 × 10^3 V/m, directed to the left
65. 2.77 × 10^6 m/s

CHAPTER 20

1. 0.025 A
3. 8.6 A
5. 82 Ω

7. **(a)** 7.9×10^5 C
   (b) 350 A
9. **(a)** 4.7×10^{13} protons
   (b) 17 C°
11. 0.0050 (C°)$^{-1}$
13. 1.64
15. 9.9×10^{-3} m
17. 189 Ω
19. 360 °C
21. $L_{\text{tungsten}}/L_{\text{carbon}} = 70$
23. 6.0×10^2 W
25. 8.9 h
27. **(a)** 13 V **(b)** 37 V
29. 250 °C
31. 1.77 A
33. **(a)** 9.0×10^2 W
    (b) 1.8×10^3 W
35. $92
37. 2.0 h
39. 32 A
41. **(a)** 145 Ω **(b)** 74 V
43. 21.7 V
45. 32 Ω
47. **(a)** 28.9 V **(b)** 16.7 W
49. 5.3 Ω
51. 64 resistors
53. **(a)** 4.57 A **(b)** 1450 W
55. 240 Ω
57. **(a)** 3.6 Ω
    (b) 33 A, breaker will open
59. 3.58×10^{-8} m^2
61. 1.0×10^2 Ω
63. 6.76 Ω
65. 4.6 Ω
67. 2.2 W
69. **(a)** 0.750 A **(b)** 2.11 A
71. 0.054 Ω
73. 30 bulbs
75. 0.054
77. **(a)** 0.38 A **(b)** 2.0×10^1 V
    (c) Point B
79. 33 A
81. 0.75 V, the left end
83. 0.94 V, point D
85. 3.43×10^3 Ω
87. 0.450 Ω
89. **(a)** 30.0 V **(b)** 28.1 V
91. **(a)** 12.0 μF **(b)** 3.0×10^{-4} C
93. 4.69×10^{-4} C
95. 1.54
97. 11 V
99. 1.80 V
101. 4.1×10^{-7} F
103. 0.15 s
105. 1.7 μF

107. 3.5×10^4 A
109. 22 A
111. 25 V
113. 6.00 Ω, 0.545 Ω, 3.67 Ω, 2.75 Ω,
     2.20 Ω, 1.50 Ω, 1.33 Ω, 0.833 Ω
115. 140 W
117. **(a)** 1.2 Ω **(b)** 110 V
119. 9.7×10^2 kg
121. C_0

CHAPTER 21

1. 75.1° and 105°
3. **(a)** 6.50×10^{-2} T
   (b) 9.10×10^3 m/s
5. 4.1×10^{-3} m/s
7. 51°
9. **(a)** Due south
   (b) 2.55×10^{14} m/s^2
11. **(a)** The answer is a drawing.
    (b) 2.6×10^{-2} T
    (c) 2.7×10^{16} m/s^2
13. **(a)** 7.2×10^6 m/s
    (b) 3.5×10^{-13} N
15. **(a)** 4.3×10^2 m
    (b) 7.8×10^5 m
17. 1.5×10^{-8} s
19. 0.0904 m
21. **(a)** $\theta = 0°$ **(b)** 0.29 m
23. 0.13 T
25. 8.7×10^{-3} s
27. 9.6×10^4 m/s
29. 8.1 N
31. 0.96 N (top and bottom sides), 0 N
    (left and right sides)
33. 57.6°
35. 53°
37. **(a)** Left-to-right
    (b) 1.1×10^{-2} m
39. 14 A
41. **(a)** 24 A·m^2
    (b) 4.8 N·m
43. 0.023 N·m
45. **(a)** 170 N·m
    (b) Increase
47. 2.6 N
49. 1.2×10^{-5} A·m^2
51. 8.0×10^{-5} T
53. 1.9×10^{-4} N·m
55. **(a)** Down
    (b) 3.1×10^{-4} T
57. 190 A
59. 6.8 A, opposite to the direction of the
    current in the inner coil
61. 1.04×10^{-2} T
63. 1.13×10^{-4} T, downward

65. **(a)** 1.1×10^{-5} T
    (b) 4.4×10^{-6} T
67. **(a)** The answer is a proof.
    (b) The answer is a proof.
69. 8.1×10^{-5} T
71. 8.2×10^{-27} N·m
73. 1.3×10^{-2} T
75. 1830
77. 140 V/m, directed toward the bottom
    of the page
79. 0.50 m, on the side of wire A that is
    away from wire B.
81. 9.3×10^{-24} A·m^2
83. I_3 is directed out of the paper, $I_3/I = 2$

CHAPTER 22

1. 150 m/s
3. **(a)** the driver's side
   (b) 2.0 m
5. (Rod A) emf = 0 V, (Rod B) emf =
   1.6 V and end 2 is positive, (Rod C)
   emf = 0 V
7. 3.2 A
9. **(a)** 3.3 m/s **(b)** 4.6 N
11. **(a)** 5.6×10^{-3} Wb
    (b) 0 Wb
13. 70.5°
15. **(a)** 7.3×10^{-4} Wb
    (b) 0 Wb
    (c) 4.7×10^{-3} Wb
17. (Two triangular ends) 0 Wb, (Bottom
    surface) 0 Wb, (1.2 m × 0.30 m
    surface) 0.090 Wb, (1.2 m × 0.50 m
    surface) 0.090 Wb
19. 2.8×10^{-3} V
21. 8.6×10^{-5} T
23. 4.8 s
25. 0.459 T
27. 0.045 V
29. (Figure 22.1b) right to left, (Figure
    22.1c) left to right
31. **(a)** clockwise
    (b) no induced current
    (c) counterclockwise
    (d) no induced current
33. **(a)** clockwise
    (b) clockwise
35. no induced current
37. **(a)** 2.4 Hz **(b)** 15 rad/s
    (c) 0.62 T
39. 4.0×10^5 turns
41. 0.150 m
43. 230 V
45. 102 V
47. 9.3×10^{-3} H

49. 220 turns
51. 0.094 A
53. 1.5×10^9 J
55. 2.3×10^{-6} H
57. 1.6×10^{-5} A
59. 1/4
61. 1/12, step down
63. 7.7 W
65. (a) 7.0×10^5 W
 (b) 7.0×10^1 W
67. (a) 0.65 V (b) 0.11 A
69. (a) left to right
 (b) right to left
71. 9.27×10^{-7} Wb
73. 12 V
75. 3.6×10^9 N/C
77. (a) 3.6×10^{-3} V
 (b) 2.0×10^{-3} m^2/s, shrunk
79. (a) 9.59 Ω (b) 95 V
 (c) 8.9 A

CHAPTER 23
1. 126 Hz
3. (a) 13.0 μF (b) 0.61 A
5. 5.00×10^{-2} s
7. (a) 0.28 A (b) Increase
9. 8.0×10^1 Hz
11. 160 Ω
13. 0.44 A
15. 0.17 V
17. 83.9 V
19. 0.819
21. (a) 0.925 A (b) 31.8°
23. (a) R and C
 (b) $R = 49.7\ \Omega$, $X_C = 185\ \Omega$
25. 270 Hz
27. (a) 51.8 V (b) -4.21 A
29. (a) 352 Hz (b) 15.5 A
31. 9.8×10^{-2} W
33. 3.1 kHz
35. (a) 1.3×10^{-3} H
 (b) 8.7×10^{-6} F
37. (a) 2.94×10^{-3} H
 (b) 4.84 Ω (c) 0.163
39. (a) $4/\sqrt{3}$ (b) $1/\sqrt{3}$
41. (a) 2.00 μF (b) 0.77 A
43. 123 V
45. 176 mH
47. (a) 113 Ω (b) $+21°$
49. 8
51. 3.11×10^3 Hz and 7.50×10^3 Hz

CHAPTER 24
1. 4.1×10^{16} m
3. 6.8×10^{-11} F
5. The answers are in graphical form.

7. 1.4×10^{17} Hz
9. 1.25 m
11. (a) 1.09×10^6 Hz
 (b) AM
13. 3.7×10^4 wavelengths
15. 150 m
17. 0.015 m
19. 0.24 s
21. 8.75×10^5 times
23. 3.8×10^2 W/m^2
25. 0.07 N/C
27. (a) 183 N/C
 (b) 6.10×10^{-7} T
29. The answer is a proof.
31. 4600 W
33. 920 W
35. (a) receding
 (b) 1.8×10^6 m/s
37. (a) 6.175×10^{14} Hz
 (b) 6.159×10^{14} Hz
39. (a) 0.550 W/m^2
 (b) 3.7×10^{-2} W/m^2
41. (a) 0.82 (b) 0.18
43. 14 W/m^2
45. 20 analyzers
47. (a) 3 (b) 9
49. 1.3×10^6 m
51. 602 W/m^2
53. 71.6°
55. increase by 9.3°
57. 3.93×10^{26} W
59. (a) 5.30 N/C
 (b) 1.77×10^{-8} T

CHAPTER 25
1. 55°
3. 14°
5. 10°
7. (a) 30° (b) 30°
9. 1.73
11. (a) Image distance is 3.0×10^1 cm
 behind the mirror.
 (b) Image height is 5.0 cm.
13. (a) Image distance is 16.7 cm behind
 the mirror.
 (b) Image height is 6.67 cm.
15. 10.9 cm
17. (a) Convex
 (b) 24.0 cm
19. (a) 2.0×10^2 cm
 (b) -6.3 cm
 (c) Upside down
21. (a) $+62$ cm (b) $+0.35$
 (c) Upright (d) Smaller
23. (a) -1.8 m (b) Virtual
 (c) 0.19 m

25. (a) R (b) -1
 (c) Inverted
27. -2
29. $+42.0$ cm
31. (a) The answer is a proof.
 (b) The answer is a proof.
33. $+74$ cm
35. $+32$ cm
37. $+22$ cm
39. $+46.7$ cm
41. 80.0 cm, toward the lens
43. 33.7°

CHAPTER 26
1. 2.00×10^8 m/s
3. 0.5411
5. 2.0×10^{-11} s
7. 1.40
9. (a) 43° (b) 31°
11. 40.8°
13. 1.92×10^8 m/s
15. 2%, decrease
17. 1.19 mm
19. The answer is a derivation.
21. 3.23 cm
23. (a) 11.3 cm (b) 33.8 cm
 (c) larger
25. 1.54
27. 1.45
29. 5.1 m
31. $n_B/n_A = 0.766$, $n_B/n_C = 1.53$
33. 1.52
35. 32°
37. 25.0°
39. The answer is a proof.
41. 0.35°
43. (Red) 44.6°, (Violet) 45.9°
45. (Red) 52.7°, (Violet) 56.2°
47. (a) -24 cm (b) 1.6
49. 61.1 mm
51. (a) -24 cm (b) 3.0
53. (a) $d_i = -75$ cm, $m = 2.5$
 (b) $d_i = -75.0$ cm, $m = 2.50$
55. (a) -7.90×10^{-3} m
 (b) -3.44×10^{-2} m
57. 48 cm
59. (a) 4.52×10^{-4} m
 (b) 6.12×10^{-2} m
61. The answer is a proof.
63. -5.6 cm
65. (a) 4.00 cm to the left of the diverging
 lens
 (b) -0.167
 (c) virtual
 (d) inverted
 (e) smaller

67. (a) 19.6 cm **(b)** 0.87 cm
69. (a) 18.1 cm **(b)** real
(c) inverted
71. (a) 46.0 cm **(b)** 43.0 cm
73. (Right eye) $-$ 0.20 diopters,
(Left eye) $-$ 0.15 diopters
75. $-$ 0.14 mm
77. (a) 11.8 cm **(b)** 47.8 cm
79. (a) 31.3 cm **(b)** 2.43 m
81. 3.7
83. (a) 6.9 cm **(b)** 3.6
85. 6.3 cm
87. 0.64 rad
89. 0.81 cm
91. 0.435 cm
93. 0.261 cm
95. -2.0×10^2
97. (a) the 1.3-diopter lens
(b) 0.86 m **(c)** -8.5
99. (a) -194
(b) -7.8×10^{-5} m
(c) 1.94×10^6 m
101. $d_i = 18$ cm
103. (a) 33° **(b)** 32°
105. -220 cm
107. 0.011 m
109. 2.46×10^8 m/s
111. (a) 0.775 m **(b)** 0.780 m
113. (a) converging
(b) farsighted
(c) 96.3 cm from the eyes
115. (a) -10.0 cm **(b)** 0.500
117. (a) converging
(b) $d_o = 2f$
(c) $d_i = 2f$
119. 0.333 m
121. -181
123. (a) 22.4 cm **(b)** 28.4 cm

CHAPTER 27

1. Constructive
3. (a) 0.68° **(b)** 1.4°
(c) 2.0°
5. 6.0×10^{-5} m
7. 0.0309 m
9. 487 nm
11. 207 nm
13. 102 nm
15. 198
17. 115 nm
19. 440 nm
21. (a) 0.21° **(b)** 22°
23. (a) 24° **(b)** 39°
25. 490 nm
27. 0.012 m
29. 0.447

31. (a) 2.59 m **(b)** 1.58 m
33. 14 000 m
35. 0.52 m
37. 3.2×10^3 m
39. (a) 1.22λ **(b)** Shorter
41. (a) 37° **(b)** 22°
43. 5.90×10^{-7} m
45. 4.0×10^{-6} m
47. 24°
49. 3/4
51. (a) 4.2×10^{-5} degrees
(b) 59°
53. 660 nm
55. 183 nm and 366 nm
57. 0.0254 m
59. (a) 0.36 m **(b)** 1.1 m
61. (a) 7.9° (violet), 13° (red)
(b) 16° (violet), 26° (red)
(c) 24° (violet), 41° (red)
(d) The second and third orders
overlap.
63. 1.95 m

CHAPTER 28

1. 2.4×10^8 m/s
3. 3.5×10^{-8} s
5. 0.15 rad/s
7. 5.57 s
9. 530 m
11. 3.0×10^6 m
13. 1.3
15. 6.0 light years
17. 3.0 m $\times$ 1.3 m
19. 2.91×10^4 kg·m/s
21. 1.80×10^8 m/s
23. -2.0 m/s
25. 5.0×10^{-13} J
27. 3.72×10^{-12} kg, the liquid water has
the greater mass
29. (a) 1.0 **(b)** 6.6
31. The answer is a proof.
33. 0.31 c
35. -0.23 c
37. 42 m
39. 1.1 kg
41. 2.60×10^8 m/s
43. 3.6×10^5 pennies
45. -0.406 c
47. (a) 4.3 y
(b) The twin who travels at 0.500 c is
older.

CHAPTER 29

1. 310 nm
3. 7.7×10^{29} photons/s
5. 6.3 eV

7. 1.26 eV
9. 73 photons/s
11. 9.4×10^9 photons
13. (a) 2.1×10^{24} photons
(b) 32 molecules/photon
15. 5.1×10^{-33} kg·m/s
17. (a) 2.124×10^{-24} kg·m/s
(b) 2.096×10^{-24} kg·m/s
19. 4.692×10^{-24} kg·m/s
21. 9.50×10^{-17} m
23. 6.6×10^{-27} kg
25. 7.77×10^{-13} J
27. 7.38×10^{-11} m
29. 1.10×10^3 m/s
31. 1.5×10^4 V
33. 8.0×10^{-6} m/s
35. 1.9×10^{-20} kg·m/s
37. $-0.0289° \le \theta \le +0.0289°$
39. 2.6×10^{-28} m
41. (a) 4.50×10^{-36} m/s
(b) 7.05×10^{27} years
43. 3.6×10^{-9} m
45. 1.2×10^{-36} m
47. 42.8

CHAPTER 30

1. (a) 6.2×10^{-31} m^3
(b) 4×10^{-45} m^3
(c) 7×10^{-13} %
3. 6.88×10^{-15} m
5. -8.7×10^6 eV
7. 434.1 nm
9. 4.41×10^{-10} m
11. (a) 7458 nm **(b)** 2279 nm
(c) infrared
13. 1.98×10^{-19} J
15. -1.51 eV
17. The answer is a proof.
19. $6 \le n_i \le 19$
21. (a) $v_n = 2\pi ke^2 Z/(nh)$
(b) 2.19×10^6 m/s
(c) 1.09×10^6 m/s
(d) yes
23. $-3, -2, -1, 0, +1, +2, +3$
25. 2, 3, 4, 5
27. -1.51 eV, -0.850 eV, -0.544 eV
29. 26.6°
31. $1s^2 2s^2 2p^6 3s^2 3p^6 4s^2 3d^{10} 4p^3$
33. (a) not permitted
(b) permitted
(c) not permitted
(d) permitted
(e) not permitted
35. carbon
37. 7.230×10^{-11} m
39. 1.11×10^{-10} m

41. 21 600 V
43. The answer is a proof.
45. 1.9×10^{17} photons
47. $1s^2\,2s^2\,2p^6\,3s^2\,3p^6\,4s^2\,3d^{10}\,4p^6\,5s^2\,4d^1$
49. 91.2 nm
51. (a) 1.08×10^{-14} J
  (b) 6.75×10^4 eV
53. 3
55. $\pm 3.16 \times 10^{-34}$ J·s,
  $\pm 2.11 \times 10^{-34}$ J·s,
  $\pm 1.05 \times 10^{-34}$ J·s, 0 J·s
57. $n_i = 6$ and $n_f = 2$
59. 30.39 nm

CHAPTER 31

1. (a) $X = $ Pt, 117 neutrons
  (b) $X = $ S, 16 neutrons
  (c) $X = $ Cu, 34 neutrons
  (d) $X = $ B, 6 neutrons
  (e) $X = $ Pu, 145 neutrons
3. 8
5. 35.2
7. $^{130}_{52}$Te
9. 9.4×10^3 m
11. (a) 0.555 357 u
  (b) 9.2217×10^{-28} kg
13. 128 MeV
15. (a) 1.858 968 u (b) 1732 MeV
  (c) 7.66 MeV/nucleon
17. 1.003 26 u

19. (a) β^- (b) β^+
  (c) γ (d) α
21. $^{35}_{16}$S $\rightarrow$ $^{35}_{17}$Cl $+$ $^{0}_{-1}$e
23. 0.019 MeV
25. (a) $^{18}_{9}$F $\rightarrow$ $^{18}_{8}$O $+$ $^{0}_{+1}$e
  (b) $^{15}_{8}$O $\rightarrow$ $^{15}_{7}$N $+$ $^{0}_{+1}$e
27. 1.59×10^7 m/s
29. 1.82 MeV
31. 3.0 days
33. 19.9
35. 379 yr
37. 146 disintegrations/min
39. 3.7×10^{10} Bq
41. 7.23 days
43. 13 000 yr
45. 0.70%
47. 90.9%
49. 6900 yr, maximum error is 900 yr
51. (a) 8 protons, 10 neutrons
  (b) 50 protons, 70 neutrons
53. 8.00 days
55. (a) 3.0×10^4 yr
  (b) 3.3×10^4 yr
57. 4 782 969 electrons
59. $^{212}_{84}$Po

CHAPTER 32

1. 0.26 roentgens
3. 30 rad
5. 2.8×10^{-3} J

7. 2.4×10^4 rem
9. 3.01×10^8 rad
11. $Z = 6$ and $X = $ carbon
13. $\gamma + ^{17}_{8}$O $\rightarrow ^{12}_{6}$C $+ ^{4}_{2}$He $+ ^{1}_{0}$n
15. (a) ? $=$ proton $^{1}_{1}$H
  (b) ? $=$ alpha particle $^{4}_{2}$He
  (c) ? $=$ lithium $^{6}_{3}$Li
  (d) ? $=$ nitrogen $^{14}_{7}$N
  (e) ? $=$ manganese $^{56}_{25}$Mn
17. 13.6 MeV
19. 4 neutrons
21. 9.0×10^{-4}
23. 9.6×10^8 W
25. 2.7×10^6 kg
27. (a) 8.2×10^{10} J (b) 0.48 g
29. 1200 kg
31. 4.0 MeV
33. 1.0 gal
35. released energy $= 24.7$ MeV
37. possibility #1 $= u, d, s$
  possibility #2 $= u, d, b$
  possibility #3 $= u, s, b$
39. 140.9 MeV
41. (a) 1.9×10^{-20} kg·m/s
  (b) 3.5×10^{-14} m
43. 1.6×10^{-3} J
45. oxygen $^{16}_{8}$O
47. 3.9×10^{18} fissions/second
49. 1.1×10^{-4} kg
51. 1.1×10^6 kg

PHOTOCREDITS

CHAPTER 1
Page 1: Photofest. *Page 3*: Courtesy Bureau International des Poids et Mesures, France. *Page 4*: Courtesy National Institute of Standards and Measures. *Page 5*: James Marshall/The Stock Market. *Page 6*: Life Science Research/The Stock Market. *Page 11*: Dennis O'Clair/Tony Stone Images/New York, Inc.

CHAPTER 2
Page 25: Warren Bolster/Tony Stone Images/New York, Inc. *Page 27*: Jeanne Drake. *Page 28*: Chris Cole/Duomo Photography, Inc. *Page 29*: Benjamin Rondel/The Stock Market. *Page 30*: Chris Cole/Duomo Photography, Inc. *Page 31*: Paul J. Sutton/Duomo Photography, Inc. *Page 32*: Lori Adamski Peek/Tony Stone Images/New York, Inc. *Page 34*: Focus on Sports. *Page 36*: Chris Cole/Duomo Photography, Inc. *Page 37* (top left): Paul J. Sutton/Duomo Photography, Inc. *Page 37* (top right): Rick Rickman/Duomo Photography, Inc. *Page 37* (bottom): John Nicholson/Allsport. *Page 38*: Todd Buchanan/Matrix International, Inc. *Page 42*: William C. Byrne/Natural Selection.

CHAPTER 3
Page 59: Mitchell Funk/The Image Bank. *Pages 62–63*: Courtesy NASA. *Page 67*: Dennis Boulanger/Agence Vandystadt/Allsport. *Page 75*: Photofest.

CHAPTER 4
Page 85: Mark Tomalty/Masterfile. *Page 86* (left): Rick Rickman/Duomo Photography, Inc. *Page 86* (center): David Madison/Tony Stone Images/New York, Inc. *Page 86* (bottom right): LeBleux-Passe/Gamma Liaison. *Page 87*: Focus on Sports. *Page 94*: Tim Davis/Tony Stone Images/New York, Inc. *Page 95*: David Davies/Tony Stone Images/New York, Inc. *Page 99*: World Perspectives/Tony Stone Images/New York, Inc. *Page 101*: Lester Sloan/Woodfin Camp & Associates.

Page 103: Rich Chisholm/The Stock Market. *Page 105*: Brian Bailey/Adventure Photo. *Page 106*: Kennan Ward/Adventure Photo. *Page 109*: Tom Hanson/Gamma Liaison.

CHAPTER 5
Page 131: Andy Caulfield/The Image Bank. *Page 137*: Duomo Photography, Inc. *Page 138*: Courtesy Ringling Bros. and Barnum & Bailey Combined Shows, Inc. *Page 142* (left) and *144*: Courtesy NASA. *Page 142* (right): Courtesy NASA and Space Telescope Science Institute. *Page 146*: Robert Mathena/Fundamental Photographs.

CHAPTER 6
Page 153: Tim Flach/Tony Stone Images/New York, Inc. *Page 155*: David Madison/Duomo Photography, Inc. *Page 158*: David Madison. *Page 164*: James D. Wilson/Gamma Liaison. *Page 168*: Rick Rickman/Duomo Photography, Inc. *Page 169*: C. Sharrock/The Image Bank. *Page 173*: Dan Feicht/Cedar Point. *Page 178*: Ron Chapple/FPG International.

CHAPTER 7
Page 188: William S. Helsel/Tony Stone Images/New York, Inc. *Page 189* (top): Bryan Yablonsky/Duomo Photography, Inc. *Page 189* (bottom left): Yellow Dog Productions/The Image Bank. *Page 189* (bottom right): Dave Cannon/Tony Stone Images/New York, Inc. *Page 190*: Graham French/Masterfile. *Page 192*: Andy Washnik/The Stock Market. *Page 195*: Graham French/Masterfile. *Page 199*: Photofest.

CHAPTER 8
Page 213: A & L Sinibaldi/Tony Stone Images/New York, Inc. *Page 216*: ©1992 Roger Ressmeyer/CORBIS. *Page 217*: Ken Whitmore/Tony Stone Images/New York, Inc. *Page 219*: Ben Van Hook/Duomo Photography, Inc. *Page 220*: Peter Griffith/Masterfile.

CHAPTER 9
Page 237: Ted Wood/Tony Stone Images/New York, Inc. *Page 241*: Uniphoto, Inc. *Page 245*: Tim Defrisco/Allsport. *Page 249*: AP/Wide World Photos. *Page 253*: Uniphoto, Inc. *Page 255*: Amwell/Tony Stone Images/New York, Inc. *Page 259*: Duomo Photography, Inc. *Page 261*: Glyn Kirk/Tony Stone Images/New York, Inc. *Page 265*: ©Paul Miller.

CHAPTER 10
Page 274: Adam Jones/Natural Selection. *Page 276*: Courtesy Ringling Bros. and Barnum & Bailey Combined Shows, Inc. *Page 282*: Richard Megna/Fundamental Photographs. *Page 291*: Travel Pix/FPG International. *Page 295*: Robert Mathena/Fundamental Photographs. *Page 297*: Everett Johnson/Leo de Wys, Inc.

CHAPTER 11
Page 307: John Warden/Tony Stone Images/New York, Inc. *Page 311* (left): Michael Melford. *Page 311* (right): Alan Carey/Photo Researchers. *Page 313*: Michael S. Nolan/Natural Selection. *Page 314*: Jim Richardson/West Light. *Page 315*: Richard Megna/Fundamental Photographs. *Page 319*: Phil Prosen/The Image Bank. *Page 322*: Rick Rickman/Duomo Photography, Inc. *Page 323*: Brian Bailey/Tony Stone Images/New York, Inc. *Page 324*: Michael Rosenfeld/Tony Stone Images/New York, Inc. *Page 327*: Photofest. *Page 331*: Gary Gladstone/The Image Bank. *Page 332*: Pascal Rondeau/Tony Stone Images/New York, Inc. *Page 335*: Paul McCormick/The Image Bank.

CHAPTER 12
Page 346: Greg Vaughn/Tony Stone Images/New York, Inc. *Page 350* (top right): Courtesy Omega Engineering, Inc. *Page 350* (bottom): Science Photo Library/Photo Researchers. *Page 351*

(right): Courtesy Dr. Sumio Uematsu, Johns Hopkins Hospital. *Page 351* (bottom): Courtesy NASA. *Page 353* (left): AP/Wide World Photos. *Page 353* (right): Richard Choy/Peter Arnold, Inc. *Page 360*: Simon Bruty/Tony Stone Images/New York, Inc. *Page 362*: Mark J. Thomas/Dembinsky Photo Associates. *Page 365*: Tony Stone Images/New York, Inc. *Page 373*: Larry Ulrich/Tony Stone Images/New York, Inc.

Chapter 13
Page 381: T. Davis/W. Bilenduke/Tony Stone Images/New York, Inc. *Page 383* (left): Courtesy NASA and The Johnson Space Center. *Page 383* (right): Noel Quidu/Gamma Liaison. *Page 385*: Gerard Fritz/Tony Stone Images/New York, Inc. *Page 389*: Pierre Du Charme/*The Ledger,* Lakeland, Florida. *Page 390*: John Callahan/Tony Stone Images/New York, Inc. *Page 392* (top): Gary Milburn/Tom Stack & Associates. *Page 392* (bottom): Frans Lanting/Minden Pictures, Inc. *Page 394*: Courtesy NASA.

Chapter 14
Page 401: Marc Chamberlain/Tony Stone Images/New York, Inc. *Page 403*: Courtesy Smithsonian Institution. *Page 405*: Ken Karp. *Page 406*: Courtesy Richard Zare, Stanford University. *Page 407*: Mark M. Lawrence/The Stock Market. *Page 409*: Ray Nelson/Phototake. *Page 418*: Andy Washnik.

Chapter 15
Page 423: Tony Craddock/Science Photo Library/Photo Researchers. *Page 424*: Dan Ham/Tony Stone Images/New York, Inc. *Page 426*: Courtesy NASA. *Page 435*: Courtesy NASA. *Page 448*: Bruce Forster/Tony Stone Images/New York, Inc.

Chapter 16
Page 458: Kim Westerskov/Tony Stone Images/New York, Inc. *Page 460*: Wada Nitsuhiro/Gamma Liaison. *Page 467* (top): Photo by Ron Garrison, courtesy Zoological Society of San Diego. *Page 467* (bottom): Courtesy Merlin D. Tuttle, Bat Conservation International. *Page 472*: Brian Kenney/Natural Selection.

Page 477 (left): G. Marche/FPG International. *Page 477* (right): Howard Sochurek/The Stock Market. *Page 479*: Simon Bruty/Tony Stone Images/New York, Inc. *Page 483*: Peter Rauter/Tony Stone Images/New York, Inc. *Page 484*: Courtesy Kurt Hondl, National Severe Storms Laboratory, Norman, OK.

Chapter 17
Page 494: Barros & Barros/The Image Bank. *Page 499*: Andy Washnik. *Page 503*: Courtesy SONY Corporation. *Page 504*: Bob Krist/Tony Stone Images/New York, Inc. *Page 506*: James L. Amos/Peter Arnold, Inc. *Page 507*: Richard Megna/Fundamental Photographs.

Chapter 18
Page 521: Peter Menzel. *Page 524*: Charles D. Winters/Photo Researchers. *Page 525*: Courtesy BMW of North America, Inc. *Page 530*: Courtesy Louis Scudiero and J. Thomas Dickinson, Washington University. *Page 532*: Uniphoto, Inc. *Page 543*: Paul Souders/©Corbis.

Chapter 19
Page 557: Courtesy Christopher Johnson and Robert MacLeod, University of Utah. *Page 559*: Tom McHugh/Photo Researchers. *Page 560*: Eric Sander/Gamma Liaison. *Page 575*: Bruce Curtis/Peter Arnold, Inc. *Page 576*: Hubatka/Phototake.

Chapter 20
Page 585: Ed Pritchard/Tony Stone Images/New York, Inc. *Page 587*: Bill Brooks/Masterfile. *Page 589*: Michael Rosenfeld/Tony Stone Images/New York, Inc. *Page 590*: Courtesy Apple Computers. *Page 593*: Courtesy Kinetic Corporation. *Page 607*: Yellow Dog Productions/The Image Bank. *Page 616*: RNHRD NHS Trust/Tony Stone Images/New York, Inc.

Chapter 21
Page 627: George Lepp/Tony Stone Images/New York, Inc. *Page 629*: Yoav Levy/Phototake. *Page 631*: Peter Arnold, Inc. *Page 632*: Gary Schultz/The Wildlife Collection. *Page 638*: Lawrence Berkeley

Laboratory/Photo Researchers. *Page 642*: Dennis Budd Gray. *Page 652* (top): Charles Thatcher/Tony Stone Images/New York, Inc. *Page 652* (bottom): Howard Sochurek/The Stock Market. *Page 654*: Jean-Loup Charmet/Science Photo Library/Photo Researchers. *Page 656*: Courtesy Los Alamos National Laboratory. *Page 658*: Courtesy Deutsche Bundesbahn/Transrapid.

Chapter 22
Page 669: Niels Van Iperen/Retna. *Page 681*: Courtesy JVC Company of America. *Page 688*: Tom Campbell/Tony Stone Images/New York, Inc. *Page 697*: Telegraph Colour Library/FPG International.

Chapter 23
Page 708: Mark Wagner/Tony Stone Images/New York, Inc. *Page 718*: Frank Cezus/Tony Stone Images/New York, Inc. *Page 721*: Courtesy Fisher Research Laboratory. *Page 726*: Lawrence Livermore National Laboratory/Science Photo Library/Photo Researchers. *Page 727*: Courtesy IBM Corporation.

Chapter 24
Page 732: Fritz Prenzel/Tony Stone Images/New York, Inc. *Page 733*: Nadie Mackenzie/Tony Stone Images/New York, Inc. *Page 736*: Jurgen Vogt/The Image Bank. *Page 738*: Courtesy Astronomical Society of the Pacific. *Page 741*: Courtesy Anglo-Australian Telescope Board. *Page 744*: Eric Meola/The Image Bank. *Page 750*: Diane Schiumo/Fundamental Photographs.

Chapter 25
Page 759: Mitch Reardon/Tony Stone Images/New York, Inc. *Page 762*: Yoav Levy/Phototake. *Page 764*: Dan McCoy/Rainbow. *Page 766*: Hank Morgan/Science Source/Photo Researchers. *Page 767*: James Lehman. *Page 769*: Ken Karp. *Page 770* (top): Courtesy Pontiac Motor Division, General Motors Corporation. *Page 770* (bottom): Paul Silverman/Fundamental Photographs. *Page 772*: Courtesy Hughes Danbury Optical Systems, Inc. *Page 774*: Jim and Cathy Church/FPG International.

Page 777: Courtesy Sandia National Laboratories.

CHAPTER 26

Page 781: ©Lennart Nilsson, from *A Child is Born*. *Page 782*: Richard Megna/Fundamental Photographs. *Page 789*: Courtesy Schott Fiber Optics, Inc., Southbridge, MA. *Page 792*: Phillip Hayson/Photo Researchers. *Page 793* (top left): Leonard Lessin/Peter Arnold, Inc. *Page 793* (top right): Courtesy Johns Hopkins University School of Medicine. *Page 793* (bottom): Margaret Rose Orthopaedic Hospital/Photo Researchers. *Page 794*: D. Parker/Science Source/Photo Researchers. *Page 796*: Gary Yeowell/Tony Stone Images/New York, Inc. *Page 801*: Med. Illus. SBHA/Tony Stone Images/New York, Inc. *Page 816*: Courtesy Tasco. *Page 817*: Courtesy Celestron, Inc.

CHAPTER 27

Page 829: Stuart McClymont/Tony Stone Images/New York, Inc. *Page 830*: Gary Vestal/Tony Stone Images/New York, Inc. *Page 834*: From *Atlas of Optical Phenomena*, Michael Cagnet, Springer-Verlag, Berlin. *Page 836*: Andy Washnik. *Page 839*: Clark Dunbar/Uniphoto, Inc. *Page 841*: Courtesy Bausch & Lomb. *Page 843*: Peter Poulides/Tony Stone Images/New York, Inc. *Page 844*: Courtesy Education Development Center. *Page 847*: From *Atlas of Optical Phenomena*, Michael Cagnet, Springer-Verlag. *Page 849*: ©Truax/The Image Finders. *Page 850*: From *Atlas of Optical Phenomena*, Michael Cagnet, Springer-Verlag. *Page 852*: "A Sunday on La Grande Jatte" by Georges Seurat; ©1996. All Rights Reserved. Reproduced with permission and courtesy of The Art Institute of Chicago. *Page 857* (left): Courtesy Edwin Jones, University of South Carolina. *Page 857* (right): James King-Holmes/OCMS/Science Photo Library/Photo Researchers.

CHAPTER 28

Page 864: The Einstein Memorial Sculpture (©Robert Berks, 1978) at the National Academy of Sciences. Photograph by Diana H. Walker, courtesy National Academy of Sciences. *Page 867*: Ulli Seer/Tony Stone Images/New York, Inc. *Page 870*: Courtesy Johnson Space Center and NASA. *Page 876*: Bill Marsh/Photo Researchers. *Page 879* (left): NASA/Mark Marten/Photo Researchers. *Page 879* (right): Dr. Leon Golub/Photo Researchers.

CHAPTER 29

Page 888: Andrew Syred/Tony Stone Images/New York, Inc. *Page 891*: John Lund/Tony Stone Images/New York, Inc. *Page 896*: Courtesy NASA. *Page 899*: Secchi-LeCaque/Roussel-UCLAF/CNRI/Science Photo Library/Photo Researchers. *Page 900* (top): From *Physics Review* 73:527 (1948), by Wollan, Shull and Marney. *Page 900* (center): Courtesy Edwin Jones, University of South Carolina. *Page 901*: Courtesy Akira Tonomura, J. Endo, T. Matsuda and T. Kawasaki, *Am. J. Phys.* 57(2): 117, February 1989. *Page 906*: Courtesy Central Scientific Co.

CHAPTER 30

Page 910: Lawrence Livermore National Laboratory/Science Photo Library/Photo Researchers. *Page 913*: Courtesy Bausch & Lomb. *Page 916*: Courtesy A.I.P. Niels Bohr Library, Margrethe Bohr Collection. *Page 919*: John Warden/Tony Stone Images/New York, Inc. *Page 933*: Charles Gupton/Uniphoto, Inc. *Page 934* (top): Hank Morgan /Photo Researchers. *Page 934* (bottom left): Mehau Kulyk/Photo Researchers. *Page 934* (bottom right): Photo Researchers. *Page 938* (left): Courtesy Australasian Medical Publishing Company. *Page 938* (right): Fritz Hoffmann/The Image Works. *Page 941*: Ronald R. Erickson/Media Interface Ltd.

CHAPTER 31

Page 947: Courtesy HCE Larsson and PC Sereno. *Page 951*: Steve Kaufman/Peter Arnold, Inc. *Page 954*: ©Centre Jean-Perrin Clermont-Ferrand/CNRI/Phototake. *Page 959*: Courtesy Elekta Instruments, Inc. *Page 964*: Patrick Aventurier/Gamma Liaison. *Page 965*: Sygma Photo News. *Page 968*: P. Loiez/CERN/Science Photo Library/Photo Researchers.

Chapter 32

Page 973: Wellcome Dept. of Cognitive Neurology/Science Photo Library/Photo Researchers. *Page 978*: Tony Stone Images/New York, Inc. *Page 986*: Courtesy Lawrence Livermore Laboratory. *Page 987*: Science Photo Library/Photo Researchers. *Page 989* (top): Will & Deni McIntyre/Photo Researchers. *Page 989* (bottom): National Institutes of Health/Photo Researchers.

INDEX

SI Units

Quantity	Name of Unit	Symbol	Expression in Terms of Other SI Units	Quantity	Name of Unit	Symbol	Expression in Terms of Other SI Units
Length	meter	m	Base unit	Pressure, stress	pascal	Pa	N/m^2
Mass	kilogram	kg	Base unit	Viscosity	—	—	$Pa \cdot s$
Time	second	s	Base unit	Electric charge	coulomb	C	$A \cdot s$
Electric current	ampere	A	Base unit	Electric field	—	—	N/C
Temperature	kelvin	K	Base unit	Electric potential	volt	V	J/C
Amount of substance	mole	mol	Base unit	Resistance	ohm	Ω	V/A
Velocity	—	—	m/s	Capacitance	farad	F	C/V
Acceleration	—	—	m/s^2	Inductance	henry	H	$V \cdot s/A$
Force	newton	N	$kg \cdot m/s^2$	Magnetic field	tesla	T	$N \cdot s/(C \cdot m)$
Work, energy	joule	J	$N \cdot m$	Magnetic flux	weber	Wb	$T \cdot m^2$
Power	watt	W	J/s	Specific heat capacity	—	—	$J/(kg \cdot K)$ or $J/(kg \cdot C°)$
Impulse, momentum	—	—	$kg \cdot m/s$	Thermal conductivity	—	—	$J/(s \cdot m \cdot K)$ or $J/(s \cdot m \cdot C°)$
Plane angle	radian	rad	m/m	Entropy	—	—	J/K
Angular velocity	—	—	rad/s	Radioactive activity	becquerel	Bq	s^{-1}
Angular acceleration	—	—	rad/s^2	Absorbed dose	gray	Gy	J/kg
Torque	—	—	$N \cdot m$	Exposure	—	—	C/kg
Frequency	hertz	Hz	s^{-1}				
Density	—	—	kg/m^3				

Greek Alphabet

α		Iota	I	ι	Rho	P	ρ
		Kappa	K	κ	Sigma	Σ	σ
			Λ	λ	Tau	T	τ
			M	μ	Upsilon	Υ	υ
			N	ν	Phi	Φ	ϕ
			Ξ	ξ	Chi	X	χ
			O	o	Psi	Ψ	ψ
			Π	π	Omega	Ω	ω